Atlas of Design

Vol. 6 »

editors
Nat Case
Aaron Koelker
Josh Ryan
Tracy Tien

published by
nacis
the North American
Cartographic Information Society

Atlas of Design, Volume Six

First Printing, October 2022

Editors: Nat Case, Aaron Koelker, Josh Ryan, Tracy Tien

Published by the North American Cartographic Information Society, *nacis.org*

The type in this book has been set in Avenir, by Adrian Frutiger.

Printed in the United States of America by Shapco Printing, Golden Valley, MN.

ISBN 979-8-9866856-6-3

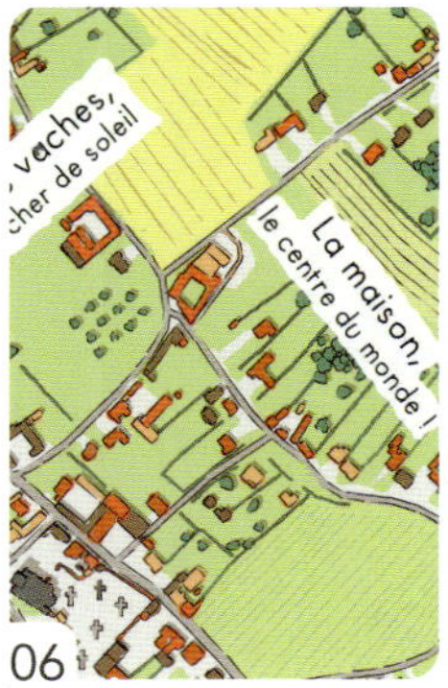

06

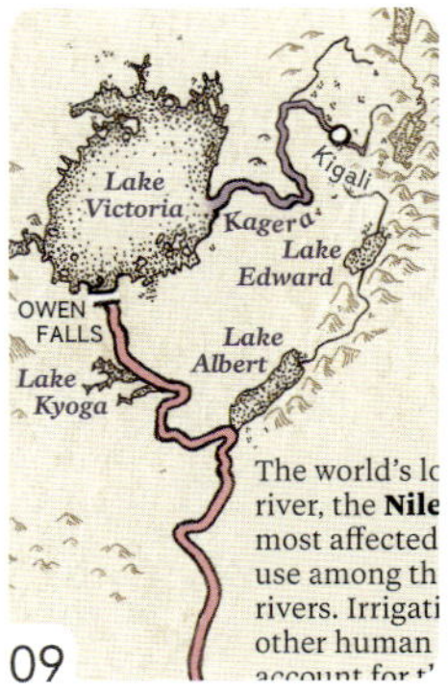

09

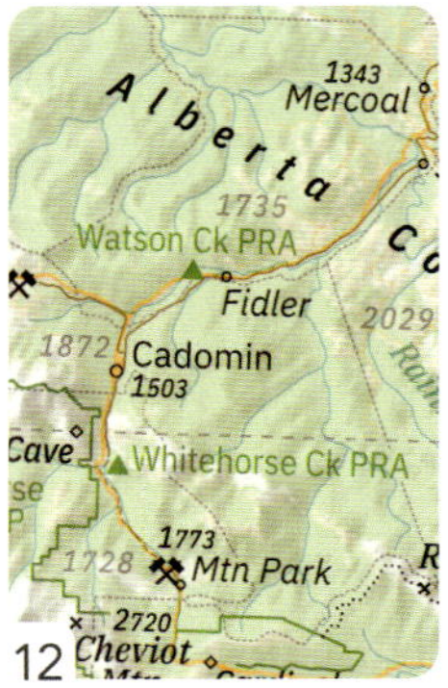

12

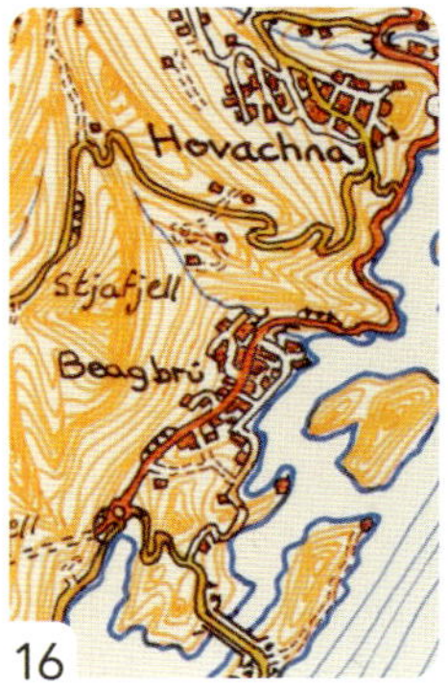

16

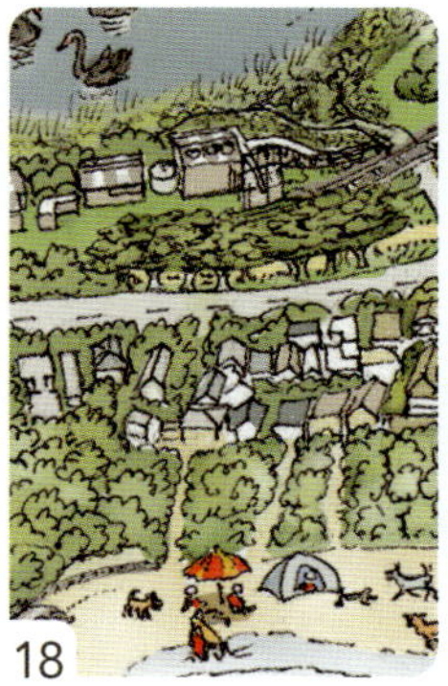
18

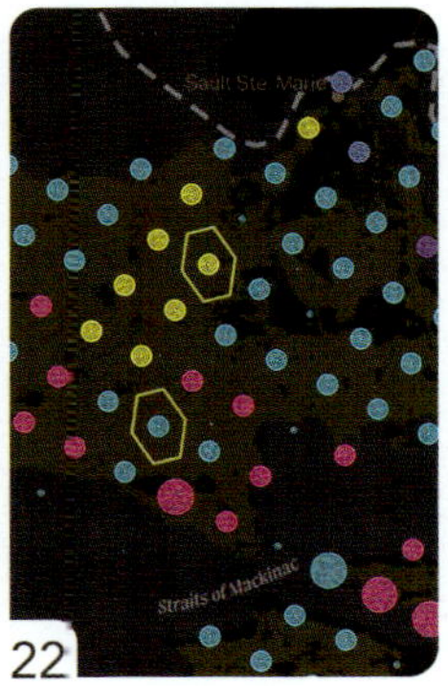

22

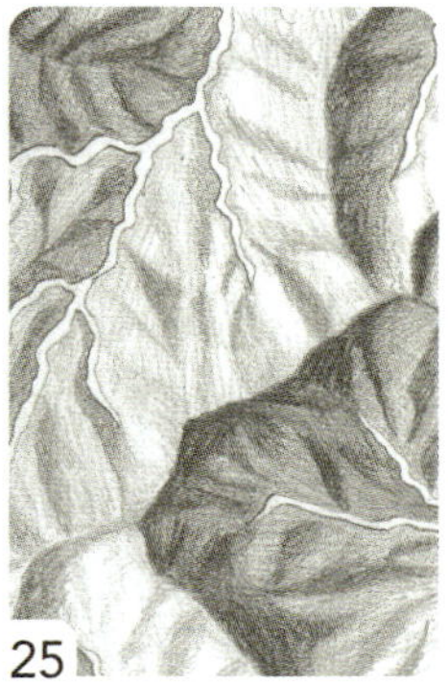
25

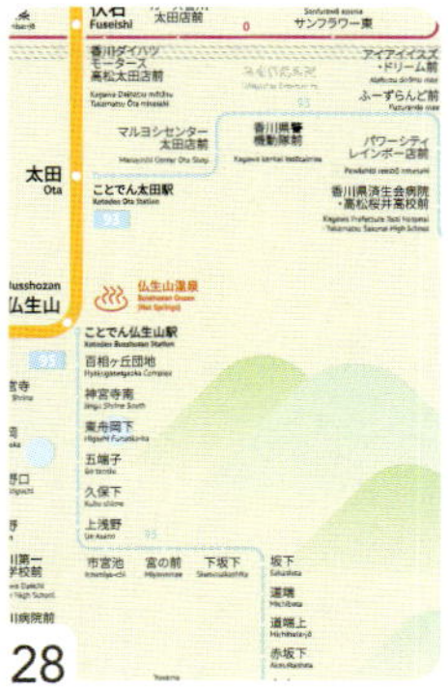
28

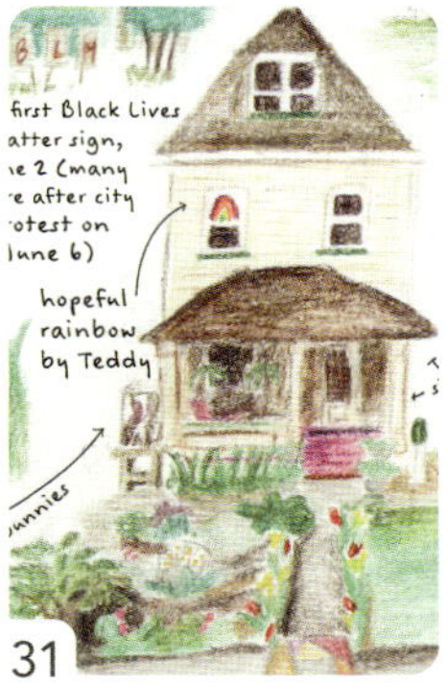

31

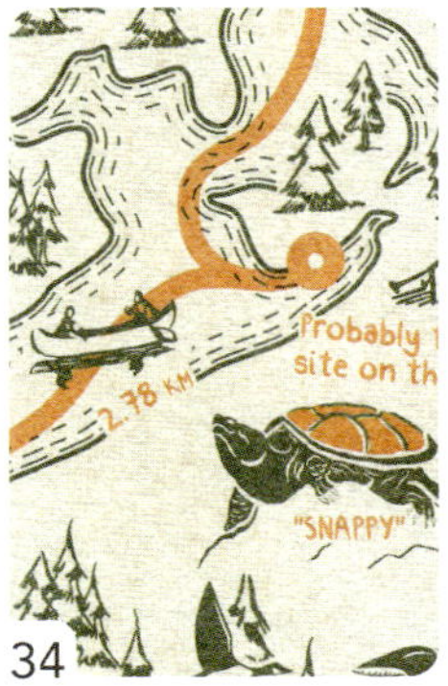

34

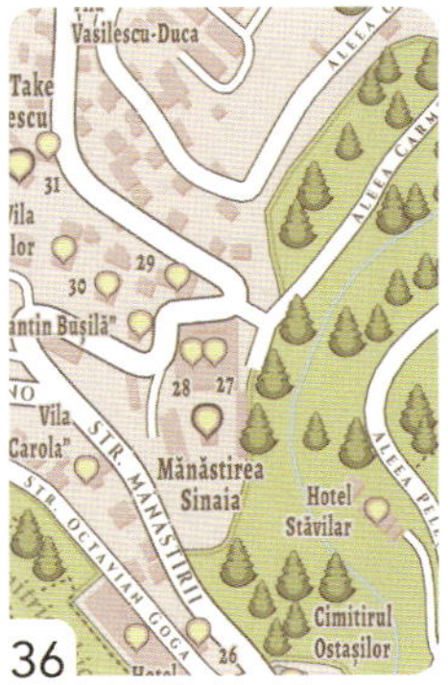

36

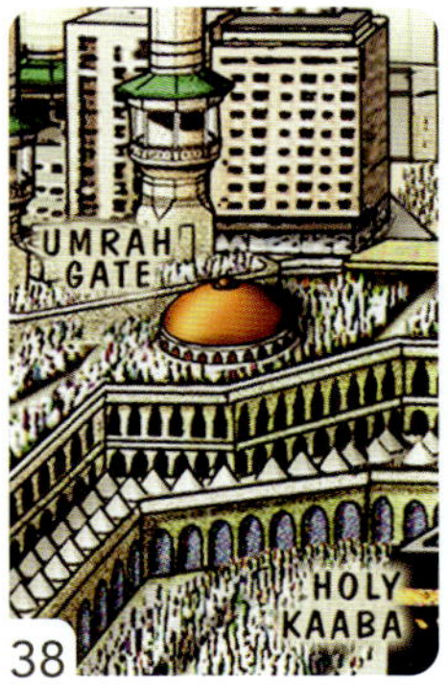

38

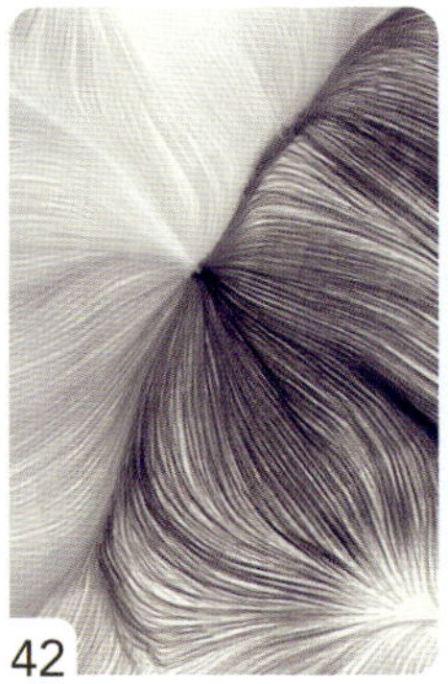
42

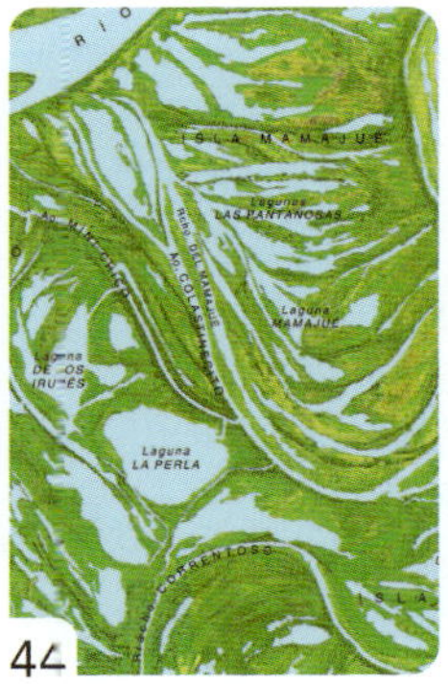

44

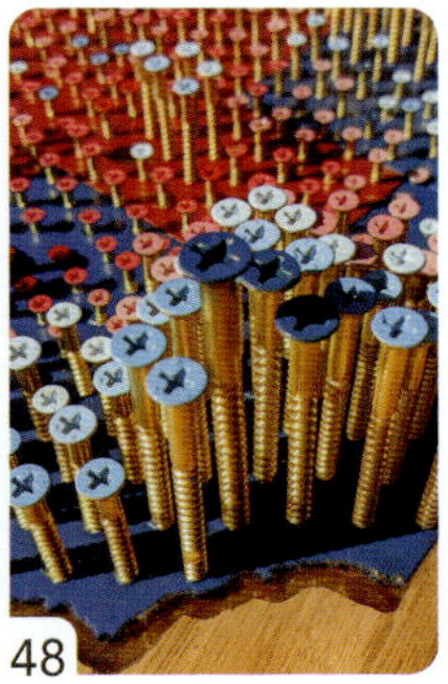
48

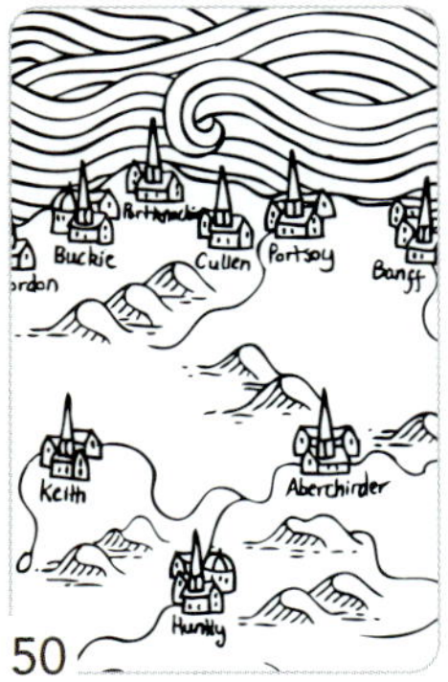

50

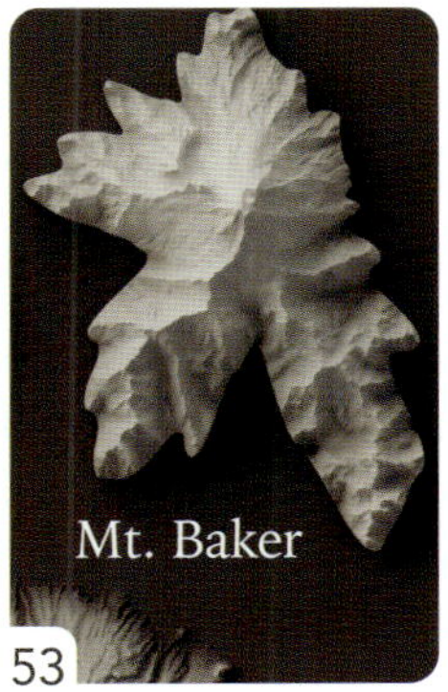

53

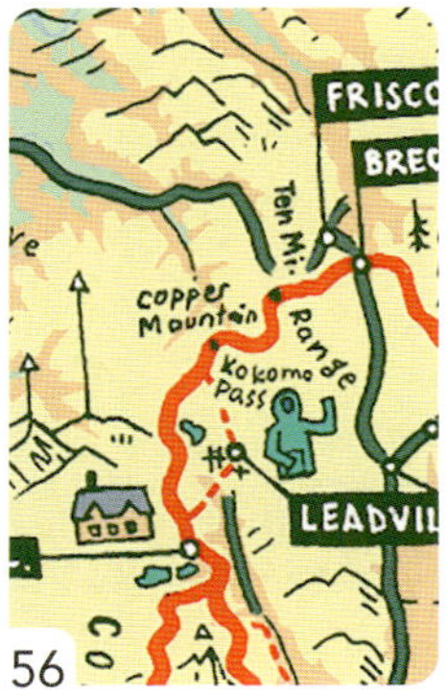

56

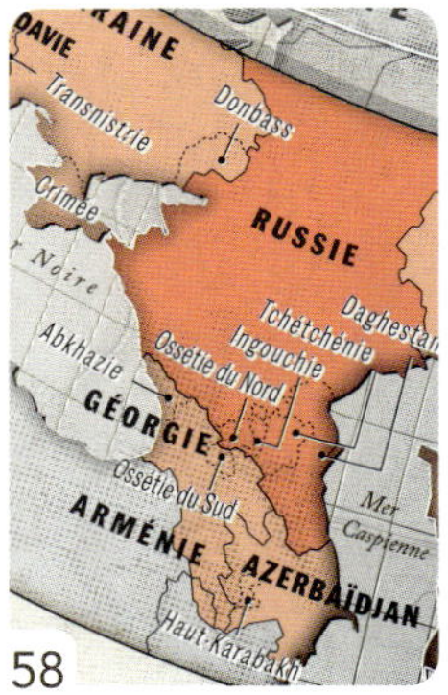

58

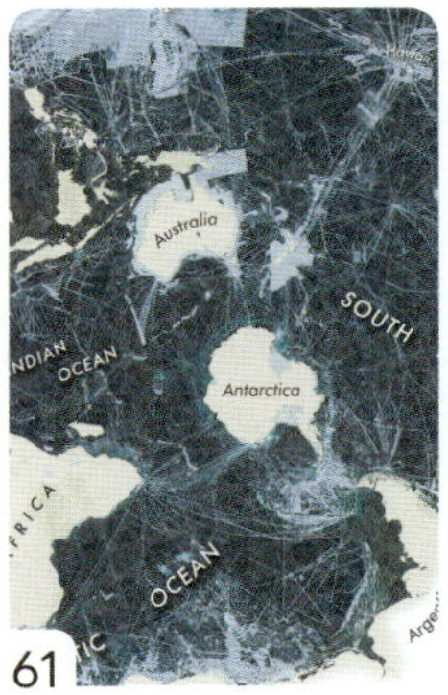

61

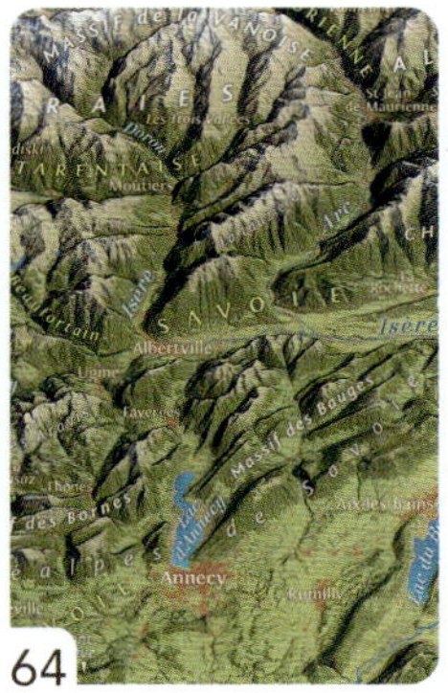

64

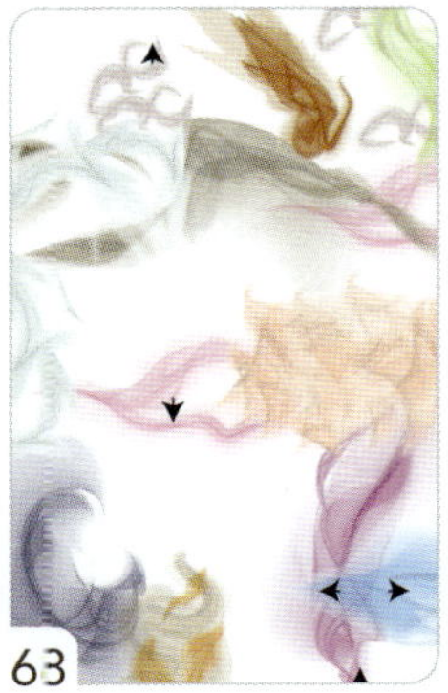
63

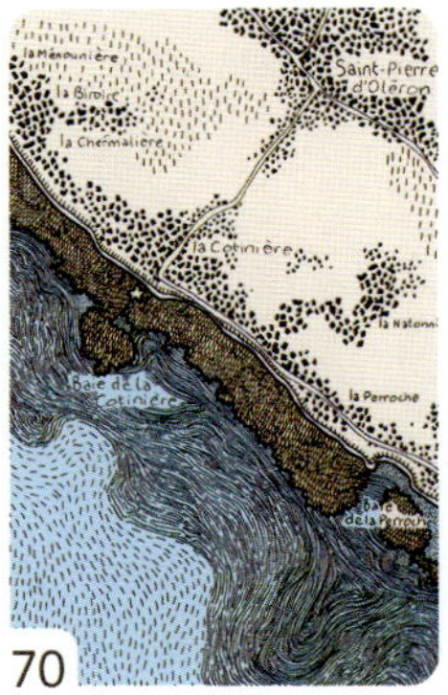

70

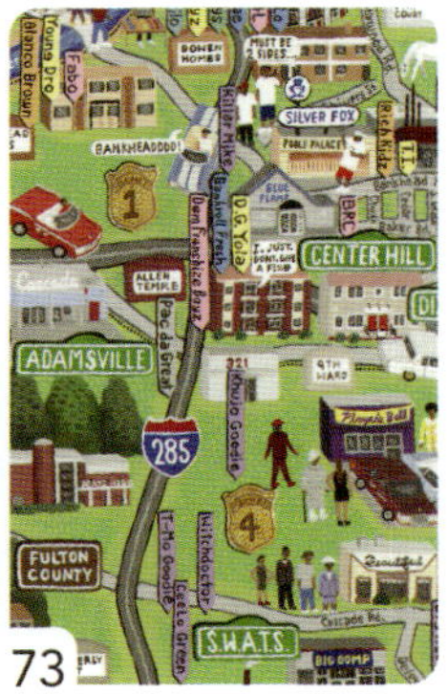

73

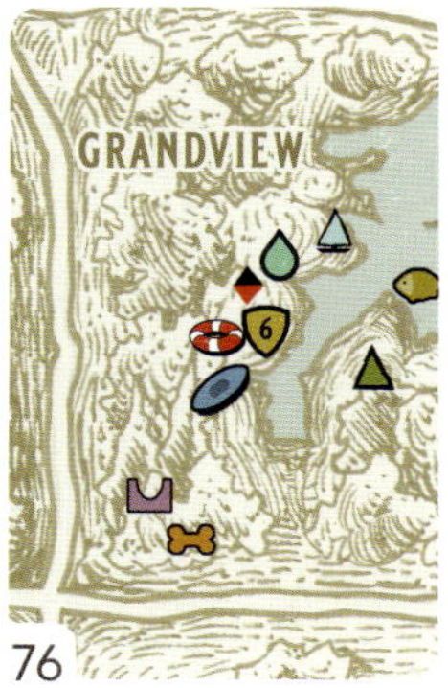

76

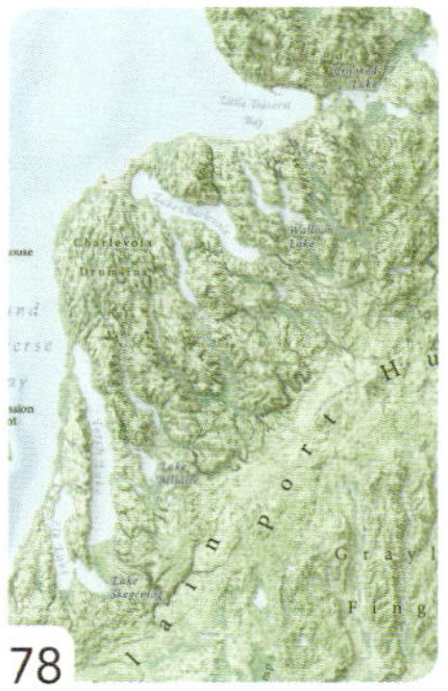
78

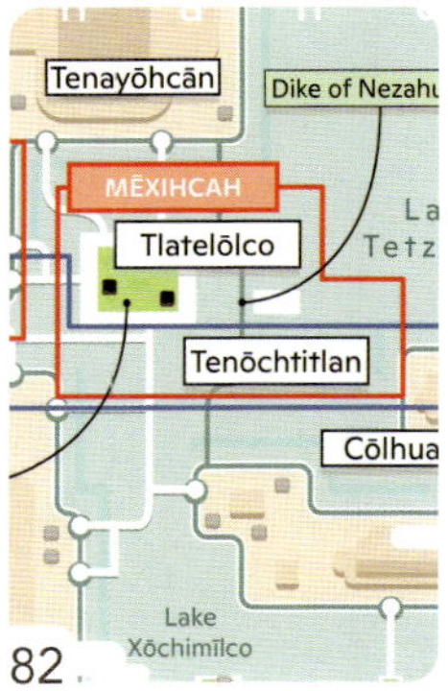

82

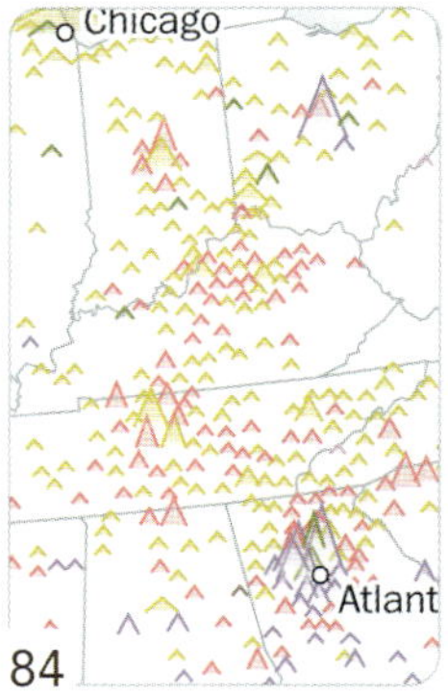

84

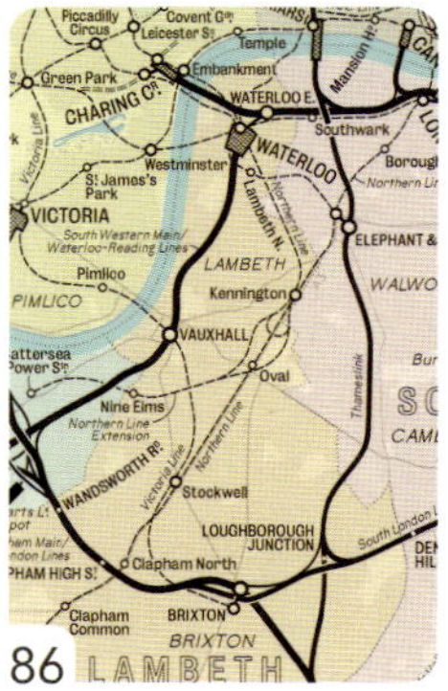

86

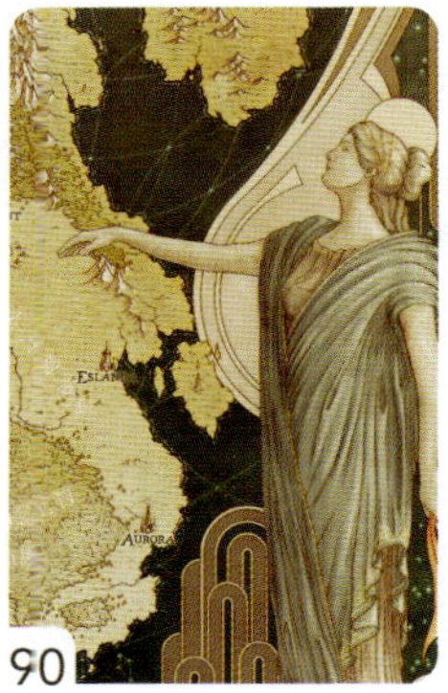
90

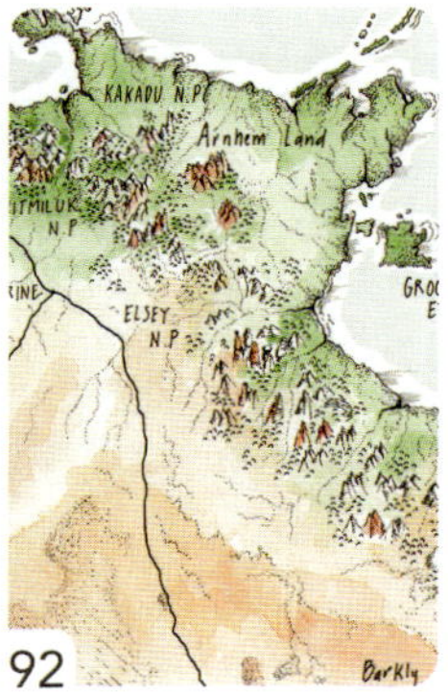

92

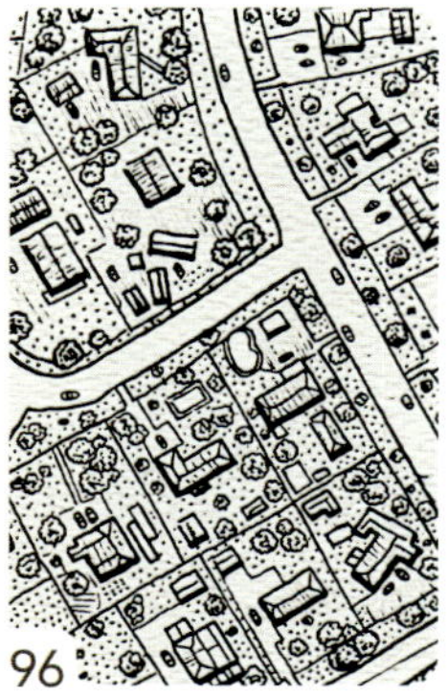
96

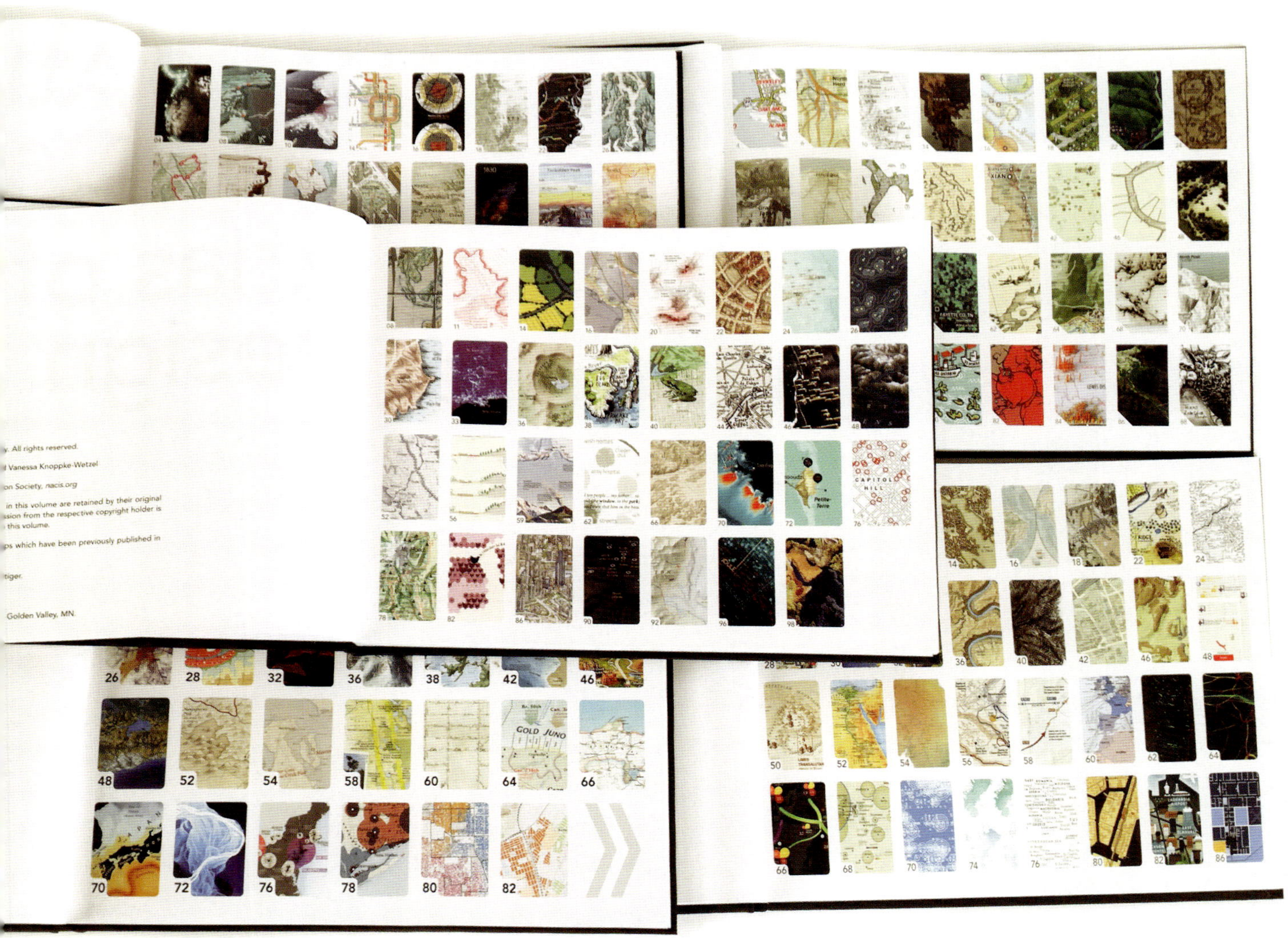

nacis

The *Atlas of Design* is a biennial NACIS publication. The North American Cartographic Information Society improves communication, coordination, and cooperation among the producers, disseminators, curators, and users of cartographic information. The organization is founded on education and information sharing, as well as promoting cartographers' works through publications such as the *Atlas*.

NACIS hosts an annual conference for professionals and enthusiasts from government, commercial, and not-for-profit groups along with academics, scholars, map/GIS librarians, artists, technologists, and students interested in the world of maps and geographic information. To learn more about NACIS visit nacis.org.

A Note from the Editors

Despite Paper Shortages® and Supply-Chain Issues™—here it is: the *Atlas of Design, Volume 6* (on satin text dull finish, prepurchased in April).

This book continues to reflect the labors of love that sustain our cartographic community, from the contributors who created the astonishing map works herein; to the judges, translators, and editors who wrestled it into book form; and to you, the readers, who continue to support this ongoing project. The Atlas ventures into uncharted territories (well, they're charted now . . .) with each iteration, and we are heartened by the evolving generosity of all our volunteers.

What would such an atlas be without maps that know no bounds? We thank all the map practitioners who took the time and care to submit an entry; we cherished each moment we got to spend with the creations. We had a feast to pick from, and could easily have made two volumes out of the contributions we waded through. To ensure that this Atlas *does* have bounds, we owe thanks to the judges: Aly Ollivierre, Armelle Caron, Daniel Huffman, John Krygier, Jordan Engel, Juliane Cron, Kenneth Field, Madison Draper, Meghan Kelly, Tim Wallace, Tom Patterson, and Vicky Johnson-Dahl.

We are grateful to our translators for literally providing us with words we cannot adequately express: Sarah Bennett, Naeema Al Hosani, Tibi Chelcea, Laurent Jégou, Ramón Perez, Adrián Blanco Ramos, and Johannes Kröger. Our copyeditors, Lys Weiss and Jeffrey K. Weiss of Post Hoc Academic Publishing Services, took the contributors' words and brought them into line so this feels like a single volume with many distinct voices.

We offer immense gratitude to our immediate editorial predecessors—Brooke Marston, Caroline Rose, Nat Case, and Vanessa Knoppke-Wetzel—who tossed this treasured hot potato to us; especially Nat, who stayed on to guide a cohort of what would otherwise have been significantly more confused editors. And we are aware of being a part of a longer chain, going back through seven more editors who have been in our position in various combinations over five previous volumes.

We thank our compatriots in the cartographic universe (including those beyond North America) at the North American Cartographic Information Society (NACIS) for their vision in supporting the Atlas. Our partners at Eastview and our volunteer forwarders, Hans van de Maarel and Amy Griffin, have helped us get all the atlases out to our readers, even when they get stuck sideways in the Suez Canal. And thanks to our print partners at Shapco Printing for their superb attention to detail in delivering a physically beautiful product.

Last but not least, to our witting (or, if you are gifted a copy, unwitting) readers: we hope you find a place among these maps, and in turn create ones that are yet to be imagined.

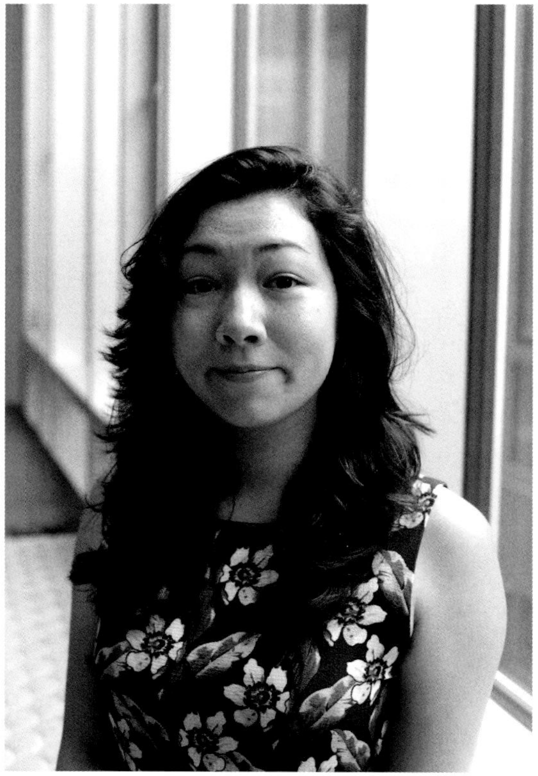

Introduction by Eleanor Lutz

Most days, I'm making charts and maps at work as a visual journalist. I enjoy the bar charts and line graphs, but I think there's something really special about a map. Maps immediately put you, the reader, into the same space as the story, in a way that charts just can't.

It can be any map: from a simple drawing designed in a few hours for a breaking news story, to an intricate arrangement of bird migrations across a continent. No matter the size and scale, if you zoom far enough out from the cropped frame, there you are — sometimes just a few miles from the story, sometimes half a world away or on the other side of the galaxy. With a little imagination, you can expand the boundaries of any map to include yourself.

I think that's why maps have the curious potential to make us think differently about where we are, and who we are. We can't escape being a part of every map, and a new way of visualizing space can change the way we think about our own place in the world. My favorite maps are the ones that do exactly that, and help us explore a different way to see.

A few years ago, I traded maps through the mail with one of the owners of an antique map shop. I didn't quite know what to expect from the trade, but when I opened the package, it contained the most bizarre and fascinating map I'd ever seen.

The map had no land, no water, and no air. Instead, it showed 133 clocks, arranged in concentric circles around one large clock stopped at noon. Each clock was inscribed with the name of a city, and arranged in order of distance away from Washington, DC—the city marked on the largest central clock. The map was called "A Diagram Exhibiting the difference of time between the places shown & Washington," and it was from an era before standard time zones were used across America.

According to the map, noon in Washington, DC, corresponded to 8:58 a.m. in San Francisco, 12:12 p.m. in New York City, and 11:23 and 30 seconds a.m. in Indianapolis. Readers could use the map to determine the complex time differences between their home and faraway places. It was a fascinating solution to a bygone problem, and it made me realize that I'd lived my whole life accepting modern time zones as an undisputed fact of life.

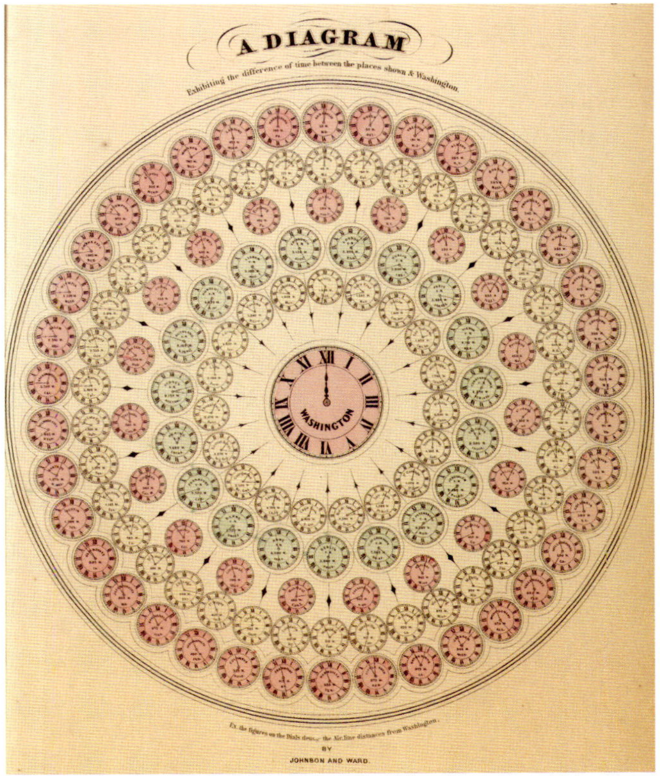

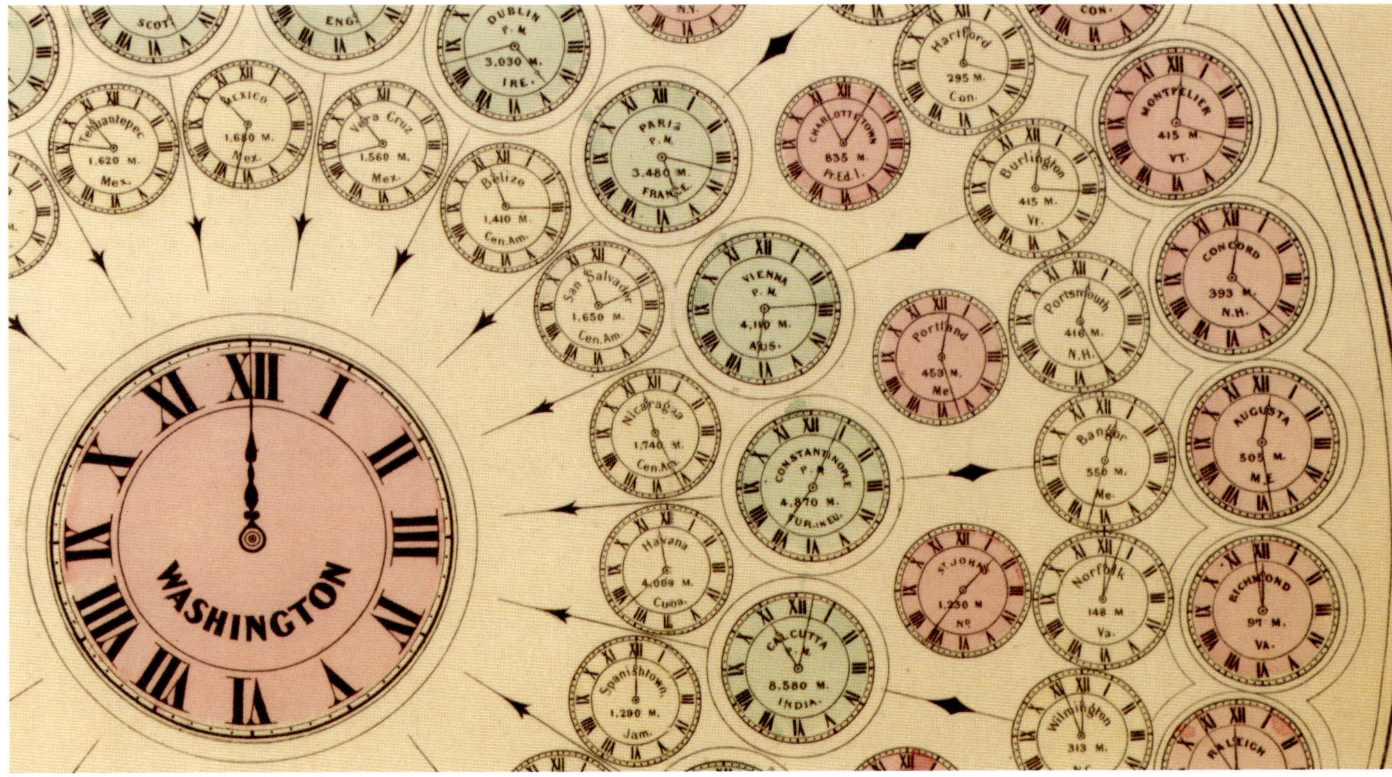

You might say that a collection of clocks, even if arranged by distance, isn't a map at all. After all, even Johnson and Ward themselves decided to call their publication "A Diagram" and not "A Map." Which poses the question: what is a map, exactly? and can something still be a map if the authors intended it to be something else?

Defining words is a slippery exercise, and I don't think a single correct answer to the question exists. We could probably quibble forever about the intricate subtleties and never come to a solid agreement. But for me personally, a map is simply something you can use to find yourself—literally or figuratively.

So street diagrams definitely count as maps, because they help you see where you stand in the physical world. Johnson and Ward's clock diagram also counts as a map, because you can use it to find yourself in time instead of space. And even imaginary maps may help you explore your own inner landscape.

When I was a little girl, I had a poster with a map of Tolkien's Middle Earth. This map certainly had no practical function. Its only purpose was to take you along on fanciful journeys, and to be honest, I wasn't even a huge fan of Tolkien. But for me the map was special because my family moved from place to place constantly. No matter how often my mailing address changed, I could always find myself on the imaginary landscape, comfortably anchored in the same place. So, was this a map? Can a fictional place be real enough to count? I think so.

Of course, just because something counts as a map doesn't mean it's necessarily a useful map, or a good one. Once you satisfy a map's basic need for a purpose, it's the style and design sprinkled on top that can elevate a plain boiled map into a geographic feast. I think it's the serendipitous spark between reader and cartographer that defines a truly great map. It's up to the cartographer whether to serve a crisp and fresh minimalist map, or a spicy irreverent one. And it's up to the reader to decide what they're in the mood to digest that day.

Illustration from *Johnson's new illustrated (steel plate) family atlas, with descriptions, geographical, statistical, and historical.* New York : Johnson and Ward, 1862.
Source: Library of Congress Geography and Map Division.

Carte de mon monde confiné,

« dans un rayon maximal d'un kilomètre autour du domicile » - Calais

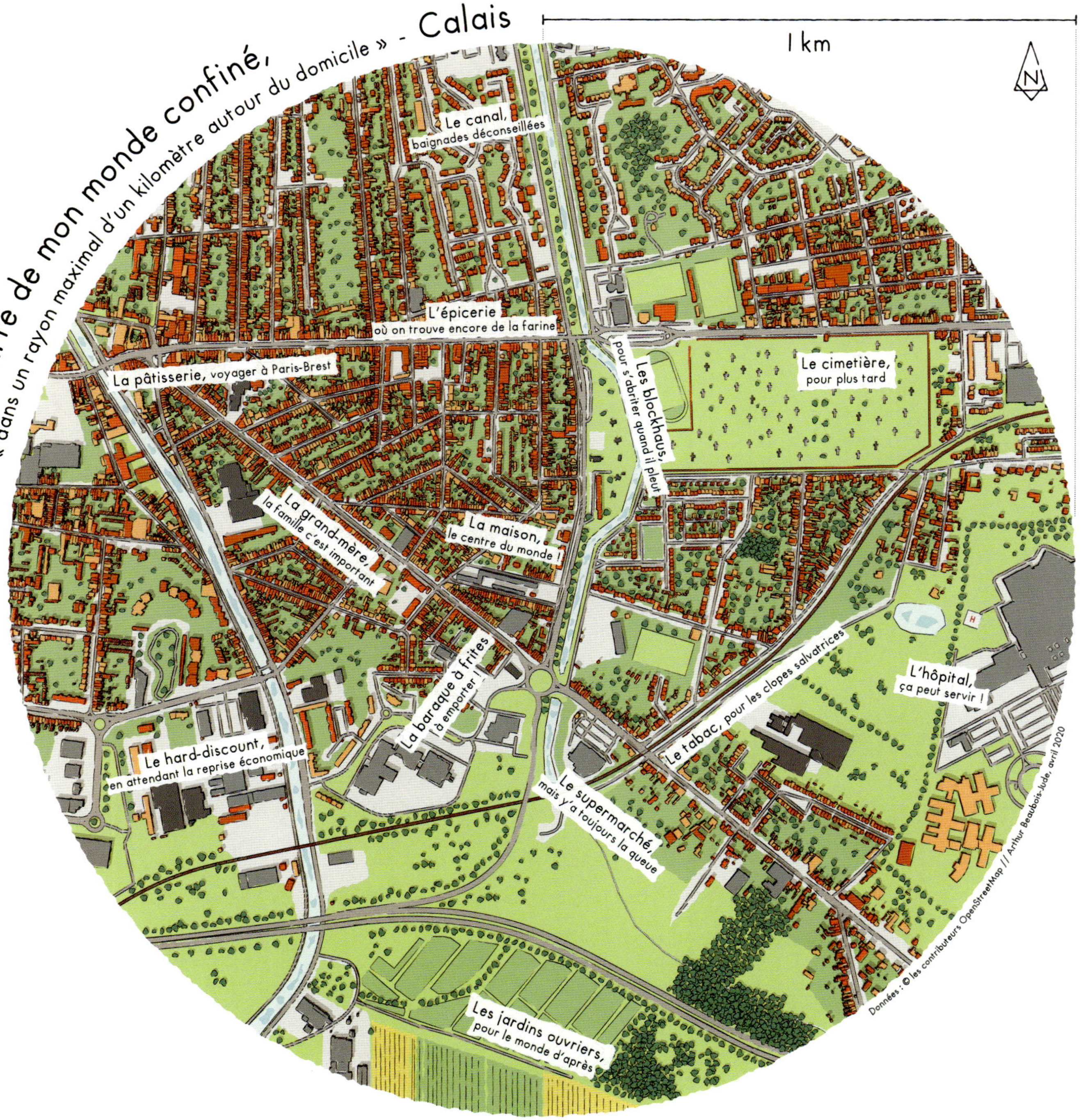

Le monde confiné de Mathilde et Benjamin,

« dans un rayon maximal d'un kilomètre autour du domicile » - Nomai

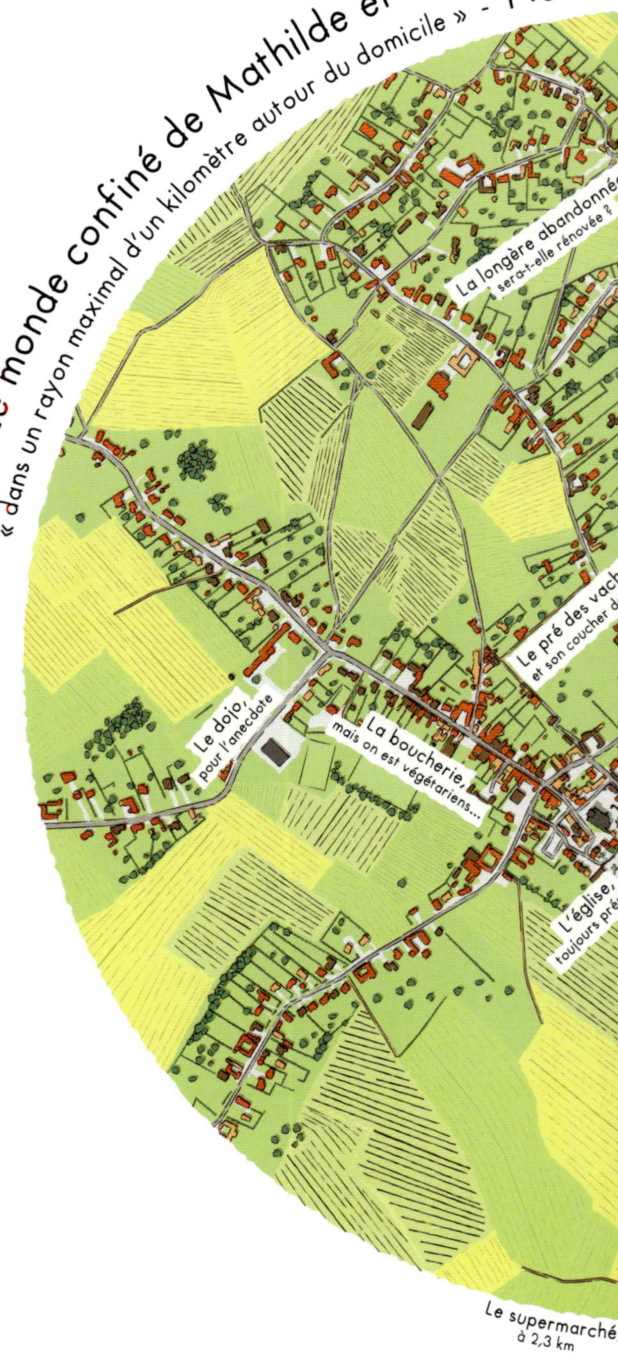

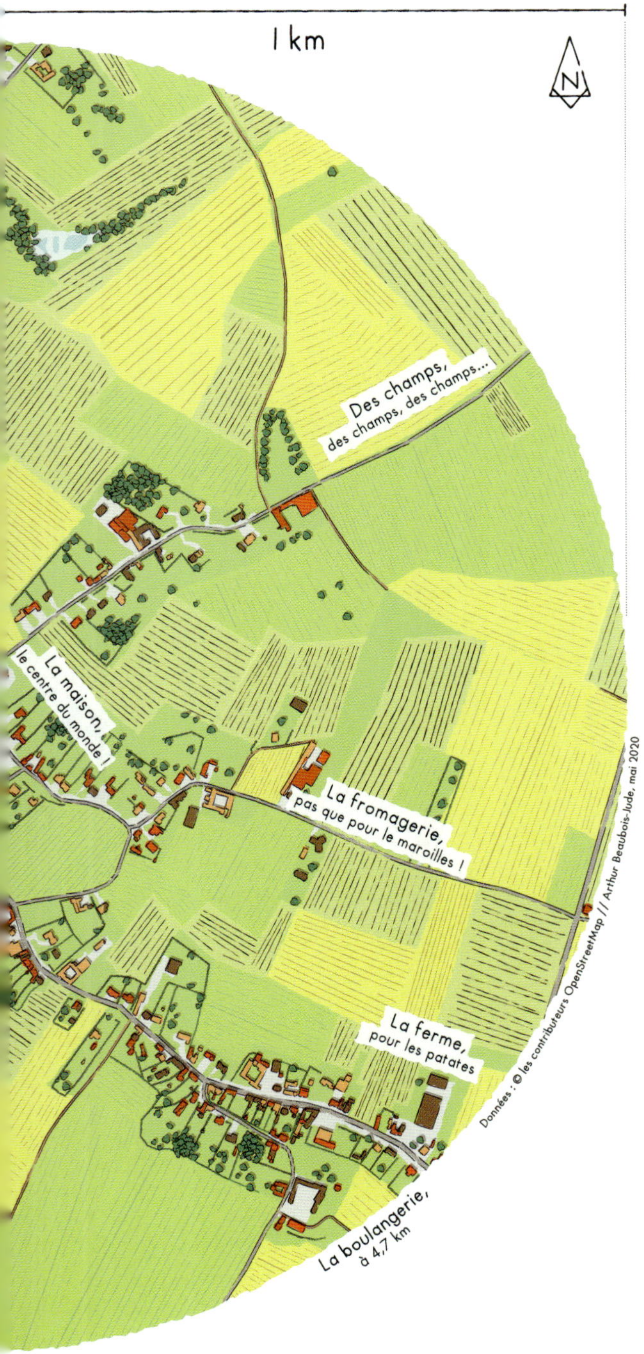

Le monde confiné de Maria,

« dans un rayon maximal d'un kilomètre autour du domicile » - Paris

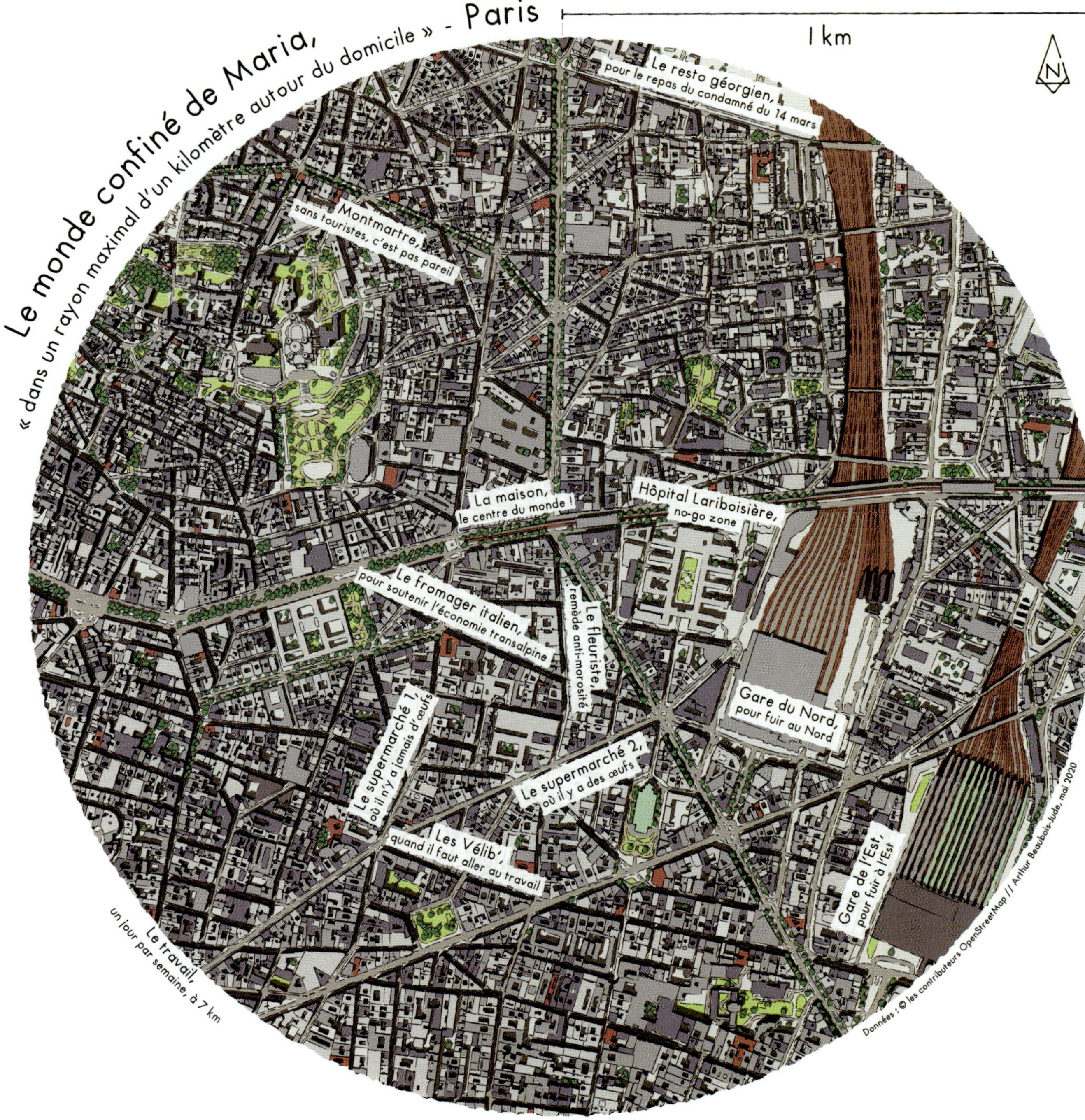

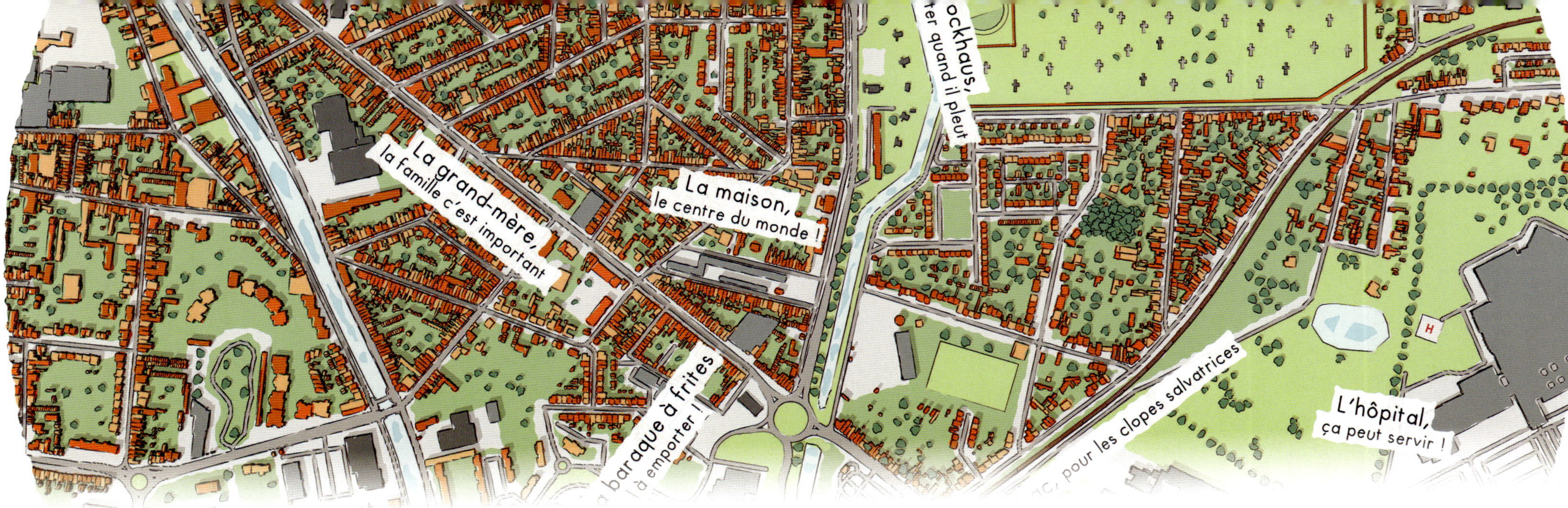

« Carte de mon monde confiné / Map of My Confined World

Arthur Beaubois-Jude

This series of emotional maps was constructed in a very particular context: during the first confinement due to the COVID-19 pandemic in France, when we were forbidden, without professional or imperative reason, to go beyond a maximum radius of one kilometer around our residence. This simple rule, very arbitrary, explains the circular shape of the map, which defined a small, individualized mappa mundi according to the location of one's home (our new center of the world). After making a map centered on my own home, I adapted it, with the help of friends who were similarly confined to their homes, to make two other versions of areas with urban fabrics very different from mine.

This series of maps was the culmination of a stylistic attempt to rediscover the spirit of "handmade" cartography—but with the help of a computer. I started with OpenStreetMap data, which I explored, amended, and reworked graphically in all their aspects.

Indeed, I sometimes feel a certain frustration, and perhaps even a kind of "complex," in carrying out my cartography by means of a computer, with all its facilities and automatisms, which our close ancestors did without, instead doing everything by hand (in accordance with a romantic vision of the profession). One problem with computer support is that everything is too clean, too perfect, even soulless, and it gives me a certain nostalgia for the styles of yesteryear. Another problem is the risk of falling into imitation.

So my goal was to try to break this perfection, by deliberately applying "graphic flaws." Rather than imitate, I sought a style that wanted to be optimistic and help to put things into perspective: in a period of anxiety, it was very natural for me that this series of maps, both in substance and in form, should have a naïve aspect, a little funny and playful at the same time.

The design was ultimately like "self-therapy," and I think other people recognized themselves a little in this exercise, since they also were being subjected to these unprecedented spatial constraints. Should we call it "Map Therapy"?

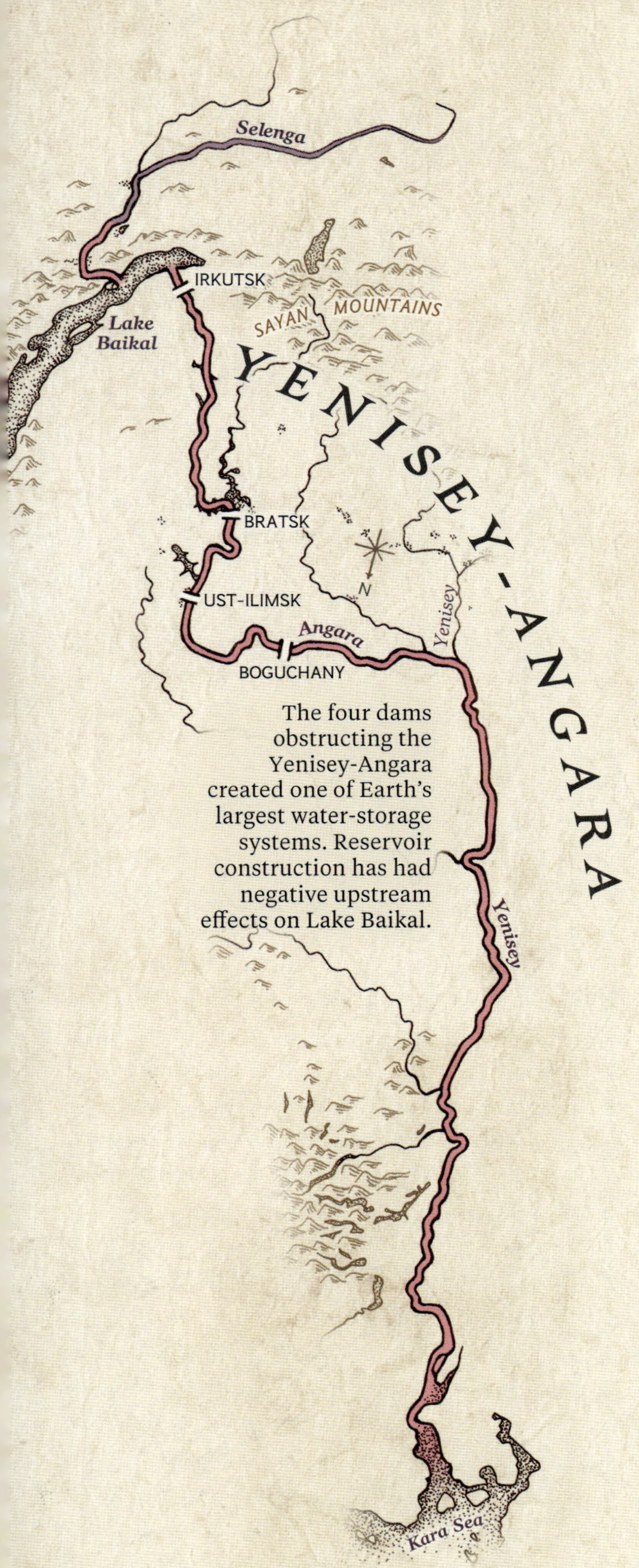

Setting Rivers Free »

Heather Gabriel Smith and Christina Shintani, National Geographic

In this atlas for *National Geographic* magazine, we wanted readers to dive into the physical and human geographies that shape and impact the health and connectivity of the world's largest rivers. What do long, free-flowing rivers have in common, and how have humans disrupted the connectivity of so many?

The idea of this atlas was born from a rich and extensive dataset. Our goal was to communicate the big-picture story told by those data, and to invite readers into the story with powerful and beautiful visuals. We accomplished this through two early decisions: to show only a tiny fraction of the data, and to draw the map by hand.

We limited the data from millions of line segments to the ten longest river backbones. For each of these river reaches, we communicated only their connectivity status (impacted, good connectivity, and free-flowing). A major challenge was designing the page to fit the ten longest rivers from four continents. Finding an optimal arrangement was not a simple task, but ultimately, arranging the rivers like puzzle pieces allowed us to show each river flowing in the same direction, at the same scale, and without crowding. Each river map is oriented with its source at the top of the page, rather than north. This promotes a reading direction that matches the flow of the rivers, a flow that becomes increasingly impeded by dams toward the bottom of the page.

We avoided a datacentric aesthetic by drawing the rivers by hand, and including only data that could fit along the river (connectivity status and location of dams). The maps and illustration were drawn in black ink and scanned before being colored and cleaned digitally. With so much detail on each river, it was important to create a strong visual hierarchy with minimal color. River connectivity is the only thing on the page that is given color other than black and earth tones, which provide important contextual information. Together, these design decisions allow the reader to absorb the most important information in an approachable way, and invite them to explore the geography that these rivers flow through, and the obstacles they face.

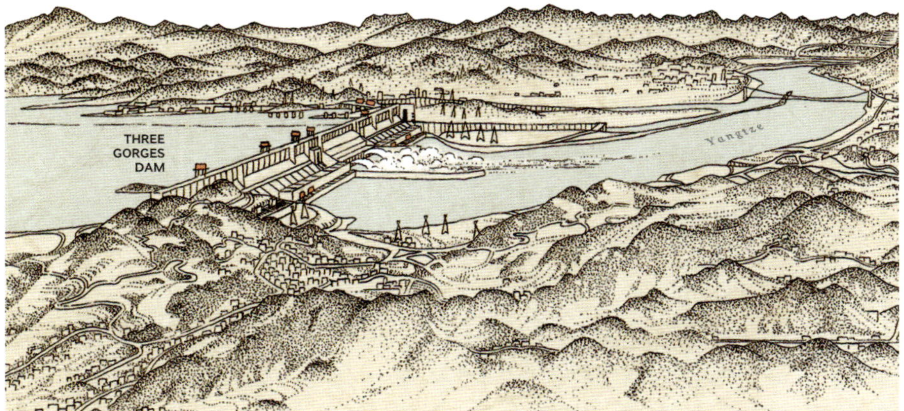

SETTING RIVERS FREE

BY **HEATHER GABRIEL SMITH** AND **CHRISTINA SHINTANI**

LONG, FREE-FLOWING RIVERS are increasingly rare. These serpentine giants should support entire ecosystems and allow the unobstructed movement of energy, materials, and wildlife in their waters and in the surrounding landscape. But humans have been harnessing their immense power for centuries, building so many dams and reservoirs, and so much other infrastructure, that now only 37 percent of the world's rivers over a thousand kilometers long (621 miles) still run free. Human impacts include dams trapping sediment so that it can no longer naturally shape the river and deltas, and regulation of reservoir water levels to reduce flow to a trickle in some areas. Just how much these obstructions disrupt connectivity can vary throughout a river, but understanding their many consequences is crucial to restoring these ecosystems.

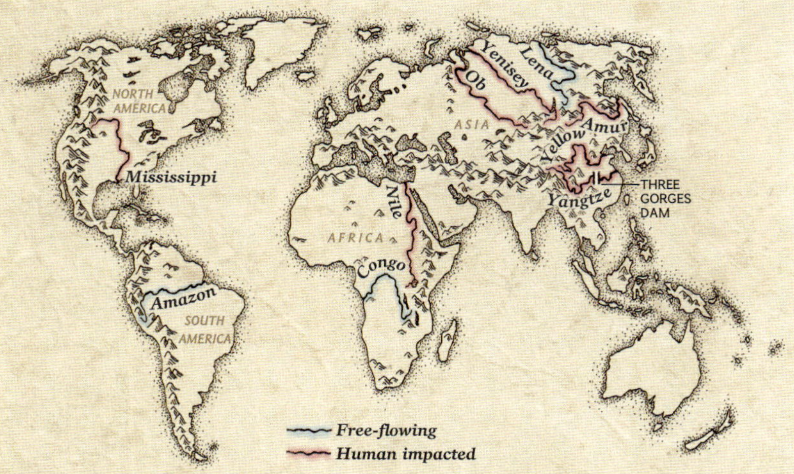

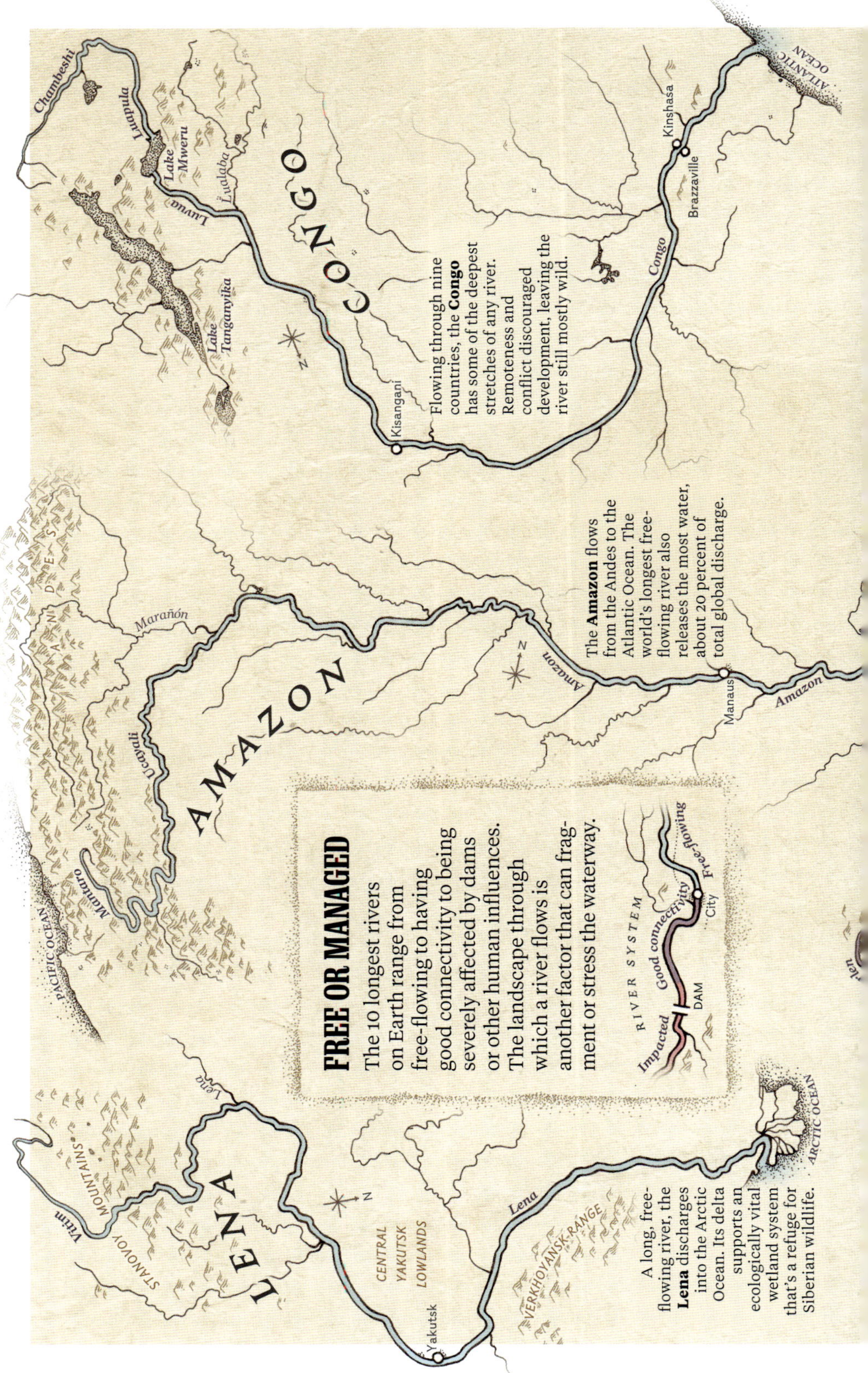

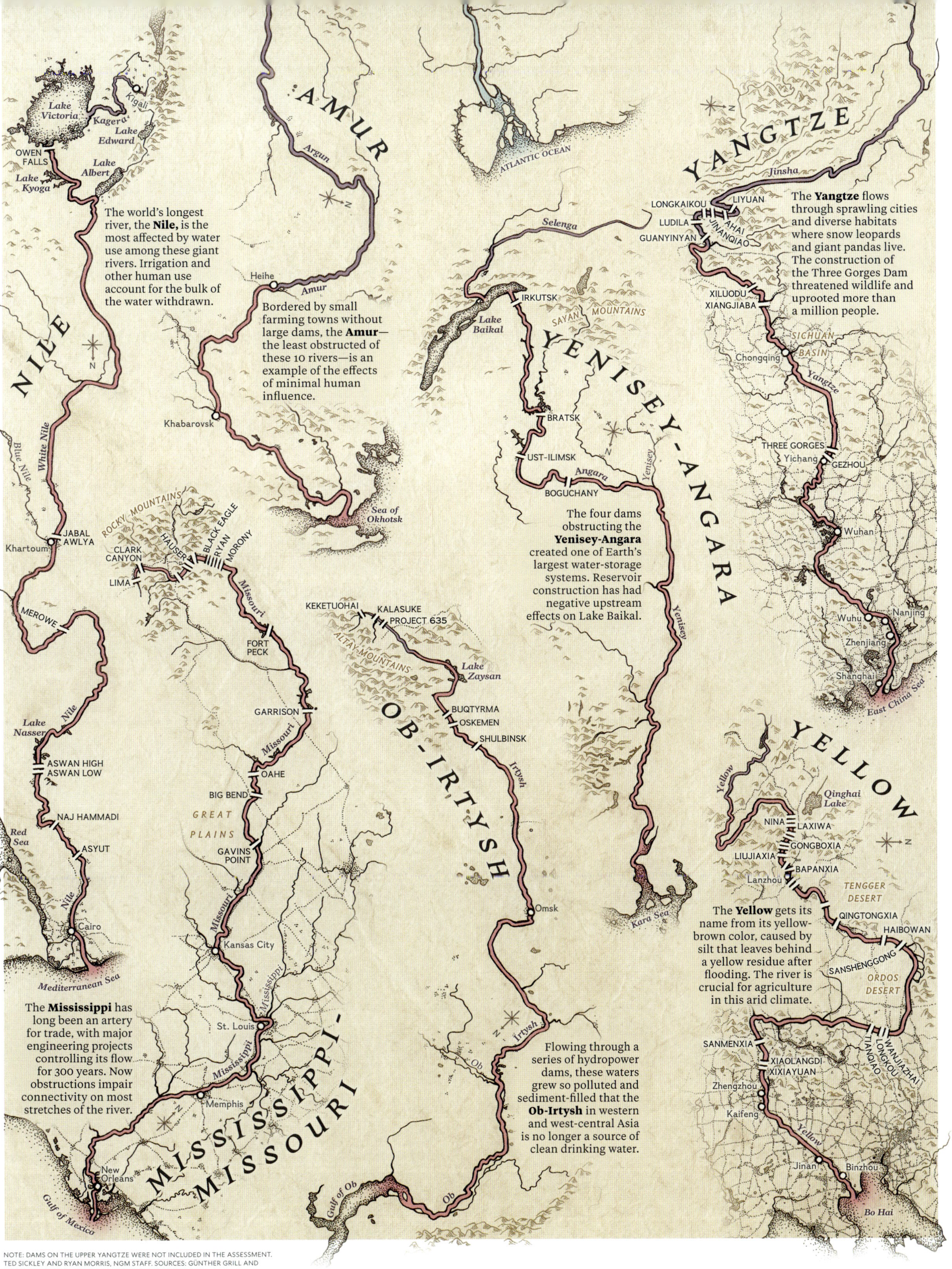

NOTE: DAMS ON THE UPPER YANGTZE WERE NOT INCLUDED IN THE ASSESSMENT.
TED SICKLEY AND RYAN MORRIS, NGM STAFF. SOURCES: GÜNTHER GRILL AND BERNHARD LEHNER, *NATURE*, MAY 2019; WALTER IMMERZEEL

Alex McPhee's Province of Alberta »

Alex McPhee, Prairie Heart Maps

I am Canadian, which means I am nearly indistinguishable from an Upper Midwesterner. But here's one of the minor details that sets us apart: the world's second-largest country does not have a very famous cartographic tradition. For this I blame our mercenary habit of heading south as soon as a prestigious job can be located in the United States. (Yet another thing that map designers have in common with Hollywood actors.)

"McPhee's Alberta" is meant to be an excellent Canadian map. This means it borrows from the American tradition, a lot. I owe a great deal to David Imus, whose "Essential Geography" series showed me that one-man passion projects can be commercially viable. Dave has a really important ideological point, which is that we can only save cartography from its Eurocentric origins by doing a better job ourselves. He also gave me the most useful practical tip of my career: make sure the label always physically resembles the feature it's matched to. That includes color, weight, and even the emotional properties of the typeface.

Now here are some more of the details that make this map uniquely Canadian. For the first time on any commercial map, we have the Numbered Treaties—the legal basis for western Canada's existence. County offices can exist outside of named localities, because our local government system doesn't follow US rules. The Athabasca Oil Sands, the Rat Control Zone, and the World's Largest Beaver Dam are some other prominent features that you'll only find in Alberta.

The last thing I have to say about Canadian culture? We are just as dry and sarcastic as the British. I won't give away every little joke in this map, but I'm very proud of using an intentionally disturbing wall of text to mark a restricted bombing range.

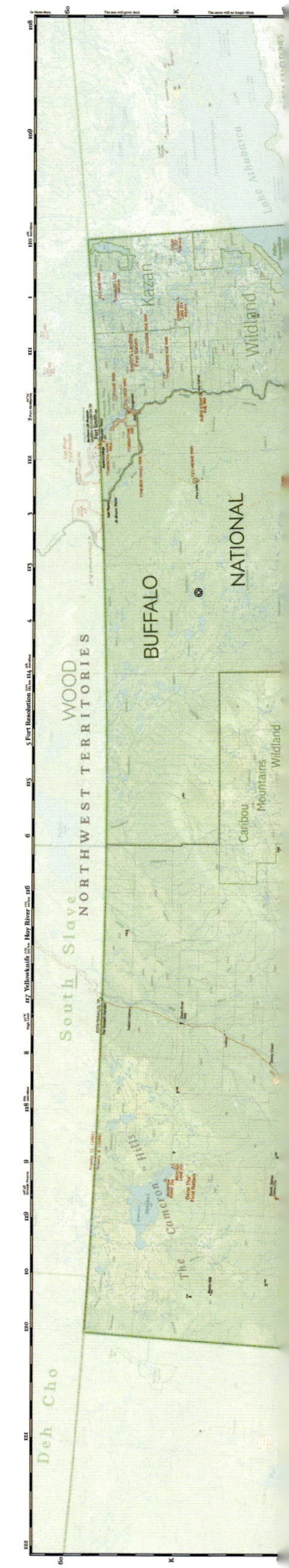

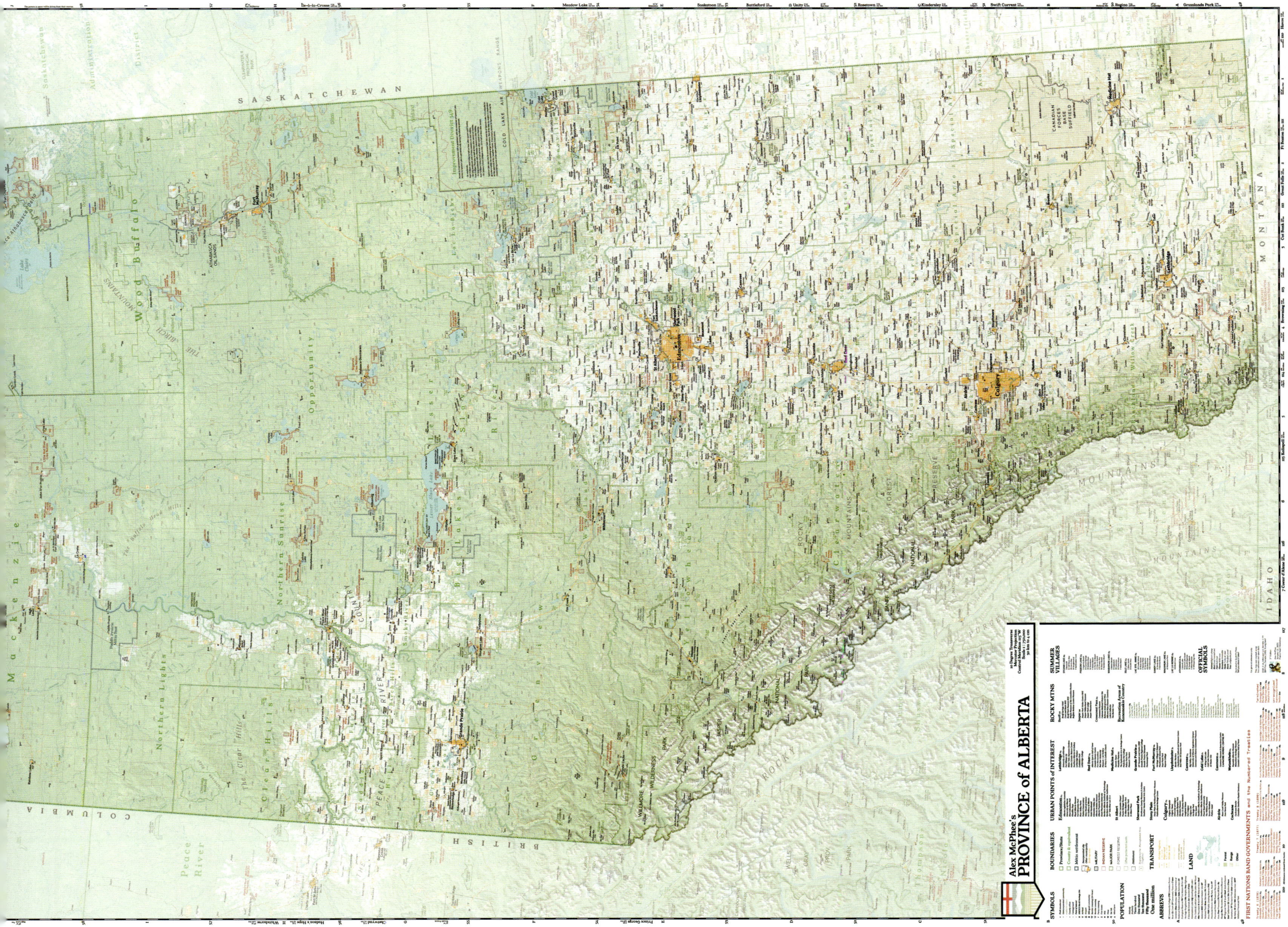
Alex McPhee's
PROVINCE of ALBERTA
SASKATCHEWAN
MONTANA
IDAHO
BRITISH
COLUMBIA
Wood Buffalo
Opportunity
Northern Sunrise
Northern Lights
Mackenzie
Clear Hills
Edmonton
Calgary
Grande Prairie
ROCKY
MOUNTAINS
SYMBOLS
BOUNDARIES
URBAN POINTS of INTEREST
ROCKY MTNS
SUMMER VILLAGES
POPULATION
TRANSPORT
LAND
ABBREVS
OFFICIAL SYMBOLS
FIRST NATIONS BAND GOVERNMENTS and the Numbered Treaties

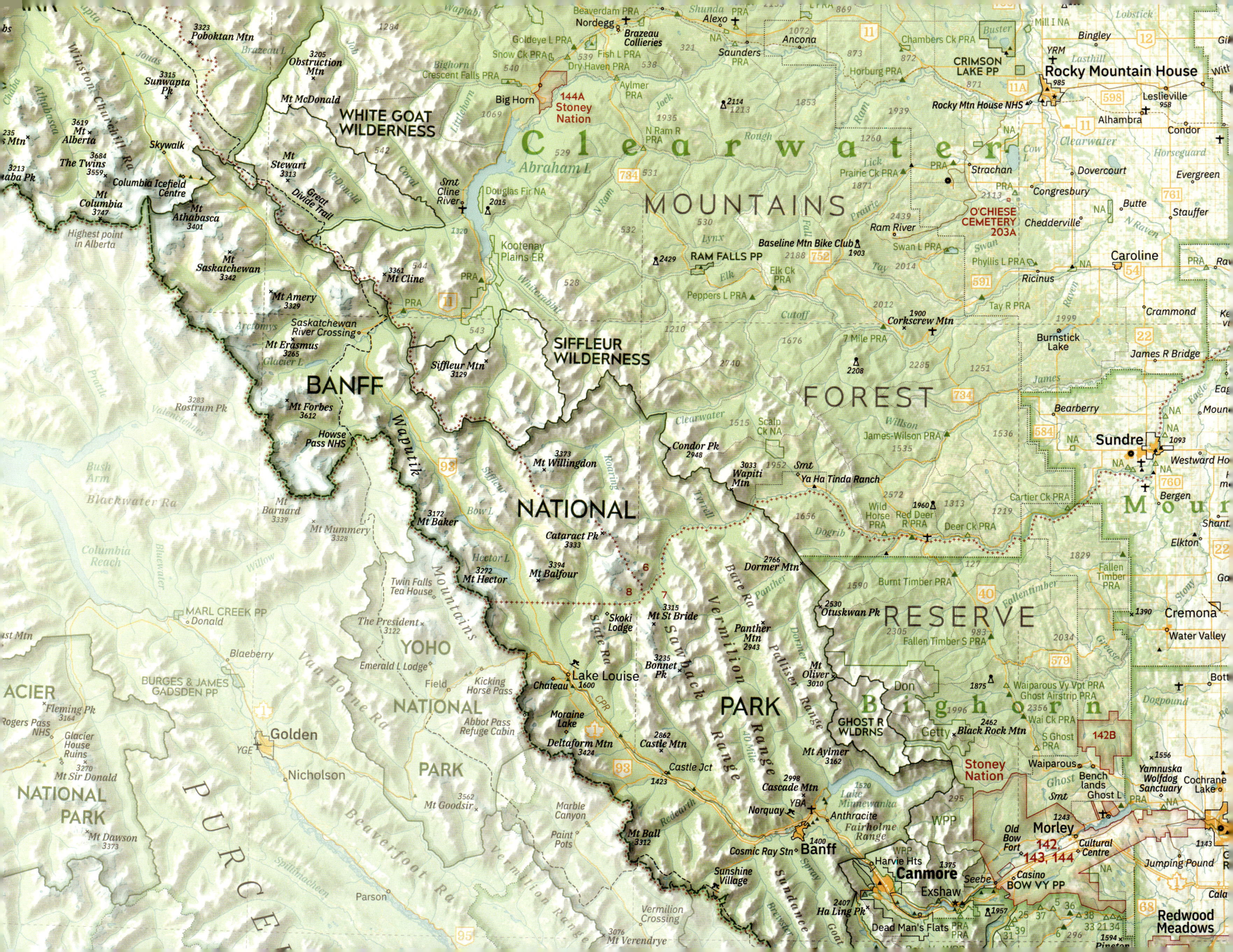

Pobokton Mtn
3323
Obstruction Mtn
3205
Sunwapta Pk
3315
Mt McDonald
WHITE GOAT WILDERNESS
Mt Alberta
3619
Skywalk
The Twins
3684
3559
Columbia Icefield Centre
Mt Stewart
3313
Great Divide Trail
Mt Columbia
3747
Mt Athabasca
3401
Highest point in Alberta
Smt
Cline River
Douglas Fir NA
2015
Kootenay Plains ER
Mt Saskatchewan
3342
Mt Cline
3361
Mt Amery
3329
Saskatchewan River Crossing
Mt Erasmus
3265
Glacier L
BANFF
Siffleur Mtn
3129
SIFFLEUR WILDERNESS
Rostrum Pk
3283
Mt Forbes
3612
Howse Pass NHS
Bush Arm
Blackwater Ra
Mt Barnard
3339
Mt Mummery
3328
Columbia Reach
Mt Willingdon
3373
Mt Baker
3172
NATIONAL
Cataract Pk
3333
Mt Hector
3272
Mt Balfour
3394
Twin Falls Tea House
MARL CREEK PP
Donald
The President
3122
YOHO
Blaeberry
Emerald L Lodge
Field
Kicking Horse Pass
BURGES & JAMES GADSDEN PP
NATIONAL
Abbot Pass Refuge Cabin
Fleming Pk
3164
Rogers Pass NHS
Glacier House Ruins
Golden
YGE
Mt Sir Donald
3270
Nicholson
NATIONAL PARK
PARK
Mt Goodsir
3562
Marble Canyon
Paint Pots
Mt Dawson
3373
PURCELL
Parson
Vermilion Crossing
Mt Verendrye
3076
Beaverdam PRA
Nordegg
Brazeau Collieries
Shunda
Alexo
Goldeye L PRA
Snow Ck PRA
Fish L PRA
Dry Haven PRA
Saunders PRA
Ancona
Crescent Falls PRA
Big Horn
144A Stoney Nation
Aylmer PRA
Horburg PRA
Chambers Ck PRA
CRIMSON LAKE PP
Rocky Mountain House
Rocky Mtn House NHS
Bingley
Leslieville
Alhambra
Condor
Clearwater
Abraham L
N Ram R PRA
Prairie Ck PRA
Strachan
Dovercourt
Evergreen
Congresbury
MOUNTAINS
O'CHIESE CEMETERY 203A
Butte
Stauffer
Chedderville
Ram River
Baseline Mtn Bike Club
Swan L PRA
RAM FALLS PP
Elk Ck PRA
Caroline
Phyllis L PRA
Ricinus
Peppers L PRA
Tay R PRA
Crammond
Corkscrew Mtn
7 Mile PRA
Burnstick Lake
James R Bridge
FOREST
Bearberry
Condor Pk
2948
Scalp Ck NA
James-Wilson PRA
Sundre
Westward Ho
Wapiti Mtn
3033
Ya Ha Tinda Ranch
Cartier Ck PRA
Bergen
Wild Horse PRA
Red Deer R PRA
Deer Ck PRA
Elkton
Dormer Mtn
2766
Burnt Timber PRA
Fallen Timber PRA
Otuskwan Pk
RESERVE
Cremona
Water Valley
Mt St Bride
3315
Skoki Lodge
Panther Mtn
2943
Fallen Timber S PRA
Bonnet Pk
3235
Mt Oliver
3010
Lake Louise
Chateau
Don
Bighorn
Waiparous Vy Vpt PRA
Ghost Airstrip PRA
Wai Ck PRA
PARK
Moraine Lake
Deltaform Mtn
3424
Castle Mtn
2862
GHOST R WLDRNS
Getty
Black Rock Mtn
142B
S Ghost PRA
Castle Jct
Mt Aylmer
3162
Stoney Nation
Waiparous
Bench lands
Ghost L
Yamnuska Wolfdog Sanctuary
Cochrane Lake
Cascade Mtn
2998
YBA
Norquay
Anthracite
Fairholme Range
Morley
Old Bow Fort
142, 143, 144
Cultural Centre
Mt Ball
3312
Banff
Cosmic Ray Stn
Harvie Hts
Canmore
Seebe
Casino
BOW VY PP
Jumping Pound
Sunshine Village
Exshaw
Ha Ling Pk
Dead Man's Flats
Redwood Meadows
Sawback Range
Vermilion Range
Palliser Range
Slate Ra
Waputik Mountains
Van Horne Ra
Beaverfoot Ra
Sundance

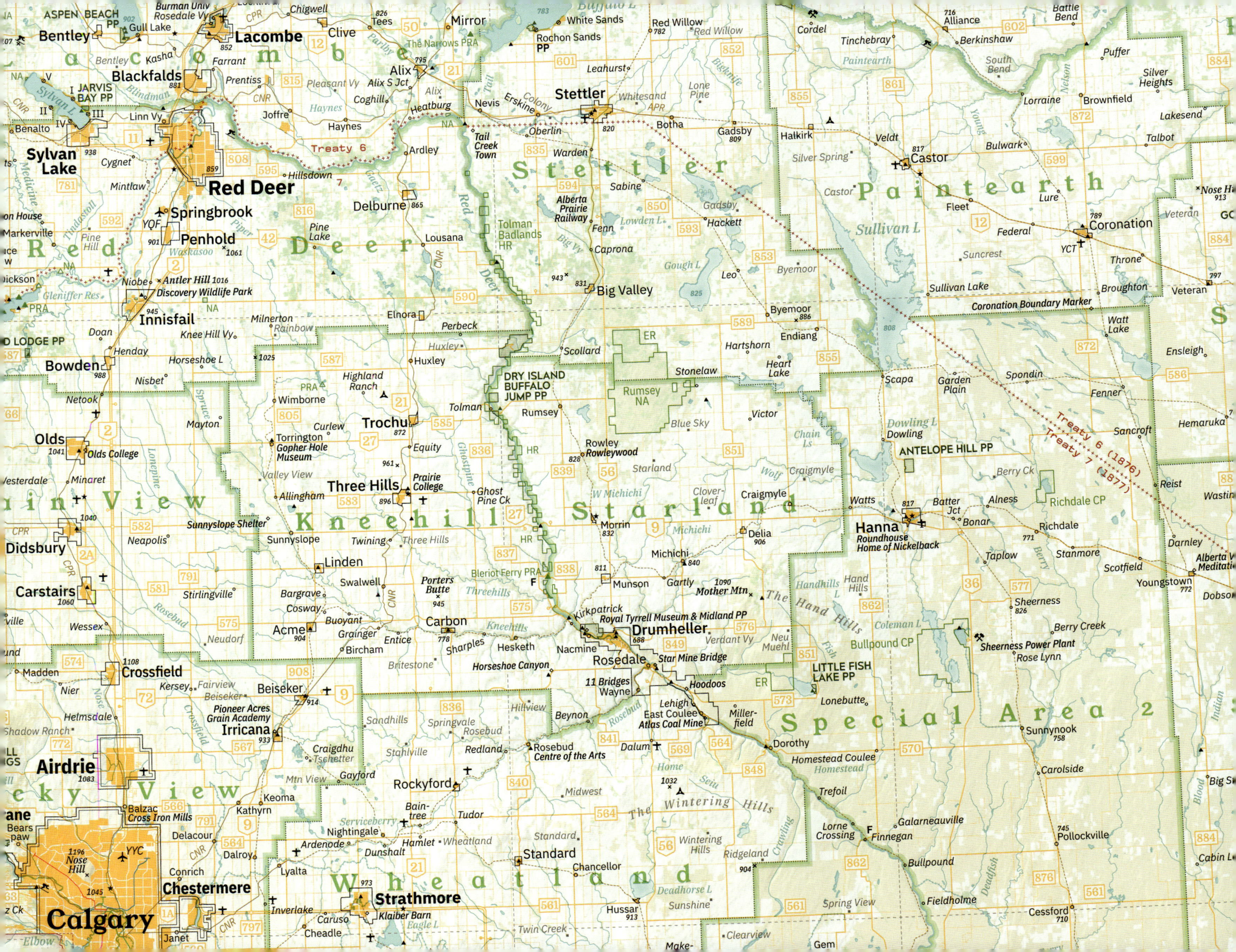

ASPEN BEACH PP
Bentley
Gull Lake
Burman Univ
Rosedale Vy
Lacombe
Chigwell
Clive
Tees
Mirror
White Sands
Red Willow
Cordel
Alliance
Battle Bend
Tinchebray
Berkinshaw
Puffer
Rochon Sands PP
The Narrows PRA
Blackfalds
Farrant
Prentiss
Alix
Alix S Jct
Leahurst
Paintearth
South Bend
Silver Heights
JARVIS BAY PP
Sylvan L
Linn Vy
Joffre
Pleasant Vy
Coghill
Heatburg
Nevis
Erskine
Stettler
Whitesand
Lone Pine
Lorraine
Brownfield
Lakesend
Benalto
Haynes
Treaty 6
Oberlin
Botha
Gadsby
Halkirk
Veldt
Castor
Bulwark
Talbot
Sylvan Lake
Cygnet
Red Deer
Hillsdown
Ardley
Tail Creek Town
Warden
Silver Spring
Mintlaw
Springbrook
Delburne
Stettler
Sabine
Castor
Paintearth
Nose Hill
Alberta Prairie Railway
Gadsby
Lure
Veteran
Markerville
Penhold
Pine Lake
Lousana
Tolman Badlands HR
Fenn
Hackett
Fleet
Coronation
Sullivan L
Federal
Red Deer
Caprona
Suncrest
YCT
Throne
Antler Hill
Discovery Wildlife Park
Niobe
Big Valley
Gough L
Leo
Byemoor
Sullivan Lake
Broughton
Veteran
Coronation Boundary Marker
Innisfail
Milnerton
Elnora
Byemoor
Endiang
Watt Lake
Doan
Knee Hill Vy
Perbeck
Hartshorn
Ensleigh
LODGE PP
Henday
Huxley
Scollard
Heart Lake
Bowden
Horseshoe L
Nisbet
DRY ISLAND BUFFALO JUMP PP
Stonelaw
Scapa
Garden Plain
Spondin
Rumsey NA
Netook
Wimborne
Highland Ranch
Rumsey
Fenner
Tolman
Mayton
Curlew
Trochu
Blue Sky
Victor
Chain Ls
Dowling L
Dowling
Sancroft
Hemaruka
Olds
Olds College
Torrington
Gopher Hole Museum
Equity
Rowley
Rowleywood
Treaty 6 (1876)
Treaty 7 (1877)
ANTELOPE HILL PP
Minaret
Starland
Craigmyle
Berry Ck
Valley View
Prairie College
Three Hills
Ghost Pine Ck
Allingham
Mountain View
Kneehill
Starland
Clover-leaf
Craigmyle
Watts
Batter Jct
Alness
Richdale CP
Reist
Wastina
Hanna
Roundhouse
Home of Nickelback
Bonar
Richdale
Sunnyslope Shelter
Neapolis
Sunnyslope
Twining
Morrin
Michichi
Delia
Darnley
Didsbury
Taplow
Stanmore
Scotfield
Linden
Swalwell
Michichi
Gartly
Mother Mtn
Youngstown
Alberta Meditati
Carstairs
Stirlingville
Bargrave
Porters Butte
Threehills
Bleriot Ferry PRA
Munson
Handhills L
Hand Hills
Sheerness
Dobson
Coswav
Kirkpatrick
Royal Tyrrell Museum & Midland PP
The Hand Hills
Berry Creek
Wessex
Acme
Buoyant
Carbon
Kneehills
Drumheller
Verdant Vy
Coleman L
Bullpound CP
Sheerness Power Plant
Neudorf
Grainger
Entice
Sharples
Hesketh
Nacmine
Rosedale
Star Mine Bridge
Neu Muehl
Rose Lynn
Bircham
Britestone
Horseshoe Canyon
LITTLE FISH LAKE PP
Madden
Crossfield
Kersey
Fairview
Beiseker
Hoodoos
11 Bridges
Wayne
Lonebutte
Nier
Beiseker
Pioneer Acres Grain Academy
Sandhills
Springvale
Hillview
Lehigh
East Coulee
Atlas Coal Mine
Millerfield
Special Area 2
Helmsdale
Irricana
Rosebud
Beynon
Rosebud
Sunnynook
Shadow Ranch
Craigdhu
Tschetter
Stahlville
Redland
Rosebud Centre of the Arts
Dalum
Dorothy
Homestead Coulee
Homestead
Carolside
Airdrie
Rockyford
Mtn View
Gayford
Home
Seiu
Trefoil
Rocky View
Keoma
Midwest
Kathyrn
Balzac
Cross Iron Mills
Serviceberry
Bain-tree
Tudor
The Wintering Hills
Galarneauville
Bears paw
Nightingale
Lorne Crossing
Finnegan
Pollockville
Delacour
Ardenode
Dunshalt
Hamlet
Wheatland
Standard
Wintering Hills
Ridgeland
Crawling
YYC
Nose Hill
Dalroy
Standard
Chancellor
Bullpound
Conrich
Lyalta
Wheatland
Deadfish
Cabin L
Chestermere
Strathmore
Deadhorse L
Sunshine
Spring View
Fieldholme
Calgary
Inverlake
Kliaber Barn
Hussar
Cessford
Caruso
Eagle L
Janet
Cheadle
Twin Creek
Clearview
Elbow
Make-
Gem

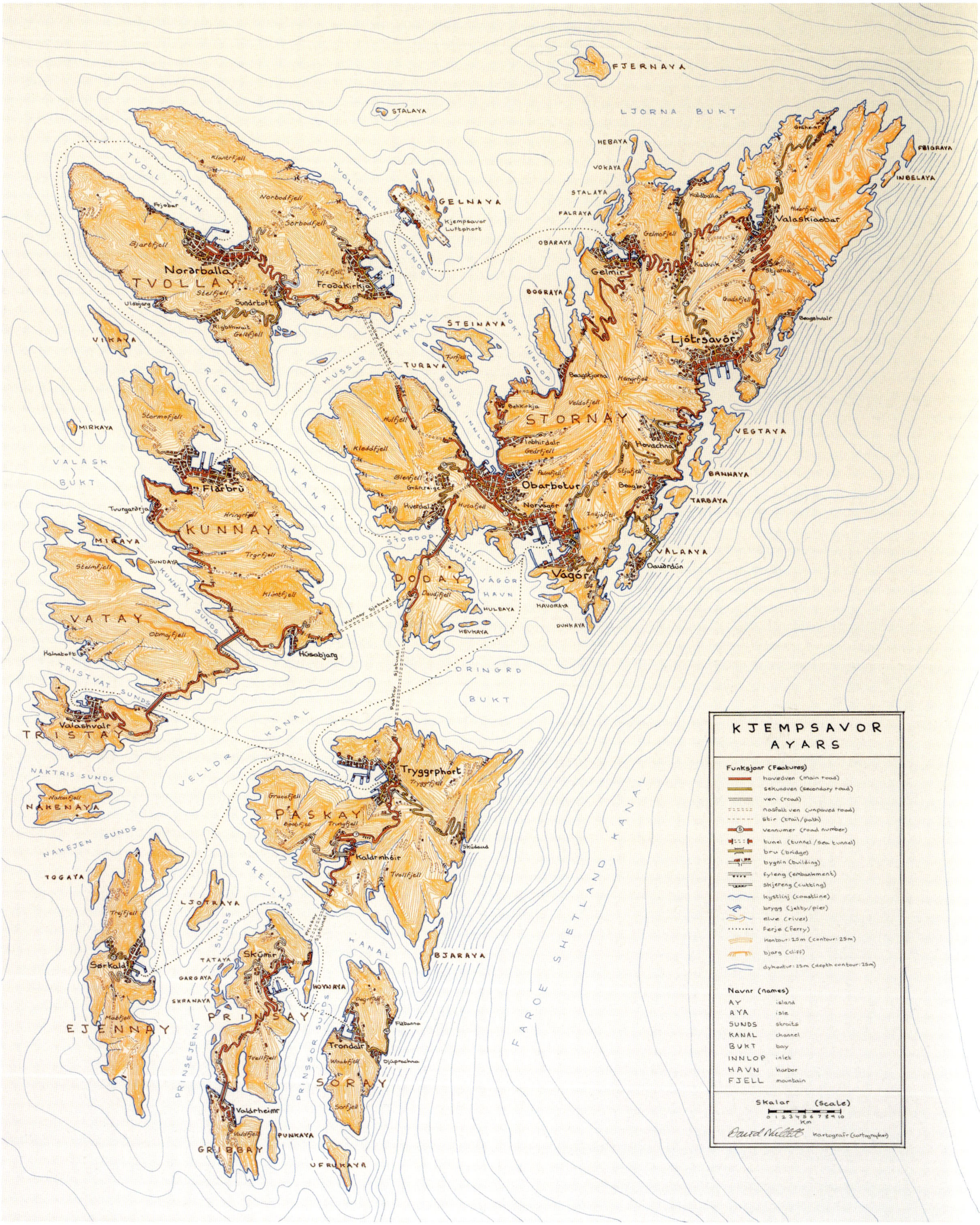
KJEMPSAVOR
AYARS
Funksjonr (Features)
hovedven (main road)
sekundven (secondary road)
ven (road)
nasfalt ven (unpaved road)
stir (trail/path)
vennumer (road number)
tunel (tunnel/sea tunnel)
bru (bridge)
bygnin (building)
fyleng (embankment)
skjereng (cutting)
kystlinj (coastline)
brygg (jetty/pier)
elve (river)
Ferje (Ferry)
kontour: 25m (contour: 25m)
bjarg (cliff)
dykontur: 25m (depth contour: 25m)
Navnr (names)
AY island
AYA isle
SUNDS straits
KANAL channel
BUKT bay
INNLOP inlet
HAVN harbor
FJELL mountain
Skalar (Scale)
0 1 2 3 4 5 6 7 8 9 10
Km
Kartografr (cartographer)
FJERNAYA
STALAYA
LJORNA BUKT
HEBAYA
VOKAYA
FBIGRAYA
INBELAYA
GELNAYA
Kjempsavor Luftphort
TVOLLGELN SUNDS
TVOLL HAVN
TVOLLAY
Norarballa
Frodakirkja
Sundrtoft
VIKAYA
STALAYA
FALRAYA
OBARAYA
BOGRAYA
Gelmir
Valaskiaobar
Ljótrsavör
STEINAYA
NOKT INNLOP
TURAYA
BOTUR INNLOP
STORNAY
VEGTAYA
BANNAYA
TARBAYA
VALAAYA
Dauðrdún
Obarbotur
Vágór
HAVORAYA
DUNKAYA
MIRKAYA
VALASK BUKT
RIGHDR KANAL
HUSSLR KANAL
STORDOP SUNDS
DODAY
VÁGÓR HAVN
HULBAYA
HEVNAYA
Flárbrú
KUNNAY
MIRAYA
SUNDAYA
KUNNVAT SUNDS
VATAY
Húsabjarg
DRINGRD BUKT
TRISTVAT SUNDS
Valashvalr
TRISTAY
VELLDR KANAL
NAKTRIS SUNDS
NAKENAYA
NAKEJEN SUNDS
TOGAYA
Tryggrphort
PASKAY
Kaldrmhóir
FAROE SHETLAND KANAL
BJARAYA
LJOTRAYA
SKELLMR SUNDS
TATAYA
GARGAYA
SKRANAYA
Skúmr
PRINSAY
HOVNAYA
Sørkald
EJENNAY
PRINSEJENN SUNDS
PRINSSOR SUNDS
Trondar
SORAY
Valdrheimr
GRIBBAY
PUNKAYA
UFRUKAYA

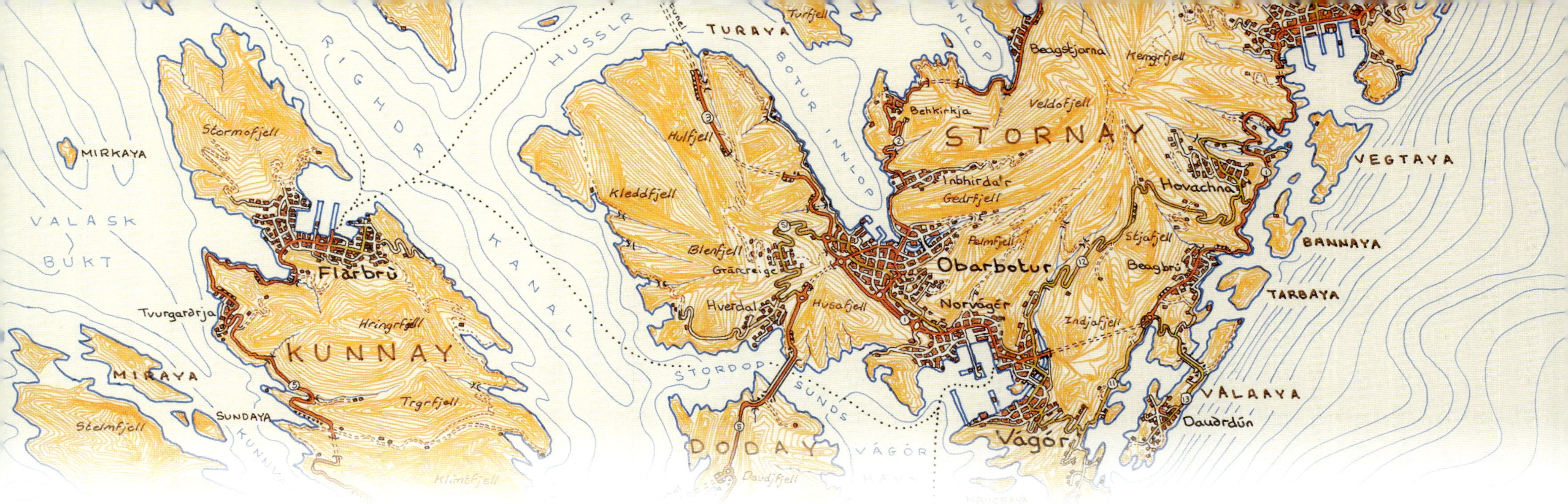

« Kjempsavor Ayars

David Nuttall, Artimaps

This hand-drawn, plausible fictitious map was inspired by the history, landscape, and people of the Faroe Islands. I have been making maps of fictional places since the age of five, and I get inspiration from the real world. For this map, I read about a new undersea tunnel in the Faroe Islands and became fascinated with the islands, their culture, and connectivity. I wanted to create a "what if" map that suggested what a fictional island-nation might look like if seamounts on the edge of the Faroe-Shetland Channel formed a landmass above the water.

In order to create a realistic setting, I researched language, culture, legends, history, engineering, and landscapes in the North Atlantic. I wanted to have a plausible setting and civilization that had a distinct background, economy, culture, and reasons for features depicted on the map. The language is semifictional, a mix of Old Norse and Old Scots, and hints at the imagined history of this landscape and the roughly translated country name "Land of the Giant Sheep."

Using pen on board, I created this map on the fly, drawing the coastline and water features first, and then the contour elevation lines, before adding the primary place-names and features. Without the ability to delete anything or use halo effects, some minimal planning for the main features is required, such as leaving gaps in the coastline for bridges. The contemporary design of the map highlights the elevation and how human development has merged with the rugged terrain. The connectivity of roads, tunnels, and ferries features prominently. This was a major factor in the color choices for the contours, roads, and buildings. I wanted the focus to be on the islands, so decided against large, filled water areas.

I want the map to work as art, but also as something that draws the viewer into the details to study the names and features of this fictional landscape—all while imagining it could be real.

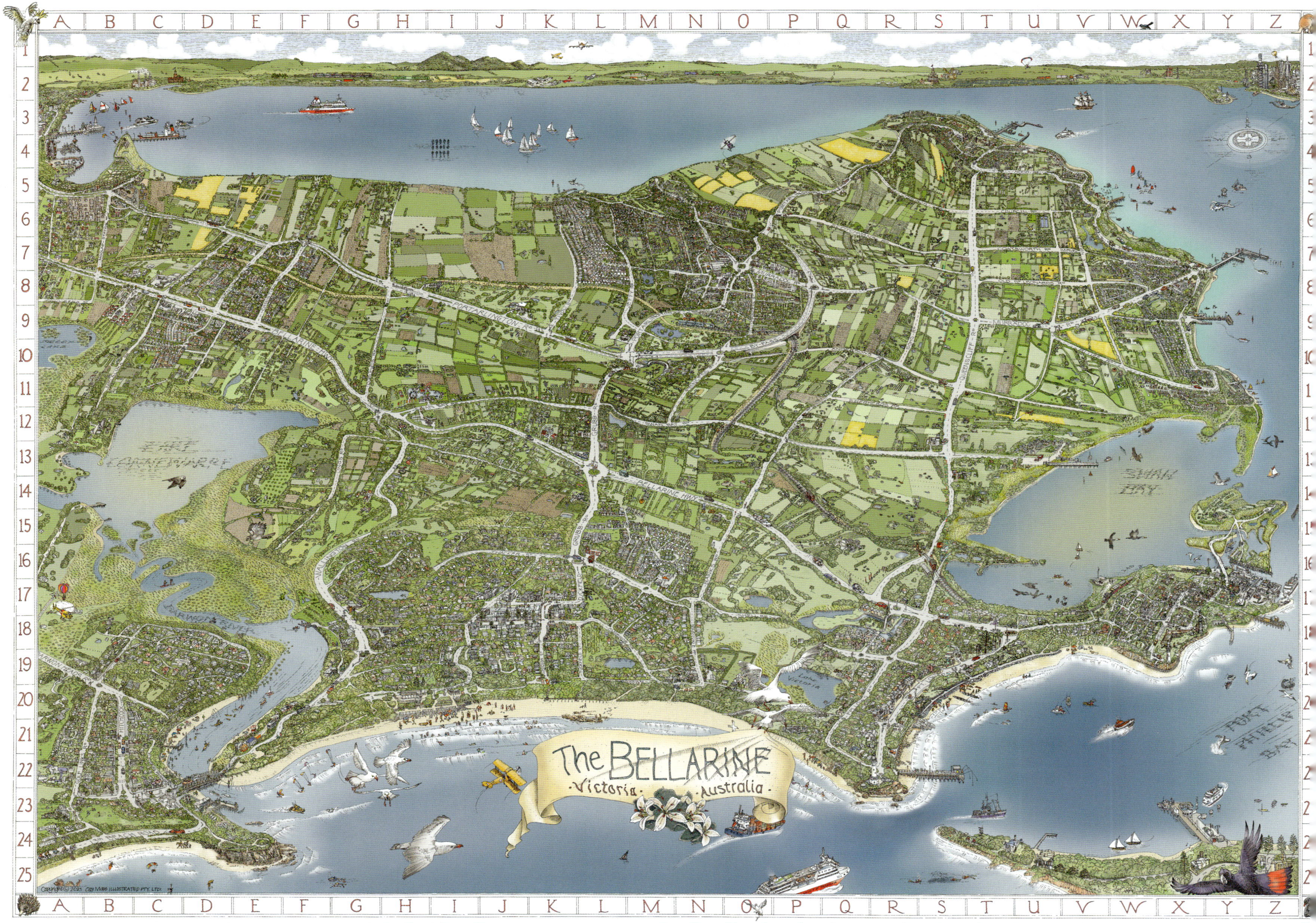

The BELLARINE
Victoria
Australia
SWAN BAY
PORT PHILLIP BAY
LAKE CONNEWARRE
BELLARINE HWY

《 The Bellarine Map

Melinda Clarke, Deborah Young-Monk, and Sean Rodwell, City Maps Illustrated Pty Ltd

The Bellarine Peninsula is on Wadawurrung Country, located southwest of Melbourne, Victoria, Australia. Melinda moved there twenty-six years ago and has always dreamed of creating a pictorial map and historical document showcasing this beautiful part of the world.

Deborah (illustrator) has also enjoyed vacations here for most of her life and is equally familiar with the Bellarine.

Working as a team, they began by studying published and online data, and laying down the geographical outline and roadways. Over the three years of research and illustration, they incorporated the detail, illustrating the beautiful coastal vistas and everything in between.

The perspective, scale, and content are critical to the map's design. Character elements, such as wildlife, people (and their activities), boats, and other interesting features discovered during the research, inject interest into the map and creatively make the illustration an engaging project for the artist in what otherwise seems endlessly detailed and daunting.

When some detail is included in the foreground, it can be "read" and accepted as oversized but not out of character. The birds and the various aircraft can indicate height above the surface and give the drawing more depth and interest as the viewer adjusts their perception of space simply because the object is drawn in a certain way, at a specific size. Having that kind of freedom, playfulness, and artistic license in such a big and challenging project is very engaging.

The original line drawing was scanned, a high-resolution TIFF file created, then changes and additions were incorporated. Sean (digital colorist) researched photos and satellite imagery to ensure the accuracy of coloring, which took a total of five months to complete.

While ensuring the project's end vision is achieved, our process considered the artistic integrity of each contributor, and helped us maintain a high level of interest and enjoyment.

This approach transfers to the map's quality and, ultimately, the viewer's enjoyment. There is so much to discover that only a dedicated viewer will see every detail.

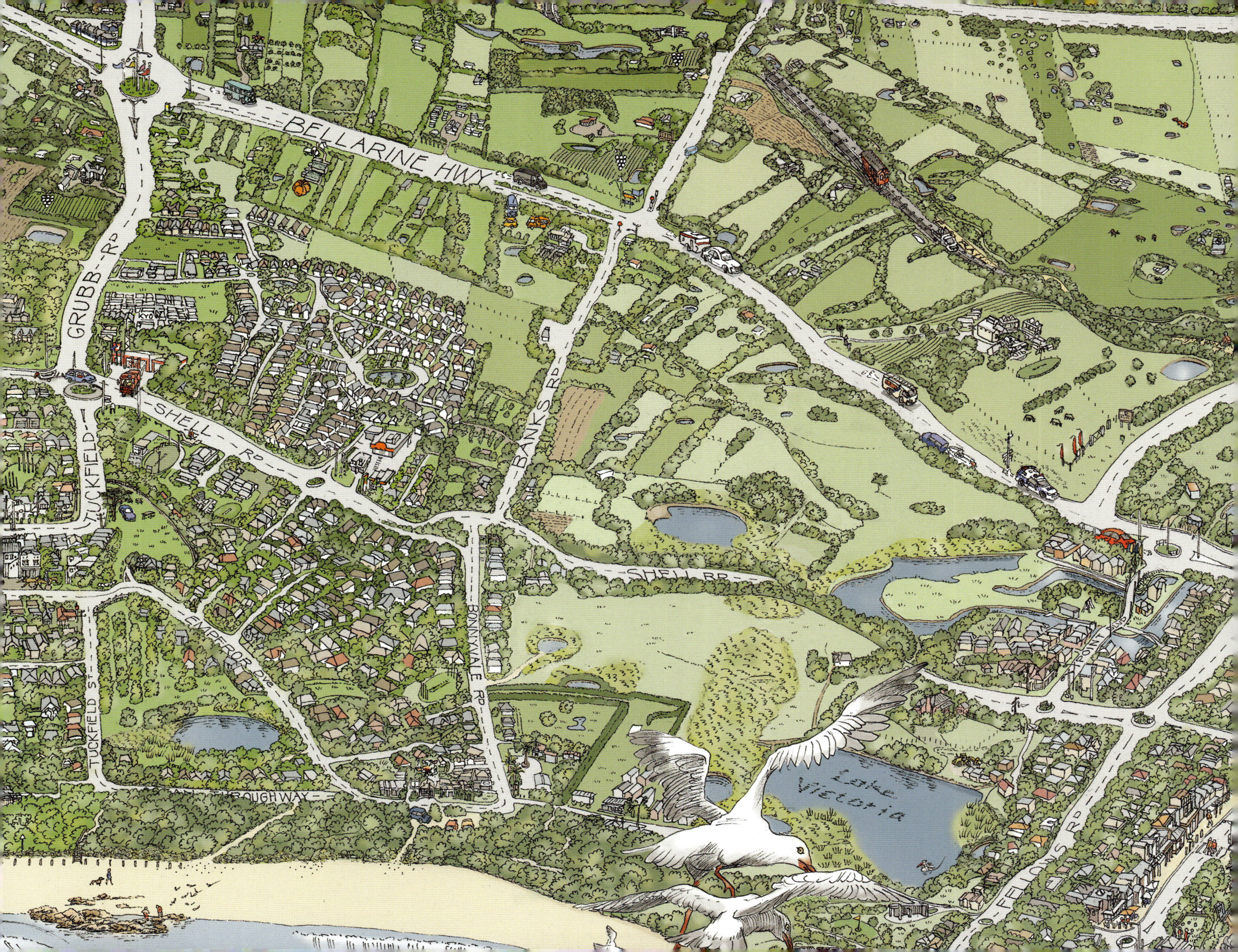
BELLARINE HWY
GRUBB RD
SHELL RD
BANKS RD
TUCKFIELD
TUCKFIELD ST
EMPEROR Dve
BONNYVALE RD
ROUGHWAY
SHELL RD
Lake Victoria
FELLOWS RD

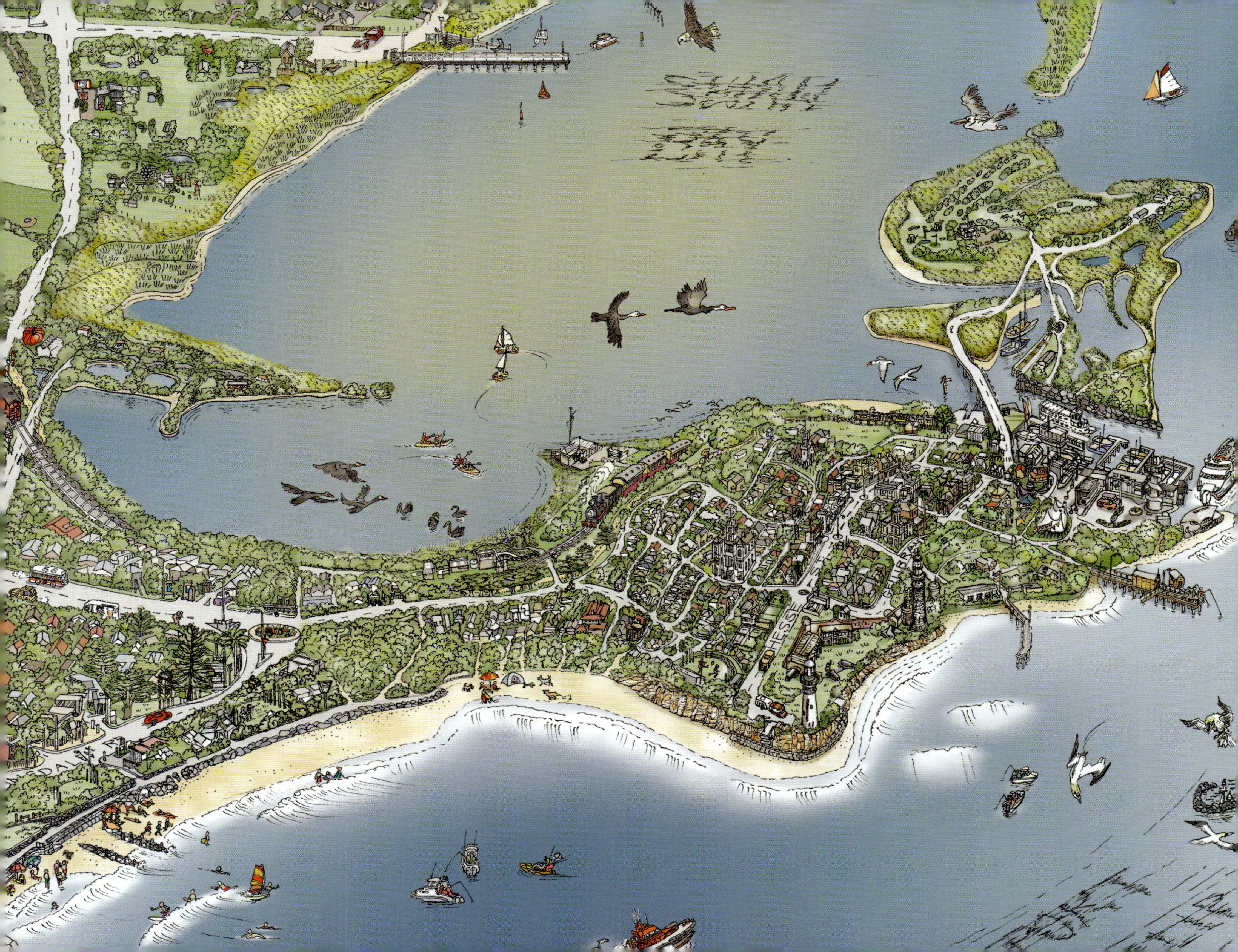
SWAN BAY
HESSE ST

The best designs, however, often contain something the viewer doesn't expect.

Slivers of an Ancient Forest »

Jake Steinberg

The northwoods are as fundamental to the cultural identity of the Upper Midwest as lakes and mosquitoes. When I was growing up in the Twin Cities, each summer promised trips north to cabin country, which meant long hours in the car, traveling roads flanked by dense timberland. Admittedly, Minnesota's forests are rarely breathtaking; the evergreens seldom tower overhead, and the understory is often a shrubby thicket. But I found grandeur in the forests' impenetrability. A wolf or moose could emerge from the brush and just as suddenly vanish like a phantom. What other secrets did the forests hold?

I found the answer in visiting a stand of 300-year-old pines at Itasca State Park. This remnant revealed what the forest had once been: a place where conifers reached more than 100 feet into the sky; where little light penetrated to the mossy understory; decaying logs formed microclimates and prevented erosion; and abundant biodiversity supported the Indigenous Ojibwe people. Loggers destroyed most of this original forest between 1890 and 1920. In truth, the woods I thought I knew weren't wholly different from the fields of corn and soybeans that supplanted the state's prairies.

I made "Slivers of an Ancient Forest" to tell the story of what the northwoods used to be, as well as what they might become. I took after *National Geographic* posters, especially their emphasis on embedding a narrative into the map. Research revealed no shortage of interesting vignettes. Exhaustive clearcutting of the forest left no tree spared without cause. An ancient grove of hemlocks in Wisconsin, known as Cathedral Pines, only stands today because the local lumber magnate's wife once held her Bible study there.

I aim to evoke a map's subject matter in each element of its design. In this case, that meant styling the graduated symbols that represent average tree age to look like tree rings—tapping into a natural cue viewers are poised to recognize. I likewise used leaf illustrations in the legend. The best designs, however, often contain something the viewer doesn't expect. Thus, I used vibrant color hues to capture the eye and invite the viewer to see the northwoods in a new way.

SLIVERS OF AN ANCIENT FOREST

The northwoods of the Upper Midwest were once covered in forests of fairytale proportion. Centuries-old pines towered a hundred feet above a biologically rich understory where little light penetrated. Reflecting on his first sight of the forest in 1890, Minnesota lumberman Richard Louis Griffin proclaimed, "I will never forget that sight or the impression it left upon my mind, as I stood there gazing upon this wonderful forest—inexhaustible, enough to last for ages as I thought at that time. Yet, within the course of a very few years not to exceed fifteen, this great forest was laid bare, leaving only a few scattering stands of pine in patches here and there." By 1920, lumber companies had exhausted the northwoods and moved on to the Pacific Northwest. Only 1% of these original forests remain. However, the forest is growing back. Previously clear-cut land is now protected from logging, but threats from climate change and invasive species threaten the forest's recovery.

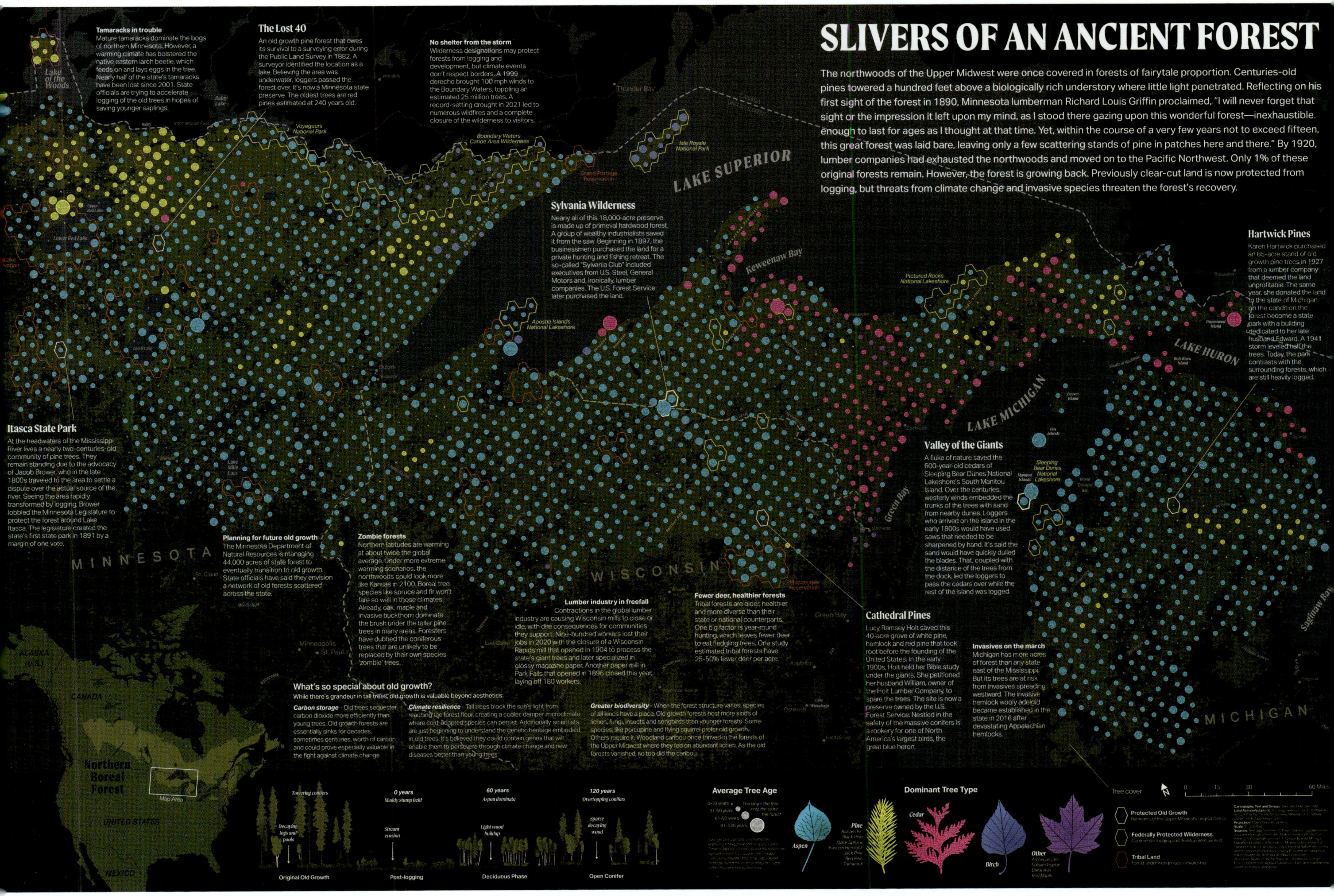

Tamaracks in trouble

Mature tamaracks dominate the bogs of northern Minnesota. However, a warming climate has bolstered the native eastern larch beetle, which feeds on and lays eggs in the tree. Nearly half of the state's tamaracks have been lost since 2001. State officials are trying to accelerate logging of the old trees in hopes of saving younger saplings.

The Lost 40

An old growth pine forest that owes its survival to a surveying error during the Public Land Survey in 1882. A surveyor identified the location as a lake. Believing the area was underwater, loggers passed the forest over. It's now a Minnesota state preserve. The oldest trees are red pines estimated at 240 years old.

No shelter from the storm

Wilderness designations may protect forests from logging and development, but climate events don't respect borders. A 1999 derecho brought 100 mph winds to the Boundary Waters, toppling an estimated 25 million trees. A record-setting drought in 2021 led to numerous wildfires and a complete closure of the wilderness to visitors.

Sylvania Wilderness

Nearly all of this 18,000-acre preserve is made up of primeval hardwood forest. A group of wealthy industrialists saved it from the saw. Beginning in 1897, the businessmen purchased the land for a private hunting and fishing retreat. The so-called "Sylvania Club" included executives from U.S. Steel, General Motors and, ironically, lumber companies. The U.S. Forest Service later purchased the land.

Hartwick Pines

Karen Hartwick purchased an 85-acre stand of old growth pine trees in 1927 from a lumber company that deemed the land unprofitable. The same year, she donated the land to the state of Michigan on the condition the forest become a state park with a building dedicated to her late husband Edward. A 1941 storm leveled half the trees. Today, the park contrasts with the surrounding forests, which are still heavily logged.

Itasca State Park

At the headwaters of the Mississippi River lives a nearly two-centuries-old community of pine trees. They remain standing due to the advocacy of Jacob Brower, who in the late 1800s traveled to the area to settle a dispute over the actual source of the river. Seeing the area rapidly transformed by logging, Brower lobbied the Minnesota Legislature to protect the forest around Lake Itasca. The legislature created the state's first state park in 1891 by a margin of one vote.

Planning for future old growth

The Minnesota Department of Natural Resources is managing 44,000 acres of state forest to eventually transition to old growth. State officials have said they envision a network of old forests scattered across the state.

Zombie forests

Northern latitudes are warming at about twice the global average. Under more extreme warming scenarios, the northwoods could look more like Kansas in 2100. Boreal tree species like spruce and fir won't fare so well in those climates. Already, oak, maple and invasive buckthorn dominate the brush under the taller pine trees in many areas. Foresters have dubbed the coniferous trees that are unlikely to be replaced by their own species 'zombie' trees.

Lumber industry in freefall

Contractions in the global lumber industry are causing Wisconsin mills to close or idle, with dire consequences for communities they support. Nine-hundred workers lost their jobs in 2020 with the closure of a Wisconsin Rapids mill that opened in 1904 to process the state's giant trees and later specialized in glossy magazine paper. Another paper mill in Park Falls that opened in 1896 closed this year, laying off 180 workers.

Fewer deer, healthier forests

Tribal forests are older, healthier and more diverse than their state or national counterparts. One big factor is year-round hunting, which leaves fewer deer to eat fledgling trees. One study estimated tribal forests have 25-50% fewer deer per acre.

Valley of the Giants

A fluke of nature saved the 600-year-old cedars of Sleeping Bear Dunes National Lakeshore's South Manitou Island. Over the centuries, westerly winds embedded the trunks of the trees with sand from nearby dunes. Loggers who arrived on the island in the early 1800s would have used saws that needed to be sharpened by hand. It's said the sand would have quickly dulled the blades. That, coupled with the distance of the trees from the dock, led the loggers to pass the cedars over while the rest of the island was logged.

Cathedral Pines

Lucy Ramsey Holt saved this 40-acre grove of white pine, hemlock and red pine that took root before the founding of the United States. In the early 1900s, Holt held her Bible study under the giants. She petitioned her husband William, owner of the Holt Lumber Company, to spare the trees. The site is now a preserve owned by the U.S. Forest Service. Nestled in the safety of the massive conifers is a rookery for one of North America's largest birds, the great blue heron.

Invasives on the march

Michigan has more acres of forest than any state east of the Mississippi. But its trees are at risk from invasives spreading westward. The invasive hemlock wooly adelgid became established in the state in 2016 after devastating Appalachian hemlocks.

What's so special about old growth?

While there's grandeur in tall trees, old growth is valuable beyond aesthetics:

Carbon storage - Old trees sequester carbon dioxide more efficiently than young trees. Old growth forests are essentially sinks for decades, sometimes centuries, worth of carbon and could prove especially valuable in the fight against climate change.

Climate resilience - Tall trees block the sun's light from reaching the forest floor, creating a cooler, damper microclimate where cold-adapted species can persist. Additionally, scientists are just beginning to understand the genetic heritage embodied in old trees. It's believed they could contain genes that will enable them to persevere through climate change and new diseases better than young trees.

Greater biodiversity - When the forest structure varies, species of all kinds have a place. Old growth forests host more kinds of lichen, fungi, insects and songbirds than younger forests. Some species, like porcupine and flying squirrel prefer old growth. Others require it. Woodland caribou once thrived in the forests of the Upper Midwest where they fed on abundant lichen. As the old forests vanished, so too did the caribou.

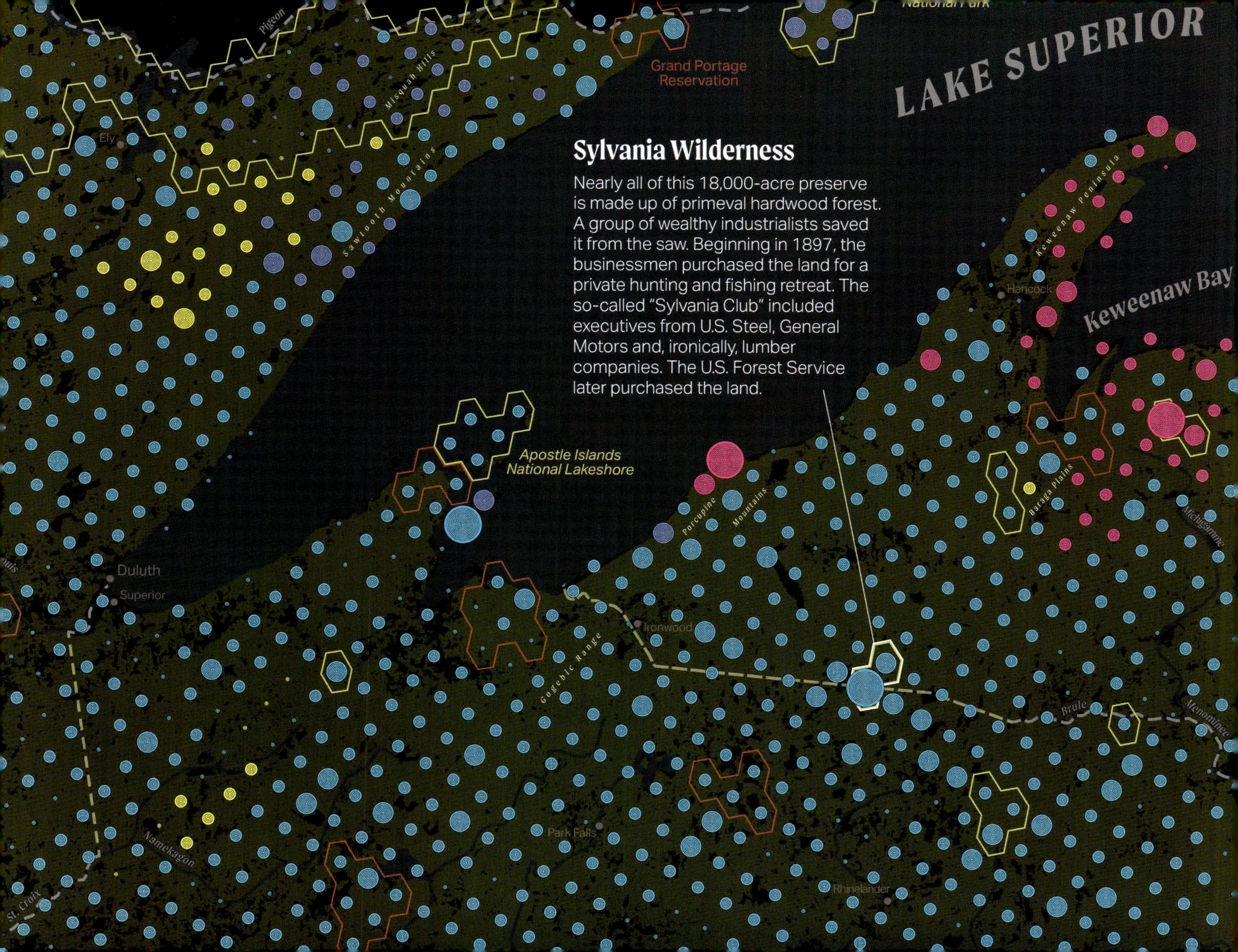

Sylvania Wilderness

Nearly all of this 18,000-acre preserve is made up of primeval hardwood forest. A group of wealthy industrialists saved it from the saw. Beginning in 1897, the businessmen purchased the land for a private hunting and fishing retreat. The so-called "Sylvania Club" included executives from U.S. Steel, General Motors and, ironically, lumber companies. The U.S. Forest Service later purchased the land.

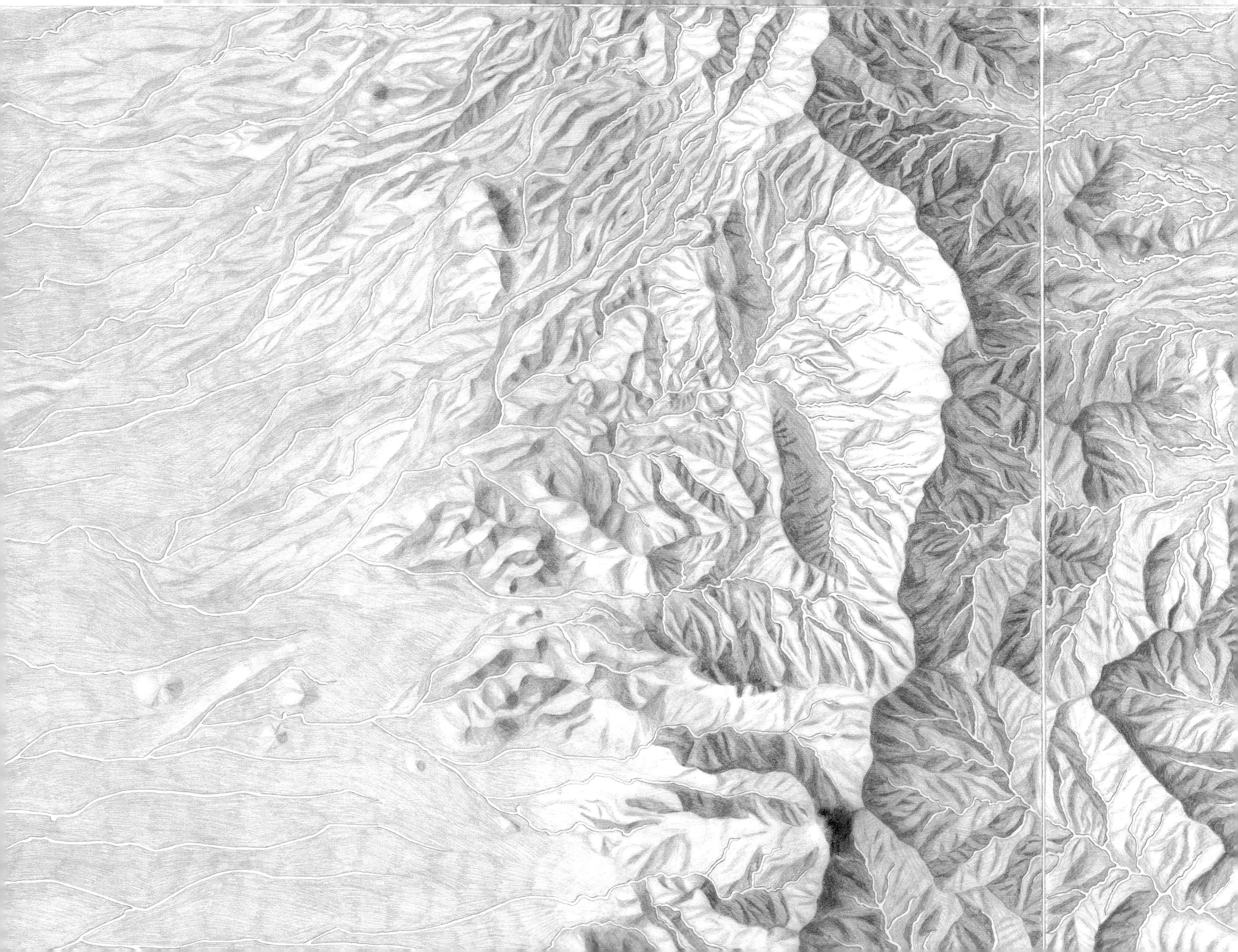

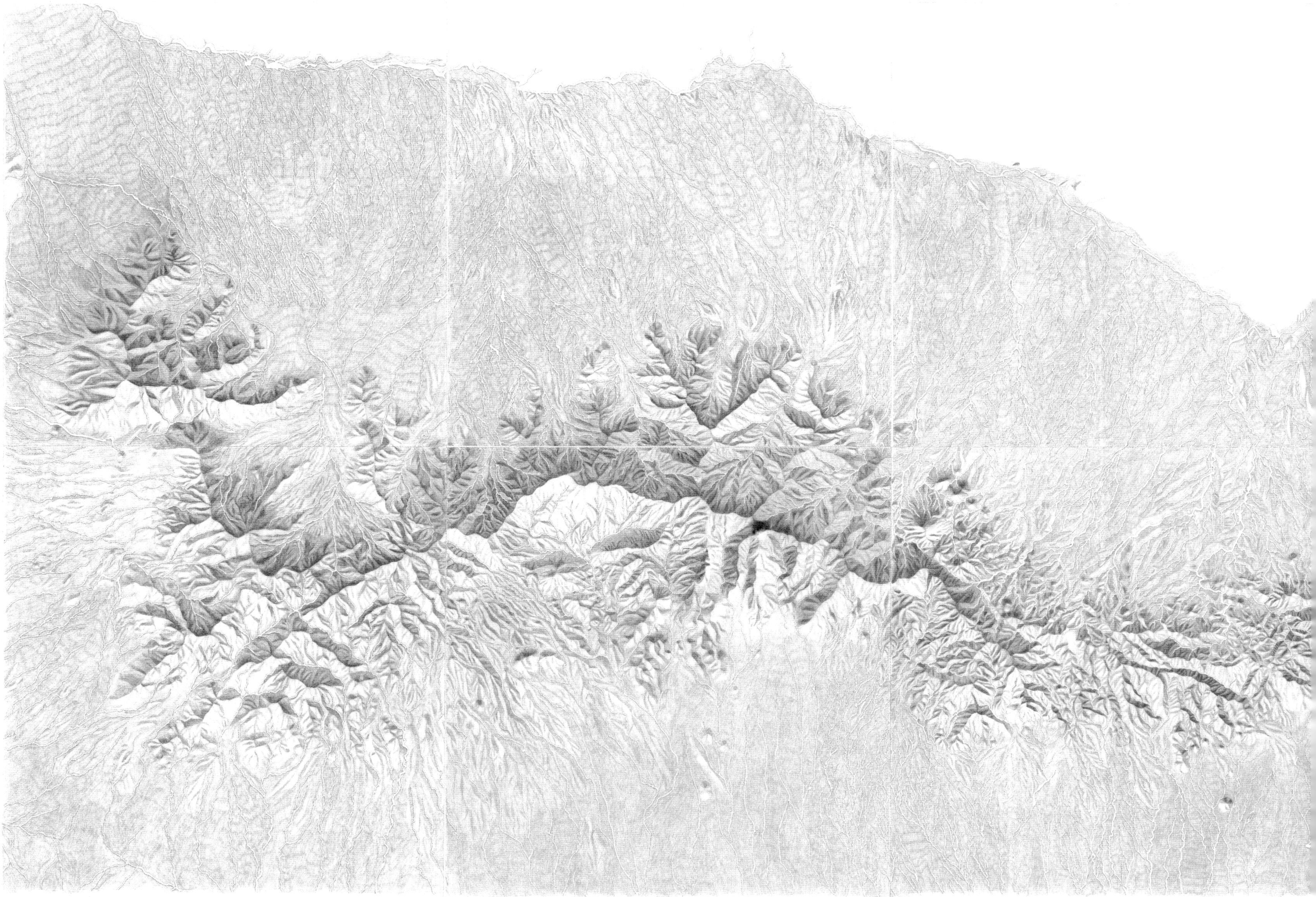

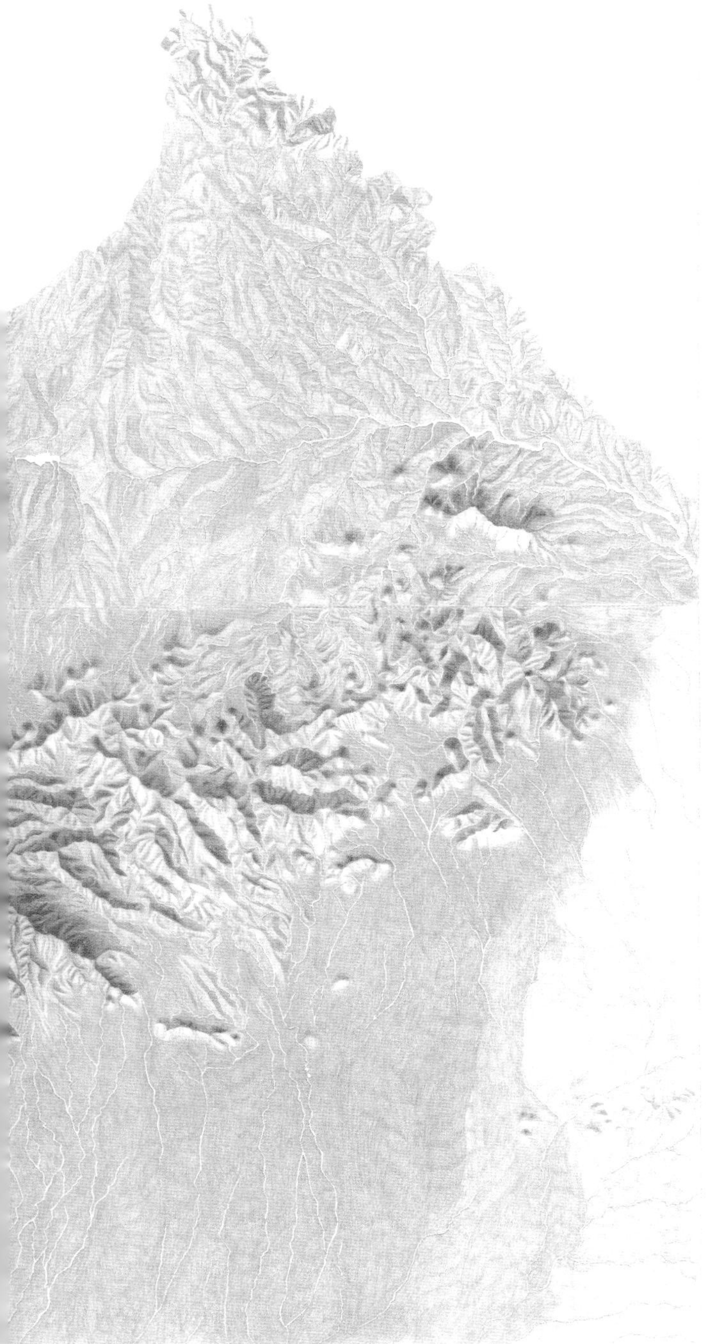

《 I'itoi's Place: A Landform Portrait

Zoé Edgecomb

I made this map as a cartographic gesture toward decolonization. It depicts a mountain range within the lands of the Tohono O'odham (Desert People). Known widely as Baboquivari, the range today bears a name that is a Spanish derivation of the O'odham name for its prominent central peak, Waw Giwulk.

In 2017, the Tohono O'odham Indian reservation gained international attention due to federal plans to build a wall along the US–Mexico border. A 75-mile stretch of this structure cuts through traditional O'odham lands, creating a barrier affecting both human and other-than-human communities. In my efforts to understand this contested place, I noticed the serpentlike figure of the range while exploring through Google Earth. In later conversations with O'odham friends, I realized that the people of these lands see a series of places, not a singular image. It is home to I'itoi, the Creator. Experiencing the land with feet on the ground, you know where I'itoi's cave is, you know the region that is considered his garden. With modern cartographic techniques and aerial imagery, the range becomes a figure. The same technologies that have defined boundaries and mapped resources for exploitation can also reveal a geological formation that transcends borders, languages, and economic systems.

With this map, I tried to deconstruct the US Geological Survey format: I downloaded PDFs from TopoView and, with the help of a light table, applied the shaded-relief technique developed by Swiss cartographer, Eduard Imhof, to define the form. I allowed the mark-making on the western side of the range, currently part of the Tohono O'odham reservation, to flow to the edge of the paper. The eastern extents stop roughly where the flow of water changes direction.

It will take a long time to shift the paradigm of the colonialist conception of land from a resource to be managed and exploited, to a system that is alive. This is what I thought of as I spent hundreds of hours with pencils, paper, and base maps from USGS and INEGI, the national map of Mexico. The border is invisible, and the power of the landform itself emerges.

The drawing has been offered to the Tohono O'odham Nation as a gift.

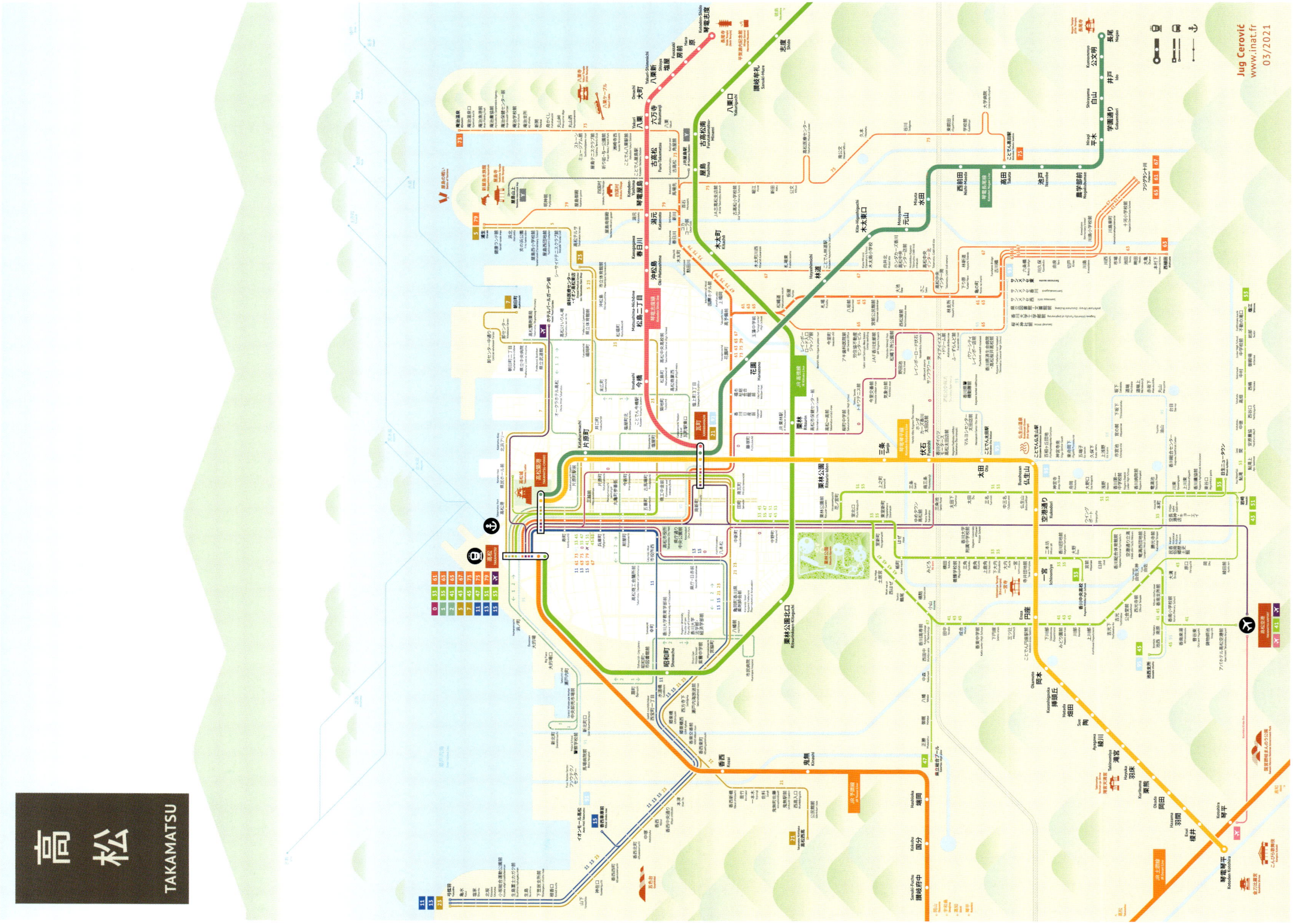
高松
TAKAMATSU
Jug Cerović
www.inat.fr
03/2021

Landscape is ... arranged in successive planes to create depth and a perspective without focal points. «

« Takamatsu Public Transport Map

Jug Cerovic

In Takamatsu, streams and bus lines run downhill toward the inland sea. The sea is sprinkled with islands, hills are sprinkled with temples.

A transit map is the image of the network, it is the image of the city, and when the setting is outstanding, you want the map to match it.

On the central, vertical axis of the layout I placed the downtown area, encased in a symbolic white disc, where cityscape is on scale and shown in great detail with all streets. This part of the map can be used right away as a pedestrian wayfinding map. Five thick rail lines radiate from downtown terminals toward the countryside. Beneath them bus lines radiate toward the hinterland, color-coded by function, with each color corresponding to a different sector. Lines that use a common corridor are bundled into a thicker line to denote cumulative frequency. Ferries are the third available transport mode and connect outlying islands to the main harbor. Thus is the information hierarchy of this transit map, laid out in successive layers for a progressive information acquisition, one transport mode at a time.

Landscape is likewise arranged in successive planes to create depth and a perspective without focal points. Hills in the foreground, then river valleys, the shoreline, the sea, the islands, the mountains in the background, and finally the sky. Custom pictograms stand for every landmark or point of interest, with a different silhouette for every single temple. For convenience all stations are labeled in both Japanese and English.

I want a transit map to be at the same time a functional diagram and a work of art. Here the map is at once a two-dimensional plan perfect for journey planning and network exploration, and a three-dimensional landscape painting that conveys the unique flavor of the depicted land: neat, legible, beautiful.

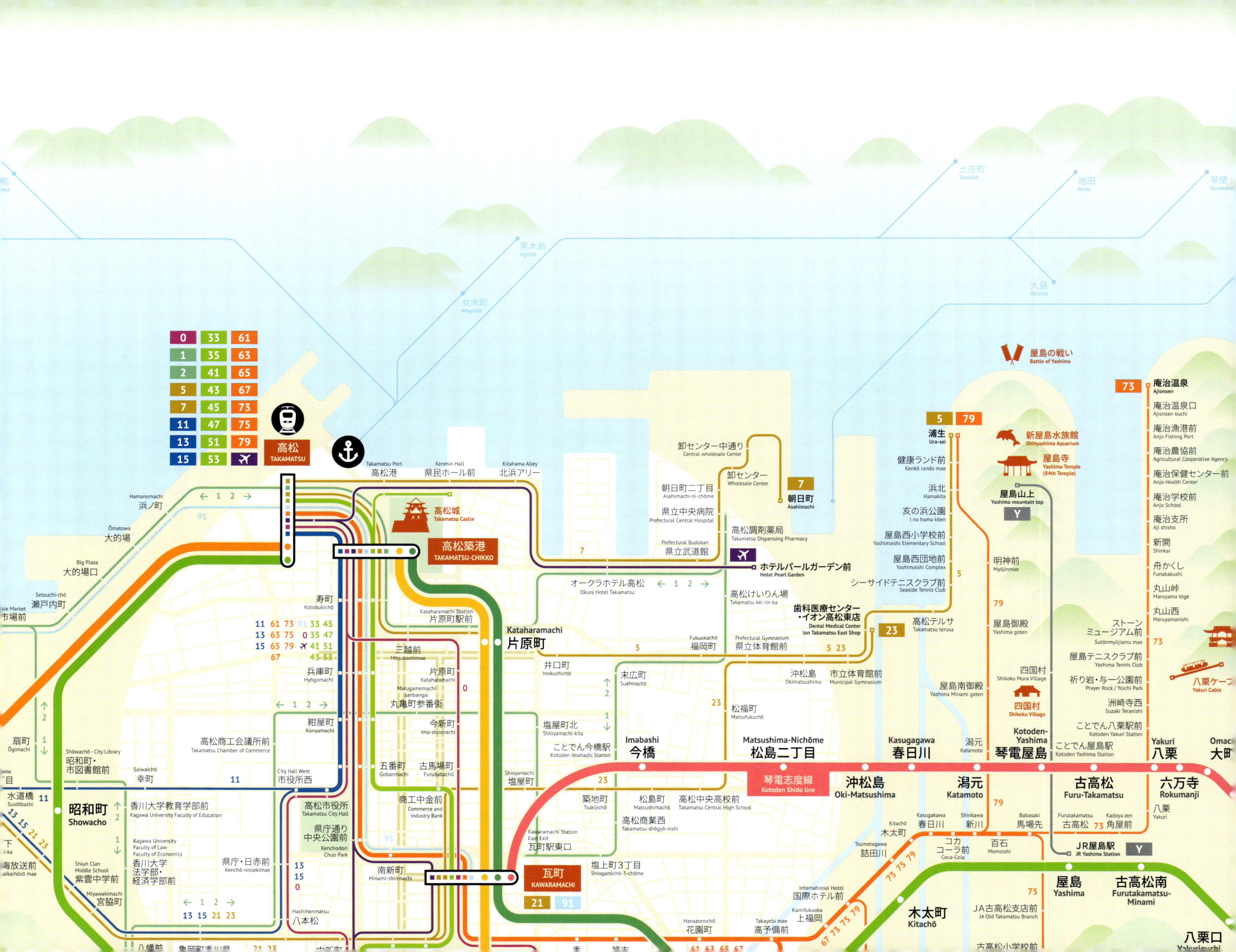

高松
TAKAMATSU
高松築港
TAKAMATSU-CHIKKO
瓦町
KAWARAMACHI
琴電志度線
Kotoden Shido line
高松城
Takamatsu Castle
屋島の戦い
Battle of Yashima
新屋島水族館
Shinyashima Aquarium
屋島寺
Yashima Temple
(84th Temple)
屋島山上
Yashima mountain top
四国村
Shikoku Village
八栗ケーブ
Yakuri Cable
Takamatsu Port
高松港
Kenmin Hall
県民ホール前
Kitahama Alley
北浜アリー
Hamanomachi
浜ノ町
Ōmatowa
大的場
Big Plaza
大的場口
Setouchi-chō
瀬戸内町
寿町
Kotobukichō
Kataharamachi Station
片原町駅前
Kataharamachi
片原町
三越前
Mitsukoshimae
兵庫町
Hyōgomachi
片原町
Kataharamachi
丸亀町参番街
紺屋町
Konyamachi
今新町
Ima-shinmachi
高松商工会議所前
Takamatsu Chamber of Commerce
Shōwachō - City Library
昭和町・
市図書館前
Saiwaichō
幸町
City Hall West
市役所西
五番町
Gobanmachi
古馬場町
Furubabachō
Shioyamachi
塩屋町
扇町
Ōgimachi
水道橋
Suidōbashi
昭和町
Showacho
香川大学教育学部前
Kagawa University Faculty of Education
高松市役所
Takamatsu City Hall
商工中金前
Commerce and Industry Bank
県庁通り
中央公園前
Kenchodori
Chuo Park
Kagawa University
Faculty of Law
Faculty of Economics
香川大学
法学部・
経済学部前
県庁・日赤前
Kenchō-nissekimae
Shiun Clan
Middle School
紫雲中学前
Miyawakimachi
宮脇町
南新町
Minami-shinmachi
Hachihonmatsu
八本松
井口町
Inokuchichō
末広町
Suehirochō
塩屋町北
Shioyamachi-kita
ことでん今橋駅
Kotoden Imahashi Station
Imabashi
今橋
築地町
Tsukijichō
Kawaramachi Station
East Exit
瓦町駅東口
塩上町3丁目
Shiogamichō-3-chōme
オークラホテル高松
Okura Hotel Takamatsu
卸センター中通り
Central wholesale Center
卸センター
Wholesale Center
朝日町二丁目
Asahimachi-ni-chōme
県立中央病院
Prefectural Central Hospital
Prefectural Budokan
県立武道館
高松調剤薬局
Takamatsu Dispensing Pharmacy
朝日町
Asahimachi
ホテルパールガーデン前
Hotel Pearl Garden
高松けいりん場
Takamatsu kei rin ba
歯科医療センター
・イオン高松東店
Dental Medical Center
Ion Takamatsu East Shop
Fukuokachō
福岡町
Prefectural Gymnasium
県立体育館前
沖松島
Okimatsushima
市立体育館前
Municipal Gymnasium
松福町
Matsufukuchō
Matsushima-Nichōme
松島二丁目
松島町
Matsushimachō
高松中央高校前
Takamatsu Central High School
高松商業西
Takamatsu-shōgyō-nishi
沖松島
Oki-Matsushima
浦生
Ura-sei
健康ランド前
Kenkō rando mae
浜北
Hamakita
亥の浜公園
I no hama kōen
屋島西小学校前
Yashimaishi Elementary School
屋島西団地前
Yashimaishi Complex
シーサイドテニスクラブ前
Seaside Tennis Club
高松テルサ
Takamatsu terusa
屋島南御殿
Yashima Minami goten
明神前
Myōjinmae
屋島御殿
Yashima goten
四国村
Shikoku Mura Village
Kasugagawa
春日川
潟元
Katamoto
Kotoden-
Yashima
琴電屋島
潟元
Katamoto
Kitachō
木太町
Kasugawa
春日川
Shinkawa
新川
Babasaki
馬場先
Tsumetagawa
詰田川
コカ
コーラ前
Coca-Cola
百石
Momoishi
International Hotel
国際ホテル前
Kamifukuoka
上福岡
Hanazonochō
花園町
Takayobi mae
高予備前
木太町
Kitachō
JA古高松支店前
JA Old Takamatsu Branch
屋島
Yashima
古高松南
Furutakamatsu-
Minami
JR屋島駅
JR Yashima Station
古高松
Furu-Takamatsu
Furutakamatsu
古高松
Kadoya zen
角屋前
六万寺
Rokumanji
八栗
Yakuri
Yakuri
八栗
八栗口
ことでん屋島駅
Kotoden Yashima Station
ことでん八栗駅前
Kotoden Yakuri Station
洲崎寺西
Suzaki Teranishi
祈り岩・与一公園前
Prayer Rock / Yoichi Park
屋島テニスクラブ前
Yashima Tennis Club
ストーン
ミュージアム前
Sutōnmyūjiamu mae
庵治温泉
Ajionsen
庵治温泉口
Ajionsen kuchi
庵治漁港前
Anjo Fishing Port
庵治農協前
Agricultural Cooperative Agency
庵治保健センター前
Anjo Health Center
庵治学校前
Anjo School
庵治支所
Aji shisho
新開
Shinkai
舟かくし
Funakakushi
丸山峠
Maruyama tōge
丸山西
Maruyamanishi
男木島
Ogichō
女木町
Megichō
大島
Oshima
土庄町
Tonoshō
池田
Ikeda
草壁
Kusakabe
0 33 61
1 35 63
2 41 65
5 43 67
7 45 73
11 47 75
13 51 79
15 53

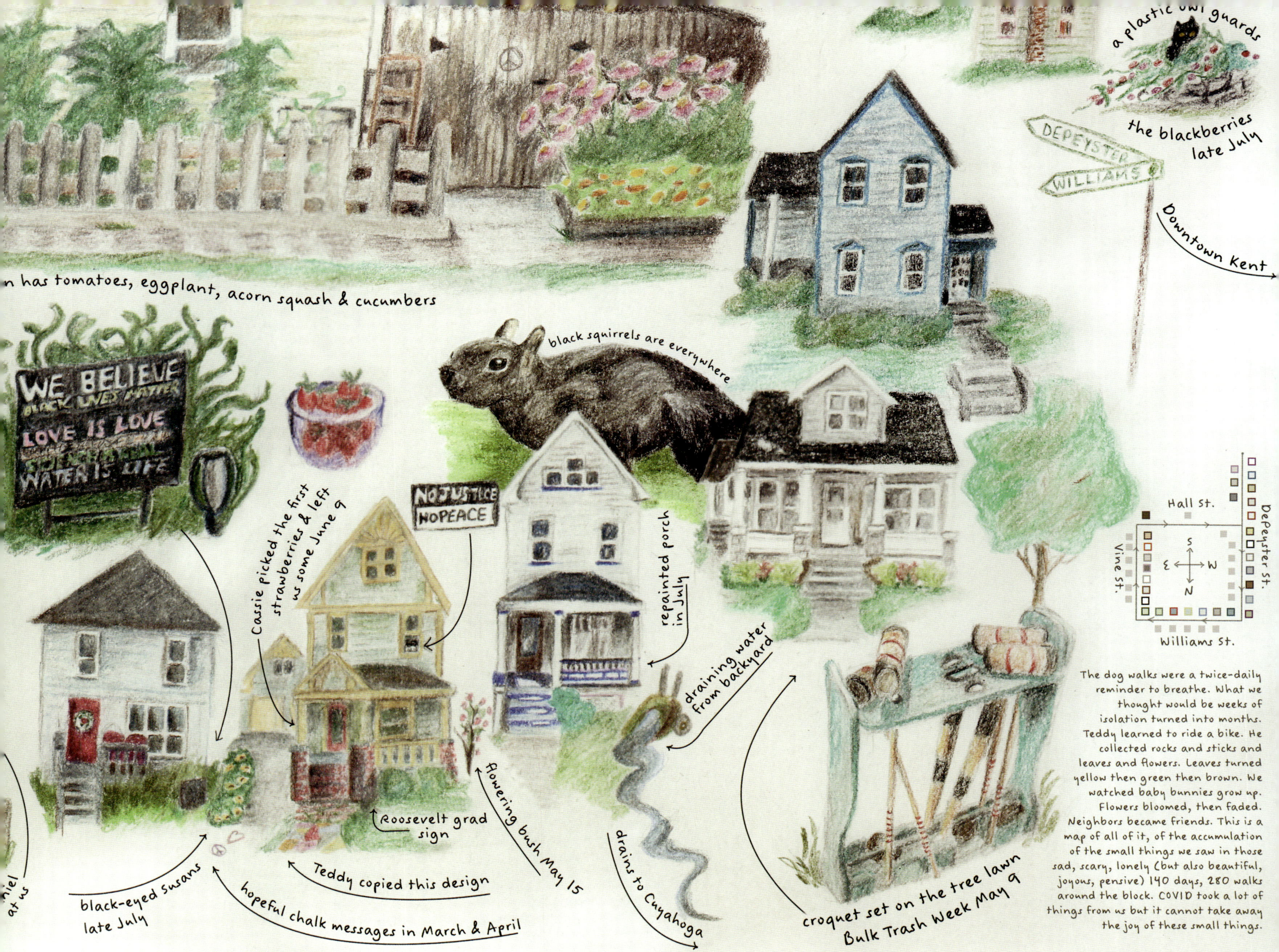

The dog walks were a twice-daily reminder to breathe. What we thought would be weeks of isolation turned into months. Teddy learned to ride a bike. He collected rocks and sticks and leaves and flowers. Leaves turned yellow then green then brown. We watched baby bunnies grow up. Flowers bloomed, then faded. Neighbors became friends. This is a map of all of it, of the accumulation of the small things we saw in those sad, scary, lonely (but also beautiful, joyous, pensive) 140 days, 280 walks around the block. COVID took a lot of things from us but it cannot take away the joy of these small things.

Jen Mapes, September 2020

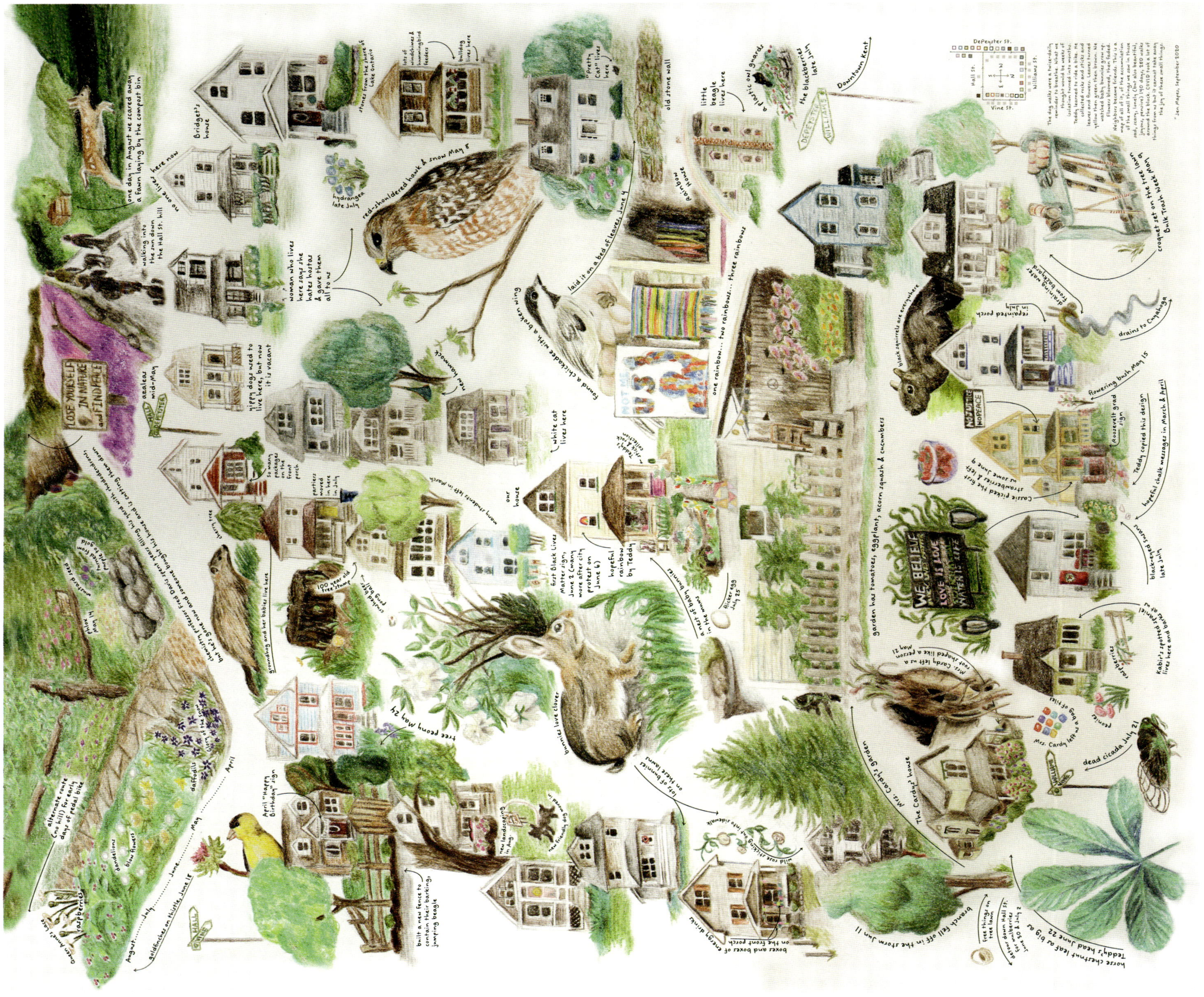

one day in August we scared away a fawn laying by the compost bin
no one lives here now
Bridget's house
stones from the shore of Lake Ontario
lots of windchimes & hummingbird feeders
bulldog lives here
"Pretty Cat" lives here
old stone wall
little beagle lives here
a plastic owl guards the blackberries late July
Downtown Kent
DePeyster St.
Williams St.
Hall St.
Vine St.
hydrangea late July
red-shouldered hawk & snow May 8
walking into the sun down the Hall St. hill
woman who lives here says she hates hostas & gave them all to us
found a chickadee with a broken wing
laid it on a bed of leaves, June 4
Rainbow House
one rainbow... two rainbows... three rainbows
croquet set on the tree lawn
draining water from backyard
drains to Cuyahoga
repainted porch in July
black squirrels are everywhere
azaleas mid-May
LOSE YOURSELF IN NATURE and FIND PEACE
yippy dogs used to live here, but now it is vacant
new hammock
flowering bush May 15
Roosevelt grad sign
Teddy copied this design
hopeful chalk messages in March & April
Cassie picked the first strawberries & left us some June 9
white cat lives here
Teddy's rock collection
so many packages on the front porch
partiers moved in here in July
many students left in March
our house
first Black Lives Matter sign June 2 (many more after city protest on June 6)
hopeful rainbow by Teddy
a nest of baby bunnies in the mulch
flicker egg July 25
garden has tomatoes, eggplant, acorn squash & cucumbers
WE BELIEVE
LOVE IS LOVE
WATER IS LIFE
blackeyed susans late July
raspberries
Mrs. Cardy left us a root shaped like a person
groundhog and her babies live here
100 year old tree stump
bunnies love clover
tree peony May 24
lots of bunnies on these lawns
Mrs. Cardy's garden
The Cardy's house
Mrs. Cardy left us a bag of tiles
dead cicada July 21
horse chestnut leaf as big as Teddy's head June 22
branch fell off in the storm Jun 11
boxes and boxes of energy drinks on the front porch
new friendly dog moved in
new landscaping in Aug
April "Happy Birthday" sign
built a new fence to contain their barking, jumping beagle
goldfinches on thistle June 18
August ...... July ...... June ...... May ...... April
alternate route (no hill) for early days of pedal bike
dandelions
blow flowers
daffodils
raspberries

The design aims to connect these memories to this community, to our home—to time and place. «

« Around the Block

Jennifer Mapes

This map was an art project with my son, Teddy, who was four years old when the pandemic began. His day care closed for five months, so walking around the block and drawing together helped pass the time. The walks brought an appreciation for my neighbors and neighborhood.

As a geography professor and cartographer, I've watched Teddy's spatial awareness carefully. Children's thinking starts very large scale (zoomed in), and so his first maps were of rooms in our house, then the route to preschool, showing landmarks from a child's eye: a curve in the road, the bumpy part of the trail, and the "waterfall" (fountain) on campus. Accordingly, I took this approach when designing this map: highlighting memories that we shared, small details from the landscape that we might use to identify a point in time (a fallen branch after a storm) or space (the house with the energy drinks on the front porch).

The map's design is built around place-based storytelling: Remember the day we saw the hawk at Bridget's house, and then there was a snow squall? Remember when Cassie left us strawberries? The design aims to connect these memories to this community, to our home—to time and place. The layout of this map was a challenge. As the legend indicates, our actual route doubled back on our street after taking a clockwise route around a single block. My design solution here was to illustrate both sides of the street and move this half-block into the center of the square block. This placed our house at the center of the drawing, as it is at the center of our mental map of our neighborhood. At this scale we don't usually think about north and south. The choice to locate north "down" on the map was subconscious and likely related to downtown being our most frequent destination. When we write, we start at the top and work to the bottom. Similarly, when our family walks, we start at our house "top" and work our way down, or in this case, around the block and back again.

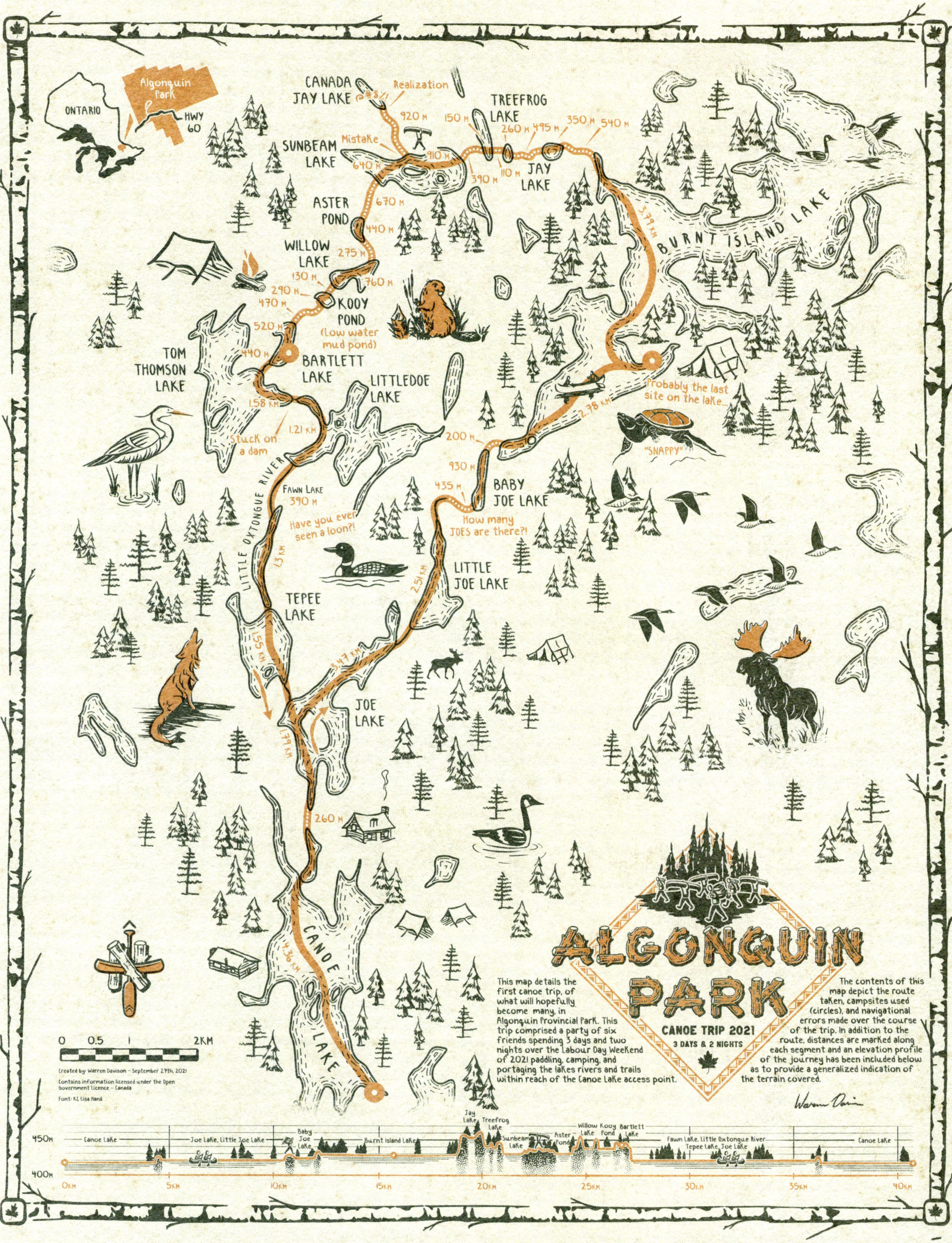
ALGONQUIN PARK
CANOE TRIP 2021
3 DAYS & 2 NIGHTS
This map details the first canoe trip, of what will hopefully become many, in Algonquin Provincial Park. This trip comprised a party of six friends spending 3 days and two nights over the Labour Day Weekend of 2021 paddling, camping, and portaging the lakes rivers and trails within reach of the Canoe Lake access point.
The contents of this map depict the route taken, campsites used (circles), and navigational errors made over the course of the trip. In addition to the route, distances are marked along each segment and an elevation profile of the journey has been included below as to provide a generalized indication of the terrain covered.
ONTARIO
Algonquin Park
HWY 60
CANADA JAY LAKE
Realization
920 M
TREEFROG LAKE
150 M
260 M
495 M
350 M
540 M
SUNBEAM LAKE
Mistake
640 M
910 M
390 M
110 M
JAY LAKE
3.39 KM
BURNT ISLAND LAKE
ASTER POND
670 M
440 M
WILLOW LAKE
275 M
130 M
760 M
290 M
470 M
KOOY POND
(Low water mud pond)
520 M
TOM THOMSON LAKE
440 M
BARTLETT LAKE
LITTLEDOE LAKE
Probably the last site on the lake...
2.78 KM
1.58 KM
1.21 KM
Stuck on a dam
200 M
"SNAPPY"
930 M
435 M
BABY JOE LAKE
LITTLE OXTONGUE RIVER
Fawn Lake
390 M
Have you ever seen a loon?!
How many JOES are there?!
1.3 KM
2.51 KM
LITTLE JOE LAKE
TEPEE LAKE
1.55 KM
3.47 KM
JOE LAKE
1.79 KM
260 M
CANOE LAKE
4.36 KM
0 0.5 1 2KM
Created by: Warren Davison - September 27th, 2021
Contains information licensed under the Open Government Licence - Canada
Font: KG Lisa Hand
450M
400M
Canoe Lake
Joe Lake, Little Joe Lake
Baby Joe Lake
Burnt Island Lake
Jay Lake
Treefrog Lake
Sunbeam Lake
Aster Pond
Willow Lake
Kooy Pond
Bartlett Lake
Fawn Lake, Little Oxtongue River
Tepee Lake, Joe Lake
Canoe Lake
0KM
5KM
10KM
15KM
20KM
25KM
30KM
35KM
40KM

These injected a healthy dose of vintage Canadiana into the aesthetic. «

« Algonquin Park Canoe Trip

Warren Davison

Algonquin Provincial Park is located within southeastern Ontario, nestled between Georgian Bay and the Ottawa River. The park contains swathes of deciduous and coniferous forests, rolling hills, and rocky ridges all punctuated by thousands of lakes, ponds, rivers, and streams, making it an ideal landscape to explore by canoe.

This map proudly documents one such canoe trip. While preparing for this trip, I pored over modern recreation maps abundant with information and great cartography. I chose to make an artful departure from this topographic style with my map, and attempt to convey the essence of backcountry camping and share the experiences of this particular trip.

To this end, I strove to mimic visual qualities from aesthetic touchstones, such as vintage camping brochures, local historic maps, and musty field guides. I kept topographic features within the base map to a minimum. Where features were absolutely required, I generalized them to reduce their precision, achieving a smoother and looser illustrative style. In lieu of topography, I opted to describe the terrain using an elevation profile. Although the terrain is not the most dramatic, I liked how the profile visualized the distance, elevation, duration, and conveyance method of the various segments of the trip as one continuous line. This illustrated chart–like depiction added both charm and data metrics to the map.

The minimal base map afforded space to adorn the route with illustrations of animals, camping ephemera, and specific moments from the trip. These injected a healthy dose of vintage Canadiana into the aesthetic. The illustrations, and the map itself, were created using a limited, two-color palette that is reminiscent of press printing. To completely convince the viewer that this map was indeed a tangible object from another era, I applied a rich paper texture and included subtle misalignments in the application of the ink.

My hope for this map is that it evokes a sense of nostalgia for those friends who were a part of this trip and that it might rekindle fond memories of time spent in nature: days spent paddling calm lakes, meals by the campfire, and nights under the stars.

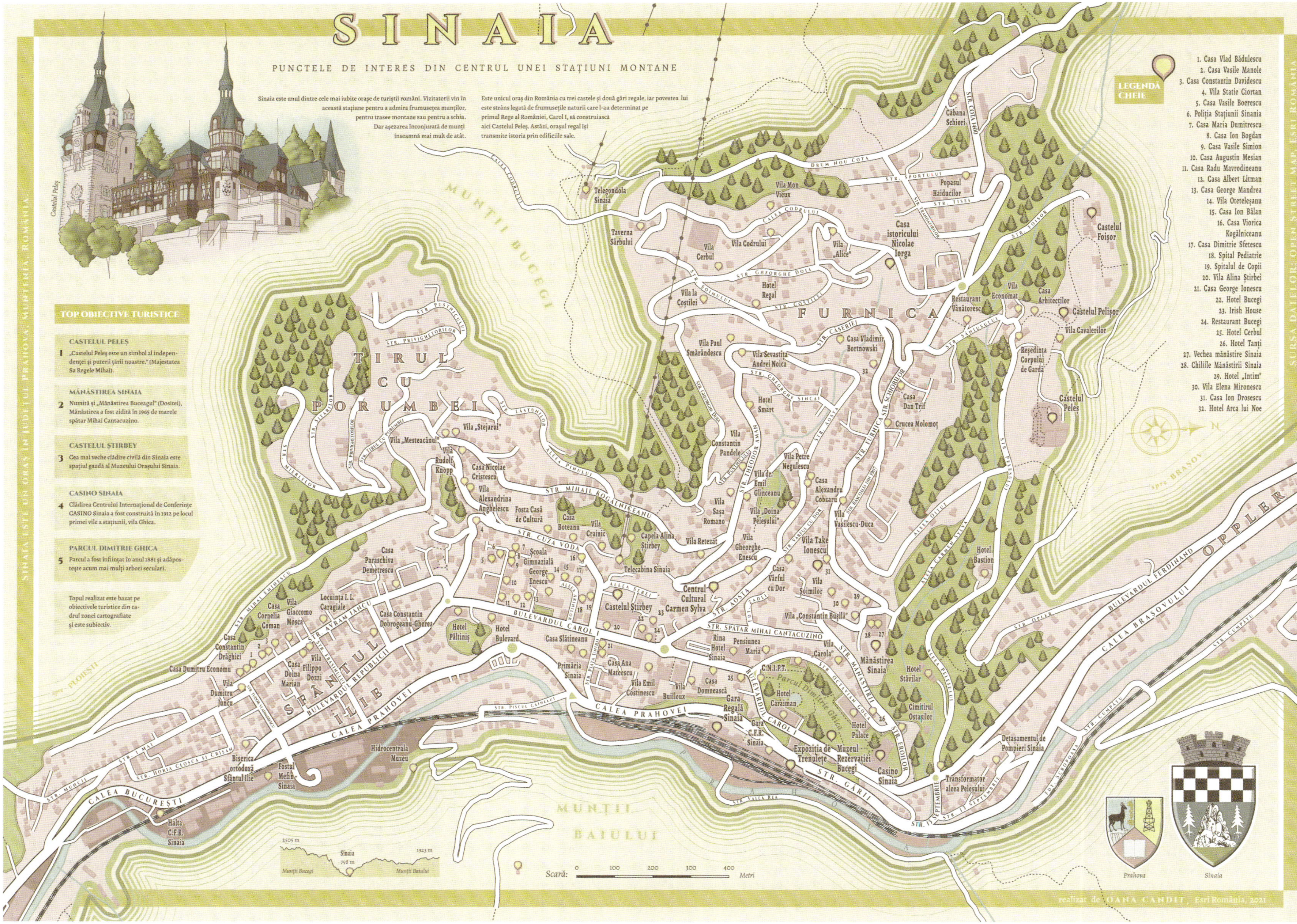
SINAIA
PUNCTELE DE INTERES DIN CENTRUL UNEI STAȚIUNI MONTANE
Sinaia este unul dintre cele mai iubite orașe de turiștii români. Vizitatorii vin în această stațiune pentru a admira frumusețea munților, pentru trasee montane sau pentru a schia. Dar așezarea înconjurată de munți înseamnă mai mult de atât.
Este unicul oraș din România cu trei castele și două gări regale, iar povestea lui este strâns legată de frumusețile naturii care l-au determinat pe primul Rege al României, Carol I, să construiască aici Castelul Peleș. Astăzi, orașul regal își transmite istoria prin edificiile sale.
Castelul Peleș
SINAIA ESTE UN ORAȘ ÎN JUDEȚUL PRAHOVA, MUNTENIA, ROMÂNIA.
TOP OBIECTIVE TURISTICE
CASTELUL PELEȘ
1 „Castelul Peleș este un simbol al independenței și puterii țării noastre." (Majestatea Sa Regele Mihai).
MÂNĂSTIREA SINAIA
2 Numită și „Mănăstirea Buceagul" (Dositei), Mănăstirea a fost zidită în 1965 de marele spătar Mihai Cantacuzino.
CASTELUL ȘTIRBEY
3 Cea mai veche clădire civilă din Sinaia este spațiul gazdă al Muzeului Orașului Sinaia.
CASINO SINAIA
4 Clădirea Centrului Internațional de Conferințe CASINO Sinaia a fost construită în 1912 pe locul primei vile a stațiunii, vila Ghica.
PARCUL DIMITRIE GHICA
5 Parcul a fost înființat în anul 1881 și adăpostește acum mai mulți arbori seculari.
Topul realizat este bazat pe obiectivele turistice din cadrul zonei cartografiate și este subiectiv.
LEGENDĂ CHEIE
1. Casa Vlad Bădulescu
2. Casa Vasile Manole
3. Casa Constantin Davidescu
4. Vila Statie Ciortan
5. Casa Vasile Boerescu
6. Poliția Stațiunii Sinania
7. Casa Maria Dumitrescu
8. Casa Ion Bogdan
9. Casa Vasile Simion
10. Casa Augustin Mesian
11. Casa Radu Mavrodineanu
12. Casa Albert Litman
13. Casa George Mandrea
14. Vila Oteteleșanu
15. Casa Ion Bălan
16. Casa Viorica Kogălniceanu
17. Casa Dimitrie Sfetescu
18. Spital Pediatrie
19. Spitalul de Copii
20. Vila Alina Știrbei
21. Casa George Ionescu
22. Hotel Bucegi
23. Irish House
24. Restaurant Bucegi
25. Hotel Cerbul
26. Hotel Tanți
27. Vechea mănăstire Sinaia
28. Chiliile Mănăstirii Sinaia
29. Hotel „Intim"
30. Vila Elena Mironescu
31. Casa Ion Drosescu
32. Hotel Arca lui Noe
SURSA DATELOR: OPEN STREET MAP, ESRI ROMÂNIA
MUNȚII BUCEGI
MUNȚII BAIULUI
TIRUL CU PORUMBEI
FURNICA
SFÂNTUL ILIE
Telegondola Sinaia
Taverna Sărbului
Vila Cerbul
Vila Codrului
Vila Mon Vieux
Vila „Alice"
Casa istoricului Nicolae Iorga
Cabana Schiori
Popasul Haiducilor
Castelul Foișor
Castelul Pelișor
Castelul Peleș
Vila Cavalerilor
Reședința Corpului de Gardă
Restaurant Vânătoresc
Vila Economat
Casa Arhitecților
Casa Vladimir Bortnowski
Casa Dan Trif
Crucea Molomoț
Hotel Regal
Vila la Coștilei
Vila Paul Smărăndescu
Vila Sevastița Andrei Noica
Hotel Smart
Vila Constantin Pandele
Vila Petre Negulescu
Casa Alexandru Cobzaru
Vila Emil Glinceanu
Vila „Doina Pelesului"
Vila Vasilescu-Duca
Vila Sașa Romano
Vila Retezat
Vila Gheorghe Enescu
Vila Take Ionescu
Casa Vârful cu Dor
Vila Șoimilor
Vila „Constantin Bușilă"
Vila „Stejarul"
Vila „Mesteacănul"
Vila Rudolf Knopp
Casa Nicolae Cristescu
Vila Alexandrina Anghelescu
Fosta Casă de Cultură
Casa Boteanu
Vila Crainic
Capela Alina Știrbey
Telecabina Sinaia
Școala Gimnazială George Enescu
Casa Paraschiva Demetrescu
Castelul Știrbey
Centrul Cultural Carmen Sylva
Locuința I. L. Caragiale
Vila Giaccomo Mosca
Casa Cornelia Coman
Casa Constantin Drăghici
Casa Constantin Dobrogeanu-Gherea
Hotel Păltiniș
Hotel Bulevard
Casa Slătineanu
Primăria Sinaia
Casa Ana Mateescu
Vila Emil Costinescu
Vila Builloux
Casa Domnească
Rina Hotel Sinaia
Pensiunea Maria
C.N.I.P.T.
Hotel Caraiman
Parcul Dimitrie Ghica
Vila „Carola"
Mănăstirea Sinaia
Hotel Stăvilar
Cimitirul Ostașilor
Hotel Palace
Hotel Bastion
Gara Regală Sinaia
Gara C.F.R. Sinaia
Expoziția de Trenulețe
Muzeul Rezervației Bucegi
Casino Sinaia
Transformator aleea Peleșului
Detașamentul de Pompieri Sinaia
Casa Dumitru Economu
Vila Dumitru Juncu
Casa Doina Marian
Vila Filippo Dozzi
Biserica ortodoxă Sfântul Ilie
Fostul Mefin Sinaia
Hidrocentrala Muzeu
Halta C.F.R. Sinaia
BULEVARDUL CAROL I
CALEA PRAHOVEI
CALEA BUCUREȘTI
CALEA BRAȘOVULUI
BULEVARDUL FERDINAND
OPPLER
STR. MIHAIL KOGĂLNICEANU
STR. CUZA VODĂ
STR. SPĂTAR MIHAI CANTACUZINO
BULEVARDUL REPUBLICII
STR. GĂRII
STR. 13 SEPTEMBRIE
DRUM NOU COTA
STR. COTA 1400
spre PLOIEȘTI
spre BRAȘOV
N
Sinaia 798 m
1505 m
1923 m
Munții Bucegi
Munții Baiului
Scară: 0 100 200 300 400 Metri
Prahova
Sinaia
realizat de IOANA CANDIT, Esri România, 2021

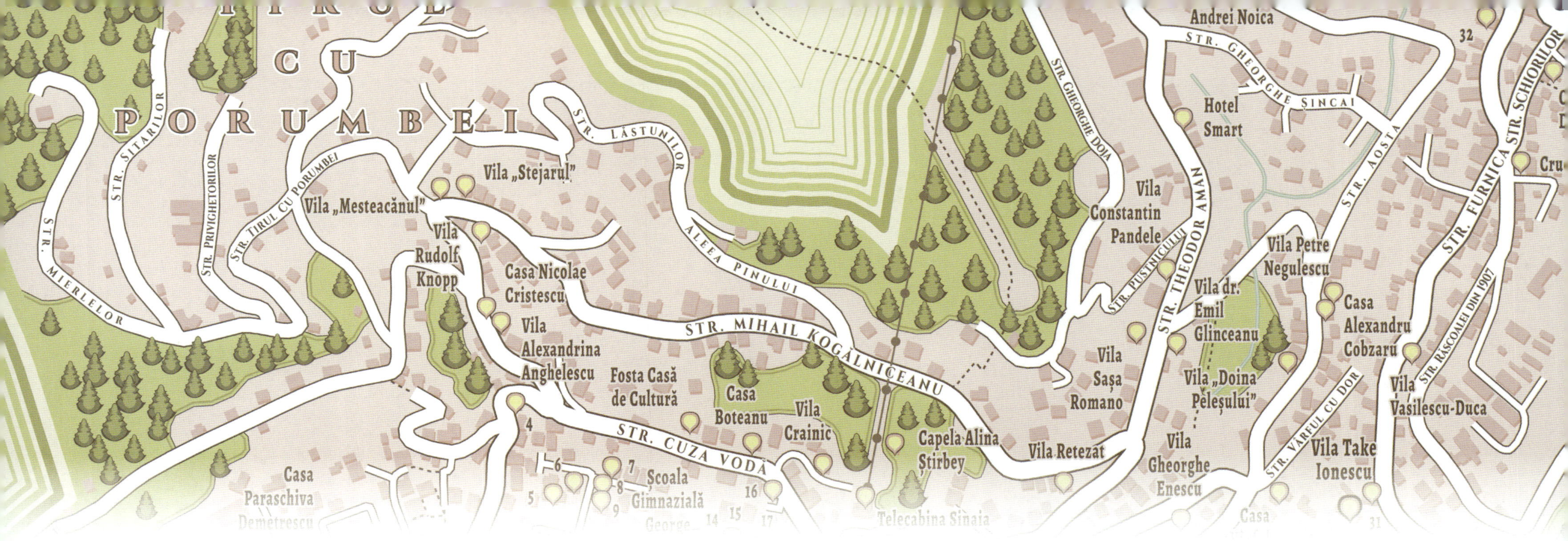

« Sinaia (Romania): Points of Interest in the Center of a Mountain Resort

Oana Candit, Esri Romania

From the very beginning the beauty of nature gradually determined the construction of the city of Sinaia, Romania, and the three castles that stand as a sample of greatness. The edifices of the city drew my interest and led me in transposing the story of the place into a map whose purpose is to be hung in a frame.

Peleș Castle is by far the most impressive building of the city, its main symbol, and it represents the first piece of this map's puzzle. Therefore, the colors in the main illustration are chosen as a theme that also reflects in the other elements. At the same time, the yellow and gold used to fill the background and points of interest were meant to remind viewers of the royal character of the settlement.

To be able to preserve this place's purity and express its historical character, I found it necessary to keep a well-balanced saturation of colors and simple shapes for the visualizations that surround the built space of the resort. On the map, the city is contoured by green lines of different widths, which show the continuity of the arboreal vegetation to the mountains. The representation of the fir trees has the purpose of enhancing the city area in the viewer's eyes.

To the future owner, the map will reveal itself as a guided tour through Sinaia's hidden gems, each building representing a new chapter in the discovery game. The inspiration behind my perception of tourist routes came from the urban games that are generally used on smartphones.

The map is different due to its north orientation, which offers a better perspective in viewing and tracking objectives. The illustration of the area outside the map frame indicates that the story and cultural treasure continues beyond the center of the mountain resort and is waiting to be explored.

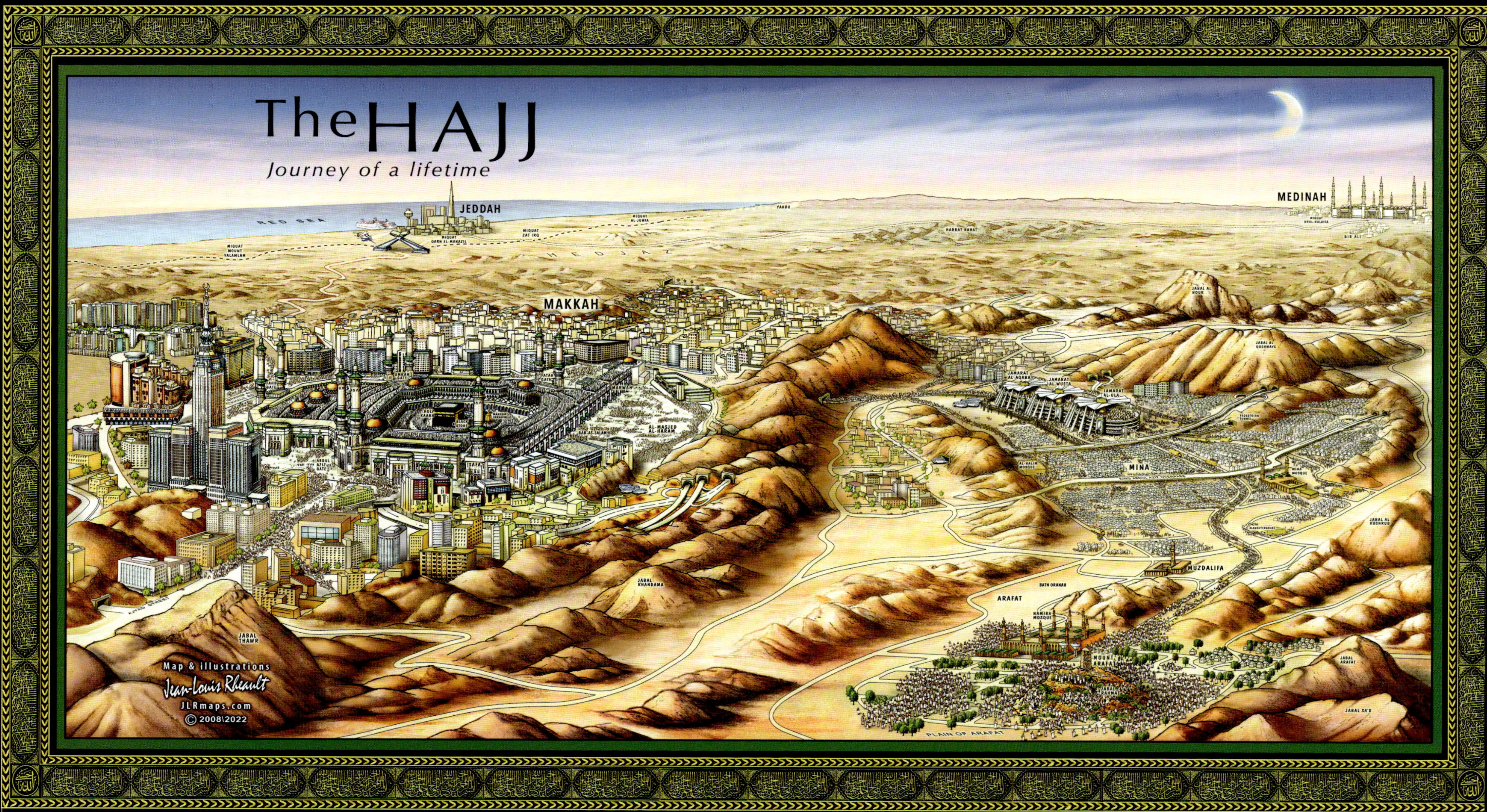

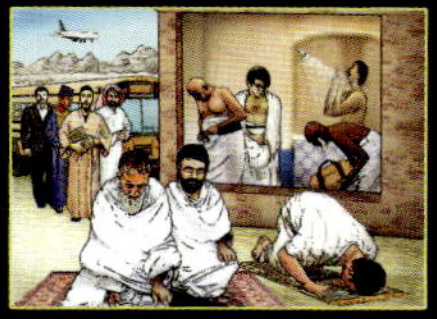
El- Ihram

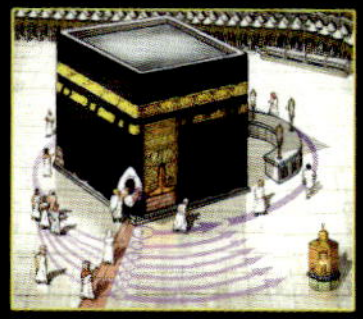
Tawaf Al Koudoum

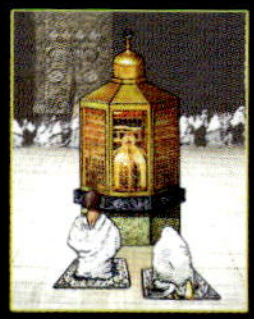
Makkam Ibrahim

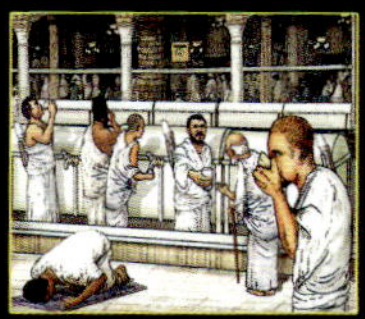
Bir Zemzem

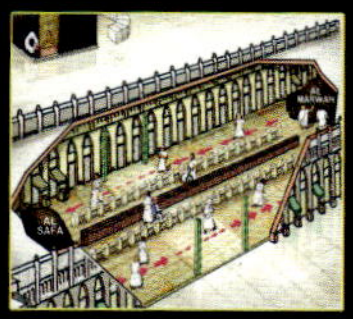
Sa'ee

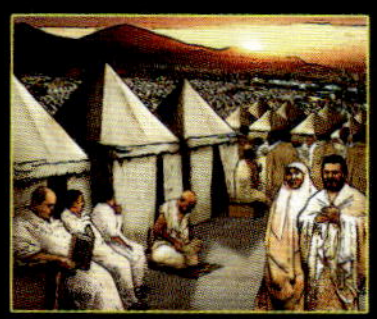
Mina

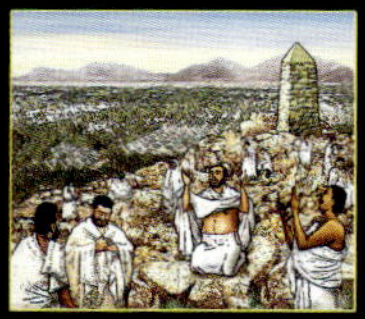
Arafat

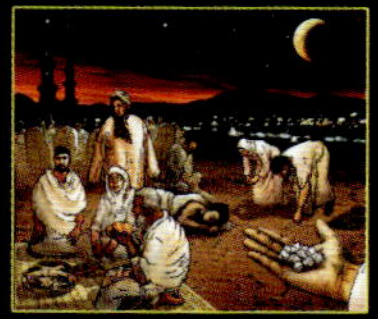
Muzdalifah

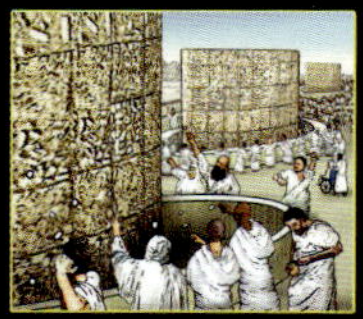
El Akabat

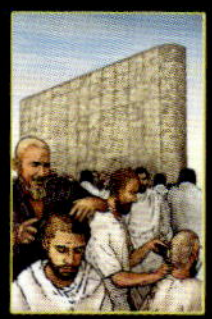
El Halk

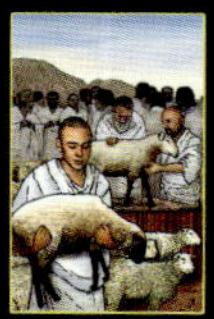
Sacrifice

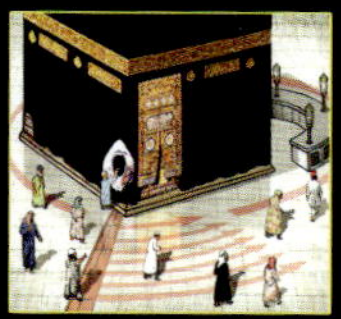
Tawaf El Ifada

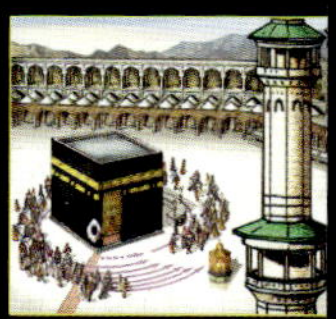
Tawaf El Wadaa

This surrender of individual identity... creates a powerful, life-changing experience and a unique mapping undertaking. «

« Le Hajj, Voyage d'une vie / Hajj, Journey of a Lifetime

Jean-Louis Rheault

The Hajj is the name of the pilgrimage to Mecca that Islam requires every Muslim to do once in their lifetime. This happens in a territory called the Hejaz that includes the holy cities of Mecca and Medina.

The first challenge of this map is that one can't enter this area without being a Muslim. I suppose I might have done like the comic-book hero Tintin and dressed for the part, but my days of such adventures have long passed.

The main design challenge is that Mecca is not a regular city like the urban views you see on a 3D view of Google Maps. It has always been a small, dusty town in the middle of a harsh and barren desert landscape with very imposing mega-structures and a huge "tent city" built for the millions of visiting pilgrims who come every year. Imagine making a map of the small town of Woodstock, New York—but at the moment of the 1969 concert!

My maps are always more about telling the story of a place than about making single-scale representations. The story of Mecca is unique in that Muslims come from everywhere on the planet, and as they enter the area, their first step is to bathe and all dress in the same plain white garb. Whether someone is a member of Brunei royalty or a Mongolian biker, all social, national, and ethnic distinctions are left behind. This surrender of individual identity into an egalitarian collective, all of whom perform the same steps over a few days, is what creates a powerful, life-changing experience and a unique mapping undertaking.

This is why the illustrations of the thirteen steps of the Hajj are prominently featured as an essential part of the map.

Another puzzle was how to illustrate crowds. I created novel custom digital brushes of tiny pilgrims which, as I much appreciated, are all dressed the same. However, along the way, an infuriating computer glitch scattered my crowds all over the canvas and buildings, and that was a real martyrdom to clean up!

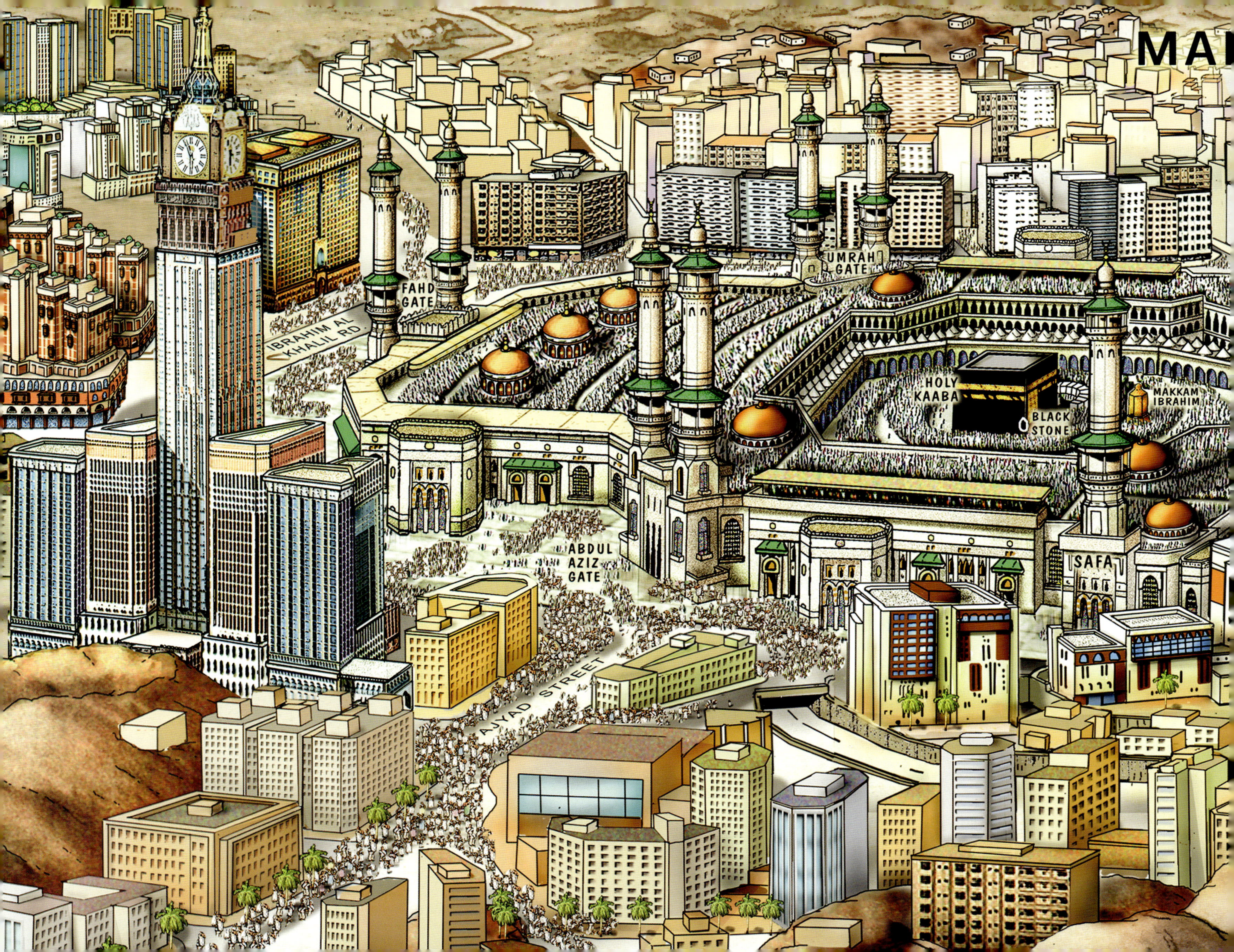

MA
FAHD GATE
IBRAHIM AL KHALIL RD
UMRAH GATE
HOLY KAABA
MAKKAM IBRAHIM
BLACK STONE
ABDUL AZIZ GATE
SAFA
AJYAD STREET

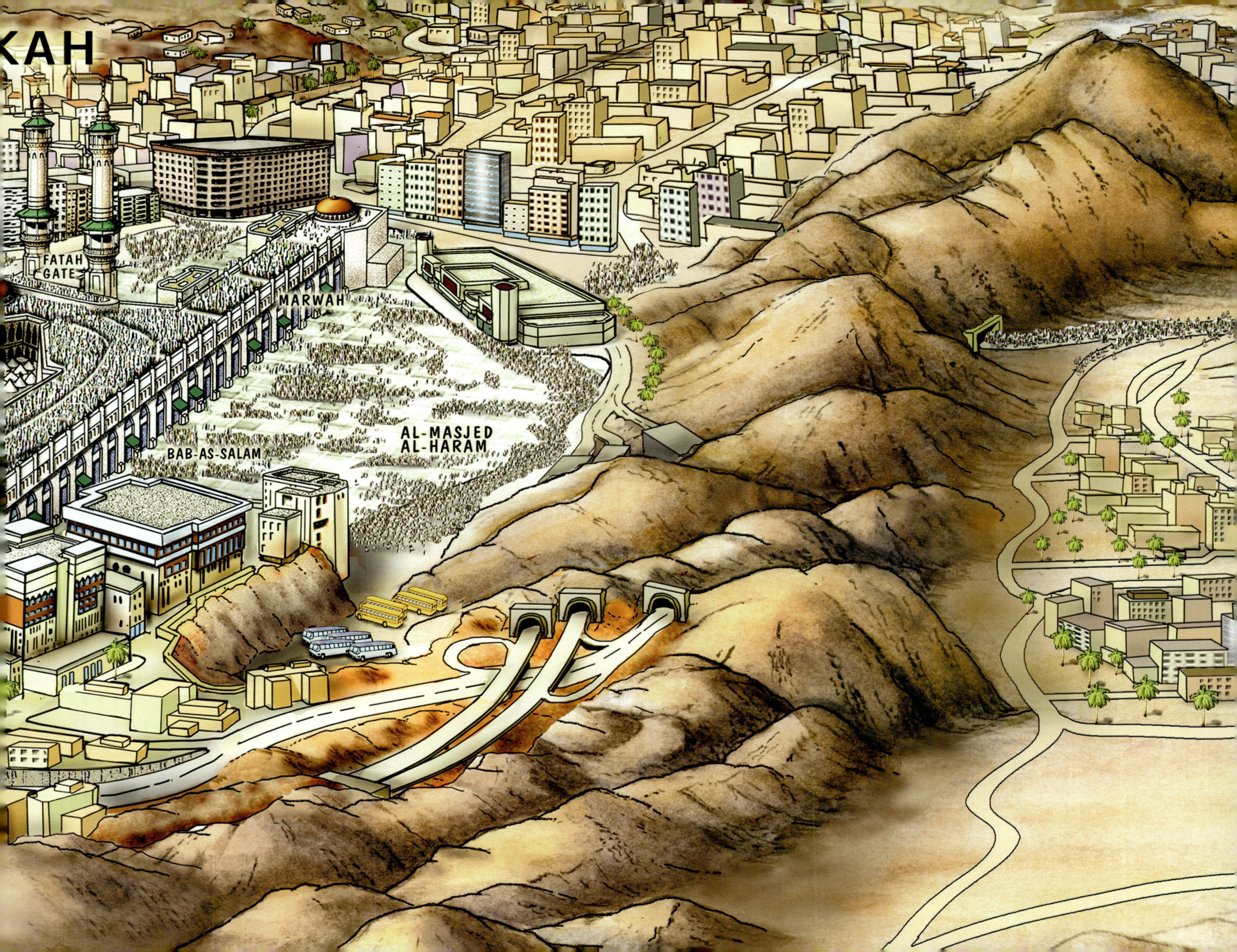
KAH
FATAH GATE
MARWAH
BAB-AS-SALAM
AL-MASJED AL-HARAM

The

Presidential Range

of the

White Mountains

New Hampshire

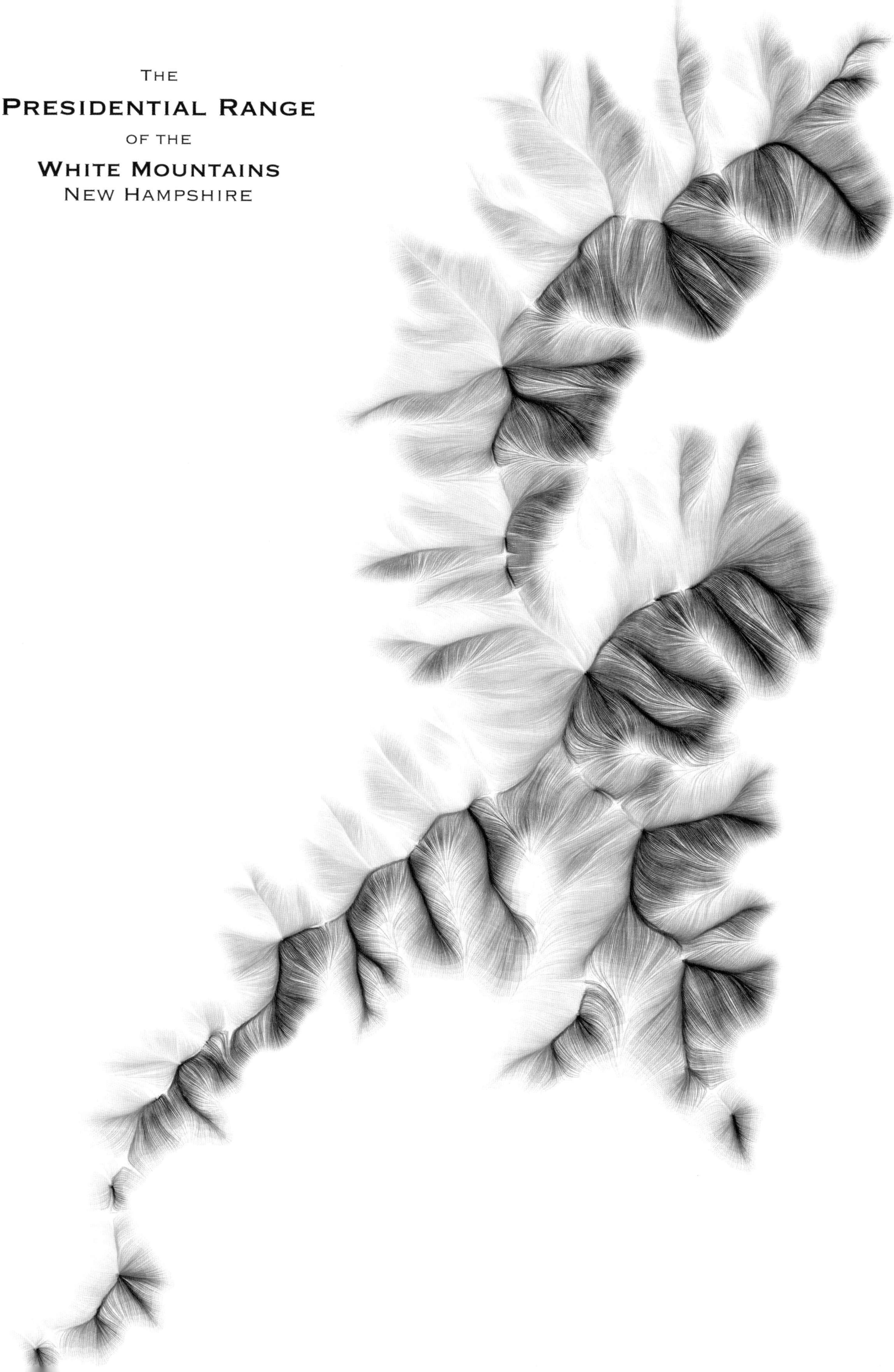

《 The Presidential Range

Andy Woodruff

The Presidential Range in New Hampshire's White Mountains, anchored by Mount Washington, is home to the highest mountain peaks in the northeastern United States. While it's obvious from certain places on the ground, the imposing yet intricate shape of these mountains can be lost in standardized terrain maps calibrated to much higher elevations. As the White Mountains have become a familiar and beloved region during my years living in New England, they've also become my cartographic practice ground, where any learning exercise doubles as an attempt to show off what the area has to offer. That is the story of this map.

The map's origins lie in a self-imposed challenge to write code to automatically render hachures, one of my favorite cartographic relief techniques. Hachures—short parallel strokes along elevation contours, indicating the direction of slope—are rare in modern cartography but are an eye-catching feature of many maps from the previous few centuries. I try to wield code as a tool for design, and while writing code allows a cartographer to do many new and exciting things, I like to use it to do some old and beautiful things, too. Thus, hachures have been the technological side-quest of my career.

Although I have yet to succeed at code for halfway decent hachures, the endeavor has led to this map, essentially hachures run amok. It's a map of flow, each line following the terrain downhill, not unlike what proven flow analysis algorithms might produce, but developed mostly with an eye for cartographic aesthetics. The goal was a map that conveys enough topographic detail for a good impression of this land, while retaining a natural, organic appearance. This feather-like map approaches that goal by stripping away information from earlier experiments with the technique, abandoning animation and color for a monochrome image of the highest parts of the mountain range. My hope is that a map like this can be meaningful to those who know the place and alluring to those who don't.

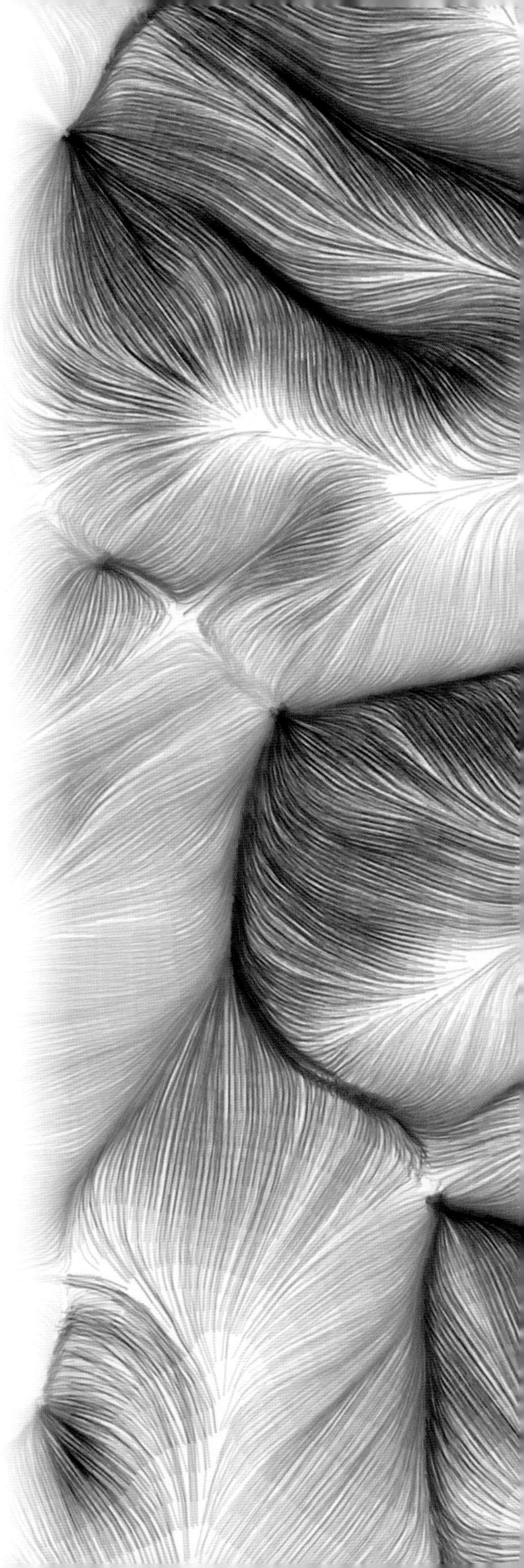

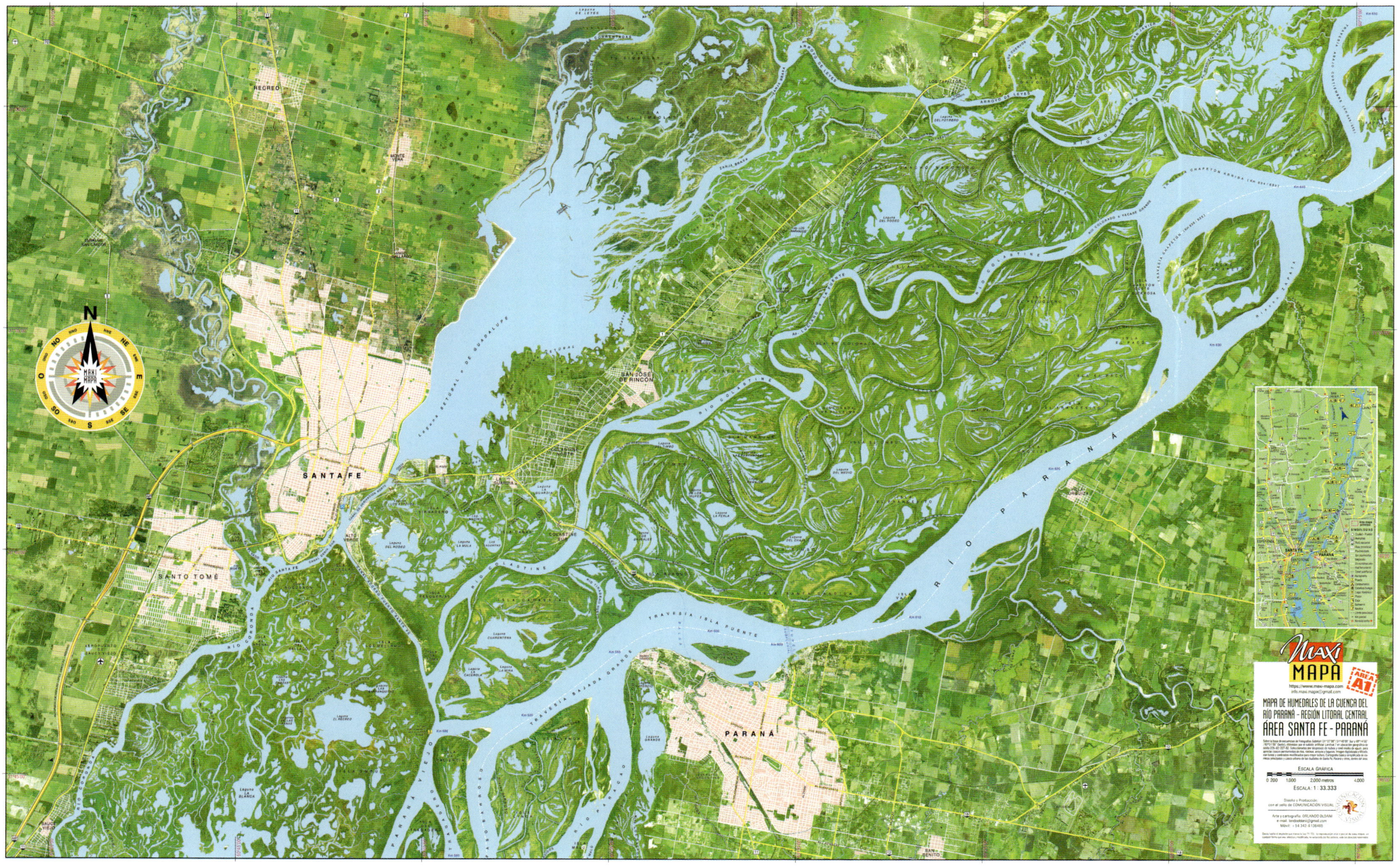
N
NE
E
SE
S
SO
O
NO
RECREO
MONTE VERA
SANTA FE
SANTO TOMÉ
SAN JOSÉ DE RINCÓN
PARANÁ
RÍO PARANÁ
RÍO COLASTINÉ
TRAVESÍA ISLA PUENTE
PARANÁ VIEJO
Maxi MAPA
ÁREA A1
https://www.maxi-mapa.com
info.maxi.mapa@gmail.com
MAPA DE HUMEDALES DE LA CUENCA DEL RÍO PARANÁ - REGIÓN LITORAL CENTRAL
ÁREA SANTA FE - PARANÁ
ESCALA GRÁFICA
0 200 1.000 2.000 metros 4.000
ESCALA: 1 : 33.333
Diseño y Producción: con el sello de COMUNICACIÓN VISUAL
Arte y cartografía: ORLANDO OLDANI
e-mail: landooldani@gmail.com
Móvil: +54 342-6136485

...living elements, such as vegetation... cannot be reduced to mere geometries, just as the landscape can't be contained in distinct classifications. «

« Wetlands in the Paraná River Basin, Central Littoral Region

Orlando Roberto Oldani

The Paraná River is the second-longest river in South America, after the Amazon. The Paraná River is 4,000 km long and has a basin of 2.8 million km^2 that occupies the southeastern part of Brazil, Paraguay, and the north and center of Argentina.

In its upper course, it runs boxed in with steep slopes and great unevenness through the Brazilian plateau. In its middle course, already in Argentina, it flows along a fault that separates two plains and the water is opaque and less rapid, presenting the majestic look of a great river of plains. The ravines on the left bank are high, while those on the right bank are low and flooded, widening into imposing wetlands, with streams and brooks running parallel to the direction of the river, and draining into it. In its lower course, the river continues southward to its mouth into the wide Río de la Plata.

I took advantage of my knowledge of graphic design and photo editing to develop these detailed maps, as I am passionate about these wetlands and I was motivated by the lack of documentation about them.

The landscape plays the primary role, so this map focuses on nature as it represents living elements, such as vegetation, that cannot be reduced to mere geometries, just as the landscape can't be contained in distinct classifications. Therefore, in the design of the maps, I incorporated satellite imagery as the prevailing element combining all the qualities expressed. I selected clear and clean imagery stripped of clouds, improved the tones and contrast digitally and corrected them chromatically using a single tone for the water. I completed the outline of the courses of the streams covered by vegetation or clogged by sand. I used these orthophotos as the background of the map, overlaying other elements, such as the simplified cartography of main roads and urban areas, to achieve the goal for its correct interpretation: clarity and accuracy of information.

I am proud to have designed this map, which is part of a seven-map series, with the goal that it will be useful for those who enjoy the river.

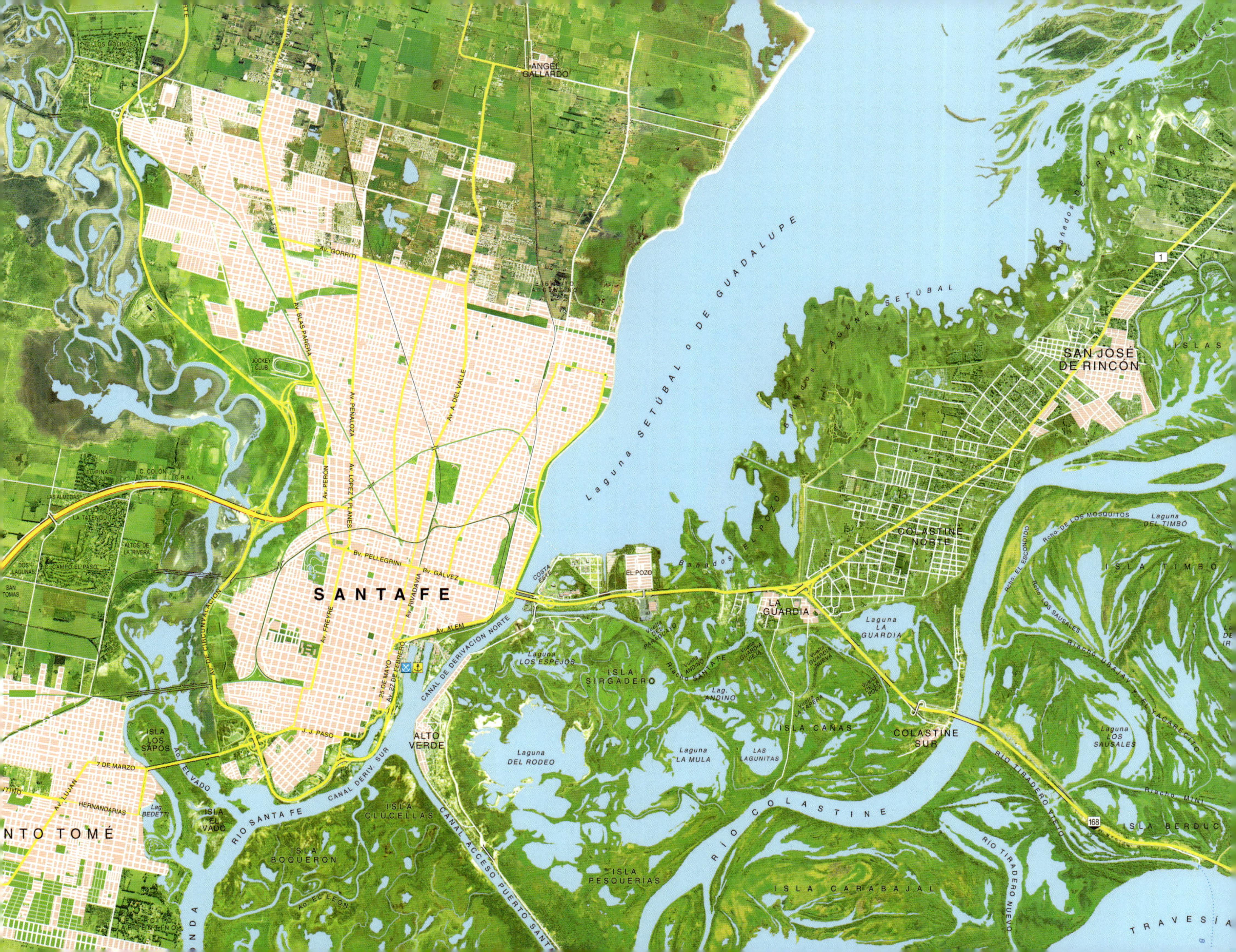

SANTA FE
ANGEL GALLARDO
Av. GORRITI
Av. BLAS PARERA
Av. PEÑALOZA
Av. A. DEL VALLE
JOCKEY CLUB
Av. PERON
Av. LOPEZ Y PLANES
Bv. PELLEGRINI
Bv. GALVEZ
Av. RIVADAVIA
Av. FREYRE
Av. ALEM
25 DE MAYO
Av. 27 DE FEBRERO
J. J. PASO
Av. DE CIRCUNVALACION
COSTA
EL POZO
Laguna SETÚBAL o DE GUADALUPE
LAGUNA SETÚBAL
Bañados del Rincón
SAN JOSÉ DE RINCÓN
ISLAS
COLASTINE NORTE
LA GUARDIA
Laguna LA GUARDIA
Laguna DEL TIMBÓ
Rcho. DE LOS MOSQUITOS
ISLA TIMBÓ
Rcho. EL ESCONDIDO
Rcho. LOS SAUSALES
Riacho UBAJAY y EL YACARECITO
Laguna LOS SAUSALES
Riacho MINI
ISLA BERDUC
TRAVESÍA
RÍO TIRADERO VIEJO
RÍO TIRADERO NUEVO
RÍO COLASTINE
COLASTINE SUR
ISLA CARABAJAL
ISLA CAÑAS
LAS LAGUNITAS
Laguna LA MULA
Lag. ANDINO
Laguna DEL RODEO
ISLA SIRGADERO
Laguna LOS ESPEJOS
Riacho SANTA FE
CANAL DE DERIVACION NORTE
ALTO VERDE
CANAL ACCESO PUERTO SANTA FE
ISLA PESQUERIAS
ISLA CLUCELLAS
ISLA BOQUERON
Ao. EL LEON
CANAL DERIV. SUR
RIO SANTA FE
ISLA EL VADO
Ao. EL VADO
Lag. BEDETTI
ISLA LOS SAPOS
7 DE MARZO
HERNANDARIAS
Av. LUJAN
NTO TOMÉ
SAN TOMAS
DOS LAGUNAS
CAMPO EL PASO
LA TATENGUITA
LAS ALMEDAS
C. COLON
PINAR
ALTOS DE LA RIVERA
1
168

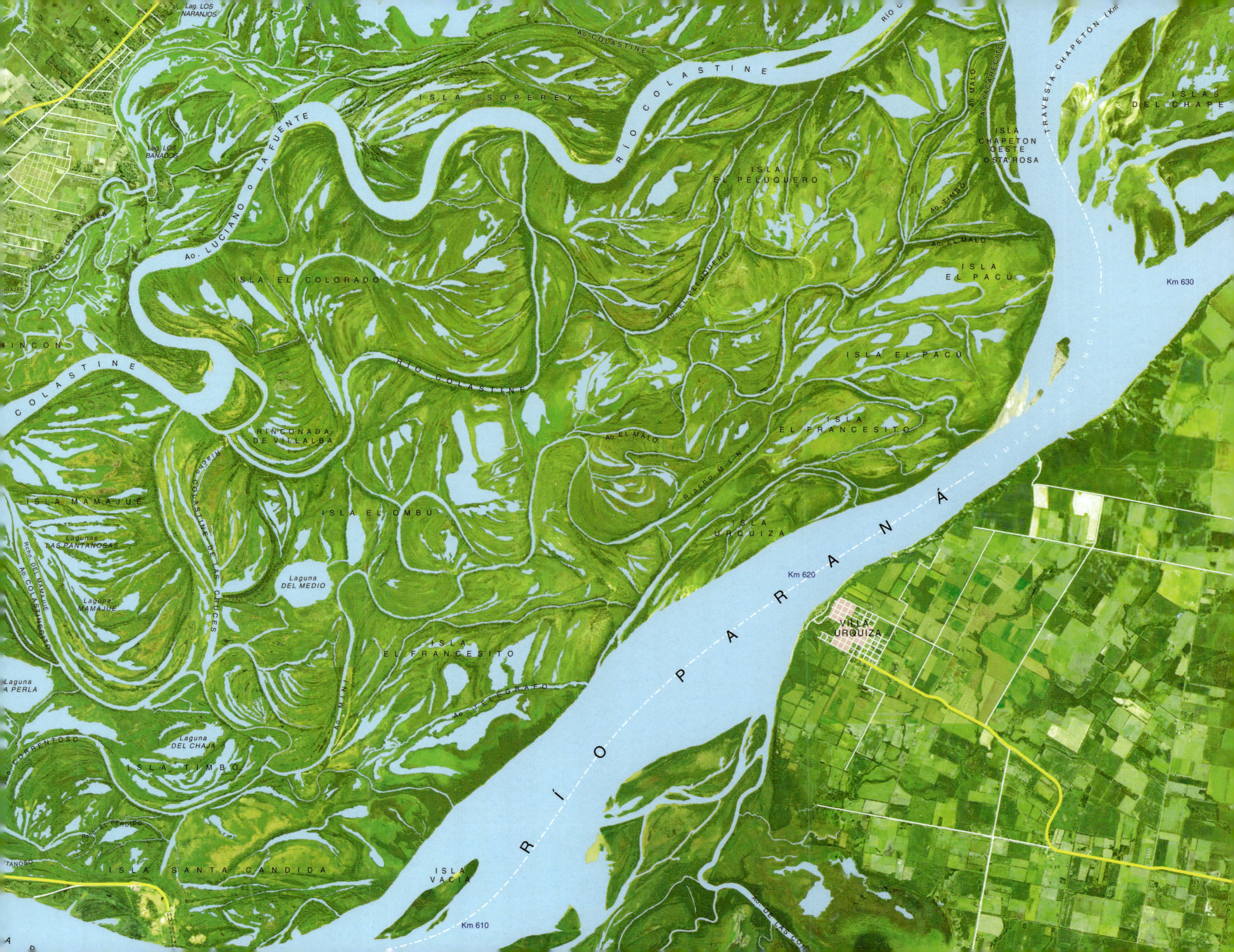

Lag. LOS NARANJOS
Ao. COLASTINE
RÍO COLASTINE
ISLA SOPEREX
Lag. LOS BAÑADOS
Ao. LUCIANO o LA FUENTE
ISLA EL COLORADO
RINCON
COLASTINE
RÍO COLASTINE
RINCONADA DE VILLALBA
Riacho COLASTINE DE LAS CRUCES
ISLA MAMAJUE
Lagunas LAS PANTANOSAS
Rcho. DEL MAMAJUE
Ao. COLASTINECITO
Laguna MAMAJUE
Laguna DEL MEDIO
ISLA EL OMBÚ
ISLA EL FRANCESITO
Ao. TELEGRAFO
Ao. MINI
Laguna DEL CHAJÁ
ISLA TIMBO
Ao. CORRENTOSO
Rcho. EL PERDIDO
ISLA SANTA CANDIDA
ISLA VACIA
Km 610
RÍO PARANÁ
Km 620
VILLA URQUIZA
ISLA URQUIZA
Riacho MINI
Ao. EL MALO
ISLA EL FRANCESITO
ISLA EL PACU
ISLA EL PACU
Ao. EL MALO
Ao. TIMBO
ISLA CHAPETON OESTE o STA.ROSA
ISLA EL PELUQUERO
Ao. EL PELUQUERO
TRAVESÍA CHAPETON 1Km
ISLAS DEL CHAPE
Km 630
LIMITE PROVINCIAL

Screwed: 2020 Presidential Election »

Kenneth Field

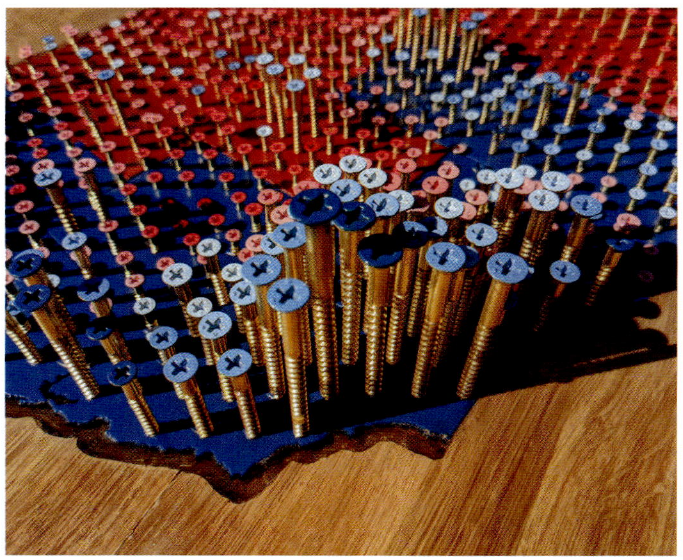

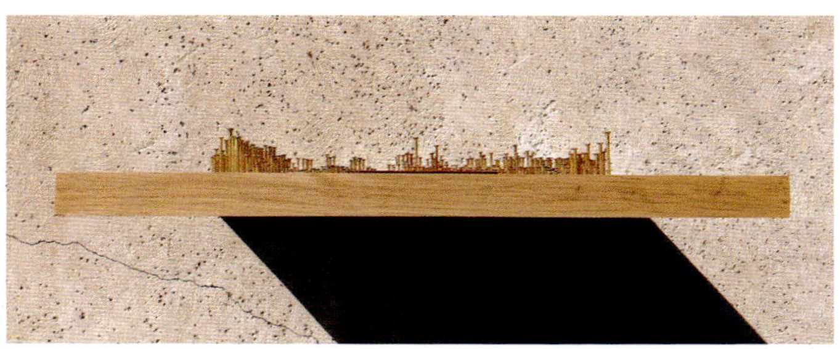

Hanging in the Tiffany & Co. store on Fifth Avenue in New York City is a portrait of Charles Lewis Tiffany, done as a 3D bas-relief consisting of tightly packed, gridded screws. The screw depth and color (in shades of gray) create a three-dimensional terrain of Tiffany's head. Upon seeing this for the first time, I thought it would make an interesting cartographic sculpture, and I consigned it to my memory bank of inspiration.

Fast-forward to the 2020 US presidential election, and my mind wandered back to the Tiffany portrait. Others have used a similar technique for Digital Elevation Models (DEMs) but not, as far as I knew, for a thematic map. I wanted to design and create a piece of "cARTographic" work, and use the process as a catharsis during a turbulent time in US politics. It provided me with many hours of enjoyment during the pandemic lockdown, as I used power tools to physically craft a map rather than clicking a mouse.

I'm not entirely sure there's a true "need" for this sort of cartographic product, but the endeavor, and the result, supports the longstanding use of maps as art. They are made to delight, and enjoyed for what they are, rather than the more usual reasons we look at and use maps.

The map functions, in the sense that there is a color-coded base map of the US states indicating the winning party in the election. The screw gauge and height represent voter turnout. Each screw head is painted in a shade of red or blue, with darker shades representing a higher share of the vote, and there's an epoxy resin–filled vignette for the coastline and lakes. It's a multivariate, thematic map that just happens to be made out of wood, paint, and screws as opposed to pen and paper, or pixels. The title plate added a playful take on the state of US politics during this turbulent period.

This map is the end-product of a somewhat serendipitous design process from an initial chance interaction with a piece of art that inspired me, but one that required detailed planning. It harks back to predigital cartography, when there wasn't an easy-to-use "undo" button.

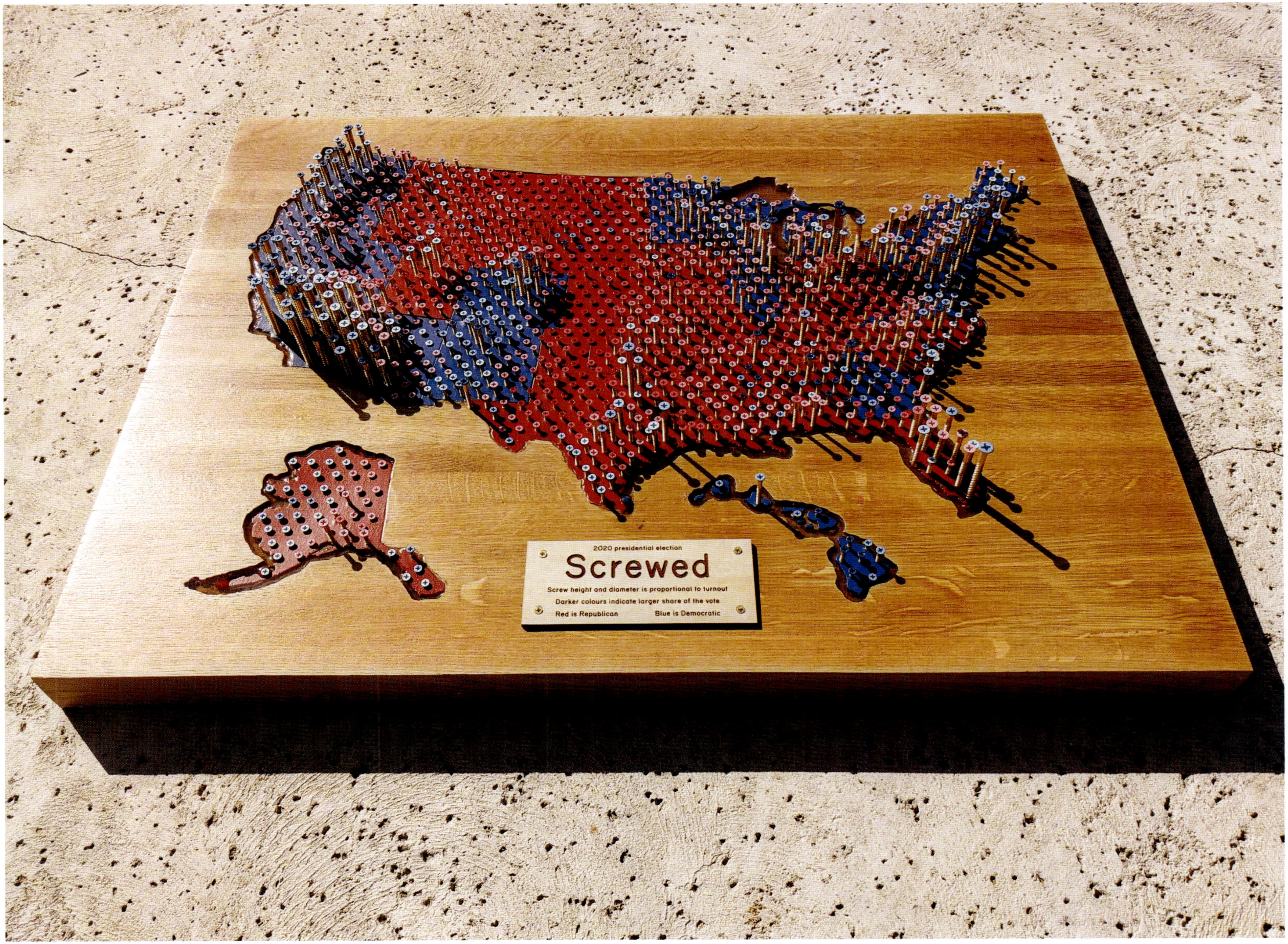
2020 presidential election
Screwed
Screw height and diameter is proportional to turnout
Darker colours indicate larger share of the vote
Red is Republican
Blue is Democratic

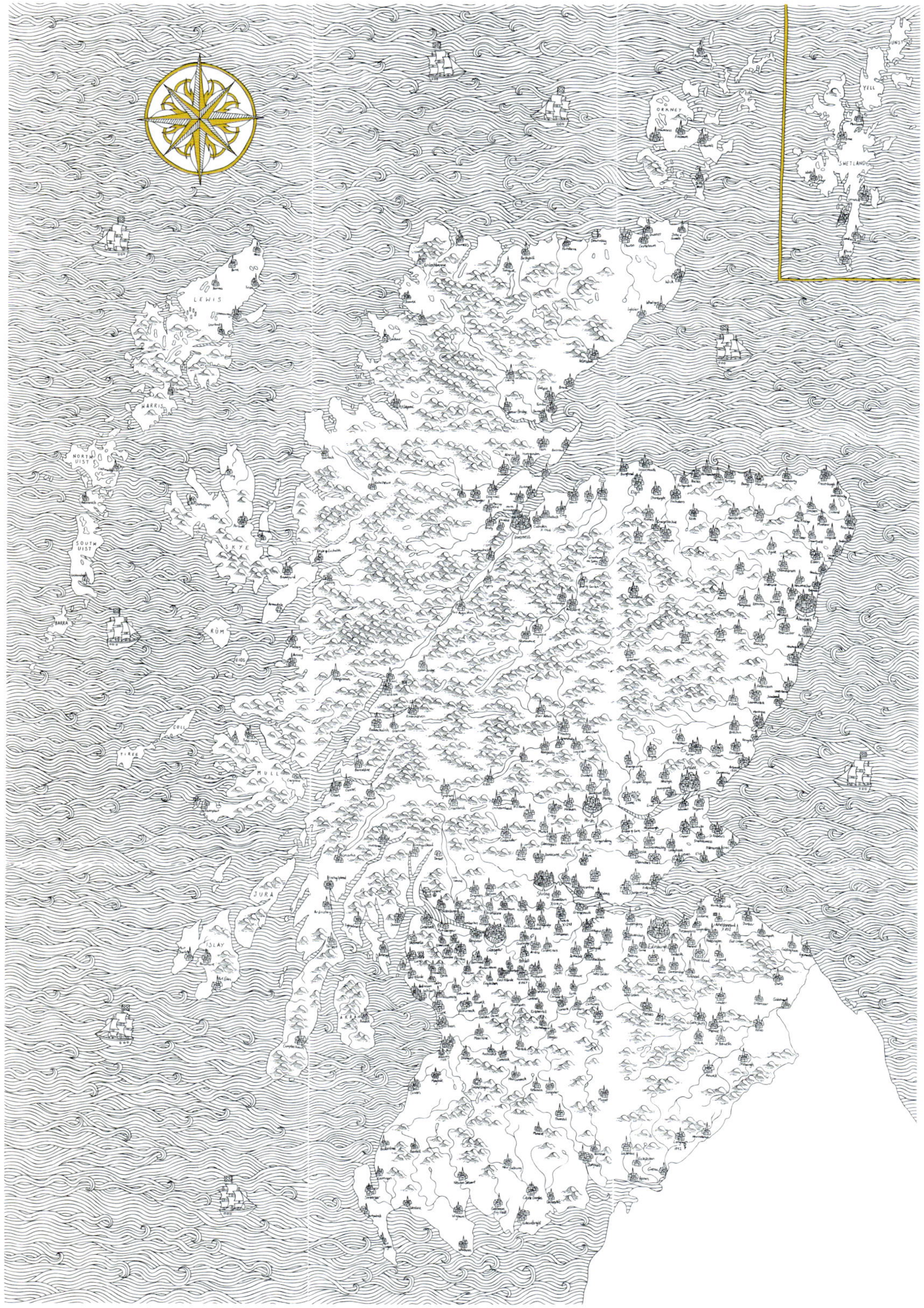

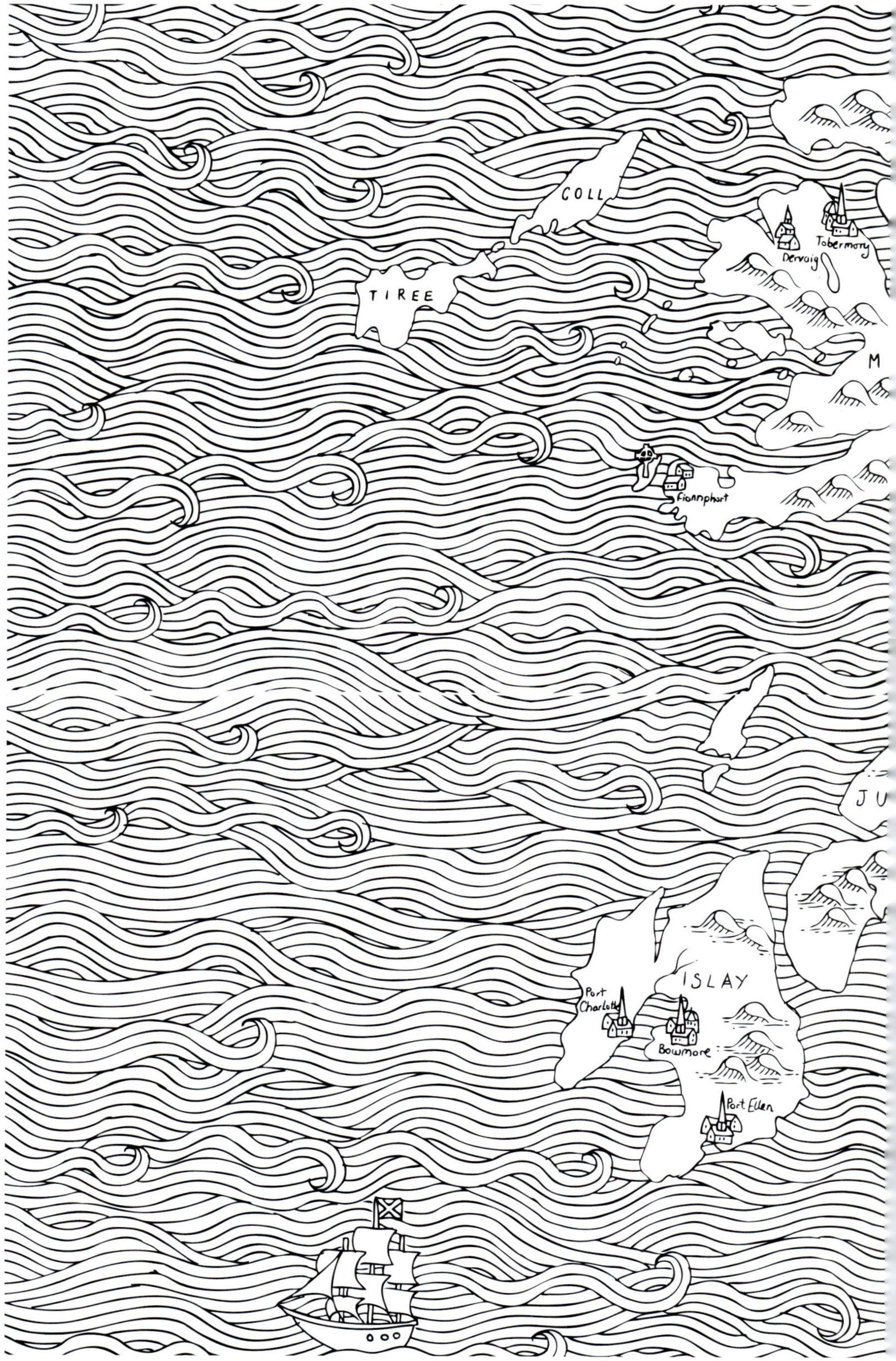
COLL
TIREE
Dervaig
Tobermory
Fionnphort
ISLAY
Port Charlotte
Bowmore
Port Ellen

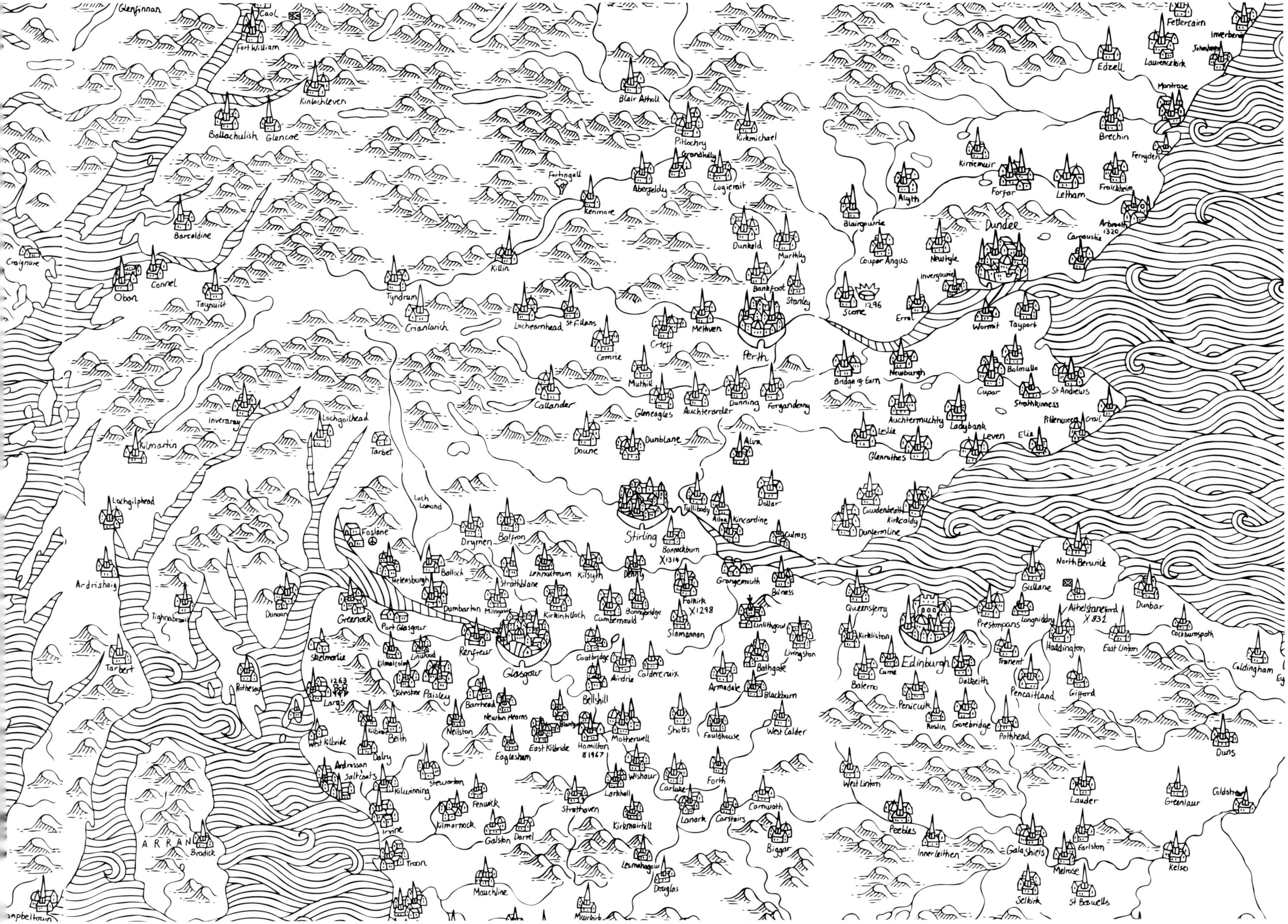
Glenfinnan
Caol
Fort William
Kinlochleven
Blair Atholl
Fettercairn
Inverbervie
Johnshaven
Edzell
Laurencekirk
Montrose
Ballachulish
Glencoe
Pitlochry
Kirkmichael
Brechin
Ferryden
Grandtully
Fortingall
Aberfeldy
Logierait
Kirriemuir
Forfar
Letham
Friockheim
Alyth
Kenmore
Barcaldine
Blairgowrie
Dundee
Arbroath 1320
Carnoustie
Dunkeld
Murthly
Coupar Angus
Newtyle
Craignure
Connel
Killin
Invergowrie
Oban
Taynuilt
Tyndrum
Bankfoot
Stanley
Scone 1296
Errol
Wormit
Tayport
Crianlarich
Lochearnhead
St Fillans
Crieff
Methven
Comrie
Perth
Bridge of Earn
Newburgh
Balmullo
Cupar
St Andrews
Muthill
Callander
Gleneagles
Auchterarder
Dunning
Forgandenny
Strathkinness
Auchtermuchty
Ladybank
Inveraray
Lochgoilhead
Pittenweem
Crail
Kilmartin
Tarbet
Doune
Dunblane
Alva
Leslie
Leven
Elie
Glenrothes
Loch Lomond
Dollar
Lochgilphead
Tullibody
Alloa
Kincardine
Cowdenbeath
Kirkcaldy
Dunfermline
Culross
Faslane
Drymen
Balfron
Stirling
Bannockburn X1314
Ardrishaig
Helensburgh
Balloch
Lennoxtown
Kilsyth
Denny
Grangemouth
Bo'ness
North Berwick
Strathblane
Falkirk X1298
Gullane
Athelstaneford X832
Dunbar
Tighnabruaich
Dunoon
Greenock
Dumbarton
Milngavie
Kirkintilloch
Cumbernauld
Bonnybridge
Linlithgow
Queensferry
Prestonpans
Longniddry
Port Glasgow
Renfrew
Slamannan
Kirkliston
Haddington
East Linton
Cockburnspath
Tarbert
Skelmorlie
Kilmacolm
Linwood
Glasgow
Coatbridge
Airdrie
Caldercruix
Livingston
Bathgate
Edinburgh
Tranent
Coldingham
Rothesay
1263
Largs
Johnstone
Paisley
Barrhead
Bellshill
Armadale
Blackburn
Balerno
Currie
Dalkeith
Pencaitland
Gifford
Newton Mearns
Kilbirnie
Beith
Neilston
Blantyre
Penicuik
Roslin
Gorebridge
Pathhead
West Kilbride
Dalry
Eaglesham
East Kilbride
Hamilton 1967
Motherwell
Shotts
Fauldhouse
West Calder
Duns
Ardrossan
Saltcoats
Stewarton
Wishaw
Carluke
Forth
West Linton
Kilwinning
Fenwick
Strathaven
Larkhall
Lanark
Carstairs
Carnwath
Lauder
Greenlaw
Coldstream
Irvine
Kilmarnock
Galston
Darvel
Kirkmuirhill
Peebles
Biggar
ARRAN
Brodick
Troon
Lesmahagow
Innerleithen
Galashiels
Earlston
Kelso
Melrose
Douglas
Mauchline
Selkirk
St Boswells

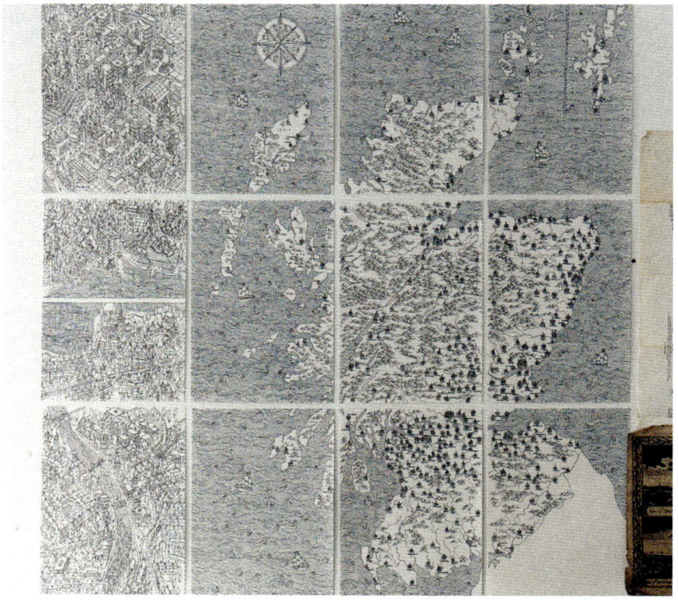

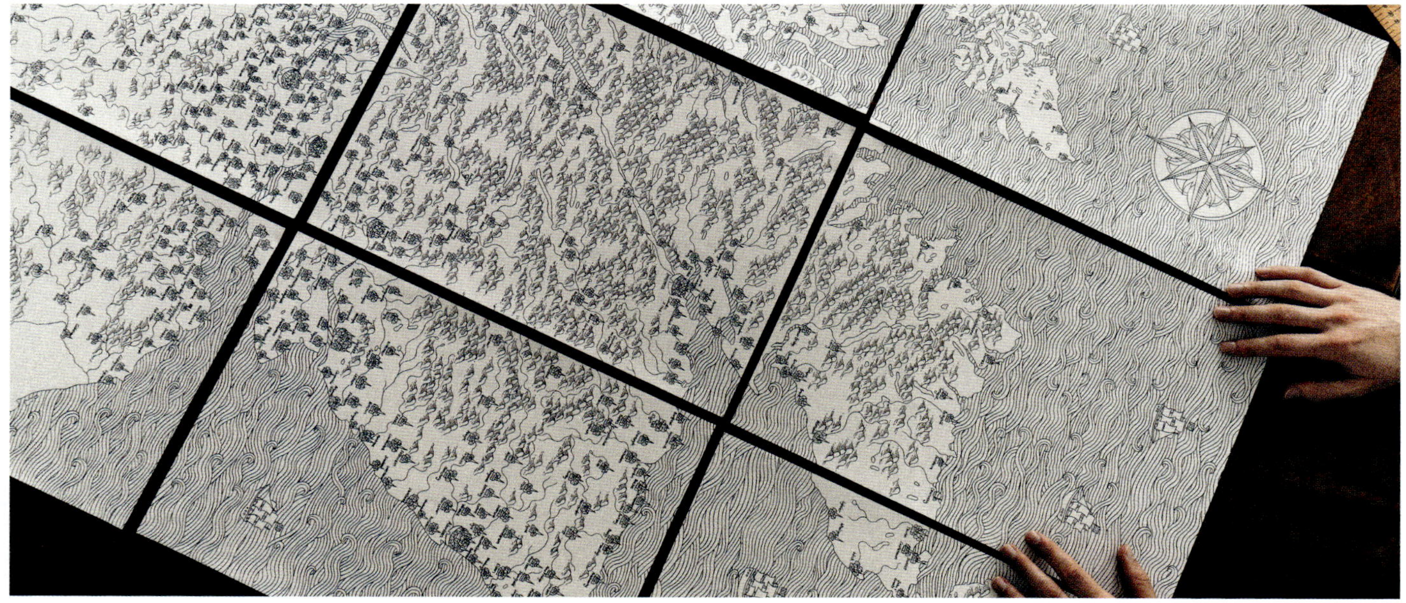

« Atlas of Scotland

Andrew Redmond Barr

To me, map design is an opportunity to present the history and culture of a place in a unique and imaginative way. I had worked on a range of Scottish cultural projects before, but this time I wanted to attempt something that would give a more expansive overview of the whole of Scotland, something that would delve deeper into the fabric of the land and reveal one of the world's oldest nations in a whole new light.

It was at that point that I started using hand-drawn mapping as a basis for telling Scotland's story, and I set to work on an "Atlas of Scotland." It was certainly the biggest creative challenge I had ever set myself.

The maps are drawn in simple black and white, with occasional flashes of gold. I took inspiration from the way gold is used symbolically in old Scottish poetry, for example, when Robert Burns (Scotland's national poet) likened the value of humankind to "the gowd" (gold).

Over many months a picture of Scotland started to emerge in segments, and I began to see how it would all work when laid out in the pages of an atlas. I decided to represent settlements over a certain size with small clusters of houses around a kirk (church) spire, which is how most towns and villages were first established. Scotland's seven cities are represented by larger clusters with multiple spires, as well as city walls and recognizable local landmarks.

Some settlements are also marked with dates representing important moments in Scottish history, including the births of literary figures, political events, and decisive battles. A Saltire (the flag of Scotland) flies atop Ben Nevis, Scotland's highest mountain. A Celtic cross stands on the Isle of Iona, representing Scotland's ancient spiritual heartland. I've also marked the standing stones of Callanish on the Isle of Lewis, and the ancient tree at Fortingall, which is thought to be one of the oldest living things in Europe.

I hope this atlas will encourage people to see Scotland in a new light, and reaffirm a sense of the country's distinct place in the world.

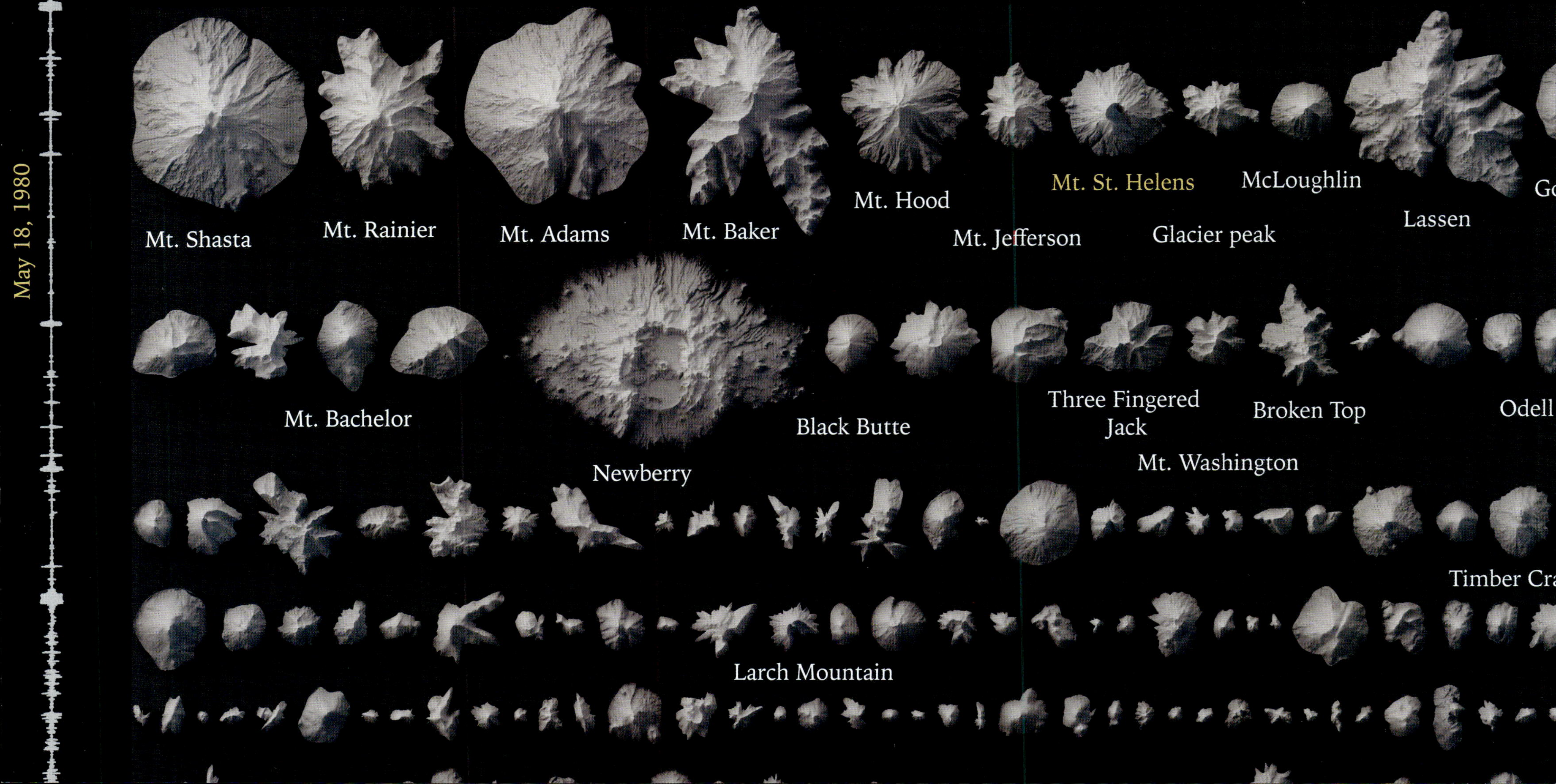
May 18, 1980
Mt. Shasta
Mt. Rainier
Mt. Adams
Mt. Baker
Mt. Hood
Mt. Jefferson
Mt. St. Helens
Glacier peak
McLoughlin
Lassen
Mt. Bachelor
Newberry
Black Butte
Three Fingered Jack
Mt. Washington
Broken Top
Odell
Timber Cra
Larch Mountain

Each edifice tells a unique story of construction and erosion.

Three Views of Cascades Volcanoes »

Leif Karlstrom and Daniel O'Hara

This map was conceived as part of a scientific study to understand patterns of volcanic activity associated with the Cascadia subduction zone in the United States. Our research goal was to estimate the total volume of volcanoes erupting over the last 2.6 million years (the Quaternary period of Earth history), and to identify how edifices at the surface relate to structures inferred in the subsurface. We found roughly 3,000 individual volcanoes, and taught a computer how to extract these landforms automatically from satellite-derived Digital Elevation Models.

Our study was a success in that it generated new insights into the volcanic history of the Cascades. But we were so impressed by the distribution of volcanic landforms—far more extensive than one might guess based on the dozen or so stratovolcanoes dominating the Pacific Northwest skyline—that we began a more playful examination of the topographic catalog. Inspired by nineteenth-century atlas illustrations depicting "comparative heights of mountains and lengths of rivers," we wanted to generate a similarly aesthetic representation, but one based on modern high-resolution data. The result is our poster, "Three Views of Cascades Volcanoes."

The centerpiece is a height-ordered sequence of volcanoes presented like a coin book. One sees immediately that large stratovolcanoes and shields are far outnumbered by small cinder cones and domes, with an incredible variety of shapes. Each edifice tells a unique story of construction and erosion. To further highlight these differences, we present a perspective view of edifices clumped as if in a single mountain mass, along with a geographically accurate representation with the volcanoes extruded from a regional topographic map.

A final design element was making human connections. For those who experienced it, the 1980 Mount St. Helens eruption was a defining life event, one that also fundamentally changed the shape of the volcano. We include a seismogram (data recording the ground shaking) from this eruption as a map border. We also worked to identify common Native American names for major Cascades volcanoes in the legend, recognizing the deep impact these landforms have had on generations of humans in the Pacific Northwest.

Volcanoes of the Cascades

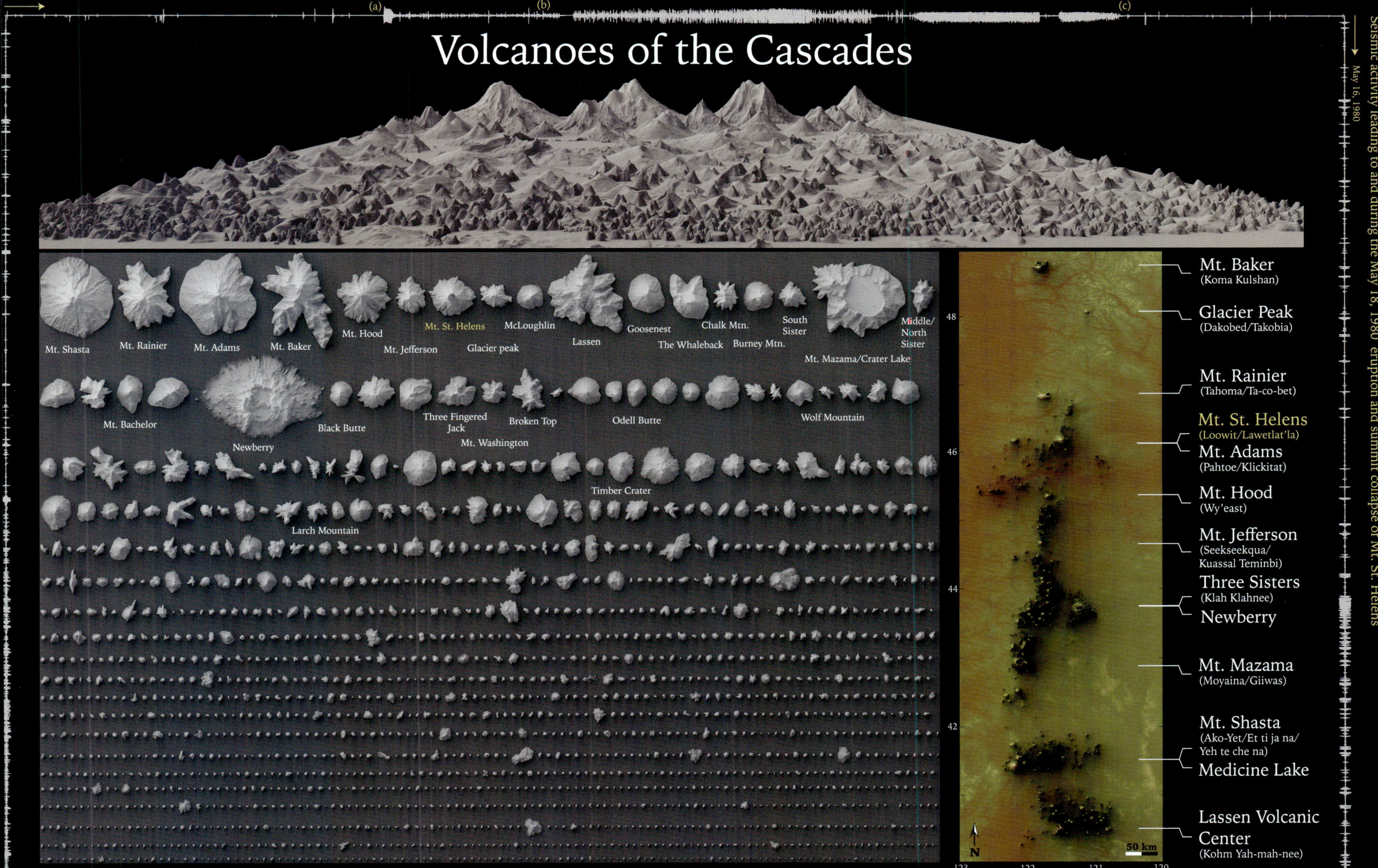

Three views of 2105 Quaternary (<2.6 Ma) volcanoes in the Cascades arc, USA, using 10m resolution satellite-derived digital elevation models. Volcano edifice outlines come from O'Hara et al., (2020), https://doi.org/10.1130/G47706.1. Top: Perspective view of clustered volcanoes. Bottom left: Height ordered, from top left. Bottom right: geography, native American place names come from local indigenous languages. Border: re-digitized seismogram from Mt. St. Helens (starting upper right, read clockwise) May 16-18 (UTC), 1980, at stations APE and MTM. Label (a) Landslide/start of cataclysmic eruption, (b) Onset of Plinian eruptive phase, (c) End of Plinian eruptive phase. For methods and list of volcanoes, visit http://www.volcanolisteningproject.org

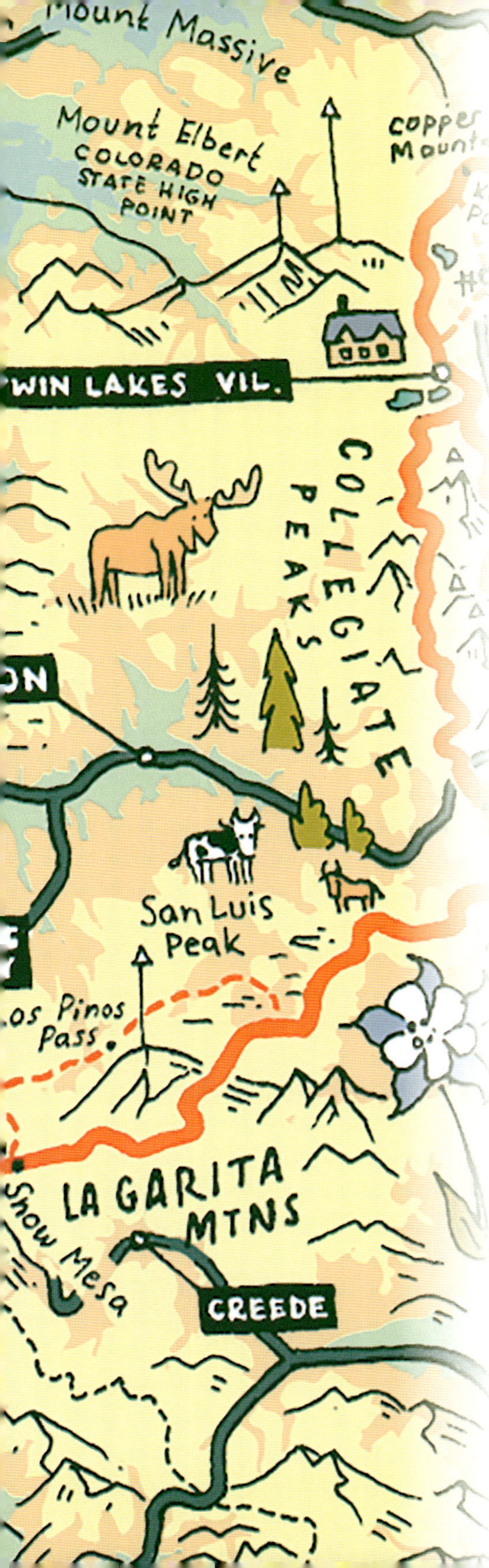

> *...the illustrations were drawn from experiences many hikers that year had in common: encountering a whole valley full of angry cows...*

Colorado Trail »

Madeline Grubb

The Colorado Trail is a rugged and beautiful 500-mile strip of dirt that traverses eight mountain ranges, meanders along five major river systems, and wanders through countless wilderness areas and a patchwork of public lands at an average elevation of 10,300 feet. Thru-hiking has always been one of my biggest passions, so I jumped at the opportunity to hike the Colorado Trail in the summer of 2021. Halfway along the trail I accidentally smashed my phone. After that, I was left with my thoughts and the landscape to keep myself entertained. I dove headfirst into getting better at flower identification, trying to sketch the mountains while I was sitting out storms, and taking in as much of the trail as possible. This map is my attempt to convey the joy of moseying through the wildflower-filled meadows of the high alpine, phone free. I felt that an illustrative map, rather than a digitally accurate one, was the best way to capture the stories and feelings that come along with the trail experience, and an approach I hope to use on other maps of long-distance trails in the future.

The towns shown are all "trail towns," important for resupply as you travel through on foot. Many of the illustrations were drawn from experiences many hikers that year had in common: encountering a whole valley full of angry cows before the La Garitas and so many moose while going through the Collegiates.

I wanted to create a map that other thru-hikers and bikers could see their own adventure in. All of the linework was done in pen and ink and the colors drawn in digitally. I hiked during a terribly wet monsoon season, and much of my time was spent under rainy skies. The muted shades of gray, teal, and browns were translated into the map's background landscape and elevation. But in those same moments the trail was also dotted with bright pops of color: sunsets, wildflowers, and the new growth of plants and trees. I tried to conjure these moments out with brighter colors for contrast.

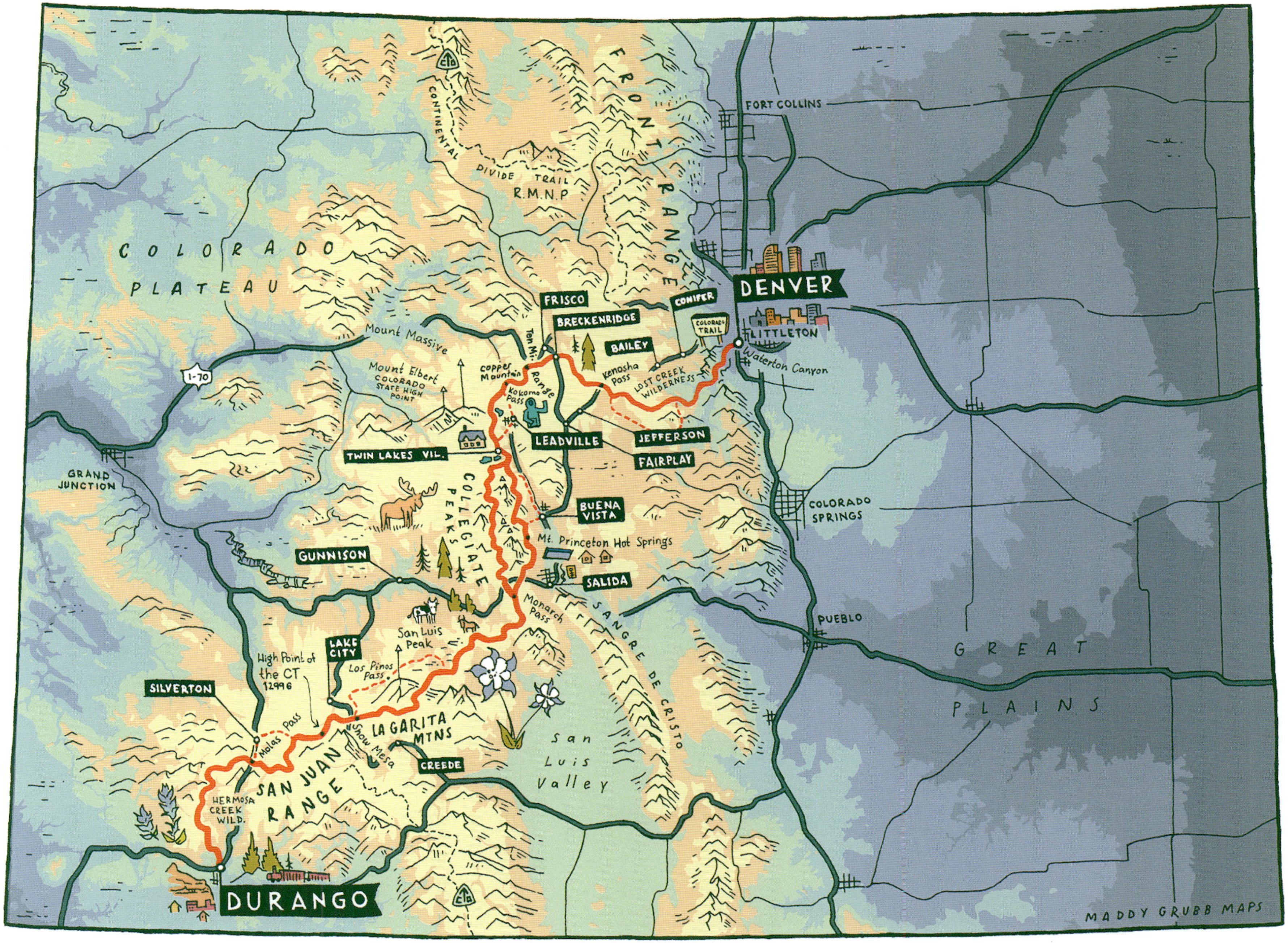

COLORADO TRAIL

500 Miles from Denver → Durango

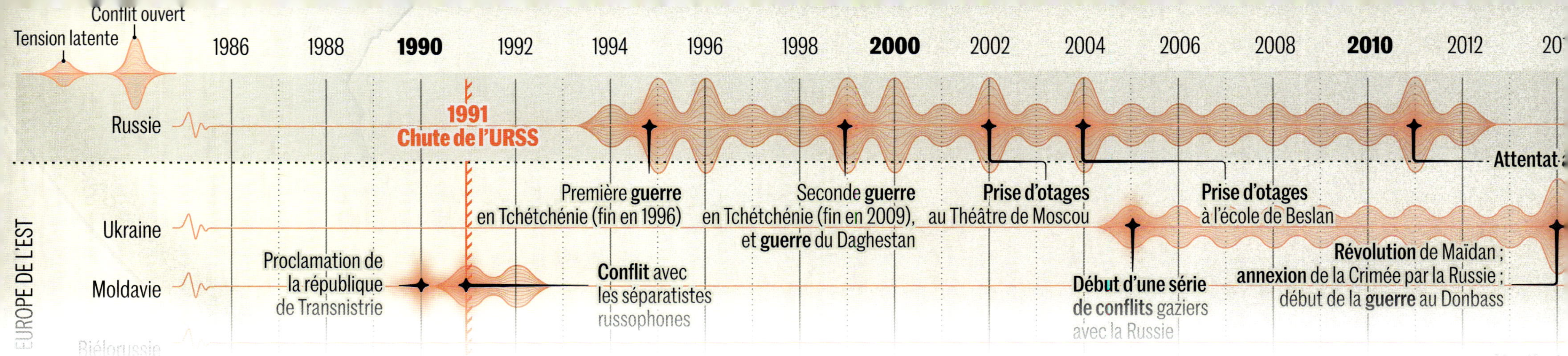

Thirty years of tension and conflict »
Contested borders, ethnic violence, Russian aggressions… The shock waves of the fall of the USSR

Xemartin Laborde, Lucie Rubrice and Delphine Papin, Le Monde

This cartography and data visualization page is part of the geopolitical section of the French daily *Le Monde*. Thirty years ago, the USSR (Union of Soviet Socialist Republics) disappeared, and this page is an opportunity to show the numerous troubles that followed this historical event: armed conflicts, massacres, insurrections, violence, in which almost all the former Soviet republics are concerned.

Historical maps are always an opportunity to explore the graphic styles of the past. Here we had fun re-creating a graphic atmosphere reminiscent of the propaganda maps drawn from the 1920s to the 1960s. We based our project on the work of three American cartographers who drew true works of art: Richard Edes Harrison, cartographer for Fortune and Life magazines; Robert M. Chapin, cartographer at Time magazine; and Erwin Raisz. These three cartographers have always been concerned with providing a visual narrative, with immersing their readers in their maps. The visual effects, the confusing projections, the contribution of didactic illustration make their cartographic plates models to follow.

In our page, we could have used a classical flat projection, but we wanted to give the visual impression of a planetary empire that crosses an entire hemisphere. The orthographic projection, which respects the spherical aspect of the Earth, seemed to us the best option. To play with the roundness of the planet and to amplify the didactic and analytic side of our cartographic page, we had the idea of proposing three zooms on the major zones of concern, emphasizing the impression of extraction, as if we were cutting and removing bits of skin from an orange. Other devices give the map a vintage feel: the blackened side of the continents, the blurring toward a faded yellow, the ice cap to recall the Cold War era, the halftone effect, the use of two-color printing combined with a photo grain, and the small flag of the North Pole are all references to maps from the 1960s. Finally, we wanted to propose an original data visualization by playing on the notion of shock waves with an aesthetic reminiscent of the 1960s.

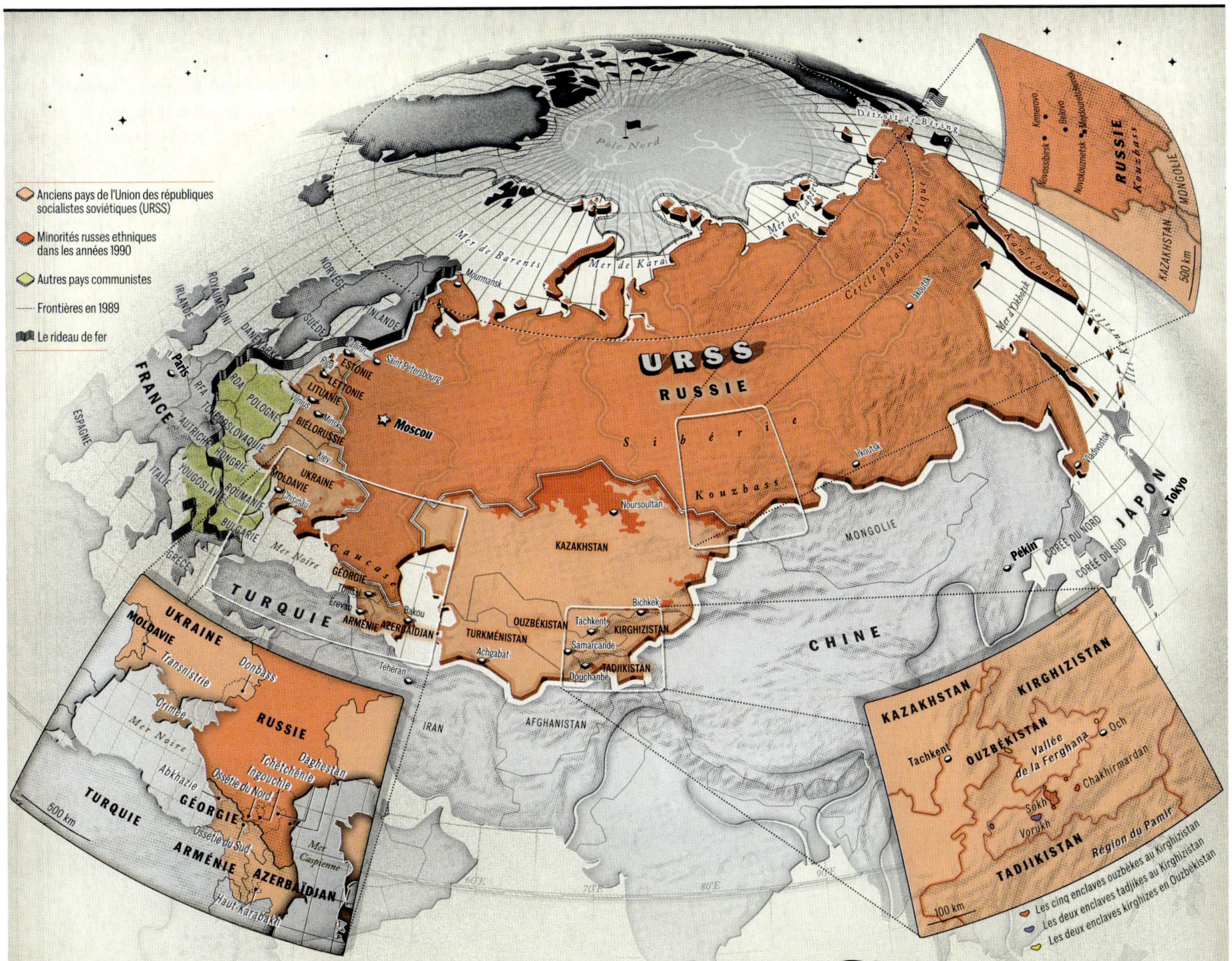

Trente ans de tensions et de conflits

Frontières contestées, violences ethniques, agressions russes... Les ondes de choc de la chute de l'URSS

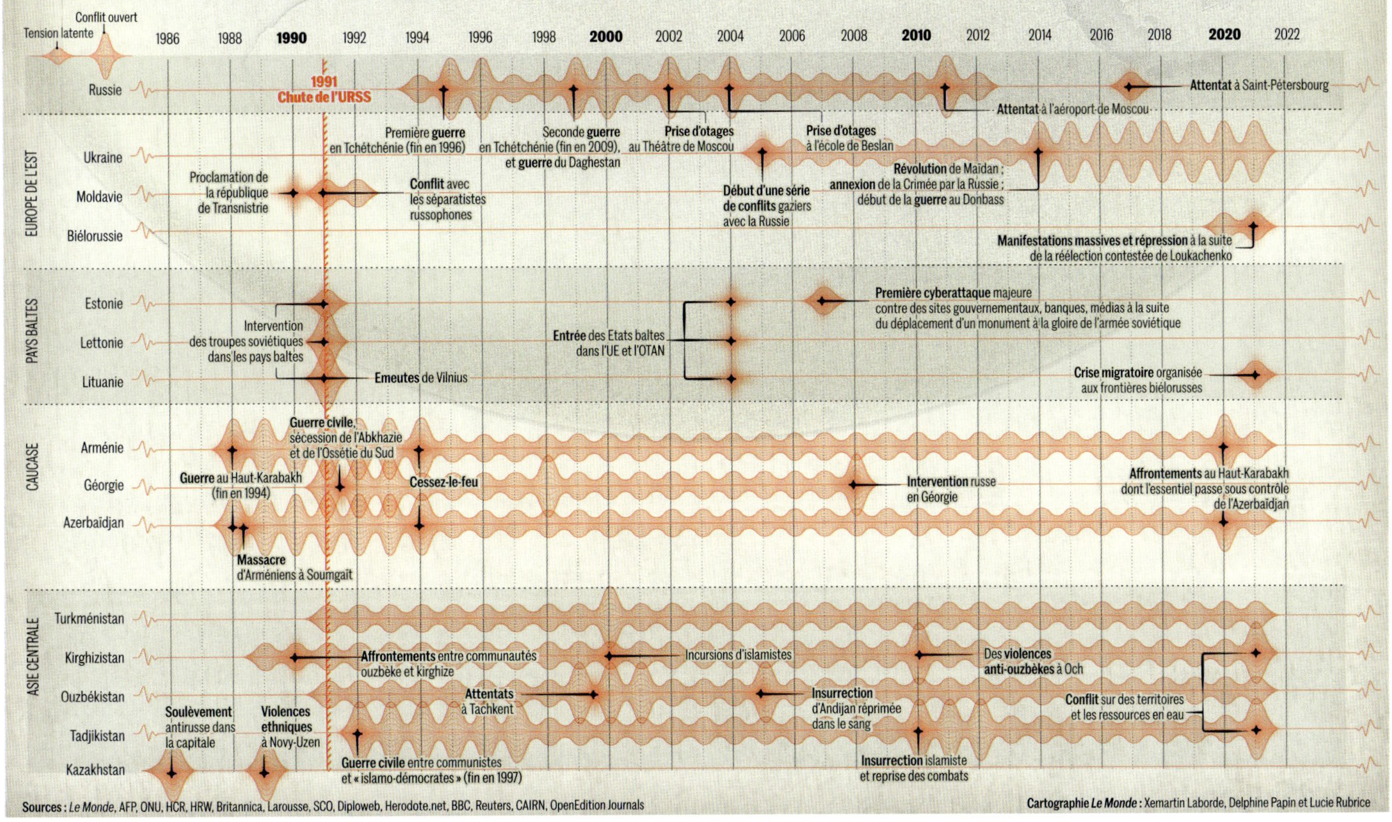

Sources : *Le Monde*, AFP, ONU, HCR, HRW, Britannica, Larousse, SCO, Diploweb, Herodote.net, BBC, Reuters, CAIRN, OpenEdition Journals

Cartographie *Le Monde* : Xemartin Laborde, Delphine Papin et Lucie Rubrice

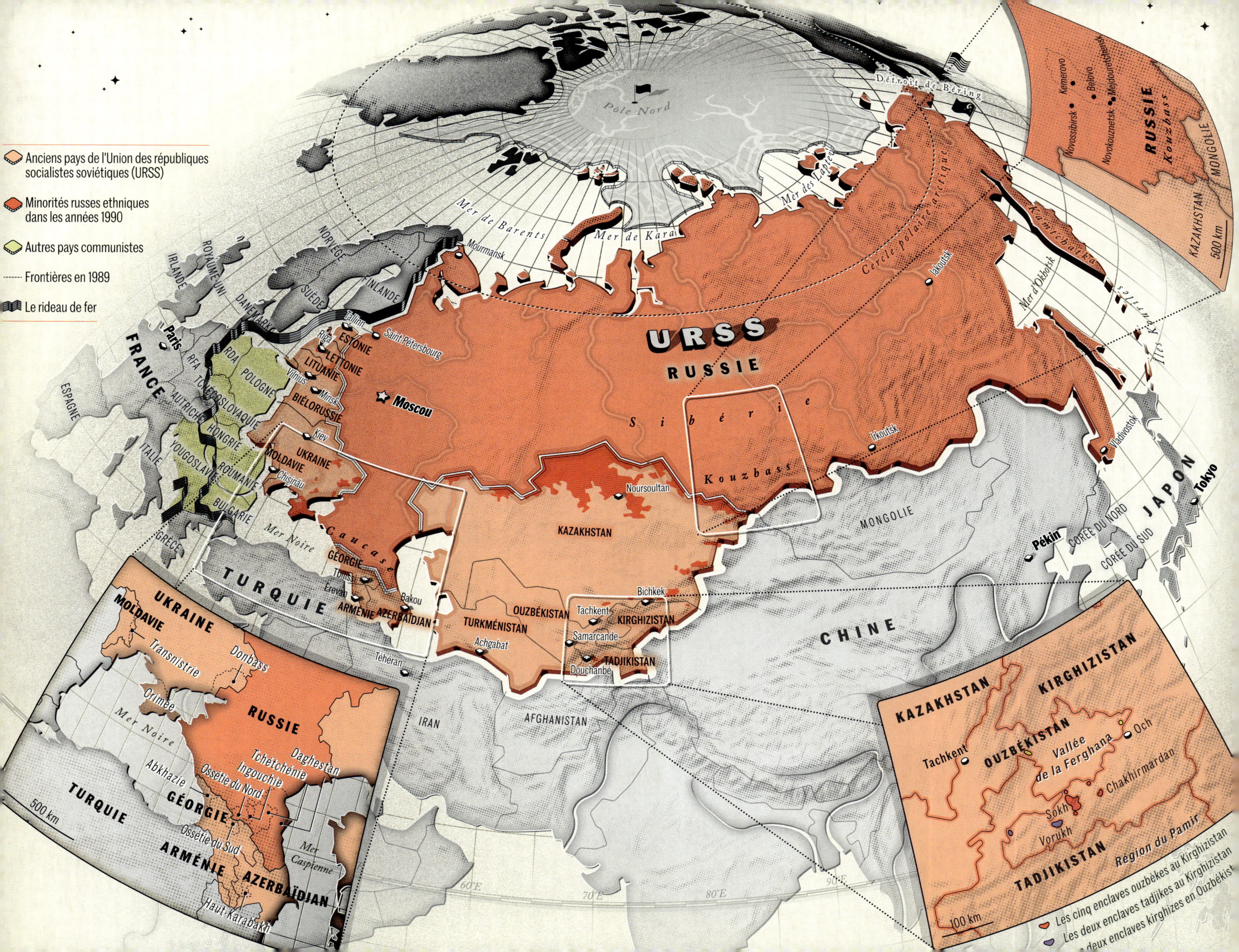
Anciens pays de l'Union des républiques socialistes soviétiques (URSS)
Minorités russes ethniques dans les années 1990
Autres pays communistes
Frontières en 1989
Le rideau de fer
URSS
RUSSIE
Sibérie
Kouzbass
Moscou
Saint-Pétersbourg
Mourmansk
Iakoutsk
Irkoutsk
Vladivostok
Pôle Nord
Détroit de Béring
Mer de Barents
Mer de Kara
Mer des Laptev
Cercle polaire arctique
Kamtchatka
Mer d'Okhotsk
Kouriles
Tallinn
Riga
ESTONIE
LETTONIE
LITUANIE
Vilnius
Minsk
BIÉLORUSSIE
Kiev
UKRAINE
MOLDAVIE
Chisinau
Caucase
GÉORGIE
Tbilissi
Erevan
ARMÉNIE
Bakou
AZERBAÏDJAN
KAZAKHSTAN
Noursoultan
OUZBÉKISTAN
Tachkent
Samarcande
TURKMÉNISTAN
Achgabat
KIRGHIZISTAN
Bichkek
TADJIKISTAN
Douchanbe
Téhéran
IRAN
AFGHANISTAN
CHINE
MONGOLIE
Pékin
CORÉE DU NORD
CORÉE DU SUD
JAPON
Tokyo
TURQUIE
Mer Noire
GRÈCE
BULGARIE
ROUMANIE
HONGRIE
YOUGOSLAVIE
ITALIE
AUTRICHE
TCHÉCOSLOVAQUIE
POLOGNE
RDA
RFA
FRANCE
Paris
ESPAGNE
DANEMARK
SUÈDE
NORVÈGE
FINLANDE
ROYAUME-UNI
IRLANDE
60°E
70°E
80°E
90°E
Kemerovo
Belovo
Mejdouretchensk
Novossibirsk
Novokouznetsk
RUSSIE
Kouzbass
MONGOLIE
KAZAKHSTAN
500 km
UKRAINE
MOLDAVIE
Transnistrie
Donbass
Crimée
Mer Noire
RUSSIE
Daghestan
Tchétchénie
Ingouchie
Ossétie du Nord
Abkhazie
GÉORGIE
Ossétie du Sud
TURQUIE
ARMÉNIE
AZERBAÏDJAN
Haut-Karabakh
Mer Caspienne
500 km
KAZAKHSTAN
KIRGHIZISTAN
Tachkent
OUZBÉKISTAN
Vallée de la Ferghana
Och
Chakhirmardan
Sokh
Vorukh
TADJIKISTAN
Région du Pamir
100 km
Les cinq enclaves ouzbèkes au Kirghizistan
Les deux enclaves tadjikes au Kirghizistan
Les deux enclaves kirghizes en Ouzbékistan

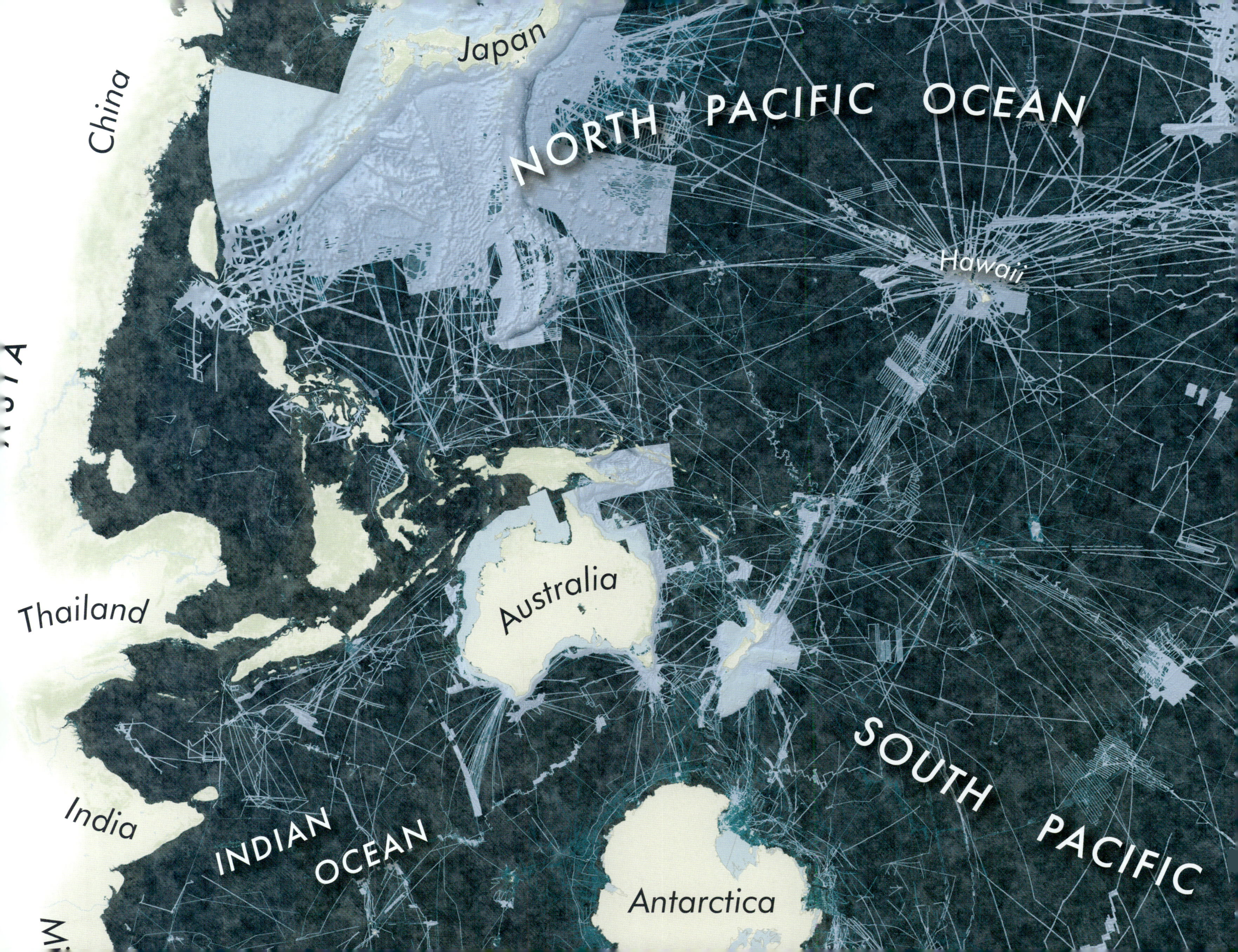
Japan
China
NORTH PACIFIC OCEAN
Hawaii
Thailand
Australia
India
INDIAN OCEAN
Antarctica
SOUTH PACIFIC

Murky Waters

■ Unmapped areas of the ocean floor by the Seafloor 2030 project

GEBCO 2020 seafloor dataset

Unmapped Seafloor by Territory

Percentage of Exclusive Economic Zone (EEZ)

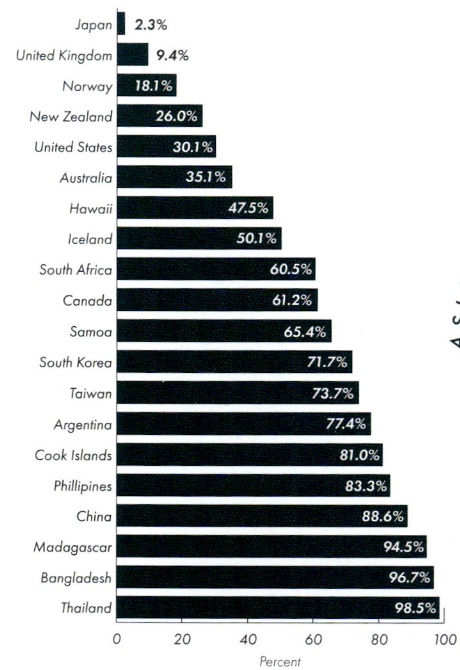

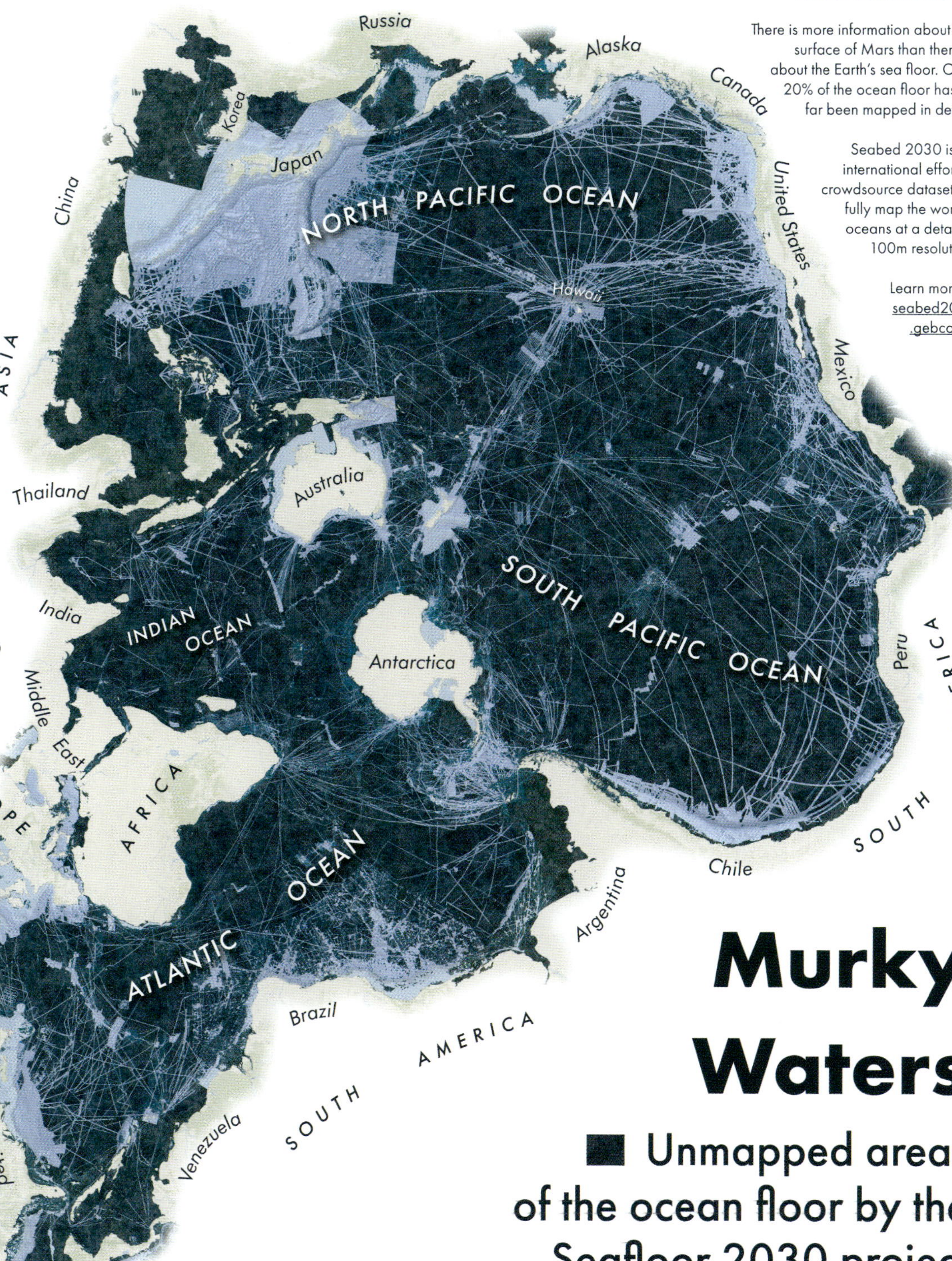

About Seabed 2030

There is more information about the surface of Mars than there is about the Earth's sea floor. Only 20% of the ocean floor has so far been mapped in detail.

Seabed 2030 is an international effort to crowdsource datasets to fully map the world's oceans at a detailed 100m resolution.

Learn more at seabed2030.gebco.net

THE MAP KIWI
Maps and cartography by Andrew Douglas-Clifford

Map by Andrew Douglas-Clifford / The Map Kiwi
andrewdc.co.nz

Data: GEBCO / Seabed 2030, MarineRegions.org, ESRI

...large areas of the planet we call home are not well known to us. «

« Murky Waters: Unmapped Areas of Seafloor

Andrew Douglas-Clifford

The oceans of the Earth are vast, covering more than 70 percent of the surface of the planet. Many people see Google Earth, with all of its imagery and bathymetry information, and think that in the age of satellites and detailed geographic information, there must be very few places on the surface of the Earth left to map.

However, unlike the dry areas on land that have been well mapped for centuries, the ocean's murky depths remain largely unmapped in detail. To date, satellites have been able to take indirect, gravity-based measurements of the sea floor, but this provides only a crude picture.

The Seabed 2030 project aims to complete a full and detailed map of the world seabed by collating many sets of direct measurements from ship surveys into a single bathymetry dataset.

When this map was created in 2020, only 20 percent of the ocean floor had been mapped in detail, and much remains murky and unknown. The map explores what areas are yet to be mapped by taking the negative of known surveys. It is displayed using the Spilhaus Projection, which is often dubbed the "projection for fish." The projection is well suited for this ocean-centric view of the world, which is practical for such data. But it also evokes a different, perhaps startling perspective—revealing that large areas of the planet we call home are not well known to us.

Beyond showing us the mammoth task the Seabed 2030 project has ahead of it this decade, the map also reveals patterns of how human civilization interacts on the surface of the sea. Thin ribbons of journeying ships can be discerned across the vast ocean, connecting islands and continents. Some nations appear to prioritize exploration of their Exclusive Economic Zones much more than others. Events in recent history, such as the search area for the missing Malaysia Airlines Flight 370 in the Indian Ocean, can also be spotted.

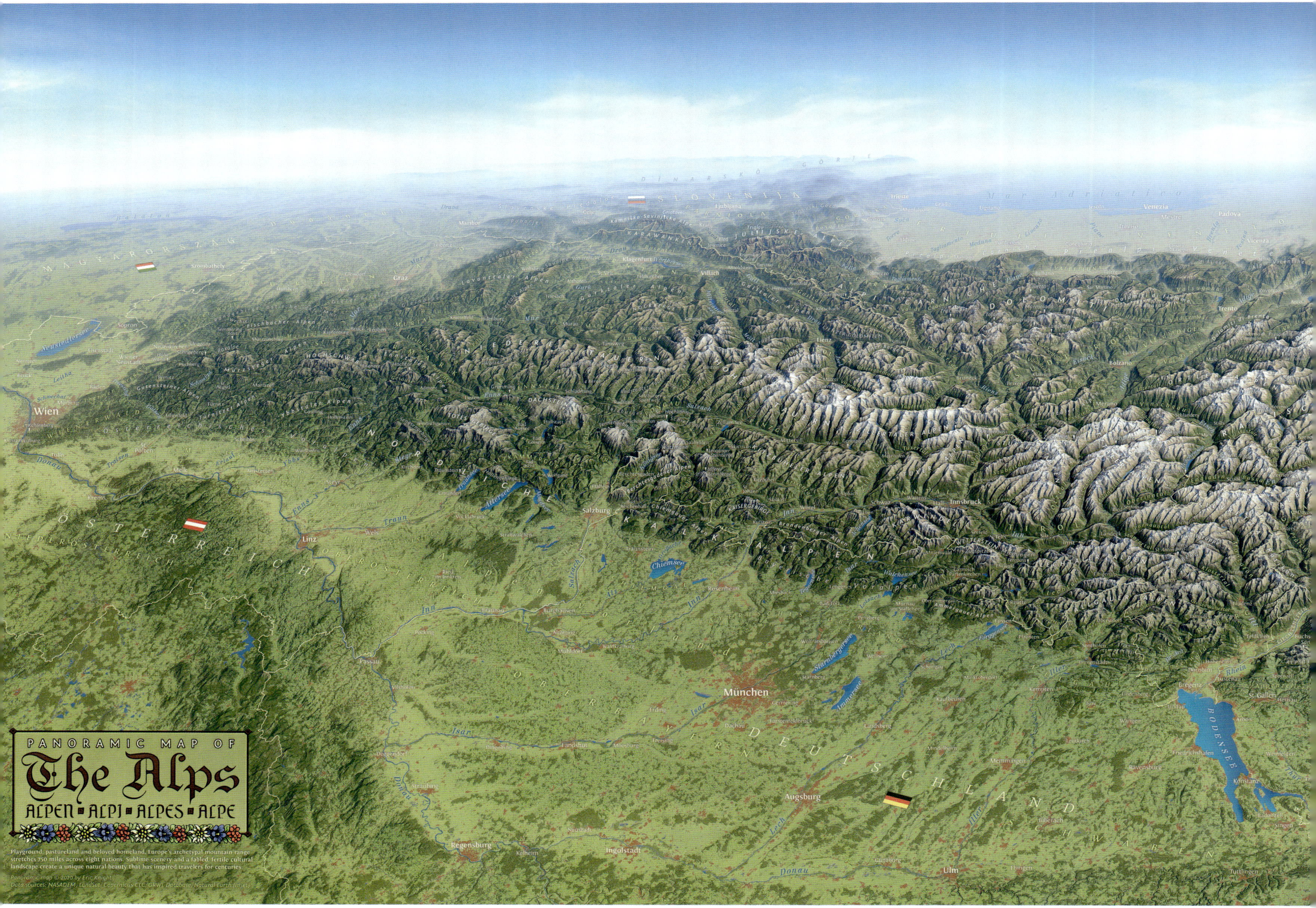
PANORAMIC MAP OF
The Alps
ALPEN ▪ ALPI ▪ ALPES ▪ ALPE
Playground, pastureland and beloved homeland, Europe's archetypal mountain range stretches 750 miles across eight nations. Sublime scenery and a fabled, fertile cultural landscape create a unique natural beauty that has inspired travelers for centuries
Panoramic map © 2020 by Eric Knight
Data sources: NASADEM, LandSat, Copernicus CLC, GRWL Database, Natural Earth (inset)
Wien
Linz
Salzburg
München
Augsburg
Regensburg
Ingolstadt
Ulm
Innsbruck
Graz
Villach
Klagenfurt
Ljubljana
Venezia
Padova
Bolzano
Trento
Chiemsee
Donau
Inn
Isar
Lech
BODENSEE
ÖSTERREICH
DEUTSCHLAND
MAGYARORSZÁG
SLOVENIJA

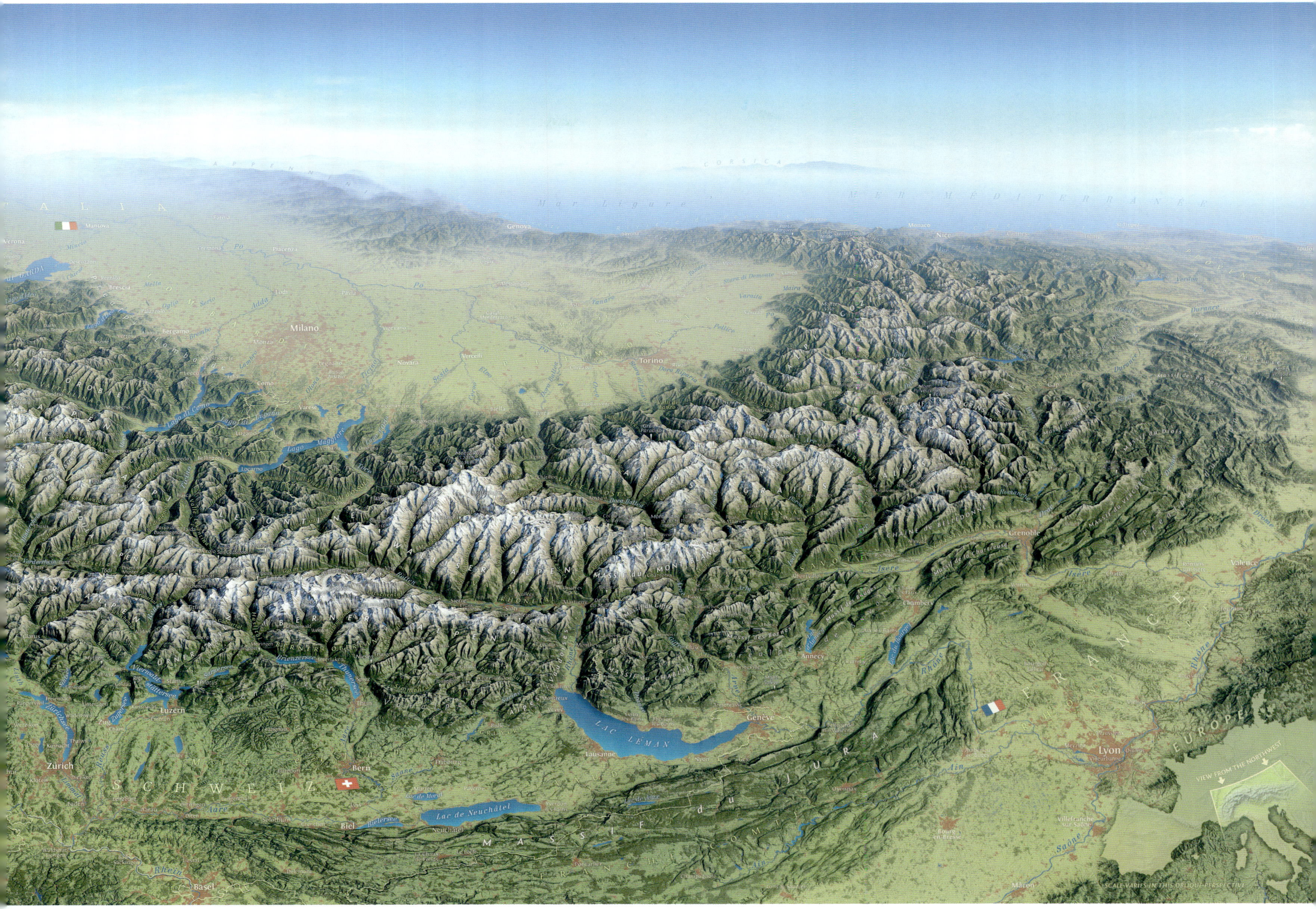
CORSICA
Mar Ligure
MER MÉDITERRANÉE
Genova
Monaco
Nice
Verona
Mantova
Brescia
Bergamo
Milano
Como
Novara
Vercelli
Torino
Po
Lago Maggiore
Locarno
Grenoble
Valence
Isère
Chambéry
Annecy
Genève
LAC LÉMAN
Lausanne
Luzern
Zürich
Bern
SCHWEIZ
Lac de Neuchâtel
Biel
Basel
Rhône
FRANCE
Lyon
EUROPE
VIEW FROM THE NORTHWEST
Saône
Mâcon
SCALE VARIES IN THIS OBLIQUE PERSPECTIVE

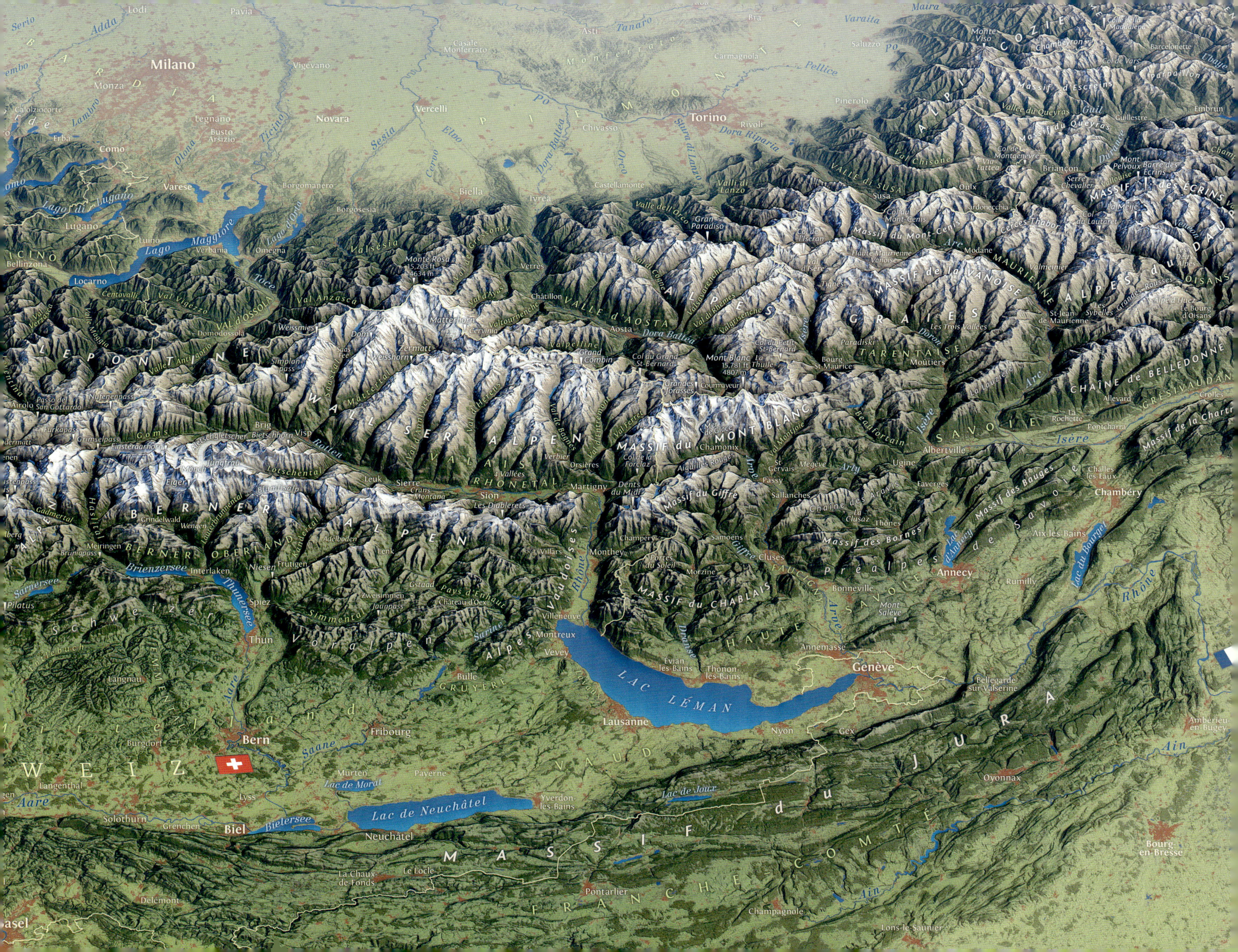

Milano
Monza
Lodi
Pavia
Adda
Lambro
Como
Varese
Lago di Lugano
Lugano
Lago Maggiore
Locarno
Bellinzona
Novara
Vercelli
Biella
Ivrea
Torino
Chivasso
Asti
Tanaro
Casale Monferrato
Carmagnola
Saluzzo
Pinerolo
Po
Sesia
Ticino
Dora Baltea
Dora Riparia
Pellice
Varaita
Maira
Monte Rosa
15,203 ft
4634 m
Valsesia
Omegna
Verbania
Domodossola
Val d'Ossola
Val Anzasca
Simplon pass
Matterhorn
Zermatt
Weisshorn
Saas
Visp
Brig
Leuk
Sierre
Sion
Martigny
Rhone
Aosta
Valle d'Aosta
Châtillon
Grand Combin
Col du Grand St-Bernard
Gran Paradiso
Valli di Lanzo
Valle di Susa
Susa
Bardonecchia
Col du Mont Cenis
Col de l'Iseran
Massif du Mont-Cenis
Modane
Massif de la Vanoise
Haute Maurienne Vanoise
Maurienne
St-Jean de-Maurienne
Paradiski
Tarentaise
Moûtiers
Bourg St-Maurice
Mont Blanc
15,781 ft
4807 m
Courmayeur
Grandes Jorasses
Col du Petit St-Bernard
La Thuile
Massif du Mont Blanc
Chamonix
Col de la Forclaz
Megève
Sallanches
Massif du Giffre
Samoëns
Morzine
Massif du Chablais
Massif des Bornes
Annecy
Lac d'Annecy
Massif des Bauges
Albertville
Ugine
Faverges
Savoie
Isère
Chambéry
Aix-les-Bains
Lac du Bourget
Rumilly
Massif de la Chartreuse
Chaîne de Belledonne
Allevard
Massif des Écrins
Briançon
Oulx
Guillestre
Embrun
Barcelonnette
Massif du Queyras
Monte Viso
Alpi Cozie
Alpes Graies
Walliser Alpen
Alpi Lepontine
Berner Alpen
Berner Oberland
Grindelwald
Wengen
Eiger
Mönch
Jungfrau
Interlaken
Brienzersee
Thunersee
Thun
Spiez
Niesen
Frutigen
Adelboden
Lenk
Gstaad
Zweisimmen
Simmental
Château-d'Oex
Pays d'Enhaut
Les Diablerets
Villars
Alpes Vaudoises
Montreux
Vevey
Villeneuve
Monthey
Champéry
Dents du Midi
Lac Léman
Lausanne
Évian les-Bains
Thonon les-Bains
Genève
Annemasse
Nyon
Gex
Bellegarde sur-Valserine
Haute-Savoie
Préalpes de Savoie
Bonneville
Cluses
Mont Salève
Rhône
Fribourg
Bulle
Gruyère
Sarine
Saane
Bern
Burgdorf
Langnau
Emmental
Aare
Schweizer Voralpen
Pilatus
Sarnersee
Murten
Lac de Morat
Payerne
Vaud
Lac de Neuchâtel
Neuchâtel
Yverdon les-Bains
Lac de Joux
Biel
Bielersee
Grenchen
Solothurn
Langenthal
Lyss
La Chaux de-Fonds
Le Locle
Pontarlier
Delémont
Basel
Massif du Jura
Franche-Comté
Champagnole
Lons-le-Saunier
Oyonnax
Ain
Bourg en-Bresse
Ambérieu en-Bugey
Piemonte
Monferrato

Making a complex mountain landscape appear familiar and harmonious is not a simple task «

« Panoramic Map of the Alps

Eric Knight

The Alps are the ultimate subject for a panoramic map. This archetypal landscape of village church steeples, manicured valleys, sparkling lakes, and gauzy summits has inspired its very own, unique and much-loved brand of painterly cartography. So when it came time to tackle this classic subject matter, I found myself taking a few more stylistic cues than usual from old-world cartographic artists like Berann, Bieder, and Koller.

The hand-painted panoramas that inspired this map typically feature bold, outsized landforms and a sunny, nearly primary color palette. They are old-fashioned landscapes, idealized and romanticized in the best sense, with a naïve sort of beauty seldom captured by satellite imagery alone. So although this map is built from modern data sources, its details have been curated, stylized, and above all simplified to evoke the sensibilities of an earlier time.

Making a complex mountain landscape appear familiar and harmonious is not a simple task, especially when the medium is not paint but instead modern geodata with all of its accuracy and excess. Terrain and land cover must be generalized, yes, but the greater issues are more subjective and not so easily quantified. I try to develop compositions that work at multiple scales, showcasing an entire region while drawing the viewer in close to explore little scenes, using text and simple shapes to lead the eye from one place to the next. Clutter is reduced, but the perception of detail remains, and gradients of color, value, and contrast help create an illusion of scale that is both vast and comprehensive yet intimate and inviting.

Given this subjective design process, one might not expect this panorama to have morphed into a thematic map, of all things, describing loss of snow cover in the Alps for *National Geographic* magazine. To me, the fusion of panoramic map design and aesthetics with real world data and context is full of intriguing possibilities, and shows how design principles developed by landscape artists of the past have had lasting significance in the way we see the world.

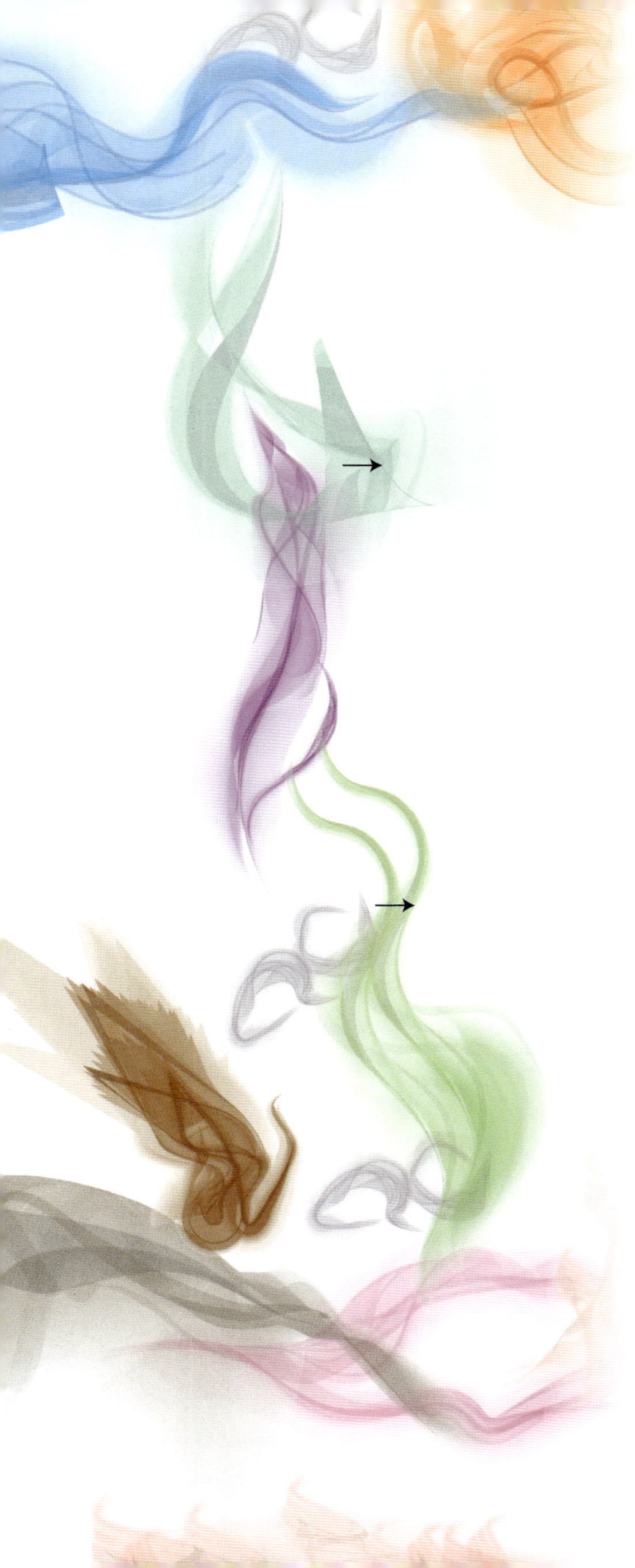

Hospital Corridor Smellscape »

Kate McLean

In my work mapping is an activity, a performed series of social explorations that enable a local population to explore their neighborhood in new ways. Conventional cartography tends to focus on what humans see; my mapping focuses instead on what humans smell. I create symbols to represent the experience of smelling, and I geolocate the symbols in the town, city, or building where they were noted. This hospital "smellmap" requires the viewer to use their olfactory imagination to conjure forth the smells: from the weak and insipid coffee floating near the hospital canteen to the many bodies in the busy entrance halls, from metallic odors of stale blood emanating from the operating theater to the sweaty, sticky, stale pong of the waiting room. Some people find it difficult to imagine smells, while for others it is easy. During the design phase, I manipulated the smells to indicate how they might waft, drift, and meander along the corridors. I removed the walls of the operating rooms and wards to further accentuate the capacity of smells to fill and hold a space of their own regardless of the boundaries we construct.

The piece was commissioned by 'Senses and Modern Health/care Environments,' an academic network exploring opportunities for collaborative thinking, making, research, and funding in the sensory history of health care environments. Early in 2020 they proposed a series of publications on the senses in hospitals: in the waiting room, the corridor, the ward, the operating theater, the laundry, the threshold. They invited me to contribute a creative, experience-based piece, to explore how we can understand the sensory environment of hospitals through arts, personal accounts, and more.

Research took the form of an open question snowballed to friends and colleagues of the Senses and Modern Health/care team members and of the designer. I simplified smell descriptions from the original longer descriptions and grouped them into similar origins. The overall map design is based on National Health Service design guidelines, including the choice of typefaces and legend layout.

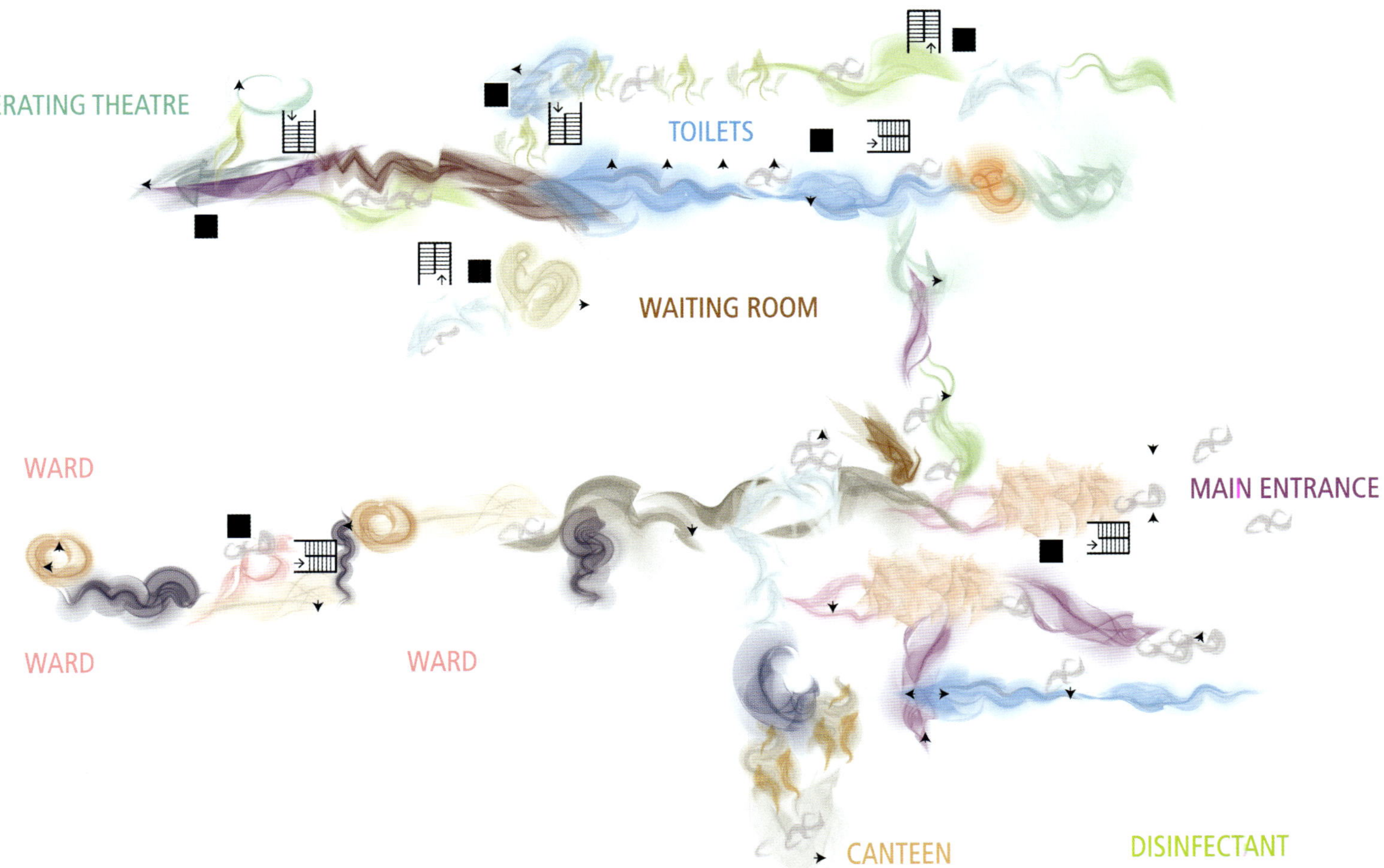

Welcome to a virtual

Hospital Corridor Smellscape

Olfactory data recalled by 27 former hospital patients, workers and visitors. February 2020.

Hospital smells are both intense and pervasive; they mingle in the corridors forming memorable but bizarre smell combinations.

KEY TO SMELL COLOURS

PEOPLE

- Nail varnish remover
- Slightly sweet and difficult to identify; floral, leftover smell of passer-by's perfume
- Old people
- Many bodies in a space, odours from all orifices and on clothing, sheets, seats…

FROM THE OPERATING THEATRE

- Chemically metallic, silver and rouge – tubing
- Metallic stale blood
- Soap in a bowl of hot water used to find a vein for chemo
- Latex gloves
- Something sterile

FROM THE WARD

- Hot wipes
- Rubber mattress
- Welcome slices of buttered toast
- First cup of tea after general anaesthetic (cosy and warm)

FROM THE CANTEEN

- Weak insipid coffee
- Heart-sinking yet also nostalgic canteen food
- Coffee and nachos

FROM THE WAITING ROOM

- A lingering smell of food in corridors (quite a meaty smell, like cottage pie)
- Burnt chicken, or toast
- Sweaty/sticky/stale odour of a waiting room
- Non-specific disinfectanty smell found particularly in hospital corridors

DISINFECTANT

- Clean, like citrus
- Bleach/disinfectant used in loos
- Disinfectant wipes (sickly sweet layered over artificial flowers)
- Disinfectant
- Purell, alcohol wipes

COMBINATION

- Mix of a floor cleaner & rubber
- Rubber, disinfectant, mince and an acrid smell that cut through from the dressings cupboards

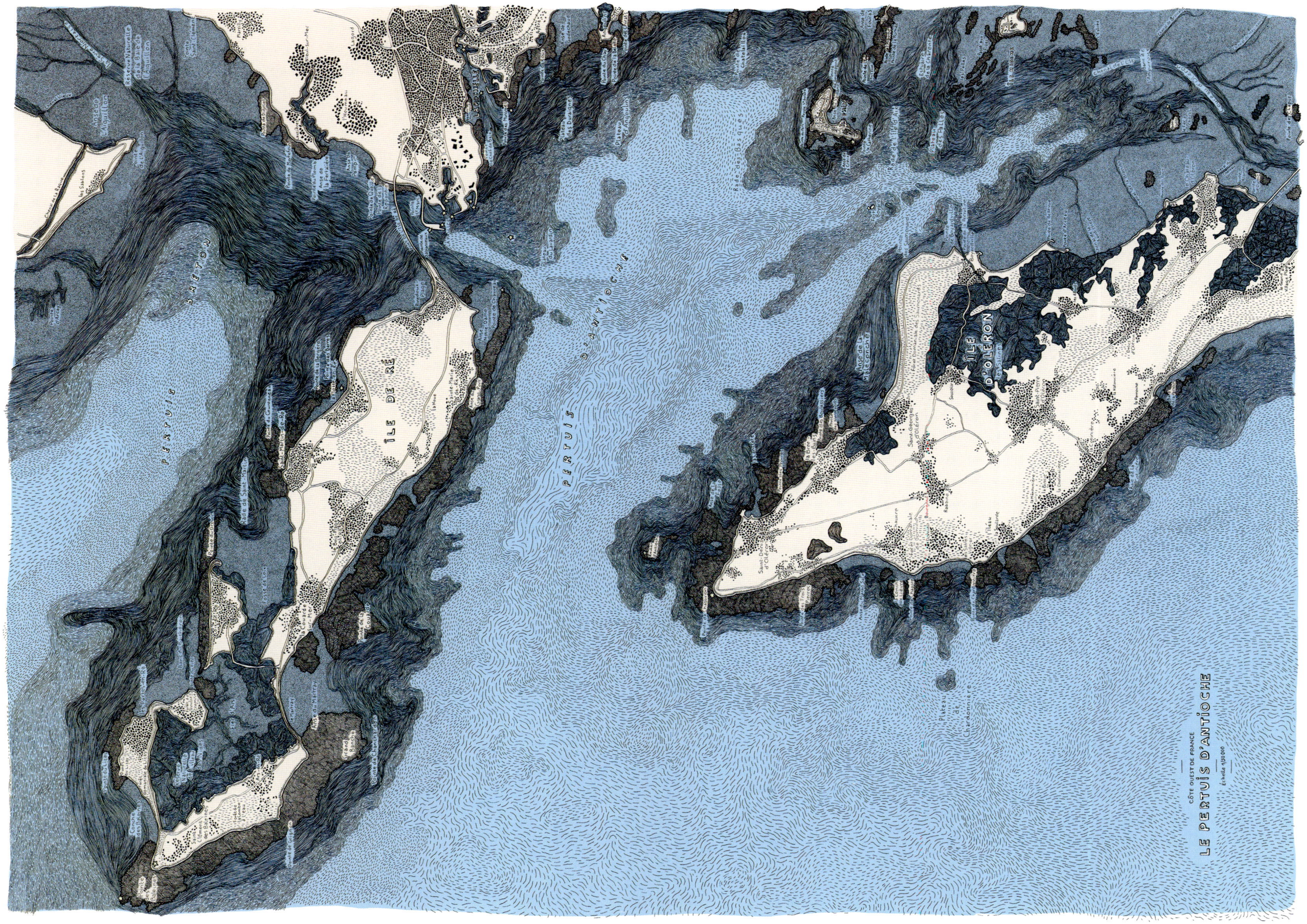
CÔTE OUEST DE FRANCE
LE PERTUIS D'ANTIOCHE
Échelle 1/50 000
ÎLE DE RÉ
ÎLE D'OLÉRON
PERTUIS D'ANTIOCHE
PERTUIS BRETON

« Le pertuis d’Antioche

Marine Le Breton

This map is an example of the nautical chart as an artistic project.

In love with strokes, lines, and dots, I create nautical charts, in the style of graphic embroidery. Inspired by the maps of the French naval hydrological service and national geographic institute, the drawing becomes the interpretation of real elements. Whether in connection with the worlds of scientists, of people from the Heritage, the Territory, of artists, and also of navigators, these charts challenge because they are at the junction of all these themes. Without forgetting the one that comes closest to the heart: the sensitive territory.

The nautical chart brings together both close and opposing themes: it is a source of fantasy (an imaginary journey), yet, at the same time, it is completely real; it uses geographical data, redrawn; it is composed of elements from Science & Technology but itself becomes an object of Art.

This project starts from reality and grows out of it. The shapes from the maps are the raw material. The artistic freedom lies in this slow and concentrated construction: an exercise that takes shape thanks to thousands of small dots, short lines, long strokes, grids, and outlines, and which has gradually become a graphic code in its own right. The heart of this work are these more or less dense tracings of the bathymetry, made by hand, using fine-tip gel ink pens, which gradually fill the sheet to the last square centimeter . . . almost.

Because I think that what remained empty must be justified.

Apart from these few rules, I allow variations: pivoting, changing the range of colors, cutting an uninteresting area, either preserving an island whole or cutting it shamelessly . . .

These nautical charts lead to discussions, and people of all ages can be captivated by them, even if the drawing can sometimes be a little difficult to decipher: not everyone knows the bathymetric system, but the presence of the seas and lands remains a reliable point of reference.

If we know the place, then the map will represent a memory, emotions, often shared. And it is no longer an art object, nor a scientific survey either—it becomes a sensitive object.

It is all of these.

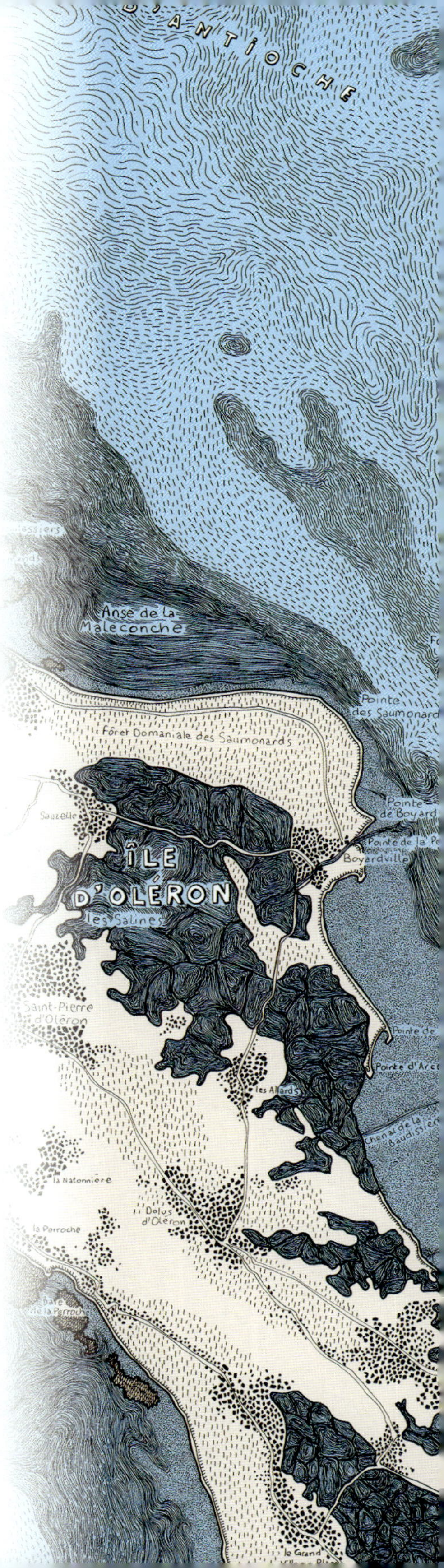

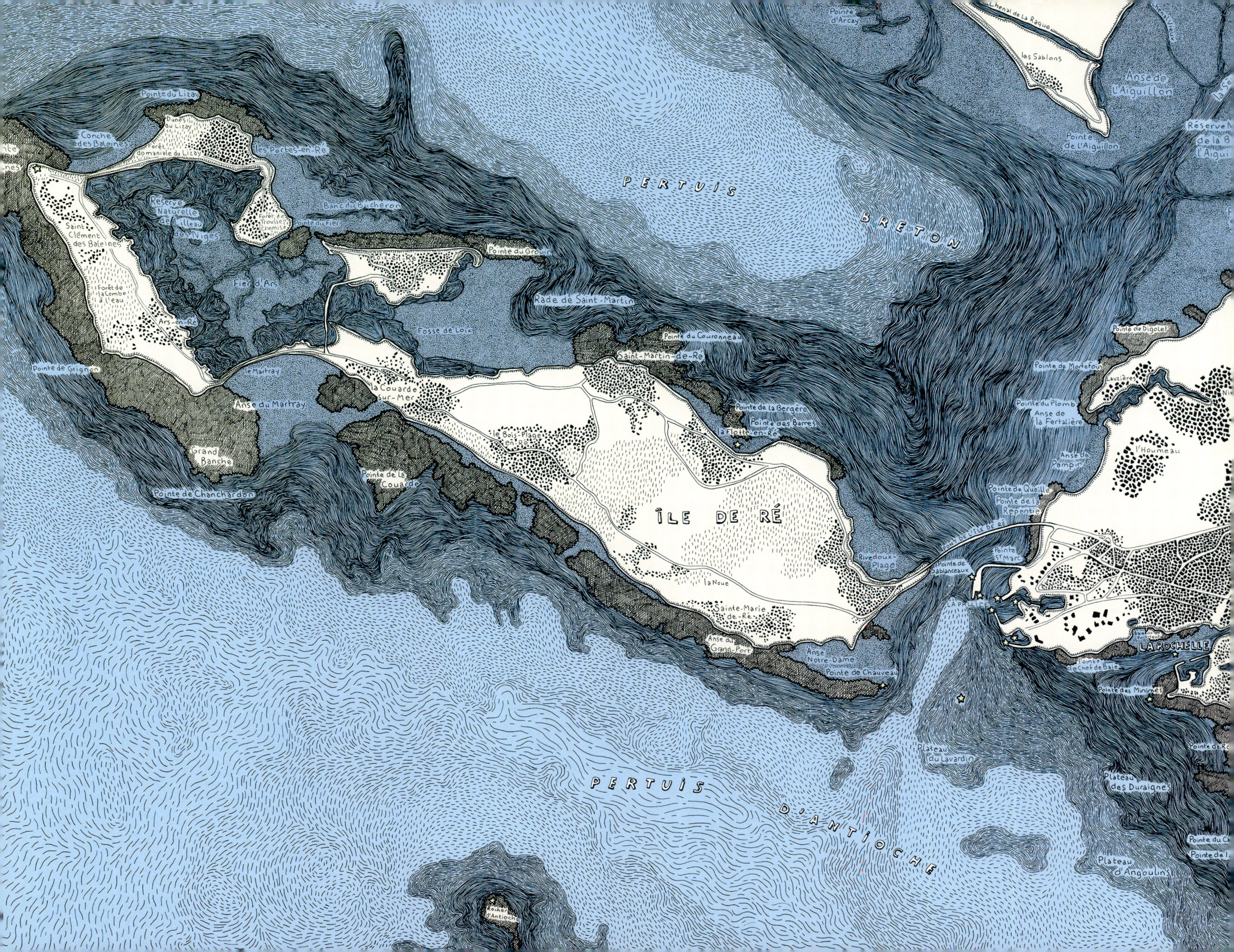
Pointe d'Arcay
Chenal de La Raque
les Sablons
Anse de L'Aiguillon
Pointe de l'Aiguillon
Pointe du Lizay
Dunes du Lizay
Forêt domaniale du Lizay
Conche des Baleines
les Portes-en-Ré
Réserve Naturelle de Lilleau des Niges
Forêt de Trousse Chemise
Banc du Bucheron
Pointe du Fier
Loix-en-Ré
Pointe du Grouin
Saint Clément des Baleines
Forêt de la Combe à l'eau
Ars-en-Ré
Fier d'Ars
Fosse de Loix
Rade de Saint-Martin
PERTUIS
BRETON
Pointe de Grignon
le Martray
Anse du Martray
la Couarde sur-Mer
le Bois-Plage en-Ré
Grand Banche
Pointe de la Couarde
Pointe de Chanchardon
Pointe du Couronneau
Saint-Martin-de-Ré
Pointe de la Bergère
Pointe des Barres
la Flotte-en-Ré
ÎLE DE RÉ
la Noue
Sainte-Marie de-Ré
Anse du Grand-Port
Anse Notre-Dame
Pointe de Chauveau
Rivedoux-Plage
PONT DE L'ÎLE DE RÉ
Pointe de Sablanceaux
Pointe St Marc
Pointe de Digolet
Pointe de Mortefoin
Lauzières
Pointe du Plomb
Anse de la Fertalière
l'Houmeau
Anse de Pampin
Pointe de Queille
Pointe de la Repentie
LA ROCHELLE
Pointe de Chef de Baie
Pointe des Minimes
Plateau du Lavardin
Plateau des Duraignes
Plateau d'Angoulins
PERTUIS
D'ANTIOCHE
Rocher d'Antioche

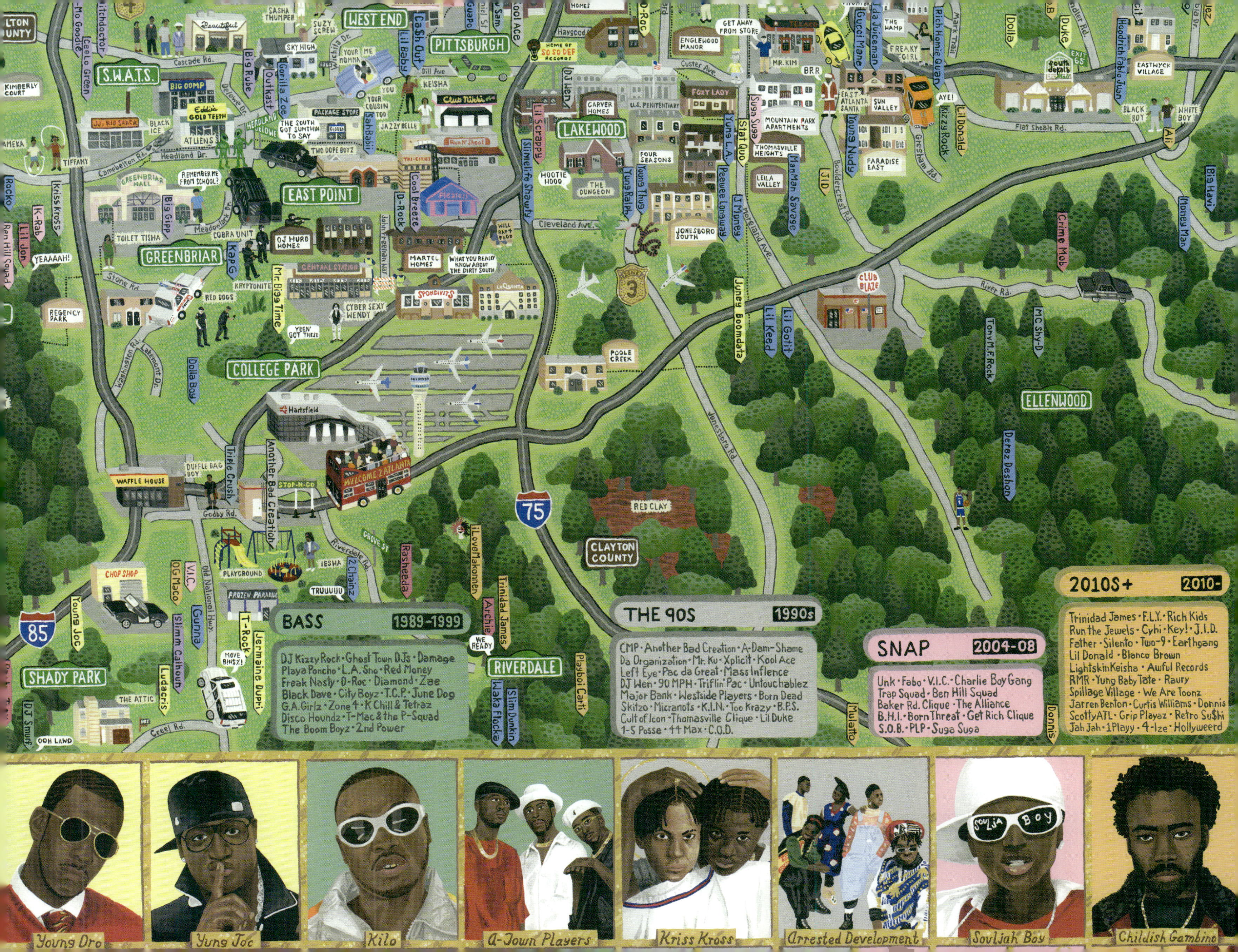
WEST END
PITTSBURGH
S.W.A.T.S.
LAKEWOOD
EAST POINT
GREENBRIAR
COLLEGE PARK
ELLENWOOD
RIVERDALE
SHADY PARK
CLAYTON COUNTY
Hartsfield
WELCOME 2 ATLANTA
WAFFLE HOUSE
RED CLAY
75
85
Cascade Rd.
Headland Dr.
Cleveland Ave.
Flat Shoals Rd.
River Rd.
Jonesboro Rd.
Riverdale Rd.
Godby Rd.
Old National Hwy.
Washington Rd.
Stone Rd.
Camp Creek Pkwy.
Custer Ave.
Dill Ave.
Moreland Ave.
Bouldercrest Rd.
Gresham Rd.
Creel Rd.
BIG COMP
ATLIENS
GREENBRIAR MALL
REGENCY PARK
CENTRAL STATION
SPONDIVITS
MARTEL HOMES
OJ HURD HOMES
CARVER HOMES
U.S. PENITENTIARY
FOXY LADY
ENGLEWOOD MANOR
MOUNTAIN PARK APARTMENTS
THOMASVILLE HEIGHTS
FOUR SEASONS
LEILA VALLEY
JONESBORO SOUTH
PARADISE EAST
SUN VALLEY
EASTWYCK VILLAGE
CLUB BLAZE
POOLE CREEK
THE DUNGEON
CHOP SHOP
PLAYGROUND
FROZEN PARADISE
THE ATTIC
OOH LAWD
KIMBERLY COURT
CLUB NIKKI
PACKAGE STORE
BASS 1989-1999
DJ Kizzy Rock · Ghost Town DJs · Damage
Playa Poncho · L.A. Sno · Red Money
Freak Nasty · D-Roc · Diamond · Zae
Black Dave · City Boyz · T.C.P. · June Dog
G.A. Girlz · Zone 4 · K Chill & Tetraz
Disco Houndz · T-Mac & the P-Squad
The Boom Boyz · 2nd Power
THE 90S 1990s
CMP · Another Bad Creation · A-Dam-Shame
Da Organization · Mr. Ku · Xplicit · Kool Ace
Left Eye · Pac da Great · Mass Inflence
DJ Wen · 90 MPH · Triflin' Pac · Untouchablez
Major Bank · Westside Players · Born Dead
Skitzo · Micranots · K.I.N. · Too Krazy · B.P.S.
Cult of Icon · Thomasville Clique · Lil Duke
1-5 Posse · 44 Max · C.O.D.
SNAP 2004-08
Unk · Fabo · V.I.C. · Charlie Boy Gang
Trap Squad · Ben Hill Squad
Baker Rd. Clique · The Alliance
B.H.I. · Born Threat · Get Rich Clique
S.O.B. · PLP · Suga Suga
2010S+ 2010-
Trinidad James · F.L.Y. · Rich Kids
Run the Jewels · Cyhi · Key! · J.I.D.
Father · Silento · Two-9 · Earthgang
Lil Donald · Blanco Brown
LightskinKeisha · Awful Records
RMR · Yung Baby Tate · Raury
Spillage Village · We Are Toonz
Jarren Benton · Curtis Williams · Donnis
ScottyATL · Grip Playaz · Retro Su$hi
Jah Jah · 1Playy · 4-Ize · Hollyweerd
Young Dro
Yung Joc
Kilo
A-Town Players
Kriss Kross
Arrested Development
Souljah Boy
Childish Gambino

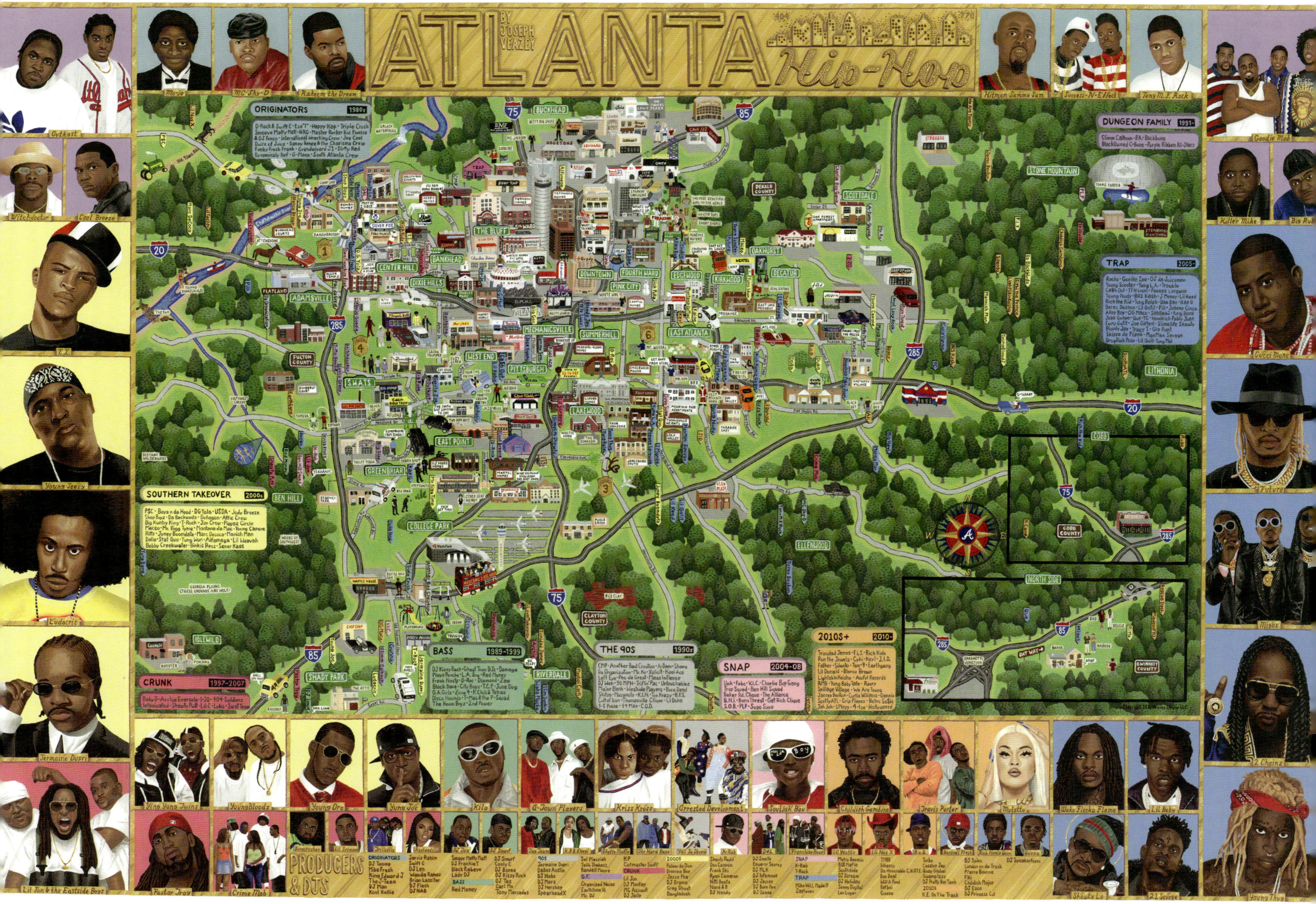
ATLANTA
BY JOSEPH VEAZEY
Hip-Hop
404
770
MC Shy-D
Raheem the Dream
Hitman Sammy Sam
Success-N-Effect
Tony M.F. Rock
Outkast
Goodie Mob
Witchdoctor
Cool Breeze
Killer Mike
T.I.
Gucci Mane
Young Jeezy
Future
Ludacris
Migos
Jermaine Dupri
2 Chainz
Lil Jon & the Eastside Boyz
Young Thug
ORIGINATORS 1980s
D-Rock & Swift C · Eze"T" · Happy Hap · Triple Crush
Smoove Matty Matt · NRG · Master Rocker Kid Finesse
& DJ Fenzo · International Wrecking Crew · Joe Cool
Duice of Juice · Danny Renee & the Charisma Crew
Funky Fresh Frank · Grandwizard J1 · Dirty Red
Dynamicaly Def · G-Mann · South Atlanta Crew
DUNGEON FAMILY 1991-
Slimm Calhoun · P.A. · Backbone
BlackOwned C-Bone · Purple Ribbon All-Stars
TRAP 2005-
Rocko · Gorilla Zoe · OJ da Juiceman
Young Scooter · Yung L.A. · Trouble
Ca$h Out · YFN Lucci · Peewee Longway
Young Nudy · 808 Kash · J Money · Lil Keed
Rich the Kid · Yung Ralph · Dae Dae · Kap G
Derez Deshon · Lil Gotit · Pill · Johnny Cinco
Alley Boy · OG Maco · SahBabii · Yung Bans
Jose Guapo · Que · YC · Hoodrich Pablo Juan
Euro Gotit · Joe Gifted · Slimelife Shawty
Bloody Jay · Tracy T · Ola Runt
Skippa da Flippa · ManMan Savage
DrugRich Peso · Lil Quill · Yung Mal
SOUTHERN TAKEOVER 2000s
P$C · Boyz n da Hood · DG Yola · USDA · Jody Breeze
Shop Boyz · Da Backwudz · Bohagon · Attic Crew
Big Kuntry King · T-Rock · Jim Crow · Playaz Circle
Maceo · Mr. Bigg Tyme · Montana da Mac · Young Capone
Rittz · Juney Boomdata · Marc Decoca · Manish Man
Dolla · Stat Quo · Yung Wun · Alfamega · Lil Weavah
Bobby Creekwater · Binkis Recs · Senor Kaos
CRUNK 1997-2007
Baby D · Archie Eversole · I-20 · 404 Soldierz
Intoxicated · Shawty Putt · Lil C · Loko · Swat Team
BASS 1989-1999
DJ Kizzy Rock · Ghost Town DJs · Damage
Playa Poncho · L.A. Sno · Red Money
Freak Nasty · D-Roc · Diamond · Zee
Black Dave · City Boyz · T.C.P. · June Dog
G.A. Girlz · Zone 4 · K Chill & Tetraz
Disco Houndz · T-Mac & the P-Squad
The Boom Boyz · 2nd Power
THE 90S 1990s
CMP · Another Bad Creation · A-Dam-Shame
Da Organization · Mr. Ku · Xplicit · Kool Ace
Left Eye · Pac da Great · Mass Inflence
DJ Wen · 90 MPH · Triflin' Pac · Untouchablez
Major Bank · Westside Players · Born Dead
Skitzo · Micranots · K.I.N. · Too Krazy · B.P.S.
Cult of Icon · Thomasville Clique · Lil Duke
I-5 Posse · 44 Max · C.O.D.
SNAP 2004-08
Unk · Fabo · V.I.C. · Charlie Boy Gang
Trap Squad · Ben Hill Squad
Baker Rd. Clique · The Alliance
B.H.I. · Born Threat · Get Rich Clique
S.O.B. · PLP · Suga Suga
2010S+ 2010-
Trinidad James · F.L.Y. · Rich Kids
Run the Jewels · Cyhi · Key! · J.I.D.
Father · Silento · Two-9 · Earthgang
Lil Donald · Blanco Brown
LightskinKeisha · Awful Records
RMR · Yung Baby Tate · Raury
Spillage Village · We Are Toonz
Jarren Benton · Curtis Williams · Donnis
ScottyATL · Grip Playaz · Retro Su$hi
Jah Jah · 1Playy · 4-Ize · Hollyweerd
BUCKHEAD
STONE MOUNTAIN
SCOTTDALE
DEKALB COUNTY
OAKHURST
DECATUR
KIRKWOOD
EDGEWOOD
FOURTH WARD
DOWNTOWN
PINK CITY
THE BLUFF
BANKHEAD
CENTER HILL
DIXIE HILLS
ADAMSVILLE
FLATLAND
MECHANICSVILLE
SUMMERHILL
EAST ATLANTA
WEST END
PITTSBURGH
FULTON COUNTY
S.W.A.T.S.
LAKEWOOD
EAST POINT
GREENBRIAR
BEN HILL
COLLEGE PARK
LITHONIA
ELLENWOOD
COBB COUNTY
NORTH SIDE
GWINNETT COUNTY
RED CLAY
CLAYTON COUNTY
RIVERDALE
SHADY PARK
IDLEWILD
CHURCH
ROOSTER
PERCIVAL
WELL UM HUH
DISTANT WILDERNESS
GEORGIA PLAINS (THESE GROUNDS ARE HOLY)
SPLASH WATERFALLS
CANDY LADY
BOWEN HOMES
MUST BE 2 SIDES...
SHAG SURFER
STEPHENSON HIGH SCHOOL
PLAYGROUND
TRUUUUU
WE READY
75
85
20
285
Ying Yang Twins
YoungBloodz
Young Dro
Yung Joc
Kilo
A-Town Players
Kriss Kross
Arrested Development
Souljah Boy
Childish Gambino
Travis Porter
Mulatto
Waka Flocka Flame
Lil Baby
Pastor Troy
Crime Mob
Bonecrusher
Lil Scrappy
Trillville
Rasheeda
DJ Jaz
DJ Smurf
Jan Jeam
K.P. & Envyi
Ghetto Mafia
The Hard Boyz
Yall So Stupid
K-Rab
D4L
Franchize Boyz
Lil Yachty
Lil Nas X
B.o.B.
Bankroll Fresh
Rich Homie Quan
Gunna
Shawty Lo
21 Savage
PRODUCERS & DJS
ORIGINATORS
DJ Toomp
Mike Fresh
King Edward J
The J-Team
DJ Man
Kool Kollie
Jarvis Rahim
Swift C
DJ Len
Wanda Ramos
Brian Lassiter
DJ Flash
DJ NAB
Smoov Matty Matt
DJ FrankieT
Black Rabeye
Lady DJ
BASS
Red Money
DJ Smurf
Cooly C
DJ Bones
DJ Kizzy Rock
DJ Taz
Carl Mo
Tony Mercedes
90S
Jermaine Dupri
Dallas Austin
DJ Nabs
DJ Mars
DJ Hershey
SpearheadX
Sol Messiah
Talib Shabazz
Randall Moore
D.F.
Organized Noize
Earthtone III
Mr. DJ
KP
Cutmaster Swiff
CRUNK
Lil Jon
DJ Montay
MC Assault
DJ Jelly
2000S
Polow da Don
Drumma Boy
Jazze Pha
DJ Drama
Greg Street
Bangladesh
Shawty Redd
Don Cannon
Frank Ski
Ryan Cameron
Nitti Beatz
Nard & B
DJ Nando
DJ Smalls
Emperor Searcy
DJ MLK
DJ Infamous
DJ Jaycee
DJ Burn One
DJ Sense
SNAP
K-Rab
T-Rock
TRAP
Mike Will Made It
Zaytoven
Metro Boomin
808 Mafia
Southside
DJ Scream
DJ Holiday
Sonny Digital
Lex Luger
TM88
Wheezy
Da Honorable C.N.O.T.E.
Dun Deal
Will A Fool
Fatboi
Goose
Turbo
Cassius Jay
Quay Global
Swamp Izzo
DJ Pretty Boy Tank
2010S
K.E. On The Track
DJ Spinz
London on da Track
Pi'erre Bourne
FKi
Childish Major
DJ Esco
DJ Princess Cut
DJ Speakerfoxxx

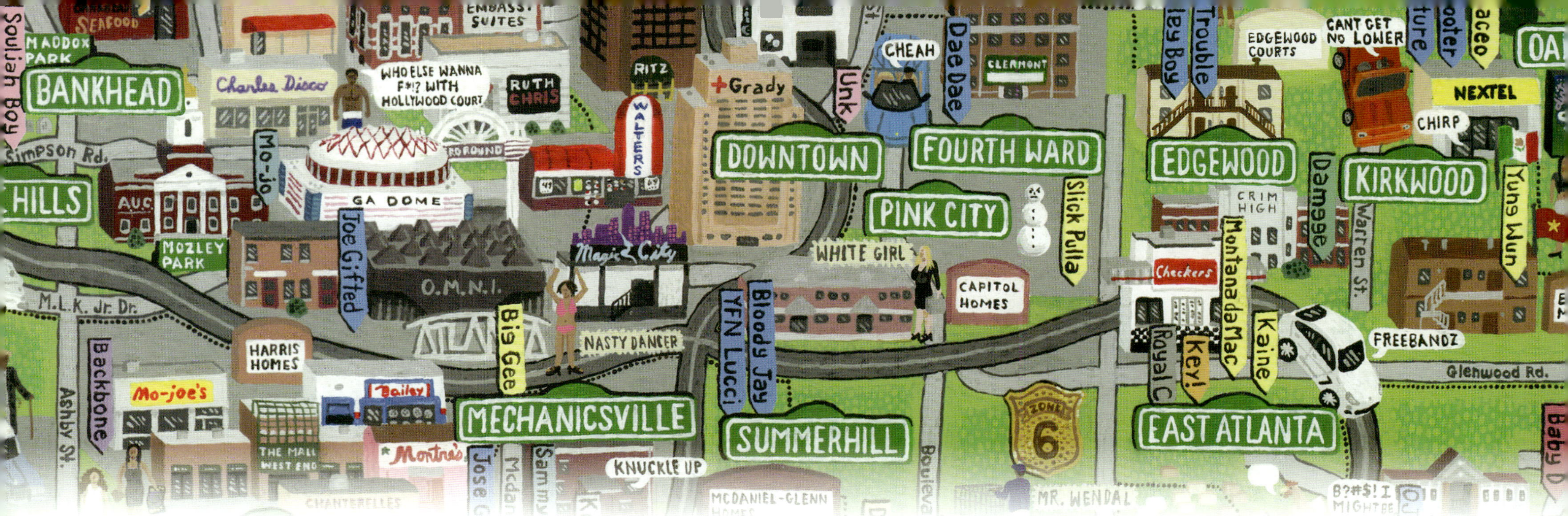

《 The Atlanta Rap Map

Joseph Veazey

"The Atlanta Rap Map" came about as a passion project, first conceptualized in 2017. As a graphic designer, I often utilize my own illustration in my work. Here, I wanted to do something self-initiated that would connect deeply with people, and be rooted in culture. Naturally I chose the topic of Atlanta hip-hop, as I have been a huge fan of this music since middle school.

Visually, my goal was to create a map that was visually stimulating, and that would reflect a sense of warmth, community, and collaboration. I tried to make it as detailed as I could possibly paint, to add a sense of wonder and awe that I feel is often left out of contemporary art and illustration. My personal style is a mix of influences including folk and outsider art, commercial illustration, graphic design, Early Netherlandish art, and others. *National Geographic* foldout maps and diagrams, illustrated local city and amusement park maps, "Springstreets" by Dan Cassaro, and the "Map of Harlem Nightlife" by Simms Campbell were big conceptual inspirations.

The content for the map was chosen after an exhaustive review of Atlanta hip-hop culture since 1980. The research took about a year and involved listening to thousands of chart-toppers, local hits, underground classics, and online interviews, as well as a few classic albums in their entirety. Every location reference in the lyrics, from neighborhoods to street names to local businesses, was collected and literally "put on the map" by being mentioned in a hit song. Artists were chosen for inclusion based on contribution to the genre, influence, popularity, and reach.

While working on the map, I often feared that it was going to turn out visually muddy. This can often happen when including a large amount of information or imagery in one piece, leaving the viewer with an amorphous blob of visuals. The gold framework and extensive forest allowed for a hierarchy between the map, portraits, header, and "Producers & DJs" sections. This ends up making the art much more approachable and acts as a guide for the viewer.

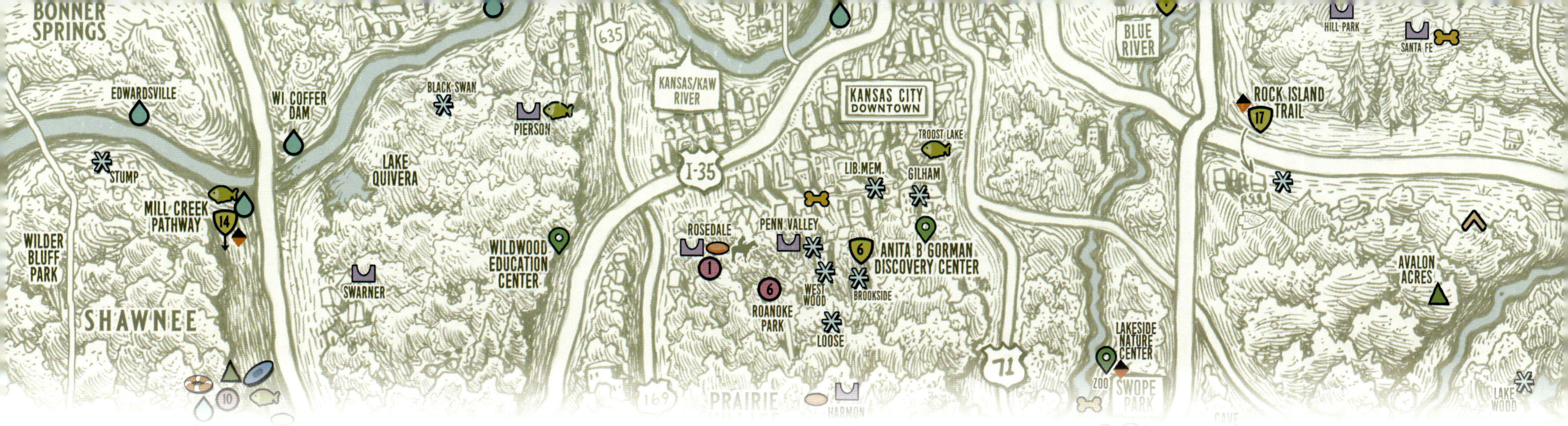

The Kansas City Map of Outdoor Pursuits »

Jeremy Collins

In the early days of pandemic-mandated social distancing, I immediately saw a need to be in the outdoors as much as the weather would allow. As I observed, others in my area did as well. I joined a few online user groups—the mountain bikers, the kayakers, the disc golfers, the general outdoor enthusiasts—and they were all asking the same things. Everyone wanted to know where to go to participate and enjoy new activities outside. As I searched around the internet, I found truncated, outdated information and had the idea to develop an outdoor pursuits map for Kansas City.

I wanted the map project to be not only aspirational and functional but also aesthetically pleasing. I spent a few months hand-drawing a base map in black ink that showed major highways and waterways. The suburbs and rural areas are shown primarily as forested, and downtown is clearly urban concrete and buildings. I hid "Easter egg" items throughout the city: animals, common domestic items, toys, ephemera, and curios. Some of them make sense. Some don't.

As the concept evolved, I sought online community input as to what people wanted to see on the map. Public input was crucial to producing something that real people actually wanted. Users pointed out smaller, more localized attractions I had missed. Perhaps the whole city wouldn't utilize these zones, but those within a neighborhood would find a level of pride in being included.

My main influence stylistically was Lewis and Clark's journals. During their expedition on the Missouri River, they camped on the river banks in Kansas City for three days. Despite their short visit, they have had an impact on local lore and pride ever since. Even though it's my home city, I tried to approach it as if discovering things for the first time, just as Lewis and Clark did.

the KANSAS CITY MAP of OUTDOOR PURSUITS & NATURE EDUCATION

CREATED BY JEREMY COLLINS
@jer.collins @jercollins.com

LEGEND: CAMPING · BOAT PUT-IN · ROCK CLIMBING · DIRT TRAIL · NON-DIRT PATH · FISHING · DISC GOLF · NATURE EDUCATION · BEACH/SWIMMING · DOG PARK · HORSEBACK · ORIENTEERING · MANAGED HUNTING · ARCHERY · SKATE PARK · SAILING/WINDSURFING · PADDLING · PADDLING+RENTALS · FIREARM RANGE · PRIMO SLEDDING · ADA COMPLIANT

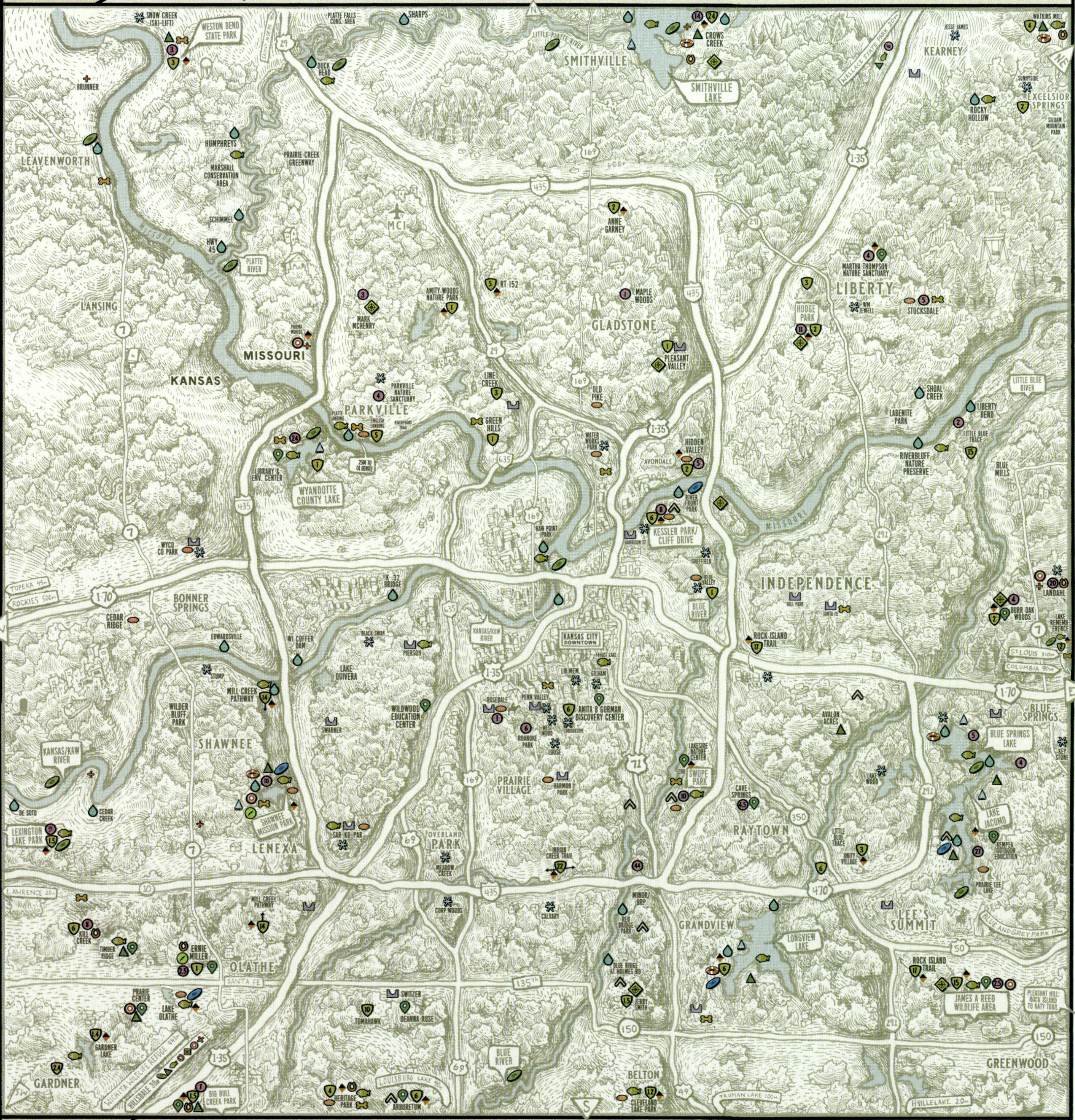

GENERAL LOCATION MAP ONLY

FOR MORE INFORMATION USE THESE WEBSITES AND APPS: ▶TRAILFORKS.COM ▶MOUNTAINBIKEPROJECT.COM ▶KCHIKER.COM ▶MDC.MO.GOV ▶URBANTRAILCO.COM ▶MOUNTAINPROJECT.COM ▶PADDLING.COM ▶ALLTRAILS.COM ▶KCPARKS.ORG ▶MOSTATEPARKS.COM

AREA HIGHLIGHTS

KESSLER PARK/CLIFF DRIVE

Location: NE of Downtown off of Paseo blvd
Description: A historic scenic byway with an elaborate disc golf course, over 10 miles of trail system and the only true roped rock climbing area in KC (which began in the 1980's). Cliff Drive is closed to vehicles Thursday-Sunday.

WYANDOTTE COUNTY LAKE

Location: NW of City Center in KCK
Description: Fabulous oak-hickory woodlands surrounding a 400 acre lake. One of the most scenic trail systems in the metro with over 20 miles of singletrack trail. Popular with mountain bikers, for fishing and rowing teams. 16 picnic areas, marina and boat rentals.

SMITHVILLE LAKE

Location: 20 miles North of the City on 169 hwy
Description: A 7000 acre reservoir with 175 miles of shoreline. Excellent camping, trails and paved paths. 26 miles of equestrian trails, a marina and boat rentals. Numerous other amenities. A great get away within eyesight of the city.

JACOMO/BLUE SPRINGS LAKE

Location: South of I70, NE Lees Summit/SW Blue Springs
Description: An 800 acre forest with two lakes, a widlifle refuge that includes a bison and elk viewing area and Missouri Town- a walk through village based on the late 1800s. The 22 mile trail system is constantly being developed and expanded.

SWOPE PARK

Location: Between 71 and 435 hwy, south of 63rd st.
Description: The pride and joy of the KC parks system. Home of the zoo, two golf courses, a fishing pond, a massive trail network, disc golf course, nature center, dog park, volleyball courts. Noted for excellent mountain biking and the bulk of the bouldering (unroped climbing) in the city.

JAMES A REED WILDLIFE AREA

Location: Southeast Lees Summit, East of 291, south of 50
Description: Twelve well stocked ponds for fishing are the highlight here, as well enjoyable paved and unpaved trails. Birdwatching, archery and horseback trails are other options.

SHAWNEE MISSION PARK

Location: West side of town, just off 435 in Shawnee
Description: Over 1600-acres including a 120-acre lake for swimming, boating and fishing, and a 44-acre dog park with beach access. Mill Creek Pathway passes through here, intersecting miles of great dirt trails built for mountain biking.

WATKINS MILL STATE PARK

Location: NE corner of map, north of Excelsior Springs
Description: A nice escape from the city but still relatively nearby. Features a 100-acre lake with a walking/biking path, fishing, and camping. Includes a preserved 19th century wool mill to explore.

MISSOURI RIVER + KC SHORELINE

Location: Winding through the Nothern end of downtown
Description: Originating in Montana, the longest river in North America passes right through downtown. KC's crowning natural jewel hosts trophy fishing in an urban environment with a few boat put-ins and ample shoreline for angling.

WESTON BEND STATE PARK

Location: NW corner of the map, near the town of Weston
Description: Perched over a notable bend in the Missouri river, this park is on the far outskirts of the map, 30 miles from the city center. There is camping, a 3-mile paved hike and bike path, a dog park and a beautiful overlook.

BURR OAK WOODS

Location: West of 7 hwy, north of I70 in Blue Springs
Description: Hiding in a neighborhood, but you'd never know it. Enjoy quiet scenic trails bordered by limestone and Burr Oak Creek. The Nature Center offers education programming utylizing a 168-seat auditorium and 3000-gallon aquarium.

ANITA B. GORMAN DISCOVERY CENTER

Location: East of the Plaza on Troost
Description: Eight acres of natural area used for education right in the heart of the city. Includes short paved and natural trails through forest, woodland, wetland, and prairie habitats. Free nature programs are offered regularly in a spacious and inspiring facility.

BIG BULL CREEK PARK

Location: Spring Hill, KS. South on I-35
Description: At 2000 acres, it is the largest park in Johnson County. Known for ever-growing mountain bike and horseback trails and non-profit group camping.

BLUE RIVER BANKS

Location: Johnson County to the Missouri River
Description: A 38-mile stream snaking its way through the city, with hiking, biking trails and occasional bouldering all along it's path. It's possible to kayak the length of the river, although unpopular due to downed trees and debris.

PARKVILLE, MO

Location: South of the airport, north of the river
Description: Quaint town that offers options for outdoor activities. There are six miles of walk and bikable pathway along the river, and 4.5 miles of beautiful dirt trail. There is a put-in for the Mo river, and plenty of shoreline for some quick casting at sunset.

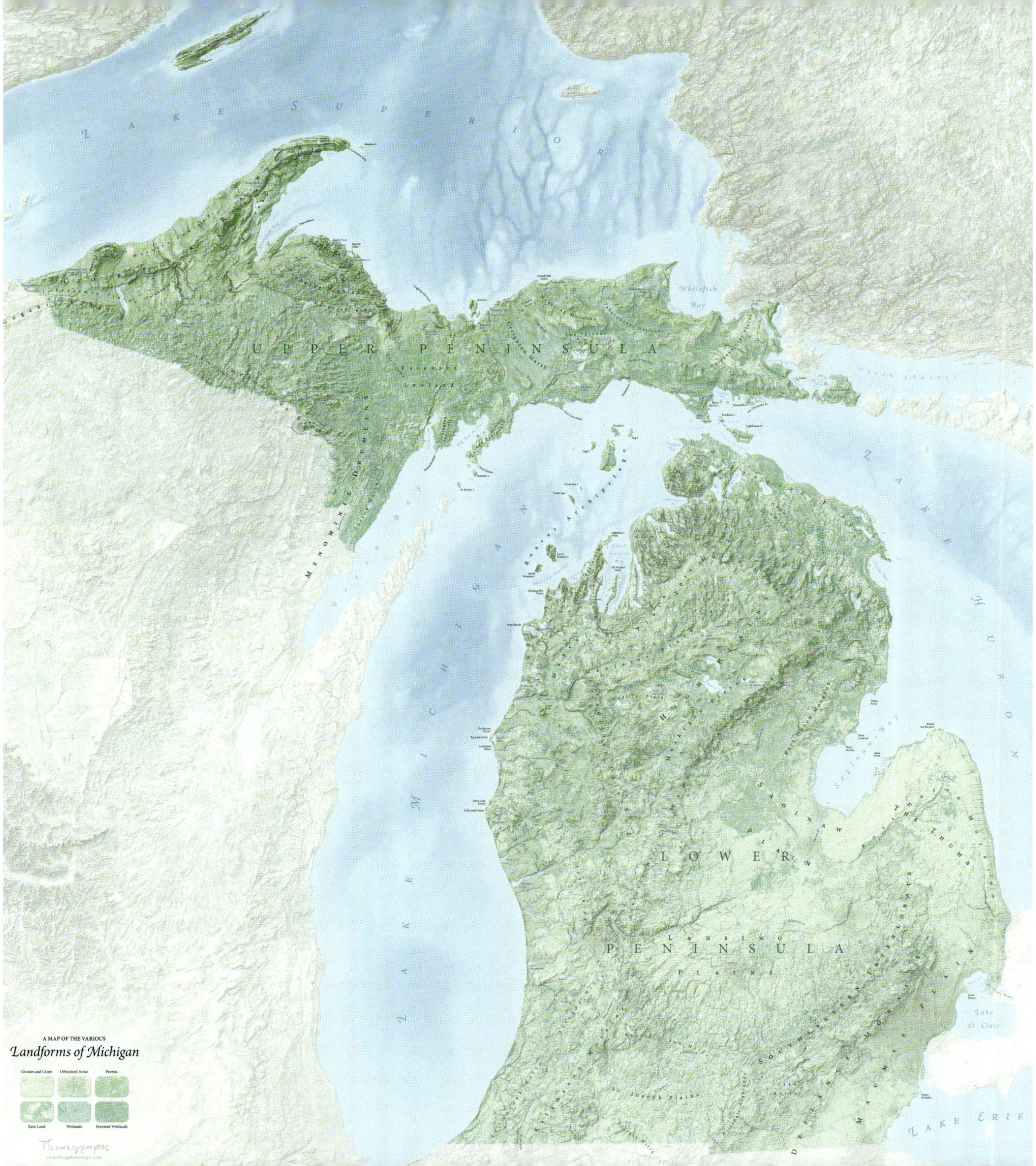
L A K E S U P E R I O R
Whitefish Bay
U P P E R P E N I N S U L A
North Channel
L A K E H U R O N
L A K E M I C H I G A N
Beaver Archipelago
Green Bay
Saginaw Bay
L O W E R
P E N I N S U L A
Lansing Plains
Saginaw Plain
The Thumb
Lake St. Clair
L A K E E R I E
A MAP OF THE VARIOUS
Landforms of Michigan
Grasses and Crops
Urbanized Areas
Forests
Bare Land
Wetlands
Forested Wetlands
Πιναкογραφος

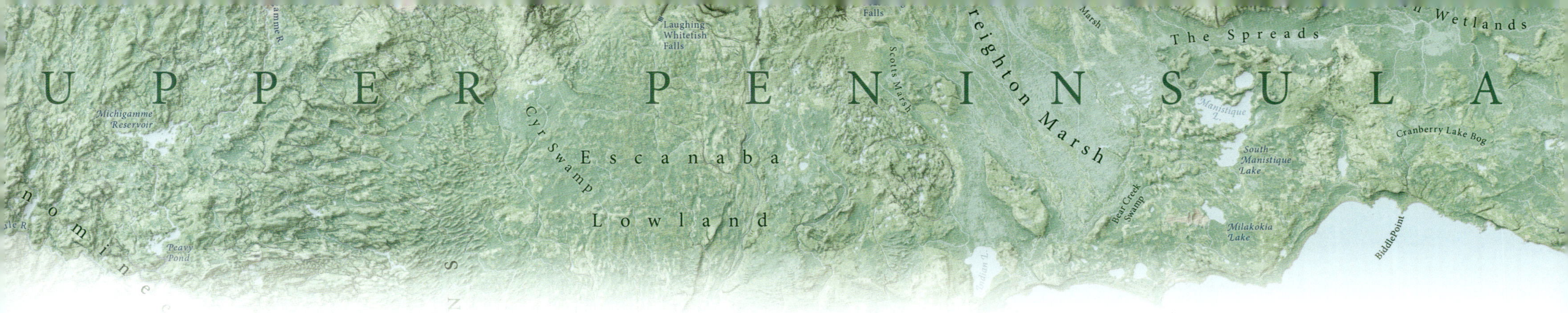

« Landforms of Michigan

Daniel P. Huffman

This is a love song about my Great Lakes home. It's the map that I wish I'd had when growing up in west Michigan, wholly unaware of being surrounded by the rich and complex topography left by ancient glaciers. Like many others, I thought of the Midwest as a fairly flat place. Once I knew better, I wanted to express the beauty of my home to others, and demand for this bit of "flyover country" the attention that it deserves. More importantly, though, I hoped that the process of scrutinizing the landscape to make this map would help me to learn more of its grammar by heart. It was a way to work through my longing for home.

I think names are key to understanding landforms. We can spot a bump on the far horizon, or find a mountain emerging from the shading on our map, but to really reify these features in our minds, I think we need to name them—to mark each as unique from all the other bumps. I devoted far more time to the labeling of this map than to any other part of the production, from extensive research that unearthed obscure toponyms, to careful (perhaps obsessive) attention to the exact spacing, curve, and position of each label.

The type is set in Minion Pro, which I admit to choosing primarily because it offered swash capitals that were elegant, without being so elaborate as to feel needlessly ostentatious for this map. I colored the labels in a way that aims to balance them with the landscape. Step away from the map and the text largely fades away, appearing once again only when you are close. The far view reveals broad shapes, while the near view gives us their identities. Underneath the text, a complex stack of shaded relief, texture shading, and land cover layers reveals fine details, dressed up in summery greens.

I hope that this map will not only give people an appreciation of a place that is special to me, but will also cause them to consider the beauty and complexity of other "flat" places.

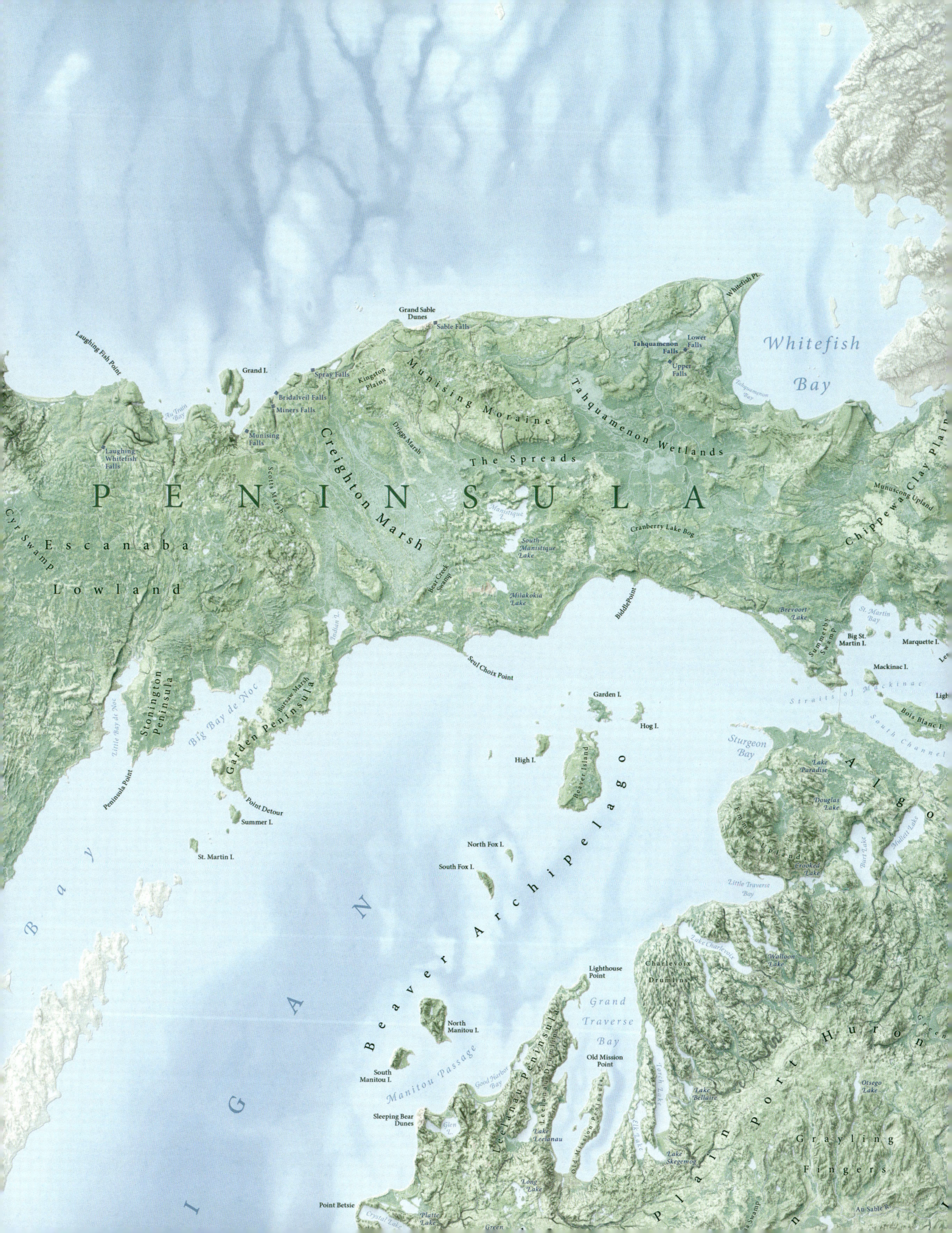

Whitefish Bay
Whitefish Pt.
Tahquamenon Bay
Grand Sable Dunes
Sable Falls
Tahquamenon Falls
Lower Falls
Upper Falls
Laughing Fish Point
Grand I.
Spray Falls
Kingston Plains
Munising Moraine
Tahquamenon Wetlands
Au Train Bay
Bridalveil Falls
Miners Falls
Munising Falls
Laughing Whitefish Falls
Driggs Marsh
The Spreads
Creighton Marsh
Scotts Marsh
PENINSULA
Manistique L.
Chippewa Clay Plain
Munuscong Upland
Cyr Swamp
Escanaba
Lowland
South Manistique Lake
Cranberry Lake Bog
Bear Creek Swamp
Milakokia Lake
Biddle Point
Indian L.
Brevoort Lake
St. Martin Bay
Big St. Martin I.
Marquette I.
Summerby Swamp
Mackinac I.
Seul Choix Point
Stonington Peninsula
Little Bay de Noc
Big Bay de Noc
Burnaw Marsh
Garden Peninsula
Garden I.
Hog I.
Straits of Mackinac
Bois Blanc I.
South Channel
Sturgeon Bay
Peninsula Point
Point Detour
Summer I.
High I.
Beaver Island
Lake Paradise
Douglas Lake
Algo
St. Martin I.
North Fox I.
South Fox I.
Emmet Upland
Crooked Lake
Burt Lake
Mullett Lake
Little Traverse Bay
Bay
N
Beaver Archipelago
Lake Charlevoix
Charlevoix Drumlins
Walloon Lake
Lighthouse Point
Grand Traverse Bay
North Manitou I.
Leelanau Peninsula
Leelanau Drumlins
Old Mission Point
South Manitou I.
Manitou Passage
Good Harbor Bay
Torch Lake
Lake Bellaire
Huron
Otsego Lake
G
Sleeping Bear Dunes
Glen L.
Lake Leelanau
Old Mission Peninsula
Elk Lake
Lake Skegemog
Port Huron Plain
Grayling Fingers
I
Point Betsie
Crystal Lake
Platte Lake
Long Lake
Swamp
Au Sable R.

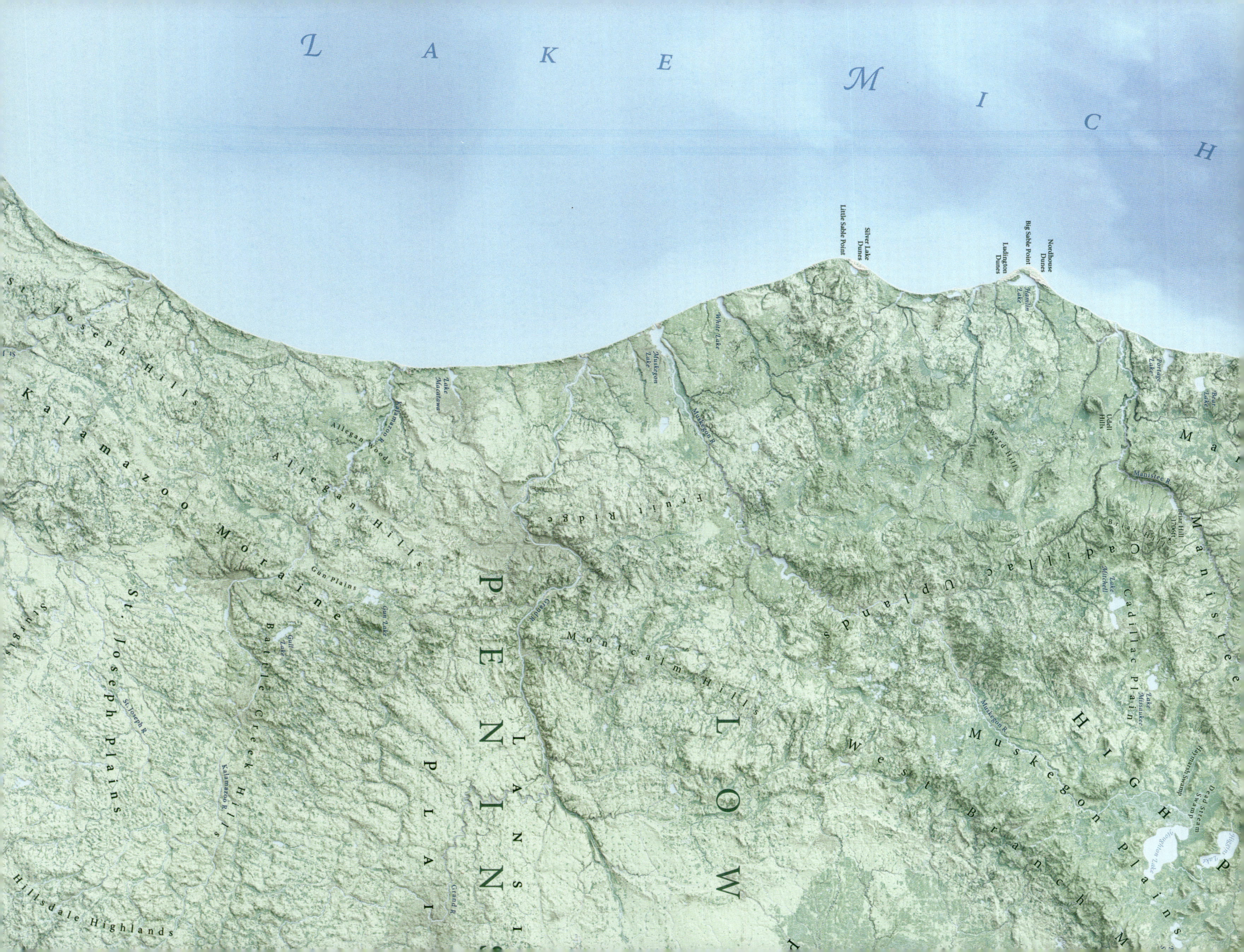

LAKE MICH
Little Sable Point
Silver Lake Dunes
Ludington Dunes
Big Sable Point
Nordhouse Dunes
Hamlin Lake
White Lake
Muskegon Lake
Lake Macatawa
Muskegon R.
Kalamazoo R.
Allegan Woods
Allegan Hills
St. Joseph Hills
Kalamazoo Moraine
Fruit Ridge
Gun Plains
Gun Lake
Gull Lake
Battle Creek Hills
Kalamazoo R.
St. Joseph R.
St. Joseph Plains
Sturgis
Hillsdale Highlands
Grand R.
PENIN
LANSI
PLAI
LOW
Montcalm Hills
Cadillac Uplands
West Branch
Muskegon Plains
Muskegon R.
Cadillac Plain
Lake Mitchell
Lake Missaukee
HIGH P
Houghton Lake
Higgins Lake
Dead Stream Swamp
Haymarsh Swamp
Manistee
Manistee R.
Briar Hill
Ward Hills
Udell Hills
Portage Lake
Bear Lake

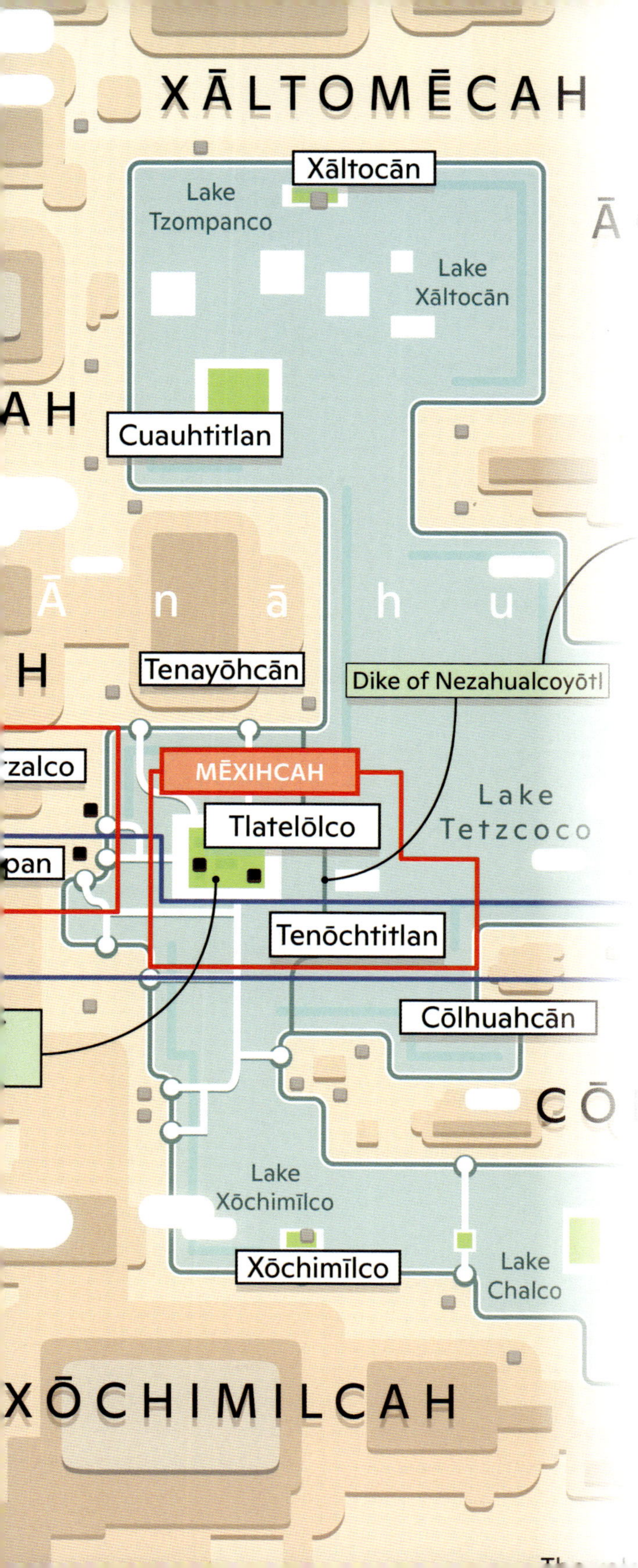

The Names and Places of the Triple Alliance »

Carl Churchill

As a descendant of settlers, I think it is a moral duty to educate myself about the history of the Americas, and especially the peoples who lived there before Europeans arrived. Often we in broader society view the history of Native peoples as an introductory section, a collection of myths and anecdotes, before the rest of the textbook. Yet the Aztec Empire, as it is now called, was just one small part of a fully fledged world, with a universe of terms and places entirely Indigenous in their creation. History that can be approached on its own terms.

This map started out as a study guide, scratched on sheets of paper as I was trying to keep track of lectures and discussions about Aztec history. I realized that there is very little data visualization (dataviz) about the Aztec world, and even less that incorporates the wealth of knowledge in historic literature outside popular history. I thought I would convert my personal journey into something others could use to appreciate this history, too—and it did not hurt that I needed an excuse to practice my Illustrator skills!

The map is centered upon Anahuac, in its original meaning as a word in the Nahuatl language: the modern-day Valley of Mexico. It is rendered in a style I have described as Pokemon-esque, intentionally reminiscent of the geometry of Mesoamerican art without direct imitation. My goal always was to evoke this world without fetishizing it. I included nearby terrain, derived from a Digital Elevation Model (DEM) of the area, but with a focus on making recognizable but simple shapes, as opposed to replicating every contour of the land in detail.

I freely moved between real places, like the island of Mexico itself, and mythical ones, like Aztlan, as much as I moved between different perspectives of illustration. It felt like an appropriate way to show that there was no hard line between mythic and real spaces in the Aztec worldview, while also adding an element of visual flair. Overall I strove toward a balance between accurate information (including many long discussions about proper Nahuatl spelling) and visual playfulness, in order to keep the history as correct as I could, without making it dry.

A GUIDE TO ANAHUAC
THE HEART OF THE AZTEC EMPIRE

What is a Aztec?
The **Aztēcah** (Aztecs) were the people who came from **Aztlān**. Historians today use it to mean the empire ruled by the Triple Alliance, but most people in it did not claim descent from the **Aztēcah**.

Aztlān
Az + (-)tlān
Egrets? at the place of (...)
The mythical northern homeland of the Aztecs.

The Origins of the Nahua
After humans emerged from **Chicōmōztōc**, some travelled to the island of **Aztlān**, before fleeing south to **Ānāhuac**.

Chicōmōztōc
Chicōme + ōztōtl + (-)c
Seven Cave Place
The origin of humanity in **Nahua** mythology.

Tōllān
A mythical city that was said to have been ruled by **Quetzalcōātl**, the feathered serpent diety who taught mankind civilization.

Pronouncing Nahuatl
Nahuatl words here should be pronounced like Spanish, except for some sounds:

Tl — Like 'clue' without the 'c'
Tz — Like 'ts' in 'cats'
X — Like 'sh' in 'shell'

Stress is on the second to last syllable.
Quetzalcoatl (Que-tzal-CO-atl)
Approximate Classical Nahuatl pronounciations used.

What is Nahuatl?
Nāhuatl, or **Nāhuatlatōlli** *(lit. the clear language)* was the language spoken by the **Mēxihcah** and the lingua franca in **Ānāhuac**. People who speak it are called **Nāhuatlācah**, or **Nahua**.

Ānāhuac
Ātl + **Nāhuac**
Water Close to (...)
The Valley of Mexico

Teōtīhuacān
A abandoned city revered by the Aztecs and others.

The Land by the Seas
Ānāhuac literally means the land by the seas. It originally meant the Valley of Mexico, but has been adopted by some today as a potential name of Indigenous origin to use instead of 'Americas'.

Āltepētl *(pl. āltepemeh)*
Ātl + **Tepētl**
Water Mountain
A individual political entity, like a city-state.

Āzcapōtzalco
The main state in the Valley before the Triple Alliance overthrew it.

The Triple Alliance
(Ēxcān Tlahtōlōyān)
The three **Āltepemeh** ruling the Aztec Empire.

Mēxihco
(island)

c. 1100
Mēxihcah
(Ethnicity)

1521
Ciudad de México
(City)

There are almost 2 million **Nahuatl** speakers today, mostly in central Mexico.

The origin of the word **Mēxihco** is unknown. Some suggest it comes from a **Nahuatl** word for 'moon'.

1527
Reino de México
(Province)

1563
Nuevo México
(Province)

1821
México
(Country)

Mexicanos
(Nationality)

1912
New Mexico
(State)

Not Nahuatl
The Aztec Empire ruled many diverse peoples, like the Totonacs, Otomi, and some Maya groups. Some have **Nahuatl** names, like the Otomi (*Hñähñu*), but they are not **Nahua**.

The volcanoes east of the valley of Mexico formed a core part of **Mēxihcah** myth.

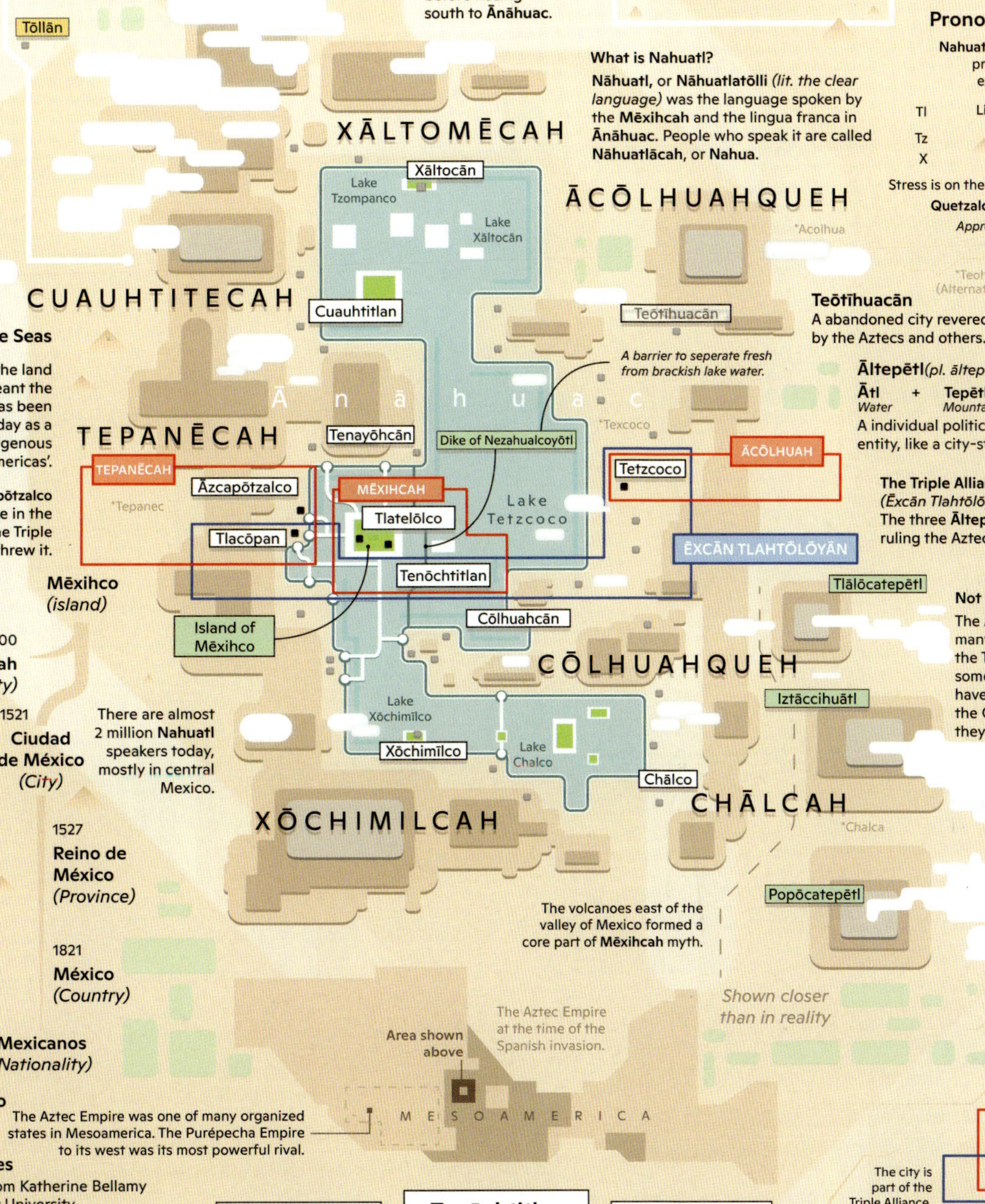

Tlacōpan
Tlacōtl + (-)Pan
Rod(s) Place on
Tlacōpan was founded by the same dynasty that founded **Āzcapōtzalco**. It joined the Triple Alliance after **Tetzcoco**. It is today the neighborhood of Tacuba.

Tenōchtitlan
Tetl + Nōchtli
Rock Prickly Pear
The famed **Tenōchtitlan** was founded on a island by the **Mēxihcah** after being barred from the cities on the lakeshore. It is today the center of Mexico City.

Tetzcoco
Te + Tzicoa + Co
People Retain Place
Tetzcoco was conquered by the **Acolhuah**, who made it their capital. The **Tetzcocah** were one of the first allies of the **Mēxihcah**. It is today the city of Texcoco.

This map represents information collected by Spanish colonial officials and Indigenous scribes primarily after the Conquest. It is not representative of a single period of time.

Data Sources
Altepemeh from Katherine Bellamy and Lancaster University.
Etymologies and definitions from the Nahuatl Dictionary, University of Oregon, David Bowles, Various sources.

Selected Primary Sources:
Crónica Mexicayotl, Códice Boturini
Códice Aubin, Códice Chimalpopoca

Map made by Carl Churchill
A special thanks to the DECM Project, Aztlan Historian, Falling Quetzal, Magnus Pharao and MajoraZ.

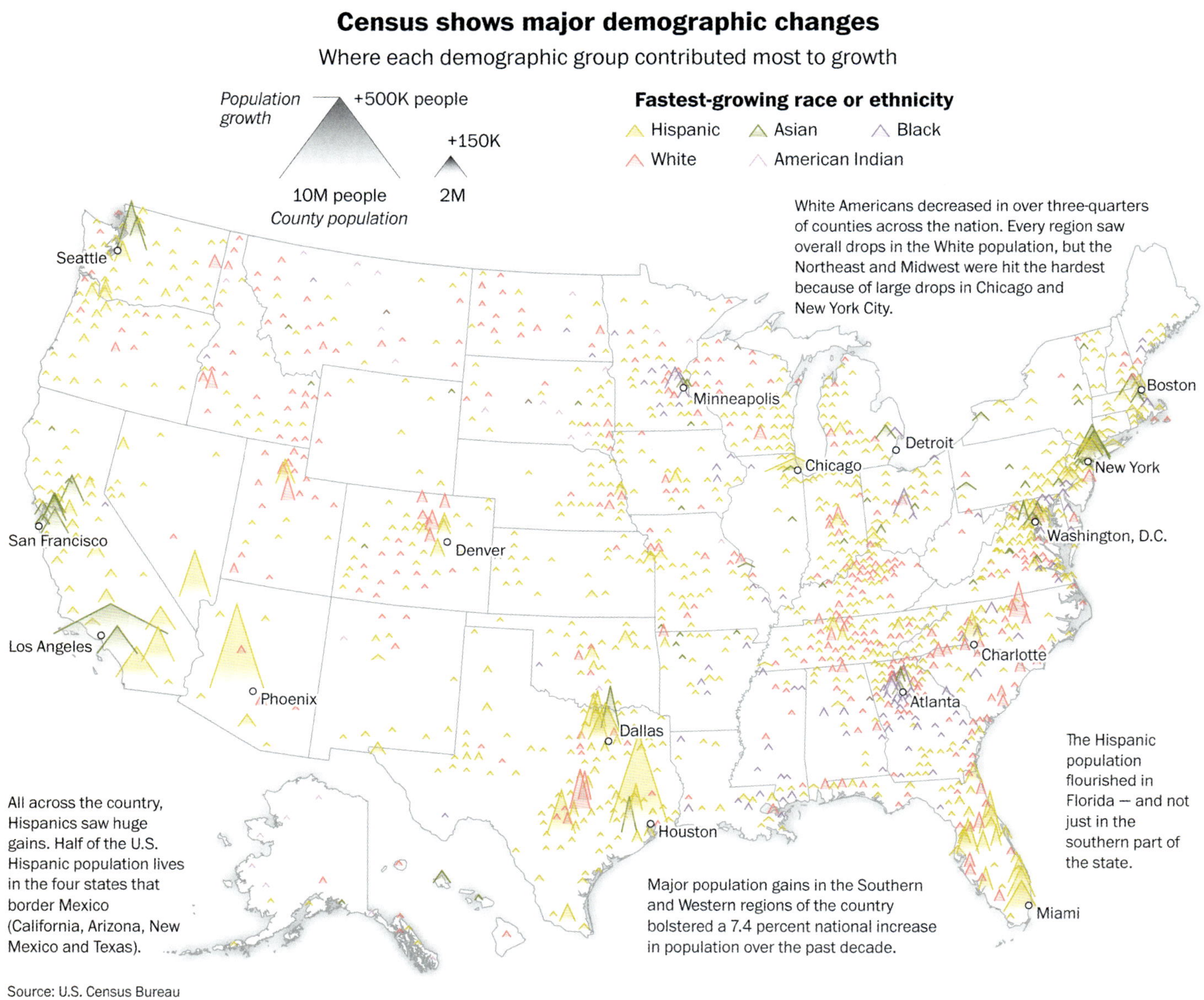

« Where each demographic group contributed most to population growth since 2010

Adrián Blanco Ramos, Zach Levitt, Brittany Renee Mayes and Ted Mellnik, The Washington Post

On the occasion of the 2020 US Census data release, we wanted to show *Washington Post* readers how America's demographics had changed in the past decade and how the collected numbers continue to shape the nation. For each of the major releases—state-level population data and county-level demographic data—we had to design maps that summarized the story for digital and print readers.

Cartographic design in a news context is about conveying a particular geographic story to the reader through a process of maximal abstraction of the real world. Every decision on a map should have a justification for how it contributes to conveying an emotion or idea related to this story.

In the county-level map, we use the metaphor of mountains to express demographic change, which offers the reader a clear and easily comparable reference. Except for borders and a few cities, we

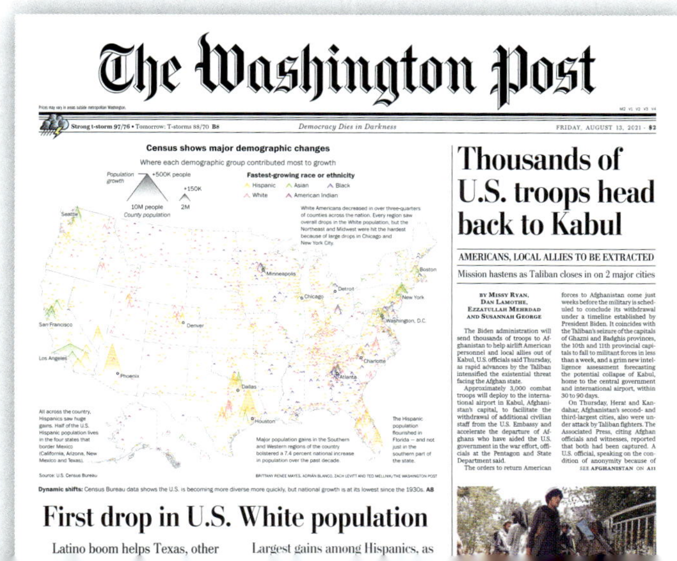
The Washington Post

FRIDAY, AUGUST 13, 2021

Census shows major demographic changes

Where each demographic group contributed most to growth

Thousands of U.S. troops head back to Kabul

AMERICANS, LOCAL ALLIES TO BE EXTRACTED

Mission hastens as Taliban closes in on 2 major cities

First drop in U.S. White population

« Railway Map of Greater London

Mike Hall

This detailed and geographically accurate plan of the entire rail-based transit network in Greater London is a product of my continuing explorations into creating maps with contemporary detail in vintage period styles, as well as my ongoing obsession with mapping my former home city. It follows an earlier, similar design for the network in Spain and Portugal.

Each element of the map takes inspiration from the style of early twentieth-century railway maps. All types of line, ranging from high-speed intercity lines to local tram lines, are represented by a hierarchy of stroke widths and styles. Every station within the area is marked by dots (and larger circles for interchanges), and details such as major termini, tunnels, depots, former lines, and new lines under construction as of November 2021 are also shown. Further details besides the railway network, such as major roads, waterways, airports, and airstrips, all add up to a comprehensive plan of the city's transport infrastructure.

Since I am not an expert in railways (though I very much enjoy travel by train), I referred to multiple online sources, such as OpenRailwayMap, as well as to historical and contemporary network maps to ensure accuracy. I'm also indebted to several people with more knowledge of railway infrastructure than I who helped me to correct errors and omissions in the initial draft.

I initially anticipated that the central area would need a separate enlargement; however, the combination of a highly legible primary typeface (Bureau Grotesque) and the print size I was working to (an A1 sheet, or 840 × 594 mm) meant I was able to display all the detail comfortably within a single image. I also made extensive use of BellTopoSans, a typeface by cartographer Sarah Bell inspired by the labeling on antique US Geological Survey maps of the late 1800s and early 1900s, which further enhanced the period look I was aiming for.

I believe that maps should be designed primarily to be beautiful, a goal that can be achieved by balancing form, color, and type within a carefully constructed composition. This map was created with these same principles in mind.

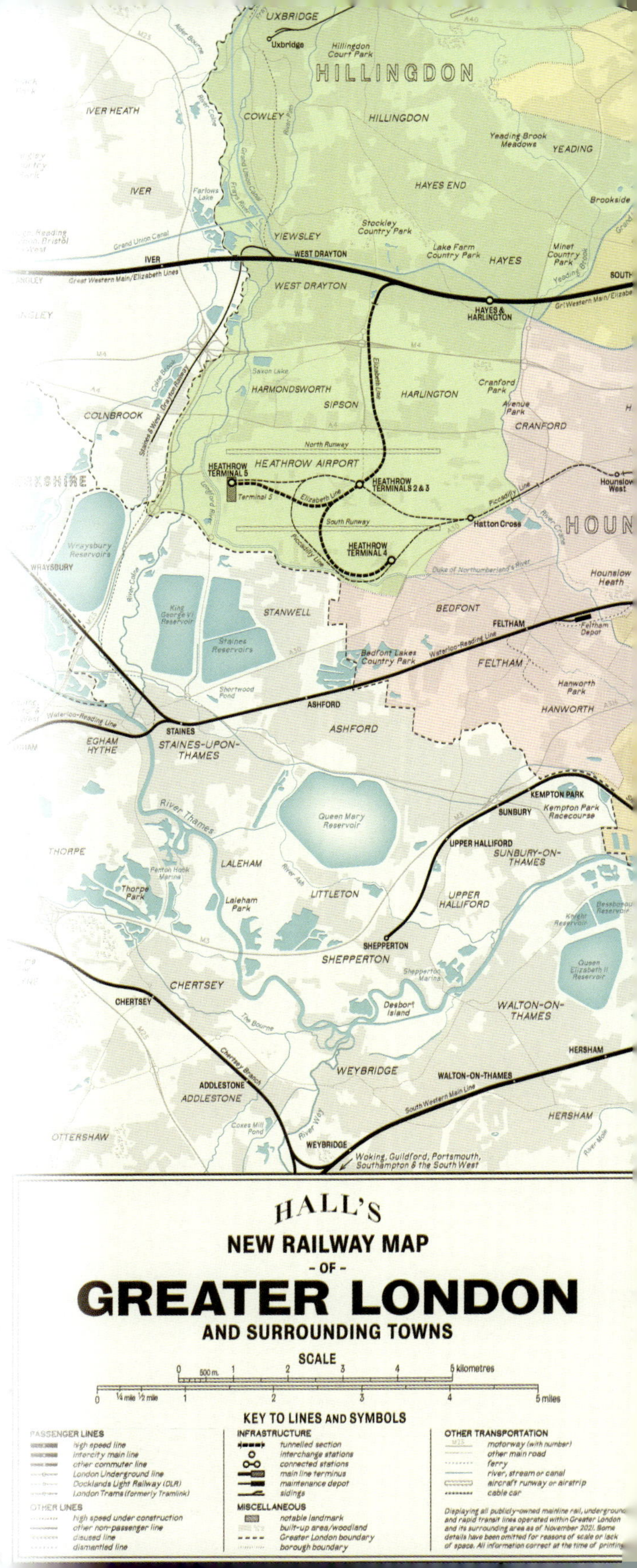

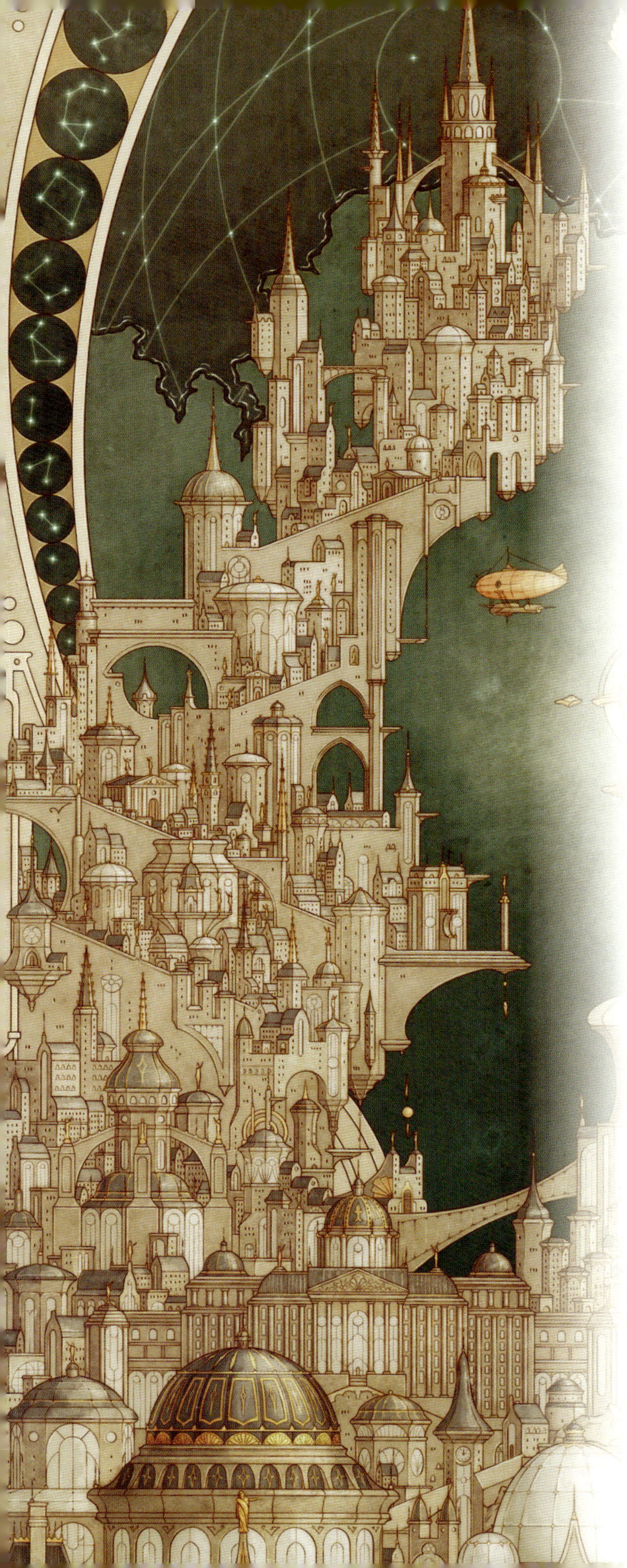

Arlys »

MistyBeee

Authors know it well: characters do as they please. As an artistic mapmaker, I can confirm it's the same for maps.

"Arlys" began as a challenge to map a country based on an imposed geography. As soon as I started exploring the world, I saw the silhouette of a woman. From that first glimpse, "Arlys" quickly became one of my most visually rich and dense works, largely driven by its own will.

The silhouette of the Lady was the guiding thread for the project, starting from the sky and naturally flowing into the verticality of various elements. This basic concept asked questions of the geography that demanded answers: for example, what is the best way to move goods from one ocean to another when the country is divided by an extensive mountain range? Airships and magic were the perfect answer, and so the land-based map was matched with a celestial chart, shaping the world with steampunk technology. This became the new basis for the ornamental style of the frame and for the surrealistic city balancing the map in the bottom left corner.

One of the greatest pleasures in fantasy and artistic cartography is worldbuilding. The map is more immersive than informative and gives viewers an idea of the landscapes they would see while traveling the country. The ornaments and unique geographic details provide clues about the cultures of the country and the possible dangers visitors might encounter, and even hint at historic and political context.

The toponymy of the map is the final touch that both asks and answers questions for the viewer. In the case of "Arlys," I chose names from the most fantastical stories of historical maps—Atlantid, Icaria, Paradis—and it's up to the reader of the map to imagine why, or how, these names are a bridge between the world of "Arlys" and our own . . .

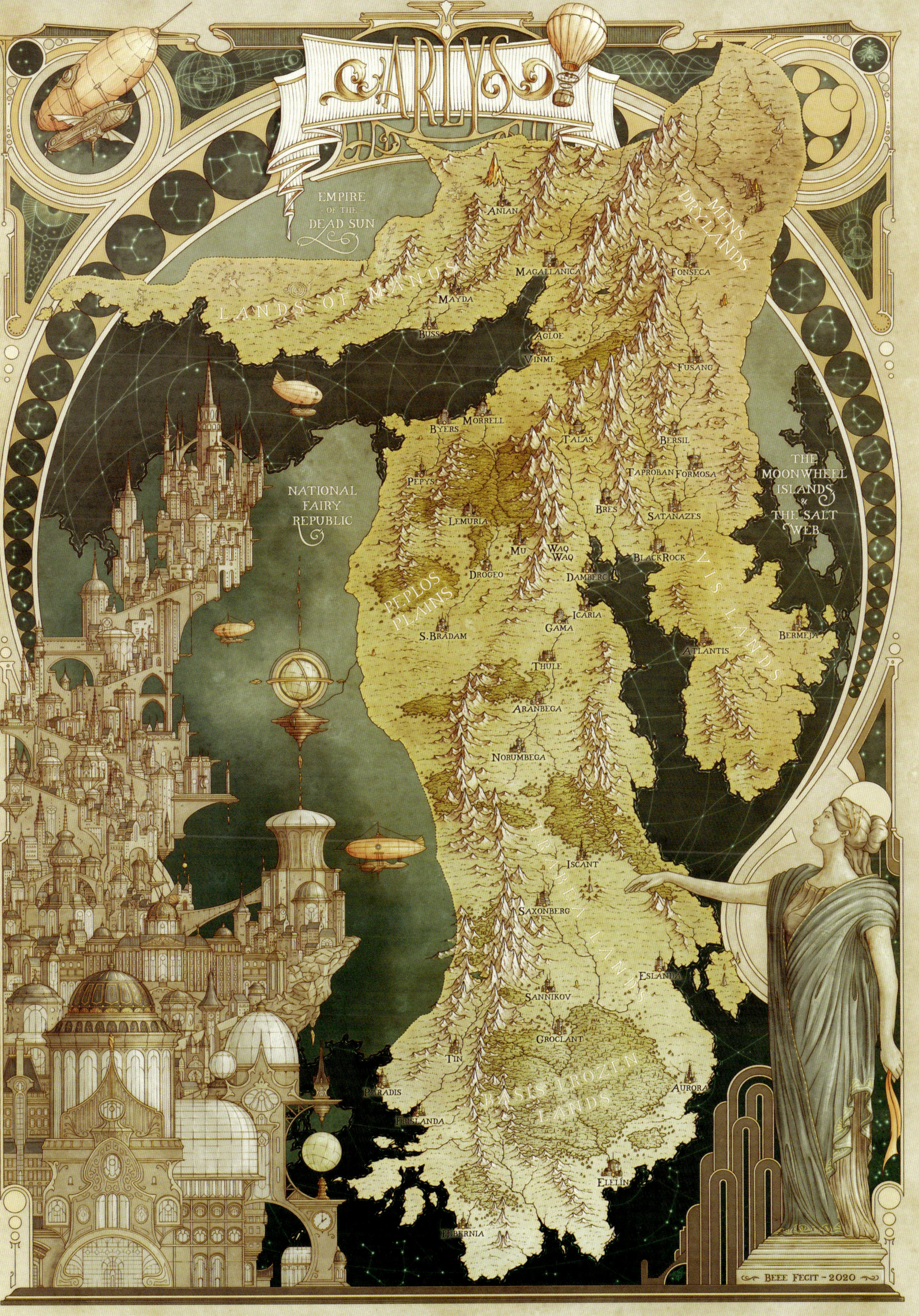
ARLYS
EMPIRE OF THE DEAD SUN
LANDS OF MANUS
MENS DRYLANDS
ANIAN
MAYDA
BUSS
MAGALLANICA
AGLOE
VINME
FONSECA
FUSANG
BYERS
MORRELL
TALAS
BERSIL
TAPROBAN
FORMOSA
THE MOONWHEEL ISLANDS & THE SALT WEB
NATIONAL FAIRY REPUBLIC
PEPYS
LEMURIA
BRES
SATANAZES
MU
WAQ WAQ
BLACKROCK
DROGEO
DAMBERG
PEPLOS PLAINS
S. BRADAM
ICARIA
GAMA
VIS LANDS
BERMEJA
ATLANTIS
THULE
ARANBEGA
NORUMBEGA
ISCANT
TRABEA LANDS
SAXONBERG
ESLANDA
SANNIKOV
GROCLANT
TIN
BASIS FROZEN LANDS
AURORA
PARADIS
FRISLANDA
ELELÍN
HIBERNIA
BEEE FECIT - 2020

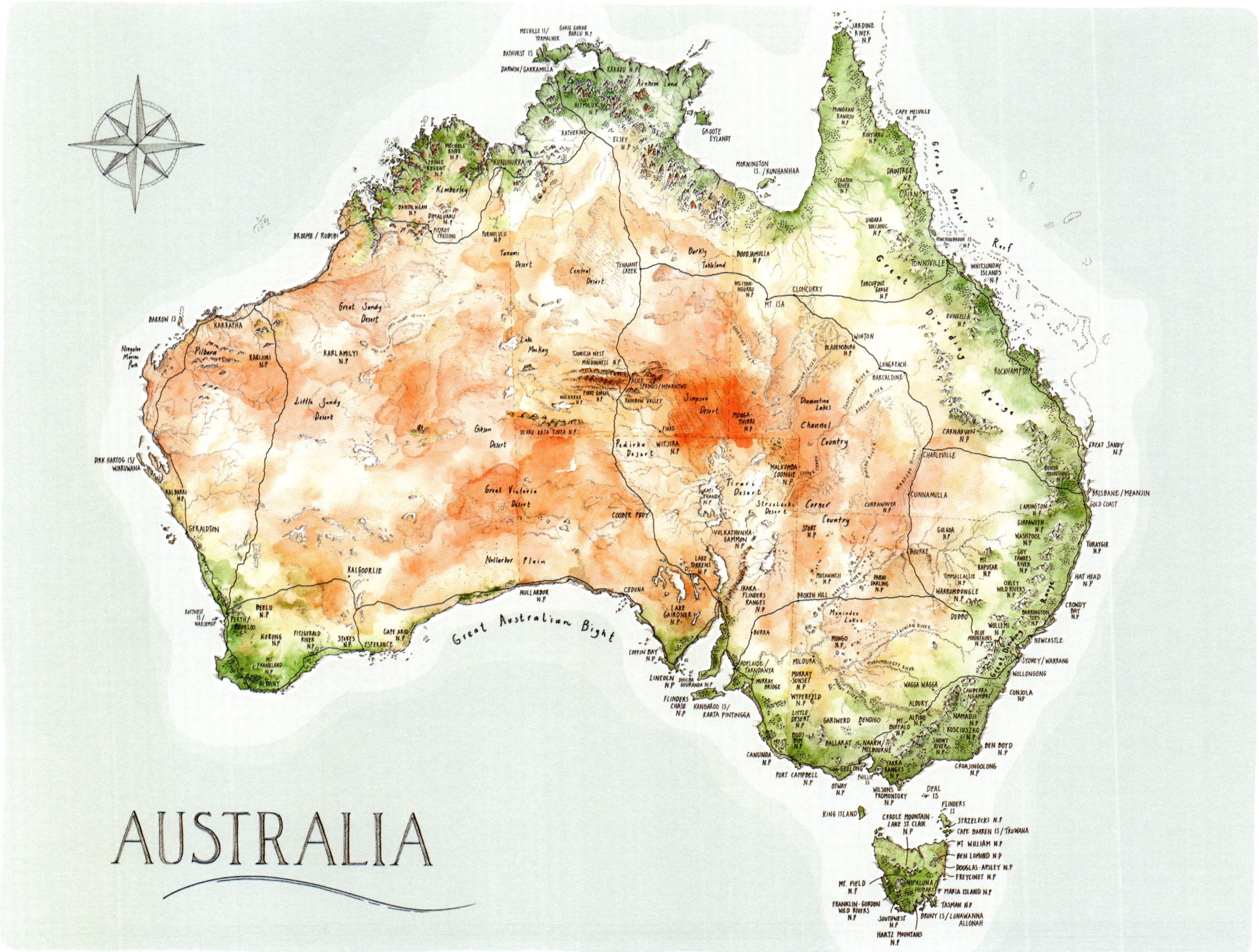
AUSTRALIA
Melville Is / Yermalner
Garig Gunak Barlu N.P
Bathurst Is
Darwin / Garramilla
Kakadu N.P
Arnhem Land
Nitmiluk N.P
Groote Eylandt
Katherine
Elsey N.P
Jardine River N.P
Mungkan Kandju N.P
Cape Melville N.P
Kinyikku N.P
Mornington Is. / Kunhanhaa
Kununurra
Kimberley
Broome / Rubibi
Fitzroy Crossing
Purnululu N.P
Tanami Desert
Central Desert
Tennant Creek
Barkly Tableland
Boodjamulla N.P
Staaten River N.P
Daintree N.P
Cairns
Undara Volcanic N.P
Great Barrier Reef
Townsville
Whitsunday Islands N.P
Great Dividing Range
Porcupine Gorge N.P
Cloncurry
Mt. Isa
Eungella N.P
Winton
Bladensburg N.P
Longreach
Barcaldine
Rockhampton
Barrow Is
Karratha
Ningaloo Marine Park
Pilbara
Karijini N.P
Great Sandy Desert
Karlamilyi N.P
Lake Mackay
Tjoritja West MacDonnell N.P
Alice Springs / Mparntwe
Rainbow Valley
Simpson Desert
Munga-Thirri N.P
Diamantina Lakes
Channel Country
Carnarvon N.P
Great Sandy N.P
Little Sandy Desert
Gibson Desert
Uluru Kata Tjuta N.P
Pedirka Desert
Witjira N.P
Charleville
Dirk Hartog Is / Wirruwana
Kalbarri N.P
Great Victoria Desert
Tirari Desert
Strzelecki Desert
Corner Country
Cunnamulla
Brisbane / Meanjin
Gold Coast
Lamington N.P
Currawinya N.P
Geraldton
Coober Pedy
Vulkathunha-Gammon N.P
Sturt N.P
Gulgoa N.P
Bourke
Mt. Kaputar N.P
Yuraygir N.P
Kalgoorlie
Nullarbor Plain
Lake Torrens N.P
Mutawintji N.P
Paroo Darling N.P
Oxley Wild Rivers N.P
Hat Head N.P
Nullarbor N.P
Ceduna
Ikara-Flinders Ranges N.P
Broken Hill
Warrumbungle N.P
Crowdy Bay N.P
Rottnest Is / Wadjemup
Perth / Boorloo
Beelu N.P
Lake Gairdner N.P
Menindee Lakes
Dubbo
Barrington Tops N.P
Wollemi N.P
Blue Mountains N.P
Newcastle
Fitzgerald River N.P
Stokes N.P
Cape Arid N.P
Esperance
Great Australian Bight
Coffin Bay N.P
Burra
Mungo N.P
Sydney / Warrang
Wollongong
Albany
Lincoln N.P
Adelaide / Tarndanya
Mildura
Murray Sunset N.P
Murray Bridge
Wagga Wagga
Canberra / Ngambri
Conjola N.P
Flinders Chase N.P
Kangaroo Is / Karta Pintingga
Wyperfeld N.P
Little Desert N.P
Albury
Gariwerd
Bendigo
Mt. Buffalo N.P
Alpine N.P
Namadji N.P
Kosciuszko N.P
Ballarat
Naarm / Melbourne
Snowy River N.P
Ben Boyd N.P
Canunda N.P
Yarra Ranges
Croajingolong N.P
Port Campbell N.P
Geelong
Phillip Is
Otway N.P
Wilsons Promontory N.P
Deal Is
Flinders Is
King Island
Cradle Mountain - Lake St. Clair N.P
Strzelecki N.P
Cape Barren Is / Truwana
Mt. William N.P
Ben Lomond N.P
Douglas-Apsley N.P
Freycinet N.P
Mt. Field N.P
Nipaluna / Hobart
Maria Island N.P
Franklin-Gordon Wild Rivers N.P
Tasman N.P
Southwest N.P
Bruny Is / Lunawanna Allonah
Hartz Mountains N.P

The individual state maps became maps of yearning, a celebration of beauty for parts of our country that we were no longer able to visit. «

« Australia

Alex Hotchin

In early 2020, as the COVID-19 pandemic reached Australia, my partner and I decided we would pack our van and take a year or two traveling our country. This became impossible when Australia's six states and two territories closed their borders to each other. We became confined first to our state, and then to our home.

So, I set out to go traveling in another way—by creating a large map of all of the places I would like to have visited. I drew each state individually, with black ink and watercolor. The tactile nature of hand on paper felt to me like I was creating my own adventure. I drew the intricate coastlines, texture, and pattern of the land in black linework first, then used the watercolor to show the diverse colors of this land. A color pattern became clear—green to the coastlines, and shades of red to the heart of the country. The patterns of deserts, lakes, rivers, mountains, and forest were of utmost importance, particularly where they traversed these now uncrossable borders.

I focused on labeling the national parks and geographic features, as this drawing was always to be about the natural places. This involved a lot of research, and I discovered places previously unknown to me. Although most Australian maps show a network of dense roads, I have reduced these to a minimum—drawing the main roads that connect the states and territories as a fine line. Thus, the road network appears frail in comparison to the vastness of this land.

The individual state maps became maps of yearning, a celebration of beauty for parts of our country that we were no longer able to visit. After I had finished drawing the last state, I combined all of my individually scanned drawings into one large map, putting Australia back together again, repairing this divided country. The texture of the sea was added, a pale color, allowing the vibrancy of the land to be the main focus. All of the joined parts made for a beautiful, and familiar single landmass. The familiarity reminding me that we are one country, and vulnerable when separated.

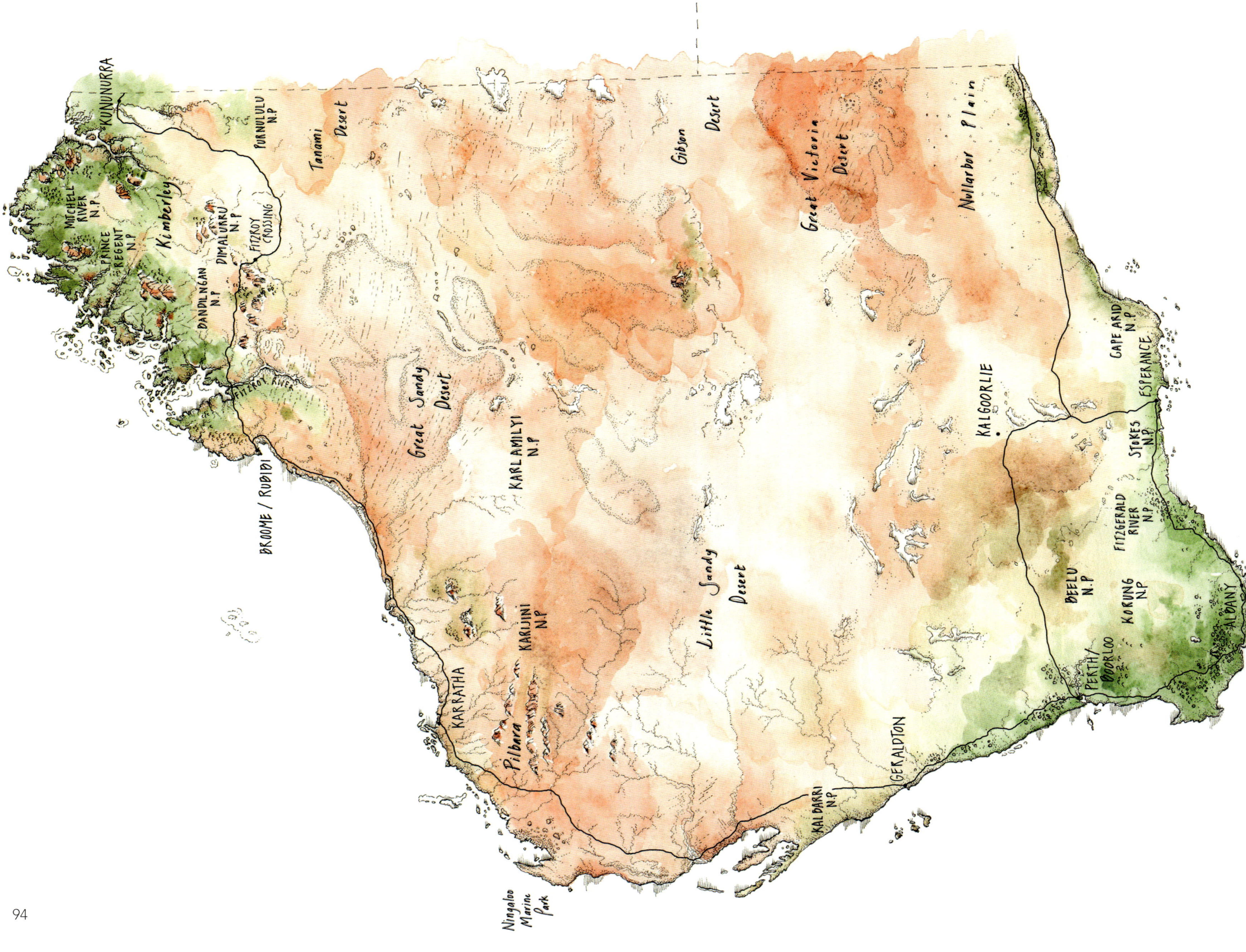

KUNUNURRA
PURNULULU N.P
Tanami Desert
Gibson Desert
Great Victoria Desert
Nullarbor Plain
MITCHELL RIVER N.P
PRINCE REGENT N.P
Kimberley
DIMALURRU N.P
FITZROY CROSSING
DANPILNGAN N.P
FITZROY RIVER
BROOME / RUBIBI
Great Sandy Desert
KARLAMILYI N.P
KALGOORLIE
CAPE ARID N.P
ESPERANCE
STOKES N.P
FITZGERALD RIVER N.P
BEELU N.P
KORUNG N.P
ALBANY
PERTH / BOORLOO
Little Sandy Desert
KARIJINI N.P
KARRATHA
Pilbara
GERALDTON
KALBARRI N.P
Ningaloo Marine Park

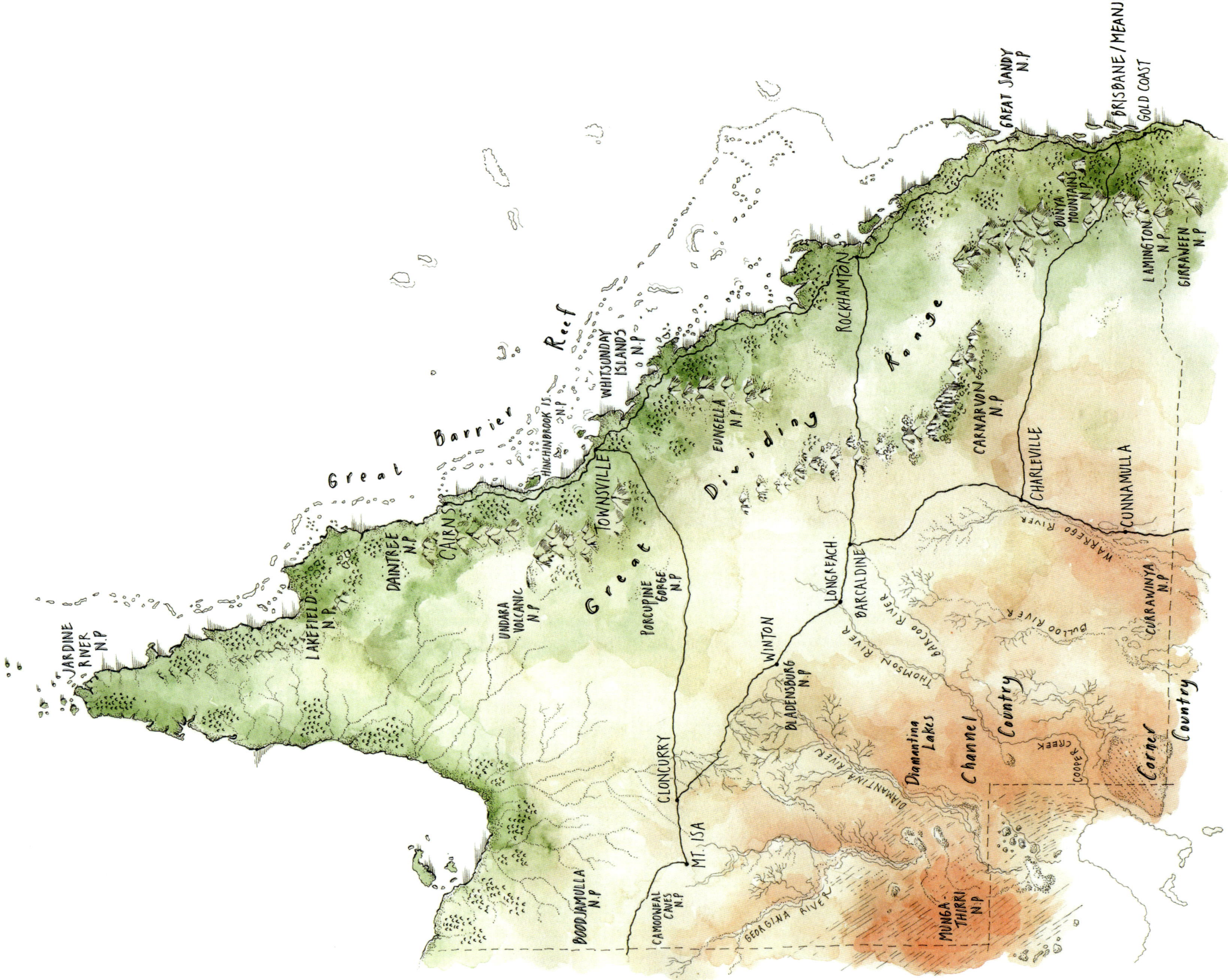
GREAT SANDY N.P
BRISBANE / MEANJIN
GOLD COAST
BUNYA MOUNTAINS N.P
LAMINGTON N.P
GIRRAWEEN N.P
ROCKHAMPTON
Range
CARNARVON N.P
CHARLEVILLE
CUNNAMULLA
WARREGO RIVER
CURRAWINYA N.P
BULLOO RIVER
Country
Corner
Great Barrier Reef
WHITSUNDAY ISLANDS N.P
HINCHINBROOK IS N.P
EUNGELLA N.P
Dividing
TOWNSVILLE
CAIRNS
DAINTREE N.P
Great
PORCUPINE GORGE N.P
LONGREACH
BARCALDINE
BARCOO RIVER
THOMSON RIVER
WINTON
BLADENSBURG N.P
Channel Country
COOPER CREEK
Diamantina Lakes
DIAMANTINA RIVER
MUNGA-THIRRI N.P
GEORGINA RIVER
LAKEFIELD N.P
UNDARA VOLCANIC N.P
JARDINE RIVER N.P
CLONCURRY
MT. ISA
CAMOOWEAL CAVES N.P
BOODJAMULLA N.P

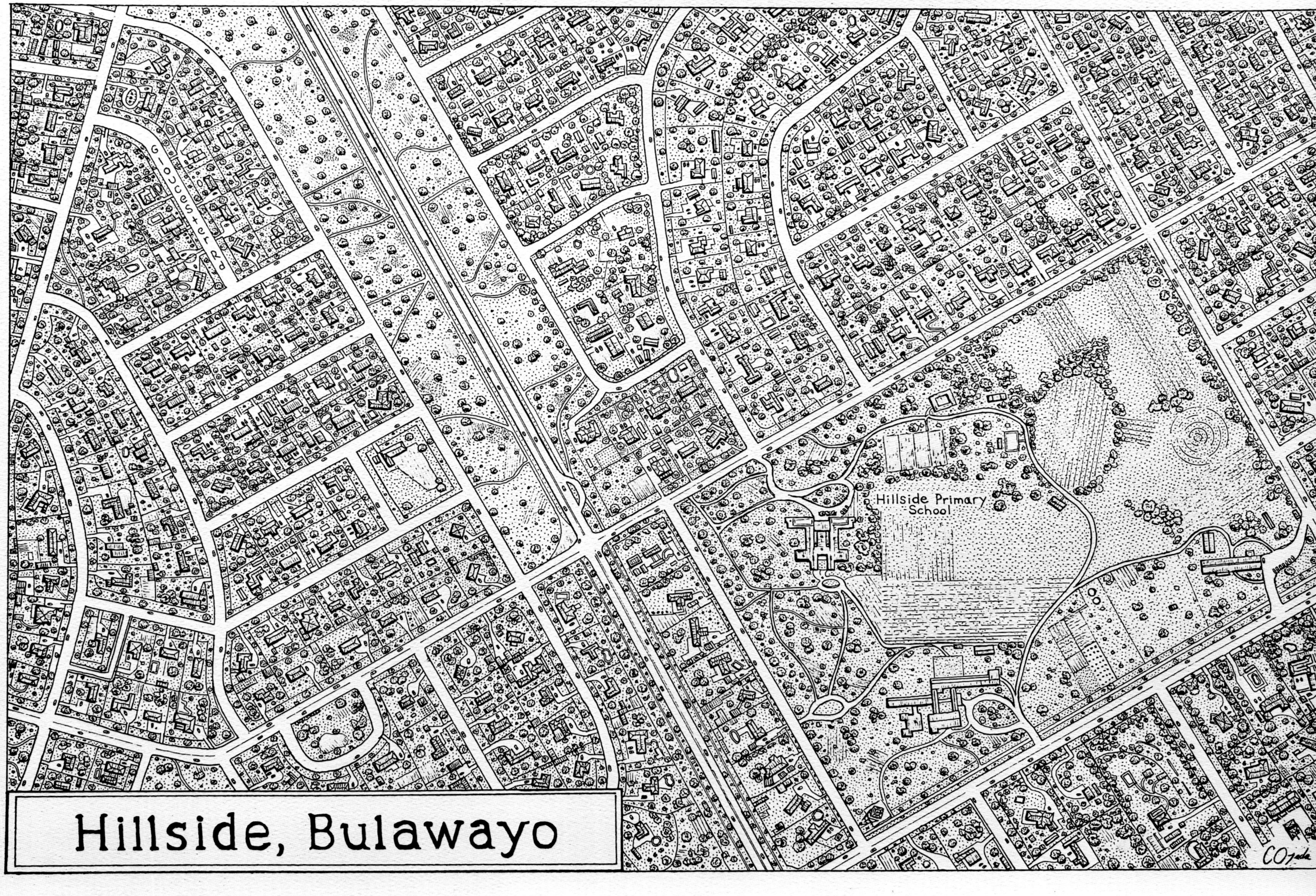
Hillside, Bulawayo
Gloucester Rd
Hillside Primary School

« Hillside, Bulawayo, Zimbabwe

Cam Ojeda

Bulawayo is the second largest city and the principal industrial center of Zimbabwe. This map focuses on one neighborhood in the southeastern part of the city, called Hillside, and it was where my uncle spent his childhood. I did this map as an art piece for him to commemorate his time there.

I have always loved geography, cities and how they formed, and maps as a way to understand places and the relationships that these places have with people and history. Maps have the power to educate and entertain the viewer, and learning about the world around us is always more enjoyable when presented in a simple visual way. For me, the more details a map had, the longer I could stare at it and be fascinated by it.

The map was done by referencing satellite images; it is hand drawn with archival ink fine-liners. This area of Africa did not have satellite images that were very detailed, and things like street views were unavailable, so I had to take some artistic liberties and do the map in the most accurate way possible but also in a way that made the piece more interesting to view. So, for example, I added cars and made most of the trees nice round shapes. Aesthetically I want to use different textures and slight shadows for depth and also have the viewer be able to look at it for a long time. When I look at maps, I enjoy all the different little details and work my way through the map, sometimes even imagining what it would be like to be down on the streets walking around. Maps are great for the imagination, and with my depictions I hope to bring about this sense of wonder and curiosity for viewers. I also wanted this map to elicit memories for my uncle, and I labeled the street he lived on and the primary school he attended, so that they are highlighted in the map and easy to reference.

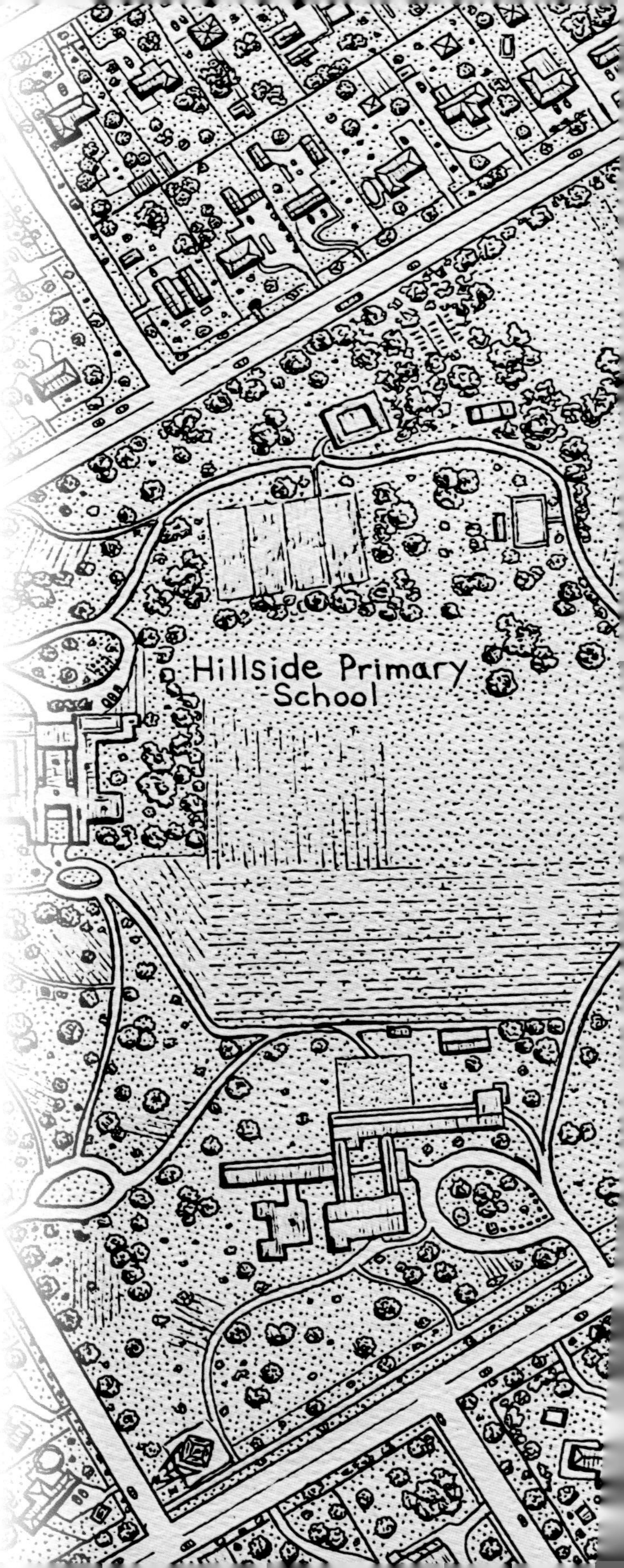

Here is this Atlas's very last page...
But rats, Volume 7 won't come for an age!
Go back and savor the details astounding...
Here's some to locate that may be confounding:

Four yellow buses all lined up to go,
A red-shouldered hawk on the same day as snow,

A fairy republic, the old Georgia Dome,
The artist's grand-mère, and Nickelback's home,

The smell of old people, the Fortingall Yew,
Two laughing fish, and marine Ningaloo,

A man with three fingers, the Clark Canyon Dam,
The Bellarine worm, and the tube at West Ham,

A leaping red dolphin, an angry white cow,
The last site on the lake (we think, anyhow).

With love,
The Editors

Wings in the Desert

A Folk Ornithology of the Northern Pimans

Amadeo M. Rea

Linguistic Consultant,
Culver Cassa

Bird Sketches by
Allan Brooks
Bob Hines
Takashi Ijichi
Amadeo M. Rea
George M. Sutton
Mike Touch

The University of Arizona Press
Tucson

The University of Arizona Press

Cataloging-in-Publication Data are available from the Library of Congress.
ISBN 978-0-8165-2459-4 (hardcover : alk. paper)

Publication of this book is made possible in part by the proceeds of a permanent endowment created with the assistance of a Challenge Grant from the National Endowment for the Humanities, a federal agency.

♾ Manufactured in Canada on acid-free, archival-quality paper containing 100% post-consumer waste and processed chlorine free.

To the memory of
Kenneth E. Hill,
connoisseur of bird books

Contents

Part 1 *Birds in Piman Culture*

Part 2 *Piman Birds*

Appendices

Illustrations *Maps, Diagrams, and Photographs*

Species Sketches

Tables

Preface

The Northern Pimans are the northernmost cluster of what I call the Northwest Mexican Ranchería Cultural Complex, a grouping excellently characterized by Thomas Hinton (1983:315–328). The Mexican Ranchería Complex consists of the Papago or Tohono O'odham; River Pima or Akimel O'odham; Mountain Pima or Oob; Ópata; Eudeve; Tarahumara or Rarámuri; Tubar; Guarijío; Cáhita (Mayo or Yoreme and Yaqui or Yoeme); Northern and Southern Tepehuán (including Tepecano); Cora; and Huichol. As a group, their cultures differ qualitatively from the Seri on the Gulf Coast, the various Yumans to the northwest, the Pueblos and Athabascans in the north, and the Mesoamerican cultures southward. Several of these tribes are now extinct as ethnically identifiable entities. Other groups still number from thirty thousand to more than fifty thousand individuals.

Members of this Mexican Ranchería Complex speak Uto-Aztecan languages and once shared a contiguous territory. At Contact, all had a ranchería settlement pattern. There were no pueblos or towns such as those among the Zuni and Hopi. The social system was egalitarian, characterized by a loose political structure, with more organized leadership as warfare demanded. Throughout the complex, there was a common subsistence strategy based on hunting, gathering, and warm-season agriculture. Most depended on summer rain (de temporal) agriculture of some sort, augmented by ditch irrigation on the floodplains of the more permanent streams. Crops grown included various cultivars of maize, several species of beans and squash, cotton, bottle gourds, grain amaranth, and chenopod.

Similar religious elements recur throughout this Ranchería Complex. There was a pattern of ceremonial drinking involving fruit (usually cactus) wine in the lowlands and corn beer at higher elevations. In addition to being religious functions, these events promoted social bonding within the group. Most, if not all, in the ranchería area also shared a symbolic deer-maize-Peyote complex. Major votive offerings, in both public ceremonies and private oration, were prayer sticks made with feathers and cotton yarn. Shamanism was practiced throughout the complex, but there was only an incipient priesthood.

All members shared a material culture best characterized by its simplicity. It included plainware ceramics, plaited basketry, and cotton spinning and weaving (horizontal looms for blankets; women with backstrap looms for sashes). Except for some stone tools such as adzes and metates, most utilitarian items could be fashioned easily and quickly. Houses were simple and usually round.

The dissimilarities apparent within this complex today are, I think, the result primarily of differing post-Contact responses to European culture. Some responses date from early in the Hispanic period, whereas others have been changes made during the twentieth century. These latter changes include larger, more defensive settlements, more complex political structures, and direct borrowings such as adobe house styles. Even implements adapted from European models, such as wooden plows and yokes, are quite uniform from one group to the next. The religious syncretism is amazingly similar throughout the complex, focusing on elaborate Lenten and Holy Week ceremonies that incorporate pascola and matachini dancers as well as locally evolved passion plays. And, of course, cultural losses during this same time period account for other differences (the present lack of plaited rectangular baskets among Upper Pimans, for instance).

This book is about those Uto-Aztecans speaking the Piman language and inhabiting a once continuous strip of the tierra caliente—the hot lowlands—from the Gila River to the Río Yaqui. All called themselves O'odham and in spite of local dialectical variations could communicate freely. The current insularity of the various subsets within lowland Pimans is the result of population decline through disease, mestizoization, acculturation, land usurpation, and the imposition of an international boundary. North of the U.S.–Mexico border, the confinement to various reservations has enhanced a sense of separatism between once contiguous groups.

The Mountain Pima are not included in this book for several reasons. Their self-designation is *Oob* rather than *O'odham*, reflecting an emic dichotomy. The Mountain Pima language is not mutually intelligible with the lowland dialects found throughout both Pimerias. It differs somewhat in phonology and, more important, in grammatical structure, being perhaps more different from these other dialects than are Spanish and Italian from each other. The Mountain Pima farm primarily by clearing hillside fields that are watered directly by summer rains rather than by manipulating runoff from arroyos or along the floodplains. Linguists estimate the separation of this upland group from lowland O'odham to be at least a thousand years (Wick Miller, personal communication, 28 June 1993). Although living near desert O'odham, Mountain Pima are separated altitudinally. They occupy the pine-oak woodlands and conifer forests of the Sierra Madre Occidental. Consequently, their cultural ecology is quite different. Their local avifauna is also distinct and deserves to be studied on its own. However, I've included cognate names for birds in the comparative linguistics sections of the species accounts, along with Tepehuán (Southern Pimans) as well as other Uto-Aztecan bird names when they were available.

I use the designations *Linnaean*, *Western*, and *scientific classification* interchangeably throughout this work for what is generally called the *biological system of binomial classification*. These four terms contrast with *folk taxonomy*, the classification system of a particular language community. Both folk and scientific systems are efforts to label the perceptual discontinuities of the flora and fauna around us. Linnaean classification systems attempt to be phylogenetic or evolutionary and may use tools such as microscopes to study fine structures or even chromosomes unavailable to the folk systematist. Folk classification usually encodes complex taxonomies of the local biota. Size is an important factor, organisms at the smaller end of the scale being classified only broadly or not at all (Hunn 1999). And folk systems are not based on an evolutionary model.

I have used the term *folk science* in this work to designate the fund of scientific knowledge of a cultural group, subsets of which would be folk taxonomy and traditional ecological knowledge or natural history. All of these sets are critical areas where human knowledge and the natural environment intersect. They are essential aspects of cultural adaptation.

Western science is typically quantified and published. (It carries more clout if it involves expensive equipment and is supported by large grants.) In contrast, folk science is observational and passed on orally. It is not overtly quantified, but I would argue that the validity of a folk scientific system is tested through time and changed as needed. We would be naive to think that cause-and-effect reasoning is something unique to Western science. The O'odham evolved a science of managing a culture in the desert that in the long term was far superior to what has replaced it: nonsustainable agriculture and sprawling urban centers. Pull the plug and cut the petroleum lifeline, and the Euro-American system of living in the arid Southwest would collapse, leaving behind a horribly degraded and much simplified environment.

Just as any study of a biological community begins with learning the taxonomy of all its constituent members, so too working out the folk taxonomy of a speech community is the basis for any ethnobiological study. Merely amassing a list of terminal taxa with their Linnaean counterparts might have sufficed fifty or a hundred years ago, but this tells us little about how a given language community cognitively organizes its biotic environment. We must explore linguistic or semantic contrast sets—that is, a group whose members are immediately included in a higher taxon (Stellar's Jay, Pinyon Jay, and so forth). But we must also work out biological contrast sets in order to map the domain of any folk taxon, which requires that we as biologists know the local flora or fauna. Does a name given for a particular specimen of *Chenopodium*, for instance, apply to all species of this genus? Are some excluded? Are weedy annual species of *Atriplex* also included? And how do these sets fit into a folk hierarchical classification system?

The development of cultural metaphors is based on shared folk science. Sadly, many translators of O'odham oral literature have failed to analyze Piman folk taxonomy and its associated folk science. Examples of this problem are given in one species account after another later in this book. My chapter evaluating documentary sources (chapter 3) may seem somewhat harsh, but the truth is that the O'odham have been sold short, particularly by later workers who have often tangled rather than straightened out the original metaphors (a result that in itself may reflect the decline in basic ecological knowledge and taxonomy among urbanites). I try to unravel these knots in the species accounts.

In common parlance, myth often takes a bad rap when contrasted with science. This is unfortunate. Myths are stories that deal with the more profound truths and aspirations of human life, whereas science probes superficials. The stories that make up the four nights of the Piman Creation Story, or Origin Myth, are of considerable depth in explaining the People's worldview, setting forth a way of life that is considered properly human. We can also distinguish other kinds of stories. Folktales are entertaining narratives, often with didactic elements. These tales are frequently etiological as well, telling how something got to be the way it is (for instance, why the Mourning Dove has a pointed tail). They are catchy, but, unlike myth, they are not very profound.

My earlier book *Once a River* (Rea 1983) and this one are complementary works, but with very different goals. The first focused on documenting scientifically, with specimens, the breeding, wintering, and migrant avifauna of the Gila River Indian Reservation and how it has changed historically. I used Pima terminal taxa as one of the tools for exploring the former status of birds along the now much-degraded middle Gila. That work included a preliminary folk taxonomy and various folk-scientific anecdotes primarily validating the folk identifications. In the present work, we come full circle, exploring now at depth the ethnographic role of birds in Piman society, using the first study as a sort of etic grid supporting the folk information. Of course, the geographic scope of the present study is far greater than the first, which was limited to the middle Gila.

I think I have almost exhausted what is to be learned from River Pima regarding their local avifauna, but in no way is this book a definitive work on Tohono O'odham folk ornithology, although I have made some contacts and, I hope, have captured the basic structure. Papago comprises six or more different dialects spread across southern Arizona and adjacent northern Sonora, Mexico. Each group may harbor differences in names of organisms and accompanying oral traditions. Some of this difference is reflected in the species accounts here, but a thorough study would constitute a good doctoral dissertation, preferably by an O'odham graduate student who knows the language. And soon! Information is vanishing at the rate of a half life per generation.

The general sounds of O'odham and their orthography are given in appendix G. Four simple rules will get you through the pronunciation of O'odham words reasonably well: (1) the accent is on the first syllable of a word; (2) the written *l* is never the sound *l* but a flap *r*; (3) the vowels are the same as in "continental" languages, except for *e*, which is a rounded, spread-lipped *u*, there being no conventional *e* sound in O'odham. And (4) covering all these rules is one umbrella rule: slow down! O'odham is a language of unhurried desert places. On most days, the wind moves casually about the desert landscape; the People talk in similar tones. The vowels especially are sensuously slow and breathy. Some are long (written double), and others are shorter, but still longer than most spoken English.

The orthography found in this work is practical, using the symbols that most readers of English and Spanish recognize. The *š*, *ṣ* or *x* is written as *sh*, and *č* or *ç* is given as *ch*. (Diacriticals may thrill the linguistic specialist, who likes to have a single symbol for each phoneme or sound, but they invite confusion among nonspecialists.) Except in direct quotes of published works, I have attempted to standardize the ten or twelve orthographies used in earlier sources for writing the Tepiman words included here. Failure to do so is not only sheer laziness, but also unfair to readers, who are led to believe differences in pronunciations exist where actually there are only different orthographic conventions. Russell's "Navitco," for instance, is "Navichu," just as Pimans say today. Failure to unscramble Russell's spelling has led to such novelties as "Huhugam" being called the "Hohokam" by most today, often with an incorrect accent to boot (as in Andrews and Bostwick 2000).

Colonial Jesuits and Franciscans compiling vocabularios did the best they could with their European orthographies, often influenced by their country of origin (Spanish, German, and Italian Jesuits recorded Pima differently). Glottal stops might be indicated by double vowels, but vowel length was ignored. We don't know if terminal vowels were voiced or unvoiced. Colonial *rh* was either the modern alveolar palatal *ḍ* or a sound that was in transition to it.

O'odham words have been italicized throughout unless they are personalized (proper nouns), as in mythological accounts, or used as place-names; then they are capitalized and set in roman type, as in Ñui, Buzzard, the character. Spanish words are not italicized.

For the most part, I have used the correct O'odham toponyms because more native speakers are themselves returning to their proper names for places rather than to the Hispanicized or Anglicized (and often badly mutilated) renditions. So you will find in the text Komaḍk not Komátke, Pisin Mo'o not Pisínemo, Haashañ Keek rather than Sacaton Flats (the accents are on the first syllables of each word). Excellent listings of place-names can be found in Saxton and Saxton 1969:186–187 and Saxton, Saxton, and Enos 1983:132–138. Because this volume is an ornithological work, the scientific name for a bird appears in the heading of the species account where there is one for that particular bird. If there is no species account for a bird, the scientific name is given at first reference to that bird in the text or the species accounts. The scientific names of plants and other animals appear in appendices B, C, and D.

The English names of birds as well as of other organisms have been treated as proper nouns for several reasons. By international convention, bird names are always capitalized in professional ornithological writing. This practice eliminates guesswork as to what constitutes a formal name and what is only descriptive. The Steller's Jay, *Cyanocitta stelleri*, for instance, is a blue jay, but it is not a Blue Jay, *Cyanocitta cristata*; not all blue-winged teals are Blue-winged Teals, *Anas discors*, as Lyndon Hargrave (1968) points out. Aside from preventing endless confusion as to what actually constitutes a name, capitalization of names conveys a dignity to the other organisms with which we share this earth, a subtle antidote to anthropocentric arrogance.

The common or English names for birds are important, just as are the scientific names. They have been standardized in various editions of the American Ornithologists' Union (AOU) checklist (1957, 1998). Changes in names are sometimes required by changes in taxonomic concept. The (belated) acknowledgement that all continental North American flickers interbreed, hence are the same species, required an umbrella name, Northern Flicker, for the entire group. The familiar name "Red-shafted Flicker" is still available for one of the subgroups readily identifiable in the field. Sometimes two populations have been found to be reproductively isolated, requiring separate common as well as scientific names, as in the case of two screech-owls. However, even when the AOU ran out of legitimate taxonomic reasons for changing names, momentum seems to have kept the process going, and many new names that have appeared are capricious at best. (The committee on nomenclature needs to be reminded of the dictum "If it ain't broke, don't fix it.") Most of these frivolous changes I have chosen to ignore.

In a number of instances, the generic names used here differ from those found in field guides and the AOU checklists. My reasons, usually anatomical, are outlined in *Once a River* (Rea 1983). At still higher taxonomic levels, the AOU has failed to untangle the polyphyletic orders Ciconiiformes and Falconiformes, in spite of the abundant evidence published on this subject. (The inertia of tradition, at least in ornithology, is sometimes more powerful than the findings of science.)

Also, I think it important to emphasize that there are no *informants* in this book. This derogatory term not so subtly implies the superiority of anthropologists who study a culture not their own. Instead, this work features Piman *teachers* who were willing to share their deep knowledge of desert plants and animals and their associated symbolic meanings as encoded in oral literature. I come as a student to the native consultant. Many of these people are cultural scholars in their own right, although usually without advanced degrees.

Nearly every conversation in this book has been tape-recorded and transcribed verbatim. In places, I have subsequently gone over passages with the consultant so that wording or meaning might be made more explicit or some point elaborated, with new or additional wording spliced in. A word or phrase has occasionally been inserted in square brackets for clarity. Otherwise, editing has been restricted to eliminating some repetitive phrases or actual divergences—changes that I also kept to a minimum. The manner of expression in English often preserved a very Piman quality of speaking. Even asides often reveal important cultural associations. Perhaps they are not asides at all, except to the more linear-thinking Westerner.

I have made no corrections in the grammar used by speakers. The language of choice for all the consultants was O'odham. In the interviews, they are usually speaking in English, their second language (far more competently than I do in my second language, I must add). Probably all learned English only when they started school. Some Tohono O'odham interviews were mostly or entirely in Piman. These interviews have been translated or at least summarized by Culver Cassa, a multilingual Pima. The all-O'odham interviews have been particularly useful in conceptual matters, bypassing the filter of English. For a few of the Piman consultants, Spanish is their second language.

Several methods were used for folk taxonomic elicitations. Whenever possible, the consultants and I tried to identify the bird in the field—the best, most accurate method. With Gila Pima, I often used specimens, either freshly collected or study skins. At one time, the O'odham evidently had more familiarity with handling birds in the flesh (see the swallow account, for instance). Sometimes just the description of field characters, behavior, and voice sufficed for identification. In many cases, particularly with aquatic and riparian species, the birds were no longer found in the local environments.

Alfred Whiting taught me a trick he used with bird elicitations among the Hopi. He cut illustrations of local species from a field guide, glued them to sheets, and arranged them in a loose-leaf binder. This method has the advantage of flexibility. As additional illustrations are encountered, they can be cut out and added, showing multiple aspects of a species. The technique also eliminates confusion caused by all the extralimital species that often appear on a page of the unaltered field guide. It has the disadvantage, when not combined with fieldwork and specimens, of confusing the consultant with regard to scale: images of a hummingbird, a kestrel, and a Red-tailed Hawk usually appear the same size on the page. Nonetheless, I used this field-elicitation notebook often, backing it up with field verifications whenever possible.

A few other conventions. I use the word *dietary* as a noun to designate all those foods that a culture considers proper food for that group to eat, whether or not they appear in a particular individual's diet. People may know that a particular plant or animal is edible, but that does not necessarily make it a part of the culture's dietary. When the phrase "the People" appears, it means the O'odham, Pimans, or true people.

I have used the more inclusive term *Piman* for Northern O'odham in general—that is, Kohadk, Tohono O'odham, Akimel O'odham, and perhaps other O'odham groups as well. It is intended to break away from the reservation mentality of separateness, an artifact imposed by outside colonial forces, and to reinforce the original sense of the People of Pimería Alta and Bajo. The designation *Pima* usually means those people living along desert streams or rivers in contrast to the Papago or Tohono O'odham. The term *Pimas* is used here for several individuals of this language community. Mountain Pima are Pimans as well, although they do not use *O'odham* as their self-designation. *Tepiman* is the even more inclusive term to designate both northern and southern Pimans, or the Tepehuán, as these latter are known. The relationships of Piman groups is further explained in chapter 1.

And finally, readers will note the many songs given throughout this book. Some people might object that some of these "songs" are of a sacred nature. But a song includes in essence music, not just words. When used in a sacred context, the music and words must be accompanied by the ceremonialist's actions and intentions. A sack of flour and a jug of water are not bread. Similarly, words printed in a sacramentary or missal are not Eucharist or communion. Nor is playing or listing to Bernstein's Mass. Although some neotraditionalists may not grasp this critical distinction, I think the singers and shamans who authorized the use of their songs in printed form did.

Acknowledgments

I owe an enormous debt of gratitude to all the Piman native consultants and teachers who have patiently endured my ethnobiological questioning (and especially my checking and rechecking points) over four decades. Occasionally it was I who endured patiently when a seemingly simple question requiring an equally simple answer expanded into a four- or five-hour response. I thank them all for their generosity. Brief biographies, inadequate to do them the justice they deserve, appear in chapter 2, "Native Consultants." Additional information on some of these remarkable people can be found in my earlier books (Rea 1997:17–28, 1998b:17–20).

Several individual Pimans should be acknowledged here. Tohono O'odham Jefford Francisco, a natural-resources manager from the Topawa district, was instrumental in running down the quail names, ultimately confirming the folk specific used by Akimel O'odham Sylvester Matthias. Larry Saraficio of Aangam village was most helpful in identifying older cowboy friends and relatives whom he thought I should interview and in setting up meetings with them. They contributed lifetimes of field experience in Papaguería. As Larry did previously for the mammal book, he again supplied most of the personal names found in appendix E.

Culver Cassa was more than just a linguistic consultant, though his name appears on the title page in that capacity. He was responsible for most of the song transcriptions and all of the new translations into English included here—no small task. Most important, Culver *thinks* mythology. He ponders meaning from the viewpoint of traditional hunters, gatherers, and subsistence agriculturalists making a living along desert rivers and at springs. We had many an energetic discussion as we tried to understand stories in a cultural context. Much of his original thinking about bird metaphors has greatly enriched this work.

The late Allan R. Phillips, my ornithological mentor beginning shortly after my arrival on the Gila River Reservation, was the first to suggest that I ask older Pimas what they knew about the birds before the disastrous collapse of the riparian community on the middle Gila. This was long before I had ever heard of ethnobiology or folk taxonomy. The direct result of his urging was my technical work on the middle Gila bird life (Rea 1983), but an ultimate spinoff is the present work.

Then there are those who patiently plowed through the pages of my manuscript. Ethnobiologist Catherine Fowler reviewed and commented extensively on an earlier draft of the entire manuscript. She also helped with Uto-Aztecan cognates. Ethnobiologist Eugene N. Anderson also read the whole thing and supplied pages of helpful and encouraging notes. Ethnobiologist Eugene Hunn read and commented extensively on the folk taxonomy chapter. Linguist and ethnobiologist Cecil H. Brown also read and commented on this chapter as well as consulted on problems that arose in the comparative linguistics sections of the species accounts. (A curious aside is that many ethnobiologists are avid bird-watchers as well, including these four anthropologists.) Philip Unitt, who took over my duties at the San Diego Natural History Museum, read all the species accounts, lending his editorial and ornithological expertise to their improvement.

Others read parts of the manuscript. Gale Monson reviewed the Golden Eagle and Crested Caracara accounts and helped resolve questions on bird distributions in southern Arizona and Sonora. (He has written books on both places.) Seymour H. Levy read the caracara account and supplied information on quail. Falconer Randy West from the Jacumba area read the chapter on bird keeping.

Still others helped with technical information on birds. The late William A. Calder, environmental physiologist at the University of Arizona, clarified details about avian hibernation. Bill Haas supplied information on owls, in particular the Long-eared Owl, helping me piece together its identification from bits of folk science. Rich Glinski, wildlife biologist with a specialization in raptors, provided feedback on the impact of Piman practices on Golden Eagle populations in the desert. Mexican vertebrate biologist Eric Mellink B. and his wife, Ana, edited the Sonoran bird names, supplying local usage. They might not approve of a few used here, but at times distinctions have to be made where there is no vernacular precedent. Avian osteologist Charmion R. McKusick supplied detailed information on the archaeological distribution and use of turkeys throughout the Southwest, as well as information on macaws.

In the documentary line, several people made significant contributions of sources I would have otherwise overlooked. J. Andrew Darling supplied me with a copy of Rinaldini's 1743 vocabulario and of Charles H. Cook's Gila River vocabulary from the late 1800s. We also enjoyed vigorous discussions of Tepiman topics. John C. Ravesloot provided me with copies of Herzog's unpublished Pima and English Creation Story transcriptions from the tribal archives. Linguist David L. Shaul brought to my attention the colonial Upper Pima vocabulario from Attí, Sonora. Alan Ferg, Arizona State Museum archivist, unearthed a number of unpublished O'odham word lists—fragments of paper tightly filed in shoe and cigar boxes—and brought them to my attention. He was also responsible for locating at the museum the historic photographs of O'odham ceremonies where feathers were used. Daisy Njoku of the National Anthropological Archives searched through mid-nineteenth-century word lists at the Smithsonian Institution looking for bird names. She and Felicia G. Pickering also researched Frank Russell photos and artifacts there. These photographs are reproduced here courtesy of the National Museum of Natural History, Smithsonian Institution.

Working in the field with friends can be exhilarating. Gary Paul Nabhan and I made a number of memorable trips to Tohono O'odham, Pima Bajo, Mountain Pima, and Quitovac Hiá ch-eḍ O'odham country. Not soon to be forgotten were such incidents as his putting our last box of matches in the ice chest (after most of the ice had melted) or the open pack of tortillas in the chest with the bloody rodent specimens (again mostly melted ice), or his driving off with my plant press still on top of the field vehicle. (He had forgotten to bring his own presses.) In spite of these mishaps, we're still on friendly terms. Various trips to Pima Bajo and Mountain Pima country were made with the indefatigable Wendy C. Hodgson, botanist of the Desert Botanical Gardens, primarily with ethnobotanical targets. I made one trip to Mountain Pima country with herpetologist turned botanist Thomas R. Van Devender and his wife, Ana Lilia Reina. We stopped every three miles to collect some plant Tom spotted, and we managed to cover in three days what Wendy traversed in six hours or less, but of course ended up with many more collections. Veterinarian friend Robert L. Nagell underwrote three trips to Mountain Pima and Pima Bajo country and assisted with fieldwork. He also contributed equipment, including computers, and a number of critical books for this project. He's convinced that ethnobiologists are at least a little bit crazy.

Two people stand out in the logistics area. The late Sally Pablo of Komaḍk village supplied refuge and companionship during the many years of my ornithological fieldwork on the middle Gila. She put up with birds being skinned on her kitchen table and with dead birds and snakes in the freezer, but put her foot down on the lizards. Sister Anne Fischer, SSND, graciously supplied working and living space at Sacaton during my numerous trips to O'odham country. She suffered through having her dining-room table appropriated for office space as Culver Cassa and I worked through the manuscript for weeks at a time.

Various individuals offered other kinds of assistance. Librarian Joyce Antorietto of the San Diego Museum of Man ran down and copied many obscure as well as not so obscure references I needed. Anita Alvarez de Williams provided references on the Yuman-speaking peoples. Carol Boardway salvaged and repaired several old but critical cassette tapes, allowing me to dub new working copies. Alaska linguist Lawrence D. Kaplan sent me a copy of Russell and West's (2003) most interesting sub-Arctic Athabascan ethnoornithology. Field biologist Robert Schmalzel arranged my first several meetings with one of his talented field assistants, Jose Enriques. Jack Schlotte traced the origin of some English bird names. Arthur C. Risser, general manager of the San Diego Zoo, recommended the artist Mike Touch. Monica Wagner at the University of San Diego assembled all the book's computer files and printed copies of the manuscript. Phil Unitt loaned study skins from the San Diego Natural History Museum for the drawings.

Then there were those who helped in less tangible ways while this book was being assembled. Culver Cassa and I spent many enjoyable evenings over the years with ethnographer and artist Barton Wright, comparing notes between Puebloans and Northwest Mexican Ranchería peoples. Wade C. Sherbrooke provided encouragement throughout the writing of all four of my books about the Pima, as did Richard S. Felger. Piman specialist Bunny L. Fontana, as always, supplied moral support and technical information. He recommended that the Southwestern Foundation underwrite the illustrations. Gary Paul Nabhan provided encouragement when spirits flagged. And surely I cannot omit Nancy J. Turner, ethnobotanist par excellence, who has enough enthusiasm to buoy up a dozen other people.

The several thousand bird specimens, on which my survey of Gila River birds was based, are housed in the Museum of Southwestern Biology, University of New Mexico. Dr. Robert W. Dickerman, acting curator, responded to numerous requests for data from the Amadeo M. Rea Ornithological Collection. For the taxonomic and distributional information on middle Gila River birds, please consult Rea 1983.

The three George Miksch Sutton sketches are used courtesy of Greta Sutton Nordstrom, Doc Sutton's last living relative. The sixteen Major Allan Brooks sketches are from *Birds of the Pacific States* by Ralph Hoffman (copyright 1927, © renewed 1955 by Gertrude W. Hoffman), reprinted by permission of Houghton Mifflin Company. All rights reserved. The sketches of Papago headgear (fig. 7.4) originally appeared in Ruth Murray Underhill's *Singing for Power: The Song Magic of the Papago Indians of Southern Arizona* (copyright © 1938, 1966 Ruth Murray Underhill). The sketch of the arrow and target, *wulivga* (fig. 8.1) is from Leonora S. M. Curtin's *By the Prophet of the Earth: Ethnobotany of the Pima* (courtesy of the University of Arizona Press, 1984). The Southwestern Foundation of Tucson underwrote the remaining illustrations by Mike Touch, Takashi Ijichi, and myself. I thank Dianne M. Bret Harte, the foundation's executive director, for her role here. Arizona State Museum photos by C. M. Wood and Edward H. Davis of Papago ceremonials are courtesy of photo archivist Susan E. Leubbermann.

I am grateful for partial financial assistance toward the completion of this book from several sources. Once again, the late Kenneth E. Hill and his wife, Dorothy V. Hill, with the assistance of Marilyn Ellis, provided an annual stipend while the writing and editing were in progress. Several other agencies provided assistance: the Wallace Foundation (Linda Wallace-Gray), the Lennie Cooper Foundation (Dr. Fred Cagle), and the Jiji Foundation (Alan Harper). The Arizona Wildlife Foundation, the Pacific Coast Conservation Alliance, and the Environmental Education Exchange administered funding. Some of the data reported here were collected incidental to ethnobotanical projects with other funding: National Science Foundation grants BNS-77-08582 and BNS-8317190.

I thank the staff of the University of Arizona Press for publishing this work, especially acquiring editor Allyson Carter, and freelance copy editor Annie Barva for her patience in plowing through the mass of linguistic data included here.

Wings in the Desert

Introduction

One hot evening in late August 1963 I arrived on the Gila River Indian Reservation. It was already dark. Having just finished college in California, I was to be stationed for the next year at the River Pima settlement of Komaḍk, teaching high school at St. John's Indian School. The following morning, during the several hours of liturgy (the Franciscans then were very monastic) I was distracted by the bird sounds that were new to me outside the cool, thick-walled adobe church. After breakfast, the tantalizing calls I was hearing proved to be the Gila Woodpecker, Inca Dove, and Cactus Wren. An exploration began that morning that would take up most of my life.

After supper the next evening, when a relentless sun slipped behind the nearby rocky mass called the Sierra Estrella, I took a walk through the maze of cultivated fields, fallow fields overgrown with mesquite, wagon roads, and irrigation ditches—and through ankle-deep dust where the fine, pale alluvial soil was churned up. An older Pima man, coming along a ditch bank, stopped to talk to me. Back at the school, I told my high school principal, Franciscan father Walter Holly, about this encounter. "Amadeo, you must have a way with Indians! These Pimas don't talk to *any* stranger, not even another Indian, but particularly not a white man." It was a good beginning.

In addition to teaching classes that year (and four years thereafter, as it turned out), I had an ornithological research schedule ahead of me: to document with specimens the local breeding avifauna, the wintering avifauna, and, most interestingly, the migrants that passed spring and fall through this relatively little collected portion of Arizona. Almost every week there was something new to learn. And of course, there were mammals, reptiles, myriad invertebrates, and, most important, plants to collect and identify. Many of my Indian students, from various tribes throughout the Southwest, began helping with the task of collecting and preparing specimens of all kinds.

It was a bird, the Le Conte's Thrasher, that really hooked me into staying on to teach after my first year at Komaḍk was over. During the first few months' fieldwork, I had collected a number of specimens that proved to be among the first half-dozen records for the desert or even for the entire state. Professional ornithologists put me in touch with Dr. Allan Phillips in Mexico, who was at the time proofreading galleys for a book he had written with Joe Marshall and Gale Monson, *The Birds of Arizona* (1964). Dr. Phillips began studying my specimens, inserting into the book's galleys some records that were particularly significant. This, too, was a good beginning for four months. And by the way, he wrote back to me: "Could you try to work out the local status of a very pale and elusive thrasher that is seldom seen?"

During the next several years, a manuscript on the reservation bird life began to take shape. The ecological environments in which I was working had been radically altered within the lifetimes of the oldest Pimas living along this section of the middle Gila and lower Santa Cruz Rivers. By the time of my arrival, there was no surface water in these rivers, and the cottonwoods and willows of the stream sides were long gone, as were the fish and the riparian birds that once made this a home. It was Allan Phillips, again, who suggested that I inquire of the older Indians what they remembered about the former bird life along the river.

The first person I asked about birds was the gardener at the school, Sylvester Matthias. Although this was a convenient starting point, little did I appreciate then the floodgates I had opened. Our working association would span the next two decades. Sylvester probably knew more plants and animals than any other single person in his generation. He even knew the names of some organisms that could no longer be identified and perhaps he had never seen in his lifetime. But they were elements of Pima traditions, sometimes remembered only because they were mentioned in legends.

Conversations with others followed. Most were Pimas who were born in the early years of the twentieth century and who were from the western villages, the Kuiva O'odham, Downstream People. Here the Gila River had surface flow for nearly half a century longer than it did among the Ta'i O'odham, Upriver People, of the eastern villages. The People here still had first-hand information about beaver and muskrats, many species of fish, aquatic insects and plants, and various waterbirds such as kingfishers and colonial nesting herons. A few still knew the names and habits of birds such as the Yellow-billed Cuckoo and Summer Tanager that once nested in the cottonwoods and willows of the former riparian woodlands.

I eventually worked among the Ta'i O'odham of villages upstream. I introduce them in chapter 2. Although considerable ethnobiological information was still included in song and story as well as in oral history, most upstream Pimas no longer remembered much of the riparian biota. Folk taxonomy and folk science had largely fallen victim to environmental degradation.

In time, Gary Paul Nabhan and several others of us ethnobiologists made explorations among the Tohono O'odham, or Papago, as they were then known to outsiders. The focus of one of our studies was the Quitovac oasis settlement in northern Sonora (Nabhan and associates 1982; Rea 1978; Reichhardt and associates 1994). Here we learned what we could from village patriarch Luciano Noriega. Delores Lewis of Ge Oidag (Big Fields, Arizona) accompanied us on some of these trips. And for years Frank Jim of Hodi Shon Vo'o (or Hodai Shon Wo'o, Charco 27) in the western part of the main Papago Reservation was a friend and patient teacher.

Later I was to learn more about birds from Simon Lopez of Santa Rosa, Vincent Josemaria of Pisin Mo'o, Leonard Juan of Aangam, Jose Enriques of Queen's Well, Joe Lopez of Vaya Chiñ, Jose Garcia of Tohono O'odham Santa Cruz village, and Hiá ch-ẹd O'odham Delfina Cameron originally from Darby Wells. Biographies of these individuals are given in chapter 2. They all had some insight to share about their personal experiences of nature. Some told of folk science, some of traditional sickness and healing, others of birds in song, story, or ceremony.

Once when I was taking a linguistics course, my Pima friend Sally Pablo, looking through my workbook of indigenous texts, said, "Hey, this one is Pima! I can read it." I checked the source: Southern Tepehuán. I had never heard of them. Her sentence word order was different, and some of the words were different as well, but she had no problem with the narrative. The name for these people was different, but clearly they were Pima by language. Sometime later Sally joined me in San Blas, Nayarit, Mexico, where I was teaching a field ornithology course, and we made the return trip together to see if we could find the "lost" Pima we had heard about from Edward (Ned) Spicer and others. We looked at likely places in Sinaloa without success. In Bamoa, where hundreds of Pima Bajo had relocated with Cabesa de Vaca in 1540 and again in 1615 and 1616, we found a barrio with the distinctive earmarks of Piman material culture, but no one who acknowledged speaking Pima. Perhaps we would have found some individuals who did so had we had more time to get acquainted. The Tepehuán, though, were more in the interior foothill country we didn't explore.

Some years later we were put on the track of Pima speakers living on the middle Río Yaqui in the state of Sonora. We went to Tónichi, crossed the river on the ferry, and stayed the night with a helpful Mexican, Ramón Baldenegro, who gave us the names of the last two speakers. The next day we struggled downriver a dozen miles through a route that would have been much easier to walk than drive. We were welcomed at Ónavas by don Pedro Estrella and later by doña María Córdova. We soon discovered that with a little reversal of the palatalization process, Upper Pimans could communicate quite well with Lower Pimans, at least with those here in the lowlands.

Between 1978 and 1988, when Pedro died, I made sporadic trips to Ónavas several times each year, often accompanied by one of the Arizona Pimas. Most of our fieldwork focused on ethnobotany, collecting plants in the thornscrub and tropical deciduous forest, but we made notes on birds and whatever other animals we encountered as well. I now regret that I didn't investigate the local avifauna with Pedro far more thoroughly than I did. But this was salvage ethnobiology, and there were many bases to cover.

We eventually made friends as well among the Mountain Pima living in rancherias along the Sonora-Chihuahua border, mostly in the outlying area of Yepáchic. Here again our collections focused on ethnobotany and agriculture, with some work on fauna as opportunity permitted. But these upland people were a bird of a different feather, if I may be pardoned the expression. They called themselves "Oob" rather than some variant of the name "O'odham," and they spoke a Piman language that my Arizona O'odham friends Danny Lopez and Culver Cassa had difficulty unraveling. And these uplanders lived in pine-oak woodlands and coniferous forests rather than down in the tierra caliente, as they called the desert. They farmed directly from the summer rainfall rather than by manipulating the water from arroyos or periodic river overflow. They were a different people.

The people included in this book are the Northern Pimans, those living throughout southern Arizona and northern and

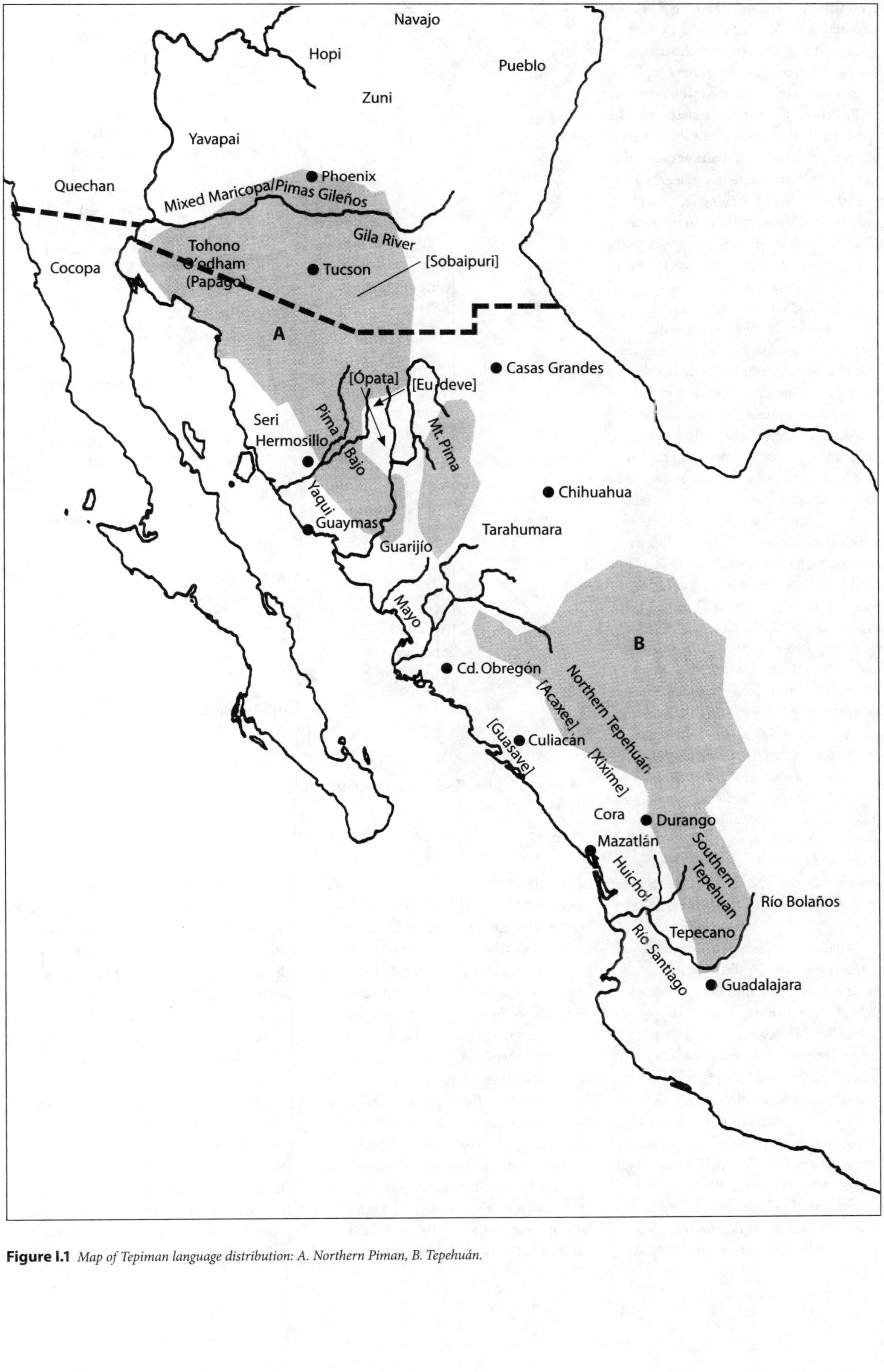

Figure I.1 *Map of Tepiman language distribution: A. Northern Piman, B. Tepehuán.*

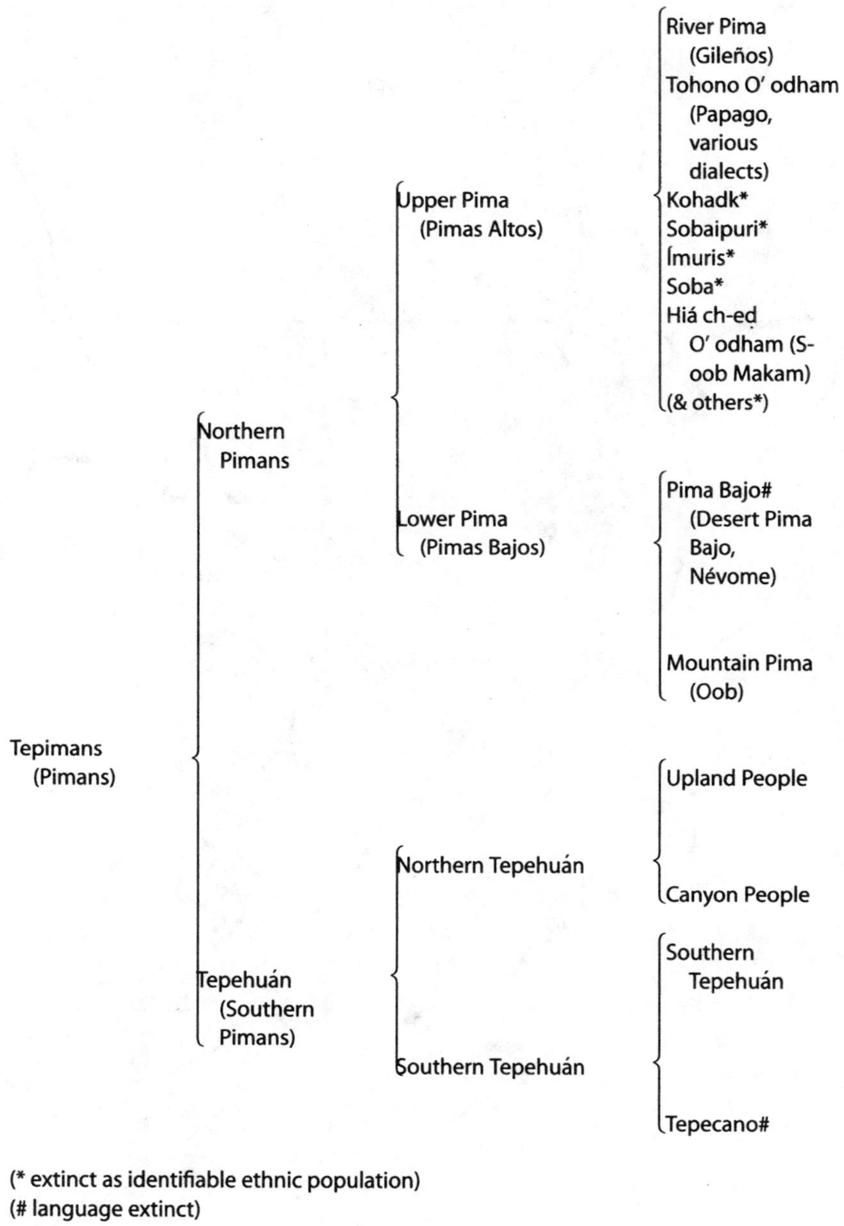

Figure I.2 *Relationships of Piman subgroups.*

central Sonora and once making claim to a contiguous arid territory of the Greater Southwest (fig. I.1). They call themselves *O'odham*, a term signifying "the People," in contrast to other humans who aren't really people in the truest, fullest sense. The word may be used polysemously, almost vicariously, for other Indians or even non-Indians. These tribal people as a group shared no central governing figure and even the historic authorities in the local ranchería may have been largely a response to the Hispanic need to interface with a hierarchical structure.

In the O'odham world, most social interaction was within one's own dialect group, and there may once have been about a dozen of these groups among Northern Pimans. The situation was very fluid. Whole villages might pack up and move to a distant location where there were better resources or security from raiders. Spaniards labeled the Pimans of the San Pedro and upper Santa Cruz Rivers the Sobaipuri, but it is unknown what their self-designation was. Franciscan historian Fr. Kieran McCarty called them Akimel O'odham. They have not survived as a recognizable ethnic group. The People who lived along the Gila call themselves *Akimel O'odham*, "River People." They somewhat derisively and pityingly referred to the People living away from permanent water sources as the *Tohono O'odham*, whereas the Spaniards called these nonriverine People the *Papago*. This cluster includes six or seven dialect groups. In 1986, the Papago tribal council officially elected to be called by their O'odham name. The Kohadk, originally living on the Santa Cruz River south of the River Pima, now occupy several villages at the northeast portion of the Tohono O'odham Reservation. (See fig. I.2.)

As native peoples began to be confined to reservations under different Anglo-American agencies, the labels *Pima* and *Papago* began to encourage feelings of separatism. However, as this study and the earlier ethnobiologies show, all these Desert People shared a basic culture in spite of minor regional variations in language and ceremonies. So the inclusive terms *O'odham* and *Pimans* are often used here to cover all the Upper Pimans or even all the northern lowland Pimans. Where there is a need to be more specific, *Akimel O'odham* refers to the Pimans living on the Salt and Gila Rivers, whereas *Tohono O'odham* signifies those people lacking a river and *Hiá ch-eḍ O'odham* those people originating from the west and southwest of the designated Papago Reservation. The terms *Piman* and *O'odham* may also refer to the language spoken by all these people. Sometimes *Papago* is used for the half-dozen or so dialect groups collectively.

Equally artificial is the Hispanic division of Northern Pimans into Pima Bajo and Pima Alto. This separation reflects a temporal and geographic division of the Jesuit mission endeavor rather than any intrinsic cultural difference. Missionaries were in Pimería Baja as early as 1619. As noted, this southern division included two distinct Piman groups: those of the hot lowlands, then called the Névome, and the Mountain Pima, among whom missionary activity began in 1673 (Pennington 1980:55–57). The field of Pimería Alta to the north was opened by Fr. Eusebio Kino and his associates in 1687. Use of the inclusive term *Pima Bajo* is no longer justified as an ethnographic label. I refer to the modern lowland people as desert or lowland Pima Bajo. Those in the uplands are Mountain Pima.

With the U.S. acquisition of southern Arizona by the Gadsden Purchase of 1853, most O'odham in Pimería Alta found themselves within yet another country. In time, these Upper Pimans forgot that they were just the northern extension of Piman country stretching nine hundred miles southward along the inner coastal plain and western flank of the Sierra Madre Occidental. Actually, Piman country begins on the Río Santiago in northern Jalisco and ends at the Río Gila. There is one break, the Taracahitan Bridge, where the Mayo, Yaqui, Guarijío, and Tarahumara form a transverse band across the Tepiman Corridor. Geographic proximity can be deceiving, though. As already noted, Mountain Pima and lowland or desert Piman languages are no longer mutually intelligible. But Río Yaqui Pimans could communicate freely with their northern relatives three hundred miles to the north.

The lowland or desert Northern Pimans covered in this work—that is, the Akimel O'odham, the Tohono O'odham, and the lowland Pima Bajo of the thornscrub and tropical deciduous forest—are no longer contiguous populations flung out across the tierra caliente from the Gila River to the Río Yaqui as they once were. Large reservations with thousands of bilingual speakers occur on the Arizona side of the border, with a few in villages along the adjacent northern fringe of Sonora. Elsewhere, Northern Pimans seem to have vanished as identifiable ethnic communities except at the pueblo of Ónavas. The last two speakers of O'odham here died in 1986 and 1988. Before they did, I was fortunate to have studied with them, too, sometimes accompanied by Gila Pimas.

My informal classes with these wonderful Piman people have spanned over forty years since that first hot night in 1963 when I arrived on the Gila. I've gone from being a full-time teacher of young O'odham to being a full-time student of their culture. Most of the old ones who really knew folk taxonomy and folk biology of birds and other organisms have since gone. It's now time to write my final examination on the birds of the Northern Pimans.

Part 1

Birds in Piman Culture

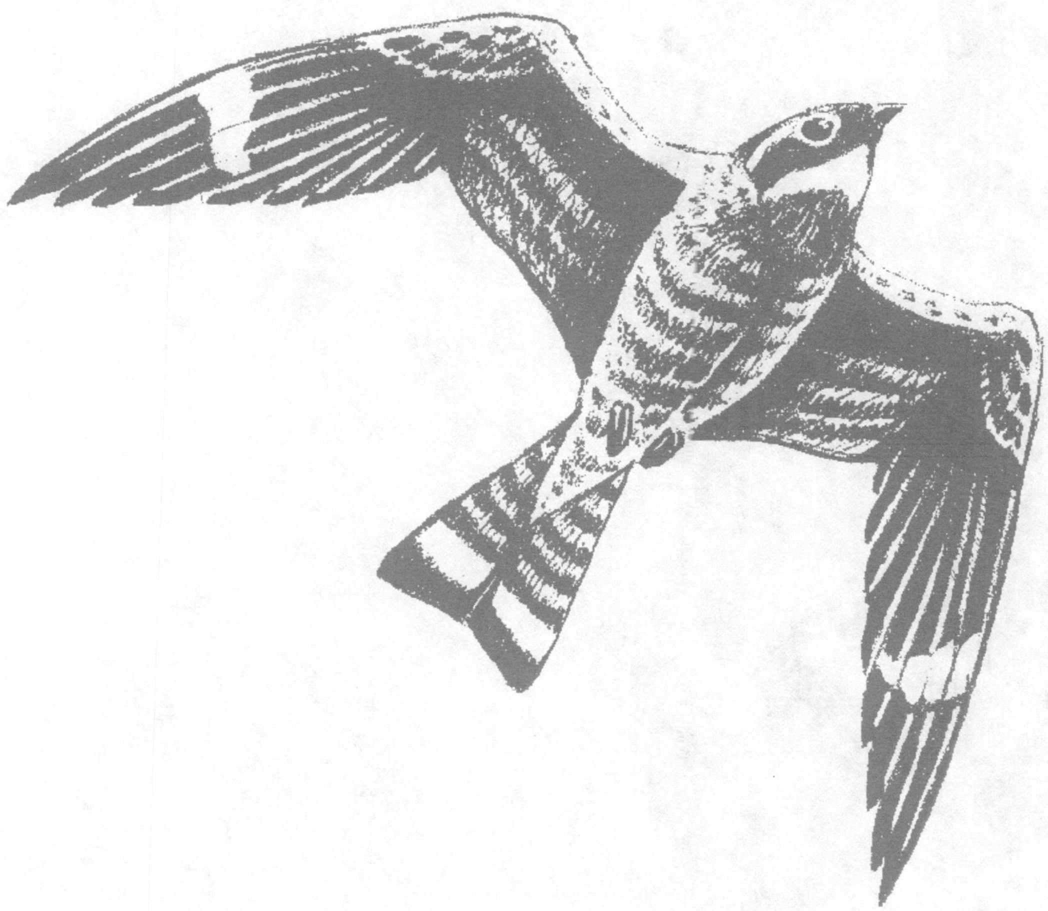

1

Northern Pimans and Their Environment

The blue-gray Estrellas tower above us, cutting the sun off early. It is only a bit after five on this early October day. I have just returned from the annual fiesta in Magdalena, Sonora, a regional celebration that once drew even these Northern Pimans that far south. Carmelita sits on a dilapidated chair to the side of her outdoor fire pit, turning thin wheat tortillas over a rusty disk blade *komal*. I take in the familiar smells of Pima country: chips of mesquite wood thrown onto the coals, fresh flour *chemait* (tortilla), and the surrounding Velvet Mesquite trees with their distinctive odor intensified by yesterday's rain. Carmelita talks slowly, in a low voice, her cadenced tones revealing her Papago connections.

Her adobe mud house is flanked by corral fences to one side, an adobe tack building to the other, with an entrance drive opposite the main house. The enclosed yard is deeply rutted by tire tracks and shaded by large old gray tamarisks. Ropes, lariats, and bridles hang casually from mesquite trees, and several saddles straddle a corral wall, with sheet plastic shielding them from the recent rain. A Screwbean, an enormous Soaptree Yucca, and several mulberry trees stretch above the low buildings. Outside this central hub, old fields, new fields, and dirt irrigation ditches dominate the surrounding flat land. Beyond this immediate sphere of human activity, mesquites, Saltcedars, and huge saltbushes reclaim wildness.

While Carmelita makes tortillas, I note the birds I see or hear from where I sit. Two Say's Phoebes call plaintively as they make buoyant sorties around the buildings: *hevel moos*, grandchildren of the Wind. A flicker calls powerfully from somewhere out in the mesquite, but I don't know which, a Red-shafted that winters here or a Gilded that breeds out in the Saguaro stands. Both are called *kuḍat*. Both come into the fields during the colder months, searching for ants on the open ground. A few Mourning Doves are picking up spilled grain. Several months earlier this ranch was overwhelmed with White-winged Doves as well, calling out distinctly, slowly, their Piman name, *ó-o- ko-koi*. But this late in fall they all have gone south. A number of tiny Inca Doves are almost invisible on the dried gray mud of the yard as they scuttle about searching for the tiny scattered seeds of amaranths and chenopods. They, too, keep reminding us of their O'odham name, *gugú*.

Either a Northern Cardinal or a Pyrrhuloxia whistles from a dense mesquite, but I can't see which. A few decades earlier only the red bird was here, but the elegant more slender gray-and-pink bird began showing up from the south and has now made its home along the middle Gila. I spot a lone Turkey Vulture heading south along the river to some roost in an old tamarisk grove, where it will be joined by dozens of its comrades. Odd how the two outcasts of the Anglo-American world, the vulture or buzzard and the Coyote, are central figures throughout Piman mythology and indeed are numbered among the four primal beings—Vulture forming the contours of the mountains so the water will flow into rivers, Coyote making out of clay the People who will then inhabit this land.

A couple of Cactus Wrens cavort around the outbuildings, through an old parked car that has seen the last of its travels, calling in churlish, grating notes. *Hokkaḍ* they're known here. They are on familiar terms with a few chickens, *chuchul*, foraging in the yard and nearby fields. Carmelita comments from her *chemait* fire that she doesn't kill her chickens, but the Coyotes try.

A gray-and-yellow bird flies into a mulberry tree and with a vigorous call identifies itself as a Western Kingbird, *chukukmal*. Meanwhile, a tiny gray bird with a yellow head is working through the bushes at the corral edge, a Verdin, its powerful calling all out of proportion to its size. In Piman thought, this *giisobī* is believed to be a great shaman, maybe because it is loud and solitary. The Verdin is soon followed by a pair of Black-tailed Gnatcatchers, even smaller, never beyond hearing range of each other. A loud ringing *churrúp-churrúp* announces the arrival of a Gila Woodpecker, named for the river just a minute's flight away. The *hikvig* not only takes up residence in these farmsteads, but finds a prominent place early in the Piman Creation Story and retains its sacred status even among contemporary O'odham who still know their culture—and their wildlife.

This is one of the working ranchos in the Pima ranchería of Santa Cruz, nestled in an arm of land between the Gila and the lower Santa Cruz as these rivers flow northwest until clearing the end of the Sierra Estrella. At the northwest end of the range, the Salt joins the Gila, and the combined water of all three desert streams heads southwest on its course to the Colorado River and the Gulf of California.

The classic desert landforms are well defined here: mountains, skirting bajadas, and alluvial bottomlands. The flatlands support a mosaic of habitats, most of them now badly degraded or entirely destroyed. The once great mature mesquite bosques are now dead mesquite cemeteries, their remnant stumps spread like grave markers across miles of dry land. Second-growth, uniform-age mesquite thickets have sprung up in some abandoned fields, especially between Santa Cruz village and Maricopa Colony, where the water table is higher. Nowhere is there mature growth.

The grasslands, some described by itinerants as being miles long, have vanished with overgrazing and the lowering of the water table. In their place, saltbush flats now cover many acres, dominated by Desert Saltbush, Seepweed, scraggly mesquites, and various species of wolfberry. Interspersed are thickets of Arrow-weed, Pickleweed, true Greasewood, Four-winged Saltbush, and Quail Brush, depending on salinity, moisture, and drainage.

The riparian woodlands, essentially what once defined River Pima country, have likewise vanished as water tables dropped and the narrow river channels were turned into broad expanses of bare sand washed in by floods. Dams upriver from the Pima settlements now divert what remains of the surface flow. The river channels are now almost pure stands of impenetrable Salt-cedars, an invasive Old World shrub or tree that thrives in postimpoundment conditions. What few individual cottonwood and willow trees survive today are found along canals and at a few pump ponds and sumps. The lagoons, marshes, oxbows, and open water that Hispanic travelers marveled at are gone, together with the many fine fish, aquatic birds, and other wildlife they once harbored. The demise of vegetational communities has been documented elsewhere (Rea 1983, 1997) and is almost too painful to review again. The local desertification of Gila Pima country is a microcosm of what has occurred and continues to develop in much of the world's xeric habitats.

The ecotone, that green edge where desert habitats abut wet woodlands, is species-rich, both in flora and fauna. Such oasis ecotones once snaked across much of the Sonoran Desert of southern Arizona and northern Sonora, as well as southward where the desert merges gradually into thornscrub and tropical deciduous forest in southern Sonora. This desert-woodland ecotone hosted many aquatic and riparian birds that we can only marvel at: herons, egrets, kingfishers, cuckoos, swallows, chats, tanagers, and warblers.

The bajadas are the skirts of rock and soil detritus that have eroded from the mountains and washed downward to their bases. They lie above the alluvial flatlands and have coarser, much better drained soils. Saguaro stands, cholla thickets, Ocotillos, bursages, White Ratany, and Little-leafed Paloverdes clothe much of the bajadas of Piman country. These piedmonts are dissected by arroyos, dry washes with sandy floors that resemble dried creeks. They have their own characteristic vegetation, the so-called xeric riparian, with Desert Ironwood, Canyon Bursage, Desert-lavender, Chuparosa, and a host of other plants.

Finally, there are the mostly basaltic mountain ranges themselves. Their vegetation includes the dominant species of the bajadas as well as species such as Desert Agave, Bush Penstemon, Desert Hackberry, Elephant Tree, Arizona Rosewood, and even native perennial grasses. Relictual communities, in particular ones cradled in higher north-facing slopes, harbor many botanical novelties that were more widespread in the Pleistocene (see D. Brown 1978; Rea 1997:56–66). These slopes once were home to White-tailed Deer herds. Desert Bighorn survive tenuously in some larger ranges despite human disturbance.

The area from the middle and lower Gila southward to the middle Río Yaqui, a distance of about three hundred miles, was once inhabited by a contiguous people whose self-designation was "O'odham" (fig. 1.1). The Spaniards called most of them "Pima." As is usually the case, the outsiders' name has stuck. At the southern border of this area was the so-called Taracahitan Bridge, a wedge of non-Piman but Uto-Aztecan peoples, speaking Tarahumara, Guarijío, Tubar, and Yaqui-Mayo. Below that was another great expanse of Piman speakers whom the Spaniards called the Tepehuán. As Hispanic colonizers progressed northward in Mexico, encountering new groups, they were aware that the Tepehuán and the northern Pimans spoke variants of the same language, but for reasons no longer known, they used two different names for these two major segments. To recognize this basic unity, linguists today often refer collectively to Pimans as a group from the Gila to the Río Santiago in northern Jalisco: the Tepimans.

When "discovered" by Jesuit father Eusebio Kino and Lt. Juan Mateo Manje in the 1690s, the River Pima, or Gileños, were occupying more than forty villages strung out along the Gila from above the Salt-Gila

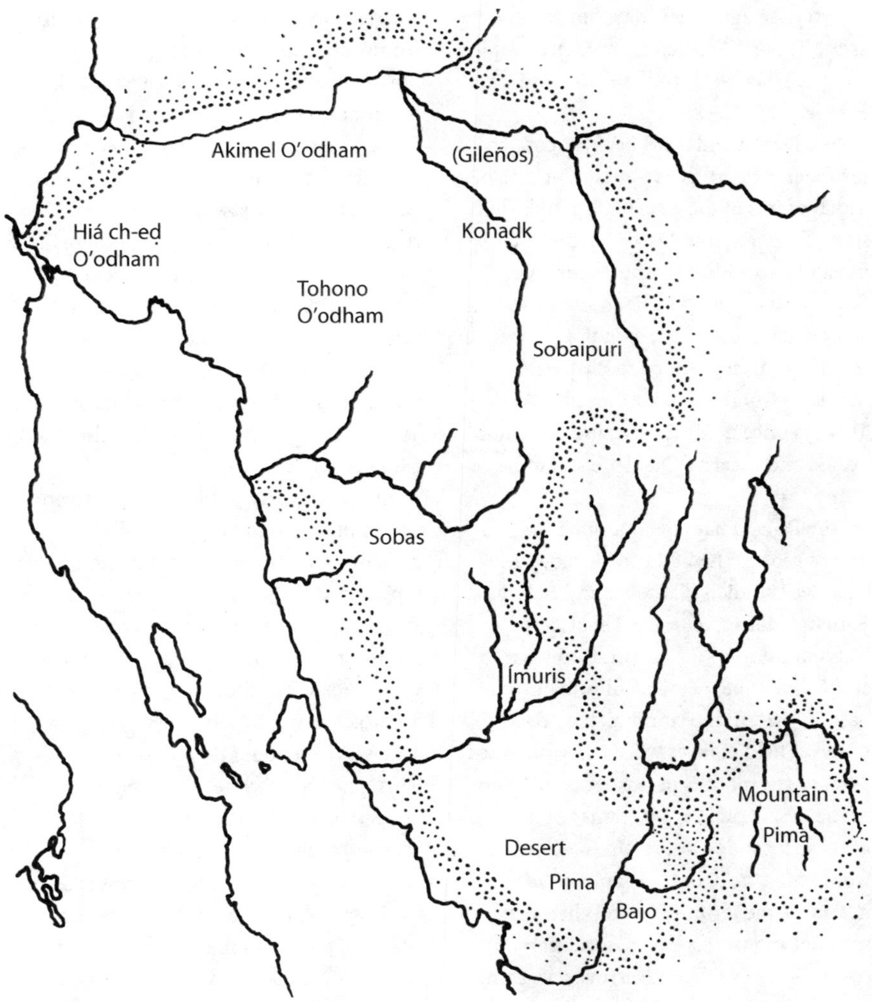

Figure 1.1 *Map of Northern Piman groups.*

confluence almost to the Colorado River. Details are few, but the settlements seem to have been more consolidated then, perhaps as a response to Athabascans, specifically the Western Apache, invading from the north about the time Europeans began exploring from the south. Most of these settlements disappeared early in the historic period, only those on the middle Gila surviving. Freed of Apache threat almost two centuries later, the Pima dispersed (once again?) to family clusters in small ranchos close to their fields and livestock.

All the family groups with their ranches made up a ranchería–a named settlement that used a common canal from the river and shared duties in ditch maintenance, communal hunting, defense, and ceremony. Daily communication was maintained within the group by the *tobdam*, hunt leader or village crier. During the next three centuries, the number of settlements on the middle Gila varied from four major and two minor ones to ten or twelve total. The lack of water in the early 1870s sent the Pimans and their allies the Maricopa scattering about the area in search of suitable farming localities.

The Piman Ecosystems

We know more about the cultural ecology of the Upper Pimans, the northern half of the Northern Pimans, because their culture survived as a functioning system into the twentieth century (Castetter and Bell 1942; Fontana 1983; Hackenberg 1983; Rea 1997:42–46; Underhill 1936). In contrast, throughout most of Sonora, Piman communities disappeared due to population declines, direct land usurpation, and loss of cultural identity through mestizoization. Upper Pimans can be divided into two rather well-defined groups on the basis of cultural ecology and relative availability of resources: (1) the Akimel O'odham (River Pima), or River People; and (2) the Tohono O'odham (Papago), or Desert People.

River People, the Gileños, relied on a rather broad subsistence base. About half their resources came from the gathering of wild foods. For the Pima, the most important resource was the Velvet Mesquite, followed by the Screwbean. The pods of these two tree foods were stored in prodigious quantities. Other major wild crops included Saguaro, cholla buds, wolfberries, and greens or potherbs of various kinds. The hunting of small game (rodents, jackrabbits, cottontails, quail, doves) and trapping of small birds (at least in winter) supplied much of the animal protein in the dietary. Fish ("*Real* Pima food!" commented one of my O'odham teachers) came from the rivers and from the canals when they were drained in winter. Large game was relatively unimportant.

The agricultural system was finely tuned to the constraints of the harsh desert environment. The Akimel O'odham practiced double cropping. In the spring, when the snowmelt in the mountains raised the water in the river and the danger of frosts had passed, the Gileños planted their warm-season crops on the islands and the lower terraces along the river. Whether they irrigated by canals aboriginally is unknown. The Creation Story says they didn't, and Hispanic documents do not mention ditch irrigation until sixty years after Contact (Lloyd 1911:120; Wilson n.d.; Winter 1973; see also Bahr and associates 1994:130–136). At some point, the River People dug miles of canals and ditches to bring water to small fields enclosed by brush fences and living fencerows.

The crops raised included at least two species of beans (teparies and limas), several species of pumpkins and squash, seven or eight varieties of maize (corn), cotton, grain amaranth and chenopod, and White-seeded Devil's Claw. Most important, all these crops were locally adapted land races or cultivars selected through generations for quick maturation and for tolerance of heat, low soil moisture, and alkalinity.

Around midsummer, following the mesquite and Saguaro harvests, fields were readied for replanting. When the monsoon rains fell in the uplands and the river rose the second time, there was a second planting of warm-season crops, called the *shopol eesabik*, "short planting." This planting had to be in at the latest by mid-August (as determined by the position of the sun's setting on the Sierra Estrella, the Pima calendar). With luck, the second crops would mature before the first killing frosts of fall.

As a result of Fr. Kino's introduction

of winter wheat and several cool-season legumes to Pimería Alta, a third planting was superimposed on the River Pima annual cycle. Wheat required additional technology: oxen and plows for preparing larger fields, sickles for harvesting, and horses for stationary thrashing. Perhaps extensive ditch irrigation was introduced at this time as well, allowing larger fields farther from the river.

What can be said with certainty is that the Pima lived in plenty. Older Pimas told anthropologist Ruth Underhill (1936) that they kept sufficient provisions on hand to see a family through one bad year. If one crop should fail, others didn't. If the mesquite mainstay, for instance, failed in midsummer, there would nevertheless be a fall crop to harvest locally or in surrounding areas. If one type of corn failed, there were plantings of half a dozen other varieties. If the cholla bud harvest was bad, several less desirable species were likely to produce (and even one that kept its fruits on from year to year).

This abundance meant that the River People had a surplus to share with Tohono O'odham friends and relatives who traveled north to help with the harvests. Freely bartered were items to which the River People had less access, such as salt, yucca roots and leaves, agave fibers, deer fat and hides. Lacking electronic media, the Desert People also brought songs, dancing, and pageants to perform for their riverine hosts. We must not underestimate this abundance. When the U.S. military, Forty-niners, and others began passing through the Pima villages in hordes on their way to California, the Pima and neighboring Maricopa responded by selling thousands of pounds of surplus wheat and other commodities. There was no market economy at the time; the Pima were not waiting for White itinerants to come flooding westward suddenly; this surplus was simply the standard provision for surviving in the desert.

Not all Northern Pimans had a rich desert-riparian ecotone on which they could depend. Spaniards, on the lookout for places to establish presidios and missions, places to bring together indigenous people where they could be "reduced" to Spanish citizens, noted that in large areas of northwestern Pimería Alta the populations were poor and naked with seemingly few available natural resources. The Spanish labeled these people the Papago, their country Papaguería. Today they are known as the Tohono O'odham, or Desert People, and the Hiá ch-eḍ O'odham, or Sand People.

The Desert and Sand Peoples' country for the most part lacks permanent streams. Much of it is in the Arizona Uplands division of the Sonoran Desert, higher than most of River Pima country. Here one finds great expanses of Creosote Bush flats and Saguaro-paloverde stands. One range, the Baboquivaris, supports chaparral and Madrean woodland communities. Wild Turkeys once made their home here, and Scaled and Mearns' Quail are still to be found.

The Papago have been termed the *two-village people*. Their cultural ecology has been well studied (Castetter and Bell 1942; Fontana 1981:19, 1983; Hackenberg 1983; Nabhan 1982, 1983; Nabhan, Hodgson, and Fellows 1989; Underhill 1939:90–112, 1940b, 1951). In the summer, they concentrated at field villages where they prepared fields at the mouths of washes coming out of the desert ranges. The names of many of these localities still retain the O'odham word *oidag*, field, or, as Spaniards wrote it, *oita*, as the terminal part. When the summer monsoons began, local thunderstorms quickly filled the arroyos, washing leguminous tree leaves and animal droppings down to the little fields below. Small brush water spreaders slowed the torrent so it would drop this nitrogen-rich detritus along with water into the prepared soil. The People would then immediately plant quick-maturing maize, teparies, and squash in the fields. With a second good soaking in a few weeks, they might pull off a harvest. In addition to the cultivated plants, there were greens and various other wild crops, in particular tree legumes, to be harvested in this season.

When the cold season arrived, the people returned to their well villages in the uplands, where water could be found throughout the winter and game (including Mule Deer, White-tail, Pronghorn, and Bighorn) could be hunted. Some villages regularly visited their Pima cousins on the Gila to help with the harvest, returning with sacks of wheat and other items they did not have locally, such as mesquite meal, Arrow-weed shafts, cotton, and cultivated devil's claw.

The long stretch between Papaguería and land still claimed by Pima Bajo on the Río Yaqui is mostly Sonoran Desert, characterized by Saguaros, mesquites, Desert Ironwood, and other tree legumes. In the north, directly south of Nogales, are the magnificent oases, green with standing water, marsh plants, and cottonwoods. Various place-names reveal former Piman occupation: Ímuris, where the river still runs; Puente Babasic, the place where the *vavish*, Yerba Mansa, grows; Tasicor, named for the Javalina; and so forth. Scattered throughout the great flatland vegetation from Santa Ana to Hermosillo one finds traces of more tropical thornscrub species: Tree Morning-glories, Tree Ocotillos, Palo Chino, and Guayacán.

In the last forty or fifty miles or so approaching the middle Río Yaqui, the last Saguaro is left behind, and thornscrub and topical deciduous forest begin in earnest. When the wet season begins with the first monsoons, the countryside bursts into a velvet green. In a little more than three months, the vegetation is again bare and leafless. Throughout the dry season, brilliant splashes of white, pink, yellow, red, or deep blue mark the blooming times for various tropical tree species.

This area is technically not desert country, if we define desert as having ten inches or less of annual rainfall. A station just south of Ónavas records more than twice that amount (Pennington 1980:121). But most of it (sixteen of twenty-four inches) falls during the three months of July through September. Temperatures, both highs and lows, match those expected in the actual Sonoran Desert. Arboreal cactus (Hetcho and Organ-pipe) grow abundantly on hillsides.

In the river bottoms where desert Pima Bajo still live, one finds numerous trees that are familiar from Upper Pima country: Velvet Mesquite, Desert Ironwood, hackberry, Blue-green Paloverde, Mexican Paloverde. The list of shared shrubs and smaller plants between the two Pimerías could go on for pages. Also, many of the same birds or closely related species make their homes throughout both desert scrub and thornscrub habitats. The common denominator is hot dry conditions for long stretches of the year.

We have far less information about the cultural ecology of Pimans living over much of Sonora. But the locations of many early historic large Piman settlements such as San José de Pimas, Tecoripa, Suaqui, and Ónavas show that they depended on *ak chiñ* agriculture utilizing summer monsoon rains. Some of these systems are still in place. Only now the method is called "de temporal." The only Pima Bajo community for which there exists a detailed cultural ecology is Ónavas, thanks to the work of geographer Campbell Pennington (1980).

Ascending the foothills out of the tierra caliente, one arrives at the pine-oak woodlands of the Sierra Madre and settlements of the Mountain Pima. As already explained, these people occupy a different ecological zone, have a different self-designation, and speak a distinctive Piman language. Their folk ornithology merits a book of its own.

Present conditions, both ecological and cultural, are vastly different from what they were at Contact. Perhaps this change is nowhere better illustrated than with the River Pima. Here the tight bond between ecological integrity and cultural, psychological, and physical health is well illustrated.

The River People, of course, depended on a river for their fishing, hunting, and agriculture—and to a considerable extent for their gathering as well. Their cultural metaphors hinged on local plants and wildlife. Songs sang of the local flora and fauna, orations imaged the desert creatures, and stories spun out from familiar places and biota. And the river depended on a healthy watershed for its very existence. So absolutely essential was this hydrology that of the four primordial beings of the Piman Creation Story, Ñui, Turkey Vulture or Buzzard, had no other job to do except to create the system of mountains, arroyos, and rivers so that water would be available on earth throughout the year and not just run off quickly to the sea as it did when the earth was first made. (Some archaic mythologist had a wonderful understanding of a functioning watershed.)

But watersheds in arid and semiarid regions are fragile entities. In post-Contact times in this region, their grasslands were overgrazed by cattle; woodlands were cut for timbers; cordwood and charcoal were made for mining enterprises; beaver were trapped out ruthlessly, allowing their water-management system to fall into disrepair; and inappropriate agricultural methods and road building introduced erosion. The result was a reduced water-delivery capacity in the uplands and an increasing magnitude of flooding and silting in the lowlands where the Pima lived (Dobyns 1978, 1981; Rea 1983, 1997:47–55).

To make matters worse, Hispanic and Anglo pioneers settled upstream on the desert rivers, appropriating for themselves most or all of what surface water remained. The irony is that it was the effective Piman defensive military enterprise that kept hostile Apaches at bay and thus allowed the settlement of Mesa, Phoenix, Florence, and other surrounding non-Indian communities. By 1870, the peoples inhabiting the first reservation established in Arizona (1859) found themselves no longer able to carry out their subsistence agriculture. Within a few years, the starving People turned to cutting their own major wild resource crop—mesquite—to sell as firewood in nearby non-Indian towns.

Not only the biota was affected. Healthy ecosystems and healthy people go hand in hand. Lost as well was a varied diet high in soluble fiber, vitamins, and essential nutrients. It was replaced by trading-post and government commodities such as finely milled flour, simple sugars, and lots of lard. Glycemic indexes in the blood stream soared. The hard physical work of hunting game, cultivating fields, manually grinding maize and wheat, and pounding mesquite pods by necessity became rare occupations. In time, the twin scourges of obesity and adult-onset diabetes would claim more than half the reservation population older than thirty-five (Brand and associates 1990; Cowen 1990; Knowler and associates 1978, 1981, 1983, 1988, 1990, 1991; Nabhan 1997; Nelson and associates 1988; Swinburn and associates 1991).

Physician and physical anthropologist Aleš Hrdlička (1906:40) wrote of the Akimel O'odham at the height of their desperation, "Primarily the Pima are agriculturalists. . . . They say that formerly they hunted and fished and gathered many clams in the Gila, but owing to the settlement of the adjacent country, which began about twenty-five years ago, and the consequent diversion of the water for irrigation farther up the river, most of the land and aquatic animals and birds as well as the clams, have disappeared."

Many birds that the People once sang about or told stories about became mere shadowy memories among the oldest Pimas: Belted Kingfishers, Yellow-billed Cuckoos, Blue Grosbeaks, Summer Tanagers, egrets, and herons. Younger people began to sit in air-conditioned offices working at computers or at home in front of the television, seldom venturing out into the remnants of habitat in their surroundings. They can now name few of the seventy-five birds or 240 plants that their great-grandparents once knew intimately.

Changes elsewhere in Pimería may be less dramatic, but have been damaging as well. Like many River Pima, the Tohono O'odham man has become the quintessential cowboy. Much of Papago country is overgrazed, invaded by thorny shrubs and Chain-fruit Cholla dispersed by cattle. Traditional foods have declined, and beef fat has invaded the diet. In an earlier day, the packrats, jackrabbits, and deer supplied precious little fat. Today trading posts push soft drinks like they were going out of style. Obesity and diabetes, virtually unknown a century ago, are taking their toll here as well.

It is harder to access the changes in the thornscrub and tropical deciduous forest of lowland Pima Bajo country, where I know much less about the plant communities and their history of desertification. Heavy cattle grazing must surely be an important component of the modern landscape and probably has been so for centuries. No doubt overgrazing has taken its toll here as it has in the north. Many medicinal and food plants are now difficult to find. An occasional straggling cottonwood or willow is all that remains of the riparian woodlands that once graced this still perennial stream. In recent years, there has been a radical shift from near-universal processing of home-grown maize to make the daily supply of tortillas to high-glycemic-index milled wheat flour, a purchased commodity.

Not all is unmitigated ecological and medical disaster among Northern Pimans everywhere, however. A medical study initiated among the Mountain Pima to monitor the occurrence of diabetes among them proved to be almost a failure: the people in the Sierra Madre worked too hard and ate too little of the wrong foods. There were great distances to be covered between little rancherias and few stores with junk foods. There were no televisions to sit in front of, and few area residents had electricity (Ravussin and associates 1994). Except in sawmill towns, most people still depend on subsistence agriculture.

Piman Culture and the Avifauna

Piman imagination seems to have soared on the wings of birds, more so than with any other group of living things. The cultural significance of birds is out of proportion to their nutritional contribution in daily life and to their biomass in the arid lowlands. Of course, the Sonoran Desert and the thornscrub and tropical deciduous forest into which it transitions southward are species-rich environments, both in plant and animal life. But the O'odham seem to have chosen birds in particular as the primary bearers of their metaphoric traditions.

The list of birds that figure in mythology and minor tales covers a broad sweep of the avifauna, with no correlation to either size or abundance: Turkey Vulture, Sandhill Crane, coot, duck, Common Turkey, Mourning Dove, Scarlet Macaw, Yellow-billed Cuckoo, Greater Roadrunner, Great Horned Owl, Western Screech-Owl, hummingbird, Belted Kingfisher, Gila Woodpecker, Verdin, Northern Mockingbird, Curve-billed Thrasher, Northern Cardinal, Blue Grosbeak, White-crowned Sparrow. Undoubtedly, others can be added to this list as more stories come to light.

At the very beginning of Piman origins, three culture heros joined Jeweḍ (Jeveḍ) Maakai, Earth Shaman, the first being, to help him create the world we now know. The first of these was Ñui, Vulture, who shaped the desert landscape with his slotted wings, making water flow possible. His counterpoint, or perhaps opposite, was Ban, Coyote. Every Piman born to an O'odham father with a Piman patriarchal lineage belongs to either the Buzzard or the Coyote moiety.

Additional birds appear as powerful metaphors in orations, another genre of Piman literature: Loggerhead Shrike, Northern Mockingbird, Blue Grosbeak, Barn Owl, raven, Northern Oriole, Turkey Vulture, Curve-billed Thrasher, roadrunner, Golden Eagle, and Prairie Falcon. They take their place with clouds, rainbows, algae, bulrushes, cattail, cane, and dragonflies of various colors—a watery assemblage for the Desert People.

Long series of sung poems are the prized possessions in Piman culture. These poems were given in dreams to their original singer, usually by some animal, often a bird. Thus, certain song groups originate from Gila Woodpecker, Belted Kingfisher, hummingbird, turkey, Lesser Nighthawk, Scarlet Macaw, and others. Birds sometimes appear in the poem's imagery itself: Turkey Vulture, American Coot, duck, Killdeer, heron, sandpipers, gull, Red-tailed Hawk, Northern Harrier, Golden Eagle, Prairie Falcon, American Kestrel, Gambel's Quail, White-winged Dove, Scarlet Macaw, roadrunner, Great Horned Owl, screech-owl (?), poorwill, nighthawk, hummingbird, Gila Woodpecker, Say's Phoebe, swallows, raven, Verdin, Black-tailed Gnatcatcher, Rock Wren, Canyon Wren (?), Northern Robin, mockingbird, Blue Grosbeak, and Northern Oriole. There must be many more. Perhaps all birds named in the Piman lexicon appear somewhere in song. Birds are good to sing about.

Moral exemplars, particularly as taught to children, included heron, Killdeer, roadrunner, Ladder-backed Woodpecker, Verdin, mockingbird, and oriole. Some of these models were negative, as examples of how not to conduct life, whereas others were positive.

Spirit power was once brought primarily by Turkey Vulture, Golden Eagle, Prairie Falcon, Great Horned Owl, and raven coming to a person in a dream. Some of these birds may still today visit the more culturally attuned Tohono O'odham. And every O'odham who still subscribes to the Piman cosmology knows that certain animals were endowed at the beginning of creation with a special nature—a sacredness, some say—that makes them dangerous. Violation of them might cause *kaachim mumkidag*, staying sickness. The birds so endowed include Turkey Vulture, ducks, Golden Eagle, Prairie Falcon, Peregrine, Gambel's Quail, roadrunner, Great Horned Owl, Gila Woodpecker, and perhaps Northern Mockingbird and a few others.

Although the O'odham carefully noted the progress of the sun at its rising and setting, marking off the summer and winter solstices, they also depended on other environmental cues to mark the seasons. Birds that were important indicators of seasonal changes included the Northern Harrier, ducks, White-winged Doves, roadrunner, and Northern Robin. All but the roadrunner are migratory.

Birds have graced the O'odham landscape with their names, in particular those rocky upthrusts from the desert floor. Some of these prominences are named for Turkey Vulture, Wild Turkey, Scarlet Macaw, Great Horned Owl, nighthawk, raven, and White-crowned Sparrow. Bird words, too, have found their way into some plant names: Turkey Vulture, Mourning Dove, Common Ground-Dove (or Inca Dove?), Scarlet Macaw, Western Screech-Owl, Great Horned Owl, raven, blackbird, and hummingbird. Overhead in the night sky, there are constellations of quail and nighthawk. Even people, in particular men, have received bird names as their own appellation (see appendix E).

Human survival in the desert depends, directly or indirectly, on rain. Ceremonial life focuses on this all-important need. Several birds were considered rain birds in Piman thought: Golden Eagle, Prairie Falcon, Turkey Vulture, turkey, hummingbirds, swallows, and probably the Red-shafted Flicker and cardinal. There was other help also. The hunter searching for game may be assisted by the raven. The shaman divining the cause of illness always uses two (or four) long primaries of the Golden Eagle or Turkey Vulture.

Some birds were hunted or trapped, supplying meat to a diet rich in vegetable resources. Other specific birds were sought for their feathers. But, most important, birds were metaphors. The intimate details of avian biology, encoded in O'odham folk science, were woven into the poetic imaginings of these desert people. Birds were wings for their thoughts and spiritual aspirations.

2

Native Consultants

Genes and cultures have something very important in common: both are repositories of coded information essential to adaptation and survival. More than this, genetic information and cultural information co-evolve in the sense that directions and latitudes set in one sphere inevitably shape constraints and choices in the other sphere.

—Virginia D. Nazarea, *Cultural Memory and Biodiversity*

The bulk of the information included in this book comes from O'odham individuals, usually older members of their communities, who shared their intimate knowledge of their desert biota with me. Most of these people I have known for twenty or more years. They have been friends and longtime teachers. When I came to O'odham country to work in the summer of 1963 and began my ornithological studies in the desert, some people called me U'uhig O'odham, Bird Man or Bird Person. My research provided a bridge of mutual interest. In time, I gradually learned that what these people knew about their environment, or what is now termed traditional ecological knowledge (TEK), was just as important and surely as fascinating as the Western taxonomic and distributional information I was pursuing (see Rea 1983).

These people were interesting in themselves as well as marvelous and patient teachers. The sessions were rewarding experiences in human discourse. I now recall some episodes with a certain nostalgia. The times Sylvester Matthias and I would drive across the desert in silence for fifteen or twenty minutes until something piqued his interest and he would comment on it. Or the many times I would pull into George Kyyitan's yard, and he would appear shortly thereafter with an arm full of orange extension cords for my tape recorder, ready to settle down for serious work. One morning when I arrived at Leonard Pancott's house, he was just finishing frying a couple of eggs. He left the pan on the stove, and we began talking. About four hours later, I suggested that he ought to eat his breakfast. "They can wait," he said, and we continued on for several more hours.

Perhaps these people were edified that someone would be interested in what they had to tell, often information not being passed on to subsequent generations more interested in TVs, VHSs, DVDs, and PCs. Anglo-American society is not the only one that is losing contact with the natural environment. This schism is affecting Native American cultures as well.

Most sessions were tape-recorded, and the information transcribed here is in the consultant's own words rather than being filtered through my rewording according to what I thought the person said. This approach preserves not only each individual's idiosyncratic speech, but also the cultural associations a person made while discussing a bird or some other organism. I occasionally restructured a particular conversation to provide a more logical sequence, or I combined several sessions on the same topic for completeness. Sometimes I went back over a transcript with the person, giving him or her opportunities to clarify or elaborate on points we had discussed. Anything I have inserted for clarification appears in square brackets.

All the consultants were speaking English or Spanish as their second language. Each had a different way of speaking. In a few cases, interviews were entirely in O'odham; they were translated by Culver Cassa. As noted in the preface, there are no *informants* in this book, a term rightfully offensive to native peoples who choose to share their fund of knowledge with an outsider. These people are rather teachers and professional consultants. All willingly cooperated in an effort to put their way of seeing the desert and living in it on record for future generations.

One thing stands out as I flip through decades of transcribed interviews with Pimans: their utter frankness. If someone knew a folk taxon, we talked about it. If they had heard the name but didn't personally know the species itself, they said so. If they had secondhand information about it, they usually gave the source. "I think . . ." was a common phrase with people who were unwilling to make statements about what they were not certain. And if they had never heard of a species, they were quite up front in saying so promptly.

A number of the consultants were *mamakai*, shamans, though I have not identified them as such for their privacy. Others were healers, those who sing the songs used in healing sessions. Their generous sharing of information has considerably enriched our understanding of birds in Piman culture. It is a curious phenomenon, but there seems to be an inverse relationship between cultural knowledge and secrecy. So much for neotraditionalists!

As it turns out, most of the people I worked with were cattlemen. The relatively younger O'odham (those born from the 1920s onward), have spent much of their time outdoors, becoming aware of the flora and fauna of the desert firsthand. With a changing culture, women increasingly spent more time with indoor activities, learning less of the biota. They themselves were the first to point out this problem. Those People born in the 1930s and particularly in the 1940s or later generally had a compromised knowledge of folk taxonomy. Younger generations of River Pima, especially, have little knowledge of the natural environment, much like their non-Indian neighbors.

At an earlier time, however, folk taxonomy appears to have been universally shared by everyone in the community—at least this is my impression based on the oldest people with whom I worked. Everyone once knew *giisobĭ*, the tiny Verdin; the great *ba'ag*, Golden Eagle; *dahiwua*, the Kangaroo rat; as well as *juḍum*, the bear. Associated folktales and folk science might be more developed among some individuals than among others, but even then traditional ecological knowledge seems to have been well spread throughout the communities before young people were taken out of their cultural settings and sent away to schools.

But does this erosion in cultural knowledge make any difference, really? Times are changing, and cultural change is inevitable. Perhaps. However, Pimans' loss of touch with their environment is systematic of a worldwide problem of aggressive homogenization. While we pride ourselves on technological innovation—the apex of some sort of social Darwinian model—the world is losing species at an alarming rate, far in excess of natural extinctions and evolution; half the languages of the world spoken only a few generations ago have been lost; and much of the genetic diversity of agricultural crops developed over the past ten thousand years has vanished. We have more and more of less and less. And the world is becoming a more boring place.

Who were my consultants? A bit of background helps place them temporally and geographically (see fig. 2.1). More about them becomes evident from what they tell us about birds in the species accounts.

River Pima

Sylvester Matthias (1913–1997) was the grounds gardener at St. John's Indian School, Komaḍk, Arizona. He had a reputation among the friars for an astonishing memory about local history and genealogy. He was the first one I asked about birds. I immediately found out that he was a storehouse of information, but we had to work out a language problem before we could proceed: he knew the birds by their Piman names, whereas I knew them only by their English and scientific names. So we began correlating our terminology. After work when he had time, he would come to my laboratory, where by now there was a sizable collection of bird study skins. He was my teacher for the next quarter century, not only about birds, but about all the biota and its connection with Piman culture.

Sylvester oftentimes knew the names for birds that no longer occurred in the desert once the rivers dried up. He would describe them, where and how they nested, what they sounded like. We would look at specimens, pictures, and travel around the country searching out live material. We eventually identified most of the plants and animals he knew, but some remain mysteries (see appendix F and Rea 1997:appendix B, 384–389). Many times Sylvester remembered plants and animals that no other Pima along the river knew. On various occasions, the bird names he provided have been verified among the Pima Bajo in Sonora and from vocabularies compiled by colonial Spanish Jesuits in the seventeenth and eighteenth centuries.

In later years, Sylvester would occasionally fly over to my place in San Diego, where we could work at leisure on tape-recording his vast fund of information. During the early part of the day while I was at the museum, he would work on masks and other Piman artifacts, and in the afternoon when I returned he would be ready with new information and with additions to the previous day's recordings. He was responsible for the overall organization of plants in my River Pima ethnobotany (Rea 1997). Unfortunately, we never attempted a similar arrangement of the emic hierarchy of mammals and birds, and no one else seemed comfortable doing so.

Early on I also began asking *Joseph Giff* (1907–1982) about the local biota. He, too, was from Komaḍk on the Gila River. Joe was a natural linguist and historian as well as a person well grounded in Piman oral literature, both sung and narrated. In his long career, he had served as policeman, tribal council member, and lieutenant governor in addition to his usual roles as cowboy, farmer, and carpenter. Joe had a

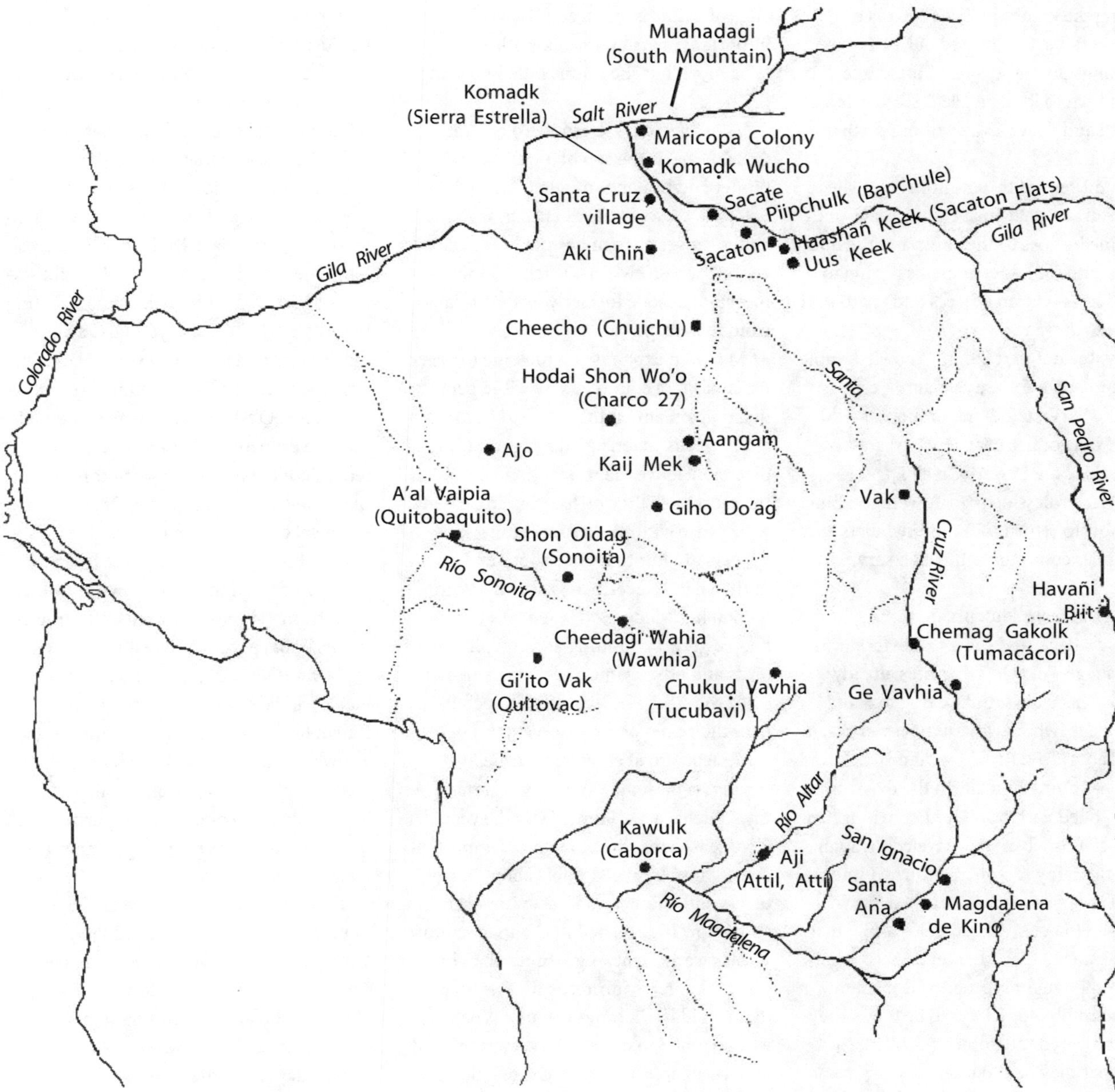

Figure 2.1 *Map of some Piman communities, Pimería Alta, discussed in the text.*

sense of humor as dry as some Arizona rivers. His daughter reported that one time when Joe ordered steak at a restaurant and the waitress asked him how he would like it, he replied, straightfaced, "Ah, *fried*. Yes, I think I'll have mine fried." After we had worked on mammals for years, I learned from state game personnel that there were White-tailed Deer in the upper parts of the nearby Sierra Estrella. Next time I saw Joe, I said, "You never told me there were *siikĭ* in the Estrellas." His prompt reply was, "You never asked."

Joe's wife, *Ruth Giff* (1910–2005), was a basket weaver as well as a farmer, an occupation once shared by Pima women as well as men. She was born at the village of Santa Cruz, not far from Komaḍk, and remembered many of the plants and animals from the desert streams. Two of her basket materials came from the river—willow and cattail; the third, devil's claw, she grew. She was also an expert at processing many crops from the desert, such as mesquite pods, ironwood seeds, Saguaro, and various wild greens.

When certain questions arose that Ruth and Joe were unsure how to answer, they would say, "We'll have to ask the old people." And they would. I regret that I did not attempt to do this myself. There were people living in the 1960s who had been raised in the last part of the 1800s, when ecological conditions here were much better and the wildlife flourished. When these people died, great stores of folk science vanished with them.

Early on *Sally Pablo* (1934–2000), Joe and Ruth's daughter, helped me get my collecting permit for the research on the reservation from the tribal governor. At the time, open areas had been designated for non-Pima hunters seeking doves and quail. One day I was searching for migrant warblers, vireos, and flycatchers in one of the old tamarisk groves, armed with my old .22 rifle and bird shot. A tribal policeman came by.

"You're not supposed to be hunting here" he warned.

"I'm not hunting," I said truthfully, "I'm collecting specimens."

"What's that?" he asked.

I was puzzled as how to respond, not

knowing the names in Pima for any of these migrants, but blurted out, "Oh . . . little things up in the tops of these trees." It was his turn to look puzzled. He just shook his head and drove away without further comment.

When I related this event to Sally, she said, "You see what you've done now? The next time he goes to the clinic and the doctor asks him for a specimen, he's going to go out in that clump of trees and try to find out what you were after."

Sally was not particularly knowledgeable about birds; her forte was plants—especially the harvest and preparation of wild foods. Many of her contributions are included in *At the Desert's Green Edge* (Rea 1997). Her professional goals were teaching Pima how to return to traditional diets and leading her community through service on the tribal council.

The road from Phoenix to Komaḍk, the Catholic village, and to its Presbyterian sister village, Gila Crossing, was already paved in those days. But getting to Santa Cruz village, which was on the lower reach of the Santa Cruz River, by car or truck was somewhat more difficult. In the dry season, the ruts filled with powdered alluvium that rose in clouds when one traveled through, and in the rainy season they turned into a sea of slippery mud. But tucked away in this ranchería of scattered fields and mud houses I was to discover some charming people who taught me a great deal about Piman ethnobiology. Or maybe it was they who discovered me, someone willing to learn what they had to offer.

Rosita Brown (1923–1977) and *David Brown* (1917–1981), for instance, became my good friends. Rosita was a basket weaver as well as skilled in making *haak chu'i*, parched wheat pinole. They farmed and had a thorough knowledge of local natural history and its connection with Piman culture. Later they left their field house and took up residence at nearby St. Catherine's Church, the hidden architectural gem of the reservation. Not only did they teach, but they made me more aware of what it was that I should learn.

Carmelita Raphael (born 1921) farmed and ranched in Santa Cruz village. We met her at the beginning of chapter 1. Her mother was Gila Pima, but her father was from Al Jeg, Menager's Dam, a settlement near the border that is a mixed Tohono O'odham and Hiá ch-eḍ O'odham village (fide Delfina Cameron). Carmelita's knowledge of plants is great, and she was an important contributor to my ethnobotany (Rea 1997), but she claims little knowledge of birds.

Theodore Peters (born 1926) of Santa Cruz village is the son of a well-known hunter and healer, S-mogich, Manuel Peters. Theodore's life as farmer, woodcutter, and cattleman has put him in close contact with New York Thicket, a mature mesquite bosque formerly just upstream from the village.

My avian explorations took me to other parts of Pima country as well. The people living upstream on the Gila had been born after serious watershed damage and water diversion above the reservation began. For the most part, they no longer had personal experience with the herons, egrets, kingfishers, and fish that once abounded on the middle Gila. Nevertheless, much environmental knowledge remained because these people grew up hunting and trapping wild birds and other game and were cattlemen.

At the village of Bapchule (Piipchulk), near the center of the reservation, I became acquainted early on with *George Kyyitan* (1912–1996), father of several of my students at St. John's. Over the years, George became one of my most important consultants, knowledgeable about both plants and animals and, most important, about how they related to O'odham culture. When nearly eighty, he retired. Then he was always ready for recording sessions that would last for many hours. Whatever we happened to discuss, he always seemed to have an appropriate Pima song that he would call up at the moment. Some were about birds; others were about mammals, plants, or other subjects (see Rea 1997, 1998b). I never found out how many songs he really knew, but like other old singers, they probably numbered in the hundreds.

Irene Hendricks (1924–2003) of nearby Casa Blanca had a relatively extensive knowledge of Piman folk taxonomy and confirmed the identifications of many plants and animals. On several occasions, she went into the field with Culver Cassa and me, helping to collect plant specimens and sharing her knowledge of natural history.

Although *Francis Vavages* (1906–1995) lived in Sacaton, the reservation agency, when I knew him, he had been raised in Haashañ Keek, Saguaro Standing, or Sacaton Flats as it is known to outsiders. A tall, wiry man in his eighties when I met him, he was a tough person to run down because he was always off somewhere on foot. He began a sort of cottage industry in Piman crafts, specializing in flutes. He supplied important cultural information, both folk science and ceremonial and mythic associations, for various plants and animals.

Clemencia Leos (1929–2007), the daughter of singer Zachary Pablo, has been a longtime friend since the days I taught several of her children in high school. Visiting her was a delight. I always enjoyed her quick mind and masterful vocabulary. One day when I mentioned that I had seventy-six Akimel O'odham bird names, she said that she probably didn't know more than ten. I doubted this, so we went through them, and she scored twenty. This situation is typical of the erosion of River Pima folk natural history that occurred between her parents' generation and her own. Nevertheless, she made some unique contributions to this book, particularly with bird songs.

Leonard Pancott (1919–2004) of Haashañ Keek did not know a great many Piman birds, but, having spent his life as a cowboy from his youth, he had good knowledge of the range and changes in local ecological conditions, in particular those related to the dropping of the water table. His descriptions of Pima hunting methods parallel those of Joe Giff, Sylvester Matthias, George Kyyitan, and Francis Vavages—all somewhat older men. He contributed significant information on the identification and cultural significance of those birds he did know.

One day Leonard got going on the neotraditionalists of the tribe, trying to be Pimas but not knowing their culture. "And they go around with long hair, thinking they're Indian." He paused. "But then, maybe I'm just jealous," he said, pushing back his Western hat to reveal a bald head.

Leslie Pasqual (born 1927) of Progressive Colony near Sacaton was still an active cattleman at age seventy-seven, working nearly dawn to dusk. When George Kyyitan and I were doing a tribal survey of traditional sites, George took me to Leslie's ranch to see how he might help us. The two men leaned over the corral fence and talked horses. All in O'odham, of course. I was so fascinated by Leslie's musical speech and careful enunciation that I focused closely on what he was saying. When the old men stopped talking, I asked, "Well, did you ever get your bay mare back?" Leslie turned to me, surprised, saying, "You speak Pima, don't you?" I had to tell him truthfully that I didn't, but that I knew just enough to follow the conversation.

Some years later when I still had some bird questions, nearly all my consultants had died. So one evening I headed down the dusty canal bank road to Leslie's place. The mesic jungle of growth where I had found a Parula Warbler and pair of Yellow-billed Cuckoos one day some thirty-four years earlier was long gone (Rea 1983:164). Leslie, after a day in the saddle, opened the tailgate of his pickup, and we sat down for some serious talking. He knew the old ways. I paid him a number of visits, and he helped to clarify the various bird questions I thought I would never resolve.

Perhaps we were kindred spirits. He is a true Luddite, living without electricity, hauling water for his horses, cattle, and household needs. He lives this way by choice. An electrical transformer and pump are not far away. His sandwich house (puddled mud jacal) is neatly plastered and whitewashed.

Culver Cassa (born 1968) is of mixed Gila River Pima and San Carlos Apache descent, but he was raised in Pima country (Sacaton) and is a natural linguist. He is one of only a few people who can listen to a Pima song, write it out, and translate it into spoken Pima and English. One time he asked if he could use an adjective adverbially because the Pima in the song said an animal was doing something "pinkly." His help has been invaluable.

If these skills aren't enough, Culver is also a master of Piman mythology. He knows what storytellers have said and where. More important, he sees the underlying meaning and metaphor, and how they relate to a desert culture's worldview and sometimes to human endeavors in general. For instance, he sees as a basic theme of Piman mythology and ritual the attempt to mediate the oppositions of individuality (epitomized by winter activities) and the needs of community and cooperative activities (such as farming in summer). Some of his original thoughts stemming from our discussion of stories are included in the species accounts.

Tohono O'odham

Jose Enriques (born 1938) was born in Kaij Mek, Burnt Seed or Santa Rosa, but his family moved soon thereafter to Queen's Well, a new settlement in the northeastern part of the main reservation, where he still lives. Although a very traditional person, Jose is a storehouse of ecological information as well. He punctuates his narrations about O'odham folk taxa and lore with comments on related matters such as slope exposure, substrate, vegetational setting, and geology. Working alongside Western biologists on various resource-management projects, he has been a teacher as well as an avid learner. A gentle person, he is always ready to share his vast information about natural history.

I made only a few visits to talk with *Vincent Josemaria* (1925–2003) of Pisin Mo'o. It was a long drive from Sacaton to his village, and he was often out on roundups. He lived with his sister, and they had no phone, so I would take my chances at catching him. None of the O'odham I had worked with up to the point when I met him knew songs for curing Prairie Falcon sickness, and I thought perhaps he would. He did, and he gave me permission to record and use them in this book.

Much of our conversations were in O'odham, with Culver Cassa doing the translations. But Vincent wasn't above using some English now and then. I once asked Culver to ask him if he knew any buzzard songs. Vincent promptly shot back to me in English, "Do you think I know everything?" Actually, he had learned a great deal about natural history as a cowboy and contributed much valuable information about birds. More important, he knew cultural relations to birds. And, yes, he did know a song for healing Turkey Vulture sickness. Vincent died of cancer, and I never had the good fortune of being in the field with him to learn more firsthand. He was a generous teacher.

Delores Lewis (?1921–2005) was a knowledgeable farmer from the village of Ge O'idag (Big Fields). He was one of the valuable teachers Gary Nabhan had while doing his doctoral research. He knew natural history well. Delores made a number of trips with Gary and me into northern Sonora, where I was able to supplement book elicitations with field identifications, which I lacked with most other Tohono O'odham consultants. A quiet, gentle person who observed much, he made many numerous contributions to this study.

My one visit to Vaya Chiñ to talk to *Joe Lopez* (?1920–2004) was after he was bedridden, although I had met him and his wife, Eva Marie (Ida), years earlier when I taught a number of their children at St. John's. Joe had been in government service as well as being a local cattleman, and his children were top-notch students. Delores Lewis was his uncle. When I went to his house, his brother answered the door saying, "Hodai Mo'o is waiting for you."

"Stone Head?" I asked. That was his Indian name, which he offered to sell to me for a horse or a couple of cows.

"What do you want to talk to me about?" he asked in his unceremonious way.

"I'm trying to finish my book on O'odham birds, and I hear you old cattlemen know a lot about birds," I told him.

"We're cowboys, not bird-watchers," he shot back. There was one important question I wanted to know and that was whether Tohono O'odham distinguished any folk specifics of quail. I struck gold on that one!

Frank Jim (1916–1998) was a longtime friend. I taught a number of his children at St. John's Indian School. Later, when I became a serious student of ethnobiology, he became a major teacher. His home village was Hodai Shon Wo'o (Charco 27) in the western part of the main Papago Reservation. Although his grandfather was Tom Childs, this was by an earlier marriage, not by the Sand Papago/Hiá ch-eḍ O'odham woman with whom Childs is usually associated (Childs 1954). Frank said he himself was not a shaman even though he was thoroughly familiar with *kaachim mumkidag*, staying sickness, and fetishes and was a participant in many types of *wusota*, healing sessions involving singing. Also he was quite knowledgeable about ceremonialism. He was a major contributor to my work on Piman mammals (Rea 1998b).

Frank treated me like I was one of his sons. When the family lived at Roll in western Arizona, their place was a good halfway point in my drives between San Diego and River Pima country. One day he told me, "In case we are not home whenever you come through, up here is the key to the house. Eat whatever is there. And here is the key to the freezer; put whatever you hunt in there." He said that because he knew I would be returning to the museum with lots of bird specimens.

Frank had many contacts in the Tohono O'odham community. One day I asked him if he might try to find me a clay bean pot, *ha'a*. They are hard to come by and expensive if you do find one. "I have an extra one that's too small for our family I'll give you." He explained exactly how to cure it with beef tallow before it is used. "You'll have to get an axle hub, you know, from a wagon, to stand it over the fire," he said, perfectly straight-faced.

"Frank! Where do you think I'll find a wagon hub in San Diego?"

"Well, I'll look around for one. Pick it up on your way back home." A week later, on my return trip, my prize bean pot was on the kitchen table, sitting on its iron ring. I still cook my beans in it.

I drive into the compound where *Simon Lopez* (born 1938) lives in Santa Rosa village, also known as Kaij Mek, Burnt Seed. He is busy painting a new outhouse between their mud house and the horse corral. I park. He and a youngster sitting facing me give no indication they are aware of me, though I park a couple meters away from them. When I get out and say something, Simon's face lights up with a warm smile, a psychological embrace. "Your Christmas card came. I was very interested," he says in the highly inflected speech characteristic of his village. Culver Cassa, Simon, and I stand around the hood of my pickup for the next three or four hours as we go over this book's manuscript. He adds so much that I must spend the entire next day making entries.

Simon explains that there hasn't been a roundup for several months now, and there are many calves that need to be trapped and branded. (Frequently I miss him because he is on roundup.) But today I have caught him on a leisurely day when we can work. The child silently takes the brush from Simon's hand and finishes the painting.

Simon's knowledge of folk science comes from years on the range as a cowboy, as was the case with so many others who contributed to this work. He has an inquisitive mind about his surroundings, which his contributions here show. He is also from a family with several generations' experience in shamanism and curing *(wusota)*. They are subjects he delights in discussing.

My single interview with *Leon Juan* (1924–2003) of Aangam was conducted entirely in O'odham, with the assistance of Culver Cassa and Lawrence Saraficio. Although Aangam is the next village just north of Santa Rosa, a different dialect is used there, so evident in cadence and inflection. Leon was a traditionalist, saturated in ceremonial knowledge. He was the one who directed younger O'odham every four years in the changing of the Ocotillo fence that surrounds the Shrine of the Four Children at nearby Santa Rosa, singing the songs and reciting the required orations.

Juanita Norris (1917–2007), a resident of the Ak Chin Reservation, was born in Paso Redondo but raised in Hikiwañ, the Tohono O'odham Reservation. She claims the Huhu'ula as her linguistic affiliation. Like Rosita Brown, she was knowledgeable about traditional ways such as making *haak chu'i*, wheat pinole. She helped with my plant and mammal books and went over all the birds she knew for this one.

Lawrence Saraficio said I should talk to his uncle *Jose Garcia* (born 1938), a lifelong cowboy with extensive knowledge of wildlife and shamanism. After several years, we finally got together at Santa Cruz Tohono O'odham village on a summer day, sitting in the back of the cool adobe church to talk about birds. We had only about an hour, and there were many interruptions, but we covered some important points on folk taxonomy and staying sickness.

Hiá ch-eḍ O'odham

The Hiá ch-eḍ O'odham, or Sand Papagos, as they have long been known (Areneños in Spanish) are descended from bands of more nomadic O'odham living west and southwest of the Papago proper. They were never granted a reservation and in fact were considered by some anthropologists to be extinct as an ethnic group (see Nabhan, Hodgson, and Fellows 1989). They proved to be dispersed, but certainly not gone. Sprinkled throughout the southeastern half of the lower Colorado River drainage, some had intermarried with other O'odham or Hispanics. I worked with two people of this dialect group.

Luciano Noriega (?–1993?) of Quitovac oasis village in northern Sonora was our principal teacher when Gary Paul Nabhan, I, and others were studying the natural history and folk ecology of this natural desert oasis (Nabhan 1982; Nabhan and associates 1982; Rea, Nabhan, and Reichhardt 1983; Reichhardt and associates 1994). The *wiigida* ceremony is carried out in this village each summer. I never learned when he was born, but he died between the summer of 1992 and the summer of 1993. When we were working there, we assumed he was around eighty.

Before the oasis was "improved" by a government project, it was a dense cattail- and bulrush-grown marsh. Luciano farmed from the springs, raising vegetables and flourishing orchard trees. These lush, cool habitats attracted myriads of birds, particularly in summer, so Luciano lived in a birdwatcher's paradise.

Although I worked with Luciano a number of times on birds and other organisms, I never worked out an exhaustive folk taxonomy, even of birds, with him. I have included those folk taxa from my notes that I am reasonably sure of, but others that I did not collect are given in Nabhan and associates 1982, table 6.

When I would visit Maricopa Ralph Cameron and discuss the names of organisms appearing in their Creation Story, his wife, *Delfina Marquez Cameron* (born 1933), would sometimes add comments from Hiá ch-eḍ O'odham. She was quite well informed about this group and spoke that dialect. She gave me a number of bird names and other information. She was from Do'ag Wucho (Darby Wells), near Moik Vahia (Ajo), Arizona. Her mother, Manuela Orosco, was from the Quitobaquito family that was ultimately ejected from A'al Vaipia and Juñ Kachk, Bates Well, when these oasis settlements became a national monument (see Nabhan 1982:89–97).

Delfina pointed out that Hiá ch-eḍ O'odham use the *v/w* distinction like the Kohadk and Akimel O'odham dialects. (She said *"vipismal"* and *"wusota."*) Gary Nabhan and I had noticed this distinction years earlier with Luciano.

"And they say we talk faster, also."

"Is that true?" I asked.

"I think it *is*."

Lowland Pima Bajo

By the time Akimel O'odham Sally Pablo and I finally found the Pima Bajo settlement of Ónavas, there were only two speakers of the language remaining. Both had worked with linguists John Robert Cornell and Ken Hale and with geographer Campbell Pennington before they worked a decade with me, an ethnobiologist.

Pedro Estrella Tánori (died 1988) was the first we met. He promptly invited us into his unplastered adobe brick house to talk Pima. Although he and the River Pima lived some three hundred miles apart and the speech communities had been separated for several centuries, it was soon apparent that these long disjunct dialects of northern Pima were readily intelligible to one another despite several major sound shifts. Earlier place-names in Arizona show that the Pima Alta once spoke like Pedro Estrella. Around the latter half of the eighteenth century, Upper Pimans began to palatalize their language (for instance, "Tuc-son" became *chuk shoñ*).

Don Pedro, as his fellow villagers called him, enjoyed fieldwork, and we scoured the local thornscrub and tropical deciduous forest on numerous trips. My work with Pedro was largely focused on ethnobotany. We collected many herbarium specimens and whatever other information we stumbled across, including avian folk taxa and natural history. I now wish we had focused more on birds. (When it comes time to write, one discovers the great gaps in the data.) After three centuries of close Hispanic contact, however, most information on mythology and ceremonialism had been lost.

I was never in the field with *María Córdova Badache* (died 1986), interviewing her instead at her home, using my elicitation notebook for birds and my daily collections of botanical specimens. I spent considerably less time with her than with don Pedro, but her input was an important check on his responses. And there were things that María remembered that Pedro didn't. Where Pedro was impish, delighting in everything we encountered in nature, María was a tiny, serious woman, utterly precise with anything she said in Pima. Pedro did not read, whereas María was a native cantor at the semana santa celebrations, singing Latin from the psalter.

I never learned when either of them was born. When I first met Pedro in 1978, he said he was eighty, but about ten years later he said he was one hundred. María seemed to have a more precise knowledge of her age, living until about her mideighties. She said she was a year or two older than Pedro.

3

Documentary Sources

As we explore the world of myth, we should remember that we are journeying not through a maze of falsehood but through a marvelous world of metaphor that breathes life into the essential human story: the story of the relationship between the known and the unknown, both around and within us, the story of the search for identity in the context of the universal struggle between order and chaos.
—David Adams Leeming, *The World of Myth*

Piman folk ornithology has undergone a vast erosion in the past seventy-five years or so. Increasingly, as people have less contact with the outdoors, even though they live in the desert, their folk taxonomy has dwindled to almost nothing. We must now rely on two sources of information: the knowledge still held by only a few people, usually very old men and women who learned the names and lore of plants and animals in their youth, and documentary sources left by those in the past who had enough appreciation for this knowledge to set it in writing. This chapter deals with the written sources.

The quality of the documentary sources varies according to the skills of those transcribing the information. Some narratives, for instance, have been transcribed with exquisite precision. Others are so distorted that the original story line can be seen only dimly through accretions of non-Piman cultural baggage.

Another problem is that some authors of documentary sources did not use any systematic orthography in recording Piman terms. Rather, they wrote words idiosyncratically as if in English, perhaps with a few hyphens thrown in between syllables for appearance's sake *(wah-choo-kook-kee* for *vajukukĭ*; *bives-cheel* for *ba'ivchul*; *how-app-ahl* for *haupal)*. These words are usually decipherable.

Likewise, there has been great variation in writers' abilities to identify Piman folk taxa and render them correctly in English in the recorded texts. Some moderns have made only a feeble effort at best toward biological accuracy. How can we understand a story when the *giisobĭ* (Verdin) becomes a species of flycatcher seldom found in the desert, the *vakav* (Northern Harrier, or Marsh Hawk) becomes a pelican, and, worse yet, the *kadgam* (Yellow-billed Cuckoo) switches biological kingdoms and is given in text as a plant?

Even worse, some documents give only a vague English gloss ("crane," "alligator," "tiger"), leaving us to guess what the original Pima term might have been. (This problem has been solved sometimes by finding an O'odham who still remembered the song or the narrative and could clarify the ambiguity.) Perhaps the greatest problem of all in understanding mythology is when authors use a general English term (for example, *deer, owl, lizard, weed*) where no such equivalent word exists in the native language. For instance, many of them lump at least five kinds of birds that O'odham talk about, each with its own name, into an English category called *owl*, a collective noun that does not exist in O'odham.

The purpose of this chapter is to identify the primary documentary sources of O'odham literature and to evaluate their authenticity and accuracy with regard to folk taxonomies. Specific misidentifications are dealt with at the ends of the species accounts.

Oral Literature

Much of what we know on any deep level about the symbolic roles of birds in Piman culture comes from oral literature such as songs, orations, and stories. Some of this oral material has been collected and published, either in English or bilingually. The richest of these oral genres is the Piman Creation Story, or Origin Myth, a loose collection of stories, interspersed with songs and occasional orations, that was officially narrated in each community during the four nights surrounding the winter solstice in late December. (Individual episodes could be and were told outside that context at other times of the year.) Poisonous animals, in particular the rattlesnake, were not to hear the stories. This taboo was more rigorous among Puebloan peoples such as the Hopi and Zuni.

In midwinter, when the sun reached its southernmost point on the horizon and seemed to tarry there for four days before starting its return trip north (as noted by the one in the village who closely tracked it), someone from the community took a handful of ground corn to the local storyteller as an official invitation for him to remember ritually the mythic origin and history of the People. In the evening, both young and old gathered in the large communal round house for the all-night event. No one dared fall asleep. During the day, people loitered and dozed, preparing for the continuation during the next night. This time was sacred for the community.

The four-night story began panoramically with the creation of the world by Jeweḍ Maakai, Earth Shaman, and the subsequent history of the three other Original Beings (Ñui, or Buzzard; Ban, or Coyote; and the People's Se'ehe, or I'itoi, also known as Elder Brother) and the People. Stories were episodic. Some episodes touched on mythic themes such as the Flood found in other people's origin stories around the world (Dundes 1988). Others were more particular to the desert itself, sometimes based on historic events and their interpretation and embellishment.

Encoded in the stories were covert messages that are usually difficult for us to tease out today. The subtle accounts completely lack overt didacticism. A story might obliquely remind listeners about starvation foods, for instance. Because real starvation might not come every generation, including this information in story had survival value (Minnis 1985:35).

Variations occurred between narrators, particularly as to which events and songs were to be included in a given canon. According to Sylvester Matthias, there were once four versions of the Creation Story on the Gila, recalled by four different narrators: Mel Cheedagĭ (Blue/Green Run) of Bapchule; Atosha (Loincloth) of Santa Cruz; Komal Hogĭ (Thin Leather) of Santan; and Ko'oi Mo'o (Rattlesnake Head) of Old Komaḍk. (One might be tempted to see in this number of narrators the sacred number four or even a reflection of the four Gospels; I think rather that the versions originate from the four large Contact-period Piman settlements on the middle Gila, each preserving its own tradition in spite of community mergings and dispersals during post-Conquest times.)

Two questions are critical here to our study of ethnobiology. First, how authentically recorded is any particular version? And second, how accurately have the plants and animals been identified in English? In the original narrative context, the People brought to the telling a shared environmental knowledge that was quite sophisticated. This knowledge included a communal folk taxonomy of hundreds of different plants and animals and, what is more, an intimate understanding of the biology of the birds and other animals forming much of mythic structure. Each animal appearing in the episodes, from the Jaguar to the tiny Verdin, carried a symbolic nuance much as the hawk, dove, bee, or pig does with us today. For most stories, knowing the critter and understanding its metaphor, as the original hearers once did, is essential—utterly sine qua non—to understanding the story.

Pimans living at the northern Tepiman frontier along the middle and lower Gila shared settlements with Yumans for at least three centuries. Although belonging to different language families, they borrowed stories from one another, just as the newly arrived Athabascans did so freely from the Pueblos elsewhere in the Southwest. Even when told in Pima, Yuman stories retain certain psychological, cultural, and aesthetic characteristics that earmark their origin. For instance, cremation rather than inhumation is a clue to the Yuman origin of an episode. Their stories include more animal characters, especially waterbirds. Stories of Piman origin portray clearly human characters, although Coyote stories are an exception.

Longer Creation Stories

To my knowledge, no creation storytellers survive today. Only a few of the oldest Indians told me about sitting up in their younger days during the longest winter nights to hear the episodes. Now just individual events are recalled, and they are usually out of context of the overall fabric of the Origin Story. Fortunately, a number of documentary sources, published or in manuscript, exist that preserve parts or all of the Origin Story. The skills of those who have translated and recorded the story or parts of it have varied greatly.

In the midsummer of 1903, J. William Lloyd of New Jersey and Pima educator Edward H. Wood rendezvoused in the vicinity of Sacaton on the Gila River Reservation. The objective was a team project: the elderly storyteller Komal Hogĭ, Thin Leather or Buckskin, a monolingual Pima, began narrating the Creation Story, which bilingual Wood translated, while Lloyd recorded everything by pencil in English, as faithfully as possible, except for a few passages of "almost Biblical plainness of speech on family matters . . . expurgated . . . for prudish Caucasian ears" (Lloyd 1911:9).

The effort, spliced with Lloyd's observations on the Pima, their country, and oral literature, resulted in the full-length book *Aw-aw-tam Indian Nights* (Lloyd 1911). Except for the interspersed songs, all of the account is here, including the formal orations, which sometimes duplicate narrative text. This is a delightful work, lacking only the Pima-language version.

In spite of being new to the Sonoran Desert, Lloyd was sensitive to the local biota and attempted to record correctly the plants and animals that occurred in the legends. In published excerpts from his field notes (which so far as I know have not been preserved), he wonders about the identification of the Bendire's and Curve-billed Thrashers and describes the behavior of a pair of gnatcatchers—all this in the days before the invention of convenient field guides.

Another account, provided in Frank Russell's Pima ethnography and also given by Thin Leather, was recorded around the same time (1901–1902). It was likely translated by Jose Louis Brennan, a skilled Papago linguist, who later lived and died on the Gila River Reservation (see Fontana 1975). This is not a line-by-line account, the episodes being rather freely summarized in English. There is no accompanying Pima version. Some of the songs are included in the story text and in a later section in Pima with interlinear translation (Russell 1908:272–282). Some episodes of the whole story appear in Lloyd but not in Russell and vice versa, in spite of their common source.

In 1920, ethnomusicologist Frances Densmore visited various Papago villages, recording Piman oral literature. Although Densmore's ostensible purpose was to collect O'odham songs and analyze their structure, while doing so she also documented much of the cultural context of the songs. For instance, she recorded large segments of the Creation Myth (Densmore 1929:15–28). Some episodes appear to be linear translations, whereas others are more like summaries, along the lines of what Frank Russell (or Jose Louis) did, and a few stories are reduced to a mere sentence or two. In comparison with the work by Russell, Lloyd, George Herzog, and Julian Hayden, Densmore's narratives appear to be authentic and sensitively transcribed. Her identifications of animals, however, are rather weak.

In the summer of 1927, anthropologist Ruth Benedict took down an extensive transcription of River Pima creation myths, shorter tales and segments, and orations. Her primary sources were Thomas Vanyiko and William Blackwater, both upriver Pima. Vanyiko (Evanico) was born around 1863 (Kroeber and Fontana 1986:87). Johnson Azul was her translator. This rich collection was finally edited and made available by Donald Bahr many years later (Benedict 2001). Benedict took down her stories only in English, occasionally inserting a Pima word when she and her translators didn't know the English equivalent. This is fortunate. Although Benedict had a good sense of narrative, she did not have a good ear for Piman words (writing *maaka* for *maakai*, *Aankap* for *Aangam*), and she didn't know the local Sonoran Desert biota well (giving "Mountain Goat" for Mountain Sheep or Desert Bighorn). But there is usually enough of the correct word to make the connection. Bahr's footnotes, however, are another matter with respect to folk taxonomy. He includes many of Dean and Lucille Saxtons' misidentifications, repeats some of his own from his earlier works (such as giving "Black Phoebe" for Verdin), and introduces a few new ones.

Two summers later, in 1929, George Herzog collected another Pima version of the Creation Story from narrator Thomas Vanyiko, who was then older than eighty. Paul Azul served as interpreter. They went over texts and songs during shorter trips in 1933 and 1936. According to Culver Cassa, this version, like Densmore's, contains considerably more Yuman inclusions than either Lloyd's or Russell's version. This document is particularly valuable because it was recorded verbatim in Pima with a parallel, word-for-word English translation. Herzog was a precise linguist who captured every Pima nuance. However, he was a poor ethnobiologist, and his English renderings of Piman folk names are sometimes farfetched (for example, Red-shafted Flicker is given as a thrasher, and Northern Harrier is called a crane). Certain ambiguities in English, such as kinship terms between characters, are clarified in the Pima original, as are the misidentified folk taxa.

The youngest documentary version of the Pima Creation Myth is that told by Juan Smith, translated by William Smith Allison, and transcribed verbatim in English by Julian Hayden (Hayden 1935) at Snaketown in 1935. Juan Smith was a traditional storyteller, a Catholic; Allison an ordained Baptist deacon; and Hayden a non-Indian archaeologist. Because this transcription was a relatively recent version, the effects of cultural erosion are evident. Juan Smith confessed that there were parts that he had forgotten (or never learned well). Allison, for his part, could not resist the temptation to preach while translating, and the resulting document is saturated with such basically non-Piman concepts as sin, wickedness, punishment, trouble, commandments, male superiority, female moral inferiority, and so forth. However, some wonderful segments appear here that are not found in any of the other recordings.

Donald Bahr has edited the Hayden-Smith manuscript and made it available to a wider audience under the title *The Short Swift Time of Gods on Earth: The Hohokam Chronicles* (Bahr and associates 1994). Bahr has done us a service by teasing out Allison's editorial sermonizing from Smith's actual text and by printing these parts separately in the page margins. However, he misidentifies all but the most obvious plants and animals appearing in the story. This misidentification is most blatant with regard to birds. Of all the excellent works published on Arizona's avifauna during the past one hundred years, not one is referenced in the book. Even a book where almost all the Piman folk taxa are identified by Pima consultants (Rea 1983), which was reviewed by Bahr in 1983, is unmentioned. Also, Bahr inexplicably changed Hayden's River Pima renderings of words into a dialect of Papago or Tohono O'odham.

Sometimes included with the actual Creation Epic is a long, complex story of twin boys who are sons of White-clay Eater Woman and their connection to two daughters of Oam Ñui, Yellow Buzzard (Benedict 2001:25, 97–122; Densmore 1929:54–80; Herzog [1929]). Because a reed flute plays a central role in much of the narrative, this story is often referred to as the Flute Story. According to Culver Cassa, it is a Yuman story that has been incorporated into the Piman canon. Ralph Cameron confirms that it is a Maricopa story. The full version of the story can be found in Spier 1933:367–396. Culver sees the Yellow Buzzard tradition as one of sadness and depression, but that of White-clay Eater Woman as one of happiness and the realization of potentials. He notes other polarities such as experience versus inexperience, older versus younger. "It's about us, men."

However, the story as told by the O'odham has a Piman overlay. Everything the boys do is cyclic. The boys experience a sense of futility (east) at each stage of the story, but the mother (west) reinforces the sense of opportunity. The boys discover their sense of virility and powers of procreation. (The Maricopa call her West Woman;

the Pima call her White-clay Eater Woman or sometimes Nasia.)

After a complex journey involving many birds and other critters, the girls find the boys. The union of the younger pair is consummated. Although temporarily there is laughter in their dwelling, Peregrine Falcon is called in by Oam Ñui to destroy the house and kill the boys (Herzog [1929]; see also Benedict 2001:82–87, 114.)

Incomplete or Fragmentary Documentary Sources

In addition to the major mythical sources that appear to be reasonably complete renditions of a storyteller's repertoire, other documentary texts include one or several individual episodes from the Origin Story canon. (It seems it was up to the individual narrator to establish his own tradition of what should be "canonical" stories, bounded only by what he could narrate within the four-night time frame.)

In 1774, Franciscan father Pedro Font (1931:34, 37–41) transcribed some fragments of episodes related by the governor of Uturituc (later known as Hejel Jek, Open Place). "It all reduces itself to fables, confusedly mixed with some Catholic truths," the priest remarked. The earnest Indian sensed the Spaniards' smirks and refused to continue the story the next day. The friar's attitude was poor soil for planting sacred myth, and Font's jumbled and confused fragments tell us little more than that certain stories already existed at this point in time. Still, they mention the hummingbird and Coyote as well as the dog that talked at the Flood.

The most distorted version of the Creation Story is that provided by Captain F. E. Grossmann in 1873 (407–410). Virtually every sentence contains statements at odds with authentically recorded versions. Whoever provided the narrative must have been an untrammeled prevaricator who understood Pima poorly. In addition to the facts being wildly incorrect, the narrative structure of the story is laced with such non-Piman concepts as prophets, kings, thrones, royal houses, and a reigning queen. Even the historical inclusions contradict other carefully transcribed accounts. As the Gila River agent from 1869 to 1871, Grossmann worked to improve the conditions of the Pima in the years when they were first faced with a lack of water (Wilson n.d.). However, his venture into mythology was not among his better achievements.

Historian Hubert Bancroft (1889:78–80) and archaeologist Adolph F. Bandelier (1892:462–463) also recorded Creation Story fragments. Bancroft's and Bandelier's renditions have distortions similar to those found in Grossmann's document. Bancroft credited his account to "five chiefs of that nation, and their very intelligent and obliging interpreter, Mr Walker, at San Francisco, in October 1873." Bandelier opened his account, "Mr J. D. Walker, an old resident of the vicinity of Casa Grande, who has been to me personally an excellent friend and valuable informant, told me this tale." Rather clearly, these three jumbled accounts (Bancroft 1889; Bandelier 1892; Grossmann 1873) all derive from the same source, who did more fabricating than translating.

In an anthology of myths, Charles Skinner (1896:2:215–218; see also Edmonds and Clark 1989:64–66) included an apocryphal Pima Flood Story but gave no indication of its source. It is so inauthentic that one might imagine it originating from Walker also.

About the time Russell and Lloyd were working on the Gila, the indefatigable American Indian photographer Edward S. Curtis visited the reservation and recorded five or six episodes (Curtis 1908:2:3–4, 14–23). These episodes are reasonably accurate summaries of the story themes, although not linear translations. There are some errors. For instance, Handsome Man, not the great shaman's daughter, gave birth to the baby that started the Flood. However, his work also includes two interesting avian references. Kuḍat, Flicker, is the boy who kicks the wooden ball that impregnates the mother of Ho'ok the monster, and Vishag, Peregrine Falcon, is the final manifestation of Ho'ok's spirit. Judging by his photos, Curtis seems to have worked in the Komaḍk region, a bastion of conservatism, and his accounts may reflect the school of storyteller Ko'oi Mo'o, Rattlesnake Head, later known as Joseph Head.

Almost simultaneously, archaeologist Jesse Walter Fewkes (1912:45–52), making excavations at Casa Grande ruins, enlisted the aid of Thin Leather for telling those stories relating to the local ruins. At least seven stories are given, six being among those that Thin Leather included in the (presumably full) canon he related to Lloyd. Fewkes's stories lack the careful sentence by sentence recording that Lloyd undertook, and his accounts sometimes stray from the Lloyd and Russell versions in significant ways. Comparing the three texts, one can easily spot where misinterpretations arose in translations. In fact, Fewkes (1912:42n.) acknowledged the problem at the time: "Some of the Pima told the writer that his interpreter had made mistakes in interpretation, so that what is give here can be regarded only as approximations to truth." Well put. Nevertheless, several avian reference are found only in Fewkes's renditions, including a hummingbird story.

Ruth Underhill (1939:10–11, 1946:8–12) offered two skeletal summaries of a Papago or Tohono O'odham version of the Creation Story. The first outlines just the origin, migration, and conquest episodes, but includes some points found nowhere else, to my knowledge. The summary in the second outline is somewhat more extensive.

Saxton and Saxton's *Legends and Lore of the Papago and Pima Indians* (1973) is a loose anthology of tales as well as episodes from the Creation Story based on various sources, nearly all Tohono O'odham. Most of the texts were collected by various workers earlier in the twentieth century. This anthology is valuable because each page is printed bilingually. Although the Saxtons are not strong in the English identifications of Piman folk taxa, the original can be checked in the parallel O'odham texts on the same pages.

J. Alden Mason (1921) obtained a version of the Tohono O'odham migration legend from Abraham Pablo of Santa Rosa village in 1919. The early part was recorded in O'odham, but the bulk of the text was transcribed in English through a translator. Only the English is published. Both the Eagle Man and the emergence stories are included together with the origins of Ho'ok and the Children's Shrine.

In the spring of 1935, Gertrude Hill (1940a, 1940b) collected several Papago stories from the village of Santa Rosa. These stories were told by the then-elderly Tohono O'odham Angelita Lopez and translated by a Pima woman. The short collection includes the story of Eagle Man and Ho'ok as well as the origin of tobacco and parrots (macaws).

Harold Bell Wright compiled an anthology of tales in a work called *Long Ago*

Told (Huh-Kew Ah-Kah): Legends of the Papago Indians (1929). Most of these tales are not from the Origin Myth and seem to have a more southern Papago tenor, suggestive perhaps of Hispanic influence. Wright (1929:viii, x) said that his role was only "to select, piece together, eliminate tiresome repetitions, harmonize confusing elements, and to make clear broken sentences and obscure construction." He further claimed, "I have added nothing. I have made no attempt to interpret." Still, in arranging these stories for English readers, the compiler introduced changes, probably more substantial than he acknowledged. The collection contains some interesting tales in which birds figure importantly. Where the birds could be identified, I have included these tales in the species accounts.

Leonora Curtin's (1949:133–136) River Pima ethnobotany includes several extremely attenuated stories, such as references to the roadrunner, the unidentified flat-headed bird *(komalk mo'okam)*, mockingbird, and nighthawk (given as "Whip-poor-will").

There are two small collections of stories by River Pima authors. The first, by George Webb (1959:90–115), contains some summaries of the primary episodes of the Origin Story. The erosion of traditional knowledge is already evident here (compared to the Russell, Lloyd, Benedict, Herzog, and Densmore documents). The second collection, by Anna Moore Shaw (1968), has been more strongly modified by the author to serve as Sunday school stories usually with a moral, making them more cute than authentic.

Ethnohistorian Bernard L. Fontana collected an important version of the Ho'ok (witch) story from Ascensio Anton of Cheedagĭ Wahia, or Pozo Verde, Sonora. The story was told in Papago, translated into English by Danny Lopez, a Tohono O'odham, and taken down by Fontana (1981:26–29). Fontana was also instrumental in obtaining a segment of the Tohono O'odham Creation Story from Frank Lopez (1980) with the assistance of Albert Alvarez and Ofelia Zepeda. It appears in Larry Evers's *The South Corner of Time* (1980:128–149). There are a few translation glitches (for example, *uus gaat*, "bow," became "gun"; *ho'idkam*, Desert Ironwood, metamorphosed into a cactus), but the segment reproduced is given in both languages.

Songs

Songs constitute another great body of Piman oral literature. They are short, tight poetic utterances, usually four lines long (sometimes three or five), almost haiku-like except for more words and more (rather than fewer) syllables than spoken Piman. Words are transformed to match the rhythm and beat of the music to such an extent that many speakers of O'odham, even some singers, are unable to discern the meanings of the songs. Take this example, first in song language, then in spoken (rather than sung) Pima, and finally in English:

> Sung Pima: kokomagi mumuval/ melonge hiosige tasho na gamhiyo ke noḍa
>
> Spoken Pima: koomagĭ muuval/melok hiosig todshag am ii k noḍa
>
> English: Gray fly/Ocotillo flower foam drinks and becomes dizzy.

In the first line, we see that song language has done more than merely extended vowels (as, for instance, in Gregorian chant), but has radically altered the structure of individual words as they would be spoken. For instance, *melhog* or *melok*, Ocotillo, has become *melonge*. Part of the beauty of the poetry comes from how the music, which we cannot appreciate here, and the modified word structure are correlated. The preeminent scholar in analyzing the structure, content, and function of Piman songs is Donald Bahr (Bahr and Haefer 1978; Bahr, Giff, and Havier 1979; Bahr, Paul, and Joseph 1997).

Cracking the nut of Piman sung poetry requires several steps before a final English version is reached. After the sung words have been written, they are reduced to spoken O'odham, and then the individual words are translated and put into meaningful English that, it is hoped, reflects some of the beauty of the original sung poetry. Culver Cassa worked out most of the songs in this book, even those originally published in a more complicated orthography by Russell (1908:272–338). Russell's more than two hundred songs were probably recorded by Jose Louis Brennan, a native O'odham speaker (Fontana 1975:xiv).

Most songs are in long series, sometimes called cycles, of up to four or five dozen individual sung poems. These series originate in dreams and are given by some animal power or other source over a period of time. (A few series come from other sources, such as the Jiosh Keli, God the Father; *hejel memḍdam*, automobile; and *chechpa'avĭ*, loose women.) Some song series that have been enumerated by Sylvester Matthias, Herb Narcia, and George Kyyitan include the following (those series marked with a dagger are apparently no longer extant; that is, the sequence is no longer among the repertoire of a current singer):

- *aaḍo*, Scarlet Macaw (Salt River)
- *ba'ag*, Golden Eagle (maybe Santa Rosa)
- *ba'ivchul*, kingfisher (Bapchule)
- *cheedagĭ baikam*, "blue tail–has" = young Western Whiptail Lizard
- *chechpa'avĭ*, loose women (common)
- *gigidval*, swallows (Salt River, Kaij Mek [Santa Rosa])
- *hejel memḍdam*, automobile
- *hevaichuḍ*, Blue Grosbeak (†?) (Santa Cruz Pima village?)
- *hikvig*, Gila Woodpecker (Komaḍk, Don [Daniel] Thomas)
- *jiawul, ñiawul*, devil (common)
- Jiosh Keli, Old Man God (Juan Manuel, Haashañ Keek [Sacaton Flats])
- *kahagam*, unidentified bird, possibly alternate of cuckoo (†)
- *ñepoḍ*, nighthawk (Blackwater)
- *okokoi*, White-winged Dove (Aangam)
- *shashañ*, Red-winged Blackbird (Adolph Juan, Komaḍk)
- *totoñ*, ant (Salt River)
- *totova*, turkeys (†)
- *vajukukĭ*, oriole (Roy B. Hill, then Blaine Pablo, Sacaton Flats)
- *vatopĭ*, fish (†)
- *vipismal*, hummingbird (Jiosh Keli [Juan Manuel], Sacaton Flats)
- *vak chechetopĭ*, dragonfly (U'usig village [abandoned] opposite Bapchule)

Songs sometimes exhibit what Bahr, Giff, and Havier (1979:247) call internal splitting—that is, the opposition of subject or action or both in paired verses of a song. We find this splitting often in songs about birds: eagle walks, falcon runs; eagle flies in a circle, falcon flies in a straight line.

Birds are important sources of Piman

sung poetry. The one who teaches the songs is a shadowy animal spirit helper, not an individual biological bird. This distinction probably parallels staying sickness and shamanism where ultimately some generic sense or spirit essence of the animal is offended and must be appeased rather than a specific physical manifestation of the animal itself (see chapter 5). The songs frequently are not even about the source animal or person. For instance, only one of the more than fifty recorded oriole songs actually images this bird (Bahr and associates 1997).

The one who received songs in dream encounters, for instance, from the kingfisher-person, would be called a *ba'ivchul namkam*, kingfisher meeter. Bahr and his associates (1997:xiii, 134–138) liken song acquisition to shamanic experiences. Once the recipient privately mastered a song, he could sing it publicly. If he practiced the song earlier where someone could overhear it, Joe Giff noted, it would vanish. Most songs in a series are historically associated with one key person who could teach his songs to others, particularly those who would accompany him in singing. A different person might occasionally dream a song in a series dreamed by another and add it to the original sequence, as in the case of the Sandpiper Song that Sylvester's father dreamed.

Many of the song sequences, in particular the longer ones, are sung for social dancing. Other songs are used in particular ceremonies. Of special interest in this book are the songs sung during a *wusota*, or healing session for curing sicknesses originating from specially endowed birds. Curing songs are more likely to feature the animal itself, alluding to its appearance and behavior. *Wusota* series are typically short, and some of the included songs may make no reference to the offended animal.

The songs recorded in this work usually depict a bird, telling a brief story about it or making a symbolic comparison:

> A black swallow
> Hangs far away
> And a black cloud
> Hangs there.

During the course of my fieldwork, Piman consultants offered songs as they discussed some species, just as they did with plant songs and mammal songs (Rea 1997, 1998b). Sometimes when a question arose as to whether a bird or other animal was a source of staying sickness, a song used in a *wusota* was given as evidence. Today many of these songs have been lost, so the ones generously given here are treasures. (Incidentally, for an excellent introduction to Native American song, see John Bierhorst's *A Cry from the Earth: Music of the North American Indians* [1979].)

Orations

Orations or ritual speeches are another form of Piman oral literature. Major sources of orations are Russell 1908:339–389 (ten speeches with Pima interlinear and free translations); Underhill 1946 (English only); Underhill and associates 1979 (a large bilingual anthology accompanying seven public native rites); Bahr 1975 (three bilingual variants of one oration); Lloyd 1911 (seven speeches given only in English); and Benedict 2001 (nearly twenty speeches, in English only). In a way, these speeches are much like psalms in that they are memorized and delivered verbatim, often in a liturgical setting. Unlike psalms, Piman orations were not intended for singing, but rather for public recitation.

Orations are difficult. George Kyyitan remarked of a Mockingbird Speech, "It's hard to understand. Pretty hard to understand. It's pretty hard to put it up [that is, to deliver or recite it]." This comment from a native speaker and singer! Lloyd and his translator struggled through Thin Leather's speeches, enthralled by their wildness and mysticism. The speaker identifications and references are often ambiguous, but these very ambiguities, notes Culver Cassa, are part of the aesthetics of this genre: they create puzzlement and demand focus of the hearer's attention. They are not casual.

The structure of a speech usually begins with an introductory section addressed to the listeners present, followed by a long body describing the journey and activities (and hoped for results) of some unidentified person and then by a brief conclusion again addressed to the company present: "And so my relatives . . . your worthy sons and daughters, knowing nothing of the starvation periods, have been happy. The old men and the old women will have their lives prolonged yet day after day by the possession of corn."

One group of orations describes the Piman, or Vupshkam, conquest of the Vipishaḍ, Short People, the original inhabitants of the middle Gila and lower Salt River valley. They contain considerable avian metaphor. Culver Cassa notes that these orations subsequently were used as war speeches, associating the People with the original conquest by the Emergenti, imaging what they hoped to accomplish on offensive expeditions into enemy country. As a form of magic, they describe in the past tense what they hope will be the outcome.

Underhill and associates (1979) give particular attention to the ceremonial or ritual context of the many orations they document. Here we are concerned mostly with the avian metaphors that are in the speeches. In fact, the orations are a major key to learning how a particular bird serves as Piman metaphor—for example, the bloodiness of macaw and the raininess of eagle—opening the way to understanding other literary genres in which the same species appears.

Lexicographic Works

Two important Piman dictionaries date from the colonial period. Both include various ornithological terms. The first is the Névome vocabulario, probably compiled during the 1660s at the desert Pima Bajo village of Ónavas, Sonora, by the Jesuit priest Baltasar de Loaysa. (This is the same pueblo where the last two speakers of the language, María Córdova and Pedro Estrella, lived.) The holographic document has been carefully edited and more conveniently rearranged by Campbell W. Pennington (1979).

The second is a vocabulario from Attil or Attí (Addi), a mission in northern Sonora—that is, Pima Alto country. It seems to have been compiled by 1774, perhaps the work of Franciscan father Antonio Benz. Less than half the leaves survive, preserved in the Bancroft Library (cited here as the Attí vocabulary [1774]). This work is particularly important because it is the only known lexicon of Upper Pima, and it was compiled before the palatalization of the language in the North (*ni, ne, nu* > *ñi, ñe, ñu; tu, te* > *chu, che; s* > *sh*, and so forth). This feature and the Contact placenames in the North show that Upper Pima phonology was little if at all different from Lower Pima until later in Spanish colonial times. The vocabulario is important also because it includes a number of folk taxa with their Spanish equivalents, which were sometimes more biologically precise than

English terms. Details are discussed in the individual species accounts, usually in the comparative linguistics section.

Several modern dictionaries of O'odham are available. Although they include folk taxa or the Piman names for plants and animals, they seldom explore or map the *domain* of a word with any precision. Saxton and Saxton (1969) and Saxton, Saxton, and Enos (1983) have produced works that focus essentially on Papago dialects. Many plant and animal folk taxa are included in these works, but their English identifications are sometimes weak, as already noted. The authors seem not to have consulted the works of biologists and ethnobiologists working in Pimería. Mathiot's (1973, n.d.) two-volume Papago dictionary is far more detailed. Her English glosses of plant and animal names reflect more of an effort to identify correctly the local desert biota.

The National Anthropological Archives at the Smithsonian Institution have several Piman word lists. There is an unpublished lexicon of River Pima gathered by F. Grossmann in 1877 (146 pages, typed, double columns). Rev. Charles H. Cook's lexicon is somewhat shorter, containing some 3,600 English to Pima words and 2,700 Pima to English entries. There is internal evidence that he and Grossmann, both from Germany, shared lists. At least some of their folk taxonomy is identical. Some of the entries shed light on how words were used at that period of Piman history. Two other early Arizona residents of the Gila River Reservation left shorter vocabularies: Ammi White (MS 625a-c) and John Walker (MS 623-b). These vocabularies contain four and two bird names each (see goose account).

Franciscan father Antonine Willenbrink served on the Gila River and Ak Chin Reservations in the early twentieth century. He was well known to the O'odham I worked with, who said he learned the language so well as to be indistinguishable from a native speaker. His *Notes on the Pima Indian Language* (1935) has word lists including more than thirty bird names, identified in English with variable accuracy. Curtin (1949:24) reproduced a key to his orthography in her Pima ethnobotany.

Archivist Alan Ferg drew my attention to various shoe and cigar boxes of word cards he uncovered at the Arizona State Museum. One had been assembled by William Kurath, an anthropology professor, before his death in March 1948. Another was compiled by linguist Kenneth L. Hale (n.d.b). Both sets give Papago words. A third set was by ethnographer Jean B. Johnson on Lower Pima, elicited at some unknown date. The words are from lowland Pima, not Mountain Pima (then included with Pima Bajo), as indicated by the name for chicken (*totor* rather than *supon* or *shupon* of the sierra). Each set contained some valuable nuggets, including some bird names.

Let me add a cautionary note. Although older O'odham, in particular those born early in the twentieth century, usually had an intimate knowledge of the local flora and fauna, they were understandably less familiar with the English names for these organisms. Consequently, certain misnomers have crept into English translations, such as *greasewood* for Creosote Bush, *prairie-dog* for one of the local small ground squirrels, and *nightingale* for the nighthawk. If the original O'odham word is given in the source, the English referent can be worked out.

Another recurrent problem is the use of a term in English that implies a broad domain in cases where the O'odham usage is actually more specific and precise. Examples are "owl" for *chukuḍ*, the Great Horned Owl; "hawk" for *vishag*, Prairie Falcon; and "deer" for *huai* or *huawĭ*, the Mule Deer, in distinction to *siikĭ*, the White-tailed Deer. In numerous cases, the O'odham were more careful biologists than those who wrote about them! Their precision is lost in these faulty translations.

Ethnographic Works

In addition to mythology and dictionaries, various ethnographic works capture, to a greater or lesser degree, aspects of O'odham ethnobiology. I continue to be amazed at the breadth and thoroughness of Frank Russell's work on the River Pima. This young Harvard Ph.D., in the terminal stages of tuberculosis, spent seven months at Sacaton beginning in 1901, dying shortly after. As Fontana (1975) noted, Russell had the talented help of a Papago field assistant, Jose Louis Brennan, virtually a coauthor, who compiled the massive sections on mythology, oration, and song that compose a significant portion of the text. I suspect that Russell's wife, whose name is mentioned only once as "Mrs Russell" in a footnote, played another key role in assembling this monumental work. Much of Russell's ethnobiological information is included in the sections "Plants Used for Food," "Medicinal Plants," "Animals Used for Food," and "Domestication of Animals" (1908:69–86).

During the first half of the twentieth century, Ruth Murray Underhill produced a series of in-depth works on the Papago or Tohono O'odham. They include *The Autobiography of a Papago Woman* (1936), *Singing for Power: The Song Magic of the Papago Indians of Southern Arizona* (1938), *Social Organization of the Papago Indians* (1939), and, most important for our understanding of ceremonialism and birds, *Papago Indian Religion* (1946). I have drawn extensively on these works. Underhill was not strong on biological identifications in English, and she (or her editors) unfortunately did not see fit to include O'odham words, which would facilitate the subsequent identifications of Piman folk taxa.

A considerable amount of information is to be gleaned from Castetter and Underhill's (1935) slim work on the ethnobiology of the Papago. Although its identifications of plants and animals are not always satisfactory, the accompanying O'odham names permit species verification. Another work from this period, *Pima and Papago Agriculture* (Castetter and Bell 1942), includes some passing references to birds.

Both birds and mammals are important in Piman shamanism and healing. The classic study of this subject area is that by Donald M. Bahr and his three O'odham associates, *Piman Shamanism and Staying Sickness (Ká:cim Múmkidag)* (1974). Although based on studies with a single shaman consultant, the work is solid. I have had the key points of this work repeatedly confirmed by both Akimel and Tohono O'odham who have been intimately involved in shamanism and healing. The caveat, as noted earlier, is that Bahr takes a cavalier attitude toward biology. When not misidentified outright, folk taxa that in O'odham have specific meanings and connotations for culturally significant reasons are often rendered in broad English categories such as "deer," "owl," and "rabbit." Bahr collaborated with Underhill in another important work, *Rainhouse and Ocean: Speeches for the Papago Year* (Underhill and associates 1979), an exposition of the Tohono O'odham ceremonial cycle.

The standard work on the Pima Bajo proper—that is, the lowland or Desert

Pima Bajo—is Campbell Pennington's *The Material Culture*, volume 1 of *The Pima Bajo of Central Sonora, Mexico* (1980). This work is written from the viewpoint of a geographer and economic botanist and is not intended as an ethnobiology. Nevertheless, it includes numerous folk taxa, although the bird life is often misidentified.

Jesuits, from the time of first contact until their expulsion from the New World in 1767, were in Tepiman country of northwest Mexico. Various ones who were stationed in Piman country wrote relaciones, or documents, describing the people among whom they worked. The "German Jesuits were better prepared scholastically and scientifically than most of their Spanish colleagues," notes Henry Beers (1979:362), hence their more extensive writings. Some relaciones contain verbal snapshots of hunting activities, bird keeping, and feather use. Others describe the country before the devastating effects of overgrazing and other anthropogenic abuses.

Fr. Luís Xavier Velarde was Fr. Eusebio Kino's successor in Pimería Alta. He wrote of the Sobaipuri Pima on the upper Río Santa Cruz in 1716 (Gonzáles 1977; L. Velarde 1931). Swiss father Philipp Segesser (1945) worked among Pimas in northern Sonora and adjacent southern Arizona (Guevavi [Ge Vavhia, Big Well] and Bac, both on the Santa Cruz). He penned a relación in 1737. Fr. Jacobo Sedelmayr (1939, 1955) left behind drafts written in 1744 and 1746 of his work and travels in Sonora, including firsthand descriptions of the middle Río Gila.

Juan Nentvig (1971, 1980) was stationed among Ópata (or Eudeve?) and Upper Pima after 1749. His *Descripción geográfica . . . de Sonora*, long known as the *Rudo ensayo*, was written in 1762. The Pradeau and Rasmussen translation (1980) is riddled with biological and other errors; Viveros's Spanish version is recommended. German father Joseph Och (1965) was at San Ignacio, an Upper Piman mission on the Río Magdalena (1756–1758). His translated relaciones appear as *Missionary in Sonora* (1965). Another contemporary German, Fr. Ignaz Pfefferkorn, was stationed at several Northern Pima missions, including Atí (1756–1762) and Ge Vavhia (1762). His book-length relación is available in English as *Sonora: A Description of the Province* (1949).

Some colonial Jesuits left rather scholarly grammars and dictionaries to aid in missionary activities. Those with word lists often include bird names in the indigenous languages correlated with Spanish. Two of these works on Pima are described earlier. The Italian father Benito Rinaldini (1743) worked among the Northern Tepehuán, principally at Nabogame of southern Chihuahua. His vocabulario was compiled between 1724 and around 1743 (see Pennington 1969:38–39 n. 57). It is the source of various Tepehuán cognates used in the comparative linguistics sections of this book.

Various other works deal with the Northern Pimans, but for the most part they do not touch on ethnoornithological matters.

4

An Overview of Avian Folk Taxonomy

Every culture has a folk classification system that tells how its members cognize the biota of their environment. Basic to an understanding of a people's classificatory system is the concept of folk taxonomic *rank* (Berlin 1976). A name is not just a name; it has one or more positions in a conceptual hierarchy. For the most part, we learn about classification (either during our own childhood or later while studying another language) through encounters with folk generics: those names that attempt to partition the discreet realities perceived in the biological world. These names are the units that people talk about mostly. Sometimes these units are subdivided into folk species and (much more rarely) folk varieties. Folk generics are generally gathered into a few named or unnamed categories that are called *life forms*. The number of life-form categories varies, but, among animals, "birds" is often one named category. Finally, in this shallow taxonomic tree, these life forms may be gathered into one of several kingdoms, generally plant or animal, although these kingdoms are unlabeled in many languages. Kingdoms are also called "unique beginners." Rank, then, is the position in the hierarchy at which a name is used.

By looking at large numbers of languages in the world, representing different subsistence levels, ethnobiologists have discovered general patterns in the evolution of the folk biological hierarchy (C. Brown 1984). The encoding sequence in a language begins with folk generics, the most basic perceptual units of nature. These generics are followed in a rather predictable sequence by folk specifics and varieties (if any), by life forms, and finally by unique beginners (Berlin 1972). In looking at how Northern Pimans name birds, we consider these categories in this chapter in the order in which they appear developmentally rather than hierarchically.

In an ethnobiology, we must keep in mind that we are comparing two different systems of classification. Almost any study of a culture's folk science must first investigate its folk taxonomy. This system is compared with a so-called scientific or biological classification as an etic grid (Hunn 1975a). This Western taxonomy is based on presumed evolutionary (phylogenetic) relationships, as best as they can be inferred by morphology and other tools. One goal of Western taxonomy is to label all biological organisms universally by means of the Linnaean binomial system. In Linnaean taxonomy, every known organism is named by genus and species. The system is also hierarchical, every organism being assignable to a kingdom, phylum, class, order, family, genus, and species.

A folk taxonomic system, in contrast, is local, selectively naming at some level the organisms in the immediate environment, both native and introduced. It is nonuniversal and nonevolutionary. It is constrained by such factors as the size of the organism and its cultural significance or salience. The native speaker may have neither the need nor the equipment to distinguish species of sprawling spurges or kinds of white-footed mice. However, both folk and scientific taxonomy are based on perceptual discontinuities in the natural world: prickly-pears are not chollas, cotton rats are not packrats, and Northern Harriers are not Red-tailed Hawks.

Biologists untrained in folk science are often content with merely collecting terminal taxa from some language community. Their implicit goal is to discover how well native speakers' taxonomy measures up to "scientific" or Linnaean taxonomy. To do so, they usually compare terminal taxa. A terminal taxon is the final label that any particular biological discontinuity may be called in a lexicon. It is not further subdivided. But cognitively the result is a mixed bag of variable rank: most labels are typically folk generics, but also included in a collection are names for any folk specifics and folk varietals that the language encodes. These names in themselves do not tell us how the people organize their perception of nature. A native system is arranged hierarchically, as is a Western or evolutionary system.

Just as it is essential for the biologist or ecologist conducting any investigation to work out the Linnaean taxonomy of the organisms treated in a scientific study, it is likewise necessary to correlate a culture's folk taxonomy with Western taxonomy in order to appreciate either that culture's traditional knowledge or its metaphors. There are numerous examples of Piman song, myth, and other genre suffering at the hands of clumsy or lazy translators.

The Folk Genus

All cultures have the folk genus as the most numerous category in the hierarchical system that classifies the plant and animal world.[1] These categories are the ones most readily correlated with the discontinuities seen in nature. They are the categories first learned by children of small-scale societies. They are the most frequently spoken about and named during elicitations. Folk genera generally exhaust the recognized organisms of the local biota (Berlin 1992:52–101; Berlin, Breedlove, and Raven 1973, 1974; Brown 1984; Hunn 1977; N. Turner 1988). Appendix A lists the most frequently recognized Northern Piman bird names alphabetically rather than hierarchically.

At first sight, terms for most Piman folk generics of birds appear to be nonanalyzable; that is, they consist of a single word or lexeme incapable of being broken down etymologically. Examples of nonanalyzable lexemes in English are *pine, oak*, and *eagle*, and in Piman *huk, toa*, and *ba'ag*. Among some 230 River Pima plant names, there is a core group of about 60 primary nonanalyzable lexemes. Partially analyzable names contain a morpheme (a minimal meaningful unit) that can be factored out. An example is *vakoañĭ*, heron/egret; *va* is a water-marker particle found in the names of many aquatic or riparian organisms (*vaamog*, mosquito; *vavukĭ*, raccoon; *vak*, bulrush). Although nonanalyzable names are considered to be the oldest in a language group, partially analyzable lexemes are probably quite ancient as well and for the most part are not consciously recognized as compound terms by native speakers. For instance, of thirty-two Northern Piman folk generic names of mammals, twenty-eight are nonanalyzable, two apparently are partially analyzable (bat, Desert Bighorn), and two are analyzable (Round-tailed Ground Squirrel, kangaroo rat) (see Rea 1998b). Of these names, only the latter two are regularly glossed by native speakers, perhaps because they are based on verbs (to stand upright, to sit down) regularly used in modern speech.

In contrast to these two categories (nonanalyzable and partially analyzable), analyzable names are usually descriptive, and the components are recognized by native speakers. Many names that the Pima coined for post-Contact plant acquisitions are descriptive. A few examples are *chuchul maamaḍ*, "chicken babies" (pear); *keli baasho*, "old man's chest" (casaba melon); *s-vepegĭ iibadkam*, "red fruit–it-has" (strawberry), and *gogoks viipiḍ*, "dogs' testicles" (plums, prunes). (See Rea 1997:392–399 for a more detailed discussion of Piman plant neologisms.) An example of an analyzable bird folk generic is *baasho s-vegĭ*, "chest red," or robin (see table 4.1).

Because of their antiquity, nonanalyzable and partially analyzable folk generics may provide clues to a language's area of origin (Fowler 1972, 1983; Friedrich 1970). However, descriptive names, such as *ba-hidaj u'uhig*, "Saguaro fruit bird," House Finch, are more likely younger additions to a lexicon, labeling biological species more recently encountered or acquired. Folk specifics are likely to be more recent lexical acquisitions as well.

Pan-Piman Folk Generics

Many plant, mammal, and bird names are broadly distributed throughout Tepiman country, but how extensively is often difficult to determine. Accurate plotting of these pan-Piman names is hampered by the general lack of folk taxonomic data from the southern half of the region—that is, among the various dialects of Northern and Southern Tepehuán. The few names known from these southern groups have been gleaned from oral narratives or from studies of linguistics and material culture. We can compare folk generic names with more confidence between the desert Pimans (that is lowland Pima Bajo, River Pima, and Tohono O'odham) and Mountain Pima (table 4.2).

In spite of the incomplete data, we still know that a number of bird names are definitely pan-Tepiman; that is, they occur both north and south of the Taracahitan divide. These folk generics include vulture, heron/egret, hummingbird, macaw, roadrunner, kingfisher, turkey, Harris' Hawk, large falcon, kestrel, eagle, raven, swallow, blackbird, and mockingbird. An additional dozen northern Piman folk generics may be pan-Tepiman as well, but data are lacking from Tepehuán dialects. And due to referent shifts between Northern and Southern Pimans, a name may label different organisms in different speech communities.

Of these pan-Tepiman names, that of the raven is curious. The *kokon* term is used by all groups except those inhabiting Pimería Alta—that is, the Akimel and Tohono O'odham, who call the bird *havañĭ*. However, the widely used southern name, according to Culver Cassa, appears to be preserved in a character's name, Kokoñip, in a northern mythic episode (see raven species account).

Another group of bird names is at least pan-Piman—that is, occurring throughout Northern Piman country, including Mountain Pima country. This group includes the names for screech-owl, coot, and flicker. Pima Bajo, unlike the River Pima (Gileños) in the north, seem not to distinguish coots from ducks at the folk generic level. The

Table 4.1 O'odham Avian Folk Generics

aaḍo (macaw, Scarlet, Military) N
ba'ag (Golden Eagle) N
baasho s-vegĭ (Northern Robin) D
bahidaj u'uhig (House Finch) D
ba'i chukul (Horned Lark) D
ba'ivchul (Belted Kingfisher) N
bichpoḍ (Abert's Towhee; Canyon Towhee) D
biit keishnam (Bendire's Thrasher) D
chechlogam (*unid. bird, mythical or real?) N?
chehegam (Ladder-backed Woodpecker) N
chehopĭ, also *vuhigam* (Common Ground-Dove) O
chev chiñkam, also *keli huch*, syn (Crissal Thrasher) D
chevel u'uhig (Black-necked Stilt) D
chivi-chúuch, sivi-chúuch, piwi-chúuch (Killdeer) N
chukuḍ (Great Horned Owl) N
chukukmal (Western Kingbird) N
duuk vaidam (Pima Bajo) (Coppery-tailed Trogon) D?
ee'eḍ vahuḍam (Barn Owl) D
gihodag (Band-tailed Pigeon) O?
giidval, pl. *gigidval* (swallows, generic) N
giisobĭ (Verdin) N
go'ogov (Logger-head Shrike) N
gugu (Inca Dove) O
haiñ uushab, haiñ julshab (*unid. waterbird, probably American Avocet) D
haupal (Red-tailed Hawk) N or ±
havañ (raven), N
heet kawuḍk (Summer Tanager) D
hevachuḍ, hevaichuḍ (also var.) *heewasjel* (Blue Grosbeak) N
hevel moos (Say's Phoebe; Ash-throated Flycatcher) D
hialak (goose; Snow Goose) O
hikvig (Gila Woodpecker) N
hokkaḍ, hogkaḍ (Cactus Wren) N
hoohi (Mourning Dove) O
jujupul (*unid. waterbird) N
kaachkĭ hugidag, (also) *kaachkĭ baasho u'uhig*, syn (gull) D
kadgam (Yellow-billed Cuckoo) N
kakaichu, kakachu (Gambel's Quail) O
kokoho (Burrowing Owl) O
kokon, koko'on (raven [Pima Bajo]) N
kookoḍ (Sandhill Crane) N
koologam (Nuttall's Poorwill) N (?)
koomagĭ u'uhig (Le Conte's Thrasher) D
kuḍvich, kuḍvik (Curve-billed Thrasher) N
kuḍat (Northern Flicker) N
kuigam (Phainopepla) D
kuukvul, kuukul (Western Screech-Owl) O (or ±)
kulisgam, kukulisgam (*unid. waterbird, avocet or White-faced Ibis?) N
kusijim (Crested Caracara) N
ñepoḍ (Lesser Nighthawk) D
ñui (Turkey Vulture) N
oám u'uhig (*unid. "yellow bird"; Yellow-breasted Chat?] D
okokoi (White-winged Dove) O
o'oḍopiwua, (also) *kauk oo'okam, chechetopiwua*, syn (Lark Bunting) D
shadav, shalav (Black-throated Magpie-Jay [Pima Bajo]) N?
shashañ (Red-winged Blackbird; all blackbirds?) N
shuudagĭ maamaḍ (sandpipers—Least, Western) D
shuug (Northern Mockingbird) N
sipok (Northern Cardinal) N
sisikĭ (American Kestrel) N
taḍai (Greater Roadrunner) ±
tamtol (White-crowned Sparrow) N
tavir (parrot, Amazona? [Pima Bajo]) N (also *s-ñeokdam u'uhig* D)
tobavĭ (Harris' Hawk) N(?)
tosiv (Western Meadowlark) N
tova (Common Turkey) N
unmu'uhig, u'unam u'uhig (*unid. jay or dream bird?) D
vachpik (American Coot) D
vajukukĭ, vajakukĭ (Bullock's Oriole) N
vakav (Northern Harrier, Marsh Hawk) N (also *s-heepĭ wuadam* syn D)
vakoañ (heron/egret) ±
vapkag, vapka'ig (ducks [and geese?]) ±
vavas (Rock [and Canyon?] Wren) ±
vipismal (hummingbird–Costa's, Black-chinned mostly) ±
visag, vishag (Prairie Falcon) N

N = nonanalyzable lexeme
± = partially analyzable lexeme
O = onomatopoeic name
D = descriptive (analyzable) lexeme
syn = folk synonym
*unid. = unidentified, see appendix F, unidentified birds

Gileños, living at the marshy confluence of two or three desert streams and various intermittent washes, not surprisingly once had an extensive taxonomy of waterbirds now largely forgotten (see appendix F, unidentified birds).

A large group of bird names was common at least to the desert Pimans (Tohono and Akimel O'odham) and lowland Pima Bajo, although seemingly unknown farther south. In addition to the folk generics already treated, they include crane, White-winged Dove, Mourning Dove, Common Ground-Dove, duck, quail, Red-tailed Hawk, Gila Woodpecker, flicker, and nighthawk. Any Northern Piman speaker, encountering someone from another O'odham dialect throughout a three-hundred-mile north-to-south range would have little trouble discussing the local avifauna. The list was probably once far more extensive than we can now document. But even with linguistic erosion, speakers throughout this area shared more than two dozen bird names.

Once again, folk taxonomic studies among Mountain Pima and especially of the various dialects of Tepehuán should round out the picture of pan-Piman or pan-Tepiman names considerably.

Is there a southern bias in bird names? Do the nonanalyzable and only partially analyzable folk genera provide any clues to an ancestral home for Tepimans? The known pan-Piman avian folk genera (table 4.2) are geographically and altitudinally widespread species that have broad ecological tolerances. In this regard, bird names are unlike plant names, which suggest that the evolution of proto-Tepiman occurred in an ecotone between desert thornscrub and foothill habitats rather than in the low desert itself. Stated differently, presumed archaic plant names indicate a non–Sonoran Desert origin for Pimans, whereas bird names do not point to any specific elevational or geographic origin.

Folk Species of Birds

As noted, the folk genus is the basic unit in folk taxonomies and language. English examples are gull, hawk, rat, deer, fly, oak,

Table 4.2 Pan-Piman Avian Folk Generics

Biological Name	O'odham	Pima Bajo	Mountain Pima	Northern Tepehuán	Southern Tepehuán (Tepecano)
vulture	*ñui*	*nui*	*nuli*	*nuí*	*nuuí*
hummingbird	*vipismal*	*vipsumaal*	*vipish, vipishi*	*vipishi*	?
macaw	*aaḍo*	N: *arho, varo*	*varo*	?	*vaar*
crane	*kookoḍ*	N: *kokoḍ*	?	?	?
heron/egret	*vakoañ*	N: *vakoani*	*vakuun [vor, vora]**	*vakoni*	*vakoon*
White-winged Dove	*okokoi*	*okokoi*	?	?	?
Mourning Dove	*hoohi*	*hoohi*	*[vohokoli?]*	?	?
Common Ground-Dove	*chehopĭ*	*tehop*	?	?	?
coot	*vachpik*	*vakaik?*	*vaki* *vakiig*	?	?
duck	*vakag* *vapkaig*	*vakaik*	?	?	?
roadrunner	*taḍai*	*taḍa*	*tarpui, tepuí*	*taragai*	?
kingfisher	*ba'ivchul*	?	*baivkuli*	*baivkali*	?
quail	*kakaichu*	*kakatu*	*[tesoli, tesholi]**	*[vaivoli]*	*[súrui]*
turkey	*tova*	*tovo*	*tova, tovo, tov*	*tova*	*[pipir]*
eagle	*ba'ag*	*ba'aag*	*ba'aga*	*baagai*+	*ba'aa*+
Harris' Hawk	*tobavĭ*	N: *tobavi*+	*[tob, tobi]*+	*tobavi*+	?
Red-tailed Hawk	*haupal*	N: *auparhi*	?	?	?
large falcon	*vishag* *visag*	*visaag*	*vishagi, viishuki*+	?	*pichaag*+
kestrel	*sisikĭ*	N: *sisika*	?	*shishikamu li*+	?
screech-owl	*kuukvul*	*tukul, tuukul*	*kukvulyi*	*kukuvuli*	*tekkur*+
Great Horned Owl	*chukuḍ*	*tukuḍ*	*tukur, tukura*	*tukurai*	*tukur,* *tukuur*+
flicker	*kuḍat*	*dugdat*	*kurta*	*kuratu*	?
Gila Woodpecker	*hikvig*	*hiikvig*	?	?	?
Ladder-backed Woodpecker	*chehegam*	*tehegam*	?	?	?
nighthawk	*ñepoḍ*	*nepoḍ*+	?	?	?
raven	*[havañ]**	*kokon*	*kokon, kokoñi*	*kokoñi*	*kokoon*
mockingbird	*shuug*	*shuug*	?	?	*suug*
swallow	*giidval,* [pl.] *gigidval*	*giigdoval*	*gigikolyi*	*pipidamuli* (?)	*gidaawal*
blackbird	*shashañĭ*	*shashan*	*shashan, shashani*	?	?

Note: See individual species accounts for details and variants, including colonial forms.
N = colonial Névome (modern Pima Bajo unknown)
+ = name cognate but biological referent different or uncertain
[ ]* = noncognate form

pine, and fir. These units are often non-analyzable lexemes, as those just enumerated. But not always. Mosquito, flycatcher, nighthawk, and pineapple are folk generics that are analyzable. Folk genera often correspond to biological genera.

The folk genus may be further subdivided into folk species. English examples are Western Gull, White-tailed Deer, Ponderosa Pine, and Monarch Butterfly. The Piman classification of plants has few examples of folk species, and these examples, as to be expected, are mostly among cultivated plants. A notable exception among wild plants is wolfberry *(Lycium)*, where Akimel O'odham regularly distinguished four folk species, three of them marked. (Their country harbors perhaps the highest species diversity of wolfberries in the world, not all of which are distinguished by names in Pima.) Only two folk species occur among mammals: *toha chuuvĭ* for the Antelope Jackrabbit and *chuk chuuvĭ* for the Black-tailed Jackrabbit.

In contrast to plants and mammals, O'odham distinguish a number of birds at the folk-specific rank. Of sixty-eight folk generics among Upper Pimans (Akimel and Tohono O'odham), at least seven or eight (10.3–11.8 percent) are polytypic—that is, divided into two or more folk species (table 4.3). This calculation discounts folk synonyms and alternate names, discussed separately later.

With Piman birds, in each case except one the unmarked folk generic also serves at a lower level of contrast as the folk specific (called the type specific). A parallel example in Western geography would be Baja California, referring to the entire peninsula at one level, but the state of Baja California contrasting with the state of Baja California Sur. (There is no state Baja California Norte just as there is no state East Virginia.) In the O'odham avifaunal lexicon, the excep-

Table 4.3 Avian Folk Species among Upper Pimans

Folk Genus	Type-Specific Folk Species	Biological Species
ñui (vulture)	*ñui*	Turkey Vulture
	ñuipa	Black Vulture
vakoañ (heron)	*toha vakoañ*	egrets
	koomagĭ vakoañ	Great Blue Heron
	tash cheenam vakoañ	Green Heron
	chuággiakam vakoañ	Brown Pelican
ba'ag (eagle)	*ba'ag*	Golden Eagle
	vakoañ ba'ag	Osprey
visag, vishag (falcon)	*visag, vishag*	Prairie Falcon
	cheedagĭ visag, cheedagĭ vishag	Peregrine Falcon
kakaichu (quail)	*kakaichu*	Gambel's Quail
	s-hadam kakaichu	Scaled Quail
	olas (o'ok) kakaichu	Mearns' Quail
chukuḍ (large owl)	*chukuḍ*	Great Horned Owl
	chukuḍ ka'amaḍ	Long-eared Owl
giisobĭ (Verdin)	*giisobĭ*	Verdin
	s-chuk mo'okam giisobĭ	Black-tailed Gnatcatcher
vavas (small wren)	*(koomagĭ?) vavas*	Rock Wren
	(?)*oam vavas*	Canyon Wren

tion is "heron/egret," where the unmarked *vakoañ* stands only for the generic, and all species are obligatorily marked. There is no focal species (type specific) of heron that is understood in an unmodified context. This situation is like that for mammals, where both Piman species of jackrabbits are marked. With the vulture, eagle, falcon, quail, large owl, and Verdin, the overwhelming focus is on the species that is unmarked. Few native speakers born after around 1920 recognize the marked forms; these forms have essentially vanished from the lexicon.

Possibly two species of small wren were marked: one being the *oam vavas*, "yellow/brown wren," that is, the Canyon Wren; the other the *koomagĭ vavas* or *s-chehedagĭ vavas*, "gray (or) blue wren," indicating the Rock Wren. Certain songs make this distinction. But the distinction in general folk taxonomy needs to be verified with someone who knows both species. So far I have not found anyone who does.

Marking sometimes serves to add a new linguistic item to the native lexicon, and with time the marking may shift from the new entity to the originally unmarked term as cultural salience shifts to the new object or species, a linguistic phenomenon called *marking reversal* (see Berlin 1972:82–83; Witkowski and Brown 1983). For instance, the Banana Yucca growing in Desert Piman country was originally called *hovij*. When bananas were encountered, they became known to the Tohono O'odham as *hovij chuigam* (Banana Yucca it-looks-like), and the native plant was called *O'odham hovij*, the People's Banana Yucca). As market-bought bananas became increasingly available, they came to be called just *hovij* in contrast to the now marked *O'odham hovij*, the low-salience native species.

I know of no examples of marking reversal among Piman names for birds. However, there is one case where the biological referent of a folk generic has completely shifted: *aaḍo*, originally the Scarlet Macaw, is now used for the domestic peafowl (a green bird), the macaw no longer being a bird in general Piman consciousness. So far as is known, no marking was ever involved. This may be a simple referent shift.

There is a possible example of a referent shift in the Piman name for chicken, *chuchul* (which would be *tutul* outside of Pimería Alta). The word may once have been used for a native gallinaceous bird, the chachalaca, and been subject to a marking reversal in post-Contact times. This needs to be investigated among Northern and Southern Tepehuán dialects, where the native chickenlike bird occurs. Just beyond the edge of the chachalaca range, the people of Ónavas call the chicken *totol*, with roots in the reconstructed Proto-Aztec **tootol*,[2] turkey (Fowler 1983:232). At the extreme southern end of the Tepiman corridor, Mason (1917:328, 330, 333) found *tovwa, tovua* being used for "hen" (apparently the chicken), but *pippir* for "turkey." Presumably this latter term is a referent shift from the native to the introduced domestic bird, but where *pippir* comes from is unknown to me. It sounds onomatopoeic. At any rate, *totol, tutul*, and *chuchul* are cognates, and all refer to some galliform species.

Folk Synonyms

Several birds among the River Pima have two or more names, or what I have called *folk synonyms*. These names do not appear to be peculiar to different dialects, as seen occasionally between Tohono O'odham groups. Or if they once were, they have since been learned and used across dialectical boundaries. Nor are they merely variants of the same word, as we see, for instance, in the name for Killdeer *(chivichúuch, sivi-chúuch, piwi-chúuch)*.

These words are not examples of overdifferentiation—that is, of giving separate names to different sexes or age groups of an organism. I know of but one apparent case of true overclassification among Upper Pimans: the juvenile and adult whiptail lizard are known by different names.

The Crissal Thrasher is known both as *chev chiñkam*, "long bill-it-has," and *keli huch*, "old man's nail," descriptive names. I found both names in use among Kuiva O'odham (Downstream People) and some Ta'i O'odham. The first is perhaps the more frequently used. I doubt this is a case of a collapsing taxonomy, with names formerly assigned to several different species being telescoped to the more common species. There are four common thrashers breeding on the Gila and five folk names. The most elusive of these thrashers, the LeConte's Thrasher, has a name I have tentatively confirmed.

The Common Ground-Dove has two legitimate synonyms in O'odham, both known on the Gila, *chehopĭ* and *vuhigam*. Various individuals I worked with knew both names for the birds. These synonyms are traceable to early colonial times in the south, and both seem to be onomatopoeic (see discussion in species account). The Mearns' Quail, found in the oak woodlands of the Baboquívari Mountains, is known to some as *olas kakaichu* and to others as *o'ok kakaichu*. Its restricted range within the desert might explain why. Different language communities encountering it gave it different names.

The Lark Bunting seems to have the most names in the Akimel O'odham lexicon: *o'oḍopiwua, s-banmakam, kauk (kavk) oo'okam*, and *chechetopiwua*. Some speakers used only one of these names, whereas others (probably most) were familiar with two or more of the synonyms. It is perhaps no surprise that this small bird was an important winter food resource, passing through Pima country most abundantly during the spring hunger period. Once wheat was introduced as a winter crop, spilled grain became an important attractant to itinerant flocks of buntings. (The equally or perhaps more important small and easily trapped bird in the dietary, the White-crowned Sparrow, has but a single O'odham name, however.)

The two raven names—*kokon* in the south, *havañ* in the north—are not strictly folk synonyms in that they do not occur within the same speech community. At least among modern speakers, one name replaces the other geographically. Eugene Hunn (personal communication, ca. 2003) suggests calling this case an *allotropic synonym*, while calling a true folk synonym (one within the competence of a single speaker or community) a *sympatric synonym*. These terms are borrowed from biogeography.

The folk generic *hialak* is also something of a mystery. It has been called "goose," perhaps a Snow Goose. Its relationship to *vapkaik*, duck, remains unresolved.

A number of birds are known by names that are not really their formal names, but rather widely known alternate or poetic names. For instance, the Northern Harrier has two Piman names. *Vakav*, a nonanalyzable lexeme, is its real name. It may also be called *s-heepĭ wuadam*, a reference to its metaphorical role of bringing down the cold from the north in the fall and spreading it over the Pima fields.

The loudest and most widespread species of thrasher in Pima country, the Curve-billed Thrasher, called *kuḍvich*, is also known as the Soft-Feather-Nested Shaman (Viigĭ Koshkam Maakai) in the Flood episode, but this mythic label seems never to be used as its ordinary species name.

The name *s-banmakam* for the Lark Bunting seems to be an alternate or poetic name for a bird already well supplied with folk synonyms. The term *s-banmakam* might also be applied to a greedy dog or even a human. However, in most contexts older speakers would probably understand it as referring to this bird.

The four gull names—*kaachki-tab u'uhig, kaachkĭ hugidag, kaachkĭ baasho u'uhig, shuudagĭ baasho u'uhig*—are descriptive of a bird found at the shore. Trips to the Gulf of California to gather salt were abandoned generations ago, so the People no longer have much opportunity to see gulls. The first three names seem to be the best Sylvester Matthias could remember. Only a few people I worked with knew names for gulls.

Unique Beginner (Kingdom)

O'odham seems to be a more advanced language in that it has both an inclusive plant term and an animal term, categories that tend to be encoded rather late in linguistic development (Berlin 1992:20–31, 190–195; C. Brown 1979, 1981, 1984). The term *ha'ichu vuushdag*, literally "things that grow up/emerge" for River Pima (*ha'ich mo vuushañ* among Santa Rosa Papago, fide Mathiot 1962, 1964), includes most of what botanists would call plants, both wild and cultivated, but apparently excludes mosses, liverworts, lichens, and of course algae (see diagram of Sylvester Matthias's arrangement in Rea 1997:88). *Ha'ichu doakam* is the reciprocal counterpart for animals, or "things with life" (from *doak*, "is alive"). Joseph Giff said it was "something animal living," succinctly summarizing it as "all moving things with life." George Kyyitan said the same. But *ha'ichu s-doajkam*, Joe said, "means 'something wild,' that is, untamed, afraid of humans, shy. But we don't use this word much." It is not a contrastive term, but a characteristic of certain *ha'ichu doakam*. Mathiot (1962) notes that items in the plant category are mostly aggregate or mixed aggregate-individual nouns, never individual nouns, whereas animate items are more often individual nouns.

But would *ha'ichu doakam* include O'odham or humans? "Naw," said George, "because has different ways of doing things." Tohono O'odham Frank Jim agreed. With Pima Bajo at Ónavas, I was told that the concept of *doákam*, "animal," includes fish, mammals, birds, bats, and domesticated animals, but not people.

Mathiot (1962:158), working with the Totoguañ dialect of Papago, gave the following breakdown, which clarifies a seeming contradiction:

ha'ichu doakam (living things)*
  hemajkam (people)
  u'uhig (birds)
  ha'ichu doakam ("animals")*

Here we see the asterisked term used polysemously, first to include all animate beings, then at another level for "animals" in contrast to people and even birds. Pilcher (1967) uses *ha'ichu doakam* inclusively for plants and animals.

The unique beginner for animals is old within the Piman language. It was attested in the 1660s in Névome Pima: "[A] todo [ser] viviente [le llaman] *doacama* [*doakam(a)*]" (All living things are called *doakam[a]*) (Pennington 1979:120). So we know that the term *doakam* is not a response of modern speakers trying to accommodate the European concept labeled by the word *animal*. (There seems to have been no corresponding word for plants, or at least none was noted in the colonial document.)

U'uhig as Life Form

In the development of languages, an inclusive "bird" word tends to be one of the early life forms linguistically encoded into a culture's lexicon (C. Brown 1979, 1982, 1984). For O'odham, this word is *u'uhig* or *u'uvig*.[3] One writer studying Piman folk taxonomy characterized *u'uhig* as animals that can fly, therefore eliminating the chicken (Pilcher 1967). However, people I worked with emphasized instead the characteristic of having feathers. By way of illustration, Joe Giff explained, "A *chuchul* [chicken] is still *u'uhig*, but generally you don't call it that, rather *chuchul*. Just like you don't call a *ñui* [vulture] an *u'uhig* very often, even though it *is*." He explained that adding the word *u'uhig* to a bird's actual name occurred mostly in song, where more syllables for line extension were needed.

Sylvester Matthias gave this illustration: "There used to be an Ostrich farm east of Tolleson on Fifty-first [Avenue, in Phoenix]. They [Pima men] saw it when they went *kuushad* or *kuunam* [on communal mounted hunts or fire drives]. They called those Ostriches *ge'echu u'uhig* [literally 'big/huge birds']. *U'uhig* includes anything with feathers, including geese, roadrunner, turkey, and *chuchul*." His examples are instructive. Although the roadrunner can fly, it does so laboriously, not high and seldom very far (although I once watched one in Pima Bajo country sail hundreds of feet down a barranca slope, then give its characteristic call upon landing). When Anglo farmers brought Ostriches *(Struthio camelus)* to the Phoenix region in the 1890s, the Pima had no trouble assigning these flightless creatures to the life form "bird." (For more on this Ostrich endeavor, which even some Sacaton Pimas got into, see Balfour 1984; Barney 1955; Colley 1979; and Murray 1939.)

The Anomalous Bat

The bat poses a problem in many folk taxonomies of the world because it seems to combine traits of both birds and mammals (Hunn 1977:288–289). One day in our studies, I asked Sylvester Matthias, "How about *nanakmel*, bat? Did you say it was or wasn't in the bird family?" He typically referred to higher groupings of plants or animals as "families." He became quite animated as he thought about this problem. "Yeah! That's another question. Yeah . . . a bat. . . . I wouldn't describe [categorize] that, because I never heard. It *flies* like a bird. All we know is [it is] *nanakmel*." Another time he said he wouldn't include *nanakmel* in *u'uhig*, "bird," but said it could be in *ha'ichu doakam*, the next higher category, "animal." And still another time Sylvester wondered aloud if the bat might metamorphose like a butterfly from some nonflying form. His criterion for inclusion in the life form "birds" was possession of feathers, regardless of ability to fly.

George Kyyitan, however, said *nanakmel* has wings, so it's an *u'uhig*, although he had heard it is related to *nanhagiu*, mice. Later, when sorting mammals into related groups, he commented on the bat that "it's half *u'uhig*, half *nahagiu* [mouse]" (see resulting figure 2.1 in Rea 1998b).

Luciano Noriega of Quitovac was more definitive. A bat flew over us at the oasis. Luciano called it *nanakam*. I asked if it was a form of bird or what. He replied that it was not a form of *u'uhig*, but like a ratón, mouse, because its mouth, ears, and hands are like a those of a ratón. His criteria are morphological, having nothing to do with flight. I didn't ask how feathers might fit into his definition.

I asked Theodore Peters his opinion. "*Nanakmel* comes from the mice—face almost like mice, *nanhagiu*. It lives in a hole in the mountain."

"How about *chuchul* [chicken]?" I asked.

"*Chuchul* is an *u'uhig* because it has feathers."

"How would you classify angels?"

"*Daam-kaachim o'odham*. They got wings, too, but they're not *u'uhig*." The word he used means "heaven people," heaven being compounded of two words meaning "above, spreading out" (Mathiot 1973:238).

Likewise I asked Jose Enriques, "What is a *nanakmel?*"

"It's a bat."

"Is it an *u'uhig?*"

"No."

"What are its relatives?"

"I don't know. It looks like rats [mice], *nanhagiu*, with a tail and [he motioned to indicate arms and membranes], like Batman. And *fur*."

"What about *chuchul?*"

"It's a bird. I don't know why he doesn't fly, though, but it's got the same wings."

At Vaya Chiñ, I posed the same question about the bat. Joe Lopez said, "I don't know. It's not like a bird." His half-brother Ramon added, "It can fly but it has skin similar to mice. They got little claws like a rat. I'm pretty sure it's got teeth."

The anomalous biological position of the bat was culturally encoded into Piman dream metaphor. The Great White Bat, Ge'echu S-toha Nanakmel, might appear to a boy in his dreams to confirm his ambiguous gender identity as a *uikwuaḍ*, transvestite (see Rea 1998b:129–130).

There is also one mythological episode that deals with the bat's anomalous position. In the fire quest tale told to Harold Bell Wright (1929:37–42), first a number of humans then several birds are sent to get fire from the sun for nighttime use, each one failing. The council decides to send a night bird. Ultimately, Nanakmel, the Bat, succeeds, returning with a clap of thunder and a bolt of lightning that ignites a tree. The next day the People begin to look for Nanakmel. "They looked all around. At last they found him hanging all limp to a tree. And poor Bat had not one feather left on his body. He was burned down to the skin. And he was so very much ashamed that no one could coax him to come out in the daytime." Before this he was "covered with nice gray feathers." Although this story explains some of the anomalous qualities of bat as having once been a bird, it doesn't account for its rodentlike characteristics. It was these morphological attributes, rather than the act of flying, that the people I asked found most important.

How Are Birds Named?

Unlike the folk taxonomy of plants, but more like the folk classification of mammals, relatively few Piman avian folk taxa are descriptive (16 percent). The list includes several sets of synonyms:

haiñ uushab, "broken mouth/bill," unidentified wading bird (American Avocet?)

vachpik, "dives under the water, submerges," American Coot

chevel u'uhig, "tall bird," Black-necked Stilt

shuudagĭ mamaḍ, "water babies," sandpipers

kaachk baasho u'uhig, "sea chest [beach] bird," gull

ee'eḍ vahuḍam, "blood squeezer," Barn Owl

ñepoḍ, (something cigar shaped), nighthawk

hevel moos, "Wind's grandchild," Say's Phoebe

ba'i chukul, "black throat," Horned Lark

biit keishnam, "it stepped on feces," Bendire's Thrasher

chev chiñkam, "long bill it has," Crissal Thrasher

keli huch, "old man's nail," Crissal Thrasher

koomagĭ u'uhig, "gray bird," Le Conte's Thrasher

kuigam, "mesquite owner," Phainopepla

baasho s-vegĭ, "chest red," Northern Robin

heet kawuḍk, "red ochre round hill," Summer Tanager

bichpoḍ, "feather-crested head," Abert's, Canyon Towhee

s-baabanmakam, "Coyote-like eaters," Lark Bunting

kauk (kavk) oo'okam, "hard bones (or back) it has" Lark Bunting

chechetopiwua, makes a characteristic sound (?), Lark Bunting

bahidaj u'uhig, "ripe Saguaro fruit bird," House Finch

oam u'uhig, "yellow bird," unidentified, perhaps Yellow-breasted Chat or goldfinch

u'unam u'uhig, "enclosure bird," unidentified

duuk vaidam, "rain+(?)," Coppery-tailed (Elegant) Trogon

Of these designations, two names are more allegorical than descriptive: the names for Say's Phoebe and Phainopepla. The phoebe, master of the air, is a "woman's daughter's child," and the woman is Wind, a personification for Pimans that is much more than a meteorological phenomenon. The Phainopepla, at home among the mistletoes festooning mesquites, is the owner or master of the all-important mesquite bosques.

Throughout world languages, the name of a bird may describe or at least suggest its call. English examples of such onomatopoeic names are "Killdeer" and "poorwill." Onomatopoeia is strongest among Piman dove names. Anyone familiar with dove calls should have no difficulty in guessing that the name *hoohi* belongs to the Mourning Dove, *okokoi* to the White-winged Dove, *gugú* to the Inca Dove, and *chehopĭ* and *vuhigam* to the Common Ground-Dove. (The terminal *-gam* of *vuhigam* ascribes what it is attached to as a characteristic quality, so this is the *"vuhi"* dove.) Owl names, too, are onomatopoeic. *Kokoho* describes one of the Burrowing Owl's calls quite exactly. And *kuukwul* may be in imitation of the Western Screech-Owl's "bouncing ball" song. (Culver Cassa suggests another possible derivation; see species account.) The name for the quail, *kakaichu* or *kakachu*, is taken from one of its most characteristic calls.

There are no Spanish or English loan words among Piman names for native birds except for the occasional substitution of *paado* among southern Papago for "duck," for which a legitimate O'odham name, *vapkag*, already exists. A similar bastardization occurs also among mammals where the original name for the Collared Peccary or Javelina, *tasi'ikol*, has frequently been replaced with *mischiñ kooji*, both words being loans from Spanish (see Rea 1998b:233–234). Several Spanish loans occur, especially among Papago, for nonnative species: *palóoma* and *papalho* (*paloma*, pigeon); *pilgíido* (*periquito*, parakeet) (Mathiot 1962, n.d.; Saxton, Saxton, and Enos 1983:50). Even *paado* may have crept in originally as a name for the domesticated duck.

There is one Yuman loan word among Piman bird names, *hialak*, goose (see species account for details). In fact, it is the only Yuman loan that I have discovered in O'odham folk taxonomy. Presuming a southern ancestral home for Pimans, it is surprising that plants and animals encountered in a northward expansion resulting in close association with Yuman speakers such as the Maricopa did not result in more borrowed names. Yet northern species such as beaver and Muskrat have unrelated Piman and Yuman names.

Some animal names are based on a presumed or metaphorical relationship expressed with a kinship term. One example is *chukuḍ ka'amaḍ*, the Great Horned Owl's grandchild, through the father's side. It is the Long-eared Owl, a somewhat smaller, slimmer version of the Great Horned Owl, but very different in vocalizations and behavior (it is localized and secretive). With the Say's Phoebe (or for the Ash-throated Flycatcher), *hevel moos*, the relationship is to a natural phenomenon. The bird is the grandchild of the maternal Wind.

Culver Cassa explained to me that relational naming generally involves assigning some culturally atypical or even unacceptable trait to the maternal side of the family:

> What I mean about using names of maternal kinship terms *(tatal, moos)* is that it shows how groupings of similarly appearing or behaving animals are reckoned. It seems to be based on the familiarity of a paternal familial village kinship. Everyday known things are similarly grouped because they look alike; they are placed in families. Let's say all doglike creatures (fox, wolf, dog, Coyote) would be called or reckoned as a family. This is like all people in a kinship system are grouped as *ogol, maam, vaav*, according to their father.
>
> The father's group was the only important one (*ogol, maam*, etc.). Alien blood would come into such a group through a wife. A foreign man would upset the patrilocal system and its economy. Because of this, anything odd in behavior or appearance could be attributed to the foreignness of the mother.
>
> Since some animals are classed thus (maybe unconsciously), another animal that is similar can be said to be part of its family through its odd appearance or behavior.
>
> There is something called *shee'e tatal*, wolf's maternal uncle. Since there are no dog species left [unnamed], its affinity must be sought in its odd similarity and behavior. Probably it may be a somewhat doglike creature that is in a pack, perhaps the Coatimundi. There is also a lizard called *ko'oi tatal*, "rattlesnake's maternal uncle." Desert Iguana is the animal. Something about the rattlesnake and the Desert Iguana are similar—either behaviorwise or in their life cycle—that merits its [the latter's] being rattlesnake's uncle.

Levels of Inclusion

As I noted for mammals, except for bats, skunks, and smaller rodents, Piman folk generics correspond one to one with biological species (Rea 1998b:123). Likewise, most Piman avian folk genera and all folk species correspond to biological species. We can list the few exceptions. *Vakoañ* includes all the egrets and herons (generously oversplit at the generic level by most current Western taxonomists). *Vapkag* probably subsumes all ducks and perhaps geese—in other words, one avian family consisting of many genera and species. Similarly, *vipismal* includes all genera and species of hummingbirds; the local breeding species are mostly green and differ primarily in adult male plumages. The name *giidval* covers all swallows including the martin. Some folk species may have been distinguished in this folk genus, but linguistic erosion has now blurred the picture so we cannot be sure. Finally, *shashañ* means specifically the Red-winged Blackbird for many O'odham speakers, but others would include all blackbirds, even the cowbirds, within this name. The River Pima, living where the Red-wing commonly breeds, are more likely to use the word in the restrictive sense than are the Desert People, who know blackbirds primarily as nonbreeding birds.

Upper Pimans refer to a *komalk mo'okam*, or flat-headed bird, whose head was stepped on at the first wine feast. The actual bird associated with this name varies from one person to another, depending on what story he or she has heard. For some, *komalk mo'okam* is the Canyon Wren. Others ascribe this incident to the Gila Woodpecker, Lesser Nighthawk, a kingfisher, Western Kingbird, Say's Phoebe, or Nutting's Flycatcher. At any rate, it is an mythological epithet, not a particular species' name.

Biological Referent Shifts

As speakers of a language disperse over a greater geographic area, they often retain a lexeme, a folk taxon, and apply it to a different biological species, particularly if the original biological referent is not a part of the local flora or fauna of the newly inhabited area. Numerous examples indicate how English-speaking colonists encountered new organisms in North America and called them by the names they had used for similar species in the Old World. Robin, blackbird, flycatcher, and warbler are a few avian examples. (Here each new referent is in a different family from its original referent.)

There are several examples of referent shifts in the Northern Piman lexicon. For River Pima, the name *bichpoḍ* indicates the locally abundant riparian Abert's Towhee. In Tohono O'odham country, this species has been replaced by the upland Canyon Towhee, which here became the *bichpoḍ* referent. The referent for *hevel moos* among River Pima is the Say's Phoebe. Among at least some Tohono O'odham, that name labels the Ash-throated Flycatcher. Both species occur in both areas, but the Ash-throated is much more prominent in the desert uplands of Paguería.

As already noted, a temporal shift in the biological referent of *aaḍo* occurred. The Scarlet Macaw, a bird of high cultural salience, was formerly reared by the Pima. During the twentieth century, as the original referent was forgotten, the name was shifted by many speakers to the peafowl.

What Gets Named and Why?

How thoroughly—that is, how completely—did Northern Pimans recognize and linguistically distinguish the birds of their environment? We will never know the answer to this question with any accuracy for this entire area. The known Pima Bajo folk taxonomy of birds is too fragmentary. But if we narrow the focus to Pimería Alta—that is, to Piman and Papago country—we can make some better estimates, even though there has been linguistic erosion here as well. I documented the historic avifauna of the middle Gila, home of the Gileños or River Pima (Rea 1983). Some 237 species of birds have been recorded historically at the northern fringes of Northern Piman country. (This number includes accidental, vagrant, and rare species as well as a few species new to the local avifauna.) How many of these species are included in the Piman lexicon of bird names? And perhaps even more interesting, why? We can venture some answers.

Approximately 118 biological species of birds (table 4.4) occurred fairly commonly to abundantly in Upper Piman country, especially in River Pima country with its former rivers and lagoons (Monson and Phillips 1981; Phillips, Marshall, and Monson 1964; Rea 1983; Witzeman, Demaree, and Radke 1997). One might reasonably expect to see these species in a year, at least before modern ecological deterioration and desertification. Of these 118 more common species, 55 are resident, and an additional 18 are species present only in the breeding season, for a total breeding avifauna of 73 species.[4] Some 15 species commonly pass through Piman country on spring or fall migrations, and an additional 30 species regularly stay throughout the winter.

Eliminating the Scarlet Macaw, a nonlocal captive species; the magpie-jay, trogon, and parrot, which occur only in Pima Bajo country; the caracara and one species of quail from southern Paguería, as well as two birds that may be only mythological *(chechlogam, unmu'uhig)*, we are left with 66 folk generics and 10 folk species, for a total of 76 terminal taxa, exclusive of folk synonyms.

The 76 terminal taxa and the 118 more common species are not necessarily in concordance. Numerical abundance does not mean that a species got named, nor does low density mean that a bird went unnamed. River Pima, for instance, maintained the names for at least four species not found in their immediate environment: *tova*, turkey; *gihodag*, Band-tailed Pigeon; *ñuipa*, Black Vulture; and *s-hadam kakachu*, Scaled Quail. The turkey and pigeon were encountered generally in the pine-oak woodlands at higher elevations. (The River Pima lexicon also maintained names for both pine and oak.) The other two birds were generally encountered when the Pima went on pilgrimage to Magdalena, Sonora. This brings us to 72 terminal folk taxa for labeling the local River Pima avifauna. Thus, River Pima named 61 percent of the 118 common local species. Some other named birds, although being local, occurred in low population densities, at least judging by modern conditions, which are probably anthropogenic artifacts. These birds include Brown Pelican, Prairie Falcon, Peregrine Falcon, Osprey, Golden Eagle, gulls, and cranes. Thus, the River Pima lexicon captures some both nonlocal and noncommon species, making the 61 percent number only a rough calculation of the naming process.

Still, some generalizations are possible. Most named species are locally breeding birds. People had an opportunity to become familiar with their breeding biology: feeding, singing, territorial defense, mate selection, nest building, incubation, and so forth. Generally speaking, these elements are the "stuff," the substance, of their folk taxonomy: distinctions based on a characteristic biology superimposed on a discrete morphology. The ecological salience of species that merely pass through or winter in small numbers is far lower than that of breeding birds. Eugene Hunn (1999) found a similar pattern among the Sahaptin of the Pacific Northwest and the Tzeltal and Zapotec in southern Mexico.

What is unnamed from the local riverine Piman *breeding* avifauna?

rails (presumably once in the large marshes, now long gone)

Ferruginous Pygmy Owl *(Glaucidium brasilianum)*

Elf Owl *(Micrathene whitneyi)*

White-throated Swift (included with swallows?)

Ash-throated and Wied's Crested Flycatcher *(Myiarchus tyrannulus)*

Black Phoebe *(Sayornis nigricans)*

Vermilion Flycatcher *(Pyrocephalus rubinus)*

Bell's Vireo *(Vireo bellii)*

Lucy's Warbler *(Vermivora luciae)*

Yellow Warbler *(Dendroica petechia)*

Table 4.4 Biological Status of the Common Birds of River Pima Country

Pied-billed Grebe, *Podilymbus podiceps*	R (f)
Turkey Vulture, *Cathartes aura*	Br
Black Vulture, *Coragyps atratus*	Tr
White-faced Ibis, *Plegadis chihi*	Tr
Canada Goose, *Branta canadensis*	W
Green-winged Teal, *Anas crecca*	W
Mallard, *Anas platyrhynchos*	W
Northern Pintail, *Anas acuta*	W
Cinnamon Teal, *Anas cyanoptera*	R
Northern Shoveler, *Anas clypeata*	W
Gadwall, *Anas strepera*	W
American Wigeon, *Anas americana*	W
Ring-necked Duck, *Aythya collaris*	W
Ruddy Duck, *Oxyura jamaicensis*	R
Red-tailed Hawk, *Buteo jamaicensis*	R
Harris' Hawk, *Parabuteo unicinctus*	R
Northern Harrier, *Circus cyaneus*	W
American Kestrel, *Falco sparverius*	R
Gambel's Quail, *Callipepla gambelii*	R
Green Heron, *Ardeola (Butorides) virescens*	R (f)
Great Blue Heron, *Ardea herodias*	R (f)
Snowy Egret, *Ardea thula*	Tr
Least Bittern, *Ixobrychus exilis*	R
American Coot, *Fulica americana*	R
Common Gallinule, *Gallinula chloropus*	R
Killdeer, *Charadrius vociferus*	R
Common Snipe, *Gallinago gallinago*	W
Long-billed Curlew, *Numenius americanus*	W
Spotted Sandpiper, *Actitis macularia*	W
Greater Yellowlegs, *Tringa melanoleuca*	W
Least Sandpiper, *Calidris minutilla*	W
Western Sandpiper, *Calidris mauri*	Tr
Long-billed Dowitcher, *Limnodromus scolopaceus*	W
Black-necked Stilt, *Himantopus mexicanus*	R
White-winged Dove, *Zenaida asiatica*	Br
Mourning Dove, *Zenaida macroura*	R
Common Ground-Dove, *Columbina passerina*	R
Inca Dove, *Scardafella inca*	R
Yellow-billed Cuckoo, *Coccyzus americanus*	Br
Greater Roadrunner, *Geococcyx californianus*	R
Barn Owl, *Tyto alba*	R
Great Horned Owl, *Bubo virginianus*	R
Western Screech-Owl, *Otus kennicottii (asio)*	R
Ferruginous Pygmy-Owl, *Glaucidium brasilianum*	R(f)
Elf Owl, *Micrathene whitneyi*	Br
Burrowing Owl, *Athene cunicularia*	Br
Long-eared Owl, *Asio otus*	R (f)
Common Poorwill, *Phalaenoptilus nuttallii*	R
Lesser Nighthawk, *Chordeiles acutipennis*	Br
White-throated Swift, *Aeronautes saxatalis*	R
Black-chinned Hummingbird, *Archilochus alexandri*	Br
Costa's Hummingbird, *Archilochus costae*	Br
Belted Kingfisher, *Megaceryle alcyon*	W (Br f)
Gila Woodpecker, *Melanerpes uropygialis*	R
Ladder-backed Woodpecker, *Dendrocopos scalaris*	R
Northern Flicker, *Colaptes auratus*	R
Western Kingbird, *Tyrannus verticalis*	Br
Ash-throated Flycatcher, *Myiarchus cinerascens*	R
Black Phoebe, *Sayornis nigricans*	R (f)
Say's Phoebe, *Sayornis saya*	R
Western Flycatcher, *Empidonax difficilis*	Tr
Western Wood Pewee, *Contopus sordidulus*	Tr
Vermilion Flycatcher, *Pyrocephalus rubinus*	R (f)
Horned Lark, *Eremophila alpestris*	R
Tree Swallow, *Tachycineta bicolor*	Tr
Violet-green Swallow, *Tachycineta thalassina*	Tr
Northern Rough-winged Swallow, *Stelgidopteryx serripennis*	Br
Barn Swallow, *Hirundo rustica*	Tr (Br f)
Cliff Swallow, *Hirundo albifrons*	Br (f?)
Common Raven, *Corvus corax*	R
Loggerhead Shrike, *Lanius ludovicianus*	R
Verdin, *Auriparus flaviceps*	R
Cactus Wren, *Campylorhynchus brunneicapillus*	R
Marsh Wren, *Cistothorus palustris*	W (Br f)
Rock Wren, *Salpinctes obsoletus*	R
Canyon Wren, *Catherpes mexicanus*	R
Northern Mockingbird, *Mimus polyglottos*	R
Bendire's Thrasher, *Toxostoma bendirei*	R
Curve-billed Thrasher, *Toxostoma curvirostre*	R
Crissal Thrasher, *Toxostoma crissale*	R
Le Conte's Thrasher, *Toxostoma lecontei*	R
Northern Robin, *Turdus migratorius*	W
Western Bluebird, *Sialia mexicana*	W
Black-tailed Gnatcatcher, *Polioptila melanura*	R
Ruby-crowned Kinglet, *Regulus calendula*	W
American Pipit, *Anthus spinoletta*	W
Cedar Waxwing, *Bombycilla cedrorum*	W
Phainopepla, *Phainopepla nitens*	W
Bell's Vireo, *Vireo bellii*	Br
Warbling Vireo, *Vireo gilvus*	Tr
Orange-crowned Warbler, *Vermivora celata*	Tr
Lucy's Warbler, *Vermivora luciae*	Br
Yellow Warbler, *Dendroica petechia*	Br
Yellow-rumped (Audubon's) Warbler, *Dendroica coronata*	Tr
Common Yellowthroat, *Geothlypis trichas*	R
Wilson's Warbler, *Wilsonia pusilla*	Tr
Summer Tanager, *Piranga rubra*	Br
Western Tanager, *Piranga ludoviciana*	Tr
Northern Cardinal, *Cardinalis cardinalis*	R
Black-headed Grosbeak, *Pheucticus melanocephalus*	Tr
Blue Grosbeak, *Passerina caerulea*	Br
Canyon Towhee, *Pipilo fuscus*	R
Abert's Towhee, *Pipilo aberti*	R
Lark Bunting, *Calamospiza melanocorys*	W
Brewer's Sparrow, *Spizella breweri*	W
Black-throated Sparrow, *Amphispiza bilineata*	R
White-crowned Sparrow, *Zonotrichia leucophrys*	W
Dark-eyed Junco, *Junco hyemalis*	W
Song Sparrow, *Melospiza melodia*	R
Hooded Oriole, *Icterus cucullatus*	Br (f?)

(continued on next page)

Table 4.4 Biological Status of the Common Birds of River Pima Country ***(continued from previous page)***

Bullock's Oriole, *Icterus galbula bullockii*	Br	Brewer's Blackbird, *Euphagus cyanocephalus*	W
Western Meadowlark, *Sturnella neglecta*	R	Brown-headed Cowbird, *Molothrus ater*	R
Red-winged Blackbird, *Agelaius phoeniceus*	R	House Finch, *Carpodacus mexicanus*	R
Yellow-headed Blackbird, *Xanthocephalus xanthocephalus*	W	Lesser Goldfinch, *Carduelis psaltria*	W

Br = breeding
R = resident (winters and breeds)
W = winters only
Tr = transient (spring and/or fall migrant)
f = formerly
Source: Based on Rea 1983 and Witzeman, Demaree, and Radke 1997.

Common Yellowthroat *(Geothlypis trichas)*

Yellow-breasted Chat *(Icteria virens)* (possibly named?)

Pyrrhuloxia *(Cardinalis sinuatus)* (included in cardinal name, *sipuk?*)

Song Sparrow *(Melospiza melodia)*

Black-throated (Desert) Sparrow *(Amphispiza bilineata)*

Brown-headed Cowbird (included in Red-winged Blackbird name, *shashañ?*)

House Finch (Is *bahidaj u'uhig* a folk taxon among some Tohono O'odham only?)

Perhaps the Pied-billed Grebe, *Podilymbus podiceps*, may have bred locally when the middle Gila River was lush, and it still does at nearby Picacho Reservoir. Because of its diving behavior and overall shape, it may have been included in the coot category, *vachpik*. But we will never know now because the question was never raised when people who might have known were alive. When there were old-growth mesquite bosques along the rivers, the People must have been familiar with Lucy's Warbler, Yellow-breasted Chat, and Bell's Vireos with their distinctive calls.

Unnamed common local wintering avifauna include:

grebes (included with coots or ducks?)

rails

?White-faced Ibis *(Plegadis chihi)*

Dark-eyed Junco *(Junco hyemalis)*

Ruby-crowned Kinglet *(Regulus calendula)*

bluebirds (Western, *Sialia mexicana*, and Mountain, *S. currucoides*)

Hermit Thrush *(Catharus guttatus)*

House Wren *(Troglodytes domesticus)*

Brewer's Blackbird *(Euphagus cyanocephalus)*

Yellow-headed Blackbird *(Xanthocephalus xanthocephalus)*

Seven named birds in the Akimel O'odham lexicon are winter residents only:

shuudagĭ maamaḍ, Least/Western Sandpipers

vakav, Marsh Hawk/Northern Harrier

vakoañ ba'ag, Osprey

kookoḍ, Sandhill Crane

baasho s-vegĭ, Northern Robin

tamtol, White-crowned Sparrow

o'oḍopiwua, Lark Bunting

Three of these species are very large birds with unusual behaviors that would single them out (the harrier, crane, and Osprey). The crane and Osprey, in addition, have distinctive vocalizations. The remaining four are small birds. The abundant sparrow and bunting were important winter food resources, and perhaps the robin was as well. For about ten months of the year the two peeps, or small sandpipers, are common to abundant visitors to the muddy shores of rivers or small ponds (Witzeman, Demaree, and Radke 1997). These birds are too active to be overlooked. Similarly, Brown Pelicans (and perhaps even White Pelicans) were too conspicuous to go unnoticed and unnamed, although they were irregular vagrants to Gila country.

Unnamed among the regularly wintering birds are other *Buteo* species, two Accipiters, grebes (perhaps included with coots or ducks?), sapsuckers, kinglet, terrestrial thrushes, goldfinches, migrant wrens, and other sparrows, including juncos. The nonnaming of the junco is particularly puzzling: it is abundant in winter and is easily trapped along with White-crowned Sparrows. Yet no one could give me a name for these common dooryard birds. Older Pimas denied any knowledge of the Bald Eagle. Unlike the Golden Eagle, it has no place in any Piman ceremony (in contrast to the Puebloan peoples to the north, who kept and used both species). Although about forty pairs of Bald Eagle now nest along some of Arizona's major rivers, and nests have been found south as far as the middle Río Yaqui in Pima Bajo country, this distribution may be a recent phenomenon arising from impoundments of rivers for water storage and the increased availability of rough fish (Hunt 1998; Russell and Monson 1998:62–63). Piman folk taxonomy lends support to the idea that this species was formerly very rare as a breeding species in southern Arizona.

Irregularly wintering birds are apparently unnamed: nuthatches, creepers, titmice, *Empidonax* flycatchers, warblers, vireos. And no birds that are solely spring or fall transients made it into the lexicon. The myriads of warblers, vireos, sparrows, and flycatchers that migrate through the desert, sometimes festooning blooming mesquites and paloverdes in spring, passed through unnamed.

Crested (Audubon's) Caracara is interesting in that it has a southern breeding distribution, reaching northward through the southern half of Papago country. This strikingly marked bird is both common and unwary where it occurs. Here it is labeled by the unanalyzable name *kusijim*. Northern strays, apparently once more numerous than now, were called in at least one River Pima family *oam ñui*, yellow vulture, an improvised folk specific (Rea

1983:135). At least two other southern birds, the magpie-jay and the trogon, were named in Pima Bajo. But how many more southern birds were named is unknown given the sketchy folk taxonomies here.

Commentary

Although many plants new to the local flora have received O'odham names, no new names have been coined for the birds that naturally invaded the northern fringes of Tepiman country during the twentieth century: Bronzed (Red-eyed) Cowbird, Pyrrhuloxia, Hooded Oriole *(Icterus cucullatus)* (?), and Great-tailed Grackles. We can add to this list two intentionally introduced European species: Common Starling, *Sturnus vulgaris*, and House Sparrow, *Passer domesticus*. These birds were surely noticed. For instance, Sylvester Matthias brought me the nest of a Hooded Oriole he found while trimming fan palms. He was familiar with the black horsehair nests of the Bullock's Oriole, but the pale fiber nest of the Hooded Oriole was new to him. At Santa Rosa village, Simon Lopez was puzzled by the now abundant grackles that have colonized, wondering if they were large blackbirds or some kind of small raven. It is a characteristic behavior of northern O'odham speakers not to extend the domain of biological referents to include new species, but rather to say, when pressed, that something is merely *like* something else in the formal lexicon (see Rea 1997:392–399).

Other birds that apparently extended their ranges northward during the nineteenth and twentieth centuries took the names that O'odham to the south had called them. They include Inca Dove, White-winged Dove, Black Vulture, and Northern Cardinal (Phillips 1968).

The naming and keeping of nonnative birds is discussed in chapter 6. Of these birds, the early naming of the chicken remains clouded. According to the Jesuit father Ignaz Pfefferkorn (1949:117), writing in the mid-1700s, Indians from the south where the chachalaca occurred applied its name to the chicken. Whether marking reversal was involved is unknown. Work with the Tepehuán may shed some light. Among colonial and modern terms labeling the chicken throughout the Tepiman corridor, three are cognate *(totol, tuchuli, chuchul)*, and one isn't (*takoko*).

Some bird names are localized because the birds themselves are. For instance, Papago named the Crested Caracara, Black Vulture, Mearns' Quail, and Scaled Quail, species found generally south of River Pima country. For the most part, however, more northern Tohono O'odham did not know the names for these four species. (But Gileño Sylvester Matthias knew the names for two of them.) Similarly, Pima Bajo named the trogon, Military Macaw, magpie-jay, and parrot—all species with more southern distributions.

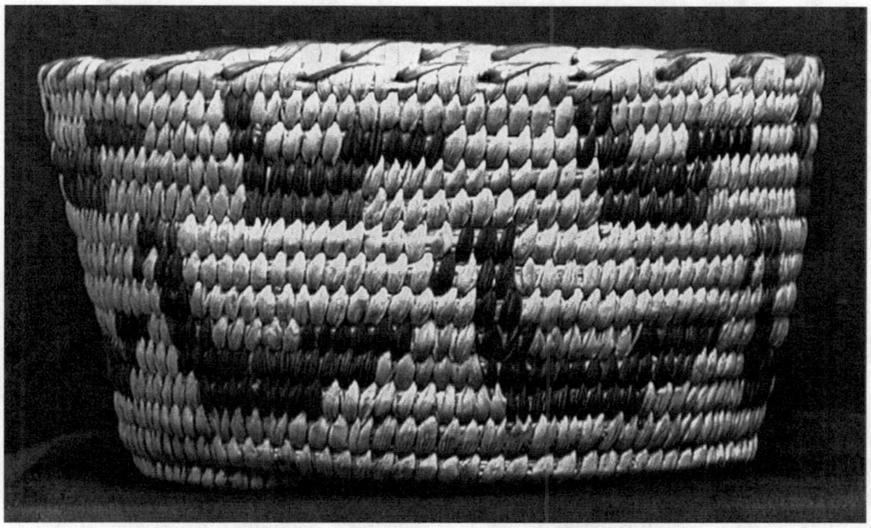

Figure 4.1 *Tohono O'odham (Papago) basket with floating waterfowl (May Collection, University of San Diego, weaver unknown, photo by Joyce Antorietto). Unlike the Akimel O'odham living in a riverine environment, the Tohono O'odham, occupying a land with very little surface water, subsumed many kinds of waterbirds into one folk taxon.*

Some generalized answers to the question of what gets named can be proposed:

1 Breeding birds are more likely to be named than nonbreeding species.
2 Wintering birds are more likely to be named than spring or fall transients.
3 Nonpasserines are more likely to be named than passerines.
4 Larger birds are more likely to be named than smaller species.
5 Among small birds, those included in the dietary are more likely to be named than those that were not.

Underdifferentiation

Conspicuously underdifferentiated are the ducks and geese, sandpipers as a family, gulls, swallows, flycatchers, sparrows, and blackbirds. To this list we can add the more secretive waterbirds (?Pied-billed Grebe, Least Bittern, all rails, Common Yellowthroat, Marsh Wren), none of which were named insofar as can now be determined. Among the Anseriformes, nearly all of which are wintering birds, no folk species at all were distinguished, with the possible exception of *hialak* (Snow Goose?). Its relationship to *vapka'ig* remains clouded. From the many scolopacids—the snipe, curlew, sandpipers, and relatives—only the smallest "peeps" were definitely designated by name. (There is a mythological reference in English suggesting that two forms were once distinguished by River Pima.) Among contemporary speakers, Akimel O'odham distinguish considerably more aquatic species than do Tohono O'odham, a direct consequence of ecological salience, few waterbirds being seen by the nonriverine Papago (see fig. 4.1).

Likewise, no folk species of swallows are definitely known, although most North American species migrate through Piman country, and two or more species may have bred locally (Rough-winged, Barn, Cliff, Purple Martin). The large towhee was named, but apparently the only small sparrows distinguished were two that were conspicuously marked with black and white and were important food resources as well. Among flycatchers, the Say's Phoebe and omnipresent kingbird were named, but none of the lesser and more difficult to discern species (Wied's Crested, Ash-throated, *Empidonax* species, pewees). Why the brilliant and confiding Vermilion Flycatcher with its conspicuous courtship flight should be omitted is a puzzle; perhaps the right person was not interviewed.

Well-differentiated groups in the lexicon include hawks and falcons, woodpeckers, doves, owls, and thrashers. In these six families, all or nearly all locally breeding species were labeled at some taxonomic level. The avocets, stilts, and various wading birds were probably well recognized in the folk taxonomy, though a number of the names cannot now be pinned to a biological species (see appendix F). Chalk that up to both linguistic and ecological deterioration.

How much of the local biota did the Akimel O'odham lexically recognize when the culture and the environment were still intact? We can say that the average person would have known at least 235 plants, 35 mammals, 75 birds, 10 fish, 20 herps (that is, snakes, lizards, and turtles), and 25 invertebrates. In other words, a command of 400 biological taxa, at some level, were readily known by name. The bulk of these names label what Western biosystematists would call biological species. This is a conservative estimate, particularly with regard to herps and invertebrates. But it should be a comfortable bottom number for the general person to have commanded.

How well does the lexicon match the local biodiversity? We have good numerical information on both the scientific and folk taxonomic taxa for three groups of organisms: plants, mammals, and birds.

At least 525 biological species of plants have been documented in the immediate vicinity of the Gila River Pima villages (Rea 1997:84–85). Some of these species are rare and restricted to local sierras, visited by few people. Others are aquatic or riparian species that disappeared from the reservation more than half a century ago. Still others are cryptic species, not readily distinguished without an optical aid. If we adjust for these exceptions, the Gila Pima linguistically recognized at least half their local, easily visible flora (183 native or naturalized plants and fungi labeled by terminal taxa). Given the cultural and linguistic erosion as well as the ecological deterioration of the past century, it is likely that Gileños once recognized three-quarters of their readily perceivable flora.

All the local mammals were named at some level, with the possible exception of the Ringtail and the seldom encountered Desert Shrew. Most Pima mammal names correspond with biological species. The exceptions are a single name for several skunk species, one name for all species of bats (parts of four families), and six names for all species of rats and mice. However, most likely each biological genus of small rodent was once labeled with a Piman folk generic name, except the grasshopper mice *(Onychomys)* that so much resemble white-footed (deer) mice *(Peromyscus)*. Further discrimination would require examination of teeth with a hand lens, equipment not in the usual inventory of the folk taxonomist constructing a taxonomy. Similarly, I think that nearly every biological species of reptile and fish was formally named. (A few of the rarer nocturnal species of snakes seemed to be unfamiliar to people I worked with, but I cannot be sure how much of this is attributable to erosion due to lack of contemporary contact.) It is only when we encounter the overwhelming diversity of Sonoran Desert invertebrates, in particular insects, that the Piman folk taxonomy appears to be inadequate. Certain names labeled whole groups, whereas others singled out conspicuous species.

Erosion of Piman Folk Taxonomy

The major consultants among the River Pima for this book and my previous books were for the most part born between 1900 and 1915 or 1920. Those born after 1920, even though native speakers, generally had a considerably reduced inventory of plant and animal folk taxa, typically less than 50 percent of what the older people knew. This fact makes me wonder how much was lost from the generation born in the last quarter of the 1800s that I never interviewed, even though I knew some. These individuals included the last of the monolingual speakers on the Gila. Anyone born on the river after 1900 could claim schooling and a good command of English, although Pima would remain the language of choice for at least several more generations.

What caused such an abrupt collapse in the biological lexicon? It was in part related to the ecological catastrophe on the middle Gila River. Surface water was seriously diminished by watershed destruction in the uplands and water diversion to Anglo and Mexican fields upstream from reservation lands. Most marsh and wetland habitat dried. The entire riparian gallery woodlands dried, grasslands vanished as perennial communities, and mature mesquite bosques died. With these habitats went their associated animals. The ecological collapse began in the 1870s and has continued to the present.

Surface water in the streams persisted longest in the western villages where the water table was highest. Here the Santa Cruz and various other washes draining northward converged. The people living at these settlements (called the Kuiva O'odham, Downstream People) remembered many aquatic plants and animals that the Ta'i O'odham had long lost from their cultural inventory, including most of the waterbirds.

Simultaneously, the O'odham passed from a hunting-fishing-gathering and small-scale agricultural subsistence economy to a wage-earning and commodity-buying economy. Traditional foods with slow-release complex carbohydrates, high soluble fiber, and less fat became unavailable or were supplanted with high-fat, high-glycemic-index foods lacking the gums and mucilages that once protected individuals against sudden surges in blood sugar levels. The heavy physical work of the former lifestyle gave way to sedentary office jobs and hours of television watching. The ultimate toll has been the twin scourge of obesity and adult onset diabetes. The ramifications of the disruption of a local ecology and the culture that had evolved in it continue to unfold. Not just bird knowledge was lost.

5

Birds, Guardians, Shamans, and Healers

In order to understand an important aspect of the relationship between Pimans and their environment, we must explore the O'odham idea of staying sickness and animal powers. Staying sickness, *kaachim mumkidag*, is a cultural phenomenon found among both Akimel O'odham and Tohono O'odham—that is, the River Pima and the Papago. It presumably was once found among other Northern Pimans as well, but centuries of close contact with non-O'odham seem to have obliterated any traces of it outside Pimería Alta.

Certain species of birds and other animals in the Piman cosmos are special in that violation of them may cause *kaachim mumkidag*. This category of afflictions does not pass from person to person as does *oimmeḍdam mumkidag*, wandering sickness. Nor can it afflict non-O'odham. Left untreated, staying sickness ultimately invades the whole body, causing death. The offense or abuse usually must be determined by a diagnosing shaman, or *maakai*. Most of the time the healing occurs in a separate ritual session, *wusota*, performed by a specialist or meeter, *namkam* (who has an association with the offended animal), accompanied by any assistants who know the meeter's songs about the species involved.

Guardians and Meeters

The social status of individuals within O'odham communities was strongly uniform with the exception of the role of *maakai*, shaman, and several hereditary offices such as the Navichu, a healing kachina, and the *tobdam*, hunt leader. However, individuals might get power that helped them in their various roles in society such as that of hunter, runner, gambler, planter, or singer in healing ceremonies. For a man, power could be elicited by killing an enemy (an Apache), killing an eagle, or going on a salt pilgrimage to the Gulf of California. Each of these activities had to be followed by a strenuous purification outside the village, lasting for a period of four to sixteen days. After that, an animal might come to the man to give him special power. These power sources have gone under various labels such as Animal Helpers, Guardian Spirits, Animal Guardians, or Animal Visitants. Most often the person who had such a helper was called a *namkam*, meeter.[1] Much of what we know of meeters and their power comes from the writings of Underhill (1936; 1938:49–52; 1939:161–162, 167–174; 1946:15–21, 266–270).

Probably any animal could come to a person to bring him power, including birds, mammals, reptiles, and even insects. The person so visited, for example, would be a *chukuḍ namkam*, Great Horned Owl meeter; *vishag namkam*, Prairie Falcon meeter; *juḍum namkam*, bear meeter; or *totoñ namkam*, ant meeter. *Ban namkam*, Coyote meeter, was a common source of power, particularly for a person belonging to the Coyote moiety. According to Underhill (1939:170), "There were certain standard Visitants, with loosely defined powers. The hawk [falcon] and the cranes [egrets] (always in a flock) gave speed in running. Coyote, turtle [tortoise], spider and many others gave curing powers. The crow [raven] gave cunning as a scout, the puma, wildcat and [Mule] deer gave hunting prowess. Luck in gambling came from a number of small birds, in one case, from the eagle."

Pima Joseph Giff explained it this way: "A person who knows all about the *chukuḍ*—its powers and its habits—is [a] *chukuḍ namkam*—means there's a *connection* between the *chukuḍ* and the person to give him great knowledge or seeing, as *kavio namkam*, horse meeter." He then added that my designation as an ornithologist would be *u'uhig namkam*, bird meeter.

The person did not seek contact with any particular animal through a vision quest, nor did he merit the encounter by good acts. The guardian freely chose one person and not another. There were four aspects of an encounter, noted Underhill (1939:264): "Power came from an animal. It came in a dream in natural sleep. It was open to anyone. Its acquisition was solitary, not a group experience." But the visitant might also come to a youth as he lay exhausted after running dozens of miles across the open desert. Particularly in the early morning before light, a parent constantly admonished a youth to show virtue so as to encourage a meeter:

> Perhaps some time in the night, something will arrive. Then if you are a good runner, you will jump up, you will run (after it). Early in the morning you will get up. Far over there (down to the west) you will sit and will listen for something. You will run over there and something which is good-luck-bringing will meet you: a coyote, an eagle, a hawk [falcon]. Then you will be a fast runner. You will be a good hunter.
>
> You should run constantly, no matter how much hardship it means; even if you are thirsty, hungry, tired. Then truly this will happen to you. You will throw yourself down repeatedly (worn out). Then there will come a hawk, an eagle, a mountain lion, a jaguar, a bear, and will tell you how to follow something alive (game) and to kill it. (Underhill 1939:162)

Not everyone who became a good hunter or runner, for example, necessarily had dream power, but it was essential for two roles: the professional shaman and the healer, a layman who conducted curing ceremonies, *wusota*. For the potential shaman, the visitor might come and encourage the person to take up this profession, teaching him his diagnosing songs. For the healer, the animal guardian might teach him a curing song or show him where to find some fetish such as a feather, claw, or unique stone to be used in the *wusota* session. For instance, Papago Maria Chona told Underhill (1936:47):

> My husband was a Coyote-Meeter. That is, when he had his medicine man's dreams, it was our comrade, Coyote, who came to him and sang him songs. It began when he went with the other men to the ocean to get salt, for that is where all our young men go when they want to meet some spirit. They gather the salt, and then they run along that great sandy place where the water is and where there are white birds flying over head. It was those white birds that had sung to my father, but with my husband it was different.

Great Horned Owl was the animal guardian who helped war shamans acquire their power. Maria Chona, discussing what her father had told her about his own war-path experience, related to Underhill (1936:13):

> All that time [in camp] the medicine man sat alone on a little hill waiting for the owl-dead [*chukuḍ*] to come to him. He had dreamed of owls himself and so he could speak with them. They always took an Owl-Meeter on to war with them to call the dead who are our spies. Soon that medicine man came back to the others and said: "Has anyone some of our old, wild tobacco?" They gave him some. Then he went back and they saw lights when he made two cigarettes. They saw those lights move as cigarettes do when two people are smoking. They heard the medicine man talking, very low, and then they saw both cigarettes thrown away. The medicine man came back and said: "The enemy are at Smooth Ground. We will come upon them at noon tomorrow." And so they did.

In the Hayden-Smith manuscript version (Hayden 1935) of the Origin Myth, various birds as well as a few other animals (snake, bear) were used by the Vupshkam (the Emergenti or Pimans) as agents in the conquest of the puebloid peoples, or "Hohokam."[2] Avian species that did so were poorwill, Verdin, eagle, "hawk" (probably falcon), Great Horned Owl, raven, and "bluebird" (*hevachuḍ*, Blue Grosbeak).

The Agents of Staying Sickness

Potential sources of staying sickness include more than forty species of animals, several plants (Sacred Datura and Peyote among them), some objects, the ocean, and even violations of ceremonial protocol. In everyday life, one hears most commonly about an animal-caused sickness. Fewer than a dozen birds among the seventy-five or so species linguistically recognized by Northern O'odham are associated with staying sickness (table 5.1), in contrast with there being about one-quarter of the named mammals so associated in the Piman folk lexicon.

Native speakers characterize these dangerous animals in various ways. Joe Giff said they are endowed with a "special nature," and Sylvester Matthias called them "sacred." George Kyyitan referred to them in Pima as having *gevkadag*, strength. They are also said to have a "way," *himdag*, that afflicts one. Because of the animal's strength or way, there is a moral relationship between the animal and an individual O'odham.

I pursued the connotations of the term *gevkadag* with George. Here's how he explained it: "Strength—*gevkadag*—or power [is] like if you have strength. Or if you had a strong mind or a strong will. Things like that. I'd think that's what it really means." Such is its application to a person. It refers to psychological rather than physical characteristics. I asked him, "You said the *ba'ag a'an* [eagle feathers] are real powerful. Is that the word you would use in O'odham for that? Is that also *gevkadag?*"

"Yeah, it is, but in a little different way. I think it is in a way of using those things, it's going to be a very hard thing. It will hurt you, it will hurt your feelings [health], or things like that come in, your sickness

Table 5.1 Birds Associated with Staying Sickness or Other Taboos

Species	Status	Distribution	Possible Reason
Turkey Vulture	S	U	primordial being; moiety mascot
Golden Eagle	S	N	metaphor of power; mythological event, origin of lustration; source of clouds
Prairie Falcon	S	N	metaphor of power; associated with eagle
Peregrine Falcon*	S	U	(same as Praire Falcon, by association?)
American Kestrel	S(?)	U	?
Gambel's Quail	S	N	?
White-winged Dove	T	U	associated with Saguaro fruiting, a major crop
Greater Roadrunner	S	U	?
Great Horned Owl	S	U	reincarnation of Piman souls
screech-owl	T(?)	U	ambiguous (Russell 1908:263)
hummingbird	T	U	survived Flood; helped measure earth to locate center
Gila Woodpecker	S	U	survived Flood; helped measure earth to find center
Say's Phoebe, Ash-Throated Flycatcher	T	U	associated with Wind, a powerful being
Verdin	T	U	survived Flood; major mythological shaman
Northern Mockingbird	S(?)	U	stabilized earth at Creation; talked during Flood

S = known agent of staying sickness
S(?) = suggested cause of staying sickness
T = otherwise tabooed with sanctions
N = northern distribution within the Tepiman corridor
U = ubiquitous within the Tepiman corridor
* The inclusion of Peregrine, in addition to Prairie Falcon, is based entirely on Densmore (1929), who throughout her work refers to "Blue Hawk" as the dangerous bird rather than to just "falcon" or "hawk." See species accounts for details.

[*mumkidag*]. That's it. I think that's the way it goes."

"But certain animals, like *chuk chuuvĭ* [Black-tailed Jackrabbit], have strength? And how do you say that strength in O'odham?"

"It's the same: *gevkadag*."

While discussing folk taxonomy and folk science, O'odham often commented on whether a biological species in question was safe or a potential source of *kaachim mumkidag*. Their interest in the question no doubt reflected my own. The first criterion was usually whether or not such-and-such had a song to be used in *wusota*; animals with a curing song were dangerous.

The offense to animals so endowed varies. It is the act in itself, rather than the actor's intention, that is significant. It involves some type of disrespect. For some, it can involve simply killing them even accidentally. But even some important game animals, such as Gambel's Quail and Mule Deer, possessed this special nature. In their case, an offense might be improper killing, butchering, or handling and disposing of the remains (for mammals, see Rea 1998b). It could also mean ridiculing the animal or even watching it die slowly; it must be dispatched as quickly and painlessly as possible. Even what belonged to an endowed animal, such as tubers stored by a gopher or chiltepins picked by a bear, must be avoided; both the gopher's "house" and a bear's tracks are dangerous. Any wood that the Gila Woodpecker has pecked or excavated into for its nest must be set aside cautiously, not burned. A carcass "claimed" by a Turkey Vulture cannot be touched by an O'odham. Feathers from a dangerous bird cannot be used until ritually purified by a *maakai*, making them safe.

Any O'odham in the role of parent, hunter, or cowboy must take precautions to avoid violations. A man whose wife is menstruating must excuse himself from the hunt. To kill any of the special animals at this time would be disrespectful to them, inviting *kaachim mumkidag*. Many sanctions fell on a man whose wife was pregnant. He must avoid all dangerous animals beginning when the couple first notice a missed period. The developing fetus rather than the parents would be the one afflicted by the violation. Sickness usually becomes manifest at or after birth or sometimes even later in the child's life. Latency might even delay the effect until adulthood. I asked Frank Jim how his wife got *kakkachu mumkidag*, quail sickness. "The same way as they do. I guess her father, *way* back. And they hadn't caught it till [now]. Just like I say, you get older, you get it bad. When you're smaller, you get it, you get cured quick."

It is perhaps difficult for Westerners to appreciate the respect that O'odham must hold for certain animals. In the past, it was part of a more pervasive pattern regulating human behavior toward the natural world. When communal hunts were announced by the *tobdam*, village hunt leader, he would never mention the targeted animals by their proper names, but had to use some oblique term that would not be offensive to the animals ("We will hunt the long-eared, white-tailed ones"). Today, traditional O'odham will never even point to a developing fruit on a vine or stalk; one must indicate it with a crooked index finger because pointing directly would give offense. One must explain to even a nondangerous plant why it is being harvested ("I am taking your branches for medicine").

The question is, Why are some animals in the Piman cosmos considered "dangerous" but not others? For instance, the Mule Deer, *huai*, is sacred, whereas the related White-tail, *siikĭ* is not. Likewise, the Black-tailed Jackrabbit, *chuk chuuvĭ*, is fraught with sanctions, whereas the Antelope Jackrabbit, *toha chuuvĭ*, is completely safe; the latter may be hunted at any time, even when one's wife is menstruating or pregnant. Among mammals, many of these dangerous ones are found only in the northernmost sector of the Tepiman corridor—that area from the Gila River to the Rio Santiago historically occupied by O'odham speakers. Both Mule Deer and Black-tailed Jacks have a northern distribution, but their counterparts, the White-tailed Deer and one of the large white-tailed species of hare are found throughout this corridor.

I have hypothesized that Northern Pimans, expanding into the northwest out of their ancestral home in the south, held the strange new species they encountered here in awe, endowing them with rigorous sanctions (see Rea 1998b:28–45; since writing this work, I have discovered that the porcupine, a species with a strictly northern distribution in Tepiman country, is a cause of staying sickness and is powerfully sanctioned). An interior southern ancestral home within a desert-sierran foothill ecotone is suggested by the O'odham non-analyzable plant names. The original area should be characterized by manzanita, pines, oaks, ash, agaves, yuccas, dasylirions, Ocotillo, a tall columnar cactus and several smaller ones, chollas, prickly-pear, mesquite. And we can also throw in a handful of aquatic and riparian plants that could come from almost anywhere.

How well does the northward expansion model explain which species of bird are agents of staying sickness? The fit is less precise than for mammals, but the sickness from several species might be explained by it. Both the Golden Eagle and the Prairie Falcon are northern species, reaching their southwestern continental limits in northern Sonora. However, the Peregrine Falcon, which seems to share all the cultural trappings of the Prairie Falcon, ranges throughout the corridor. (Unlike the folk taxonomic situation for the two jackrabbit species, the more widespread falcon is linguistically marked, *cheedagĭ vishag*, whereas the northern species is unmarked, simply *vishag*.)

The quail case is an interesting one. Various species occur in Northern Piman country, but the sanctions apply only to the Gambel's Quail, and then only to its *sivoḍa*, distinctive black topknot, lacking in the other southwestern quail species. The "dangerous" species is found throughout the arid regions of Sonora south to the Sinaloan border. If the southern origin/northern invasion hypothesis is correct, the Tepiman ancestral home would be pushed farther south than that indicated by any other "dangerous" northern vertebrate species. Most of the other powerful animals reach southward only to about central Sonora.[3]

Other factors might explain why certain other birds carry sanctions. In Piman cosmology, Turkey Vulture is paired with Coyote as two of the four primordial beings in Creation. Both are moiety mascots and are strongly associated with staying sickness. The mockingbird likewise has a key role in Creation and later in the Flood event. Indeed, Russell (1908:245) stated clearly, "After the waters [of the Flood] had gone down Elder Brother said . . . it is dangerous to harm anything that came safe through the flood." The hummingbird is a survivor of the Flood, as is the Gila Woodpecker. These key players in the Piman pantheon merited special status and respect. (One would expect the Verdin and Curve-billed Thrasher to have special status as well.) The Great Horned Owl still holds a unique position among Pimans because of its association with the dead.

It should be pointed out, however, that O'odham themselves generally offer no rationale for causality beyond the fact that certain species have been endowed with "strength" or "way" from the beginning.

Some animals are universally considered powerful or dangerous, but not all. Certain villages or districts lack certain kinds of sickness. The potential to be afflicted seems to be associated with where one comes from, not with the community where one happens to be living. For instance, Frank Jim, a western Tohono O'odham, knew the dangers from certain animals to other O'odham, such as his wife, that would prove no danger to himself. In discussing quail sickness, for instance, he said, "They have that over there, we don't have that. It don't bother us over there, but out here [Sacaton] might. They got it a little different. We have *kakkachu*, but we don't have [its] sickness."

The Diagnosticians

Historically, Piman *mamakai*, shamans, could play a number of roles in an otherwise egalitarian society. Individual shamans specialized in different fields. In war, they located the enemy in the east. They could predict (or even aid) the coming of rains for summer plantings. They were called upon to ensure good crops. They played specific parts in certain ritual events, in particular the rain ceremony, much in the manner of a Puebloan priest. They could strengthen a village's chance for winning in games. And they diagnosed the causes of disease in staying sickness and identified sorcery. According to Russell (1908:256), these diviners were in a separate class from the other categories of shamans. The role of the shaman as diagnostician is the only one that has survived significantly through the twentieth century. The functions and theories of the diagnosing shaman have been studied by Bahr and his O'odham associates (1974).

When a shaman is called to hold a *doajida*, divining session, for a sick person, he attempts to locate what organism or object or lapse is causing the patient's problem. The session is usually held at night. The *maakai* uses a gourd rattle in one hand and a matched set of divining feathers, usually from the Golden Eagle, in the other (fig. 5.1). He has a set of diagnosing songs that may last part or all of the night. He also uses tobacco smoke from rolled cigarettes and his quartz crystal to help him visualize the problem. (For less deep-seated illnesses that are not yet strong, a shorter ceremony called a *kulañmada* may serve to identify and remove some *mumkidag*.)

If successful, the *maakai* "sees" what is causing the problem. The violation may be in the patient's distant past, something now long forgotten. Or it may even be something that the person's father did during the mother's pregnancy, as noted earlier. The symptoms reported by the patient are of help. Some of the physical manifestations of violations are very specific, others broad and overlapping. There may be multiple violations, stacked like cards in a deck. They vary in strength, manifesting a confusing suite of symptoms. The shaman must determine where to begin, which must be cured first, like figuring out where to start tearing down a building. Blowing tobacco smoke over the patient's body and sucking may help to separate out various latent sicknesses so that they may be perceived individually. Then the *maakai* informs the patient of the offenses, which by now the sufferer may have forgotten.

The session with the shaman, the *doa-*

jida, may diminish or even cure the sickness, depending on its strength. Simple cases where the *mumkidag* has not yet pervaded the body may respond. The *maakai* may also be a specialist with the particular "way" involved. But usually a *namkam* or lay specialist who knows the songs of that animal, who is a meeter of that animal, must be engaged for a *wusota*, a separate healing session.

Here is how Leonard Pancott of Haashañ Keek (Sacaton Flats) explained it: "The medicine man looks at him and visualizes what causes it, as if he's 'allergic' to something. It will be later, after it's abused, that 'allergy' will appear [expressed latency]. Then they find somebody who sings the songs. [Then] have to have the song for *wuso* from what you killed or what you abused. Sing all night. Then he will get over it."

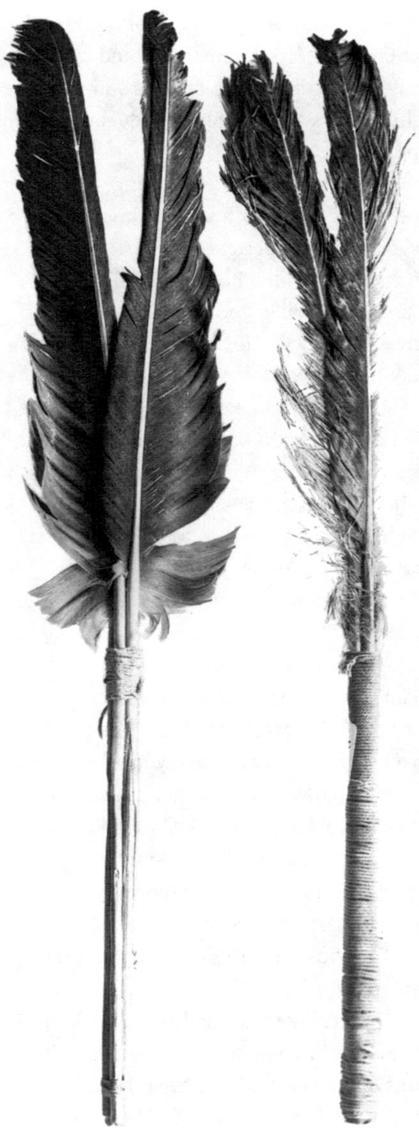

Figure 5.1 *Shaman's divining wand (courtesy of the National Museum of Natural History, Smithsonian Institution, Washington, D.C.).*

Likewise Joe Giff of Komaḍk said, "[T]hat's the way they treat any sickness that is caused by some animal or some *thing*. It takes another medicine man or somebody who will examine the patient, and he's the one that will diagnose the ailment, you see, and he will tell what it *is*, or what caused it. And then he will refer him to a man who will sing those songs, you see, and he [the patient] will go there."

Frank Jim from western Papago country said, "We go to the *maakai*. He's got the strength to find out all about that. We don't know which it belongs to. They find out what causes that [the symptoms]." Sylvester Matthias of Komaḍk explained that in a diagnosis the *maakai* might determine that "*jevho at i m-taach*, the gopher infected you; [you're] sick from gopher's *power*."

Meeters as Ritual Healers

The healing generally occurs at a special ceremony called a *wusota*. Some ceremonies are held in the daytime, but most are nocturnal events. Although the offense occurred at a particular time and place to a particular individual animal or plant, it is in some generic sense that the animal species, not the individual, has been abused and must be appeased. The curing is aimed at the generic essence.

The healer is a layman, not a shaman, because he does not "see" or diagnose the cause of sickness by meeting with his spirit helpers. His performance lacks charisma. He functions much as a pharmacist in relation to a physician, as Joe Giff put it. He is indeed a specialist in that he has been given a song or a series of them by his animal guardian, which he is now free to use and to teach to others. Unlike the shaman, he is not secretive about his songs; they are for public use. Healers frequently gave me songs about some animal I was writing about and often helped translate and explain them for publishing.

The *wusota* session, however, embodies some of the same ritual actions of the shaman's *doajida*. The curer may blow on the patient and apply some particular fetish that he has received directly or indirectly from his guardian (fig. 5.2). The fetish is standard for particular sicknesses: a horned lizard carved from paloverde wood, a Great Horned Owl feather, a Mule

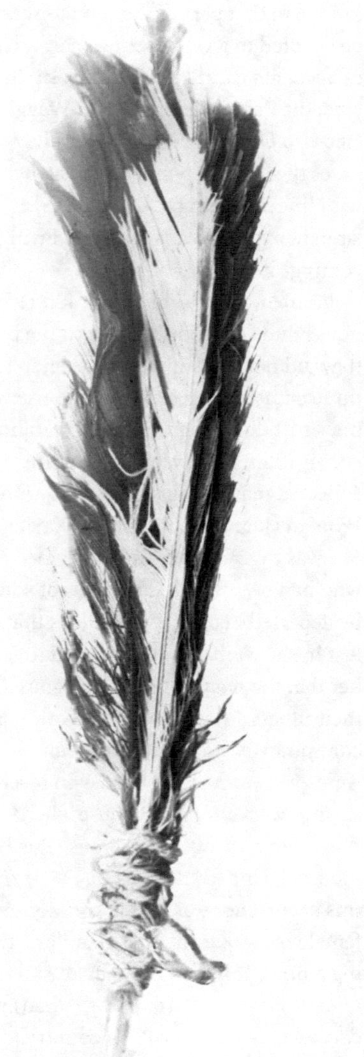

Figure 5.2 *Mockingbird fetish (courtesy of the National Museum of Natural History, Smithsonian Institution, Washington, D.C.).*

Deer tail. These actions may suffice. Typically, the *wusota* involves singing the songs acquired from his animal helper. According to Underhill (1946:287), "The [*wusota*] performance itself was public, and half the village attended to help with singing and gain a blessing." The goal of the ritual is to persuade the causal agent to quit sickening the patient. The singing is considered the ultimate cure, sufficient to rectify a correctly diagnosed affliction. Underhill (1939:175–176, 1946:288) asserted that the songs are not prayers to the afflicting animal's strength. However, Bahr and his associates (1974:97–98) consider them prayerful actions: "The idea is that the strength must ultimately be lifted by a spiritual recipient of the song."

Sylvester said that a *wusota* would begin at dusk and might last all night. The curing of *ñui mumkidag*, vulture sickness, was unique in that part of the ceremony might be devoted to a greatly shortened version of a *wuagida*, girl's puberty rite. In this case, the dance was called Ñui Wuagida (see also Underhill 1946:289). Culver Cassa notes that both ceremonies are associated with linear dancing as opposed to circular dancing. The bat is associated with the coming-of-age ceremony.

Women, too, might dream and receive power and songs for healing. In the past, they did not participate, however, in the three occasions for exposing themselves for power (the salt pilgrimage, eagle killing, or enemy killing). Densmore (1929:114–134) collected many songs originating from Owl Woman (Juana Manuel of San Xavier). She was a *chukuḍ namkam*, owl meeter, who had been taught hundreds of songs by departed people. Some clients that she lost in the *wuso* session returned to tell her that the *maakai* had misdiagnosed their illness. Women who know the songs commonly participate in the *wusota*. There were even women shamans who received their powers after menopause when they were no longer dangerous (Russell 1908:256; Underhill 1939:72, 1946:277), but this occurrence was rare. I know of several female *mamakai* still living and practicing among Tohono O'odham and another famous *oks maakai* (female shaman) now deceased, from River Pima country.

A firsthand account by George Kyyitan of staying sickness and *wusota* is somewhat atypical in that a diagnostic session, *doajida*, is lacking. George was discussing sinews and arrow making, noting the feathers that might be used, such as dove and even chicken, then said, "But [not] eagle, unless you know how to clean [purify] it, because you can't put it in your mouth. That's another one that's very superstitious to the people. And it's true, too. I know that it is true, because I got that one, *ba'ag* [eagle staying sickness]. Me and one of my friends. We were kids at that time. His grandfather used to have some of those eagle feathers. And, because, I think, they are medicine man, some from Tohono. And he came over there. And he used that one and he left [it].

"And *he* got some, this friend of mine; he got some of those little ones. And he said, 'I got some feathers and we're going to make some arrows.' 'Alright.' So we did. I split the feather and put it in our mouth, and scrape it up, clean that up. He did the same thing. Now finally, in about a week after, we had sores, *all* over. *Nothing* will cure it!" George scratched his short white crew cut vigorously, recalling this event that occurred more than half a century earlier.

"On your scalp?" I asked.

"Yeah. *Nothing* will cure it. *Nothing!* No medicine. No, any of these [things]. But you have to get somebody to do it for you, those that knows how to sing the eagle [curing songs] and use that feather on you [*wusota*]. And it [went a]way, after that. We cured that up. Because we never paid any attention to what *we* did. And what *we* used. And how *we* used it.

"Finally, *they* ask us, what we did. 'Did you do this? Or did you do that?' Then he [George's young friend] told them what we did this. 'Well, it's a *long* time. No wonder you got it all over already [they said].' He got his first. Then after that I got mine. [inaudible] And it cures [it] right there."

"Did you have to go to a *maakai?*"

"Naw. *Maakai* won't do it."

"No?"

"No. Got nothing to do with it."

"How did you get rid of eagle sickness?"

"The person got to know [how] to do all this. And he's the one that will cure it." Apparently, in this case, the causality was so clear that the *wusota* was used directly without a preceding *doajida*, or shaman's diagnosis.

"Who knew how to fix you up?"

"One—Old Man Peters. And my dad. Those are two [who knew the curing songs]. They did alright. That's how I got alright. *No* more."

"Song?"

"Yeah. Have to use a song and a feather, eagle, *ba'ag*. Because it's the one [that did it]." (Five years later George gave another account of this incident; it appears in the Golden Eagle species account.)

Several things relating to staying sickness bear noting in this story. First, the eagle feathers belonging originally to the *maakai* may not have been purified yet to render them safe for general use. Or more likely the boys were not the authorized users of such powerful objects. Second, in this particular case the use of a diagnostician, or *maakai*, was not necessary because the symptoms were obvious. So no *doajida* was held. As Underhill pointed out (1946:288), "Some symptoms seemed to be so widely known that when they appeared no diagnostician was called for, but the curing singer came in immediately." Third, the two people who sought to perform the *wusota* over the boys knew the eagle songs. One of the men was probably Keli Manol, Old Man Manuel Peters, from the west end of the reservation, a renowned healer.

Finally, the use of eagle feathers in the *wusota* ceremony was as a fetish, not as a *maakai*'s divining instrument. In this particular session, they happened to be from the same species of bird, but were probably not the primary feathers as used for the *maachwidag* (divining wand). Underhill (1946:300) noted "down" as being used as the fetish in this specific healing ceremony. Russell (1908:209, 263) wrote, "The eagle causes hemorrhage. The remedy is to sing the eagle song . . . and to pass the down of the eagle over the part. The eagle is also blamed for the lice that find refuge in the hair of the Pimas. The remedy is to blow cigarette smoke over the head." No specific Eagle Song occurs among Russell's anthology of medicine songs, and his set of eagle songs is given as circling songs. But then, songs occasionally had dual purposes.

Frank Jim told of self-healing. He had been diagnosed with eagle sickness years ago, but had experienced recurrences of it (see the Golden Eagle account). I asked him, "Now when you get that, *ba'ag* sickness, what do yo do? Do you have a *maakai* do it, or do you take care of it yourself?"

"Sometimes I take care of it myself, but if I don't feel good, I go see the *maakai*, let him do it, clear it out before I use it again. That's the reason I've got the *a'an* [feather in his medicine box]. A lot of things I get sick with it, I use it myself."

"So on some things a person can cure himself."

"If you know just exactly what it is. If you feel it, and maybe it's this one. You try the *ba'ag*. Maybe it's *huai* [Mule Deer]; you try it. You try all what you've got [have been diagnosed with], and *one* of those things will be the one. But when I feel that stuff, I know it's the *ba'ag*. And I just go all over it every evening, for four nights, four evenings. And I feel better."

"You do that with singing or without singing?"

"I don't know to sing [it]. I don't know the song, Ba'ag Song. I never did hear it. I just do it myself. About five minutes, before I go to bed. But if I don't feel better, then I see the *maakai*."

"If you knew the song . . ."

"I'll sing that for myself while I did

that. If I know it, I could sing it, it would be better, but I never did hear the Ba'ag Song. Same way with a lot of other stuff, I never did heard it. Some I heard, but I'm trying to figure it all out, you know, like deer, like horses, and something else. I try to figure out. I know some. If I figure out where to start it, then, figure out how it goes. It's pretty hard to do, but after I got away from there [his home village], I just forget it. Everybody always says, 'You know how to sing that.' Yeah. But I can't sing it by myself."

6

Bird Keeping and Rearing

The keeping of birds, primarily as a source of feathers for ceremonial use and for fletching arrows, was widespread throughout the Greater Southwest. Even the domestication of the turkey in the Anasazi and Mogollon regions seems originally to have been as a feather source, not as a food source. The ethnography of eagle keeping among the Hopi and Zuni and various other pueblos, continuing to the present day, has been well documented (Beaglehole 1936:18–19, 25; Fewkes 1900; Parsons 1939:28–29; Voth 1912). The Mohave, Cucapá, and Havasupai raised hawk and eagle nestlings until the feathers were large enough to pluck, then released the birds (Drucker 1941:131; Spier 1928: 151–152). Likewise, the Maricopa, living adjacent to the River Pima, kept eagles and hawks for arrow feathers, and supposedly vultures and "crows" as well (Spier 1933:74, 373–378). In the early 1600s, Jesuit historian Andrés Pérez de Ribas (1999:410 and n. 61) noted captive eagles among one group of Ópata. Obregón's (1928) even earlier chronicle of tame parrots and eagles in cages at Oera (Nuri), Sonora, most likely pertained to Cáhita Indians rather than to lowland Pima Bajo (Pennington 1980:23–26). Coastal Cáhita kept both kinds of birds for their feathers (Beals 1943:12–13). The River Pima annalist noted that some raiding Apaches in the mid-1800s carried with them a live vulture (Russell 1908:45). Jicarilla Apache kept young eagles for plucking (Opler 1936:207).

The history of bird keeping among Northern Pimans is not as complete as one would like. Yet with colonial relaciones, ethnological reports, and oral history, a fairly accurate picture can be pieced together. One early Jesuit (Pfefferkorn 1949:117) commented on the aptitude for raising birds and other animals among the Pima. The popularity of the chicken in colonial times, for instance, seems to have been primarily for pleasure because the Pima themselves at this time did not eat the birds or their eggs, which they sold to Spaniards.

An animal that is kept is called *shoiga* (pl. *shoshoiga*) in Piman, usually translated "pet" by native speakers. One's horses, cows, chickens, or other kept birds were *shoshoiga*. A man's cattle were just as much his "pets" as were his dogs. The term does not necessarily imply domestication, but rather a relationship between owner and animal. When Joseph Giff explained this relationship, I asked if my *ñuñui*, vultures kept in a cage, would be *shoshoiga*. "Yes," he answered. In fact, after a meal, a Pima woman would gather up the bones and say, "Here, take these to your pets," meaning my vultures. The birds of prey that Northern Pimans kept were their *shoshoiga*, even though they were not domesticated.

In general, cages were made of vari-

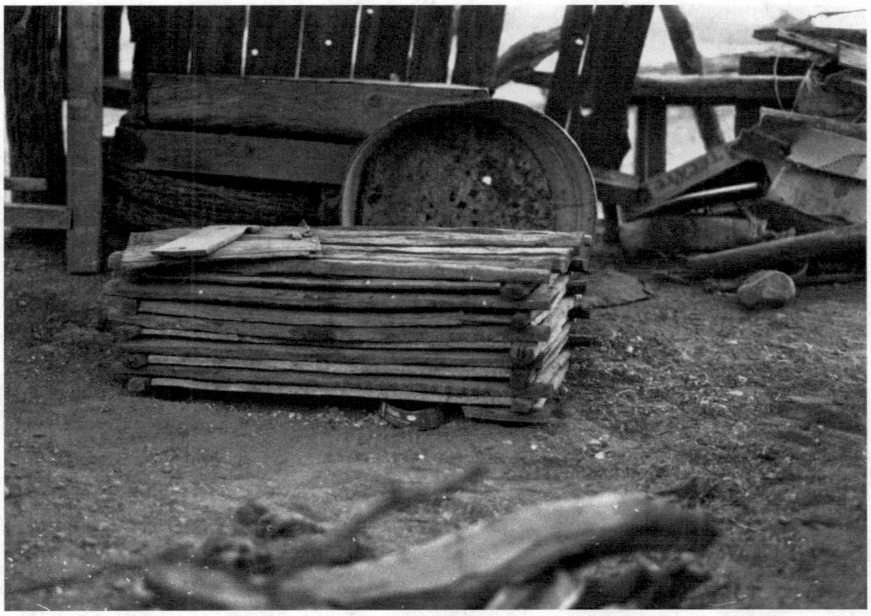

Figure 6.1 *Pima Bajo bird cage, Ónavas (photograph by Amadeo M. Rea).*

ous plants with slender stems of uniform diameter (fig. 6.1). In the far north, Arrowweed was the material of choice. In other areas, willow, Saguaro ribs, Carrizo, Batamote (Seep-willow), and Otate *(Arundinaria)* were used where available. Eagles required stronger cages made from mesquite (or willow?) branches or Saguaro ribs. No figures of Piman eagle cages survive. In any case, two sets of alternating sticks or poles were laid up log-cabin style, the overlapping ends bound with the inner bark of willow, rawhide, strips of rag, or, later, wire (see Kissell 1916:144; Russell 1908:102). The roof was made with a latched row of more rods. These smaller cages were virtually indistinguishable from bird traps.

In a legendary account, Falcon Meeter inhabits a house of Saguaro ribs that the people build for him (Benedict 2001:145). He sings this song:

My house of cactus ribs,
From the western door eagle comes
And flies back and forth.

My house of cactus ribs,
From the western door falcon comes
And flies back and forth.

When used as a trap, these structures were called *eḍpa*. The same word was used for the flat or tapestle-style trap as for the cage-style trap. However, Sylvester used the following terms for coops: *u'uhig kii*, "bird house [cage]"; *okokoi kii*, "White-winged Dove cage," and *ba'ag kii*, "eagle cage."

These structures are still to be seen in many Pima Bajo yards, where they are made of cactus or any other uniform-size sticks that may be handy. They are more often used to keep chickens away from some prized plant, such as a tree seedling, as they are to keep animals confined.

The idea of log-cabin-style confinements is widespread throughout Tepiman country and much of northwest Mexico, at least where Native Americans survive. In larger examples, such as goat corrals, the corners may not be secured so that the poles can be moved from place to place and readily reconstructed just by repiling the logs.

The primary bird species kept by Northern Pimans were macaws and perhaps other species of parrots, Golden Eagles, Red-tailed Hawks, turkeys, and later chickens. Occasionally other wild birds were kept. From time to time, Pimans may have adopted from missionaries or other Europeans such birds as pigeons, Muscovy Ducks *(Cairina moschata)*, guinea fowl *(Numida meleagris)*, Graylag Geese *(Anser anser)*, Chinese Geese *(A. cygnoides)*, and Mallards *(Anas platyrhynchos)*. Of these examples, we can be sure only of the pigeon because it is still being found in some villages.

Macaw Rearing

Beginning around A.D. 1100, macaws became an important trade item to the various cultures throughout the Southwest: Hohokam, Mogollon, Anasazi (Hargrave 1970). The primary species identified from osteological remains in archaeological sites is the Scarlet Macaw, although occasionally other species of parrots appear. This is interesting because the Scarlet Macaw is a tropical species ranging north on the east coast of Mexico to about the Tropic of Cancer (see species account). By the Medio Period (A.D. 1150–1425), Casas Grandes in northern Chihuahua seems to have been an important breeding center for the macaw trade in the Southwest. Birds of all ages, including nestlings, and eggs have been recovered here (Di Peso 1974a:267, 269, 272–273, 1974b:182–185; McKusick 1974).

It is possible that some practices similar to those at Casas Grandes were occurring late in the Classic Period along the middle Gila. One of the Vipishaḍ *vapaki*, pueblos, that the invading O'odham overcame in the western part of modern Pima country was noted for its many captive macaws ("*aadok* [*sic*]") (Benedict 2001:50). The narrative says the Vupshkam, invading Pima, took the macaws themselves, but at a neighboring settlement where many eagles were being raised they took only the supply of eagle feathers. It may seem like an insignificant detail, but it would have been much more practical to march away with some live macaws than with heavy and hungry Golden Eagles.

Northern Pimans were rearing Scarlet Macaws at the time of Contact (Kino 1919:291–292; Manje 1954:168; Pfefferkorn 1949:124; Segesser 1945; L. Velarde 1931). The reported localities were widespread in the Pimerias, including one Papago village, but with no evidence, so far as I'm aware, among Mountain Pima. The reports cover from at least 1701 to 1767. Luis Velarde's account of "muchas guacamayas que crían los pimas" (many macaws being raised by the Pimans) at San Xavier del Bac and neighboring rancherias leaves little doubt that the birds were actually being raised that far north. In these various relaciones, the stated purpose for keeping the birds was for plucking feathers. There are apparently no reports of captive macaws from the subsequent Franciscan period, but few relaciones from this period discussed ecology or natural history. O'odham appear to have been important intermediaries in the trade of macaws and their feathers to the western Pueblos (Tyler 1997:22–23).

We know nothing about the caging used for these powerful-billed birds. One Jesuit said guacamayos were kept in people's houses. Adobe pens were used at Casas Grandes. A passage in the River Pima Creation Story implies that these birds were fed seeds of devil's claw, melon, and broth made from corn (Lloyd 1911:99–101). Both wild and cultivated forms of devil's claw were maintained by River Pima, and melons (after Contact) and maize were important local crops. Various other wild seeds may have been used, such as the fruits of Graythorn. Pimas said they themselves ate only the sweet flesh, not the hard seed, of this plant (Rea 1998a:156). The same may have been true of the mesquite seed, an abundant resource. For the most part, macaws ate food that was part of the Piman dietary, so parrot raising was a luxury for a people with a stable subsistence base in the desert.

It is unknown when macaw rearing stopped among Northern Pimans. Perhaps archaeology will eventually help fill in some gaps. The Scarlet Macaw continued to be one of the most significant avian metaphors in Piman oral literature until the turn of the twentieth century. But most younger Pimans by the early 1900s were no longer familiar with these great birds, and the biological referent for their word *aaḍo* shifted to the only large long-tailed bird they knew, the peacock, then to be seen on Anglo-American farms.

Among the Pima Bajo, Pennington (1980:197) noted that a bright green *perico* was sometimes caught and kept as a caged bird, but by the time of my own fieldwork in 1978 the practice had ceased. The only local parrot on the middle Río Yaqui is the White-fronted Amazon.

Hawk and Eagle Rearing

River Pima kept both Golden Eagles and Red-tailed Hawks captive. They were probably not abundant, but likely every village maintained a few. What is surprising is that no Hispanic visitor noted this Piman practice, and of the thousands of Anglo-Americans who streamed westward through Gila Pima country in the mid-1800s, only one even mentioned this practice. William R. Goulding (n.d.), in his 1849 diary, commented that "most every family have dogs and a kind of pet eagle or hawk which is used in place of a cat."

At the turn of the twentieth century, ethnographer Frank Russell (1908::86) reported, "Eagles and red-tailed hawks were kept for their feathers. All the inhabitants of a village assisted in catching lizards and in furnishing other suitable food for the village bird. The feathers were regularly plucked for the paraphernalia of the medicine-man." He used the past tense, suggesting perhaps that he himself did not see any of the birds.

The picture from these two sources is skimpy but important. It would seem that in the half century between Goulding and Russell, fewer birds were kept. But if we turn to what the Pima themselves had to say about keeping these two birds of prey, we fill in a rather good picture of an important custom that is no longer practiced. There are at least nine firsthand accounts of raptor raising.

Ezell (1961:77), a documentary historian, broke rank and used some information supplied by "an elderly informant":

> The Pimas used to go up the Gila to where there were cliffs in which the eagles nested. One man would take a blanket and lie out all night as close to the nest as he could and just stay there still and listened to the young in the nest until he would tell how many birds there were in the nest. Then they would gather this mescal and make rope from it, and they'd make a loop in this rope. Some men would go to the cliff above the nest and they'd lower one man to the nest. He would get the young eagles and he'd be lowered on to the ground; they would never pull him up again to the top.

Ezell's Pima consultant further noted that the young birds were kept in Saguaro rib cages in the village. Their wing feathers were used for fletching war arrows and the "down" tied in men's hair.

Sylvester Matthias of Komaḍk was the first Gila Pima to tell me about hawk and eagle keeping, and our discussions of this subject spanned more than two decades. In one session, he explained, "Anyone could have an eagle as long as they could scale ropes on cliffs to reach the young. [They] pluck and distribute feathers to relatives."

"How many [Pimas] had *ba'ag* [eagles] in cages when you were a kid?" I asked him.

"Not too many. The Maricopas had them. They get the young *haupal* [Red-tailed Hawk], in cage, for their feathers."

"How long?"

"Must be several years. My uncle, Vincent Jackson's father, would hunt rodents for someone's eagle, so he would get the feathers."

"Would they pull the feathers out?"

"Yeah. Or maybe they would *fall out* like those buzzards that roost in the cottonwood trees."

"Which ones?"

"The end feathers of the wing. We call that *mamchuvdak maakai*. They will be saving it. The *maakai* will buy it. Especially that eagle."

"You know where they got eagles?"

"There's a cave [cliff] on the other end of the [Sierra] Estrellas [northwest end]. The Maricopas know it. They have to use ropes to go down to get the young eagles."

"The white- or brown-headed one?"

"The brown. And there was a *vakoañ ba'ag* [Osprey], that goes after fishes."

"Did the *vakoañ ba'ag* have a nest ground?"

"That I don't know—where they nest."

Sylvester explained that, like the *haupal*, Red-tailed Hawk, *ba'ag* were kept in a stouter cage made of mesquite limbs, with a little house inside. There were customary places for the capture of eagles. The eagles, he said, used the same nest sites on cliffs every year, and one had to use ropes to descend to the nest. All the young would be taken from the nest. A traditional nesting place was at Two Points north of Montezuma Peak in the Sierra Estrella. Theodore Peters's father, Manuel Peters, who hunted in the Estrellas, also knew of this eagle's nest in the saddle between Twin Peaks. "They'll have to scale down by ropes to get the young eagles." Sylvester noted that the Maricopas were primarily the ones to capture young eagles at this nest. Pimas also went to A'agaj Ñe'edam, Singing Antlers or Four Peaks, in the Mazatzal Mountains northeast of Phoenix, to capture nestlings, but not, according to Sylvester, to the Superstition Mountains near Four Peaks.

The eagles were held sacred to the Pima, Sylvester said. The shaman used the outer primaries for brushing away disease or "spirits" in his curing sessions. He thought that no visions or purification were involved with the capture of eagles, but this must have been part of the overall desacralization of the reservation starting in the last decades of the 1800s. At least in recent times, there was no attempt to capture the adults. Arrows fletched with eagle feathers were used solely in fighting the Apaches (and perhaps other enemies), never in hunting.

Another time Sylvester explained: "Pimas just swiped them out of the nest—in rocky places. Lowered someone by ropes. Grab a couple of young eaglets and bring them out. Go back to the same place year after year for eagles, just like herons [that is, both species reused the same nests]. Take as many as they can get hold of." Parents, he said, came to their young in the cage. "Pimas just stripped the feathers off the live eagle and the next year it would be ready for another crop. For someone who had an eagle caged, others got rodents, or even a jackrabbit so he [the owner] wouldn't have to hustle so much [getting food]. At stripping time everyone who brought food got feathers."

Francis Vavages of Haashañ Keek likewise discussed captive eagles: "They used the feathers, because I heard they're tough, won't wear out [*sic*] like any other feathers. This man from [Sacaton] Flats, he was telling me about it. He knew where there was a nest of the eagle, a big one. It's right by the Red Rock, that mountain over there. There's a cliff up there some place. And that's where his nest is. So every year he [the eagle] has little ones, two little ones, young ones. He figured that he might have some young ones again because he wanted to raise one of those things.

"So he went up there and sure enough, he found them. So he picked one up, and he brought it home to Flats. That was way back, I guess. He said after it got big, it stands about that high [a meter tall]. And they even made a cage for it, of wood, you know, closed like that, to let him stay in. And they even made a table to eat on. The same wood. Put it together, put the food or whatever they can kill, put them on there. So this man who was telling me said he used to go out and hunt birds for it. And once a month he would get paid with two feathers. [laughter] Two feathers!

"So I guess he [the eagle] got hungry one time there, and the man had gone some place. So she [his wife] decided to go in there and feed it with something else. And I guess he was already mad. So she turned her back. He jumped and clawed her from the back.

"So when the man came [back], he had to kill that *ba'ag*. That's what this man was telling me, that was feeding him. When they was small, let him go out and kill birds and feed him. And they said the way they buried this bird was they buried him like a human. They went up on that Olberg Mountain, someplace there, and buried him up there. Make a big hole there, and put him down. And put rocks over it. And he didn't know [remember] where it was. But he knew that's what they did: buried him like a human person."

"Did they take the feathers first or bury him with the feathers?" I asked.

"Just buried him with the feathers, I guess. They didn't want to . . . [not finished]."

"There's something about if your *haupal* or *ba'ag* dies," I said, "it's supposed to be bad luck?"

"Yeah, yeah, yeah. That's what they did."

Pima storyteller Anna Moore Shaw (1968:97–103, see also 1974:32) gave a story in an historical setting (probably the Salt River Reservation) of a Pima youth who seeks *ba'ag* on Tall Red Mountain (Camelback Mountain). He wanted to keep the eagles for pets. There is even a reference in the story to a real person, Chukuḍ Naank, Great Horned Owl's Ears, the renowned calendar stick keeper (see Russell 1908: plate IIb). The boy watches two eaglets, planning to take them when they are large enough. When he decides they are ready, he climbs the mountain and gets lost. He goes to sleep and in a dream the mother eagle appears to him and says, "Your people are lovers of birds and take good are of them; but a mother's love is greater. It was I who caused you to be lost on this mountain. But I'll let you go if you will promise to leave my chicks alone." The boy agrees. He calls the eaglets *napawchick (napochik?)*.

Joseph Giff said, "They kept young eagles, too [in addition to *haupal*, Red-tails]. They caught them in the mountains and kept them for pets. They raise them for pets. They have them for feathers. Maybe couple times they take the feathers off, then they turn them loose. Then they go on."

Ruth Giff added, "Grandpa had one. And we used to go to the river and catch them *vaptopĭ* [fish], for food." Joe confirmed that this was the Golden Eagle, the one with the yellowish head.

Even though the Tohono O'odham had fewer food resources, they, too, kept hawks as a source of arrow feathers. They were "housed in wicker cages and fed with rats and lizards. The tail feathers were pulled whenever they grew long enough" (Castetter and Underhill 1935:43).

Much of what applied to eagle keeping was true of keeping the Red-tailed Hawk as well. Sylvester noted that the *haupal* nests in the arms of the Saguaro. His father would use *vaapai*, Saguaro ribs, to lift the young from the nest, then pull out the feathers they needed and replace the birds. Pimas also took the young out of the nest in early May and raised them for their feathers. Cages for keeping hawks were made from Saguaro ribs. Boys would hunt food for them. The birds were fed rodents, rabbits, and chopped jerky. Those who contributed food expected to get some feathers from the caged Red-tail. The birds, he said, were not tethered in the manner used by the Pueblo Indians. Feathers were pulled out at the time of the year when the bird "sheds" (molts). Wing feathers were good for fletching arrows, but tail feathers were better. The next season they would look for more young and let the caged birds loose "because they were superstitious [*sic*]" about keeping them longer than a year or two lest they die in captivity. Apparently, even after released, captive-reared Red-tails would return for a while to be fed. "They're tame. They'll come back for food, after rodents." He thought that the River Pima discontinued the practice in the early 1900s.

What Sylvester did not say is that the nestling Red-tails produced tail feathers that were barred light and dark brown, whereas birds kept for a year would probably grow out the bright rust-colored tails of the adult plumage.

George Kyyitan of Bapchule told this story about a captive hawk:

> A long time [ago] the old people [men] just have [wear] that *atosha* [breechcloth]. And they had somebody died there. And they kept him inside there, [in] that round house, *olas kii*. And they go in there, one by one and see him.
>
> There was a *vatto* [arbor] there, and this man that lives there, he's got a pet hawk, a Red-tailed Hawk, *haupal*, pet. Every time, they have to feed him, give him something. And he does that. He said, this *haupal* was sitting up there, on top of that *vatto* there, and this man that will go in there, went in there and stooped down, I think, crawl in there. And his *atosha* was hanging down and his balls shows over here [on the side]. Then this bird saw that one, so he thought he could get them. And he went [flew] over there and caught him [from behind] with his [talons].
>
> This man was trying to chase it off.

> Ah! Ah! Ah! It caught him *right* there, right on the balls. Couldn't [wouldn't] let him go. Just keep hanging on there. He tried to get him loose and couldn't make it. Then he take one leg and just brake it up.
>
> Maybe just a joke, or maybe it's true. Francis [Vavages] told it; he knows it.

Each of these accounts helps flesh out the details of Piman eagle and hawk rearing. The birds were never domesticated: they were taken wild, then released and not bred in captivity. Known nesting places or eyries of eagles were sought year after year. People would travel considerable distances to some eyries. Either one or both nestlings might be taken. An eagle cage was large enough for a person to walk into. Caged eagles were supplied with a platform on which to eat. This may have been to help protect the bird's tail feathers from abrasion while it was eating, but this reason was never stated.

The individual who captured and kept the raptor was considered its owner, but the feathers it produced were doled out to anyone in the community who helped supply food for the captive. Most often the village youth were the hunters. Reptiles, rodents, fish, cottontails, and even jackrabbits were brought in. Feathers were almost currency in this exchange system. The feathers most sought after were the long broad flight feathers and the *viigĭ*, fluffy feathers, from the bird's crissum (undertail coverts) and thighs. Certain outer primaries of the eagle were the domain of the shaman, who would buy them.

After feathers were harvested one or more times, the birds were released, and it was considered a misfortune to have one's eagle or hawk die in captivity. But if an eagle did die or was killed (as in Francis's story), it was given all the accords of a human burial. Small circles of stones still to be seen on flats near historic eyries may represent eagle rather than human burials. (Shamans, because of their great power, were sometimes buried in places remote from the village, and the same may have been true for eagles.)

Although it seems that River Pima had largely dispensed with religious customs as part of a pattern of desacralization initiated in the last quarter of the 1800s, eagles were still considered power birds, and their feathers had to be "tamed" by a *maakai* or shaman before they were safe for general use, such as for hair decoration. George Kyyitan told a personal story of the powerful sanctions that could result from failure to observe proper restrictions (see Golden Eagle account also).

The keeping of Red-tailed Hawks was apparently surrounded by fewer sanctions because they were not high-status birds, nor were they a source of staying sickness. Culver Cassa thinks this species was of greater significance to the Maricopa and that the adjacent River Pima adapted their keeping for more utilitarian reasons: the birds supplied a ready source of feathers for hunting arrows (see Spier 1933:74). The prized bright rusty-colored tail feathers, however, do not appear until the bird is more than a year old, so to get rufous feathers the hawk would have to be kept until it molted into its adult plumage.

George's story suggests that the Red-tail may have sometimes been tethered to the ramada or arbor that was once found in every Pima yard. However, the bird in his story may have been released, but was still returning to be fed. (Its apparent owner had just died, and the bird was hungry.) This situation might explain Goulding's 1849 comparison of kept eagles and hawks to household cats.

Finally, Anna Shaw's comments, put into the mouth of the parent eagle, that the Pima were lovers of birds and took good care of them echoes Fr. Pfefferkorn's (1949:117) statement from two centuries earlier that Pima had an "inclination" toward raising birds and other animals.

There is a mythological account of an eagle quest by the twin sons of White-clay Eater Woman that reinforces many points from these ethnohistorical accounts (Benedict 2001:97–122; Densmore 1929:56–59; Herzog [1929]). When the boys are old enough—that is, after they have mastered various hunting skills and before they seek women—they set out on their quest. The boys take turns climbing to the cliff containing the two nestlings. The smaller nestling is black (actually the older bird with more dark emerged feathers), and the larger is white (the younger bird still all in its fluffy natal down). The birds are hung around the neck for the descent *(kushvi'otk)*.

During the return home, first one then the other eagle dies because of a storm that brings cold wind and rain. Each bird is buried in a hole in the earth, together with its captor's clothes and bow. (In one version, the birds are cremated.) During the night, the boys' mother resurrects the birds, taking them home. Coyote, in a rare helping mode, builds them a house (the word used here is *kii*, house, not *edpa*, trap/cage). The birds are fed lizards and other small animals that were hunted. After the feathers are plucked once or perhaps twice, the boys' two birds are released.

Perhaps the Hohokam once kept eagles, as is suggested by an event in the Pima Creation Story. When the invading Vupshkam (that is, O'odham) reach the western part of modern Pima country, they came to one Vipishaḍ *va'aki*, pueblo, where the people had many eagles for pets. They fought and overcame the settlement and took as booty the supply of eagle feathers. The *va'aki* is mentioned only in one version of the Origin Story and was adjacent to another pueblo where Scarlet Macaws were being raised (Benedict 2001:50).

There is no evidence that Pimans trained birds for falconry or flew their birds as Old World peoples do. The transition from captive birds to free individuals was assisted by feeding the released birds until they acquired hunting skills. Eagle keeping, although it had utilitarian overtones, must have been related to eagle killing. Eagle capture and eagle killing, at least among Tohono O'odham, was essentially a power quest experience for young males (see *ba'ag* species account). To what extent this was true of River Pima is unknown.

Golden Eagles and Prairie Falcons are usually paired in Piman song and oration, both being power birds, both dangerous because they are sources of *kaachim mumkidag*, staying sickness. Why were not falcons kept? Their feathers were needed for curing *vishag mumkidag*, falcon sickness. Prairie Falcon bones frequently occur among bird remains in Anasazi sites north of Pima country. There is at least indirect evidence that this more northern culture may have tried to keep falcons. I identified an archaeological skeleton of Prairie Falcon with an injured and mended wing that had been buried in a kiva floor. Underhill (1938:110, 1946:250) recorded two sets of Tohono O'odham songs that suggest that falcons as well as eagles may have been caged at some time. One of the problems in the historical record is that writers universally misidentified the Piman folk taxon *vishag*.

Underhill (1936:62–63) was told by the Papago woman Maria Chona that her son, an incipient shaman, "used to bring birds home to the house and keep them; he could speak with birds. Once he kept a hawk in our house and killed jackrabbits for it. Once I saw him coming home in the evening after he had been after the horses, and there was something shining on his shoulder. It was the eyes of an owl. 'What will you do with it?' I said. So he let it fly away. But I think he spoke with it after that, behind the mountain."

The Hopi and the Zuni, as well as other Puebloan tribes living north and northeast of the Pimans, also kept eagles and occasionally other hawks (summarized in Tyler 1979:43–49). The details are perhaps better known for the Hopi (Voth 1912). As with Upper Pimans, the birds were captured primarily for ceremonial usage. Hopi eagle keeping differed from Piman in a number of significant ways. Hopi Eagle clan and its related clans "owned" the eagles and had exclusive rights to their eyries within a fifty- to sixty-mile radius of the pueblos. The numbers kept were substantially greater, thirty-five birds being recorded for Oraibi pueblo in one year. The young birds were taken in the spring, tethered by one leg on rooftops, and fed rabbits and rodents. In July, on the morning following the Niman ceremony, closing the kachina season, the eagles were captured and strangled. The flight feathers were then plucked from the dead bird, then the carcass skinned with the remaining body feathers still attached. Finally, the eagle carcasses were buried with food offerings (blue *piki*, wafer bread) in an eagle cemetery. The sacrificed birds were sent with prayers for rain but were expected to return to earth the following spring to nest.

The pluck and release tradition seems to have been more widespread among other Puebloans (Parsons 1939:28–29). And with Jicarilla Apache, Opler (1936:207) noted that the young were plucked twice, then set free.

At least some Puebloan influences can be found in the ceremonialism of Upper Pimans, most conspicuously in the Navichu complex. The details of eagle rearing seem to differ strongly between the two groups. For both Puebloans and Pimans, eagles are powerfully associated with rainmaking. Another point of potential similarity may go unnoticed. As noted, beneath one of the traditional eagle eyries on the Gila various, small "shrines" may represent eagle burials. An explanation for these circles has been lost from memory. Dog cemeteries (Rea 1998b:190) still exist in certain Pima villages, so other animal burials would not be out of place.

Other Wild Birds

A few other wild birds were kept. On the Gila, Russell (1908:86) observed that Sonora Doves, as White-wings were often called, "were and are yet confined to log-cabin cages built up of arrowwood rods." He did not say if these birds were captured as nestlings (they probably were) or what their function was.

During my early work on the Gila, I saw a caged Black-headed Grosbeak and Northern Mockingbird. Sylvester Matthias also noted mockingbird keeping in earlier days. David and Rosita Brown reared some free-flying Western Screech-Owl young, but these cases, it seems, were exceptional. Other species may have been kept at an earlier time. Akimel O'odham and some Tohono as well told me that *o'okokoi*, White-winged Doves, were regularly raised. They were kept under the *vatto*, not inside the house, and were later released, not eaten. People just liked to have them as pets. In the old days, Tohono O'odham similarly kept "a few species of birds" as pets, but no further details were given (Castetter and Underhill 1935:43). The Yaqui and Mayo aboriginally caged mockingbirds (Beals 1943:12). The keeping of mockingbirds as pets is recorded also for the Akwa'alá (Paipai), Cucapá, Maricopa, Yavapai, and Hualapai (Drucker 1941:131), all Yuman speakers.

Pennington (1980:196–199) noted that Pima Bajo confined White-winged Doves in round or square cages made from Carrizo, Otate, or Batamote (Seep-willow) stalks. They were kept as pets and fed worms, locusts, beans, chiltepines, and wild fruits. At maturity, the birds were eaten. He noted that mockingbirds were kept because they sing well and were fed on this same diet. Northern Cardinals were also kept here for pets. Whether this was a traditional practice or part of the widespread Mexican practice of keeping songbirds confined in inadequately small cages is difficult to say.

Domestic Birds

Rock Dove, Domestic Pigeon, *Columba livia*

The common domestic pigeon, known to ornithologists as the Rock Dove, may be found feral about some northern Piman villages, in particular those settlements that have taken on a more urban demeanor. Although it flies free, it still depends on humans for food resources as well as for buildings for nesting and roosting places. The birds most often seen are the "wild" type—that is, mostly gray with a white rump and metallic green and purple head, neck, and chest.

Pigeons have been among Pimans long enough to be formally named. Sylvester Matthias, a Kuiva O'odham, Downstream Person, said, "*Paplo* are pigeons, but George Kyyitan [a Ta'i O'odham] would say *paploñ*. People ate the squabs. Can make soup out of them, but they have to boil them!" In either case, the name is a loan word from Spanish *paloma*. The word is interesting in that it is in the plural form, as indicated by the *p* terminating the first syllable. Thus, it joins a number of other words, such as *ants, leaves*, and *teeth*, that are fixed in the plural or aggregate.

For Tohono O'odham, Mathiot (n.d.:143) gives *papalho*, same as *palóoma*, both derived from Spanish. I have not heard the latter used for the bird among Akimel O'odham, but a famous Pima *maakai* (shaman) and merchant who lived across the river from Sacate was called Palóoma.

Irene Hendricks of Casa Blanca used *paplo* for the domestic pigeon, but noted that the Ring-necked Dove or Ringed Turtle-Dove *(Streptopelia risoria)* as well as other captive doves are called *jujkam hoohiga*, "Mexicans' Morning Dove."

Domestic Chicken, *Gallus gallus*

The Old World chicken has been among Northern Pimans at least since the establishment of Spanish missions in Pimería Baja in the early 1600s. (Despite claims that the chicken was in the New World before Europeans arrived, all archaeological bones so far have stubbornly refused to yield pre-Columbian dates [Carter 1971].) Whether the chicken—like various Old World cultivated crops such as melon, watermelon, and perhaps wheat—arrived in the north by indigenous trade in advance of the Spaniards is not known. The earliest record of the bird in the north appears to be in the Névome vocabulario (1660s), where gallina is listed as *totori* and *totoli*, with the annotation that the word is a mexicanismo. Three centuries later the chicken in this community was still known as *totol*.

Several accounts of Piman chicken rearing appear in early Jesuits' accounts. For a broader survey of references to chickens appearing in colonial documents pertaining to the Greater Southwest, see Schroeder 1968:106–108. In his 1737 relación, Fr. Philipp Segesser (1945:148, 151) penned this description of a Pima house: "The straw huts have a single living room where father, mother, children, dog, cat, and chickens live together, and in which fire is built." He added, however, that Pima and Papago of the time would not eat chicken meat, although his houseboys raided eggs from his chicken coop for five months before being caught. He himself raised ducks and geese at his mission, in addition to hens.

Padre Juan Nentvig (1971:83), writing in 1764, said, "Se crían en Sonora bastamente gallinas domésticas; en algunas partes también patos y guajolotes" (They raise in Sonora plenty of domestic chickens; in some parts also ducks and turkeys). He mentioned wild doves and various species of quail as well, not saying that they were raised. This reference may primarily concern the Ópata because the bird names included are in that language.

German father Ignaz Pfefferkorn (1949:117), stationed at Piman villages in northern Sonora and southern Arizona, then at an Eudeve village before the Jesuit expulsion, wrote *tuchturi* for "Spanish chicken," but did not specify whether this name was Piman or Eudeve. He raised an interesting question, however: "It is not known whether our European chickens were in Sonora, or even in America, before the arrival of the Spaniards, or whether they were first brought there by the Spaniards." After noting that the Indians had their own name for the birds, the Jesuit then proceeded with a folk linguistic argument: "However, there is in Sonora a species of native bird, called *chachalaca* by the Spaniards, very similar to our fowl. These birds are always called tuchturi by the Indians, and hence it is more probable that our chickens were unknown to the Indians previous to their introduction by the Spaniards, and that when they saw the European chickens, the Indians named them tuchturi because of their similarity to chachalacas." Though the remainder of the proposition may not particularly advance his argument, it does shed important light on the native Sonorans' responses to the two galliform birds and to bird keeping in general: "This theory could in some measure be confirmed by the fact that the Indians have always eaten chachalacas, but they dislike the taste of our chicken and its eggs. Certainly, they enjoy raising chickens, not to eat them, but to sell pullets and eggs to the Spaniards and further because they have an inclination for animal and bird raising."

This account is in keeping with Pennington's (1969:154) impression among the modern Northern Tepehuán. "Chickens," he wrote, noting their abundance, "are rarely eaten; they seem to be valued primarily because their eggs can be traded to the mestizos for corn and beans." (At Zuni pueblo around 1880, Bourke found a similar situation [Schroeder 1968].)

Among Upper Pimans, the chicken is known as *chuchul*, the word being in a fixed plural or aggregate form. Fr. Kino no doubt introduced chickens to his ranches in Pimería Alta along with cattle and horses. His goal was the sustainability of population centers. Both padres Pedro Font and Francisco Garcés remarked on chickens being raised on the Gila by 1775 (Font 1930:4:44; Garcés 1900:104; Garcés 1965:7), and they seem to have been well established, if not numerous, by the time of the California Gold Rush. J. Ross Browne (1869:110) noted that Pimas sold six hundred chickens to the military in 1861. The significance of chickens in late historical times is uncertain. At some time, they entered the Piman dietary.

I raised the question of the chicken's origin to Sylvester Matthias. "That's the question. It's not mentioned in the tradition [Origin Story, legends, oral history]. Not like *a'aḍo* [Scarlet Macaw], turkey, and woodpecker. The Spanish or the Pilgrims—that's when the *chuchul* get familiar." (By "Pilgrims," he means an Anglo introduction.)

There is some information on breeds. During Lloyd's 1903 stay among the Pima, he found "a big-footed Mexican *choo-chool*, or chicken" among the Pima of the Santan area (1911:5). At this same time, Russell (1908:86) wrote, "Until the recent introduction of large American breeds a small breed of poultry had been kept for several years. This had been obtained from the Mexicans, from whom also the manner of calling 'pul,' 'pul,' 'pul' was derived. With the new breeds came the American custom of calling 'peep,' 'peep.'"

George Kyyitan and Leonard Pancott from the middle Gila villages said that when they were young, there were only three breeds of chicken: *s-jejevk chuchul*, the Plymouth Rock; *s-toota chuchul*, white; and *s-vepegĭ chuchul*, red chicken. Downstream I asked Theodore Peters what kind of chickens he recalled from the early days. "Rhode-Island Red [was] the only kind. At night they got a ladderlike, they go in by themselves, into the *chuchul ha kii*."

"Did anything bother the *chuchul?*"

"*Baaban* [Coyotes are] the only guys that kill them—not foxes. *Visag* [Prairie Falcon] go for *chuchul*—just like a bullet—maybe kill itself. The way I heard, they use their wings to kill a chicken. *Ban* will swing its head around and put it on his back, just like he was carrying a sack. That's what people say. They always swing it, hold it in the teeth. Too heavy to drag it, too heavy to carry it in its mouth—and slow. Skunk makes hole in the egg. *Chukuḍ* [Great Horned Owl] no bother to chickens."

Discussing poultry, I asked Ruth Giff, also from Santa Cruz village, how many *chuchul* her family kept in the early 1900s. "About ten maybe. We didn't even eat the eggs. Some we saved for more chickens."

"Did anything besides the *baaban* bother the chickens?"

"Nothing, unless a *vishag*. [They had] *chuchul kii*—chicken house. Everybody had chickens."

"Did you have *totova* [turkeys]?"

"No. We had chickens. Some people do [have] *tova*; not very many."

"Ducks and geese?"

"No."

However, Francis Vavages said *chuchul nonha* (chicken eggs) were eaten. "Can you eat them when they start to form?"

"No. Just let them go," he replied.

According to Sylvester, *kokokok* is the clucking of a hen, whereas *chuchul am kuuk* indicates chicken calls. He also noted that *hihivdag* is small pox, but *chuchul hiivdag* is chicken pox, obviously a back translation of the English name for the disease. Here in the north, the rooster is called *keli chuchul*, "old man chicken," and the hen *oks chuchul*, "old lady chicken," or just *chuchul*. Mathiot (1973:214) adds *chioj chuchul* for "young rooster" and *uvĭ chuchul* for "young hen." A chick is *chuchul maḍ* (pl.

chuchul maamaḍ). Chickens were included in the category *shoiga*, generally translated "pet," the same as horse, cow, dogs, and other kept birds.

Anna Moore Shaw (1974:30), originally of Gila Crossing settlement, recalled this teaching from the old days, "If a hen should crow like a rooster, it is a bad sign. The hen should be killed to stop evil spirits [*sic*] from bringing harm to you and your family."

Village feast-day celebrations among both Tohono and Akimel O'odham might have featured a rooster pull, called *chuchul mo'o bebhe*, "chicken head pull." A rooster was buried in the sand up to its neck. Mounted men raced past it, leaning over the sides of their steeds, attempting to pull the bird out of the ground by its head. The one who succeeded then had to defend his prize in a free-for-all. "They wear just Levi jumpers because it's strong, the straps; no shirt because it tears up," noted Theodore Peters. The one who escaped finally with the head of the chicken was the winner (Joseph, Spicer, and Chesky 1949:85; Rea 1998b:108; Underhill 1946:323). So far as I know, cock fighting was unknown in Northern Piman country.

Two plant names are derived from a fanciful resemblance to the chicken. Among Pima Bajo, the blue flowers of a day-flower, a small plant in the Commelinaceae, were called *totol kakam*, "chicken cheeks." In the north, River Pima likened pears hanging in a tree to chicks, calling them *chuchul maamaḍ*, "chicken babies."

Pfefferkorn's suggestion that the Piman (or Northern Sonoran Indian?) name for the chicken was derived from the native name for the chachalaca is a logical one. If that were the case, one would expect a marking reversal, with the new item initially marked or modified by an adjective. After time, as the introduced item, in this case the domestic bird, became more culturally significant, it would take the unmarked name, and the original referent, here the chachalaca, would become modified or marked. It appears this was the case: the native chachalaca, found in the periphery of Pima Bajo country, is now known as *sha'i totol*, "brush 'chicken.'"

As noted, the two main forms of the chicken name among northern O'odham are *totol* in the south (Pima Bajo) and *chuchul* in the north (Pima Alto). (The consonant shift is to be expected, the vowel shift is not.) Cognate terms are found among other Uto-Aztecans: *totorí* (Rarámuri or Tarahumara), *tótoli* (Yaqui-Mayo), *totoi* (Mountain Pima, uncommon). Seemingly different was the colonial Southern Tepehuán *takoko*, gallina de castilla, and *takuk kulidade*, rooster (Rinaldini 1743).

Pfefferkorn's hypothesis may not be correct (or may not be the complete answer). The Nahuatl *tootolin* indicates turkey or domestic fowl. "From its basic meaning 'turkey' this has been extended to chickens and doves," notes Karttunen (1983:248). So the Nahua word may have moved into the fronteras (border regions) with the chicken itself. But what about the term *chachalaca*, which would appear to be Aztecan also? Karttunen (1983:44) writes, "That this is a back loan from Mexican Spanish is shown by the lengthening of the stressed vowel and by the word-final glottal stop, both of which are characteristic Spanish loans into Nahuatl." But then the other Aztecan names for the bird must be considered: *chachalacámetl, chachalacani*, and *chachalaquiliztli* (Macazaga Ordoño 1985).

At any rate, there was no simple transmission of the fowl with its name throughout northwest Mexico as there was with wheat (Rea 1997:337). In the extreme south of the Piman corridor, Mason recorded *tovwa* (plural *topwa*) for "hen" and *suípir* for "chicken." The first appears to be the turkey word (*tova* throughout the north). Among colonial Southern Tepehuán, Rinaldini (1743) listed the chicken as *takoko*, gallina de castilla, which he contrasted with *tova*, gallina de la tierra. Pennington (1969:154) found the chicken to be called *takúku* among Northern Tepehuán. Another noncognate term, *shupon, supona*, appears commonly among Mountain Pima. The now extinct Eudeve used *pedéko* or *bedéko*. The also extinct Tubar said *kakayá*, which might suggest the Piman quail word, but more likely is simply onomatopoeic. (Their quail was *tsokorá*.)

In summary, Pima Bajo, living at the very edge of the northern range of the chachalaca, used a *totol* word for chicken, as did their neighbors the Taracahitans. In Pima Alto, speakers all now call chicken *chuchul*, with the colonial intermediate *tuchturi*. Still other names were used by geographically adjacent Sonoran speakers.

Domestic Turkey

At least two breeds of turkey, the Large Indian Domestic (LID) and the Small Indian Domestic (SID), were widely raised among the Anasazi and other prehistoric Puebloan peoples of the Southwest (McKusick 1980, 1986; Rea 1980). The birds were initially used for feathers, but by A.D. 950–960 there is evidence from Mesa Verde that some individuals of the LID breed were being eaten. (Merriam's Wild Turkeys were hunted and may have been eaten at Mogollon Rim sites as early as around A.D. 600.) Also it appears that flocks of turkeys may have been herded into agricultural fields to help control grasshoppers and other damaging arthropods (Lang and Harris 1984:97–101; McKusick 1986:9; P. Velarde 1960:45–51, giving a widespread Puebloan folk tale; see also Cushing 1901:54–62).[1] But the remains of this species are almost totally absent from Hohokam sites. To my knowledge, no domestic turkeys have been recovered from protohistoric or historic Piman sites in Arizona.

But did Pimans ever capture and rear the chicks or incubate the eggs of the wild bird? The evidence is meager. Mountain Pima, living in the pine-oak country that is the natural habitat for the bird, gave me contradictory information. Pimas from the Nabogame-Yepáchic area in Mexico said that wild birds were taken either as eggs or chicks and raised. They had numerous birds about the rancherias, but these birds seemed to be mixed breeds, some erythristic. But at Yécora, a lower-elevation Mountain Pima settlement, I was told that chicks from wild-taken birds or eggs would soon die when kept, unlike the domestic breeds. Clearly, more needs to be learned of the details of turkey raising in the Sierra Madre.

Down in the tropical deciduous forest and thornscrub, Pima Bajo Pedro Estrella said that the turkeys at Ónavas are caught as young from wild parents in the mountains. Pennington (1980:244, 364) was told that around seventy-five years earlier domestic turkeys were maintained by Pima families in this pueblo.

One colonial reference from southern Pimería Alta observed that "[i]f [Wild Turkeys] are brought home with an idea of feeding or raising them, whether they be young or old, they refuse to eat and die from mourning in a short time," although some Spaniards managed successfully to rear wild birds from eggs brooded by a chicken (Pfefferkorn 1949:118).

Russell (1908:86), in his enumeration of domestic animals, makes no mention of turkey under the entry "poultry." I asked

Francis Vavages, "Did the Pima ever raise *totova?*"

"Yeah. Sometimes they get little ones and raise them with the chickens. They used to be one in that high place in [Sacaton] Flats. He's a Hopi [Elwood Myron, married to a Pima]. He raised turkeys, and Thanksgiving he'll sell them to people. But lot of times I've seen them, maybe just one or two among the chickens. Everybody used to have their chickens. I had some chickens for a while, there, but there's too many dogs. Dogs will kill them, too."

I posed the question to Leonard Pancott, "What about *totova?* Did anybody raise *totova?*"

"They used to. My mother-in-law used to raise them here. She lived down here. She'd raise a few. There's people who owned chickens and turkeys in them days. Not business or anything. They raise their own, a few of them."

"And what would they do with the *totova?*"

"They used to eat 'em."

"The whole birds or the eggs or both?"

"You know, I never know of anybody eating *totova nonha* [turkey eggs]. So they must never eat them. Maybe somebody's got two, three turkeys. And I think mostly it's just for Thanksgiving, and you know, holidays. Eat the young ones." (Both Leonard and Francis were referring to turkeys being raised at Haashañ Keek, Sacaton Flats, by the same family.)

Among Tohono O'odham, Joe Lopez of Vaya Chiñ said that turkeys were too delicate to raise in that country. A Papago woman at a roadside rancho in northern Sonora on the way to Caborca had a not very large white-rumped male turkey. When asked in O'odham about this bird, she explained that she had a female with eggs, but no *cría* (brood) yet (late May). They apparently breed here, along with her numerous *chuchul*.

Among River Pima, eagle and Red-tailed Hawk raising dropped out sometime in the first quarter of the twentieth century when arrow making and bow hunting were replaced with rifles and shotguns, and traditional healers became a thing of the past. I have not seen any turkeys in the villages here. Chickens continue in some villages, even Sacaton, but as more people move into urban-style housing units on the reservation, traditional animal husbandry ceases.

Tohono O'odham settlements still largely retain a ranchería ambience. Cattle raising is important; men maintain and use horses, and a few chickens are likely to be encountered about houses.

The Pima Bajo pueblo of Ónavas is remote and rural, though now equipped with electricity and television, which has changed the communal structure. Men are typically mounted, cattle ranching being an important profession. Rooster calls periodically punctuate the night, and the occasional cow or burro wanders into town, arousing a dozen or more dogs who contest its presence. In the past, I've seen a flock of turkeys here, though they are much more common within the rancherias of the Mountain Pima, wandering casually into houses searching for spilled kernels of maize. Don Pedro of Ónavas said that local turkeys are caught as young from wild parents in the sierra, but Mountain Pima gave conflicting reports on the efficacy of this practice, as we saw.

7

Feather Use

As seen in the individual accounts of the various species, birds served as powerful symbols and metaphors in Northern Piman culture. Birds figured in myth, tale, oration, song, and ceremony. Not surprisingly, the use of feathers by Pimans was once pervasive. Birds were perhaps even more important to the People as sources of feathers than as food resources. More species were sought for their feathers: Wild Turkey, Turkey Vulture, Great Horned Owl, macaw (certainly Scarlet, but probably Military at one time as well), Golden Eagle, hawk (Red-tailed and probably Harris'), falcon (Prairie and perhaps Peregrine), and flicker. Less certain was the actual use of Blue Grosbeak, cardinal, and oriole feathers to make stringed offerings. These last three were important in color directions and as symbols in oral literature. "Dangerous" feathers included those of the Golden Eagle, Great Horned Owl, Turkey Vulture, Prairie Falcon, roadrunner, quail, and hummingbird. The connections between these birds and staying sickness are dealt with in more detail in chapter 5 as well as in their individual species accounts.

Feather use might be divided into two categories, ceremonial and utilitarian, though these classes are constructed by outsiders and might not correspond to anything in the Piman mind. Ceremonial uses might well be considered utilitarian to O'odham, although utilitarian uses might not be ceremonial. At any rate, it is hard to imagine any Piman ritual in which feathers did not play a conspicuous and integral part.

Ceremonial Uses of Feathers

In the past, desert agricultural peoples lived in a precarious marginal environment. Rain meant food or starvation. Even many important wild resources, both plant and animal, depended on rain. But precipitation itself wasn't always dependable. Not surprisingly, desert peoples such as the O'odham and the Pueblos evolved elaborate ceremonial systems with ultimately a single major aim: to ensure rain.

For Upper Pimans, at least, tobacco smoke, ceremonially offered, and Saguaro wine, consumed in a complex liturgy, bring wind, thunder, clouds, and ultimately rain. Rain is blind, so he must be lead along by his brother Wind (in this case, both are male). That is why most summer rainstorms in Piman country are proceeded by violent winds. The Desert Bighorn, snakes, toads, centipedes, dragonflies, and Red Velvet Mites are rain associates. The songs for rain are poems featuring these animals. Likewise, certain birds were considered by Pimans specifically as rain birds: the turkey, eagle, Turkey Vulture, Prairie Falcon, gulls, and swallows. Feathers of most of these birds, except the swallow and perhaps the gull, played a role in rainmaking ceremonies. At least formerly, one of the functions of a *maakai*, shaman, was to sing songs and use specific feathers from his store to request rain. The choice of feathers was not random. Each bird or other animal appearing as a rain animal was based on its mythological persona.

The turkey seemed to me an unlikely bird to be associated with rain except that these birds are found in the mountains, and rain-filled summer clouds build up over these ranges. There are stories about Coyote and turkey in the pine country, one of which makes this connection explicitly (Benedict 2001:165):

> The turkeys were holding a rain ceremony, and they sang:
> Our feathers bring the clouds and it will rain.
> Our feathers bring the clouds and it will rain.

I asked artist-author Barton Wright, who works with Hopi ceremonialism, what connection there might be. He said if you hold up a turkey feather by the base, you will have a visual metaphor of a rainstorm. The uppermost part is white, symbolizing the thunderhead. Below it is the black band, the dark underpart of a summer rain cloud laden with moisture. Next there is the brown part, the sky, finely flecked with gray, the falling rain. Each turkey feather, so frequently used on Hopi pahos, prayer sticks or votive offerings, is a rainstorm. Whether there was a similar symbolism in the Piman mind is unknown. We can only note that in general Mesoamerican symbolism was broadly shared throughout this region.

The connection of Golden Eagle with rainmaking can be traced to the killing of Eagle Man in the Creation Story. Culture hero Se'ehe, or I'itoi, tells the people to watch for a white cloud over the cliff that will indicate his success over the monster. After killing Eagle Man and his son, Se'ehe plucks their feathers, both the dark ones and the downy ones, throwing them in the different directions to be the source of storm clouds. Henceforth, the *viigĭ*, fluffy feathers, of the eagle will represent rain clouds and will be an aid, together with song and tobacco smoke, in calling storms (see eagle account; also Benedict 2001:183–184; Densmore 1929:52; Underhill 1939:105–110, 1946:243–252).

For instance, around the turn of the twentieth century, Maria Chona was diagnosed with enemy scalp sickness. During her *wusota*, curing session, the singers sat around her. "Each had four down feathers from under the tail of an eagle fastened to a little stick. That is one of the most sacred things we have, those eagle down feathers, because they bring the rain; they almost *are* the rain. All the men who have Prisoners [scalp effigies in a basket] have those feathers to keep their Prisoners quiet" (Underhill 1936:21).

The Prairie Falcon is intimately associated with its larger relative the Golden Eagle, so it, too, is a rain bird. Its feathers would be used in calling rain and its name invoked in song. For instance, one song says that when eagle alights, there is thunder, and when falcon perches, there is a flash of lightning—the two precursors of rain.

Ñui (Turkey Vulture) is also a rain bird. Part of the rationale is contained in several episodes in the Creation Story. In the beginning, after Earth Maker created mountains in the desert, the water after a rain just ran to the sea. Ñui is sent to fly over the earth and with his primaries cuts out the valleys, arroyos, and tinajas that capture and divert the water so it will be available for use over a longer period of time. It is Buzzard who creates the hydrologic system of the desert.

In another episode, Buzzard steals "medicine" from Bitter Wind's house, taking it into his feathers, then distributing it over mountain tops. Consequently, there will emerge from the desert ranges the roaring of wind and the noise of thunder before storms. And the People will go into these mountains for songs and power. Apparently not all the "medicine" left Ñui's feathers. Even now the shaman may use long, slotted outer primaries of the Turkey Vulture in diagnosing illness.

And in still another story, Ñui, after killing culture hero I'itoi (Se'ehe), an act that was bad but necessary, says that he will send wind and clouds to the People. So ultimately he is promising the People rain, something good. Thus, Turkey Vulture's feathers, in particular the outer wing feathers, are powerfully associated with rain and water, as well as with curing. The dark clouds in vulture songs may rumble with thunder (see species account). Black Vulture feathers, however, have no ceremonial function.

Bahr (in Bahr and associates 1994:315 n. 15) links eagles with rain and moistness, but vultures with hot light. However, both can be used in divining and healing, and both are associated with rain, so I do not think Bahr's interpretation holds. Rather, both birds were endowed with a special status from the beginning. Oam Ñui, Brown Buzzard, speaks of the mysterious power in his feathers (see species account).

And, although my teachers never mentioned this, the People could not have overlooked a most conspicuous behavior of Turkey Vultures: when afternoon rainstorms threaten, these great dark birds return from all directions to their roosts, spread their wings and tails horizontally, and soak up the fresh rain (a behavior called passive bathing) before preening.

Red items, such as Saguaro fruit and Coral-bean flowers and seeds, are also associated with rain. Red-shafted Flicker (and cardinal?) feathers seem to have played a role in rainmaking as well, judging by a set of *omina*, prayer sticks, collected on the Gila. At an earlier time, the prized Scarlet Macaw feathers probably would have done so as well. *Aaḍo*, the macaw, originated from the blood of a *siivañ*, Puebloan chieftain, or, according to another story, from the blood of the monster Ho'ok. In some accounts, the birds are later released in the mountains to the east or southeast, whence come summer rains. The Scarlet Macaw is preeminently a blood metaphor, graphically used in song and oration about Apache killing. This event itself is associated with rain, perhaps because the Apache occupy the high country in the east, from where the summer rains come.

In her study of Upper Piman basketry, Kissell (1916:169–172) found a veritable lineup of the rain bird feathers contained in the half-dozen medicine baskets she collected: turkey to bring rain, buzzard (emergency cases for rain), "chicken hawk" (last resort for rain), and eagle; also found was owl ("mastery over certain secret forces in nature, more particularly in this case over sickness and distress due to the owl, the chief among which is fever"). The hawk was probably *vishag*, Prairie Falcon; the owl was probably *chukuḍ*, Great Horned Owl.

At Santa Rosa, Simon Lopez's ceremonial dancers carry three rain bird effigies on sticks: *toha u'uhig*, a white bird found especially at the ocean (probably a Common Egret); *giidval*, the swallow, either black or blue; and *kookoḍk*, which he called "geese" but described as big gray birds with long legs, originally cranes, although the current referent is more likely the Great Blue Heron (see photograph of the Skipping Dance on the frontis).

In addition to the shaman's wand of eagle or vulture primaries, specific feathers were used as fetishes in curing sessions, together with songs. They include the wing feathers of Prairie Falcon, as well as feathers from eagle, Great Horned Owl, and Turkey Vulture.

A shaman who was thought to be practicing witchcraft was called a *paḍ maakai*, "bad shaman." Such a sorcerer might put a curse on another person, even one living in a distant village. The sickening curse is usually from some object shot into the victim's living area or even his or her body (Bahr and associates 1974:61–64; Underhill 1946:278–279). A friendly *maakai* in a trance or divining session would determine what the offending object might be and remove it from the premises. If it were in the victim's body, the shaman would suck it out. Frequently, the object might be a feather of Great Horned Owl (or ghost) located somewhere in the house, yard, or ranch.

Making votive offerings is a worldwide practice found prominently today throughout Asia and among Native Americans. The forms are varied, but usually consist of some material object that is offered, often at a shrine, accompanied by a prayer petition. Offerings may include prayer flags, a stone or bead, a coin, water, some ground meal, or the burning of incense, tobacco, or candles. Feathers are one of the most important votive offerings traditionally used by native peoples of the Greater Southwest, from the pahos of the Hopi to the prayer arrows and "God's eyes" of the Huichol.

An event in the Piman Origin Epic illustrates various types of votive offerings. When Corn Man and Tobacco Woman quarrel, Tobacco abandons the People, going to the west, the dryer region. But she herself is a votive offering needed to induce wind and clouds and ultimately rain. Without her, now, there is no precipitation. After two years of drought, Corn Man and his sidekick, Pumpkin Man, unable to grow, leave the People also, going toward the east, where there is more rain.

In this predicament, the People send Verdin, a powerful shaman, with four votive offerings to entice her back: a shaman's crystal, a Hohokam slate palette, eagle tail feathers, and *viigĭ* (soft feathers) from an unspecified bird (Lloyd 1911:219). In another version of this story, when Corn Man leaves the Pima after Tobacco Woman's anger at her gambling losses, Corn expects the people to coax him back with votive offerings of *viigĭ* and a bead (Bahr and associates 1994:87; Hayden 1935).

The *viigĭ* are the fluffy feathers of the eagle or hawk, plucked from the bird's thighs, lower belly, or undertail coverts. (They are often but incorrectly called "down feathers.") Underhill (1946:25) noted that these prized feathers were "the offering *par excellence*, placed with a fetish, deposited at a shrine, or worn at a ceremony." Their value derived from their resemblance to rain clouds. Once they had been ritually purified, together with the person who had plucked them from the bird, they could be traded to others for their personal ceremonial use.

The *viigĭ* gave its name to one of the major Piman ceremonies, the *viigida (wiigida)*, also called the harvest festival or the prayer stick festival. Among Akimel O'odham, it was referred to simply as Navichu from the main impersonator in the event. These feathers were offered as "food" to scalps kept in scalp baskets. They were also offered to the stone hearts of the great water monster from Quitovac and renewed annually at the summer ceremony.

Komal Hogĭ, or Thin Leather, the River Pima storyteller, gave Russell a set of four rain songs. The second and longest song is a growth song, describing the waving corn tassels and squash leaves about the singers. The singers are spending the night among the new plants, "singing them up." "The darkness of evening falls as we sing before the sacred âmĭna." The prayer sticks compose the altar set up in the fields. The song concludes, "The white light of day dawn yet finds us singing, while the squash leaves are waving" (1908:332).

These *omina*—prayer sticks, healing sticks, or medicine sticks (Mathiot n.d.:465)—were among the most sacred objects kept by a shaman. They were usually made of short sections of Arrow-weed, pointed at one end, with feathers of certain birds attached to the upper ends with homespun cotton string, then bound into a bundle of four or more individual sticks. The sticks were often plain, but might be painted blue, green, or yellow (fig. 7.1).

It is hard to appreciate the importance of these prayer sticks, which fell into disuse among the River Pima long ago. But one oration about I'itoi (Se'ehe), following his resurrection, describes the sheer power the *omina* once evoked:

> He examined the sticks, but none suited him;
>
> He eyed along the river, that green snake, which he had made, and found the sticks that pleased him.
>
> He cut those arrow-weeds, he found there, into four pieces, and blew smoke over them.
>
> And out of them came sparks of light, that almost reached the Opposite World, the World of the Enemy, where things are different.
>
> And when he saw the light from the sticks he smiled within himself;
>
> He was so pleased he had found the sticks that suited him. (Lloyd 1911:136)

The oration continues, telling how culture hero I'itoi then endows these original prayer sticks with power, finally depositing them where Sun stops to drink at the end of his day's journey.

Among Tohono O'odham, Underhill (1946:26) found eagle feathers to be the usual choice for attachments to prayer sticks. "Under certain circumstances turkey down replaced eagle down, but the feathers of no other birds were used." Desert People used Catclaw, *Acacia greggii*, as the wood (Castetter and Underhill 1935:76). River Pima were more diverse in feather selection. In addition to these two species, at least Red-shafted Flicker and a "blue bird" were used.

While explorer Carl Lumholtz (1912:107) was visiting the Children's Shrine at Santa Rosa, the Tohono O'odham village, he visited other local shrines: "I was shown the place where the elder men, sitting in a circle, make the small, bright blue prayer-sticks for the great harvest [*wiigida*] festival. One end is pointed, and to the other is tied a turkey feather; about two hundred and fifty of these are distributed at the feast in the interest of health, and some are buried later, to keep the fields moist" (1912:107).

Underhill (1946:25–26) provided the additional details about construction of a prayer stick among Papago: "The usual stick was unpainted, about four inches

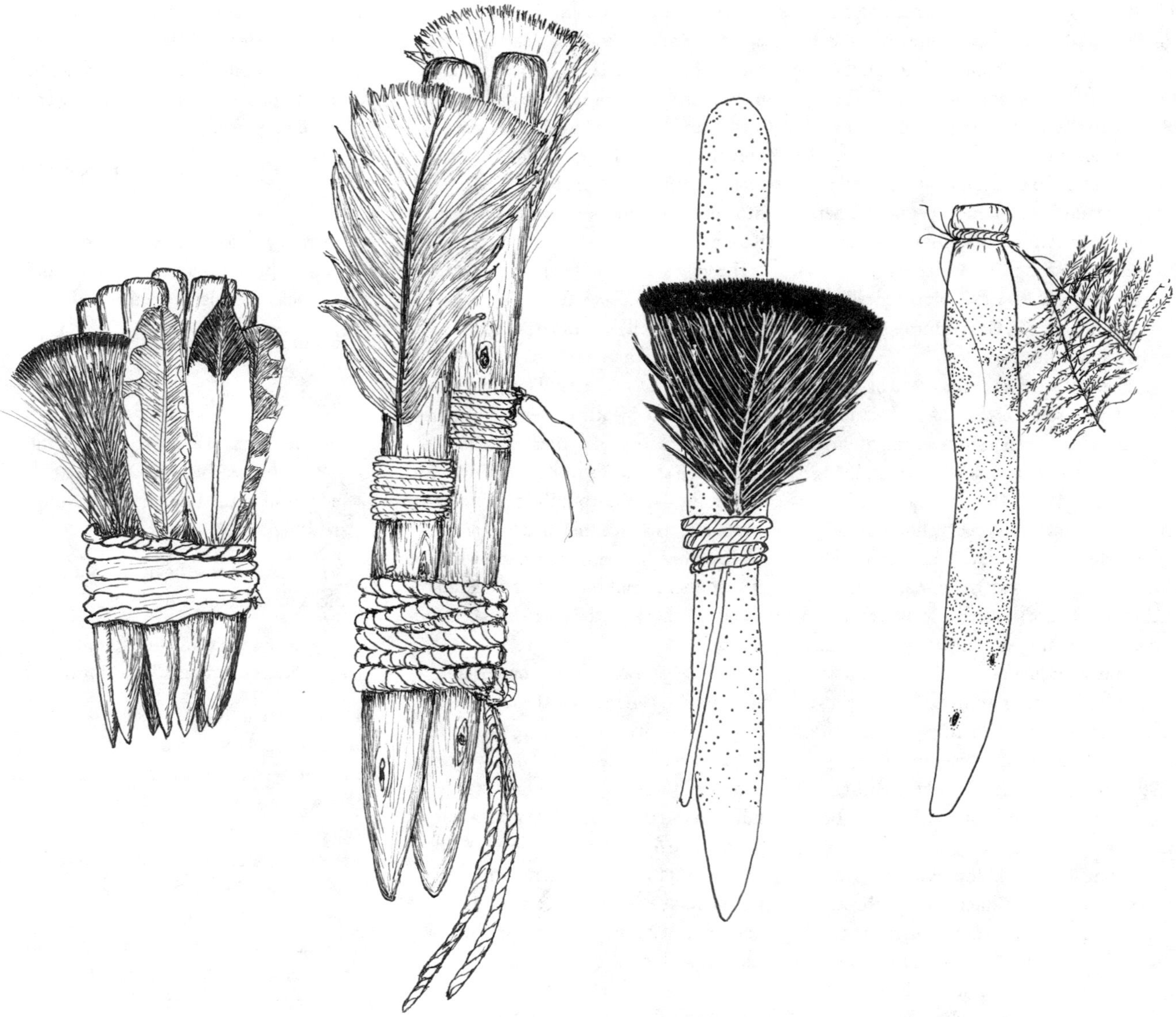

Figure 7.1 Omina *(Piman prayer sticks).* (a) *Akimel O'odham* omina *bundle (Red-shafted Flicker and turkey feathers bound with cotton yarn and cloth strips to Arrow-weed sticks, painted blue or green).* (b) *Tohono O'odham* omina *pair (turkey feathers bound with cotton twine to a stick painted ultramarine blue).* (c) *Single Tohono O'odham* omina *(turkey rump feather tied to a stick painted ultramarine blue.* (d) *Tohono O'odham* omina *(unidentified viigĭ attached with sinew and twine to a stick painted red ochre).*

long, pointed at one end so that it could be driven into the ground, and notched at the other end. One or two feathers were fastened with sinew to each end of a cotton string. The string was put through the notch at the top of the stick and tied around the neck of the stick with another string."

The prayer sticks used during the ten days preceding the *wiigida* ceremony were somewhat different. As described by J. Alden Mason (1920:15),

> The day after the council [setting the ceremonial date], the entire population . . . busy themselves cutting bundles of sticks which are left to dry for a day. The following day they are shaved and cut to a uniform size, about a foot in length, adorned with fine turkey feathers on one end and tied up in bundles of about 8 inches in diameter. These sticks are painted blue or green and represent the growing corn, the turkey feathers being the leaves. Early in the evening of the tenth day before that set for the celebration, all the bundles of sticks are gathered into one great bundle and set in the center of the feast ground.

These sticks were later distributed to each participant of the ceremony. On the day of the celebration, the *chu'i wuadam*, meal spreaders, sprinkled cornmeal on each person to keep sickness away and promote longevity. "As [the participants] near the [dancing] place there is a small pile of prayer-sticks and every adult man takes one and cleanses himself with it, putting a little corn-meal on the stick and brushing it over himself. There are only a few sticks there and as each man cleanses himself he returns the stick to the pile" (Mason 1920:15). Small wonder that the *viigĭ (wiigĭ)*, the fluffy feathers tied on the prayer sticks, have given the whole ritual its Tohono O'odham name, *wiigida*. As noted earlier, on the Gila the parallel ceremony is called just Navichu, from the name of the principal masked dancers.

There were various uses for the *omina*. In addition to being planted in the fields to ensure good crops, they enclosed an area in which men sang the growing songs. During many rituals, they were planted upright at the edges of ceremonial areas to enclose sacred space. They marked the trail that sacred dancers had to follow during the *wiigida* ceremony. At Quitovac, where the ceremony is still held each summer, when the Navichu dancer (fig. 7.2) and his retinue go out through the village on their begging trip, a prayer stick is temporarily stuck into the middle of five piles of sand where individual families leave food offerings. During the former salt-gathering pilgrimage, one might be planted on the beach. Or the men might carry them in their hands as they waded into the mysterious water to be washed by four waves. Then the prayer sticks and cornmeal were thrown into the ocean.

Bahr (in Benedict 2001:34, 186, 219) has confused these prayer sticks, *omina*, with counting sticks, *u'us hikchka*, used to track the days until the festival begins. Indeed, there are two orations where time keeping with *omina* is implied (Benedict 2001:198, 200). That they are two different items with different functions is clear from Mason's (1920:15–16) comment, "In addition to the great bundle of prayer-sticks, ten tally-sticks which are preserved from year to year are stuck in the ground near the fire and every evening one of them is pulled out by the chief of Achi on going home. This keeps the tally of the days and when only one stick remains, the celebration is held the following day." (See also Underhill 1946:138.)

Figure 7.2 *Navichu dancer (from Underhill 1938).*

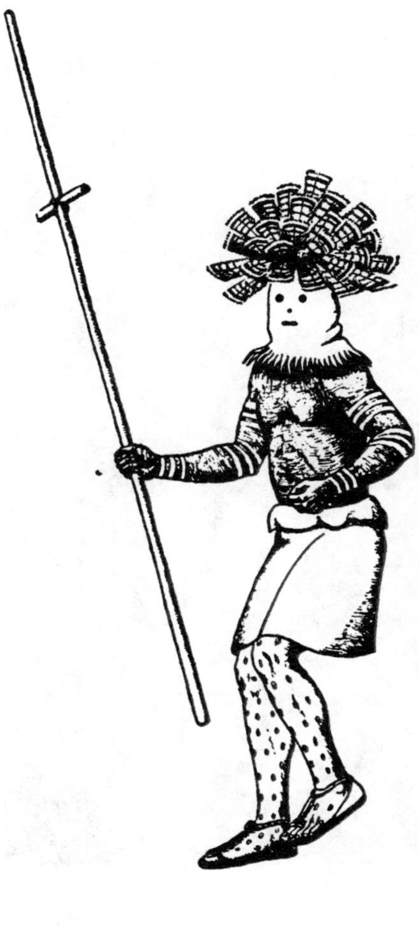

Piman orations reference several additional types of paho votive offerings with feathers. One consisted of feathers attached to string, presumably homespun cotton (Bahr 1975:40–45, 54–57, 70–71; Russell 1908:342–343). Two types—*a'an gi'aḍada*, stringed regular feathers, and *viig gi'aḍada*, stringed soft feathers (fluffies)—could consist of a single feather on a string or a series of feathers hung like Christmas tree lights. One passage suggests that whole feathered skins of orioles and swallows were used as offerings (see Bahr 1975:62–63). Prayer sticks, stringed feathers, and whole bird skins are common types of prayer offerings among the Pueblos to the north (Tyler 1979:1–6).

When the O'odham new year, beginning with the midsummer Saguaro harvest, arrives, the world must be renewed in the wine ceremony. In the "sit and drink" speeches, Culver Cassa notes, four different-colored mockingbirds remake the world. Then the cosmos knows that the world is finished; it all can start over again. Now the clouds come again, and the seasons will start over as at Creation. Now the rain cycle can start over. Finally, the thrasher from the zenith settles it off. After the Mockingbird Speeches, the "sit and drink" ceremony begins, with human participants from the four directions.

While the Saguaro juice *(navait)* is fermenting within the communal round house in preparation for the great wine feast, a prayer stick is planted before the door (Jones 1971:16). During fermentation, the *navait* is sung over by its guardians—four medicine men who attempt to locate and predict rain (Densmore 1929:155). During this rain divination, one man "[makes] motions with a little stick with a feather on the end," as Papago Maria Chona explained her own experiences to Underhill (1936:46):

> Once I fell asleep [during the wine feast ceremony]. My father-in-law was one of the medicine men standing in the center of the ring, turning his eagle feathers this way and that to catch the wind that was bringing rain, because it was dry yet, no rain had fallen. My father-in-law shook his eagle feather over me and drops of rain fell in my ear. Then I woke and sang again.
>
> The medicine men made magic all night long, while we sang. They used to go to the string of eagles' feathers hanging outside the council house and shake them. If it was to rain soon, drops of rain would come whirling off those feathers. One man made a little pile of dirt and put a bit of eagle down on top of it. Eagle down is like the clouds; it calls them. Soon water came from the dry sky and wet that eagle down, so that the pile of earth was damp. Then he took up that eagle down and sprinkled all of us dancers with it.

At the conclusion of an eagle killer's purification, his guardian and each of the assisting singers stroke his body with a prayer stick. When the songs and the purification are finished, the guardian hands the prayer stick to the killer and tells him to take a cold bath. The blessed and now safe eagle feathers are then distributed.

The insignia of the *maakai* or shaman is the wand made from a pair of matched (that is, left and right) outer primaries or pinion feathers. These feathers might be from the eagle, the most frequently used, or from the Turkey Vulture, Great Horned Owl, or "hawk" (probably Prairie Falcon). They function as an aid in diagnosing or divining (see the discussion of staying sickness in chapter 5). The wand is called *maachwidag*, meaning "learning thing" fide Bahr. Culver Cassa suggests a derivation through *maach*, divided, in reference to the physical shape of the wand. The bases of the feathers are bound with sinew or some other substance to form a handle.

Papago Maria Chona told Underhill (1936:48) that her husband, a *maakai*, "got eagle feathers from an old man and tied them on a handle, four of them, tied two and two, beautifully wound about with string." The Turkey Vulture *maachwidag* Russell (1908:187) collected from the Gila Pima is made with four feathers (see fig. 5.1).

At a shaman's death, his gourd rattle and divining plumes "were not inherited but it was thought that they should not be placed in the grave, because that would ruin their power. The power should be left loose in the world for someone to find,'" wrote Underhill (1939:189).

A song for curing *jevho mumkidag*, gopher sickness, incidentally gives us a bit of insight into how the shaman made his or her intuitive diagnosis (Russell 1908:319, reworked by Culver Cassa):

Bahange macheweina ngingikova
 yewena
bahange macheweina ngingikova
 yewena
yokovañ wua koviki ñeina totolinga
ñimui gayegayomina.

The eagle primaries four times blow,
The eagle primaries four times blow.
I turn to look and light shines into and
 crosses my heart.

Prayer sticks were used as the fetish in curing certain forms of staying sickness. Russell (1908:265–266) described one affliction called *jeeñim mumkidag* that made children fretful. Culver Cassa explained the cultural context of this sickness. Children of anyone who had an enemy scalp fetish that was not treated respectfully were subject to this disease. The *jeeñim o'odham*, the ritual curer in this case, had to be a war expedition leader or ceremonialist who had undergone enemy-slaying purification. From this ritual, he had acquired a bundle of sixteen *omina*.

The *wusota* or treatment of *jeeñim mumkidag* made use of prayer sticks tied with feathers from a blue bird and a red bird. Unfortunately, the Piman names for the two birds are not given in Russell, so we don't know the species. Thin Leather told Russell that this sickness sometimes affected the throat and caused the victim to lose flesh (1908:265–266). In this case, the *omina* were soaked in an olla, sucked four times by the patient, then rubbed across his body. The shaman's plumes, the animal fetishes of the curing singer, the eagle feathers, sea shells, and the Apache scalp won by the "ripe man" were all ceremonial property, noted Underhill (1939:92–39). These items were the most personal of all property, for no one but the owner dared touch them, she said. (In my experience, though, healers have not been reluctant to show me their fetishes and discuss their use.)

Men of a particular Piman settlement would meet at night in a communal house to discuss issues. Talking continued until a consensus was reached on any proposed action. At intervals, a cane tube filled with tobacco was lighted and passed in a circle in a formal fashion. "The ceremonial pipes of cane had bunches of little birds' feathers tied to them," wrote Lloyd (1911:89). He noted one carried by the old storyteller Komal Hogĭ, Thin Leather. The statement is puzzling because it appears to be the only reference to feather decorations on these well-known pipes. The mystery may be resolved by something that Thin Leather related to Russell (1908:107, fig. 24, 265–266) about *jeeñim* disease. During treatment, the shaman or his assistant "blows through a tube, called the tcUnyim cigarette, upon the forehead, chin, breast, and stomach of the patient. The tube has a bunch of feathers attached called a-an kiatuta [*a'an gia'aḍada*]." Komal Hogĭ was a *maakai* assistant. It would seem that the feathered pipes were those used in curing, not in communal smoking.

During various ceremonies such as the *wiigida* (fig. 7.3) and the *eḍa i meḍ* (var. *e'eḍa i voopoĭ*), the parallel run ceremony, dancers carried feathered wands in each hand that were called *iagta*. A Pima dance group still performed the *eḍa i meḍ* during

Figure 7.3 *Papago* wiigida *(Nanavichu and their attendants) (courtesy of the Arizona State Museum, University of Arizona, Tucson).*

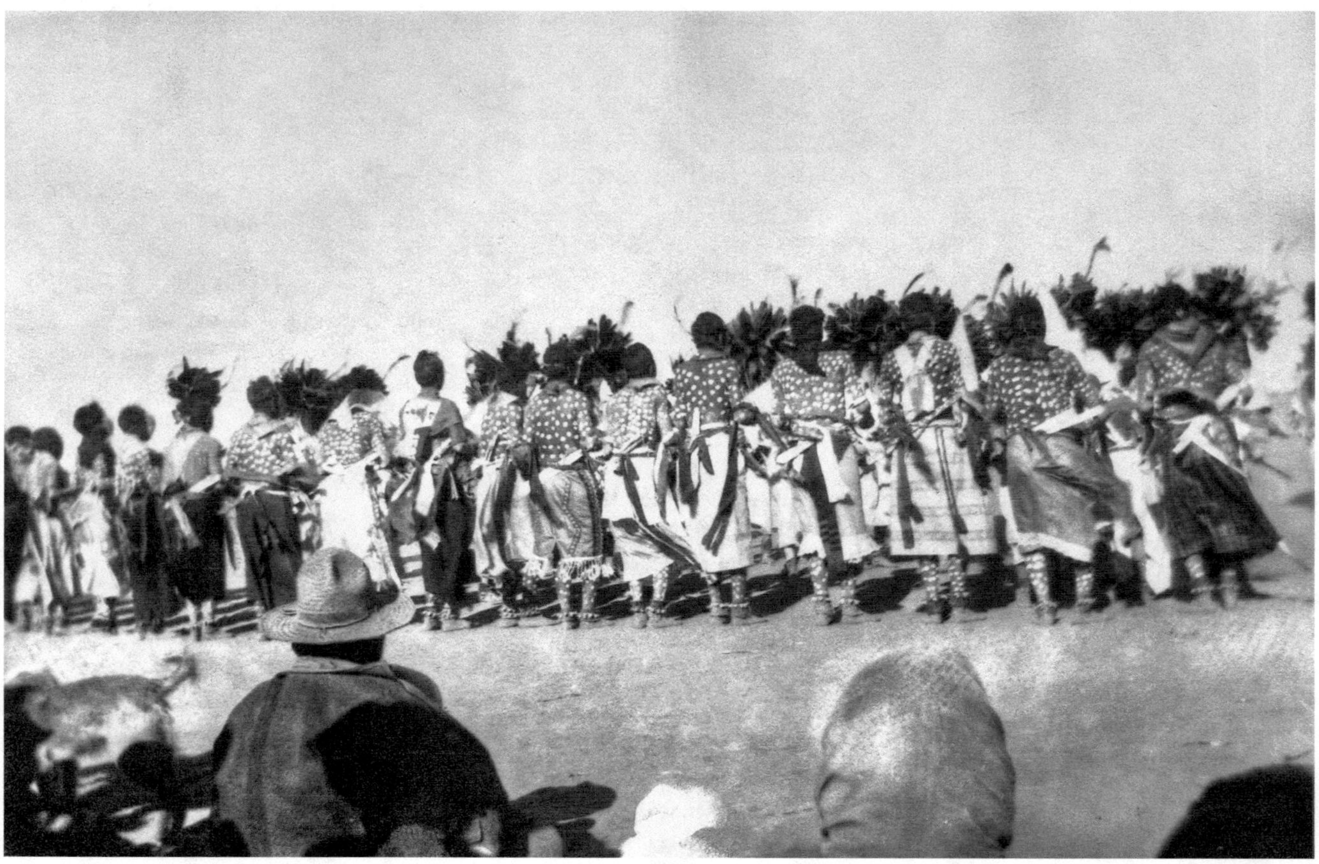

the 1960s when I lived at Komaḍk. Don (Daniel) Thomas was lead singer with the *hikvig ñeñe'i* (Gila Woodpecker songs). There were four singers with gourd rattles and two lines of dancers. The *iagta* they carried in each hand were about three and a half feet long. Sylvester Matthias said that there were also *giigdval, hevachuḍ*, and *siikĭ eḍa i meḍ* (that is, swallow, Blue Grosbeak, and White-tailed Deer parallel run dances). He also referred to the ceremony as the *e'eḍa i voopodam*. Ruth Giff added that Don also had the *tova* (turkey) *eḍa i meḍ* and someone else the *vavas* (Rock or Canyon Wren) *eḍa i meḍ*. She thought there was also a *shashañ* (Red-winged Blackbird) *eḍa i meḍ* and recalled that some dancers also carried, in addition to the sticks, a rainbow and some clouds. And on the Gila, at least, white and red crepe paper came to be substituted for feathers; the white originally must have been eagle *viigĭ*.

Upper Pimans had a kachina-like ceremonial figure called Navichu and several types of attendants called Vipiñim, Kokshpakam, Chu'i Wuadam. All were masked or had painted faces. According to Piman tradition, they are beings who originated in the north. Navichu himself wore a kilt, a belt with rattles, and a case mask with small mouth and eye openings, topped with a mass of turkey tail feathers and two longer "horns" of turkey primaries. He carried a crooked, clumsy bow and four arrows trimmed with turkey feathers. At Quitovac, the only place where the ceremony is still performed, Navichu and his assistant wear a leather mask topped with several ancient, foxed eagle feathers. In each hand, they hold long staves with several eagle feathers loosely attached to each end and the middle.

J. Alden Mason (1920:18) was told by a participant that the Nanawichu dancers "represent sahuaros, the giant cactus, and wear turkey feathers on their heads to represent the fruit and carry rattles. They also carry long poles which represent the poles with which the sahuaros are gathered." The photo by Lumholtz (1912:opposite p. 92) appears to be a mass of turkey feathers, and the text (p. 93) says, "the top is adorned with a large bunch of plumes from the turkey, hawk, and a black sea-bird." One Navichu oration makes mention of this headdress (Benedict 2001:182; Lloyd 1911:212).

One of Navichu's attendants is called Kokshpakam on the Gila ("Kokshpa," according to Mason [1920:22]). "He carried a wand tipped with blue feathers and his duty was to sing in a high piping voice, the first and last words of each song," wrote Underhill (1946:141). The kind of feathers is not stated; blue-feathered desert birds are rare. Two of these wands made by Sylvester Matthias are tipped with *Buteo* feathers, probably Red-tailed Hawk.

A hereditary Navichu from the Gila and another from Quitovac told me that the overriding purpose of this elaborate ceremony is to make the world beautiful again and ultimately to bring rain. The immediate function, however, is the curing of ailments of the legs, in particular the knees. As Russell (1908:266) reported on the Gila, after the singing of the song by a chorus of fifteen or twenty singers, the Navichu "goes to each patient and pats him with eagle feathers until he has presumably drawn out all disease. He then throws away the feathers."

This description fits a pattern in Piman culture of capturing evil. The cleansing and collecting are accomplished with a spiny plant such as Ocotillo or cholla cactus (Underhill and associates 1979:83–88). But Navichu uses the feathers as the ceremonial sponge.

Great Horned Owl feathers were used in a native "baptism" or infant-naming ceremony performed somewhere between three days and a month of birth. As Bahr and associates (1974:53) describe it, the shaman gave to each parent to drink a gruel he made from white clay and crushed owl feathers. "Then the shaman brushes their heads with an owl feather and finally he blows, spits, and presses on the top of each head." Russell (1908:187) gave a somewhat different version of this rite of purification for the River Pima child: "Putting a sacred pebble and an owl feather into a seashell containing water, the medicine-man waved an eagle feather about, while the parents and the child drank the water and ate some white ashes or a little mud." (For additional details on this rite of passage, which included the naming of the neonate, see also Russell 1908:188 and Underhill 1939:157, 1946:313. Similar elements were described for the girl's coming-of-age ceremony described by the Papago woman Maria Chona [Underhill 1936:35–36].)

The number of songs specifically mentioning Great Horned Owl feathers that Densmore (1929:105, 117–134) collected underscores the importance of this species. The songs were from Owl Woman, a specialist in *wusota* for *chukuḍ mumkidag*, owl sickness (See Great Horned Owl account.)

There are several reports placing *chukuḍ a'an*, Great Horned Owl feathers, in a special category. For instance, early in the twentieth century Russell (1908:192) found: "The owl was not so much a bird of evil as of mystery and death, and its feathers were sought for their magic potency in medicine and other ceremonies. If a lad shot one, he had to pluck the feathers from the bird before it died or the magic power of the plumes was lost." Later he noted, "If possible, the feathers were taken from a living bird when collected; the owl might then be set free or killed" (252). In describing the lustration (purification ceremony of a warrior after killing an enemy), Thin Leather told Lloyd (1911:92), "The slayer finds an owl and without killing him [it] pulls long feathers out of his wings and takes them home." These feathers were used to decorate the bag in which the scalp fetish was kept.

The gathering of molted feathers of large birds from beneath communal roosts was a well-established tradition, at least along the Gila River, where there were large cottonwoods and willows. The practice is encoded in one episode recorded in at least three mythological accounts (Densmore 1929:42; Lloyd 1911:77; Russell 1908:219). When the People have recourse to Se'ehe (I'itoi) against a too successful gambler, he instructs them to bring him all the large feathers that can be found under these roosts.

The trophy from a successful raid on Apaches might be the scalp of a killed enemy (Underhill 1946:185–210). The scalp was prepared as a fetish and kept in a special lidded basket (see Kissell 1916:170). Both the enemy-killer and the acquired article had to be purified before becoming safe enough to enter the community. The owner was obligated to "feed" this now tamed being. Among suitable offerings might be eagle *viigĭ*. Failure might result in *kaachim mumkidag*, staying sickness (see Bahr and associates 1974:289).

The scalp fetish that Lloyd (1911:92–93) described, possibly in the possession of Thin Leather, consisted of a lock of Apache hair with an attached ball of Creosote Bush "gum" (probably lac). It was kept in a buckskin bag on which were tied a feather of an owl, a feather of a "chicken hawk" (most likely Great Horned Owl and Prairie

Falcon), and soft feathers of an eagle, with a string of blue beads (Kino beads) around the neck of the bag.

Although hummingbird feathers were not actually used, they commanded respect (Russell 1908:211). If a hummingbird were inadvertently killed, its tail feathers had to be kept. If found dead, the bird had to be buried with offerings. The hummingbird's sacred status in Piman culture stems from its mythological role at the time of the Flood (see the species account).

In Piman oral literature—that is, myth, oration, and song—feathers often appear as color symbols. Examples of this ornithological pallet are: raven for black, Blue Grosbeak for blue, oriole for yellow, cardinal for red, mockingbird for white. Given the overall pallid, muted hues of the xeric landscape, it is perhaps not surprising that traditional literature used these brilliantly plumed desert birds as vivid images gracing their poetic utterances.

This example of an oration segment combines colors, plants, and bird feathers into a wet, colorful image (based on Benedict 2001:188):

> Earth Doctor sent Jevho (Gopher), his *shoiga* (pet).
> He came to the Black Spring and planted *ko'okpoḍk* (giant tule).
> He covered the surface of the water with the feathers of *havañĭ* (raven).
> He came to the Blue-Green Spring and planted *uḍvak* (cattails).
> He covered the surface of the water with the feathers of *hevachuḍ* (Blue Grosbeak).
> He came to the Red Spring and planted *mamtoḍ* (floating algae).
> He covered the surface of the water with the feathers of *aaḍo* (Scarlet Macaw).
> He came to the White Spring and planted *vak* (small bulrush).
> He covered the surface of the water with *viig* (fluffy feathers) of *ba'ag* (Golden Eagle).

Storage of Feathers

Feathers require care because they are subject to damage by moths, carpet beetles, and certain other insects. At one time, Northern Pimans kept on hand quantities of feathers, particularly from the macaw, eagle, and hawk. All men had to maintain a good supply of feathers for fletching arrows. No early commentators mentioned how they were stored until needed. The Pueblos made special feather storage containers (see Ladd 1998:123).

Underhill (1946:138) found that turkey feathers for prayer sticks were kept in plaited baskets for the *wiigida* ceremony. These baskets were probably the same as the *vasha* or *washa*, a long rectangular woven basket with a fitted lid (see Kissell 1916:170; Russell 1908:145). This type of weaving seems to have disappeared entirely from among Upper Pimans, though it is still found among Lower Pimans and other tribes to the south. Scalp fetishes and their feather offerings were kept in a *vasha*, as were the shaman's articles. "Owl feathers were always given to a dying person," noted Russell (1908:252). "They were kept in a long rectangular box or basket of maguey leaf [*sic*, *Dasylirion]. If the family had no owl feathers at hand, they sent to the medicine man, who always kept them." At Quitovac, the guardian of ceremonial objects kept eagle feathers in a sack (Jones 1971:23–24).

Utilitarian Uses of Feathers

Fletching Arrows

Arrow making was a constant activity for men and boys because arrows were broken or lost in hunts or wagers. In a way, arrows were a sort of currency, particularly in bets. Making an arrow required three basic raw materials: shafts, sinew, and feathers. "When [a man] had collected all these things," Underhill (1940b:26) reported, "he generally called a number of neighbors and an old man who knew the special songs for arrow-making. If all of them worked all day, they could make a hundred arrows and their host would feed them while working."

The best feathers for fletching were rectrices (tail feathers) and secondaries (inner wing feathers), but primaries (outer wing feathers) were also used. Eagle feathers were reserved exclusively for war arrows. Adult Red-tailed Hawk tail feathers were prized for hunting arrows because of their bright rusty color. But the barred wing feathers of the Red-tail and most likely other buteonine hawks were used as well. There is a mythic reference to a boy plucking feathers from a Harris' Hawk, implying that this dark species was also used. (The passage pairs this hawk with a Red-tail, both found dead and plucked by the boy, but does not say what the intended purpose was.)

Turkey feathers were also important in arrow fletching. Underhill (1940b:26) added "crow" (raven) tail feathers. I was told that even chicken feathers were used in later times, though I have never seen an Upper Piman arrow with chicken feathers. Young, still unskilled hunters frequently lost arrows, so a special type of arrow was made for them by simply binding two whole smaller feathers to the shaft (figs. 8.1, 8.4, 8.5b). Tail feathers of a White-winged Dove could be used on these practice arrows.

"Owl feathers were never used by the Pima to fletch arrows, for owls were of ill omen," Evanico told Woodward (1933:167). "[T]hey were of the land of the dead, and it was believed that spirits of the departed returned in the form of owls and bothered people" the old man said. Leonard Pancott, discussing arrow fletching, added another reason, "And one of the policies, or whatever you want to call it, in making arrows, is we're not allowed to use owl feathers because they say that owl feathers got ashes—they're going to get in your eyes when you shoot. Whatever that is, them owl could spark with their feathers."

Personal Adornment *(A'an Siivodag)*

Pimans' long black hair might be decorated with daubs of white paint or with feathers, particularly for special occasions. As Russell (1908:158, 163) summarized for the River Pima, "Pride of person manifested itself among the men in the care of the hair and elaboration of designs painted upon the skin. Feathers and beads were also worn in abundance." And, again, "The men ornamented their long rope-like locks with the soft breast feathers of the eagle, turkey, or other large bird." Warriors embarking on war expeditions decorated their hair with eagle feathers. There are mythological references to even small feathers from white or brightly colored birds being used for decoration.

River Pimas told me about two kinds of feather headgear, the *siivoḍ* and the *bichpoḍ*. Castetter and Underhill (1935:50) supplied more details about them from Papago usage:

> No headdress was ordinarily worn, although warriors had various caps and ornaments. The *siwoṭ* was a single upright plume, stuck in a headband

which might be woven or be merely a deerskin thong. The *pitcpoṭ* was a row of feathers around the head like a crown thrust under a buckskin band. A more elaborate bonnet was the *acki [ashk] wonam*—a buckskin cap with upright feathers sewed closely all over it with an especially high tuft on the top. There is also mention of a cap of peccary skin (*Pecari angulatus* [*Tayassu tajacu*]) fitting closely to the head with upright eagle *(Aquila chrysaetos)* or hawk [falcon] feathers (because they are the best fighters) sewed around its edges and a mask of buckskin hanging over the face and tied around the neck. Warriors might also tie a bunch of eagledown [*vopo* or *viigĭ*] to the hair at the top of the head, or some long feathers to the hair at the back.

Underhill (1938:71) showed three war headdresses made mostly of turkey tail feathers (fig. 7.4). One of these headdresses includes five equally spaced turkey primaries, the other seven. A third piece has a basal ring of back or rump feathers. The feathers appear to be attached to a basal band and project upward, much like a Plains Indian warbonnet. (There are drawings of two additional headdresses in the same illustration, but nothing diagnostic about the feathers.)

Figure 7.4 *Papago war headdresses (from Underhill 1938).*

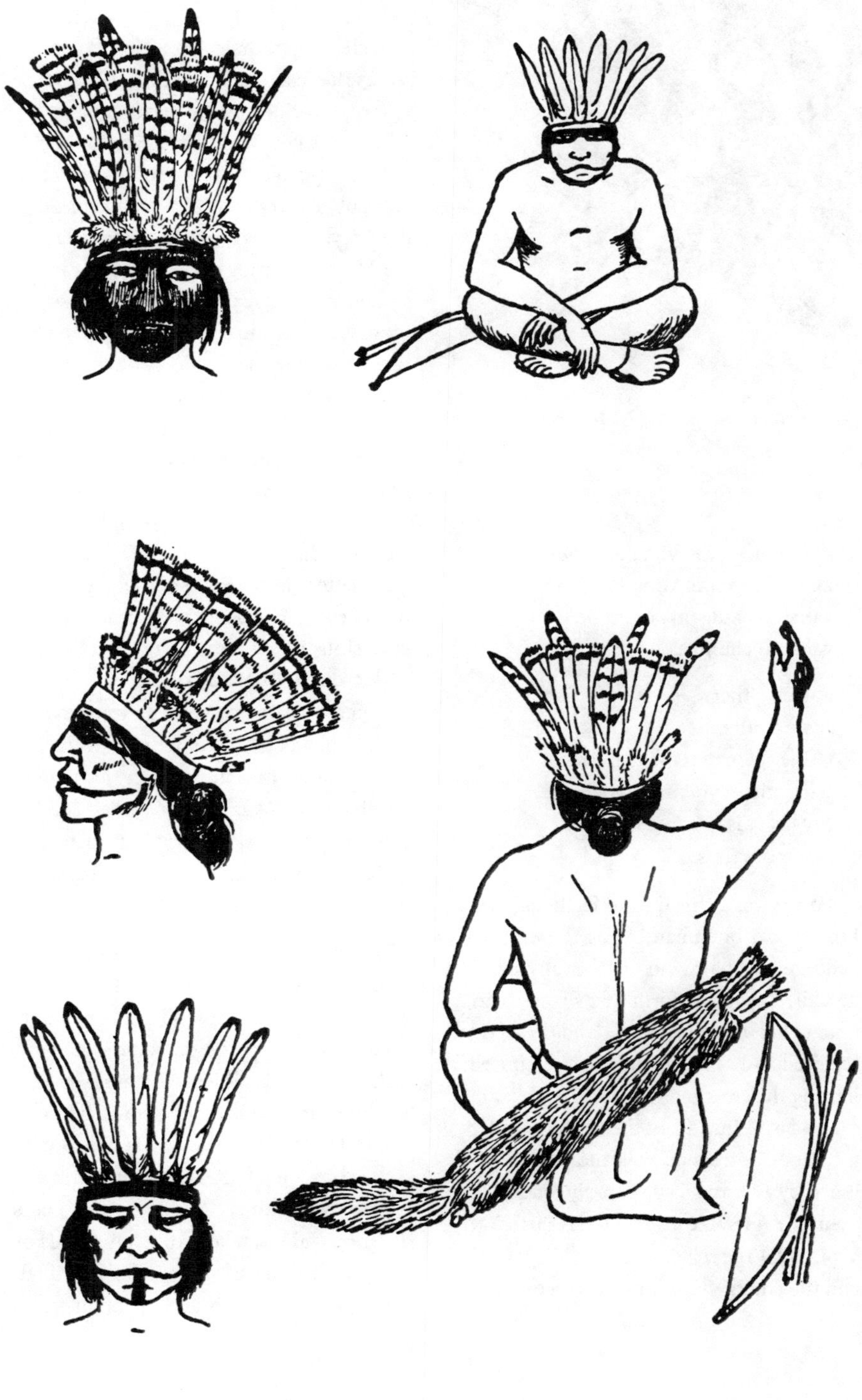

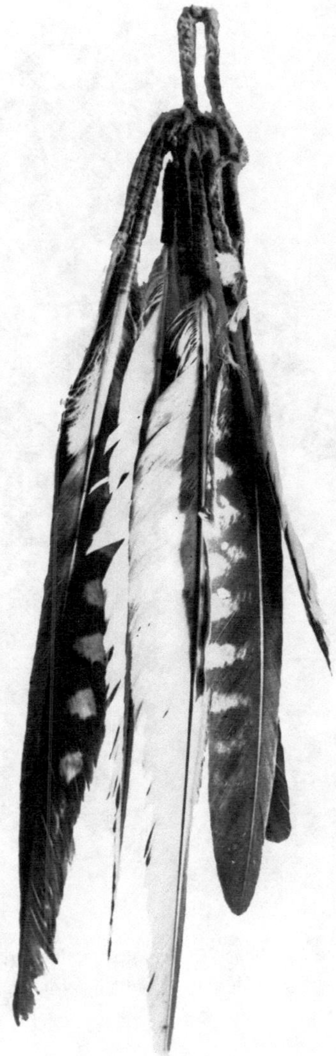

Figure 7.5 *Pima feathered hair ornament (courtesy of the National Museum of Natural History, Smithsonian Institution, Washington, D.C.).*

There were other nonceremonial headpieces. Russell (1909:115–116) collected a hairpiece attachment made of seven Prairie Falcon wing feathers (fig. 7.5) (Russell called them "hairpieces," for attaching to the hair, or labeled them "hair ornaments," noting that they may be attached to the hair to form part of the headdress). The base of each feather had been secured with sinew to a strip of cotton cloth and the strips braided together. How these hair attachments were worn, upward or downward, is not clear from his description.

Russell's (1908:116) prize collection was

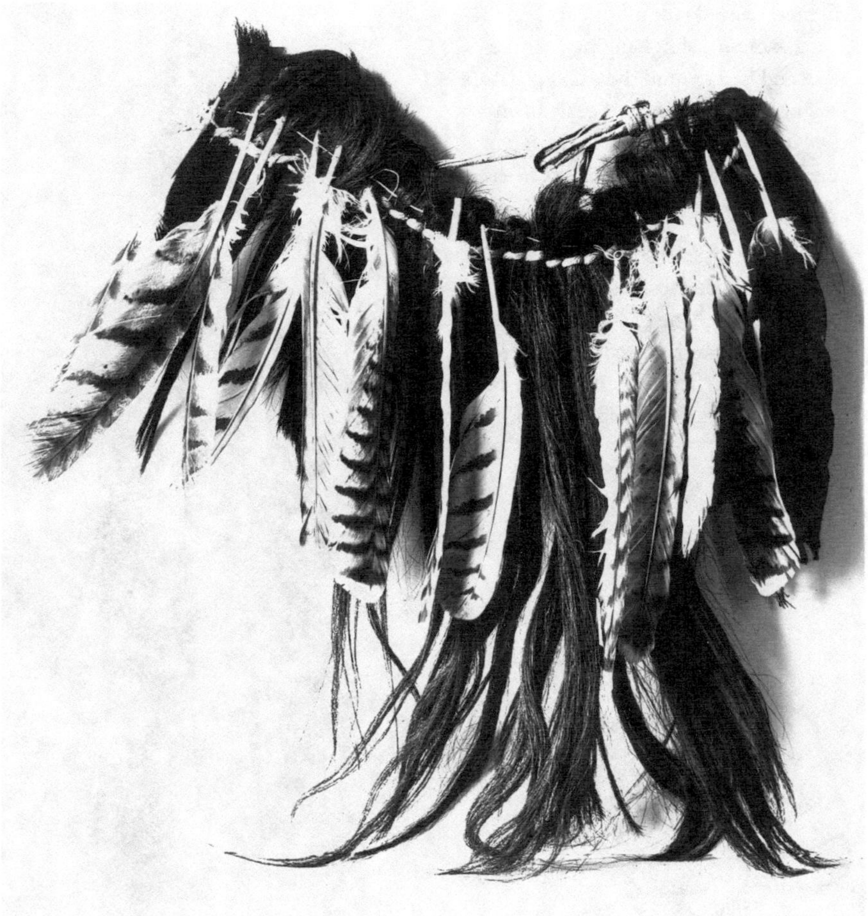

Figure 7.6 *Pima feathered war headdress (courtesy of the National Museum of Natural History, Smithsonian Institution, Washington, D.C.).*

a headdress from old Komaḍk made of Apache hair and fifteen long flight feathers from raptors (fig. 7.6):

> on the left four owl feathers symbolizing keenness of vision by night; next are three hawk, then one owl, and again hawk feathers to the number of five, symbolizing keenness of vision by day; on the right are two eagle feathers, the symbol of swiftness. Thus the wearer of this headdress possessed the courage and cunning of the hated enemy, the keen sight by day and by night of the birds that have great magic power, according to Pima belief, and the swiftness as a trailer of the king of birds, which occupies a prominent place in Piman mythology.

A song Russell (1908:316) recorded for curing Coyote sickness features a headpiece with eagle feathers :

> Bahangae yahane
> bahange yahane
> kumes hege vañ vonamhachuna a
> kumes hege vañ vonamhachuna a
> vahamo chuigama ñi imuidage
> vahamo chuinga.
>
> Eagle feathers,
> eagle feathers.
> Using them you make me a hat.
> Using them you make me a hat.
> My heart is more so like them.
> It is more thus.

Power was derived from feathers. The Piman warrior embarking on the warpath bedecked his hair with *viigĭ*. "Many wore headdresses in any form that suited them," noted Underhill (1939:131), "and these were repaired and made by the warrior himself." During the emergence of the Vupshkam, or modern Pima, from the Underworld, the People wore macaw headdresses fashioned by a shaman, supposedly rendering them invulnerable against the resident Vipishaḍ. (The magic did not always work, and the emergence narrations record losses sustained by the invaders as well as by the Puebloans who defended their fortifications.)

The successful warrior who had taken Apache scalps was esteemed. Ruth Benedict (2001:204) was told they "were often spoken of as brave men"—"[they] would pierce [the septum] of the nose and adapt the headdress of the brave, but this was never a part of the purification ceremony. The headdress was like the headdress of the Sioux but had no tailpiece."

A mythological account tells of raven feathers used in making a hair decoration. In preparing for a gambling bout between two *sisivañ*, one man takes a piece of Catclaw wood, splits it four ways, bends each end over, and attaches a "crow" feather to it, creating his hair decoration (Lloyd 1911:236). This is a Hohokam story, or at least one ascribed by the narrator to that culture. A *siivañ* was a Vipishaḍ leader of a *va'aki*, a puebloan community associated with public architecture, now in ruin. Another mythical character called Prairie Falcon Meeter, wearing a headdress of raven feathers, joins an expedition going after Apaches. Someone else in the war party wears a headdress of white feathers of some sort. These headdresses are taken aloft by a whirlwind. Falcon Meeter transforms himself into his avian mentor and brings down his headpiece (Benedict 2001:146–147).

An elderly Papago woman told Castetter and Underhill (1935:48–49) that young men in the latter part of the 1800s would come to dances wearing leggings of feathered hawk skins. These items were most likely made from Red-tailed Hawks.

Whereas feathers played an important role in Piman culture, other parts of birds were only of minor importance. Jesuit Joseph Och (1965:127), who arrived at San Ignacio Pima mission in 1756, described the initiation of young males age fourteen or fifteen as warriors:

> The induction ceremony for a new recruit is very barbarous, and accompanied by not a little bloodshed. The poor naked youth is brought before some of the old tested soldiers who must bear witness that he has sufficient heart to stand up under the test he is now put by the chief [*sic*]. [After the youth undergoes two ordeals, the initiator now] uses various talons, taken from large birds of prey, which have been carefully readied for the test through drying and disten-

tion. With these he stabs, hacks, and scratches and tears the candidate's entire body so that he bleeds all over. During all of this the recruit must stand cheerfully without writhing or turning.

Pueblo and Plains Indians made fine high-pitched whistles from the long bones of large birds such as eagle and crane. These whistles usually had ceremonial functions. I have found no evidence of them among Northern Pimans, however. In fact, use of bird bone seems to be absent from their oral tradition. Whistles used in game hunting and musical flutes were fashioned from reed or Otate (Pennington 1980:209–210; Rea 1997:102).

I have used the past tense frequently in this chapter because feather use by River Pima and Pima Bajo has all but ceased. Shamans and healers still use those feathers that are needed in their profession. But even among Tohono O'odham, who have maintained more of their traditional ceremonial life, knowledge of folk taxonomy and of uses of avian species is now greatly eroded among younger and even middle-aged generations. No longer are the folk identification, ecology, and behavior of birds and other animals the currency of thought among most younger O'odham. Without these three, metaphors collapse.

8

Hunting and Trapping, Game Preparation and Cooking

And they went out and found an *o-a-pot* [*uupaḍ*], or cat-claw tree, and cut a piece of its wood to make a bow, and they made a fire and roasted the stick over this, turning it, and they made a string from its bark to try it with; and then they found arrow-weeds, and made arrows, four of them, roasting these, too, and strengthening them; and then they went home and made a good string for the bow from sinew.
—Pima Creation Story, from J. W. Lloyd, *Aw-aw-tam Indian Nights*

Animal protein played a relatively minor role in the traditional dietary of northern Pimans living in arid regions. Most of the protein came from vegetable sources. Both the tropical thornscrub and the Sonoran Desert are rich in leguminous species bearing abundant, dependable annual crops. Mesquites, paloverdes, and Desert Ironwood were common trees throughout the region wherever there were streams and alluvial flatlands. Their seeds were gathered and stored for year-round use. Although the amount of agriculture varied between groups, all groups raised leguminous crops such as teparies, a species of desert bean, and limas. Also maize (corn), the cereal staple, when processed in the Native American manner, contributed various amino acids, as did the pseudocereal crops, amaranths and chenopods (Rea 1997:67–75). Nevertheless, all groups supplemented the daily fare with mammals, birds, and, where available, fish. With the possible exception of turkeys, animals from all these groups were wild-taken foods until the Hispanic introductions of Old World domesticated animals.

As seen in the descriptions of the Piman categories of hunting, big game supplied a relatively small amount of the animal protein (table 8.1) in those Piman communities where canal irrigation was practicable, such as along southern Arizona's streams. Socially organized hunting was overlaid on the social structure of irrigation agriculture. Here trapping and much of game hunting could be called "garden hunting"; that is, Piman agriculturalists hunted extensively those animals (both birds and mammals) that proliferated as a result of small-scale subsistence farming (Linares 1976). Small Piman fields—interlaced with ditches and fenced with living fencerows as well as thick brush fences—created ideal cover and breeding or wintering conditions for those animals the O'odham most frequently preyed on: cotton rats, packrats, cottontails, jackrabbits, Gambel's Quail, Mourning Doves, White-winged Doves, White-crowned Sparrows, and Lark Buntings. These animals, together with fish from the river and canals, were the vertebrates

Table 8.1 Weights of Piman Game Birds

Species	Number	Mean (grams)	Mean (English)
Mourning Dove	235	119	4.2 oz
White-winged Dove	30	153	5.4 oz
Gambel's Quail	249	166	5.9 oz
	727	159	5.6 oz
Wild Turkey: males	54	7,400	16.0 lbs
females	55	4,222	9.3 lbs
Lark Bunting	40	37.6	1.3 oz
White-crowned Sparrow (interior races)	100	27.0	0.95 oz
Abert's Towhee	32	47.1	1.6 oz

Note: Weights are whole birds, uncleaned.
Sources: Campbell and Lee 1953; Dunning 1993; Amadeo M. Rea, San Diego Natural History Museum and University of New Mexico collections.

that most frequently made it into the pot on a day-to-day basis. Pronghorn, Desert Bighorn, White-tailed Deer, and Mule Deer might be added only occasionally in a year.

After Contact, wheat and other Old World grains were added to the agricultural cycle as a third crop grown in the cool season. The winter crop, too, attracted prized game animals. Cotton rats reached high densities in grain, forage, and fallow fields. Great numbers of birds were attracted as well. Here is how Ruth Giff, a Pima woman who participated in the manual harvesting of wheat in her youth, described the situation: "*Hoohi, kakkaichu, o'okokoi* [Mourning Doves, quail, White-winged Doves] are around all summer. They killed them when we're working in the fields, cutting the wheat. Of course, *kakachu* is always around in the winter. *Hoohi* and *o'okokoi* would be lot of them around where they are cleaning the wheat, trashing, just sitting around. They shoot them for their meal. Make a big stew out of them." Here again, the birds most sought were those that increased as a result of Piman land-management practices, another example of garden hunting.

The Desert People, Tohono O'odham, were more marginal agriculturalists, farming essentially a single planting during the midsummer or monsoon rains. They had no rivers but developed a system of channeling the rainwaters that collected in arroyos onto the soil deltas at the lower ends of these drainages (Nabhan 1983, 1984). The method is called *akĭ chiñ*, "wash mouth" farming. Even these small fields, with their rich growth of edible greens and seedy annuals, attracted birds and mammals that the people relished. But for much of the year, Papago hunters stalked medium- and larger-size mammals out on the plains and in the desert mountains.

The Hiá ch-ẹd O'odham, Sand People, who lived in still lower and dryer country west of the Tohono O'odham, hunted a broader range of animals than any of ecologically better endowed Pimans (Nabhan, Hodgson, and Fellows 1989). They were more nomadic than their neighbors, ranging in the Pinacate region into the upper Gulf Coast, where they took crustaceans, species of fish, reptiles, mammals, and birds that were either unavailable or tabooed to Tohono O'odham and other Pimans (see Rea 1981b).

Training for Hunting

Hunting began early for the Piman boy. At a young age, he was given a miniature bow and some small arrows to practice shooting small game. A rawhide sling, at least historically, completed his hunting gear. The edible rodents and small birds, in particular sparrows and doves, would be his earliest targets. "Children hunted about the camp, generally roasting the small animals they killed on the spot. Boys might dig rats out of their holes. After the rains all children chased the birds while their feathers were too wet to fly and caught them in their hands" (Underhill 1939:159). In summer, both boys and girls searched for singing cicadas, *kokotpol*, and would sing a song while bending down the branch on which the insects rested.

Boys were expected to progress in their hunting skills from small birds to larger rodents, then to the larger lagomorphs (cottontails and jackrabbits), and finally to big game. This sequence is encoded in myth in the Flute Story where White-clay Eater Woman's twin sons take first quail, then cottontail, then jackrabbit, and finally a Mule Deer. After this, they are ready to capture eaglets (Herzog [1929]).

Another story appearing in the Ho'ok myth gives some insights into how youngsters hunted: "And one day two little boys were hunting doves after the manner of the country. They had a little kee [*kii*, house] of willows and a hole inside in the sand where they sat, and outside a stick stuck up for doves to light on. And when the doves came they would shoot them with their bows and arrows" (Lloyd 1911:110–111).

Play and hunting were combined in the activities of small bands of boys roving through the riparian woodlands, as related by River Pima George Kyyitan. "Used to be *lot* of cottonwood trees all along the river. Saturday morning, no school. Meet these children that were here, kids here. Then get our sling shot, then go over there and hunt doves in there, in those cottonwood trees, way down there, *clear* up to the end, then come right back again, on the side [of the river] where we left.

"Then we start building fire, roast them right there. Eat 'em. That's what we do. Those White-wings and others, *hoohi* [Mourning Doves].

"After that, when we get back here, noon time, we'd rest for a while. Then late[r] evening when it kind of cools down, then we come over here with our bow and arrows, then start shooting. They'd put up a cardboard box on a wire. Somebody would pull it. Run it right in front. Then we'd try to shoot. Then come right back again. That's why we know how to shoot jackrabbit, rabbit, things like that running."

"Like that *wulivga?*" I asked.

"Yeah, *wulivga*. Yeah. They used to use that one, too. But now nobody knows about bows and arrows."

The *wulivga* (fig. 8.1) was a target several inches thick and about a foot long made of some softer plant material such as Arrowweed or grass bound with willow, Screwbean, or mesquite bark (Curtin 1949:125, 131; Rea 1997:148, 1998b:64; Russell 1908:178). The target was dragged on the ground as well as tossed in the air, while the young archers took turns shooting at it. As with so much of early skills development, it was a game in which the boys bet their arrows. As Curtin remarked, "Boys at

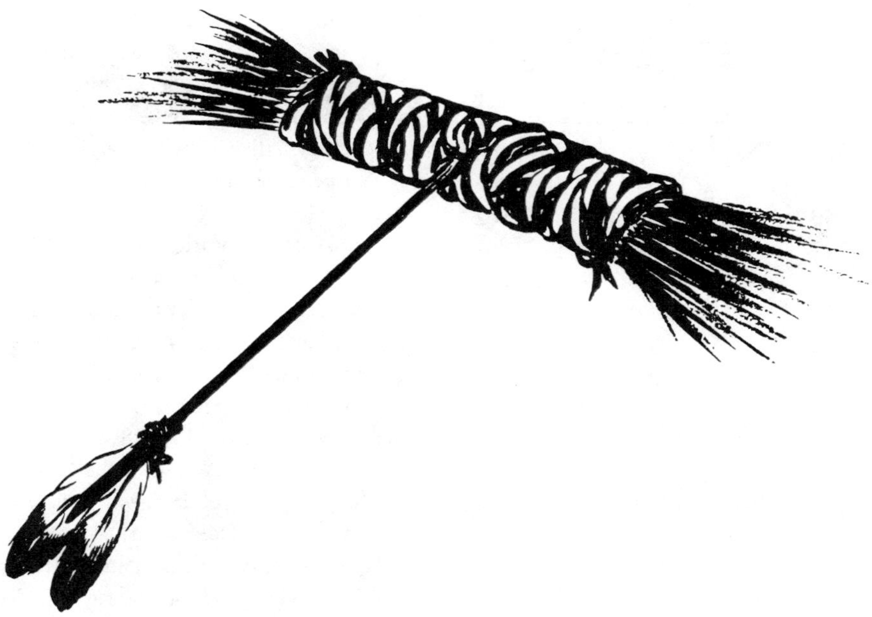

Figure 8.1 Wulivga *(target) shot with with boys' practice arrow (from Curtin 1949:125).*

this age [twelve to fourteen] were supposed to take up the duties of manhood, such as hunting and protecting the family from the enemy." On the Gila, all the older men still remembered this practice. For instance, Francis Vavages of Sacaton Flats told me, "The *wulivga* kids make out of willow. Sort of a target, and bind it with something; put it way down there and try to hit it with arrow. Sort of practice, too, when they hunt with their bows and arrows."

The end result was a skilled marksman. "The boys practice day after day, year after year, until it is not surprising that they became such accurate shots with the bow. Woe to the birds and squirrels that cross their path" (Russell 1908:178–179). About the same time as Russell, Lloyd (1911:190) wrote: "The Pimas were formerly famous for archery, and the shooting of bird on the wing, and of jack rabbits at full run while the archer was pursuing on horseback, were favorite feats." The term used for a good marksman or sharpshooter was *sheli wuidkam*. Joseph Giff explained these additional terms: *s-mu'adag*, to be good at hunting and killing (game); *huai s-muakam*, one who is good at killing Mule Deer; *s-o'oiddchugdag*, a good hunter in general. Sylvester added that hunting abilities come in the family. Some individuals are natural hunters.

As with many other tribes, the Pima boy was bound by a first-kill prohibition called *ki'omĭ* (or *kió'omĭ*). He was obliged to give it to a mentor. As Joseph Giff explained, "*Kió'omĭ* is like a disease or curse resulting from eating one's [own] first kill. Give it to an old man, one who is a good hunter. In return, receive a kind of blessing, especially to be a good hunter. Now a youngster just abstains. Don't give it to anyone special."

When a youth, Joe killed his first dove, a *hoohi*, and was excited, waving it over his head. He gave it to his father. The taboo was just for the dove. He didn't think it was for subsequent species, *tasi'ikol* (Javelina), *chuuvĭ* (jackrabbit), *tobĭ* or *toobĭ* (cottontail). Just the first kill. At least, that was all he was told. "I hope that's all, because if I kill a *huai* [Mule Deer] now, I don't want to have to give it away!"

David Brown agreed that *ki'omĭ* applied only to the first animal, of whatever species, a little boy kills. He should give it to his father or some other person to cook and eat.

Likewise, Sylvester Matthias, a third Kuiva O'odham, explained: "Anything that's a first killing, it's prohibited that he eat it—even a dove or a quail. Give it to a parent or guardian. Word for it is, *kio'omĭ—ki'omĭ* means when he eats his first killing, then he won't be a good hunter. Same way with a girl—her cooking. Same word is used [for the violation]. First *chemait* [tortilla]. Don't say anything about first planting. Young man can eat first fruits of the field. They are very strict about that *kio'omĭ*. Youngsters scared about that. A practice the Pima kept very sacred." Leonard Pancott, a Ta'i O'odham, said, "Don't eat first four killings. Give it to an old person." This practice is more rigorous, but it follows the Northern Piman ceremonial number.

From the Totoguañ dialect (Santa Rosa dialect of Papago), Mathiot (n.d.:86, 88) recorded *mu'adch* and *mu'madch* as "one's first kill" (from *mu'ma*, to kill something). She noted the prohibition in the appended examples.

Categories of Pima Alto Hunting

The one Piman word *o'oida* covers all types of hunting and fishing activities, both individual and social or communal. It might best be rendered "animal procurement." (A different verb, however, was used for trapping.) Individual game hunting was relatively uncommon among the River Pima, who had a well-organized, socially integrated irrigation system at least by the middle eighteenth century, if not aboriginally. Perhaps one individual in a ranchería would venture out periodically on his own to hunt big game—primarily Pronghorn, two species of deer, and Desert Bighorn. (Whether these hunters were the same individuals who supplied the community's needs for Wild Turkey feathers is unknown, but they would have been the ones most likely to encounter this species in the uplands.)

Here is how George Kyyitan described individual hunting or game stalking: "*Vipiam*, to get nearer them, like that *vipiam g totobĭ* [stalking cottontails]. Go all by yourself and hunt. You have to go slower and slower and get nearer. [I] heard that from someone from Komaḍk. He went after some rabbits." Mathiot (n.d.:276) defined *vipiam, vipiameḍ* as "to go hunting object." Most of River Pima hunting was in the form of communal events arranged by the *tobdam*, hunt leader, who announced village activities at dawn from his housetop. Hunting took various forms, each of them formally named.

The most widespread type of hunting was the *shaada*, a drive or surround conducted on foot. It was probably practiced throughout Piman country and among many other tribes of the continent. Runners formed a large arc a mile or two wide in some open grassy or brushy area and closed in, like a horseshoe, to where archers were waiting to make the kill. The best time for a *shaada* was after a big rain. The drivers carried bows and arrows or just pointed sticks and clubs to kill rabbits and other prey. Meanwhile, some old men gathered wood, dug a pit, and prepared the animals for immediate roasting. One old man, if skilled, could skin and gut great numbers of jackrabbits in a very short time. When the drivers reached the central area, the *tobdam* directed them to a new area to circle. If the weather was not too hot, the drives might go on all day. At the end, all the game was divided equally among the participants.

Cottontails and jackrabbits were the main game taken, but other animals might be taken as well. Pima George Webb, born around 1893, participated in the *shaada*. "On those hunts on foot, they would kill cottontail rabbits, doves, and quail. A Pima is a good runner. You should see him hunting quail. A quail only flies a short way. When it flies, the Pima is running under it. When it lands he is right there to kill it. Even old men." Sylvester from the next village downriver said, "[W]ould start early as they can. Maybe last all day. When they get [to the designated] spot, they all circle. Some [animals] get away; they follow them." Joe Giff said, "*Shai* [*shaada*] more often when it's raining or drizzling. Quail don't fly when they're wet."

Another type of communal hunting was the *kuunam*, fire drive. This surround was carried out away from villages and farming areas with their brush fences—particularly in the grasslands, the saline sacaton flats, saltbush flats, and temporarily dried marshlands. It was described not only by several of the older Gila Pima men I worked with (Rea 1998b:53–56), but by two Jesuits working in Northern Piman country in the eighteenth century, Philipp Segesser (1945:157) and Ignaz Pfefferkorn (1949:198–199).

Fire drives took principally two of the most esteemed of Piman game, *vopsho*, cotton rats (*Sigmodon* spp.), and *kokson*, packrats (*Neotoma* spp.). Cottontails and even jackrabbits might be encircled as well, but most birds would fly away.

Once Pima acquired horses, still another type of drive developed called the *kuushada*, mounted drive. This approach extended the area covered by the *shaada*, foot drive, many times over. Leonard Pancott noted that whereas the *shaada* was carried out close to the fields where game gathered, especially around clusters of mesquite trees, the *kuushada* was way out in the open country. Sylvester agreed. "They used to go way far—maybe spend the night over there—in New York Thicket." Using bows or later shotguns, hunters shot jackrabbits (the primary game) or cottontails from galloping horses.

Like the other two types of drives, the *kuushada* was under the direction of the *tobdam*. This boisterous drive was dangerous, and the leader's commands, said Leonard, were obeyed explicitly. Anyone who disobeyed was expelled from the hunt. "We do as he tells us. We're brought up to take orders. We respect him." Still, accidents happened. The 1885–1886 entry for a Gila Pima calendar-stick narrative states, "Two youths were thrown from their horses during a rabbit hunt and killed."

A fourth type of communal hunting, the *huhu'ida*, was not technically a surround. The word literally means to chase something in all directions. The *huhu'ida* was a group stalking of big game, probably originally intended to supply ritually killed deer or other animals for the postharvest ceremonials. It occurred primarily in November or December. It involved a large group of hunters who were afield at least a week, killing deer and jerking the meat.

The game sought in all these hunts were primarily rodents and lagomorphs (cottontails and jackrabbits). The taking of quail, for instance, was incidental. (For more details on these types of hunting, see Rea 1998b:48– 66.)

Still another type of social hunt was called the *wulivga* or *ha wepegida*, nocturnal hunting mainly for birds. It was practiced uniformly throughout Pima country. The word *wulivga* has two different meanings, "target" and "torch," based on a similarity in how these two items were made. Sylvester Matthias explained, "Two kinds of *wulivga*. Game for the boys—target practice. Somebody be dragging [it], somebody try to shoot it. Made out of a bark–willow bark or Screwbean bark." The other kind was the bird hunt. All the men born early in the 1900s gave me virtually identical accounts of this activity. My description here is based on 1974 interviews with Joseph Giff, supplemented with notes by David Brown and Sylvester Matthias.

After winter wheat was planted, a special type of bird hunt called the *wulivga* was held. (*Wul* means anything tied together.) This was a nocturnal hunt using a torch made of bundles of Arrow-weed two or three inches thick tied very tightly, hence the name of the hunt. One person would swing the torch to blind birds that were at roost in the fencerows or brush-piled fences, and another person would kill the birds with a stick.

Torch hunts were undertaken after days of rain. In a drizzle, Gambel's Quail especially were more flightless and would flock up under the brush. (I suspect that another reason for practicing this type of hunt in rainy weather was to avoid setting brush fences on fire, although no one mentioned this reason.) Abert's Towhees, White-crowned Sparrows, and quail were the specific objects of the *wulivga*.

In more recent times, a hunting lantern was made by cutting out the sides of a rectangular five-gallon metal can and filling it with brush. Either the turned-up sides or a piece of bailing wire served as the handle. It was swung around, providing lots of light, just as was the original Arrow-weed torch. (Still later a flashlight was used for the same purpose.)

Others also commented on this nocturnal bird hunt. At the opposite end of the reservation, Francis Vavages gave me this parallel account: "They usually do that with Arrow-weeds. They burn for a long time. They go out at night for quail. They switch [them] back and forth. They use a mesquite branch with a lot of branches to hit them. [This method of hunting is called] *wepegi[da]*. In winter time. To get their food. It's sort of mean to the poor birds. But they need to get their food. Dry Arrow-weed. Bundle—tied up, waving back and forth, *ha wepegi[d], tt o wepegi g kakaichu*. Cut mesquite. When it comes out, they hit it. They can't fly at night. A lot of times, they're doves in there, too. And those little winter birds. *Tamtol* [White-crowned Sparrow]. They only come out in winter. They're so small, I don't know what you can get out of them."

"What did they put inside?" I asked him.

"Anything fine. For a torch—they bundle up dry Arrow-weeds. They have to use green bark or fine limbs, tie it up so they won't burn. Pull the torch together. If somebody dies [at night], they bury them right away. [They] use that torch. Certain time, they have a wake."

"How long do they burn?"

"About twenty minutes. By then, maybe another torch [will be] ready."

George Kyyitan of Bapchule village also spoke of the *ha wepegid* hunt: "*Tobĭ* [cottontail], doves, quails—at nighttime when they sleep, they gather in a bunch. Big can, put wood in there, one or two carry it. *Wepegi[d]*—light at night. We usually use it on quails, too, at nighttime. When they sleep at night. They have to gather up around that bush, anything around there. You know, where they sleep. You get *big* can to put wood in there to build a fire, and one will carry that or two of them. And then they'll go around. Sometimes they use that bow and arrow. When they saw [see] where the quail will be, they shoot them right there. They get them."

"With that torch light?"

"They'll just stay there, because it's going to be very dark."

"What do you call that when you take that light out at night to hunt?"

"*Ha wepegi[d], ha wepegi[d]*, because that light will go in there."

Silvester said that the original term for this form of hunting was *wulivga*, from the torch, but it became known as *ha wepegi[d]* when the can lantern and eventually a flashlight or even headlights were used. In modern times, *wepegi* can mean "electricity" or even "movie," and *wepegidakuḍ* means "flashlight." The word is ultimately derived from lightning strike.

These various practices describe hunting among the Akimel O'odham of the Gila. Some of the communal hunts such as the *shaada* and the *kuunam* were known among the Tohono O'odham. "The hunt [*shaada*] took place in spring," wrote Castetter and Underhill (1935:42). "All the men and boys engaged formed a circle about two miles in diameter with a line of the best bowman at one point. These men remained stationary while the circle converged and drove the rabbits toward them."

Fire drives probably were carried out wherever Pimans lived in habitats that accumulated enough grasses and forbes to sustain large rodent populations, especially cotton rats, and to carry a good fire. Papago hunted roosting turkeys and Mourning Doves with arrows (Castetter and Underhill 1935:41). Otherwise, the elaborate nocturnal bird hunting, a form of garden hunting, was probably unique to the riverine peoples living in rich, mesic habitats. Was it practiced in Pimería Baja, for instance?

Probably individual hunting was more common than communal hunting among Tohono O'odham, who lived in smaller, more dispersed rancherias. Here is Underhill's (1939:77) summary:

> The Papagos hunted deer, hare and rabbit, small rodents and birds. Most of this work was done by individuals or small unofficial groups. But deer was hunted officially for the cleansing ceremony in the autumn and there were communal rabbit drives when visitors were to be feasted, especially at the time of the drinking ceremony before the rains. For communal hunts of both kinds, there was a leader who set the day, chose the locality, called the people together at dawn, appointed the beaters and made the required speech. His title was tópetam [*tobdam*], rabbiter. He passed his office down, as did the others, by instructing a younger man, usually a kinsman, and finally asking the council to ratify his appointment as the new leader.

As noted earlier, among both groups of Upper Pimans, the *tobdam*'s talk made only indirect reference to the animals to be hunted that day, out of respect for the animal. The crier's dawn speech might say that they were going to hunt the long-eared ones rather than specify jackrabbits as such. Although killing was necessary for food, it had to be done in a way as to avoid giving offense. It was thought that animals cooperated with the hunter, allowing themselves to be taken for the People's benefit.

One may wonder about the time, depth and geographical extent of some of the categories of Piman hunting here enumerated, based on accounts by Gila Pimas born early in the 1900s. Fortunately, we have at least a partial answer preserved in the writings of the ethnographically observant Jesuit Pfefferkorn (1945:205):

> Hunting is greatly to the liking of the Sonorans. It provides them with their favorite food, meat, and frequently with the skins from which they fashion their quivers and war bonnets. Finally, it gives them pleasure. For that reason they spare themselves no trouble in this activity. They go out mounted and on foot, and show themselves to be very skillful and fearless hunters. The astonishing audacity with which they attack the most ferocious beasts I have already described, as I have also described their gay mouse and rat hunts.

From this passage, we know that both the *kuushada* (mounted drive) and the *shaada* (foot drive) were practiced in early colonial times, as was the individual hunt. The rodent hunt Pfefferkorn refers to is the *kuunam* (fire drive). In fact, he penned the most detailed description of Piman fire hunting (1945:198–199; see also Rea 1998b:53–56; Segesser 1945:157). The Névome vocabulario has entries for such activities farther south as stalking and hunting using the head decoy *(mohana)* and drive *(sada)*. From these several locations and the associated terminology, it would appear that hunting techniques were rather similar throughout the range of the Northern Pimans, both Pima Bajo and Pima Alto.

Hunting Territories

The range covered by Upper Piman hunters was rather enormous, even though it was curtailed somewhat by potential encounters with enemy Athabascans from at least the late seventeenth century onward. I mapped the hunting areas most frequently mentioned by Gila River Pima born early in the 1900s (see Rea 1998b:62). They include such distant areas as Table Top Mountain, one of the most commonly mentioned locations, as well as west on the Gila to the bend north of Gila Bend. Most hunting took place within a fifty-mile radius of the Pima villages, but oral history accounts of agave gathering, coupled with hunting, took parties to the southeast as far as the Tucson region. Writing of the *kuuvid*, Pronghorn, Russell (1908:81) said, "[Pima] hunters of former centuries successfully stalked these animals upon the mesas, particularly upon the higher grassy plains to the eastward."

In 1983, Sylvester provided some details on Gileño hunting localities:

> *Kuunam*, fire drive, [was held] where lot of brush, mostly around Avondale and Tolleson—not much farming [there then]. My father told me that one time they were invited to *kuunam* up there—because [the white] people got [wanted to get] rid of their rabbits [that were] eating [their] trees [presumably orchards or citrus groves]. Also [they *kuunam*] up along Lone Butte along [the] Gila; also along Havañ Kosh [Raven's Nest, near Salt-Gila confluence], where there was just a wasteland. The Chandler area [also] was just wide open with a kind of grass called *toota muḍadkam*, "white tassels-it-has"—I don't know what that

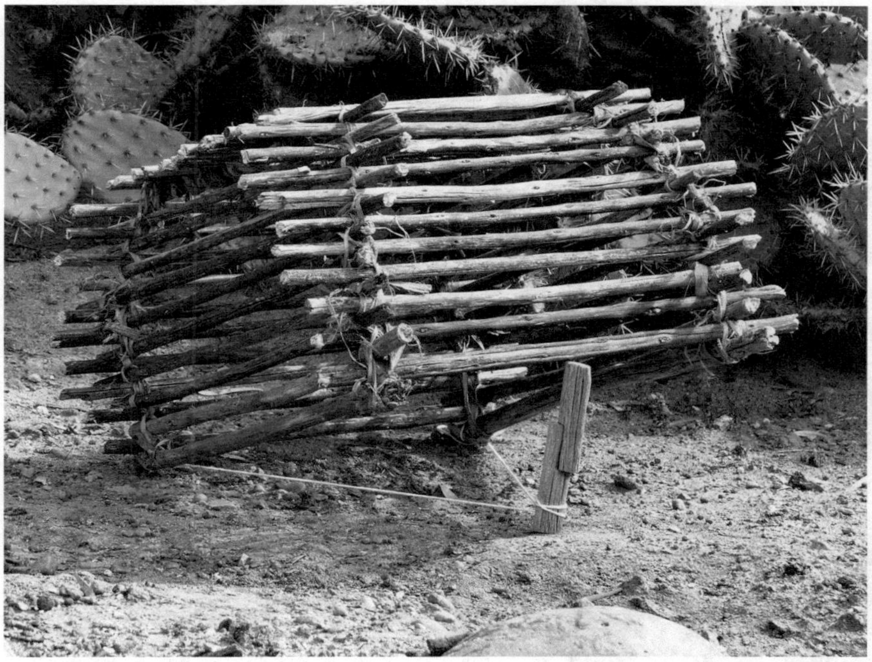

Figure 8.2 *Pima eḍpa (cage-type bird trap) (photograph by Peter Axel).*

> is, but some kind of grass [his grandmother said]. *Kuunam* would kill out mesquite—from the brush land [*sic*].

Hunting areas were determined by the individual hunter or, in the case of communal activities, by the *tobdam*. In his study of land use and tenure, W. W. Hill (1936:586–587) was given this picture of River Pima property rights: "All [Pima] lands but those assigned were free for purposes other than agriculture. There were no privately owned grazing, hunting, or gathering districts. Likewise, the wild products growing on assigned land might be harvested by anyone. 'They matured without work on the land owners' part.' However, this applied only to the products, not to the tree or bushes. No one but the owner had a right to cut down a tree, cactus, or bush that grew on his land."

A similar pattern seems to have held among the more dispersed Papago rancherias. As summarized by Underhill (1939:99),

> The men's deer hunting grounds were determined in a very loose way, like the women's gathering grounds. A men's hunting party was arranged in the council but not for relatives only, since skilled deer hunters were few. They went to certain mountains rather loosely understood to be pre-empted by their village but no informant could say what would be the consequences if they went elsewhere. "Probably no one would say anything." Then follows another example of the economy of abundance. "There weren't many hunters. There were deer enough."

Bird Trapping

Among the Pimans, at least the Gileños trapped birds regularly for food. Sylvester Matthias explained that although usually the man hunted birds, anyone could trap them, and either men or women cleaned them if they knew how. There were two types of traps, both apparently called *eḍpa* as far as anyone I asked could recall (fig. 8.2). To my knowledge, only birds were trapped by Akimel and Tohono O'odham, never mammals, though Pennington (1980) records several species of mammals taken in deadfall traps by Pima Bajo.

One trap was indistinguishable from the cage or coop (see fig. 6.1). It was made of Arrow-weeds, arranged log-cabin style, bound at the corners with the inner bark of willow, with leather or rawhide, or in later times with strips of rags or wire. The roof was either flat or domed and equipped with a handle. The trigger, called the *siishpadag*, was two pieces of Arrow-weed, each notched in an H shape, one resting on top of the other. One edge of the trap was lifted and the trigger placed beneath it, the weight of the trap holding it in place. (Sylvester made his traps so that extra sticks could be stuck into the top to increase the weight of the trap as needed. If the trap were too heavy, the trigger would not trip readily. If too light, strong birds such as quail might tip it over.) Some grain or other bait was scattered on the ground inside the trap. A noose of horse hair or string attached to the lower end of the trigger rested on the ground with the bait. Birds scratching and eating the food eventually tugged at the noose, dislodging the two-piece trigger so that the trap fell down capturing one or more birds.

This type of trap worked particularly well with quail because of their habit of scratching the ground for seeds. I once saw such a trap in Pima Bajo country, and ethnographer Thomas B. Hinton, who knew this country thoroughly, was familiar with these traps. Castetter and Underhill (1935:43) described these traps among the Tohono O'odham as a very rough cage of split Saguaro ribs. The Tohono O'odham took primarily quail, but also occasionally Mourning Doves. Delores Lewis of Ge Oidag described this trap, calling it *la'ashpakuḍ*, from the verb *la'ashp*, to trap or snare (Mathiot n.d.:24). Quail traps with the same trigger mechanism have been recorded from the Maricopa, Mohave, Kumeyaay, and Cocopa (Campbell 1999: 61–62; Kelly 1977:52; Spier 1933:72; Stewart 1947).

The simpler trap was a flat structure usually made of Arrow-weeds bound with rawhide or leather. When made flexible, without crossbars, it served as a door to the *olas kii*, round house. But when made rigid, it served two different functions: as a tapestle, or hanging shelf, called *kuukta*, in the *vatto* (arbor) or *kosin* (storage shed), and as a bird trap. Kissell (1916:140– 147) described and illustrated the technique for making this trap. River Pimas said it could be called *eḍpa* like the cage trap, although *ma'ishpaḍdag* is more likely its specific name. This word comes from *ma'ishp*, to cover something.

"When used as traps they are tilted at an angle of 20° or 30° from the ground and supported in that position by a short stick to which a long cord is attached. Wheat strewn under the trap lures small birds, which are caught when a jerk on the line removes the supporting stick," wrote Russell (1908:101). So important was this animal procurement that he added, "One of these traps was seen at nearly every house during the winter of 1901–2," a point

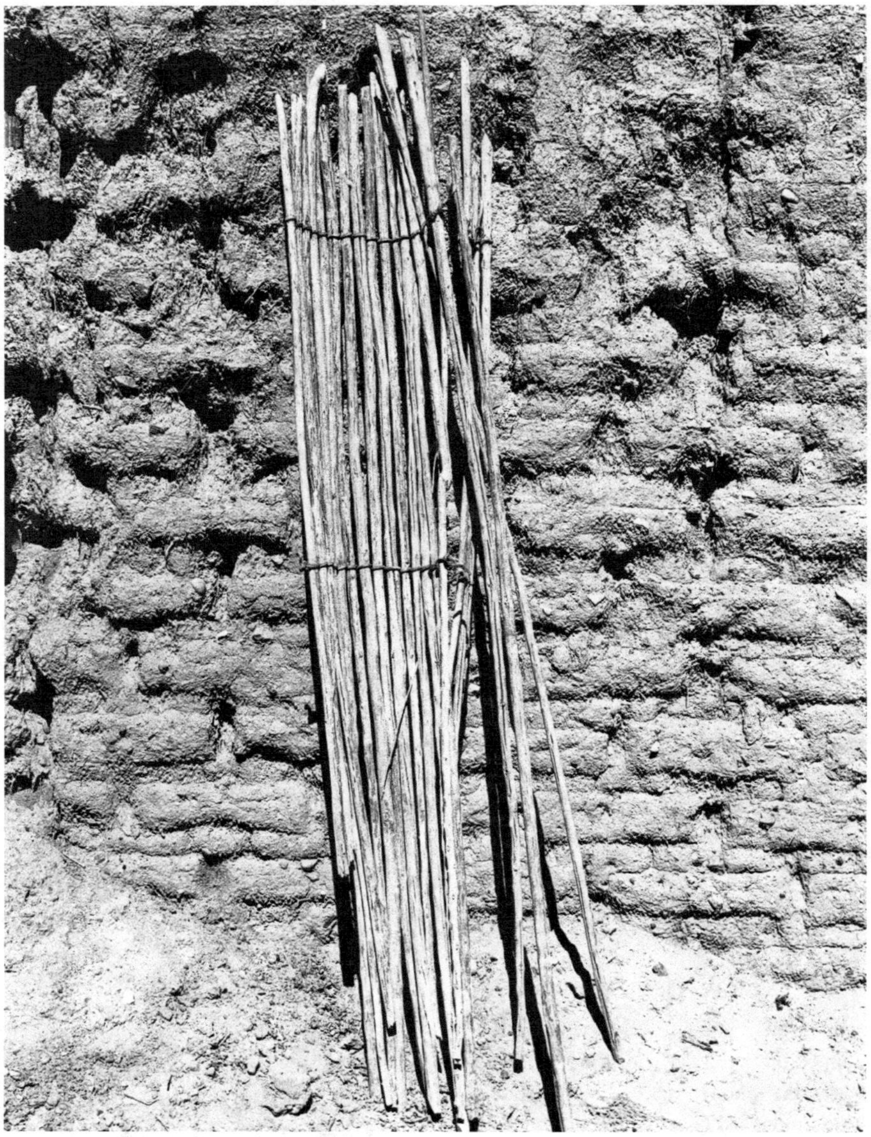

Figure 8.3 Ma'ishpaḍdag *(latticework bird trap, partial), Ónavas (photograph by Amadeo M. Rea).*

he had made earlier while discussing *tamtol*, White-crowned Sparrows (81). The flat latticework, perhaps with some added weight on top, simply fell on the birds and smothered them (fig. 8.3). The operator then gathered his or her dead birds and reset the trap for the next catch.

Many older Pimas knew these traps. For instance, Francis Vavages recalled: "*Eḍpa* I think they call it. Sort of like a trap. I know they used to trap the quail, like that. They put the Arrow-weeds together like that. Pile up dirt, and use a string, bury the string, about halfway to there and spread wheat on that thing there. And the quails will come and trip that thing, you know."

"Do the quail themselves trip the thing, or does somebody have to pull the string?" I asked.

"There's two ways of [doing it], I think. But you have to do it away from, say from that house to here. But I have to stay here if I'm going to catch those birds. I'll just keep watching them. But a lot of times they put a string there. When a bird walks in there he trips on that string and then it does that [falls]. A lot of them get away, too."

At another village along the Gila, I asked George Kyyitan, "Remember those cages they used to make for trapping quail?"

"Yeah. *Eḍpa*. Some they put it this way. With little sticks. Some they use the screen, just flat one. That's the way they trap them."

Then he described still a third type of *eḍpa*, a trap of probably more recent vintage: "Yeah, I know what you mean—*eḍpa?* Yeah, they used to [use] that one there. And they used to get quails with it, too. And sometimes they just get a screen. Or that wire fence [mesh] that uses for the house, for plastering [stucco wire]. That's what they made, too, a little trap there. And put dirt around it, there, so it would hold it down. Then, on down at the end, make a little trail there, of that wire, and put it in the ground, too. And put dirt on the sides. Then inside they could measure them, see how far, how high they are, them quails. Then, goes that way. And then, from over here, they could make some, . . . put some wheat in there, or anything they could eat. All the way down inside, to the inside. Then they go in there. Then some [go] in there, can eat in there. They don't know which way to go [out]. That's the way they get 'em, too."

He illustrated how this particular kind of trap was made from three pieces of mesh: one for the sides, one for the roof, and a third for the entrance. Like a fish trap placed in a stream, it funneled the foraging birds through a small opening into a larger compartment. Unlike mammals, which usually search for the opening through which they entered, the trapped birds would panic and try to fly up through the screened roof of the trap. So far as I know, this type of trap was not used before Pimans had access to wire mesh or hardware cloth. It had the advantage that it could continue to catch birds after the first entered. If there was enough food inside, the first bird would continue to eat, acting as a decoy attracting additional birds from a flock.

The bait and the type of trap targeted certain species that were relished for food and were abundant around Piman farms and homesteads. But traps were not always species specific. I asked Francis what birds besides quail might be trapped. "Oh, they catch *tamtol*, winter birds. Little birds. All you got to do put one in your mouth. Same way with the *bichpoḍ* [Abert's Towhee]. *Bichpoḍ* is that bird too that likes to eat—you plant melons or squash, he'll go in there. It's got that wide beak, too. They'll crack those [seeds] like those birds, those talking birds [parrots]. They'll crack those. Lot of times we'll go around there, you know, and try to plant melons. They start digging. They seem to know where the seeds are, too. All you'd see is the shells around there. Yeah."

The White-crowned Sparrow (*tamtol*) he mentioned was a target species, but the Abert's Towhee, a year-round resident, probably ended up in the pot as well. Mourning Doves could be trapped year round, but were especially abundant in winter, when northern migrants augmented local populations. Likewise, the Lark Bunting, another target species, came in large flocks wherever seeds and grains were available. Great numbers passed through the desert on their return migration in the late spring, the time of the spring stress season in the Piman annual subsistence cycle. Most other animals, such as mammalian game, were at their lowest numbers at this season.

But not all birds were welcomed. I asked George Kyyitan, "What kinds of birds besides the *kakaichu, hoohi,* and *o'okokoi* [quail, Mourning Dove, and White-wing] did they catch in those traps?"

"Any kind of bird that they really want. You know that *go'ogov* [shrike]? That's a mean one. They bite, too."

"They eat that?"

"No. *Shashañ* [Red-winged Blackbird], same thing. *Sipok* [Northern Cardinal], that's a mean bird, too."

"Did they eat those?"

"No."

Contrary to what some have thought, the People were definitely discriminatory in what species they considered legitimate game. In this conversation, doves, quail, White-crowned Sparrows, and Lark Buntings were sought as food. Incidentally trapped shrikes, cardinals, and blackbirds were released, although getting them out of a trap was a nuisance, as anyone who has handled them knows. One might have expected that at least the blackbirds would have been eaten, but they were not. (Ruth Giff said once that they did eat *shashañ*, but this appears to be a lapsus; I was unable to confirm her claim.)

There is evidence in Piman mythology of the use of hair nooses for ensnaring raptors, as seen among Puebloan peoples (see Prairie Falcon account). I found no one still familiar with this practice.

Nest Robbing

The most frequently eaten wild bird eggs were those of the Gambel's Quail, a ground-nester. Both young and grown people searched for them, particularly because a nest has so many eggs. When a nest was found, an egg was held up to the sun or broken to determine the state of incubation. Most said that if there was evidence of embryo development, the eggs were left in the nest. If fresh, the eggs were roasted on a bed of coals or boiled. There were proscriptions about washing after handling quail eggs (see species account).

Eggs of certain doves might be eaten. Most River Pima consultants recall eating eggs of both Mourning and White-winged Doves, though "these others just got two eggs," lamented one man. However, Tohono O'odham generally said that White-wings and their nests should not be molested because of their association with the Saguaro crop. It seems neither the Common Ground-Dove nor the Inca Dove was bothered, even as eggs.

Vincent Josemaria of Pisin Mo'o, a westerly Papago village, said that they would even eat nighthawk eggs when found on the ground, just as they did quail. People from other places did not mention these eggs, and Akimel O'odham said they were not bothered. In contrast, Castetter and Underhill (1935:43) stated flatly, "No bird eggs were ever used" by Tohono O'odham. Their information reached back several generations, so this practice may have been the original tradition.

Pennington (1980:194) reports similar behavior among lowland Pima Bajo: "Eggs *(nono)* of this *cordoniz* [a local species of quail] are valued as food, and it is claimed that there are generally from 10–12 eggs in a nest *(napo)*. When people are out walking they watch for evidence of quail and then search for their nests."

Lowland Pima Bajo Hunting

Among the Northern Piman groups, we know much less about the bird-hunting practices of the lowland Pima Bajo living on the middle reaches of the Río Yaqui. A sentence on colonial Desert Pima Bajo has been translated two different ways: "The Nébome also hunted the game that abounds in their montes [foothills or thornscrub], especially deer. They are very skillful in shooting these, as well as birds of the air, which are plentiful" (Pérez de Ribas 1999:391 [p. 360 of 1645 ed.]). The original Spanish says: "Tambien para sustento se valían de caza, de que abundan sus montes, en particular de ciervos, porque son muy diestros en flecharlos y aves del aire, que no les faltan" (Pérez de Ribas 1944:2:150). And Pennington (1980:193) translates the last sentence as "skilled in killing flying birds with arrows." It is unclear how much skill should be attributed to these archers.

Pennington's (1980:189–199) treatment of bird hunting at Ónavas unfortunately contains many errors, including birds that are unlikely to occur in that part of Sonora. He obviously did not get field or specimen verifications of the avian folk taxa the Pimas discussed with him, and the fine distributional work on the state's avifauna was not yet available (Russell and Monson 1998), but even van Rossem (1945) is not cited. In the individual species accounts in this book, I have tried to unravel such misidentifications, but with no remaining speakers of this dialect of Piman, many riddles continue to be unresolved.

Pennington (1980:195) describes a "pyramidal" trap that is the same as the Upper Piman quail trap: "This trap is made from sticks arranged in a pyramid and tied together so that the top of the trap will support a stone. Maize or wheat grains are placed inside, and one side of the small structure is raised slightly and supported by a stick that will collapse when a string anchored to another small stick stuck in the ground amidst the grains is disturbed."

In addition, Pennington found evidence of the figure-4 trap being used with either a crate or rock deadfall to take quail. He also found this type in use among the Northern Tepehuán of southern Chihuahua (Pennington 1969:120). Although the *eḍpa* is characteristically Piman, the figure-4 release trap may be borrowed from sierran neighbors. So far as I know, it was not used by Akimel or Tohono O'odham.

Pennington (1980:194) noted a third method of taking quail: "A long piece of string is anchored to a grain of corn that is placed in a spot where quail are known to feed. The hunter hides nearby. When the grain of corn lodges in the throat of a bird, the string is tugged sharply and the bird strangles."

In colonial times, quail seem to have been an important animal resource to the Pima Bajo, another example of garden hunting. These birds proliferate in areas with small-scale farming, particularly where fencerows provide nesting areas and shelter throughout the year. Several species of doves, most important the Mourning and White-wing, were important game birds for these people.

Hunting Equipment

The Northern Piman hunting equipment was not elaborate. Men carried a bow, *gaat* (pl. *gagt*) and hunting arrows, and boys had smaller versions of them together with a sling. Akimel and Tohono O'odham men born early in the twentieth century who supplied me with the information given here said that they hunted in their youth exclusively with these arms, not using rifles or shotguns until they were at least well into their twenties.

The choice wood for Upper Piman bow staves was Desert Mulberry, *gohi*; this tree grew only at higher elevations (3,500–5,000 feet) at the desert's edge north and east of O'odham country, mostly in lands dominated by marauding Apache. The next most prized wood, at least for Gileños, was Osage-orange, *ho'idkam koli*, a small, tough, thorny tree planted in the surrounding Anglo-American communities for fences and hedgerows. Most bows, however, were generally made from the Goodding's (Western Black) Willow, *che'ul*, locally available when the river ran. Other plants used included Screwbean, *kuujul*; Coyote Willow, *s-i'ovĭ che'ul*; and either the branch or root of Catclaw, *uupaḍ*. (Joe Giff was told that the Catclaw root made the best bow.) Even wolfberry, *kuavul* (*Lycium* spp.), is said to have been used by Papago for bow staves (Castetter and Underhill 1935:70). Lowland Pima Bajo bows were made from the tough wood of Güilonche *(Diphysa occidentalis* or *D. suberosa)* (Pennington 1980:208).

Both modern and colonial sources describing the bow say that it was from four to five and a half feet long, about two inches wide at the middle, tapering to an inch or a bit less at either end. The belly or inside was flattened, the back rounded. It was usually undecorated. Shaping and smoothing were accomplished with a rough stone and a handful of sand. Areas that showed signs of splitting were bound with wet tendons, which shrunk tightly.

Russell (1908:95) described the River Pima bow as "simple, undecorated, and not very carefully made," but this may not always have been the case. The colonial Jesuit Ignaz Pfefferkorn (1949:201, 203, 204), writing in the mid-1700s of Upper Pimans living somewhat to the south, observed, "Sonorans perform no task with greater diligence and more particular care than the fashioning of their bows and arrows. For these are the implements which sometimes procure for them a good roast and which further are their protection against the attacks of their enemies." Then with a German twist he editorialized, "Weapon making is, then, the principal and likewise the most agreeable occupation of the Sonorans, because they can engage in it conveniently while sitting still."

The Piman bow was usually a simple bow with a single arc, but it was occasionally made with two arcs (compound curve) or even slightly recurved. Shaping was achieved by means of heating in hot ashes. The shape was held by holding the bow in position with stones or between posts until it dried. A nock at either end held the bowstring. The Flute Lore Story of the twins has descriptions of bow making by the boys' mother and by Coyote, each using somewhat different techniques (Densmore 1929:59). In this particular case, Coyote did the superior job. Other information on bow making may be found in the works by Castetter and Underhill (1935:70), Rea (1998b:74–76), Russell (1908:95), and, for Northern Tepehuán, another Tepiman group, Pennington (1969:119–121).

The information assembled by Pennington (1979:52, 1980: 213–214) from the Névome vocabulario indicates that archery terms have changed little over time and that there was considerable uniformity in the form of equipment and its associated terminology from the Gila River to the middle Río Yaqui.

The bowstring, *gaat gi'aḍag*, was made of sinew, *tatai*, preferably from deer or horse, but cow served as well. "The bowstring was a two-ply cord of deer sinew which was soaked, split with the teeth and rolled on the thigh," wrote Castetter and Underhill (1935:70) for the Papago. "This hard rough cord was made rather smooth by running it through a small perforation bored in a stick of sahuaro wood *(Carnegia gigantea)*." Three-ply bowstrings were also known among Pimans. Pfefferkorn (1949:201) said the colonial bowstring was made "by twisting the intestines of various animals, and is about as thick as the quill of a raven's feather." But as is seen in several other contexts in his relación, the term *gut* has been used where tendon is clearly the material intended. The Northern Tepehuán, a southern Piman group, used sinew, leather strips, or ixtle fiber to make bowstrings (Pennington 1969:121).

Herbert Narcia said that the bow nock for attaching the bowstring was called *vaḍgaj*. In colonial times, Pfefferkorn (1948:201) wrote that "[t]he bow is bent only when the Sonoran is about to use it; otherwise it is left unstrung." This practice would have preserved the power of the bow, but it would seem that other Pimans (or at least later ones) left their bows strung.

The introduction of European firearms produced a classic case of linguistic marking reversal in hunting equipment (see Witkowski and Brown 1983). Northern Pimans call the bow *gaat*, and when the shotgun and rifle were introduced, the new item became known as *vainomĭ gaat*, "iron bow," and the traditional weapon was distinguished as *uus gaat*, "wooden bow" (pl. *u'us gagt*). More recently, as firearms became increasingly used and archery declined, the gun became simply *gaat*, and the bow obligatorily was marked *uus gaat*. Some people, unaware of the linguistic history of the phrase, have mistranslated the latter as "wooden gun." And by linguistic extension, the word *u'u* or *uush*, arrowhead, became also the word for bullet.

Arrow Making

Upper Pimans who had access to Arrowweed, *u'us kokomagĭ*, a colonial plant of high water tables and usually alluvial soils, used this plant almost exclusively, trading it to others who lived outside its range. The plant grows in extensive thickets. The stems are of uniform diameter for a yard or two, branching near the tops. With only moderate selection, abundant arrow blanks could be cut that required little or no straightening. Other Pimans had to be content with less suitable plants such as the young shoots of Catclaw or Soaptree Yucca stalks.

A complex vocabulary is associated with Piman arrows and arrow making. Sylvester Matthias, who was still making arrows when I consulted him, demonstrated the techniques and supplied the terms. According to Russell (1908:96), "The shaft length for an individual is measured and cut to reach "from tip of forefinger to nipple of breast of the maker." An arrow blank that has been cut, straightened in hot ashes while still green, then dried but not yet shaved smooth is *shelna* (pl. *sheesha*), from *shel*, "to straighten." At this point, the blanks may be bundled and hung on the house posts to await further attention.

In the next phase, the nock or notch, *uhugig* or *u'ugig*, is cut into the upper end of the shaft to receive the bow string, and the sides beveled. Frank Jim, western Tohono O'odham, called the nock *sha'aḍk*. (Joseph Giff remembered that arrows for fishing were made with the larger, heavier end down and were unfeathered.) The pile or point of the arrow, *mu'ukvuidaj*, is shaped and hardened in the fire. (The first part of this word comes from *mu'uk*, "sharp," the second from *vui*, "to face forward.") Once the blank is shaved smooth of its bark and polished with clean river sand, it is called a *hivshana*, from *hiv*, "to scrape or rub something." These items again can be bundled and hung up until needed. Some used an arrow straightener called *shelnakuḍ*, which were generally found in the ruins and were not common.

The sequence of George Kyyitan's technique differed slightly. He said, "Put the [Arrow-weed] *sheesha* out in the sun. When bark kind of wrinkled up, time to scrape it off. Straighten them out before too dry, *after* scraping."

The next steps involve the real art of arrow making: fletching and binding. To fletch an arrow is *u'umaḍad* or *a'anchud, a'anḍad*. For the Pima, choice feathers for hunting arrows were from the Red-tailed Hawk, though occasionally other large birds were used. Sylvester said that *ñui a'an*, Turkey Vulture feathers, though from a sacred bird, might be used if they were picked up at a roost or from a road-killed specimen and as long as they were not the long outer primaries, which were reserved exclusively for *mamakai*, shamans. Western Papago Frank Jim said the feathers of any bird could be used except those of *ba'ag*, Golden Eagle, and *vishag*, Prairie Falcon, which could be handled only by *mamakai*. (Presumably *chukuḍ*, Great Horned Owl, was excluded as well.) Eagle feathers were used exclusively on war arrows, after being properly purified.

Two feathers, from opposite sides of the bird, are used. The feather is split along its rachis (shaft), then one end held between the teeth and the left hand while the arrow maker shaves the pith and much of the rachis away so that the split feather will lie smoothly along the shaft. The action is much like that used by a basket maker in paring down weaving splints and in both cases is called *hivñua*. George Kyyitan said that the word *tap* means "to split the feather" (for arrows), and *tapidwua* means "splitting feathers for arrows." The outer vane from the left side of the bird must be matched with the outer vane from the right on one arrow, while the corresponding left and right inner vanes are used on a different arrow. This is necessary so that the arrow will fly straight. One cannot use two left-side feathers or the matching halves of a split feather on the same arrow.

The arrow maker soaks his bundle of dried sinews in a dish of water and puts several strands into his mouth, chewing and further softening them with his saliva. Laying two prepared feather vanes along the shaft, he binds them at their bases by a twist of now supple and gluey sinew, then rotates the arrow with both hands while still maintaining tension on the sinew with his mouth. Parting the vanes, he binds the feathers at several intermediate points, then binds the uppermost end near the nock. The terminal bindings are much broader than the intermediate bands, which keep the feather rachis from flaring away from the shaft. The excess rachis protruding from either end of the binding is now cut away with a sharp knife. Then the two attached vanes are trimmed to a narrow width (about ten to fifteen millimeters, or about half an inch). The vanes may be six or more inches long.

Finally, the arrow is finished by binding it tightly with more sinew just below the nock and above the pile at the front end, "where you put force on it; this prevents [it] from splitting," explained Sylvester. I have seen a Pima arrow where the uppermost wrapping of the feathers also served as the binding for the nock, there being little space here for the shooter's fingers. Sylvester's arrows typically had an inch and a half space, the butt, between the feather and the nock. The completed arrow is called *hapod* (pl. *haapod*).

Hunting arrows had two vanes, and war arrows typically three, but Sylvester said that even conventional arrows made to sell might have three vanes to meet buyer expectations. Lloyd (1911:190) noted, "War arrows were a yard long, with three feathers instead of two, and tipped with flint or, later, with iron. But even a wooden arrow would kill a deer."

River Pima arrows (fig. 8.4) were typically colored red. Sylvester used powdered red ochre, *heet*, when it was available. His source was Goks O'odham Dak (Do'ag) on the Salt River Reservation. Otherwise, he used moistened red crepe paper, which created a similar effect. Even the bow might be stained red. "Both bows and arrows are sometimes stained with the blood of the jack rabbit," wrote Russell (1908:96), "and war arrows may be dyed at the ends with the cochineal which makes its home on the Opuntias [prickly-pears]." These scale insects produce a brilliant purple-red color.

The color used on the war arrows probably recalled the metaphor of the Scarlet Macaw, associated with slain enemies and their blood, seen in so much of Piman oral literature (see that species account). Scalps and any other enemy trophies were harbingers of rainstorms, so welcomed by the Desert People. Bright red objects are paradigms of summer storms. The red of hunting arrows is a theme in a tale where Coyote is subjected to four temptations. (Sylvester said he did not remember the details of all four.) As was commonly done, Coyote puts up his arrows in the bet. When he succumbs to the fourth temptation, eating some stink beetles *(chukukui)*, his arrows turn into Red Racers (snakes), which climb up into a mesquite tree.

River Pima hunting arrows apparently were always simple, but the compound, or two-piece, arrow was at least known among them. In a war speech recorded by Russell (1908:360), the foreshaft of Apache arrows was called *vood*. Orations preserve older speech, so this type of arrow was probably simply forgotten on the Gila with the decline of hostilities.

Compound arrows were used by other Piman groups, however. The following information on Tohono O'odham arrows is summarized primarily from Castetter and Underhill (1935:70–72). Arrows with heads were compound—the foreshaft, *vood*, being made of Creosote Bush, *shegoi*; Desert-broom, *shuushk vakchk*; or Western Soapberry. The main shaft, *chu'idak*, might be Soaptree Yucca or Palmilla flower stalk, *takui*, and probably Carrizo or Common Reed, *vaapk*, from Akimel O'odham country. The main shaft was bored out, and the forehaft pounded in. The joint of the two shaft sections was sealed with the melted pitch from White Brittlebush, *tohavs*, then bound for the length of the joint with sinew. The projectile point was first set with this pitch, then secured with sinew (see Underhill 1951:88).

According to these authors, war arrows were fletched with the feathers of eagle or one of three genera of hawks (which are not specified, but probably buteonine). Some hawks, surely not *Accipiter* as stated, "were kept in cages and their tail feathers plucked." Turkey Vulture, turkey, and "crow" (that is, raven) feathers could not be used for war arrows, but at least the turkey feathers presumably served for hunting

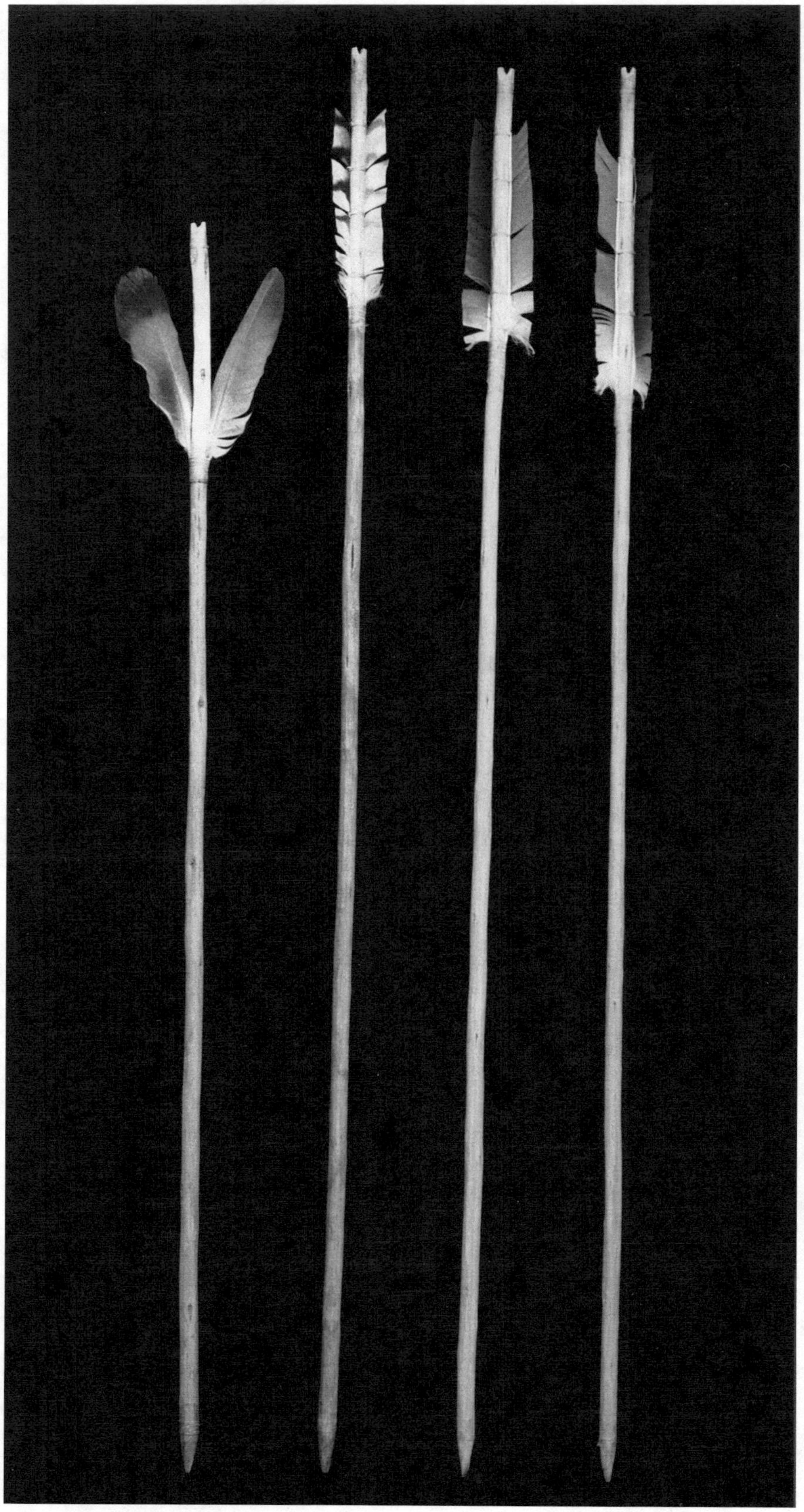

Figure 8.4 Haapod, *River Pima hunting arrows* (dodak ab ha to'a, *boys' practice arrow on left) (courtesy of Marvin Popkin).*

arrows, as is known from other sources. The vanes were set at a slight spiral on the shaft, unlike any Akimel O'odham arrows, to my knowledge.

Here is Underhill's (1938:70) description of the war arrows:

> These are not like deer arrows tipped with light stone, or rabbit arrows made of wood only. They are tipped with heavy quartz found only in four places, and having magic powers because it is [believed to be] the congealed blood from a hero's wounds. A man must have made a long journey to get it, and must have flaked it with a deer antler, over months of labor. He must have gathered reeds for the shafts in the season when reeds were fresh. He must have hard wood for the foreshafts, sinew and gum for the binding, and eagle tail feathers; also an old man who knows the arrow songs. If he has shot off all his arrows on a recent war party and since then has been tilling the ground for his kin, there is no hope. No one will give him war arrows: their possessors have too dire a need for them.

Tohono O'odham hunting arrows were simple—made of Creosote Bush or Arrow-weed from Pima country—or else compound, with Soaptree Yucca main shafts. "This made an exceedingly light arrow particularly desirable for birds" (Castetter and Underhill 1935:72). Frank Jim noted that Arrow-weed was one of the items the Tohono O'odham would collect when they came to cut wheat for the Akimel O'odham in the spring. Underhill (1951:88) illustrated both the simple and the compound arrow.

In the mid-1700s, the German Jesuit Ignaz Pfefferkorn worked with Upper Pimans at the settlements of Attí on the Río Altar and Guevavi on the Santa Cruz River, as well as with the Eudeve at Cucurpe. He usually lumped the native peoples under the term *Sonorans*. In a rather detailed relación written after the Jesuit expulsion (1767), he documented a great deal of colonial native life, including details of hunting and war equipment for an area where no ethnically identifiable Pimans survive.

War arrows, Pfefferkorn said, were compound and usually two ells (four feet) long. Three raptor feathers, about four inches long, were attached to the upper shaft. The removable thinner foreshaft, with its arrowhead, usually stayed in the wound unless it passed on through the body. Each man, Pfefferkorn noted, kept in constant supply fifty to sixty war arrows.

Colonial-period hunting arrows were remarkably like those made recently by River Pimans: "The arrows with which the Sonoran hunts game differ from war arrows in that they consist of but one piece of wood and are not armed with a flint. Instead, the arrow is sharply pointed and the point is slowly roasted in fire until it becomes strong and hard. The advantage of this kind of arrow is that it can be retrieved undamaged and can be used again after the game has been slain" (Pfefferkorn 1949:203).

Inexperienced young hunters were prone to losing lots of arrows, so they often used just the *hihivshana* or *sheesha* as unfeathered arrows. As their skills increased, they used a special type of arrow called *dodak ab ha to'a*, "attach them entire." This is the *hivshana* (pl. *hihivshana*), or smoothed arrow blank, to which two entire feathers, such as the tail feathers from a pigeon or White-winged Dove, are attached with sinew at their lower ends but not at their upper end near the nock. The feathers must be matched left and right, as in conventional fletching (fig. 8.5b), but the preparation requires only a fraction of the time. The shaft is still bound at both ends with sinew for strength. Although they look clumsy, Sylvester said that they still shoot quite accurately. (See fig. 8.1.) In a narrative, Herzog ([1929]) called the featherless arrow *ooto*.

What little is known of lowland Pima Bajo hunting equipment comes from the colonial Névome vocabulario and the modern studies by Pennington (1979, 1980). Only simple arrows had heads. Shafts were made from Bonpland's Willow, Maoto Negro (an acacia), Amapa Colorada, Palo Colorado, or Brazil. Compound arrows were made from Otate, a reedlike grass, with a short wooden foreshaft. These compound arrows never had heads. This seems to have been true for Pima Bajo in colonial times as well. However, this practice is just the opposite of what Pfefferkorn (1949:202–203) observed to the north and what Castetter and Underhill (1935) reported for the adjacent Papago, where projectile points were used on compound arrows. Pima Bajo used vulture feathers for fletching.

The finished arrow was called *u'u*, but Upper Pimans restricted the term *u'u* to the arrowhead (or, by extension, to a bullet or cartridge). The form *uush* (pl. *u'ush*) in Pimería Alta can refer either to the arrowhead or to the stinger of a scorpion or bee. Colonial Pima Bajo called the (apparently compound) arrow *huhu* (that is, *u'u*), while restricting the word *hapoda* to "cuando el palo de la flecha no es puntiagudo" (when the arrow tip is not sharp pointed) (Pennington 1979:52). As noted earlier, modern Upper Pimans now call finished arrows, which are blunt tipped, *hapod*.

Pennington (1980:209) adds, "According to the contemporary Pima [Bajo], bows and arrows were once used to kill fowl. When asked if blunt-tipped arrows were used for stunning small game so that the meat would not be damaged, the Pima noted that hunters knew how to gauge their strength and therefore blunt-tipped arrows were unnecessary."

Arrow Poisons

The powerfully toxic euphorbiaceous shrub *hínata* or Yerba-de-la-flecha, *Sapium biloculare*, occurs through much of Northern Piman country (see map in Turner, Bowers, and Burgess 1995:356). Its white, milky sap provided a fine arrow poison for many tribes, including both Northern Pimans and Seri (Felger and Moser 1985:300–301). It was also one of the primary plants used for stupefying fish. The Piman name for *Sapium* goes northward at least as far as Quitovac, Sonora. Although the species reaches its northward extension in the western Sierra Estrella, I have found no Gila Pima familiar with the tree or its use as poison. On the Río Yaqui, where the tree is common, Pedro Estrella objected to my even touching it; I had to clip and handle my herbarium specimen *(AMR 51)* between sheets of paper. Another plant in this family reported to have been used for poisoning arrows is Candilillo or Jumete, *Euphorbia collectioides* (Pennington 1980:209).

On one occasion when Sylvester had traveled with me to visit Pedro Estrella in Sonora, Pedro explained the use of this plant. Sylvester said that he had heard of rattlesnake venom being used on poison arrows as well as jackrabbit blood. "Let it stand so long that the blood will turn into poison." He said he had never heard of River Pima using any plant this way. "Sometimes people wounded [with an arrow] would die several months later, maybe from poison." However, in a Piman war oration, the word *hinaj* is used for Apache arrow poison, a word clearly associated with *hínata*, the pan-Piman term for *Sapium* (Russell 1908:360). It would seem that the information was simply lost this far north, where the shrub is so rare.

In 1737, Fr. Philipp Segesser (1945:173) noted that snakes were induced to bite liver soaked in this sap. The mixture was then rubbed on arrow points. According to what Ónavas Pima Bajo told Pennington (1980:209), poisoned arrows were not used in hunting.

Now a captain, Juan Mateo Manje (1954:98) in 1697 referred to poison arrows being used still farther north: "The Janos [who had joined the Apache on raid] related that those who were wounded by the deadly and active poison of the arrows of the [Sobaipuri] Pimas had died in agony." Manje did not explain how this poison was made. A body count of 168 was scored by the Pima warriors.

Quivers, Wrist Guards, and Other Hunting Gear

The Piman quiver is called *vogsha* (pl. *vopogsha*), a word that also designates "stepchild," which also clung to one's back. We find the word *voksa* (pl. *vopoksa*) in the ancient Névome lexicon (Pennington 1979:19).

In colonial times, according to the relación of Fr. Ignaz Pfefferkorn, the quiver was made from calf, deer, or some other wild mammal. But "tiger" skins were prized above all others. From Pfefferkorn's description (1949:108, 203), this cat is the Jaguar, the epitome of the powerful warrior in Piman thought. "The quiver is shaped like a long, narrow purse with a rounded but somewhat pointed bottom. It is no wider than is necessary for the usual number of arrows, being about seven or eight inches shorter than they are so that the Sonoran can pull them out readily. It hangs on a strap diagonally across the back from the left shoulder to the right hip. In this the arrows rest points downward. The tail of the arrow protrudes at least a handbreadth over the shoulder so that in time of need the arrow can be used instantly" (203–204).

Pfefferkorn may have had the Eudeve in mind while writing some of this description. The Caborca warriors sketched by Fr. Kino in 1696–1697 hold double-arced, recurved bows, and one has a quiver with the opening at his right hip, apparently

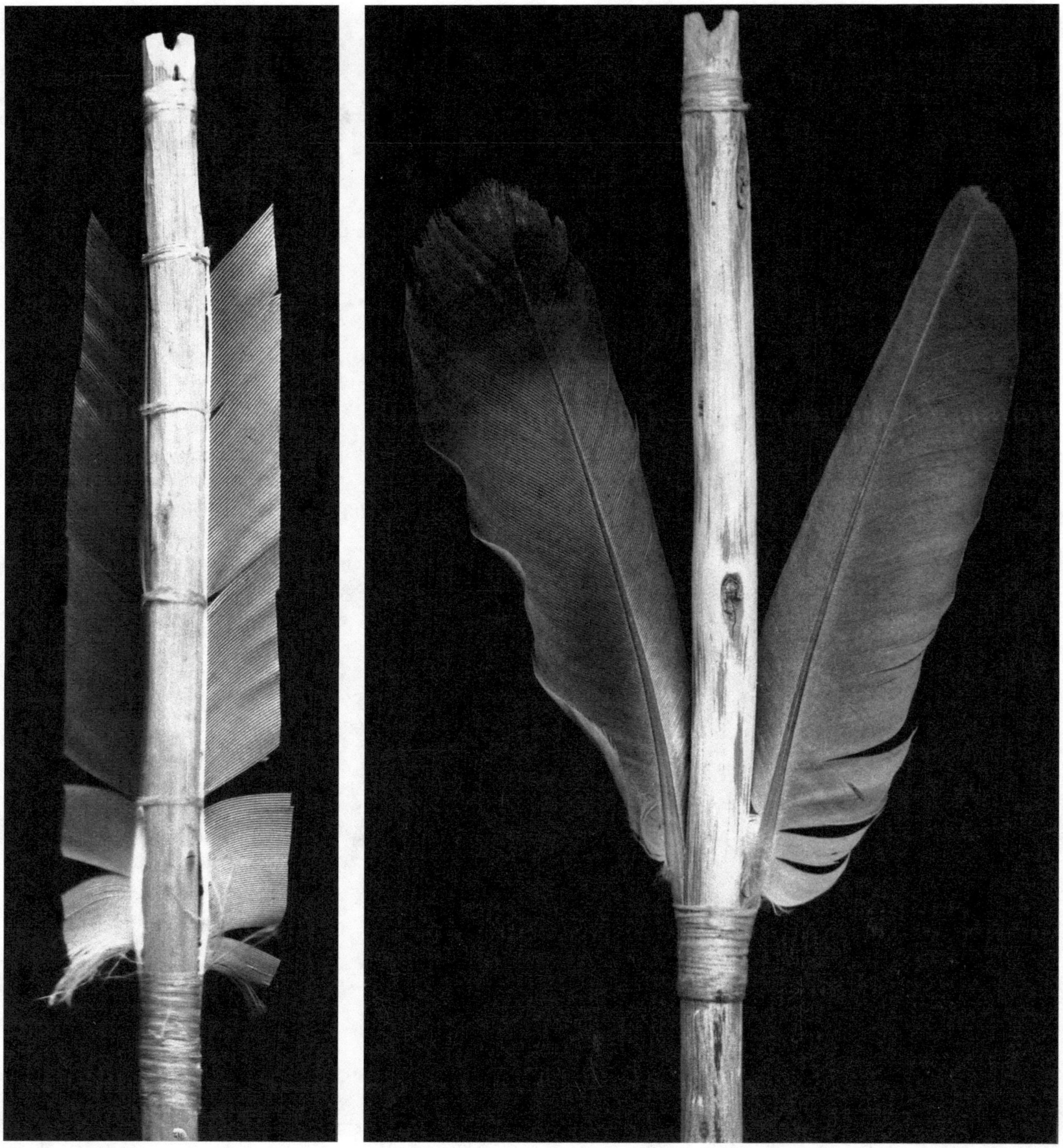

Figure 8.5 (a) *Fletching on* hapod, *hunting arrow;* (b) *fletching on* dodak ab ha to'a *(practice arrow) (courtesy of Marvin Popkin).*

slung from the shoulder (Ezell 1983:153). Underhill (1938:54, 71) showed a Papago hunter and a warrior with their quivers slung across the back of their hips, also opening at the right. The quivers appear to be made from either a fox or young Coyote skin, with the full tail hanging at the lower end, the arrows inserted in the throat region. They are attached by a belt around the front, not from the shoulders (see fig. 7.4).

Castetter and Underhill (1935:69, 72–73) found the Papago making hunting quivers from skins of such animals as Wildcat, Cougar, and Coyote. They were of rawhide, with the fur left on, the case being stuffed with grass and dried. The legs and tail were removed, except in the case of the Coyote, where the tail was left on. Larger quivers used on war expeditions were made from Pronghorn or White-tailed Deer rawhide or buckskin and were somewhat more elaborately decorated. The strap was a two-inch-wide strip of fox skin "with the hair left on to prevent chafing."

Similarly, either deer skin or the complete skin of a Gray Fox served as a quiver among historic lowland Pima Bajo (Pennington 1980:204, 359). The Gila Pima specimen collected by Russell (1908:plate XIII) is a whole Wildcat skin, with the feet still attached. The arrows are thrust into the anal end, with part of the fletching projecting, the points resting in the sewed-up head. Underhill (1951:27, 91, 97) showed similar quivers of Bobcat and one with arrows inserted down the throat of a whole fox skin.

The wrist guard or bow guard is called the *shooshba, shooshoba*, or, in colonial Pima, *sosopa* (probably *sosoba*). Wildcat skin was prized for wrist bands, Pfefferkorn (1949:106) wrote, providing "protection for the left arm against injury to the artery from the powerful snap of the bowstring." This use is even embedded in the Pima Creation Story. Handsome Man, who ultimately caused the Flood, is supplied with a wrist guard of Wildcat skin and turquoise earrings (Lloyd 1911:36). As described for the Gila Pima more recently, it was a band of leather, tied on with thongs at the two ends. With the decline of archery hunting, the *shooshoba* became an ornament worn by some men.

A hunting spear seems to have been used widely by Northern Pimans. As described by Pfefferkorn (1949:204) in the mid-1700s, it was about six feet long. "The spear's point they make almost as hard as iron in a slow fire. But it cannot properly be counted among their arms for they make no use of it in war. It is employed only on wild animal hunts." Pennington (1980:211) was told that modern Pima Bajo hunted Collared Peccary with a wooden lance. The Gileño lance described by Russell (1908:96) appears to have been a weapon. But in describing the various types of drives, elderly Pima men told me that participants might pick up any stick, such as a *vapai*, Saguaro rib, as they went along, using it for a club or spear. There were no formally fashioned rabbit sticks such as the Hopi used in their drives.

Pima boys used a sling made of leather for hunting small game, in particular doves and other birds. One collected by Russell (1908:120) has strings twenty-seven inches long. The pouch is three-by-seven inches. One made by Sylvester is similar. He said the thong should be about as long as the user's arm. The thumb goes through the noose at one end, while the other fingers hold the end to be released. I suspect that the sling is recent, although Russell (1908:120) noted, "Slings were used by Pima youth before the advent of the whites."

Sylvester called the sling *vagiwuikuḍ*, saying it is the form used by Kuiva and Salt River O'odham. George called it *valigwuikuḍ*, as Sylvester predicted for the Ta'i O'odham. "River rocks in good shape are best; used on rabbits, game birds, and to keep horses away." Tohono O'odham Frank Jim said that both a sling and slingshot were used when he was young, noting that one could "get a jackrabbit with them if it's asleep somewhere." He called it *vagivuakuḍ* (from *vagiva*, "pull and release," fide Mathiot n.d.:219). Theodore Peters of Santa Cruz called the sling *vipidwuikuḍ* and the newer slingshot, with rubber straps, *vagiwuikuḍ*. Most people recalled this latter one being used.

Loss of subsistence agriculture for Gileños due to watershed mismanagement and water diversion upstream from the villages led to a decline and ultimate loss of traditional fishing, hunting, and trapping for animal food. In time, beef largely replaced native game species, resulting in a diet very high in fats, one of the factors contributing to obesity and epidemic adult-onset diabetes. Similar trends in diet change among Tohono O'odham are producing the same medical complications.

Game Preparation and Cooking

Each hunter's kill had to suffice for the whole of the economic unit to which he belonged, which might mean from two to ten families. He butchered his deer on arrival and then called his children to carry pieces to each household, according to the number of its members. Occasionally, "the deer was all gone but the bones before he had finished," and his own family got nothing. But the gifts were, of course, reciprocated, and he had done his part toward the support of the unit (Underhill 1939:99).

I brought George Kyyitan a quail for him to show me how it is skinned. With the bird belly down and the head away from him, he split the skin of the back open from the back of the neck down through the interscapulars to the rump. Then he cut the wing feathers off at the carpels; next he cut the head off and the legs off at the heel joint. Now he could pull the entire bird's body out of the skin, leaving behind its "dress" as he called it—that is, the feathers attached to the skin. (Anyone concerned with the potential for contracting *kakachu mumkidag*, quail sickness, would pull off the head first to avoid further contact with this dangerous part.) Opening the belly, he left in the liver and heart, pulling out and discarding the rest.

I asked Ruth Giff how she cleaned quail. "Just clean it, gut it. Pull feather off. Leave the skin on. Cut the neck off. Then boil it or fry it. Eat the skin."

"Have to gut them?"

"Yeah."

She said that feathers of the *hohohi* come off easily, but ducks have to be plunged in boiling water, like a *chuchul* (chicken), to loosen the feathers. Rosita Brown said this might be done to quail as well.

There were two schools of thought on cleaning quail. Some plucked the feathers after removing the heads, feet, and wings. About an equal number of Pimans made an incision in the body and slipped the skin off with the feathers still attached.

One day talking of small game, Sylvester observed, "Boil cotton rats whole, fur and all; skin them after. Or fry in grease. Remove intestine. Just cut belly open. With quail and doves, though, you can just eat bones and all–or chew up and throw away [spit out] bones. And small fish that are cooked well—you can chew up bones and all. But big fish you have to throw out bones so they don't get stuck in throat."

Grease frying, however, was something new, picked up from Europeans and made possible by the fat of the Old World domestic animals they introduced. Pimans used the precious fat from deer for making paint and rubbing on skin that cracked in the cold desert winters. Traditional cooking was limited to broiling over the coals *(ga'a)*, pit roasting *(chuama)*, and stewing in water *(hihidoḍ)*—all ancient, unanalyzable words. Joseph Giff said that "to cook something in lard" is *manjegĭ ch-eḍ bahijid*, the word *manjegĭ* coming from the Spanish *manteca*, lard. "See, that's not even Pima," Joe noted. Small birds might be stripped of feathers and thrown on the coals to cook with the skin still on or boiled in an olla.

Avian Anatomy

Northern Pimans, in keeping with most tribal peoples with a hunting, gathering, agricultural subsistence base, had little linguistic elaboration for plant partonomy and seemed to have distinguished no parts of the flower with specific words. However, the process of pollination was fully appreciated. We see this not only in the care taken to keep the lines of wind- and insect-pollinated crop varieties pure, but in the prohibition on bothering the White-winged Dove, a major pollinator of the Saguaro cactus, a significant wild crop for the O'odham. The paucity of plant part names contrasts strongly with the abundance of animal anatomical terms, in particular those of vertebrates (see Rea 1998b:95–102). This situation reflects perhaps the ease with which the terms for human body parts can be applied to animal partonomy in general. Likewise, the daily butchering of vertebrates no doubt promoted this linguistic elaboration.

The possession of feathers ultimately is what defines "bird" in Piman thought, regardless of whether the bird can fly well or even at all. The particular species of bird provides the cultural metaphors. In ceremony and daily life, individual feathers of specific birds are the tangible symbol of the metaphor. As to be expected of a culture that prized feathers so highly, there is a rather elaborate feather vocabulary among Pimans.

The word *a'an* can mean feather or feathers as well as the whole wing of a bird. The meaning is clarified by context. A flight feather, for instance, that is used for fletching arrows is *a'an*, and the verb for "fletching arrows" is *a'anchud* or *a'anḍad*. Upper Pimans refer to the Plains Indians generically as A'an O'odham, Feather People. The polysemous use of the word *a'an* has confused even linguists. Saxton, Saxton, and Enos (1983:59) call the *a'anam totoñĭ* "a feathered ant," where what is intended is really the emergent winged ants that take flight briefly in summer.

Similarly, the term *vopo* has several meanings (see fig. 8.6 for sketches of types of feathers). It can mean the true down of a bird, such as that covering a chick; the short feathers on the body of an adult bird; and the short fur of a mammal or even the short body hair of a person. With River Pima, at least, the term has an allegorical reference to one's more distant relatives.

A specific type of feather called the *viigĭ* was greatly prized for ceremonial purposes as well as for personal decoration. These feathers are the loose or fluffy feathers found on certain large birds. Important sources were the thighs and undertail coverts or crissum of raptors. Such feathers, called semiplumes, do not have barbules interlocked into a distinct vane, except perhaps at their tips. The *viigĭ* are often incorrectly translated as "down," which they are not. The down of a bird is the *vopo*, as noted.

The axillaries, the soft feathers of the

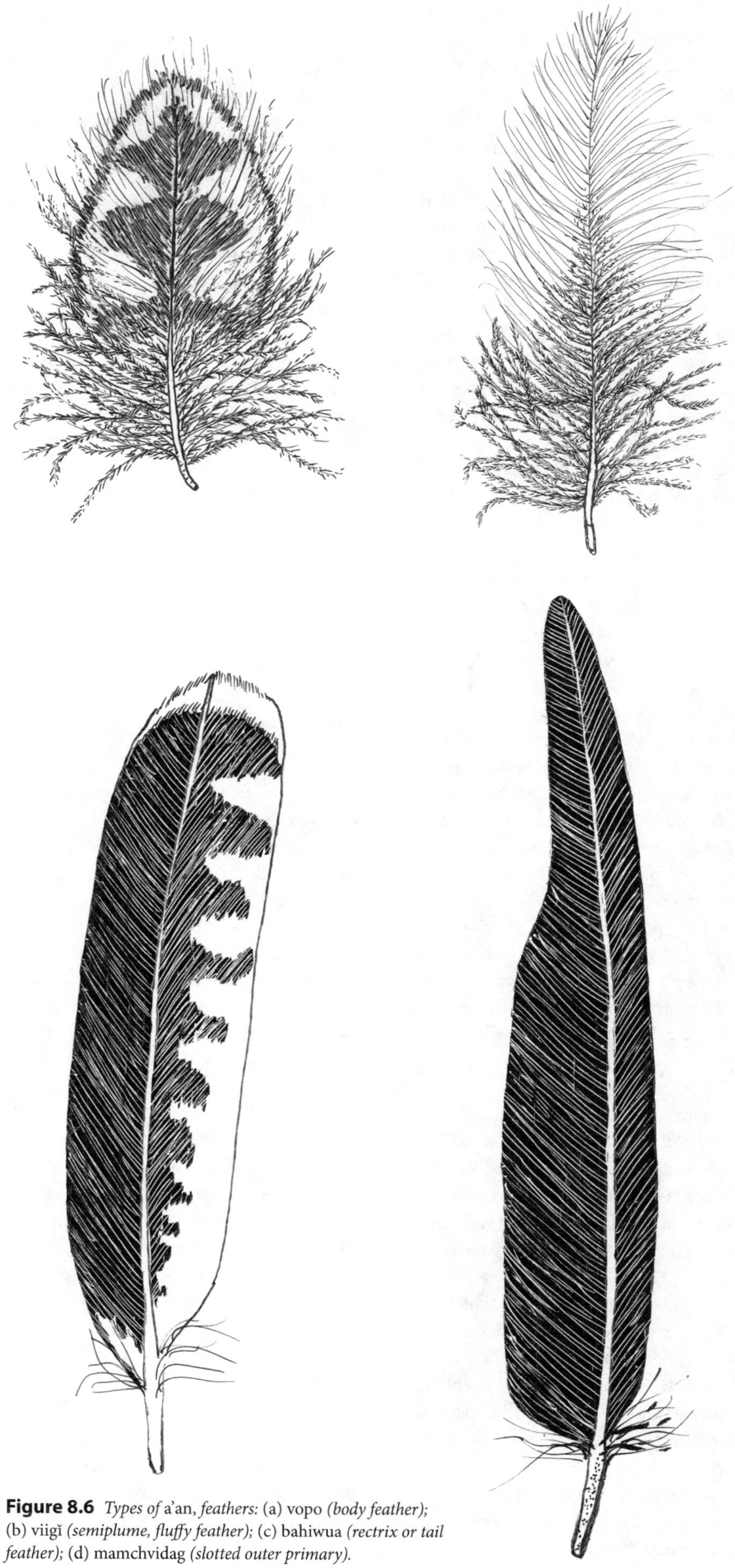

Figure 8.6 *Types of* a'an, *feathers:* (a) vopo *(body feather);* (b) viigĭ *(semiplume, fluffy feather);* (c) bahiwua *(rectrix or tail feather);* (d) mamchvidag *(slotted outer primary).*

underwing adjacent to the body, are known as *hehekaj*, from the plural of *hek*, armpit. These feathers were the prized feathers on raptors. The word does not mean "a wing feather" as implied by Saxton, Saxton, and Enos (1983). The primaries, secondaries, and upper wing coverts would be called *a'an*. An individual tail feather of any bird is a *bahiwua*, from *bahi*, the entire tail of a bird or other animal.

The outer several pairs of wing feathers of large birds such as eagles and Turkey Vultures were reserved for the use of the *mamakai*, shamans. They had a special name. Joseph Giff explained that *mamchvidaj* meant these feathers belonging to the bird, but still on it, whereas *mamchvidag* signifies these feathers already off, belonging to the *maakai*. These primaries (generally numbers seven through ten) are slotted and narrow on their outer third or so, most conspicuously on the inner web.

Several terms are used to indicate the crest or head decoration of a bird. One frequently hears *siivoḍa* (*siivoḍk* [?]), crest or topknot, such as on a quail *(kakaichu)* or cardinal. Sylvester Matthias explained, "*Siivda* are feathers stuck in the hair. When it's up here, on top, just like the *kakaichu*. But not here, on the back of the neck." One day George Kyyitan pointed out a peak west of his home in Bapchule that he called *siivdakam*: "it has a *siivda*" because "they have something on top, like a quail." He also noted that a comet is called *siivdakam hu'u*, star with a headpiece that sticks up. Mathiot (n.d.:166) defined the adjective *siivoḍk* as "tuft-shaped (said of anything shooting up and falling outward." This would apply most appropriately to the Gambel's Quail.

The other term indicating a bird's crest is *bichpdam (bichpodam, bichpoḍ* [?]). It is the source of the name for the towhee, a yard bird with a short erectile crest (see Abert's Towhee species account). While the Phainopepla and Northern Cardinal have a *siivoḍ*, the towhee has a *bichpoḍ* (fig. 8.9c).

Most words labeling the outer and inner parts of a bird are the same as those used for humans or other mammals (figs. 8.7 and 8.8). A few terms are unique to birds.

The beak of a bird is its *chiñ*, the same word that would be used for the mouth of other animals. The crop, noted particularly in quail, is the *kosaj*, or *hugij*, food sack (from *hugigaj*, its food). Theodore Peters noted this is the sack for the seeds that is discarded when one butchers. "They just put them there, but haven't grind them yet, a different place."

A bird's gizzard is *hodaigaj*. This is one instance where we can be sure the functional anatomy was understood in Piman folk science. Sylvester explained the word means "its place to grind." Although *hodai* in general means "stone" or "rock," it is used specifically for the metate or grinding stone. Just as *maakai hodaigaj*, shaman's stone, refers specifically to the crystal used in divining, a woman's *hodai* means her metate, also called *machchuḍ*. Various seed-eating birds eat grit, which serves for grinding the hard seeds in the muscular gizzard.

Another organ whose function was appreciated was the avian ovary, nestled against the kidney in the pelvis. In the

Figure 8.7 *External anatomy of a bird (quail).*

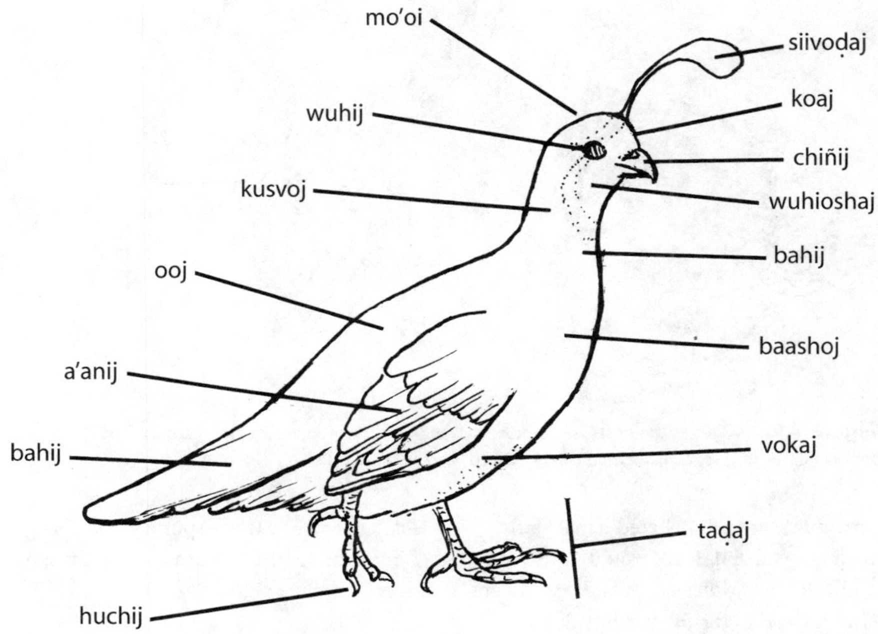

Figure 8.8 *Internal anatomy of a bird (quail).*

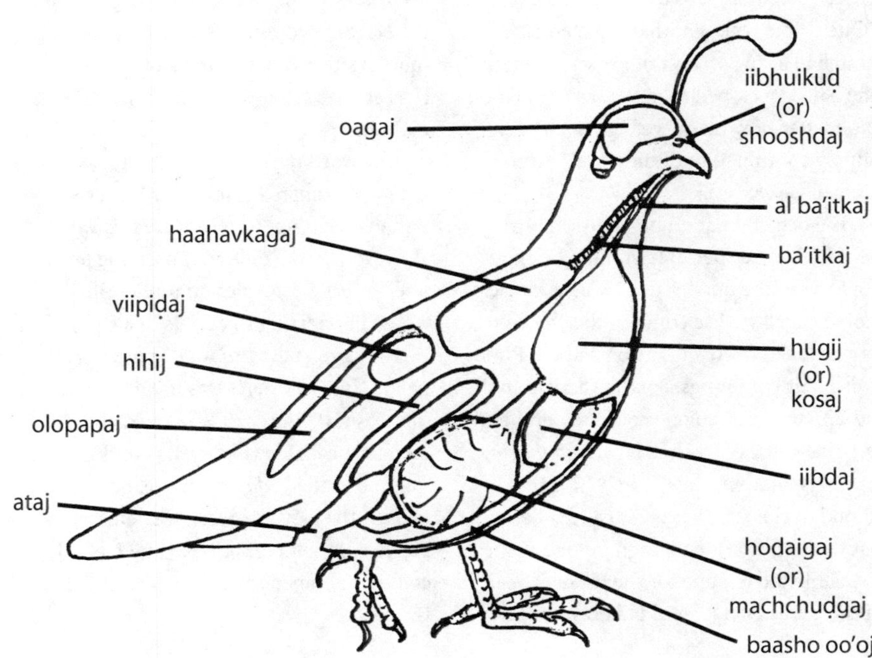

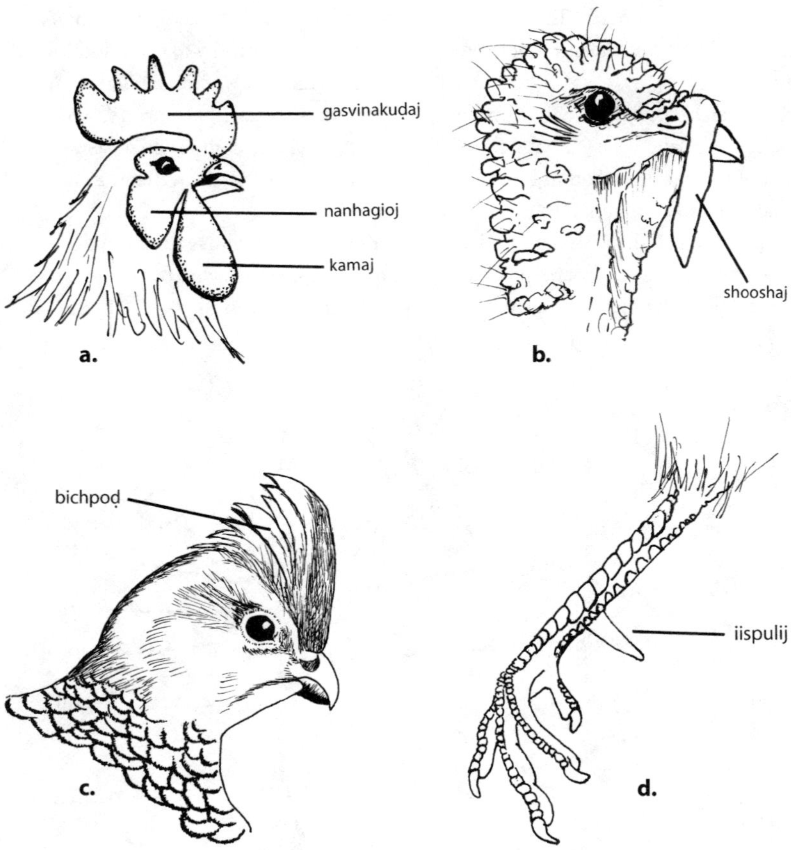

Figure 8.9 *Specialized avian structures:* (a) *chicken* (gasvinakuḍaj, nanhagioj, kamaj); (b) *turkey* (shooshaj); (c) *Scaled Quail* (bichpoḍ); (d) *turkey* (iispulij).

Mamsid is "to hatch," as in *u'uvhig ha-mamsid g e-nonha*, "birds hatch their eggs" (Mathiot n.d.:48). *Vavañig* or *vavañhig* means to be in a flock or herd, as in *totova an ge vavañig*, "the turkeys are in a flock" (Mathiot n.d.:242).

breeding season, butchered female birds such as quail and doves would have ova and yolks at various stages of development emerging from the ovary. Outside the breeding season, the ovary regresses to a scarcely noticeable flat mass. In training a novice *maakai*, the shaman "coughed up" white objects resembling mistletoe berries. He rubbed them into the aspirant's chest. "One [Pima] thought that the medicine-man had a sort of 'nest of power' wherein the balls developed as in the ovary of a hen. No matter now many were given off the supply continued undiminished," wrote Russell (1908:257).

The ornamentation of galliform birds required some special terminology. The topknot of the quail is its *siivda*, as already noted (fig. 8.7). The comb of the chicken is *gasvinakuḍ*, which is also what the Pima call a hair comb (figs. 8.9a). (There is another name, but no one could remember it.) The wattles of a chicken are *nanhagio*, earrings, or *kam*, cheek (fig. 8.9a). The snood of the male turkey is called *shosha*, mucus (fig. 8.9b). Although no one I asked remembered a name for the turkey's beard, if there was one, it would likely have been *eshpo*, beard. The tarsal spur is *iispul* (pl. *i'ispul*), the same word used for equestrian spurs (Sp. *espuela*) (fig. 8.9d).

Eggs and Nests

Upper Pimans call a bird's nest *kosh*. Colonial Névome used the word *napo* (Pennington 1979:84). A sample given in the Névome vocabulario is "baagui hu-mosuri sitatababa tu baupa ami nanapota" (Las águilas siempre hacen sus nidos en las más altas peñas). Contemporary Pima Bajo use the word *napo'o*. Pennington (1980:194, 242) wrote *napo*. Present-day Upper Pimans seem not to recognize this word. However, Shaw (1968:102) used the term *napawchick* (*napo* plus an unidentified word) in referring to eaglets in a story.

The aggregate term for egg or eggs is *nonha*. In Névome, it was *nono* (Pennington 1979:63). The shell is *kokoḍk*, and the membrane inside is *kosaj*, its sac. The yolk might be called *eḍaj*, its inside, or *nonha huḍad*. The Névome word for yema de huevo was *nono urhadi*. No one I asked could remember what the white of the egg is called.

Part 2

Piman Birds

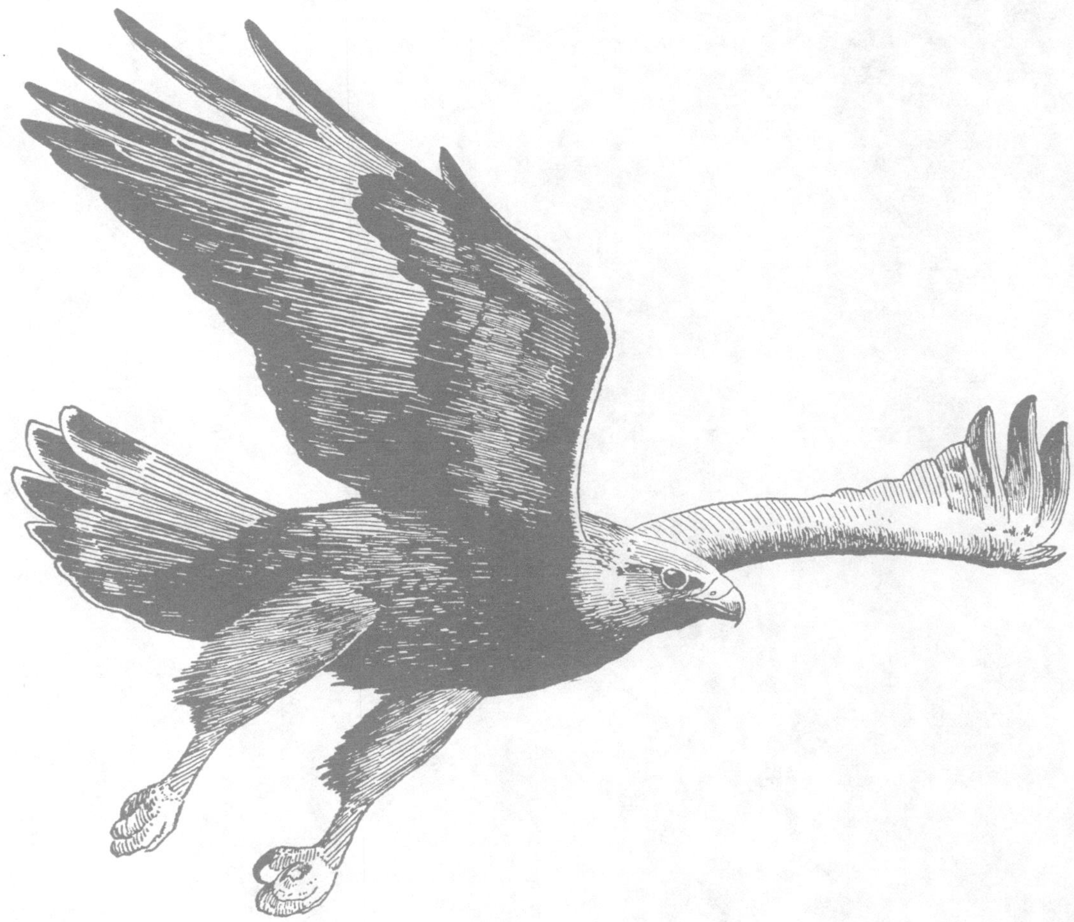

Plan of the Folk Generics (Biological Species Accounts)

This part brings together each named Piman category of bird, mostly the folk generics, except in the few cases where they are further subdivided into folk species. Each account is a terminal taxon (except for *vakoañ*, heron). All the species are included in the life form *u'uhig*, bird (see chapter 4 for details). In Piman orations, *u'uhig dadadam*, "flying birds," are most frequently contrasted with *jeveḍ meḍdam*, "earth-running creatures," the running (and crawling?) creatures of earth. This is poetic language. In practical discourse, *ha'ichu doakam* is the term for animal life, or, as Joseph Giff aptly defined it, "all moving things with life."

There is no emic sequence of birds (or of any other group of organisms). Therefore, the arrangement is roughly phylogenetic. In a few cases where birds are considered related in Piman thinking, they have been placed next to each other—for example, two folk species within a folk genus. Unfortunately, there is no one now living, to my knowledge, who can construct a hierarchy such as Sylvester Matthias did for plants (Rea 1997:88) and as George Kyyitan and Frank Jim did for mammals (Rea 1998b:22–23).

Headings

The *first* job of ethnobiology, after discovering a set of names (folk taxa) used by some language community, is to determine how these names map to Western scientific or Linnaean taxonomy. A name may represent a particular biological species (often the case with birds and mammals) or a genus or a family or even an entire order of organisms. This domain of a taxon is determined by carefully working with native speakers in the field. This first task is set forth in the headings. This map to the domain of a terminal taxon includes here both English and Mexican Spanish names in addition to the scientific names.

Each account usually begins with the River Pima (Gileño) name with its variants. The form that linguistic consultant Culver Cassa has determined as the basal variant is listed first. Plurals, where used, are identified as such. Next, on the same line and separated by a semicolon, come Tohono O'odham or Papago versions if different. Papago is a collection of a half-dozen dialects, some of which are virtually indistinguishable from River Pima (as is Kohadk). Other dialects may differ more or may have an entirely different name for the organism in question. Hiá ch-eḍ O'odham (Sand Papago) may follow. There may be no difference between the name used by these various dialectical groups. The next line is the Pima Bajo word from the lowland People of the Río Yaqui region of Sonora, Mexico (the Pima Bajo term is actually given first in the heading in a few instances where appropriate). (Where known, Mountain Pima names appear in the comparative linguistics section of an account).

The next line lists the family and order names used by Linnaean scientists. Below that in italics appears the binomial or scientific name—that is, the genus and species. The AOU Checklist Committee on Classification and Nomenclature maintains a tight rein on orthodoxy in these matters. I have not slavishly followed the AOU names and classification where my own and colleagues' taxonomic studies have suggested differently. These departures, based primarily on anatomy, especially osteology, are explained in my earlier work on Pima birds (Rea 1983).

The Linnaean names are then followed by the common English name and one or more Spanish names. English names are standardized by the AOU, but Spanish names are more variable. I have tried to stick to local usage in northwest Mexico, in particular Sonora, but many birds have only generic names. Ana and Eric Mellink have been the primary source for Sonoran usage. Where specific distinctions not in the local vernacular were needed to match the distinctions Pimans make, I have followed Birkenstein and Tomlinson (1981); Edwards (1972); Russell and Monson (1998); and Sada, Phillips, and Ramos (1987). (It seems the O'odham were often better ornithologists than the colonizers who followed them.)

Text

The first several paragraphs of each account (whether folk generic or folk specific) usually include a description of the species and a summary of its natural history, all from the viewpoint of Western ornithology. Most of this information is not found in the attenuated texts of the many fine field guides now available. These verbal sketches are intended to serve two main functions. The first is to help the reader, especially younger generations of O'odham, identify the birds in the desert and thornscrub that older O'odham once knew so well. Plumage, behavior, and habitat all aid identification, as do the illustrations. The second function is to give enough natural-history background to compare with O'odham folk science and to help readers understand the often elaborate metaphors they then built upon this intimate folk ornithological knowledge. Once we know the biology of a species, we can begin to understand the metaphors and symbolism O'odham used in their oral literature.

Everything that follows in the body of the account is the O'odham story about the bird. This is the Piman ethnoornithology. The coverage of dialectical groups is arranged roughly from north to south, with River Pima material first and desert Pima Bajo last. Far less is known about these southern people, now much acculturated. The Gileño information is rather thorough, but now quite eroded. Undoubtedly, a great deal more remains to be learned about the folk ornithology of the Tohono O'odham.

Folk science—that is, the O'odham knowledge of the bird's natural history—is generally given first. This information was proffered while consultants and I were discussing a species. Then comes information from the various Piman genres (song, myth, tale, oration) that are built upon this knowledge.

The length of an account varies greatly according to the species' cultural salience. Some accounts are thus very brief because in discussions the species was named and recognized, but was not considered of much significance. In other cases, the species loomed large in the O'odham worldview, and even what is given here does not exhaust its cultural relevance. The end of this folk section may tell how the bird's name has been incorporated into plant names or place-names.

Most ethnographic interviews have been taped and are transcribed here generally verbatim, with minimal editorial insertions placed in square brackets. All of these people were speaking English or Spanish as their second language. Material has occasionally been spliced from several different interviews with the same person. Sometimes I have taken a tape transcription back to someone, read it to him or her, then inserted any additional comments the consultant wished to add.

As noted in the introduction, names of a biological species are capitalized throughout the text proper, as is customary worldwide among ornithologists. Names used in a generic sense are not: thus, Prairie Falcon, Gray Fox, and Velvet Mesquite, but falcon, fox, and mesquite. Piman folk generics are italicized throughout, but not capitalized. When a bird's name is used as a personal name or toponym, it is capitalized but not italicized.

Finally, there may be several terminal sections with specialized information in an account ("Comparative Linguistics," "Misidentifications," "Technical Notes").

Comparative Linguistics

Many accounts close with a linguistic comparison first within the Northern Piman segment, then throughout the Tepiman groups insofar as information is available. Precious little is known of the ornithological folk taxonomy of the various Tepehuán groups. Even among the Mountain Pima, only a beginning has been made.

Time depth is provided by several Jesuit vocabularios (Pennington 1979; Rinaldini 1743) and one Franciscan vocabulario compiled in Spanish colonial times (Attí [1774]). Pennington's is Névome (lowland Pima Bajo), Rinaldini's is Tepehuán, and the Attí is Upper Piman from northern Sonora. (In contrast, the oldest extensive word list from the Gila Pima is Grossmann's 1877 manuscript.)

Finally, I have attempted to present comparative folk taxa from other Sonoran-branch Uto-Aztecan groups, insofar as they could be gleaned from published sources. Most come from lexicons or vocabularios where taxa were elicited haphazardly, usually by authors without much knowledge of the local fauna. As with almost all folk taxonomic work in the region, little or no attention was given in these materials to mapping the biological domain that a word may cover; these writings lack biological contrast sets.

Misidentification(s)

Although touted as the best-known group of organisms on earth (perhaps a questionable assertion), birds have suffered rather ignobly at the hands of ethnographers and lexicographers who have delved into Piman culture. This neglect has occurred in spite of the fact that the birds of the state of Arizona have been so thoroughly studied (Brandt 1951; Monson and Phillips 1981; Phillips, Marshall, and Monson 1964; Rea 1983; Swarth 1914; Witzeman, Demaree, and Radke 1997).

This part attempts to correct the many misidentifications that have crept into writings on the Pima, blinding Western readers to the marvelous folk taxonomies of the O'odham and particularly to the intricacies of the metaphors they have constructed based on their knowledge of the natural history of desert birds. If you don't have the right key, you can't open the door.

For those who think I may be a bit cranky about these misidentifications and missed metaphors, I invite you to read an essay by Gary Paul Nahban (2004:27–44) dealing with the problems of Pima sung poetry. Jerry Craddock's (1998) stimulating paper on translators' widespread insouciance and lamentable incompetence is also relevant to this part.

Technical Notes

Also included in the species accounts are notes of a more purely scientific nature that would be out of place in the brief introductory sketches, but that are important, nonetheless, to the taxonomy, ecology, or management of the species in discussion. For instance, the discussion of the impact of eagle capture on local desert populations seemingly has not been a topic of ethnographic concern, but is a biological question that needs to be raised. Other notes may discuss why, for instance, a genus used here may differ from what you will find in your field guides.

Species Accounts

ñui, ñuwi, ñuvi, pl. *ñuñui, ñuñuwi* *nui* (Pima Bajo)

Vulturidae (Cathartidae), Ciconiiformes
Cathartes aura
Turkey Vulture ("Buzzard"); Aura

A large dark bird effortlessly soaring through the desert sky is likely to be a Turkey Vulture, one of the most important birds in the Northern Piman cosmology. From a distance, the bird may appear black, but seen closer the plumage is very dark brown. Its tail is narrow, and its long wings show silvery gray below, with a blackish leading edge. The bird's small head is red, as are its feet, and the bill is white. (In their first winter, young Turkey Vultures are all blackish, even their beaks and feet.)

Should there still be any doubt that the bird is a Turkey Vulture, watch for the dihedral angle of the wings: they are set in a shallow V rather than being flat as in corvids, most hawks, and the other species of vulture. Turkey Vultures may be seen flying by themselves, in pairs, in small groups of a dozen or more, or even in very large flocks when they are migrating across the desert.

Turkey Vultures are scavengers, searching the desert throughout the daylight hours for animals that have died. Because their food supply is not dependable, the birds expend as little energy as possible in foraging. They fly high in the air, riding on thermals—columns of air rising from the heated earth. The slotted outer primaries, like fingers at the wing tips, help prevent stalling during slow flight. Only when a bird gets too low or must fly when there are no thermals does it resort to the more energy-consuming flapping flight.

Vultures are reported to have exceptional eyesight, with visual cells packing the retina of the eye. But their ability to smell is also acute, probably their first cue in locating food (Houston 1994; Stager 1964). The olfactory lobes of the brain are large, and so are the nares, or nostrils. The septum is perforate, allowing free passage of air. (Most other birds lack olfactory capacities.)

I once watched a Turkey Vulture foraging in Tohono O'odham country on a very windy day. It flew lower and lower in a slanting spiral of ever-tightening circles until it was below the level of the raised roadbed. The gusts were sweeping across the road where I discovered the object of the search: a roadkilled Red Racer. The bird was zeroing in using olfactory cues, not visual ones.

Except when incubating, Turkey Vultures congregate in large communal roosts, such as old cottonwood groves. After spending the night, they catch the warmth of the early morning desert sun. While basking, they spread their wings, exposing a dark fluffy patch on their lower back.

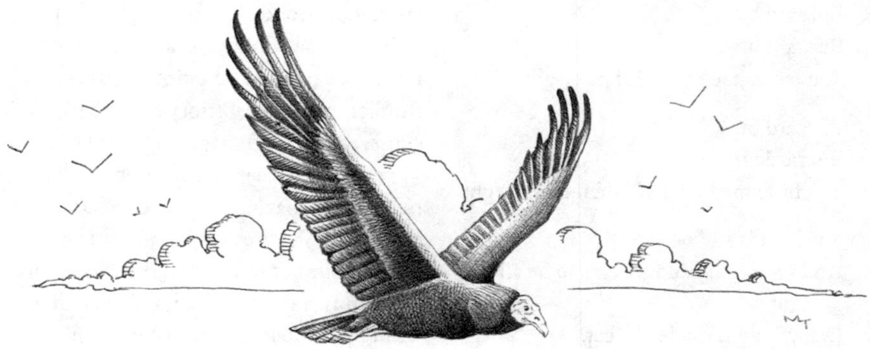

ñui, ñuwi, Turkey Vulture (Mike Touch)

This way the birds take advantage of solar energy, passive heating—rather than depending on metabolic energy from food—to raise their body temperatures, which are lowered during the night. This interval also gives the desert air time to warm, producing thermals, before the big birds, one or several at a time, sail leisurely off for the day's search.

Once on the ground, the Turkey Vulture is cautious if it is the first one arriving at a carcass. Having no talons and a relatively weak bill, it is now vulnerable. On approaching a potential meal it may make a few furtive sideways stabs with its beak before deciding that the animal is indeed dead and not just asleep or injured. Vultures arriving at a carcass on which another is already feeding may dispense with caution. At larger carcasses, vultures typically begin feeding at the eyes, genitals, and anus. Smaller animals may be eaten whole.

Unlike Black Vultures, Turkey Vultures have short legs (or tarsi). They pin down food with their inner toes while feeding carefully. The geometry of body parts is just right. Their bare head allows them to poke about without fouling their feathers with blood or gut fluids. Once its large crop is filled, a vulture runs and flaps till airborne.

Unique to New World vultures and their close relatives the storks is the behavior of urinating on their feet (the exposed scaly part of the "legs") in order to cool their body by evaporation, an avian equivalent of sweating. Although the skin of Turkey Vultures' feet is actually pink, we usually see the white accumulation of chalky urates (see technical note).

Vultures like water and will bathe in places where they feel secure. Once water soaked, a vulture is vulnerable. In Pima country, I have found them bathing in the tinajas high in the Sierra Estrella after summer rains. During the warm season, wading into shallow water prevents the buildup of excessive urates on the feet, which would reduce the cooling effect of urohidrosis. Although Turkey Vultures drink free water, there is sufficient water in their food to supply them indefinitely.

Almost all that is known to Western science about the behavior and anatomy of this marvelous bird has somehow been encoded into Piman mythology. Coyote and Buzzard (as the Turkey Vulture is often called in the American vernacular) are two of the most important beings in the Piman cosmos. Although Bear, Badger, Eagle, and Falcon are powerful figures, none looms so large in the Piman pantheon as Coyote and Buzzard.

The very first segments of the Piman Creation Story, the great epic recounted annually on the four longest nights of the year, remembers the origins of these two beings. And every listener with a Piman father has one of them as his or her mascot. Buzzard claims two clans, Maamgam and Vaavgam, the Red People; and Coyote claims two clans, Apapagam and Apkigam, the White or Yellow People. A fifth clan, the Ogolgam or Ogalim, the one that only partly emerged from the underworld, is assigned to either one animal or the other, or even sometimes to Bear.

The People refer to Ñui as *s-oam t-veemgal*, "yellow/brown moiety cousin," and to Ban or Coyote as *s-koomagĭ t-veemgal*, "gray moiety cousin." Together Buzzard and Coyote form the totality of human nature. Separately they form reciprocal parts of a contrast set. Coyote is aggressive, direct, bumbling, and mischievous. He acts first, then thinks. Vulture is quite the opposite. He is slow to act, cautious to the point of timidity, contemplative about all his actions. Quite like their real-life counterparts, Coyote is the extrovert in Piman literature, whereas Buzzard is the classic introvert. Bear is sometimes said to befriend Coyote people, whereas Mountain-lion befriends Vulture people (Underhill 1936:31). Like their companions, Bear is bold, and Cougar is cautious.

Ñui bears a special mark of his high status. An Upper Piman male might distinguish himself in battle, particularly by killing an enemy (a necessity that these generally peace-loving peoples abhorred). Such a warrior earned the right to pierce the septum of his nose and wear a badge signifying his valor. This nasal ornament (either a ring or nose plug) was called a *dakoshda*. Ñui achieved this special status: his pierced septum is still to be seen.

In the Piman Genesis, Jeweḍ Maakai, Earth Shaman or Creator, first makes Creosote Bush, then ants, and then termites *(hiopch)*, which continue the process of enlarging the earth. "Then he created a Person. He made him out of his eye, out of the shadow of his eyes, to assist him, to be like him, and to help him in creating trees and human beings and everything that was to be on the earth. The name of this being was *Noo-ee* (Buzzard). Nooee was given all power, but he did not do the work he was created for. He did not care to help Jehwertamahkai, but let him go by himself. So the Doctor of the Earth himself created the mountains and everything that has seed and is good to eat" (Lloyd 1911:28).

This origin is not merely a poetic translation. Another version from a different narrator at this point says, "He made a person. He took the shadow of his eyes and made a human being out of it. It was Oam Ñui, Yellow Buzzard. To him he talked when lonely. And he was his servant" (Herzog [1929]).[1] In a third version, Earth Doctor "held his torch behind him and cast his shadow before him. He reached out and laid hold upon his shadow. He seized it in his hand, and of this he created Buzzard" (Benedict 2001:6). Yet a fourth telling has: "Now it is Earth Doctor's turn to make something. He was sitting down, and the streams of light that came out of his eyes with which he sees everything. He reached out and got some of this light and made Buzzard out of it (Ñui)" (Hayden 1935:4;

see also Bahr and associates 1994:51–52). Each rendering emphasizes the mystical origin of this primordial being.

Then Jeweḍ Maakai made two dolls, who were the first two people, but there was no death, so they increased until there was nothing to eat, and then people began killing and eating one another. Earth Doctor did not like this, so he pulled the sky down, killing them all. He and Ñui escaped through a hole in the sky. On top of the fallen sky, Jeweḍ Maakai made another man and woman and again made a heaven and earth. This time the people began turning gray progressively earlier until they were becoming gray in their cradles. So he again made a hole in the sky through which he and Turkey Vulture escaped.

Jeweḍ Maakai now made a third world. But this time the new people smoked too much, taking up the habit progressively earlier until they began even as babes, so he again took his staff, pulled the sky down on them, and escaped with Buzzard. Then the fourth time "he created earth as it is now" (Lloyd 1911:30).

"But at first the whole slope of the world was westward, and tho there were peaks rising from this slope there were no true valleys, and all the water that fell ran away and there was no water for the people to drink. So Jehwertamahkai sent Nooee to fly around among the mountains, and over the earth, to cut valleys with his wings, so that the water could be caught and distributed and there might be enough for the people to drink" (Lloyd 1911:30–31; see also Bahr and associates 1994:51–52, 54–55; Benedict 2001:9; Hayden 1935:4).

While Vulture molds the lowlands, the narrator sings four times: "Buzzard is the bird who gives shape to the earth." Then, as the great dusky bird works on the uniform mountain masses, the narrator again sings four times: "Buzzard is the bird who gives shape to the mountains." Here we see Buzzard, flying low over the earth, gouging out the drainages with his slotted pinions or outer primaries, forming the hydrology now found in the desert. Though his overall creative contribution is not great, Vulture is responsible for the flow that not only provides water for people, animals, and wild growth, but also makes possible the several types of agriculture on which the People depend.

Four songs now celebrate the creation of the fourth world, in which we presently live (Bahr and associates 1994:52; Hayden 1935):

Buzzard bird,
Buzzard bird,
You have made the land just right.

Buzzard bird,
Buzzard bird,
You have made the mountains just right.

The land is still now
And on it everything seems to be all right
Everything is made perfect.

The mountains were shaking
But now they are still
And on them everything is perfect.

Four is the requisite Piman ceremonial number, but so far in the narrative there are only two primordial beings. Two more must be added. Once a month, male sun and female moon meet. From this union is born Ban, Coyote, who wanders around until he encounters Jeweḍ Maakai and Ñui, joining their company. "But now out of the north came another powerful personage, who has two names, See-urhu [Se'ehe] and Ee-ee-toy [I'itoi]" (Lloyd 1911:31). This latter being, though the youngest, insists on being called Elder Brother. At last the others relent and do so. (In the version that Russell [1908:208–209] obtained from Thin Leather (Komal Hogĭ), Elder Brother is the scion of Earth, who is female, and Sky, who is male. In Juan Smith's version of the Origin Story (Hayden 1935), Ñui is made after Se'ehe and Ban.)

Some time after the repeopling of the earth, Se'ehe makes a man, "the most beautiful man yet made," who in the end is responsible for the Flood (Lloyd 1911:36–50; see also Russell 1908:209–213n.; Saxton and Saxton 1973:45–61; Rea 1997:173, 275; Underhill 1946:9). Each of these four primordial beings devises a method of surviving the Flood. Jeweḍ Maakai helps some people escape into the underworld. Four small birds say they will hold on to the sky with their bills and escape. "And Nooee, the buzzard, the powerful, said he did not care if the flood did reach the sky, for he could find a way to break through." The waters rose, "but Nooee still floated in the air and intended to keep on the wing unless the floods reached the heavens" (Lloyd 1911:38, 42). Underhill (1939:126) noted, "The North Star, [is] called the Unmoving Star:—it is the stopper for the hole in the sky through which Buzzard took refuge in the flood."

After the waters subside, Ñui takes no further active role in the creative process, leaving the work of repeopling the earth to Jeweḍ Maakai, Coyote, and Se'ehe. The results are rather disappointing. Only Elder Brother makes satisfactory people. They, of course, are the O'odham.

Ñui's reticence in the creation events reflects its behavior. Ravens, caracaras, and even Black Vultures may be abroad with the first hints of daylight in their search for carrion. Turkey Vultures, however, remain leisurely on their roosts, soaking up the morning sun before dispersing across the desert.

Among several stories obtained at the Tohono O'odham village of Santa Rosa village is the account of Ho'ok, the monster woman, born of a young Piman woman who conceives when she is impregnated by means of a kickball (Hill 1940b). As a child, Ho'ok begins to injure other children with her long claws. When punished by her mother, she leaves the village. Vulture is sent to bring her back. The crying young monster refuses, singing four times: "Way over there the Buzzard, my uncle, wants me to come back." Ultimately, the people must seek the help of I'itoi to destroy this being who is eating children and even adults. The role of bringing Ho'ok back to a dance in the village so she can be intoxicated with narcotic cigarettes falls again to Buzzard. This appears to be the only version of this well-known story to establish a relationship between Ho'ok and Ñui.

Throughout much of mythic time, Elder Brother is the being most intimately connected with the Pimans. He has homes in South Mountain, the Baboquivaris, and two other desert ranges. But in time his indiscretions, particularly at the *wuagida* ceremonies for pubescent girls, become so severe that the People decide they must dispose of him (Lloyd 1911:125–131). This task poses a serious problem because even though they kill him, he comes back the next day: their power is not great enough. They must seek the aid of one of the Original Beings, his coequal.

Ñui volunteers. He calls the People together at his house (a ruin formerly west or south of Muahaḍagĭ, South Mountain—or south of Tempe) and creates a working microcosm, just as Jeweḍ Maakai had made the macrocosm, so the People can see his power, "for Juhwerta Mahki had made him to have power, tho he had not cared to use it." As Elder Brother says later in his resurrection speech, Ñui now "shaped the mountains, defined the water courses, placed the trees, and in four days completed his task" (Russell 1908:339).

After four nights of demonstrations, Ñui flies up to the roof of his house, awaiting the sun to rise, as vultures still do at their roosts. Wearing his turquoise nose ring or plug, he flies to a great height to where he can see I'itoi's heart. The next day he does the same, wearing a nose ring of sparkling shell. This *dakoshda*, worn though the pierced septum, is the insignia of a brave warrior. The following day, wearing a *dakoshda* of glittering shell, he flies up and converses with Sun about exchanging places with him, which they agree to do. On the fourth day, Vulture rises in place of the sun, shooting heat down on I'itoi with Sun's bow or gun, finally killing him.

Densmore (1929:21–22), in her quest for songs, recorded a version of the Origin Story from Vamori village in Tohono O'odham country. Here Brown Buzzard, after killing I'itoi, sings, "I have done the worst thing now in killing you, my brother, but I am going to leave your wind and your clouds [for the benefit of the People]."

Later, some say four years, I'itoi returns to life. He then goes to the world below this one, where Earth Doctor lives with some of his people. Elder Brother leads the four and a half clans up from the underworld. These Vupshkam, or Emergenti, are the People that Jewed Maakai had helped to escape to the underworld at the time of the Flood. The People emerge as a great invading army. "So they went on, slowly, camping at one place, sometimes for many days or several weeks, making their living by hunting game. And whenever they stopped they sent scouts and spies ahead to look out for the next stopping place, so that they might go ahead safely. And this went on for many years" (Lloyd 1911:149).

According to Russell (1908:227–229), when the invading army reached "the people living in the large pueblos, the ruins of which yet remain scattered about the Gila and Salt river valleys," it conquered them one after another (see also Bahr and associates 1994:207–229; Benedict 2001:41–53; Hayden 1935; Lloyd 1911:154–161). A succession of animal helpers aid the Vupshkam in their conquest. But Se'ehe had one account he wanted to settle personally: "Elder Brother told his army to capture Vulture alive. 'How can we identify him? We do not know him,' said they. Elder Brother told them to capture the warrior with white leggings; they were the distinguishing mark of Vulture. They obeyed and brought the defeated leader to Elder Brother, who scalped him; this accounts for the naked head of the vulture of to-day." Pimans may not have understood the physiological function of urohidrosis, which was discovered only recently by ornithologists (Hatch 1970; Kahl 1963), but this passage shows they were careful observers: Vulture only *wears* the white leggings—his "legs" themselves are actually red.

In the song that the invaders sing while subduing him, Vulture is hailed as *havañ ali*, "Child of the Raven!" (Russell 1908:228, 282). In Lloyd's and Hayden's versions of the destruction of the *vapaki* (pueblos), Eagle and Prairie Falcon fly up into the sky to locate Ñui, who had said he might escape there, as he had done earlier at the Flood. In Benedict's (2001:51) version, it is the "eagle people [who] found him in the air and brought him down from above" to Elder Brother.

The scalping of Ñui is no mere "and so to this day" tale, but rather a pivotal mythic event in Piman ceremonial life, at least formerly. Ñui's dancing around his own scalp is a precursor of the Scalp Dance that followed the killing of an *oobĭ* (enemy or Apache). The enemy-killer spent the next sixteen days outside the community undergoing purification, and the scalp he took stood atop a pole in the center of the settlement while postmenopausal women danced around it. A set of scalp songs traces the origin of this dance to Buzzard himself at the original dance (Russell 1908:137–138; Underhill 1938:85–92, 1946:187–189). In one of these songs, Ñui, the scalped one, speaks (Russell 1908:337):

> Yohonde kaviñik
> Yohonde kaviñik
> Kenda megali ñuhuñuive daiwuñek
> Hanovañ wegañ jewed dame tonoholi.
>
> Sandy Hill,
> Sandy Hill.
> Just then little vulture arises.
> "Here around me I shine upon the land."

The sandy or gravely knoll (O'od Kavichk) mentioned in this song still stands, having somehow escaped both road building and a housing development. It was once a location for celebrating the victory dance after a successful campaign against the Apache. The guardian in charge of the sixteen-day lustration opens his speech to the secluded enemy-slayer by invoking Oam Ñui, Brown (or Yellow) Vulture (Underhill 1946:193):

> Am not I, even I, your guardian?
> And I tell you something.
> It seems that at the east there lives a
> great yellow buzzard,
> A great shaman
> Surely he is your guardian, and, arriving
> from there,
> He reveals something to you.

Joseph Giff told the story of how Vulture kills Se'ehe almost exactly as Thin Leather had told Lloyd (1911), noting that Ñui borrowed Sun's bow and four arrows, shooting one into the pond where Se'ehe bathed. But Joe went on to explain Ñui's revenge for his subsequent scalping:

> Ñui's the one who gave nature jealousy. They scalped him; that's why he's got a red head. Then Ñui said he doesn't care what happened to him. He started singing.
>
> And there was a man who lived there. And he was a hard-working man. And he goes out and plants, cleans his fields and all that. And his wife goes to the dances, stays there all night. So he comes home hungry and cooks something, for the wife was already gone. And in the mornings, he'd go out and she would heat up what this man had cooked. And it happened for four nights.
>
> And then Ñui had this in mind, that something was going to happen. So that's where jealousy originated. The wife goes to the dance, singing and dancing, and the husband kicks her around. Before that nobody was jealous. They didn't know anything about it. That's Ñui's invention. But the way they did to him, scalped him and left him without a scalp. So he put that into people—jealousy.

A segment of an episode in the Creation Story that Densmore (1929:36–39) obtained from the Tohono O'odham village of Vamori (see also Benedict 2001:63–64) is missing from Thin Leather's version in Lloyd (1911). Here a *maakai*, shaman, in the west named Bitter Wind prepares a house for the literal-minded boy A'an Hitpag, Feather Braid. "The house was full of 'medicine,' which was said to be something like the heat vibrations that rise from the desert in the summer." The medicine was for the boy, but Oam Ñui, Brown Buzzard, arrives there first and is offended by Bitter Wind's lack of welcome. "Brown Buzzard was so angry at Beater [*sic*] Wind's words that he spread his wings against the house and took out every bit of medicine of every sort. He sang this song: Am I an eagle? My feathers are filled with mysterious power." As Buzzard left the house, he "flew over the highest and the lowest mountains and dropped into each mountain top some of the medicine that he had taken out of the house. Because of this some of the mountains became full of medicine, as the house had been. Brown Buzzard said that because of this medicine there would be a roaring of wind or noise of thunder and a shaking inside of these mountains before a storm and this would be a warning to the people."

The boy for whom the medicine has been intended arrives and undergoes a sixteen-day purification under his mentor's tutelage. On his return home, he "entered every mountain and learned songs." Ever since, the two main ways for learning songs and obtaining their power are by going to the ocean, that mysterious body of water, or to mountains where Ñui left strong medicine. Certain prominences poking up through the sea of alluvial silt across the Sonoran Desert are especially important places for power and song.

On the banks of the Little Gila, Clemencia Leos sang an Oriole Song that seems to relate to Brown Buzzard's medicine-spreading attribute. In her song, the vulture gives the singer Sacred Datura, a powerful and dangerous intoxicant, from the tips of his primaries.

> Sohama ñui si voho sho'ig ndodowa
> kotoḍoba hiosig hejel yahan kukab vasib bek n i ichu
> kuñ am iyokai nonodake kam shakali himenakaiyu
> usi chuchime iabe chume tai.
>
> Brown Turkey Vulture did me wrong:
> Datura flower at his wing tip he gives me.
> I take a drink.
> Upon finishing I'm drunk and go along staggering
> Upon the trees I am entangled.

Both in distributing the water for the People and in spreading powerful medicine, Ñui soars low around the earth and manipulates the mountains with his wings. So a *maakai* uses the vulture's outermost wing feathers, the slotted outer primaries, in divining, as he (or she) might do with the Golden Eagle.

Vulture feathers are also used in certain healings. For instance, Papago Maria Chona (Underhill 1936:62) had a son who was a *maakai*:

> Once, when a rattlesnake bit me, he cured me. I was picking beans among the weeds in our garden and, along with a handful of beans, I grabbed the snake. I had hold of him near the head but he turned his head back and bit my wrist. My son killed the snake. Then he chewed some greasewood [*sic*, meaning Creosote Bush] leaves and put them on the little hole where it had bit me. He made a mark with a buzzard's feather all around my wrist and told the poison not to go beyond that mark. Then he split the feather and tied the two pieces into a string. He bound that string around my wrist, tight, and the poison never went beyond it. Two nights it hurt me and my hand swelled, but not beyond where my son had made the mark.

Outside these major mythic episodes, there are other tales involving the Turkey Vulture. Some are merely etiological, whereas others touch on broader cultural themes. In two tales, we encounter attempted cooperation between Coyote and Buzzard.

Among Pimans there are several versions of the fire quest. In an Akimel O'odham account, first Coyote, then Roadrunner, who is finally successful, tries stealing fire from Wihom, Thunder. In one account from the Tohono O'odham, as told to Harold Wright (1929:37–42), Ñui is sent to steal fire and almost succeeds, going to the east where Sun begins his ascent. Upon Ñui's return, "they saw that his head had no feathers at all and was covered with blood. . . . Buzzard told the people how he had flown straight into the sun and how when he went to take some of it his head was so burned the feathers fell off and all his other feathers were burned black."

In a Tohono O'odham story, Vulture and Coyote cooperate in an effort to recapture Coyote's wife, who has been stolen by Eagle (Saxton and Saxton 1973:67–73; see also Shaw 1968:63–69). During his search for her, "[Coyote] found a carcass and began eating it. Suddenly Buzzard came and said, 'I know where your wife is now. I'll tell you where she is and take you there. But from now on, when you kill something, you must always remember me and leave something for me.'" Coyote rides on Vulture's back while they fly up to the upper world where Eagle has taken her. This upper world seems to be the realm of birds, though they are not named. Once again we see that Ñui has access through the sky, as in the Flood. Curiously, we never learn whether or not Coyote's wife is recovered because Coyote keeps bumbling the efforts.

Many tribes have a tale of the game animals being confined by someone, becoming inaccessible to the hunters. Pimans are no exception, and they have various renditions of this theme (Russell 1908:217–218; Saxton and Saxton 1978:232–234; Shaw 1968:43–55). In a version recounted by Gila Pima George Webb (1959:106–108), Brown Buzzard is the one who scouts out the lost game, and Coyote is responsible for their release.

There are various versions of the origin of the Saguaro, one of the most important plants to the desert Pimans. In one (Lloyd 1911:217–229), Corn Man and Tobacco Woman have a child who is a pumpkin. During the day, while the pumpkin child's parents are at work in the fields, some children, playing with it as if it were a doll, drop it, breaking its neck. In frustration, Corn Man again abandons the People. The child sinks into the ground, coming up at Gakoḍkĭ, "Crooked" (Superstition) Mountains.

The mother, Tobacco, and the maternal grandfather seek assistance in finding the lost baby. The people enlist the help of Ban (Coyote, also called Tohavs), who smells around where the child had lain, then searches the countryside.

> And he came to where the Giant Cactus was and thought it was the baby, but was not sure, and so came back and told them he could not find it.
>
> And they wanted Nooee to go, and Toehavs said to Nooee: "I did see something, but I was not quite sure, but I want you to examine that Giant Cactus."
>
> So Nooee flew around and around and examined the Giant Cactus and came back, and when the people questioned him said, "I have found it and it is already full-grown, and I tell you I think something good will happen to us because of it."
>
> And when the Cactus had fruit the people gathered it, and made *tis-win*

[*navait*, Saguaro wine], and took the seeds and spread them out in the sun.

(In a different story accounting for the origin of the Saguaro, it is Raven who is sent out to find the child-turned-cactus, discovers it in Quijotoa Mountains [Giho-do'ak], eats the fruit, and vomits it for the People on his return.)

The summer Saguaro wine feast, bringer of the monsoons, is not the only association of the vulture with rains. Kissell (1916:171) collected a medicine man's rain basket from Covered Wells that contained "a stem of the shrub ash tree with an attached buzzard feather, for emergency cases when there is great need for rain and the preceding charms [turkey feathers accompanied by chants] have proven ineffectual."

As noted earlier, Upper Pimans through patrilineal descent belonged to either the Buzzard moiety or the Coyote moiety, according to their clan. Moieties were matched against each other in games. If these groups ever regulated marriage, that function has been lost. Here is how Sylvester Matthias explained their roles: "[There were] things [Coyote people] bragged about—they oppose the Buzzard clan [*sic*]. They would say, 'Ah, the Coyote eats rotten meat.' And we'd turn around and say, 'The Buzzard eats rotten meat, too.' They're working against each other. Races and gambling against each other."

Francis Vavages concurred: "They used to tease each other—against Ñui and Ban: 'Coyote always steals chickens because he doesn't want to work,' and 'Ñui always wants to eat rotten things.'" Hair pulling seemed to be the limit to physical conflict (Parsons 1928:456). Someone told me that marriage between moieties might, in a joking manner, be cited as an excuse for marital quarrels.

Sylvester noted that these moiety assignments did not mean that Coyote or Buzzard *made* the people in the different sibs and moieties. The sib or clan names, now largely forgotten with loss of language, come from the word a child used to address his or her father (Lloyd 1911:147; Parsons 1928; Underhill 1939:31–34). Vulture people are red, have a red velvet-ant for their mascot *(shoiga)*, and were said by some to be descended from the original inhabitants of the land. Other versions of the Creation Story have all five clans emerging together in their conquest of the earlier inhabitants. Lumholtz (1912:354–355) was told that "[a]ny animal that has red about it belongs to the red people," hence to the Buzzard moiety.

There were some ceremonial associations with the moieties. Underhill (1939:31–32) noted that at the end of the sixteen-day purification for an enemy-killer, other killers blew smoke over him. If the man finishing his purification were of the Coyote moiety, the smokers held sweet grass (probably *vashai s-uuv*, nut-sedge) in their mouths, but if he were of the Buzzard moiety, they held *vavish*, Yerba Mansa. She added, "At the corn dance, which is now only a tradition, Coyote people painted their bodies with white splashes, Buzzard people with red." According to Cassa, Coyote represents individuality or winter, whereas Turkey Vulture signifies cooperation or summer.

Both the Coyote and the Turkey Vulture may cause staying sickness among Pimans, not because they were animals newly encountered by the northward-marching Vupshkam (both animals are found throughout the Tepiman corridor), but because this mammal and this bird are of utmost cultural significance among all creatures. Yet either animal could visit a person and give him power and song to cure staying sickness. "Moiety influenced a man's choice of guardian spirit and his dreaming of songs," wrote Underhill (1939:32). "A Coyote man would be very likely to dream of Coyote, although other animals would be possible. He would never dream of Buzzard, who helps his own 'partners.'"

The staying sickness caused by each dangerous or special animal is characterized by a suite of symptoms. "The vulture or turkey buzzard causes sores, especially syphilis, and sore eyes on the baby if the parent eat [*sic*] a dead animal just before the child is born," wrote Russell (1908:263–264). "The remedy is to sing the buzzard song and pass the wing feathers of the buzzard over the child." Underhill (1946:289) recorded the same symptoms among the Tohono O'odham, noting that the wing feathers were used as the fetish in a curing session. "Sometimes a dance in the formation of a girl's puberty dance [*wuagida*]" was part of the healing. (Neither Russell nor Underhill gave a vulture curing song.) Interestingly, this is exactly what Sylvester described for the River Pima. "The buzzard—they're scared of buzzard—because [of] Ñui. Oh, yes, they have a *wuagida* about *ñui*—Ñui Wuagida. Just to chase away the curse of the *ñui*. The *ñui* can put a curse—*taach [chuuda]*—on you" (see p. 49–50). He characterized the symptoms as aches or dizzy spells (seizures).

Juan Gregorio told Bahr that buzzard sickness was caused by anyone killing a vulture, picking up its feathers, or disturbing its corpse. Symptoms are sores all over the head (Bahr and associates 1974:285). Vincent Josemaria told us the same. He sang this song from a *ñuwi wusota*:

Yahan hema epova himekhu
chama vo'inoghime
si'alinge wuhuwui va kukughimehekhu
nema kainoghimekhu
chama vo'inoghime.

A feather again goes,
There it turns around,
Toward the east goes roaring.
Thus sounding going,
There it turns around.

Not only are the vulture and its feathers special (that is, sacred), but any carcass that the bird has claimed as its own cannot be touched. (The same is true of anything that the bear or gopher, for instance, has claimed; see Rea 1998b:30, 158, 209). Ruth Giff told a story that illustrates this *taatchuda*, but in this case the vulture also adopts the stricken person: "A woman ate a dead cow. And a buzzard had come already and ate the eyes out. And she was pregnant. And the baby was born without eyes. He was blind. When he grew up, the buzzard would take him—up—up—and teach him songs. The song about the train—that hadn't got here [south central Arizona] yet. I wonder what state it was [in then]—*vainamĭ voog* [the railroad]."

The blind man Ruth spoke of was a Pima singer who learned a song about the *vainamĭ voog*, "iron path/road," long before the railroad even reached Arizona. The man would be called a *ñui namkam*, Turkey Vulture meeter, because of the special role played by this animal helper. He was born blind because while his mother was pregnant with him, she violated a taboo of eating a carcass that the vulture had already claimed by first eating out its eyes. Ruth added that the Turkey Vulture itself was so strongly tabooed that it would never be eaten, not even in starvation.

The two longest wing feathers (primaries) of either the Golden Eagle or the Turkey Vulture were used by the *maakai* during the *doajida*, the diagnosis session. The two feathers were bound at their bases and held in the left hand while the shaman held a gourd rattle in his right hand to accompany his singing (fig. 5.1, left). The feathers would be held curved either up or down, according to Simon Lopez.

The birds themselves were not to be killed because they were too powerful. Feathers were collected, Sylvester said, beneath the roosts where molting birds dropped them. Sylvester used *ñui* tail or wing feathers as well as those of the Red-tailed Hawk in fletching his arrows, but then he belonged to the Coyote moiety. Although he did not say so, perhaps at one time such feathers would first have to be purified by a *maakai* before they were safe for general use.

The gathering of molted feathers is apparently an ancient practice. Near the beginning of the Eagle Man episode in the Creation Story, I'itoi "went westward and found a powerful doctor who had a daughter, and said to the father: 'I want your daughter to go around to all the big trees and find me all the feathers she can of large birds, not of small birds, and bring them here'" (Lloyd 1911:77). In another version by the same storyteller, we read, "Elder Brother sent the son to a roosting place of large birds to get feathers" (Russell 1908:219). In still another version, in response to complaints about an unbeatable gambler I'itoi instructs the people to go to where the largest birds sleep and to gather their feathers (Densmore 1929:42). The collected feathers are cut into segments, then parched to make a pinole. When the gambler drinks the feather pinole, bumps appear on his skin, then feathers emerge on his body, and he turns into a huge eagle.

Piman mythology embodies many natural traits of the *ñui*, Turkey Vulture, but Piman folk science on this bird is also extensive. Sylvester noted that the bird nests in caves. Molted feathers were gathered from under their communal roosts, usually a dead tree. If harassed, vultures will vomit on people. In Piman history, these birds were never molested or kept caged or tethered for their feathers, as were Red-tailed Hawks and Golden Eagles.

One day I return from camping in the Sierra Estrella while working on a book. I stop to see George Kyyitan, pointing out from his house the white spot near the end of the range where I had camped. George thinks for a few moments, then says he knows a song that describes a storm over the Estrellas, comparing the black clouds hanging over the mountain to buzzards. Then he sings for me this song:

> Chuchuka chevangi gan hu va himeda
> chuchuka ñuñui a wepo mamasing ga
> hu va himeda
> Meeko si kaidam o betoñime
> jewena wesiko betoñime.

The poem appears short on paper, but when the various lines are repeated and the whole piece sung through four times, it amounts to forty-eight lines. I catch the word *ñuñui*, vultures, and ask George for his translation. He pauses while he sings the song through under his breath to himself, then translates:

> "There's a black cloud going. And it looks like a black buzzard. That's why it's booming *all* over." See, when that cloud burst, or whatever you call it, it's booming all over when they sing that. See, 'cause it's something like thunder. It's the only thing it says in there. It's not very long. Yeah, not much talking there. It's a Totoñ Ñe'i, an Ant Song, from Salt River [Reservation]. But it mentions *ñui* in there and *chevagĭ*, "cloud." It's just a [circle] dance song. They sing this one *all* over. Every time they have dances.

To understand a song, one must convert the poetic language into spoken Pima, which may take considerable work. For instance, Culver Cassa renders the words of this Black Cloud Song in spoken Pima as:

> Chuchk chevag[i] gan hu va hihim
> chuchk ñuñui ha wepo maas
> gan hu va hihim.
> meeko si kaidam o betoñim
> jewẹd wesko betoñim.

And he gives this English translation:

> Black clouds are drifting over there
> Appearing like black Turkey Vultures
> Drifting over there.
> Far away it is loudly rumbling
> All over the land it rumbles.

Russell (1908:292–295) collected a set of ten swallow songs used in the Circle Dance. One of them is about the Turkey Vulture:

> Sikola imeda vañi
> kiñuñui vawepo ñiwuada'a, haiya
> damaive kachime basho vañulina
> kañeida'a
> idovaña ibeinanga chechenonge'e.

> In a circle I fly,
> I behave like a vulture, haiya.
> In front of the heavens I loiter and
> watch.
> This my breath becomes blue.

"*Ñui* is always copying somebody," Theodore Peters said. "If somebody copies somebody, they say, 'Just like *ñui!*' There was a man like that at Ak Chiñ. The women gave him the name Ñui." Culver Cassa explained this anecdote as a contrast set behind culture heroes Ñui and Ban (Coyote). Coyote rushes in impetuously and is the first to get into trouble. Buzzard has the opposite characteristics. He is so cautious that he does something after everyone else has tried it and found it safe. The metaphor reflects the two animals' real-life behavior.

One day when we were discussing reptiles, Sylvester described the dark Desert Spiny Lizard as being *s-ñuimagĭ*, colored like a buzzard. Pimans have a color spectrum of three named colors but add variants to this limited lexical palette by using *-magĭ*, "the color of"—as in purple, *s-iibhaimagĭ*, the color of ripe prickly-pear fruit.

Buzzard also lends its name to several O'odham place-names. Along the Gila River there is a Ñui Vavhia, "Buzzard Well," among the Kuiva O'odham (Downstream People, western Pima), just as there is a Ban Vavhia, "Coyote Well," among the Ta'i O'odham (Upstream People, eastern Pima). A rocky outlier in the West Sacaton Mountains is called Ñui Mehidag, "Burnt Buzzard." As with many toponyms, no one remembers any longer how the peak acquired this name. Underhill (1939:134) mentioned a Buzzard Claw Mountain (presumably Ñui Huch Do'ag) east of Papago country.

The Canyon Bursage, growing along the edges of desert washes, is called *ñuñui jeej*, "vultures' mothers," among the Tohono O'odham and at least among some Akimel O'odham. The metaphor seems now lost. Several other plants are *jeej*, "mothers," to something. Frank Jim noted that such plants are usually associated with *a'akĭ*, arroyos. Leonard Pancott, a Pima cowboy, called Canyon Bursage *ñui hehevo*, "vulture's eyelashes." (He had not heard of the plant called *huai hehevo*, "Mule Deer's eyelashes," Annual Sowthistle.)

COMPARATIVE LINGUISTICS. The name for the vulture is a pan-Piman folk generic.

River Pima say *ñúi*, and Papago *ñuvi* or *ñuwi*. Pima Bajo say *núi* (pl. *núnui*). Mason (1917) recorded *nuuí* among Tepecano (Southern Tepehuán), and Pennington (1969) recorded *nuí* among Northern Tepehuán. Among Mountain Pima, I noted variants *nuli* and *nuyi* as well as *nui* (pl. *nunui*). The colonial Névome vocabulario (Pennington 1979) gave *nunudi* for Zopilote (Black Vulture), with the annotation, "No tiene plural." However, the form as given *is* plural. Colonial Attí (Attí vocabulary [1774]) gave *nui* (pl. *nunui*) as the local folk generic (see the next species account, Black Vulture, for the discussion of folk species).

Among other Sonoran Uto-Aztecans, Aura (Turkey Vulture) is *huesó* (Eudeve), *wilú* (Tubar), *cacháwi* (Yaqui, *wiiru* fide Shaul), and *apíguiri* (?) (Rarámuri, or Tarahumara). In my own field notes for Yaqui, I have *kucháwi* for Turkey Vulture and *wíiru* for Black Vulture.

TECHNICAL NOTE. New World vultures, the Vulturidae or Cathartidae, are most closely related to the storks (Ciconiidae), and neither family claims relatives among the superficially similar herons, cranes, or hawks. Both storks and vultures lack the syrinx or avian vocal apparatus. They have similar skeletal and muscular systems, as well as similar DNA (Campbell and Tonni 1983; Emslie 1988a, 1988b; König 1982; Ligon 1967; Olson 1979, 1985; Rea 1983; Sibley and Ahlquist 1986, 1990). And both storks and vultures share a unique cooling method called urohidrosis: they urinate on their feet (the unfeathered tarsi and toes). The evaporation cools the blood, which is then circulated to other parts of the body (Hatch 1970; Kahl 1963). Although Turkey Vultures have red or deep-pink feet, the evaporated urates collect on the bare skin, coating it chalky white.

ñuipa, pl. *ñuñuipa*
s-tuk nunui (Pima Bajo)

Vulturidae (Cathartidae), Ciconiiformes
Coragyps atratus
Black Vulture; Zopilote

In many ways, the Black Vulture is quite the opposite of the more commonly seen Turkey Vulture. Its wings are relatively shorter and rounded, its tail short and broad, giving the bird in flight a somewhat batlike appearance. The toes may be seen projecting just beyond the tail. The bird is entirely black except for a whitish patch at the base of the primaries, seen from below. There is no grayish trailing edge. The wings are held entirely flat. Unless there are good thermals, the Black Vulture alternately flaps and glides, lacking the seemingly effortless buoyancy of the Turkey Vulture.

Other differences are noticeable on the ground. The Black Vulture is longer legged and longer headed. The bill is dark and slender, with narrow nares. The bare skin of the head is corrugated, hanging in a loose sack at the front of the neck, above the feather line. The short tail feathers appear to have been clipped off evenly at their tips.

Black Vultures differ from their cousins in behavior as well as in structure. Highly social, Blacks are usually found in flocks of a hundred or more and often forage as a group rather than as individuals or in pairs. Heavy in flight, they are quite active on the ground. At a carcass, they are aggressive, springing up and jostling around, with subdominant individuals waiting on the sidelines until those higher in the hierarchy are sated. Food is wolfed down, even when the bird is solitary. None of the Turkey Vulture's patient pinning down and pulling at a morsel underfoot.

Lacking the Turkey Vulture's olfactory abilities, the Black depends on its relative to locate carcasses, sometimes joining it in mixed flocks. The Black Vulture frequents places such as beaches where food is dependable. In many Latin American towns, the Black Vulture is urbicolous, throwing in its lot with humans, taking advantage of slaughterhouses, garbage dumps, and fishing docks. In open-air markets, it stands its ground against pigs and dogs that might also scavenge for discarded items.

At least in northwestern Mexico, where the two species occur together, the Black is known in Spanish by its Nahuatl-derived name Zopilote, from *tzopilotl*, and the Turkey Vulture is called Aura, from an unknown Indian source. (There are various other names for them throughout Latin America.) The Black Vulture is supposedly new to the borderlands of Pimería Alta, having pushed northward from the southern half of Sonora in the late 1800s, reaching Magdalena, Sonora, sometime between 1887 and 1925 (van Rossem 1945:52), Sells,

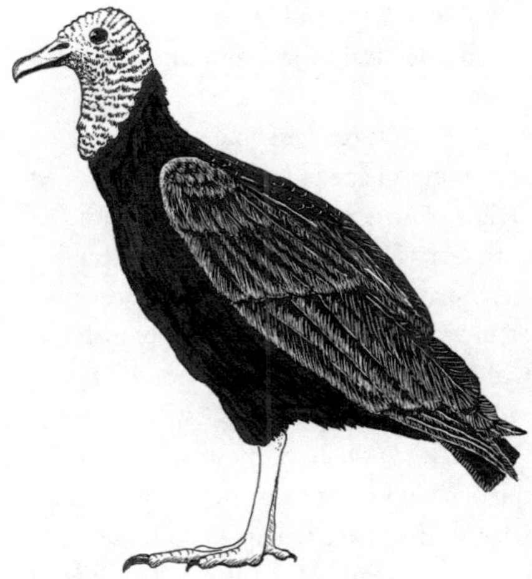

ñuipa, Black Vulture (Amadeo M. Rea)

Arizona, by 1920, Tucson two years later, and the Gila River by 1947 (Phillips 1968; Phillips, Marshall, and Monson 1964). At least this is what ornithologists think.

Folk taxonomy suggests differently. The Névome vocabulario was composed at the village of Ónavas, Sonora, Mexico, on the middle Río Yaqui, probably between 1658 and 1671. A Jesuit, most likely Padre Loaysa, recorded under the entry Zopilote, "*nunudi*, No tiene plural [*sic*]" (Pennington 1979). There is no entry for Aura or any indication of how the two local, very distinctive species were distinguished. Thus, we lack a biological contrast set for colonial lowland Pima Bajo. And, curiously, the entry as given *is* plural or aggregate, at least in form.

During my work with don Pedro Estrella, one of the two last speakers of Pima in Ónavas, three hundred years later, we encountered both species of vultures frequently. Pedro called the red-headed Aura just *nui* or, where needed for contrast, *keli nui*, "old man vulture." The Black Vulture he called *s-tuk nunui*, "black vulture." Linguist Ken Hale (n.d.a) recorded *nunui* as both Zopilote and Aura, curiously again in plural form. (Pedro's "old man" epithet may be a carryover from the Piman Creation Story, but we lack any mythology from Pimans this far south.)

In Upper Piman country, usually only the Turkey Vulture is seen. At the Pima village of Santa Cruz one November, Ambrose Juan noticed three Black Vultures flying over, calling them *s-chuk ñui*. I may have been witnessing a spontaneous coining of a name, or at least a recent naming, occasioned by the addition of a species new to the local avifauna in the past thirty years.

Black Vultures are somewhat more common in Papago country. In 1980, when Delores Lewis was afield along the border with Gary Nabhan and me, we saw both species as well as the Crested Caracara. Delores called the Turkey Vulture just *ñui*, which he described as "the one with the red head and longer wings," while calling the Black *al ñui*, literally "small vulture," the one with "narrower [*sic*], shorter wings—they have to flap—flap faster. They have black head." He contrasted both with *kusijim*, the caracara.

In many ways, Sylvester Matthias was an information pipeline to the past: he knew things, often folk taxa, that other Pimas his age had long since forgotten or had never learned. One day in 1986, after discussing the Scaled Quail in the Baboquivaris, he mentioned to me *ñuñuipá*. The first part of the word meant "vultures," and the second part he pronounced almost as an explosive *pah!* What was this? "Kind of like a buzzard, but they have a short tail. It's in the buzzard family. Across the border in Papago country [Sonora]. The Papagos call it *ñuñuipa*."

Several years later I asked him again about *ñuñuipá*. "As soon as they hit Mexico, at Sasabe. Register—get their brand on their horses. Then on to Magdalena."

"How did the Pima know that name?"

"That's the question! Must have been from the Papagos, because they're neighbors."

"How long ago?"

"About the turn of the century."

His response needs some background. Sylvester belonged to a family of traditionalists from the western or downriver part of the Gila River Reservation. Traditionalists were almost exclusively from Catholic families, minorities who clung to old ways when the reservation was rapidly becoming properly Presbyterian through the efforts of the German missionary Rev. Charles H. Cook. Catholics settled in small villages usually near a church and school. They continued to learn Piman songs and myths and to carry out Piman rituals to a certain extent. Most Presbyterians underwent cultural amnesia early in the twentieth century. (A notable exception seems to have been the Pima author George Webb of Santa Cruz.)

What Sylvester was explaining was the annual October trek to Magdalena, Sonora, for the big fiesta honoring a hybrid San Francisco Xavier (a Jesuit) and San Francisco de Assís (founder of the Franciscan order). Traditional Catholic families from Gila and Papago country would load their wagons with camping goods and trading items and head southward to northern Sonora, crossing the border at Sasabe, where they registered not their license plates, but their brands.

I have made this crossing a number of times myself in early October, heading for the festival. I would stop at this same tiny sleepy settlement to take care of vehicle registration. And always upon leaving the aduana (customs depot) with hands full of official papers, I would look up and see Black Vultures circling over town.

Some eleven years after our first discussion, Sylvester and I were at the Salt-Gila confluence, on 115th Avenue west of Phoenix. I found two freshly dead Muskrats hanging in a Salt-cedar. While taking the carcasses down, I noticed seven Turkey Vultures and two Black Vultures circling not very high between us and the nearby Estrellas. I pointed the vultures out to Sylvester. He knew both.

"I didn't know there were *ñuñuipá* here!" he said when he saw the two kinds.

"Yeah, in recent years."

"The only *ñuñuipá* you could see was just this side of the border and across the border [at Sasabe]."

I already had a Muskrat voucher specimen from Pima country, so I photographed the two carcasses and left them on the gravel for the vultures to dine on.

Nearly twenty years after Gary Nabhan and I had our first outing with Delores Lewis, I visited Delores at his home village of Ge Oidag, Big Fields. I sprang the name *ñuñuipá* on him, with no indication of what bird it might be. "You mean the black ones? Just a few come around here." I handed him my field guide. "Yeah, there it is," he said, pointing out the Black Vulture.

I found that many Tohono O'odham cattlemen were readily familiar with both the Black Vulture itself and its name. Simon Lopez knew *ñuwi*, the red-headed one, and *ñuwipa*, the black one, as well as the caracara, which he associated with them. The first two he said have always been around Santa Rosa, but the caracara that he saw occasionally he considered new to his area, and he was unsure of its name, though other cowboys had told him what it was called.

Vincent Josemaria of Pisin Mo'o knew *ñuwipa*, but wasn't sure of how to make the plural; he suggested maybe *ñuñuwipa*. He also associated the two vultures with the caracara, as did Jose Enriques. "The short-tailed vulture, they flap their wings as they fly," Jose observed, although he didn't know their name.

During her Papago studies earlier in the century, Underhill (1939:33) left a hint that the two vultures were part of the local avifauna. In discussing the position of Piman clans within the two moieties, she observed that the "Waahwkam [Vaavgam] is slave to the others and has no real partner, but looks to black buzzard who is slave to the yellow buzzard, the moiety totem." This seems a reference to the Black Vulture in contrast to the Turkey Vulture (often called *oam ñui*, "yellow [or] brown buzzard," in mythological contexts), but without the original O'odham words we cannot be sure.

After my consultations with Simon and Vincent, the Piman name was now verified in both Akimel and Tohono O'odham territory, country where the Black Vulture was supposedly of recent origin. I wondered what time depth this folk taxon might have. Almost nothing from the colonial period is documented on the Upper Pima language. A delightful exception in the Bancroft Library is a holographic partial vocabulario compiled early in the Franciscan period (ca. 1774), at the Pima village of Attí (now called Addi or Adil), fifty miles west of Magdalena. More than half the manuscript is missing, but under "Sopilote" (Zopilote) is found the entry:

> *ton nui*, pl. *ton nunui*: assi llaman a la aura. el mas proprio *nuipan* vl [vel] *nui*.
>
> *ton nui*, pl. *ton nunui*: thus they call the Turkey Vulture (Aura). The more proper [name for Zopilote] is *nuipan* or *nui*.

Although this entry is less precise than we would like, it seems to imply that there were at that time two biological species of vultures on the Río Altar, the parallel drainage west of the Río Magdalena: the Zopilote and the Aura. The former is more properly called *nuipan* in the local Pima dialect, the latter *ton nui*. The Black Vulture name is cognate with that known to Sylvester, Delores, and other Papagos two centuries later. The Turkey Vulture folk taxon is a bit puzzling. As indicated with the words *nui* and *ñui*, Piman spoken in the northern country had not yet palatalized the *n* in certain word positions. Thus, the word *ton* could mean "shiny," as it does today, or it could be *toñ*, "hot," as in current Upper Piman. There is no way to be sure.

Piman folk taxonomy therefore strongly suggests that Black Vultures have had an ebb and flow at their northern limits. They may have ranged into northern Sonora in the late eighteenth century, then receded, returning again about a century later. But the O'odham never lost their name for this bird.

One other linguistic conundrum needs to be resolved. Both the seventeenth-century Jesuit and the twentieth-century linguist Ken Hale recorded the vulture word in a plural form. Pedro Estrella usually used the plural or aggregate form, as when he called the Zopilote *tuk mo'okam nunui*, "black head–it-has (pl.) vulture." And when speaking of the Black Vulture, Sylvester almost invariably used the plural, *ñuñuipá*. Perhaps it is only an accident of elicitation. But certain Pima words are in fixed plural form: *haahag*, leaves; *totoñ*, ants; and *taatamĭ*, teeth, are a few that come to mind. "Well," as Joe Giff once joked, "those things don't often come in the singular!" The same might be said of the highly social Black Vulture.

COMPARATIVE LINGUISTICS. For variations on the folk generic name, see the preceding species account. Although the Zopilote name is written in the Attí vocabulario as *nuipan*, I wonder if the terminal *n* was really there or just written from the force of habit of writing (and hearing) Spanish words such as *pan*, bread. As noted, the attached terminal morpheme is strongly accented. Culver Cassa says it signifies "wants to be" or "trying to be." Saxton, Saxton, and Enos (1983:46) correctly identify the folk species as Black Vulture, writing it *ñuwiopa*. But the two major accents indicate two distinct words in this compound name.

I have heard several variant names for the Black Vulture: *s-chuk ñui* (Ambrose Juan, Santa Cruz Pima village), *al ñui* (Delores Lewis, Ge Oidag [Big Fields] Papago village), but they seem to have been impromptu identifications in the field. In later discussions, Delores always used the conventional name.

Among other Uto-Aztecans, the Zopilote is *tekó* (Eudeve, Tubar), *wíiru* (Yaqui), and *wirú* (Rarámuri). The Spanish name is from Nahua *tzopílotl*.

***vakoñ, vakoñĭ*, pl. *vapkoñ*; var. *vakoañ, vakoañĭ*, pl. *vapkoañ* *vakon, vakoan* (Pima Bajo)**

Ardeidae, Ardeiformes (mostly)

herons and egrets; garzas

Several species of herons and egrets are included in this folk genus, as well as the pelican (see the next four entries). I have almost always heard the word *vakoñ* or *vakoañ* used with a modifier designating a folk species. The most commonly referred to is the *koomagĭ vakoañ*, Great Blue Heron.

As with many plants and animals associated with water, this lexeme contains the Sonoran root morpheme *va*. The heron word is found in the Névome vocabulario as *vacoani* (Pennington 1979:55), where it is defined as the garza parda, which is contrasted with *s-toa vacoani*, "otra [garza] menos blanca" (a smaller white heron). In modern Pima Bajo, I got *vakoan*. The name appears as *vakóñi* in Northern Tepehuán (Pennington 1969), *vakoon* in Southern Tepehuán, and *vakoon, vapkon* with Tepecano (Mason 1917:319, 329). It was recorded as *vakoni* in colonial Southern Tepehuán (Rinaldini 1743). Among Mountain Pima at Yepáchic, I found both *vakiig* and *vakuun* and even *warra*. Pennington (n.d.) recorded *vakin*, and Hale (1964) entered *vakiñ*. River Pima may use *vakoñ, vakoñĭ*, or even *vakoañ* or *vakoañĭ*.

Saxton, Saxton, and Enos (1983:54) define *shuudagĭ u'uwhig*, literally "water-bird," as a crane, heron, or other waterbird. It may be that the *vakoan* word is missing from Papago; Delores Lewis, Joe Lopez, Jose Garcia, and Simon Lopez didn't use it, nor did Vincent Josemaria. For instance, Simon referred to the Great Blue Heron in English as "those ducks with long legs"—in O'odham, *che'echew kakiokam*, "tall/long legs–it-has." Joe Lopez called egrets *toha u'uhig*, and Jose Garcia used *koomagĭ u'uhig* for Great Blue Heron.

Culver Cassa notes that the Sobaipuri village name Bacaoncos on the upper Santa Cruz River, upstream from Guebavi (Ge Vahvia), would be Vakoañ Kosh, "Heron Nest," in modern Upper Pima. At the time of Contact, the sounds had not yet been palatalized in the north. And Spanish orthography frequently rendered the Piman *v* as *b* in place-names.

Although *vapkoñ*, herons, appear rather frequently in Piman oral literature, translators and transcribers have frequently glossed the word as "crane" (see crane account).

COMPARATIVE LINGUISTICS. The Yoeme (Yaqui) dictionary lists *vakoni* as "duck" following the Spanish loan word and an indigenous term. Surely this is heron/egret, for which there is no listing (Molina, Velenzuela, and Shaul 1999:215). Many Piman and Cahitan terms are virtually identical. Among other Sonoran-branch Uto-Aztecans, we find *watzó, wachó* (Raramuri); *batósa, batósta* (Eudeve); and *batosal* (colonial Cahita).

tash cheenam vakoañ

Ardeidae, Ardeiformes

Ardeola (Butorides) virescens

Green Heron (Green-Backed Heron); Garcita Verde, Garcita Oscura

Hiking on the lower Salt River, where it forms the north boundary of the Gila River Reservation, I become aware first of a bright yellow eye peering at me from a thicket of green willow leaning over the water, then an absurdly short tail flicking nervously. Gradually my eyes make out a long, stretched, maroon neck, sharp bill, and shaggy blackish crown and crest. A white line runs jaggedly from the chin to the chest. What seems to be a mass of dark willow leaves with pale edges resolves even more slowly into a body no longer than the skyward-pointing neck it supports. Finally I see the yellow-orange legs. As I step forward to close the few yards that separate us, this feathered apparition scrunches down and lunges off over the water, while loudly uttering an irritated *skeow*, unmistakably an avian expletive. A few trees upriver it darts again into a thick willow and vanishes.

This little fishing bird (around sixteen

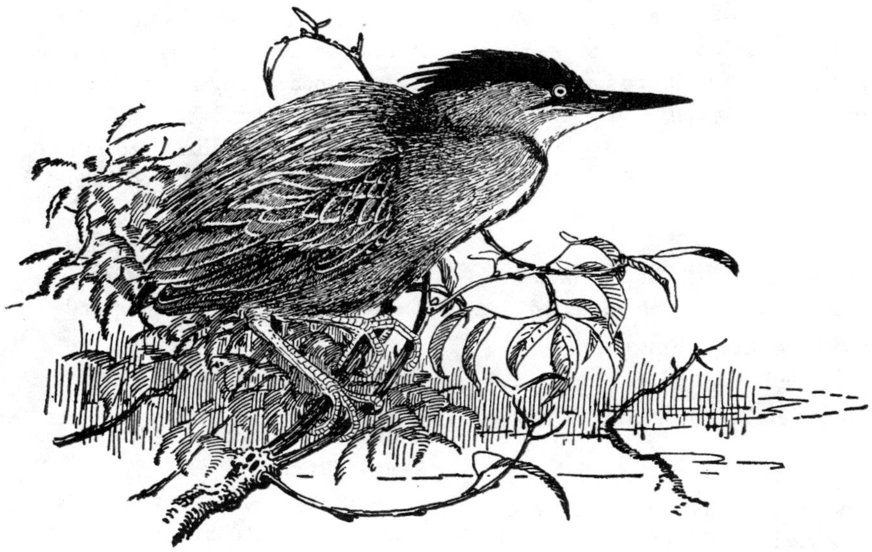

tash cheenam vakoañ, Green Heron (Allan Brooks)

to twenty-two inches long) is the Green Heron. Not much of the bird seems green, really. The erectile crown and the pale-edged wing and tail feathers are a dark slate-greenish. The elongate back feathers are powdered gray. A gray wash finishes off most of the underparts. This is the adult. The immature in summer and first winter has few elongate feathers and lacks most of the greenish tinge. In this plumage, it is dark brown above, streaked brown and white below, and the brown wings are dotted with buffy white spots.

Unlike the other herons in the area, the Green Heron is generally solitary. It prefers the marshes and stream sides that once characterized much of Northern Piman country. Here it nests and roosts in riparian thickets, particularly in willows and cottonwoods. Its nest is a stick platform, where the four or five pale unmarked blue or green eggs are placed. Young in the nest assume a bittern posture when disturbed—that is, neck, head, and bill pointing stick upright. But in adults, the "preference for horizontal postures" is characteristic (Hancock and Elliott 1978:179). Color pattern, feather texture, and behavior all serve to conceal the bird.

The Green Heron forages mostly early in the morning and again toward dusk and on into the night. Small fish are important food, but other aquatic animals are taken at small ponds and backwaters. Green Herons also feed in dryer situations such as fields and pastures where they take invertebrates, especially grasshoppers, as well as small vertebrates. They are mostly quiet birds, giving a deflected *kyew* call at times, in addition to the *skeow* note when flushed.

The Green Heron has long been gone, at least as a breeding species, from the reservation portion of the Gila, but M. French Gilman (n.d.) found it breeding annually in his yard at Sacaton early in the twentieth century. Where not molested, it may be quite confiding. It must have survived still longer in the downstream villages until the river died there, too. There are winter records for the lower Salt (Rea 1983). In summer 2003, I saw one feeding at White Horse Pass in an artifically created riparian community.

After discussing the Great Blue Heron, *koomagĭ vakoañ*, Sylvester Matthias of Komaḍk on the Gila River said, "Another, it's small. Used to nest along irrigation ditches. *Tash cheenam vakoañ*. They're small. Just build a nest like a heron. Sometimes the young ones will drop to the ground. Kind of bluish gray with yellow beak and yellow claws. Nest separately [not colonially like other herons]." He also noted its black cap. Sylvester explained that *tash* means "sun," but that *cheenam* is an old word whose meaning is no longer known.

Later I recorded this conversation with him. "*Tash cheenam vakoañ* is small, like a duck, about that color. But the beak is pointed."

"Where did it nest?" I asked.

"In the cottonwood trees. But the leafy ones, not the dead ones."

"What kind of nest?"

"Yeah, just like a heron. Gray, but beak is yellow."

"How do they sound?"

"Haaagha!"

"How do they nest?"

"Just one in a tree, maybe one in another; except the *koomagĭ vakoañ* [Great Blue Heron], whole bunch nest in one tree." He noted that the big heron nested from year to year in the same place. "And they [Green Heron] just build their nest any old way. So sometimes the young ones drop to the ground. Little ones."

"What do they eat?"

"Oh, most likely insects, like butterfly. Worms, *vopoliu* [woolly-bear caterpillars]."

On still another occasion, he said, "*Tash cheenam vakoañ* is small and nests in the cottonwood trees. They're small. They don't make no sound. They're not wild. They'll peck you if you do anything. Used to be common in the cottonwoods along the irrigation ditches."

Sylvester noted that the Green Heron came in much closer to the Piman homes than did the *koomagĭ vakoañ*, but it roosted and nested in very leafy cottonwoods. It did not roost in a dead tree down on the river as other the herons did. He said the Pima considered it a very mean bird. A Pima youngster climbing a tree once fell into a Green Heron's nest, and the parents pecked the boy. The incident happened in Sylvester's boyhood, but he could not recall the youngster's name.

No one else I worked with on the Gila recognized the name *tash cheenam vakoañ*. George Kyyitan of Bapchule said he didn't know this bird, but suggested that the name means "find the sun heron," from the word *cheegam*, "to find." The verb is apparently archaic, appearing in no contemporary dictionaries. Culver Cassa surmises that perhaps the name is an allusion to a person in the community who served as a sun watcher or calendarist who sets the dates for planting and ritual. The reference would be to the bird's behavior of freezing in a bittern posture, head and bill pointed upward, when it is detected. The sun was revered as a deity (Russell 1908:250–251). This bird looks for the sun, as if in prayer. Even today, a Pima delivering a public oration or prayer will often use both *tash* and the Spanish loan word *josh*, sometimes in tandem, in referring to God.

MISIDENTIFICATION. Sylvester had described this bird as "gray," and I incorrectly identified it as the Black-crowned Night Heron, *Nycticorax*, in a previous work (Rea 1983:146). On talking considerably more with Sylvester and listening again to our earlier tapes, I realize that I had misconstrued what he said. It became clear that he was talking about the Green Heron. Sylvester had not recognized the bird from my study skin but was impressed with the strikingly different adult and immature plumages of the night heron as shown in the field guides, where size was not evident. Green Herons also have distinctive age differences in plumage. This story serves as a caution regarding the ultimate need to have consultants in the field when at all possible to identify species because pictures and even preserved specimens can be misleading.

Also, my original notes and typed 1965 manuscript have *chúrr-nang*, but the word appears incorrectly as *cunam* in Rea 1983.

Culver Cassa and I later verified *cheenam* with Sylvester several times.

TECHNICAL NOTE. My reasons for including the various forms of *Butorides* (1849) in the genus *Ardeola* (1822), based on postcranial osteology, are given in Rea 1983:141–142.

koomagĭ vakoañ
gee vakoan, s-koomagi vakoan
(Pima Bajo)

Ardeidae, Ardeiformes
Ardea herodias
Great Blue Heron; Garzón Cenizo

Dropping its very long leg, and uncoiling its snakelike neck from an S-shaped curve, a large grayish bird with a six-foot wingspan alights gracefully on the ground or in a tree top. That this stately bird can coordinate its long bill, neck, legs, and wings is rather remarkable. The largest North American heron is also one of the most common and widely known species in this group of long-legged fishing birds. It is fully as large as a Wood Stork *(Mycteria americana)* or Sandhill Crane.

Individuals appear along desert streams, in O'odham fields, and at charcos and other shallow ponds. Here they seek fish as well as insects, amphibians, and rodents—indeed, just about any animal they can capture and swallow. Pocket gophers and cotton rats flooded out of irrigated fields are major foods where fish are scarce.

Adult Great Blue Herons are mostly bluish gray, with a broad black line over the eye, a white streak down the front of the neck with black markings, black in the bend of the wing, a black tail, and rufous thighs. The sides of the neck are washed vinaceous. Elongate feathers of the back and lower neck give the bird a loose, shaggy appearance. In the breeding season, there are several crest feathers, and the unfeathered areas of the head become more intensely colored. Young birds lack these elongate feathers, have an all-dark crown, and are somewhat duller.

Although usually solitary in feeding, these birds nest colonially, placing their large stick nests, two to four feet across, in a dead tree, if one is available. These heronries or rookeries may contain from a dozen to many dozen pairs of breeding birds. When startled, Great Blues may utter a single loud *krrauk*.

Sylvester Matthias was the first to tell me about this bird. "Across the Gila [from Komaḍk], the *vapkoañ* used to nest. There were a lot of dead cottonwoods. There were good farms there. Till that flood [of 1906]. That's where the Coöp [Colony] started. Some [Pimas] went back [upstream] to Sacate."

"Which heron?"

"*Koomagĭ vakoañ*; *s-toota vapkoañ* nest in cottonwood trees, also. Great Blue Heron nests here. They say they come nest every year [in a] dead *auppa* [cottonwood]. I saw one [colony] across the river. They're not scared of anything, too. Stand on one leg."

"What do they eat?"

"Mostly fish."

He said the nests were in one dead cottonwood over the water opposite the village. There were at least a dozen nests. The Pima did not bother the nests or young. They never found either of these herons nesting by themselves (that is, noncolonially). Pimas say if you chase a Great Blue Heron by horse, you can catch it by its dangling legs, though no one knows, Sylvester said, whether anyone actually attempted to do this. (It would be unwise,

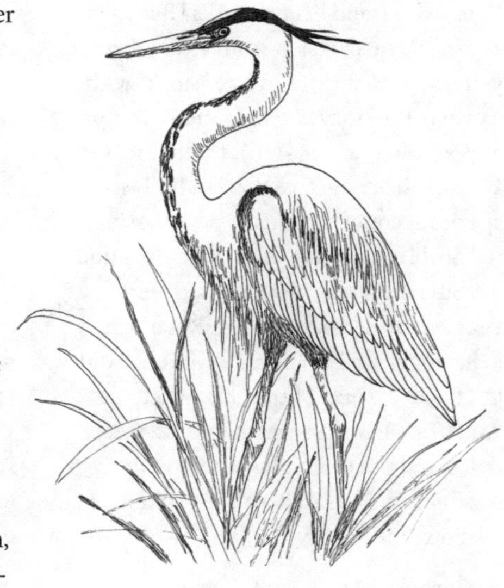

koomagĭ vakoañ, Great Blue Heron (Takashi Ijichi)

in any case, because this heron will defend itself vigorously with its dangerous bill.)

In another conversation, Joe Giff said, "*Koomagĭ vakoañ* is the big one."

"Did that one nest around here?" I asked.

Sylvester said, "They nest in the cottonwood the most."

Joe added, "Sometimes there would be a flock of them nesting in one tree."

Sylvester threw in, "Yeah, right!"

"Did the Pimas ever use those birds for anything?"

"Naw," said Sylvester.

"Did they use the same tree? Like, if they had a colony one year, would they come back?"

Sylvester, "Yeah, same [tree]."

Joe elaborated, "Use same nests. The nests are there, when they leave, they fly off. Then they come back. They are not great big pieces of wood, but they are big. They just pile them up, spread it out, and nest in there."

Joe explained that although the word *koomagĭ* is usually glossed simply as "gray," the connotation is more something indefinite, as "blurred," "dirty," or "dusty." For example, the word is also used in *u'us kokomagĭ*, the Piman name for Arrowweed.

Shaw (1968:79–81) recounted a tale about Coyote, Little Frog, a flood, and some long-legged birds she called "herons." She doesn't give their Piman name. Ruth Benedict recorded this story, calling the birds, who stand below, storks (Benedict 2001:162). In a version from an O'odham youngster, Neff (1912) wrote "stork" with a query. The story is also found in Saxton and Saxton (1973:98–100), where the bird is *kookoḍ*, the crane (see the Sandhill Crane species account). In this Papago version, the bird lands in a tree, which cranes don't do, but herons can. (But then, Desert People may not have known waterbird behavior as well as River People once did.) On the dry banks of the Gila, where the Athel Tamarisks around his corral were dying and the mesquites scarcely surviving in his fields, Leslie Pasqual told me his version of this story:

> Coyote was sitting on the debris [in the river]. You know Coyote. Some of those gray *vapkoañ* [flew over]. "Hey, friends, come and help me." They just started laughing and then turned and flew back and settled down. Ban told the *vapkoañ*, "I want you to put your beak on your [neighbor's] tail and make a bridge for me." So they did, all the way to the other bank. Old Coyote was just walking along. About half way he said, "Hey, it *sure* stinks!" "What is it that you're saying?" "It *sure* stinks!" They just flew off. Coyote just went into the river and drowned.

Herons make an appearance in another River Pima story, one with earmarks of Maricopa (Yuman) origins (Benedict 2001:106). The twin sons of White-clay Eater Woman (Nasia), on a quest for cane (Giant Reed, Carrizo) to make flutes, encounter a flock of herons, "each standing on one leg." This is supposed to be one of the frightening encounters forewarned by the boys' mother. One of the twins says, "Perhaps they will run us through with their long spears." The other brother sings:

> You are nothing but cranes [herons].
> All you do is live on fishes and things
> that live in water;
> All you do is stand on one leg.
> You are nothing.

We know from biology, as well as from Herzog's ([1929]:64–65) parallel Akimel O'odham account where the birds are specified as *vopkoañĭ*, that this encounter is about herons. The one-legged stance and the fishing behavior characterize herons, not Sandhill Cranes, which are largely vegetarian, at least in winter when they visited Piman country (Walkinshaw 1973:78–143). And herons do defend themselves with their sharp bills.

Later in the same story, the younger twin goes west (the Yuman direction of the dead) in search of his late paternal grandmother. Reaching the shore of the ocean, he is unable to cross, but a Great Blue Heron at the water's edge becomes large, spreads its wings, allowing the boy to walk across to the shore of the other side, evidently Baja California (Herzog [1929]:96–97).

The only other reference to heron in song that I have heard is in Joe Giff's Swallow Song (see American Kestrel account for the full version):

> Vakoañ u'uhig
> Si'al keekkhim
> K eda am hihinnag.
>
> Heron bird,
> East standing [dawn].
> Just then honking.

The heron's patience in its "stand and wait" hunting posture became the basis for a Pima aphorism. Clemencia Leos of Haashañ Keek (Sacaton Flats) said that when someone doesn't show up for an appointment, the person "stood up" says, "I was just standing there waiting, like a *vakoañ*."

At Quitovac in Sonora, Luciano Noriega called the Great Blue Heron we saw on the pond *kookoḍ* (which elsewhere is used specifically for the crane; see the Sandhill Crane species account). He said, "Se comen. Mucho carne in pecho" (They eat it. Much breast meat). These far western O'odham ate many kinds of animals that other Pimans eschewed (Rea 1981b).

Simon Lopez of Santa Rosa village said this "duck with long legs" sometimes visited his horse corral. As noted earlier, he called this species *che'echew kakiokam*, "long legs–it-has."

At Ónavas on the Río Yaqui, both Pedro Estrella and María Córdova called this bird *gee vakoan* or *vakon*. But don Pedro also used the more widespread name *s-koomagi vakoan*. Pennington (1980:192) was told in this area that the *vakwan*, garza morena, was killed and roasted. In this settlement, I was told it was the crane, grulla, that was eaten (see that species).

COMPARATIVE LINGUISTICS. See the preceding treatment of the folk genus *vakoñ*, heron and egret. In colonial Névome, the use of the name by itself seems to imply the Great Blue Heron as it still does among Mountain Pima. This is probably the focal species among all northern Piman speakers. Most Mountain Pima now call the bird *vool* or *vora*, although one old man at Nabogame ranchería volunteered *koomagi vakuun*.

MISIDENTIFICATION. Although both Herzog ([1929]) and Benedict (2001:120–121) incorrectly translated the word for the bird that spread its wings across the ocean as a crane, Herzog's O'odham version uses *koomakĭ vakoañ*, the Piman name for Great Blue Heron, throughout the story.

toha vakoañ, pl. *toota vapkoañ* (Akimel O'odham); *toha u'uhig* (Tohono O'odham, Hiá ch-eḍ O'odham)

Ardeidae, Ardeiformes

Ardea alba

Common Egret, Great White Egret, Great Egret; Garzón Blanco

Ardea thula

Snowy Egret; Garza Patos Amarillos, Garza Dedos Dorados

The startlingly white Common Egret is a bit smaller than the Great Blue Heron (thirty-nine inches long versus forty-six) but has a decidedly more gracile appearance, its long black legs and toes and its long yellow bill and neck being somewhat more slender. The plumage is entirely white, with lacy aigrettes or dorsal plumes in the breeding season. Common Egrets tend to be solitary, feeding both in water (salt or fresh) and in damp areas such as pastures and flooded fields, in particular alfalfa. They are stately like Great Blue Herons and also like them often hunt by standing silently until a fish or other animal approaches too closely.

The still smaller Snowy Egret (twenty-four inches long) is more compact, entirely white, with black legs and feet and contrastingly yellow toes. Breeding birds have loose nuptial plumes on the back, lower neck, and crown. More sociable, they are more likely to be seen in small groups, feeding actively in shallow water or even on land, following a tractor plowing or the water flooding across a field.

toha vakoañ, toha u'uhig, Snowy Egret (Takashi Ijichi)

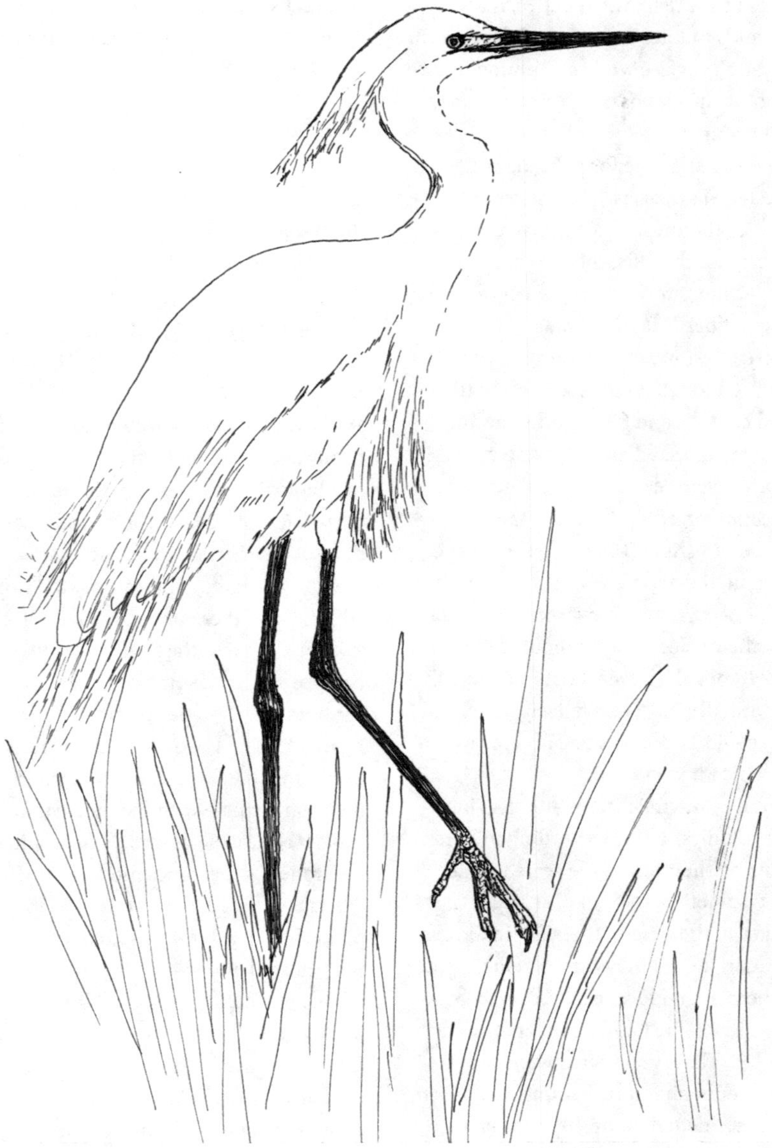

One or the other, but more likely both of these biological species are included in the Piman folk taxon *toha vakoañ*, "white heron/egret." In more recent decades, the still smaller and chunkier Cattle Egret, *Ardea ibis*, has joined the other two species of egrets in Pimería Alta, feeding in irrigated fields and pastures with livestock. It probably would be included in the folk-specific name.

This folk distinction between the Great Blue Heron and the egrets goes back at least three centuries, when we find in the Névome vocabulario a grayish heron (garza parda) contrasted with *s-toa vacoani*, a somewhat smaller white heron ("otra [garza] menos [y] blanca") (Pennington 1979:55). *S-toa vakoan* was still used by contemporary Pima Bajo speakers on the Río Yaqui. This differentiation is found today not only among the River Pima, who have had more contact with these water-birds, but also among Mountain Pima, who call the egret *toha vakiig* or *toha vakuun* (Isidro Cruz, Yepáchic). Tohono O'odham Joe Lopez and Simon Lopez and Hiá ch-eḍ O'odham Luciano Noriega called the egret *toha u'uhig*, "white bird."

Both Joe Giff and Sylvester Matthias said that some *s-toota vapkoañ* nested in cottonwood trees across the river from Komaḍk, among the Great Blue Herons, and like them making huge nests. Both the Common and Snowy Egret typically nest in mixed colonies with other species. But which egret might it have been? Common Egrets are known to breed on the lower Colorado River, more than 150 miles to the west. Snowy Egrets may be found there also, but they once bred at Picacho Reservoir, not far from the reservation (Monson and Phillips 1981) and more recently at Painted Rocks Dam near Gila Bend. Conditions at this desert oasis just downstream from the Pima villages, described vividly by the Jesuit Jacobo Sedelmayr in 1744 (1955:24), have been so radically altered that today we can only guess at their former biological richness.

The Tohono and Akimel O'odham performed a Skipping Dance called *chelkona* or *eḍa emed* as an intervillage competition (Underhill 1946:117–125). It was intended to bring rain to the host village and provide food for the guests. A line of boys and a parallel line of girls executed the steps to special songs while carrying in both hands effigies of birds, rainbows, clouds, or mountains (see frontis and figures in Fontana 1983:140). The life-size birds, with long necks and long legs, were portrayed as white with yellow bills and feet. They were said to come from the coast. Underhill called these cotton birds "cranes," but the description fits egrets much better. Even the wrap-around that the girl dancers wore to match the birds was white.

COMPARATIVE LINGUISTICS. The folk-specific distinction between egrets and other herons is not limited to riverine Pimans. We find garza blanca in colonial Cahita as *batosal* (where it is contrasted with *corobabuai*, garza parda) and *batósa, batósta* among Eudeve (Lionett 1986:117; Pennington 1981:125).

TECHNICAL NOTE. My reasons for including the genus *Egretta* (1817) in the broader genus *Ardea* (1758), based on osteology as well as on external morphology, are given in Rea 1983:143. In both internal and external structure as well as in behavior, the Common Egret is the bridge between the two genera (see also Hancock and Elliott 1978:235). For the osteological reasons for placing the species *ibis* in the genus *Ardea* rather than in the genus *Bubulcus*, see Rea 1983:146.

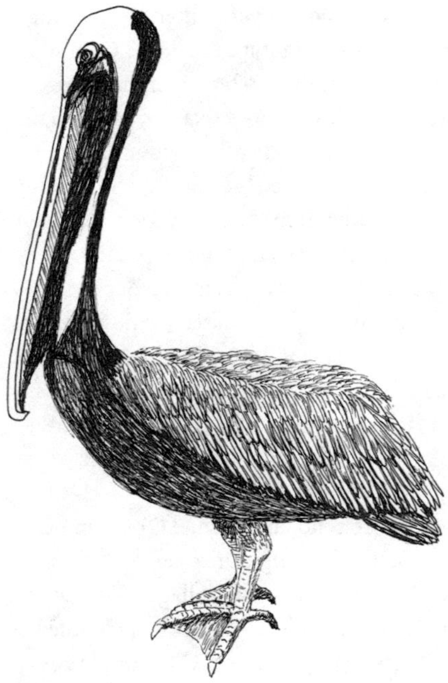

chuággiakam vakoañ, Brown Pelican (Takashi Ijichi)

chuággiakam vakoañ

Pelecanidae, Pelecaniformes
Pelecanus occidentalis
Brown Pelican; Pelícano Pardo, Alcatraz Moreno
?*Pelecanus erythrorhynchos*
?American White Pelican; Pelícano Blanco, Borregón

One doesn't usually associate pelicans with the Sonoran Desert, but two species do manage to visit Pimería Alta at times, most especially along the Gulf of California coast, but also as transients at inland points.

The Brown Pelican is a large, heavy-bodied bird with a long bill supporting a collapsible gular pouch that is not usually noticeable. Adults with contrasting head and neck patterns (yellow, maroon, and white while breeding) are found along the gulf, where they breed on islands. Documented vagrants to the interior, such as southern Arizona, have all been immatures, whose plumage is all gray brown, paler below. While fishing from the wing over the ocean, the Brown Pelican taxies just over the waves, the wing tips almost touching the water, until it spots a fish and plunges in to make the catch.

The still larger White Pelican is all white except for the black wing feathers. Its bill, gular pouch, and feet are deep yellow. Unlike its more marine relative, it breeds on islands in interior lakes and fishes from the water's surface while swimming. I have seen both species passing through River Pima country.

The Piman name for pelican, explained Sylvester Matthias, is "*chuaggiakam vakoañ*, 'a heron with a net.' We used to see them on the Gila, the big running river." *Chuaggia* is a carrying net or net bag of agave fiber (Mathiot 1973:213; Russell 1908:113; Saxton and Saxton 1973:xix) once used to tote items such as a bottle gourd canteen and food while traveling. It became the saddlebag. Sylvester said the birds were seen and described by his grandmother, who recounted their diving method of fishing. The species was never common, but apparently of somewhat regular occurrence. They were always dark ones, their color being compared to a Great Blue Heron. She never mentioned one that was white. They were relatively tame, and a Pima could walk up to one and throw it a fish, which it would eat.

Young Brown Pelicans, after leaving natal colonies, and nonbreeding individuals during their first three years wander about, sometimes stopping at larger rivers or lakes in the desert. Their destination is the ocean, where they can fish efficiently. On these inland localities, they may become quite tame (Rosenberg and associates 1991:126). These great birds were seen frequently enough in Pima country to merit the name "netted heron." When the Gila was running, with its various tributaries, the fish-loving Pima were able to spare a few fish for these odd visitors, as they did for the Europeans who first visited them (Bartlett 1854, 2:242; Kino 1919:197).

Luciano Noriega said that White Pelicans came occasionally to Quitovac oasis in Sonora, but he didn't know an O'odham name for them.

MISIDENTIFICATION. Bahr and associates (1997:127, 155) mention "pelican" in Oriole Song number 23. Pima singer Vincent Joseph translated it: "What kind of bird is it that goes low. It must be a pelican bird that goes low. Everywhere the land is made foggy. (Fog exists there)." The bird term (at least in song language) is *wakawa*, as Bahr writes it, using, as he does throughout, Tohono O'odham rather than Akimel O'odham spellings. This bird is really the *vakav*, the Northern Harrier, a more likely species for the Pima to be singing about (see that species account). In this case, the terminal *a* is added for the beat or meter of song language.

kookoḍ

Gruidae, Gruiformes
Grus canadensis
Sandhill Crane; Grulla Gris

We have Spanish vernacular to thank for firmly anchoring the Piman word *kookoḍ* to a biological species early in the colonial period. Castilian Spanish folk taxonomy distinguishes *grulla* as "crane," *garza* as "heron" or "egret," and *cigüeña* as "stork." Although all these birds are long legged, and long necked, the Spanish Jesuit who composed the Névome dictionary on the Río Yaqui in the seventeenth century directly identified two of the three in the local Piman dialect. In his vocabulario, the term cognate with the modern name *kookoḍ* appears as *cocorh*, "grulla," that is, "crane" (*rh* becomes modern *ḍ*). The grulla is distinguished from the garza parda, given in Pima as *vakoani* (the Great Blue Heron) and from "[la] otra [garza] menos blanca, *s-toa vakoani* (the smaller white egret) (Pennington 1979:55, 57). (The Jesuit also gave in Pima "*mumuguiadama* es blanca," which Culver Cassa interprets as *mumgiadam*, "bowing head forward," a rather clear reference to the Wood Stork [*Mycteria americana*]. So far as I know, no vocabulario identifies the stork in Spanish as such.)

Similarly, farther to the south, an Italian Jesuit, Benito Rinaldini, recorded the names of these two birds among the Northern Tepehuán. In his 1743 vocabulary, grulla is given as *cocore* and garza as *baconi*—that is, *vakoni*. So both the crane and the heron words are historically documented in the northern and southern segments of Tepiman.

Of the two North American species, only the more common Sandhill Crane is to be expected in the Southwest. It is a bird of stately movements that congregates in numbers in open areas such as fields or playas in winter. Its wingspan is as large as a Great Blue Heron's, but it is distinguishable by its bare scarlet crown and a "bustle" that covers the tail feathers. (Young Sandhills have a brown neck, and their heads are entirely feathered, with rufous worked into the gray body plumage.) In flight, the crane extends its neck, whereas the herons and egrets pull their long necks into an S curve shortly after taking to the air. The powerful, deep-throated trumpeting of cranes may be heard as they fly between roosting and feeding places or while they are foraging on the ground.

Perhaps at one time Tepimans had more occasions to see this elegant bird. The once extensive Pima wheat fields and the large open marshy and grassy areas formerly along the middle Gila would likely have attracted the birds. The many islands in the Gila when the river ran full may have provided the essential night roosting places. Today cranes are highly localized in a few favored places in the desert, both in Arizona and Sonora. Crane populations worldwide have tended to decrease with the onslaught of "civilization."

In more than thirty-five years of bird work in Piman country, I have never encountered cranes there. But there is no doubt that the People once knew this bird. In the story related by Saxton and Saxton (1973:98–100), Coyote is caught in a flash flood and takes refuge in a tree. Then a crane, *kookoḍ*, arrives and sings:

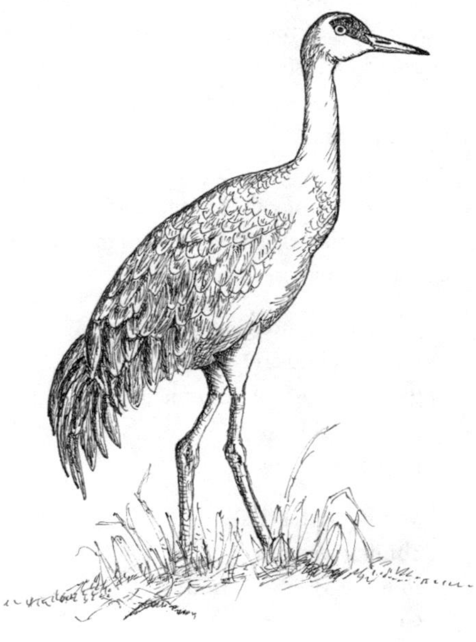

kookoḍ, Sandhill Crane
(Amadeo M. Rea)

Haa! Lulumii.
Haa! Lulumii.
Wat o komalii.
Wat o komalii.

Hey, settle down!
Hey, settle down!
It will run shallow.
It will run shallow.

I have not heard this song sung, but just reading the words with their resonating vowels suggests to me the rattling, trumpeting call of a crane. In no way does this sound resemble the seldom-heard sharp squawk of a heron.

Shaw (1968:79–81) gave a somewhat more elaborate story of Coyote causing a flood, then being saved by long-legged birds she called herons (their Piman name is not given). Coyote says derisively to one of the birds that is carrying him to safety, "*Ge u'us kakiokam*, big stick legs you have." The bird dumps Coyote into the river, and he drowns.

In his 1877 Pima dictionary, Captain Grossmann listed "crane" as *kawkurt*, a close approximation, given his orthography. The Reverend Cook (1943), another German from that time period, likewise identified *kawkurt* as crane. Modern lexicographers have preserved the association of crane with Piman *kookoḍ* (Mathiot 1973:489; Saxton and Saxton 1969:165), but with increasing fuzziness as familiarity with waterbirds has declined.

I never asked most of the older Pimas on the Gila about *kookoḍ*. The single exception was George Kyyitan, who said he thought it was something like a *vakoañ*, mentioning it in a discussion of *kukulisgam*, an unidentified wading bird (fide Culver Cassa). On the Gila, however, most folk generics for waterbirds are well defined: *vakoañ* (heron/egret), *vachpik* (coot), *vapkagĭ* (duck), *hialak* (goose), with various synthetic names for other waterbirds such as gull and sandpiper.

Sandhill Cranes are to be expected farther south along the middle Río Yaqui. María Estrella (Pedro Estrella's daughter) and Carlos Gareliz, another part Pima from Ónavas pueblo, confirmed that a few (one or two or up to as many as five or ten) gray grulla come in winter months to cornfields along the river and that the People do eat them, "but [they] are very wild, hard to get!" In contrast, Pedro Estrella had told me that the *vokoan*, heron, was not eaten. "La carne es pura negra."

COMPARATIVE LINGUISTICS. Although the O'odham word for "crane" appears constant through time and space, the English referent seems to range widely through a number of birds associated with water. I think that rather than this being a problem of the biological referent, it is a problem more of the native speaker not knowing how to gloss *kookoḍ* in English and of the naive elicitor taking the response at face value or supplying a misidentification.

For instance, J. A. Mason (1950:30) wrote "geese" (!) for Juan Dolores's *kohkoht*. However Dolores (1936), in a paper on Papago nicknames, gave *kohkohT* as "[a] tall, slim, long-legged person," but also glossed it in English as "goose." Dolores, whose first language was O'odham, was not all that up on English folk taxonomy: a goose is squat, plump, and short-legged! In an earlier paper, Dolores (1923) had defined the word as "geese, crane," as did Kurath (n.d.).

Saxton and Saxton (1973:98), in a bilingual recording of Tohono O'odham legends, identify *kookoḍ* as "crane," describing it as bird with very long legs to stand in water. But in another work (Saxton and Saxton 1969:165), they identify the taxon more broadly as "crane, goose, heron." Does it really mean all these birds, or are speakers just getting fuzzy about their concept? In a subsequent dictionary, Saxton, Saxton, and Enos (1983:33) give *kookoḍ* as simply "goose" and *kookodmad* as "gosling." But, as with Dolores, we have a clue that goose is not the referent of the native concept. These authors record the name for filaree, *Erodium*, a small weedy annual in the desert, as *kookod-oipij*, "goose's awls," or *kookod-hoho'ibad*, "goose's needles." But the shape of the filaree seed head really suggests a crane or heron, *not* a goose, as both the Linnean generic and English names indicate.

Simon Lopez also called *kookoḍ* a goose, saying it was gray, big, with long legs. It was one of his three kinds of rain birds, the others being the swallow and the *toha u'uhig*, a white bird from the ocean (gull or egret). Although he used the English word *goose*, his concept includes the Great Blue Heron.

Mathiot (1973:489) said *kookoḍ* is "either pelican, sea gull or crane." Did she mean it could be *any one* of these birds or just *one* of them? Did the consultants really intend to include all these birds in the domain of the word, or were they just confused about what to call them in English, perhaps no longer being familiar with the biological referent? We see a similar but complete shift in referent with the Scarlet Macaw, *aaḍo*, which metamorphosed into a "peacock" when Pimans became unfamiliar with the originally designated species.

Underhill, in her voluminous writings, frequently mentioned "crane" where surely she must have been referring to herons or egrets, the common birds appearing at desert streams and small bodies of water. Wright (1929:32) called his *kaw-kawt* "sea gulls." The birds in the two songs he gave most likely are gulls because they are described as white and associated with the ocean (see gull account). In a collection of Tohono O'odham legends, Saxton and Saxton (1973:98) call *kookoḍ* a crane, as noted earlier. In the Akimel O'odham Origin Story, Herzog ([1929]) gave the English word *crane* as the translation for the Pima words for both the Northern Harrier (a hawk) and the Great Blue Heron. Similarly, Benedict (2001:106), recording the same story, wrote "crane" for heron.

Although herons are somewhat common today on small ponds and charcos in the desert, few people encounter cranes there any longer. Among contemporary Tohono O'odham, both Luciano Noriega and Delores Lewis called the Great Blue Heron *kookoḍ*. Delores did not use the word *vakoañ*, the generic heron/egret word. Neither did Simon Lopez, who called these birds in English either ducks or geese. Whether they were using alternate names or telescoping various names into one folk taxon is conjectural.

COMPARATIVE LINGUISTICS. Grulla (crane) appears in several Sonoran-language lexicons. The colonial Lingua Cahita (Yaqui-Mayo) gave *Coorohe (ko'orowe)*, and the modern dictionary gives *ko'obwabsa'i* (Buelna 1890:163; Molina, Valenzuela, and Shaul 1999:209). "Stork" was *koró* among extinct Eudeve (Lionnet 1986:117; Pennington 1981:216). In modern Rarámuri, it is *go'oró* (Brambila 1983:283). All these names mimic the deep-throated rolling calls of the Sandhill Crane.

MISIDENTIFICATIONS. The reference to crane feathers in the translation of Juan Gregorio's oration seems to refer to the oriole; the word is *s-waakchu* (Bahr 1975:62–63).

My inclusion of the folk taxon *haiñ julshap (uulshab)* under Sandhill Crane was incorrect in Rea 1983:147; cranes are absent in southern Arizona during the summer (I now have it listed in appendix F, unidentified birds). This bird is most likely an American Avocet or Wood Stork.

vachpig, vachpik, vachpikdam

Rallidae, Gruiformes

Fulica americana

American Coot (Mudhen); Gallareta

A dark gray to blackish bird, looking superficially like a duck, finds its way to about any spot of permanent open water in the desert, especially if the pond provides some aquatic vegetation for cover. It congregates in numbers at ciénegas, streams, and ponds, including sewage ponds, calling loudly with a variety of semimusical clucks and chortles from clumps of cattails or bulrushes. Its whitish bill with red spots near the tip is conical, not flattened, and its greenish feet are lobed, not webbed, distinguishing it from all the ducks. This is the American Coot, popularly known as the Mudhen.

Coots dive well, both to escape and to feed. They gather seeds, vegetation, and small animals from the water's surface or below. When pressed, they will patter laboriously across the surface of the water, splashing noisily until finally taking wing. Once it is airborne, the flight is strong, and a white band shows at the back of the wing.

At one time, coots bred abundantly in the sloughs of the slow-moving middle Gila and at other oases that provided cover throughout Pimería Alta. The nest of floating vegetation, often placed in cattails, may contain a dozen or more buff-colored, spotted eggs. The ashy chicks, splashed with orange or red on the foreparts, are able to swim and dive at hatching.

River Pima, more familiar with waterbirds than were their desert cousins, distinguished *vachpik*, the Mudhen or coot, from *vapkaik* or *vapkakĭ*, the term labeling all the species of ducks.

I asked Joe Giff if he knew the Mudhen. "*Vachpik*. They can dive under water and travel under water. They don't fly up from the water—it flaps its wings *on* the water. They nest on the river. *Vach* means to dive. Something similar to a duck." He mentioned their fishy flavor, termed *s-do'i-uv*, a raw, wet, or unpleasant smell to the Pima, such as the odor of the ocean, fish, wet dog fur, or certain plants. He did not think coots were ever eaten by the Pima.

Sylvester Matthias, making the same distinction between coots and ducks, said he doubted he would have differentiated coots from Common Gallinules, *Gallinula chloropus*, if they had been present. This similar bird, also called a Common Moorhen, is somewhat more slender than a coot, has an olive cast to the upper parts, a brownish to red and yellow bill and frontal shield, and unlobed greenish yellow toes. Like the coot, the gallinule is a noisy bird, but because it requires even more emergent vegetation, it has disappeared from most of Pima country along with the river drying and water being lost. We are about a century too late to learn if this species was ever singled out in the Piman lexicon of bird names.

Sylvester and George Kyyitan were visiting the ethnobotany trail with me at the Desert Botanical Garden in Phoenix. A male and female Mallard were on the small pond. Both old men called them *vapkagĭ*. Also on the pond was a black coot that they pointed out as *vachpik*. They both said they had never heard of Pimas' eating it. The three birds were quite tame, so we had good close views for confirming the identifications.

Theodore Peters of Santa Cruz village called the coot *vachpikdam*, saying "just goes under water," adding that it was not eaten. He too contrasted it with *vapkagĭ*. Leonard Pancott of Haashañ Keek (Sacaton Flats) used the same form of the name, agreeing that Pimas never ate the coot and only occasionally the duck.

There is a song in the ant series from Salt River that elegantly pictures the placid coot (Bahr 1997:56, 76, 99):

Tondam shuudagĭ kaach
tondam shuudagĭ kaach
g eḍa vachpigdam oimmeḍ
miak o ñeid
s-ap o maasim viinog.

Shining water lies
Shining water lies
In it the Mudhen wanders
Come and look
Nice looking it floats.

Coot makes a cameo appearance in a story where Kadgam, Yellow-billed

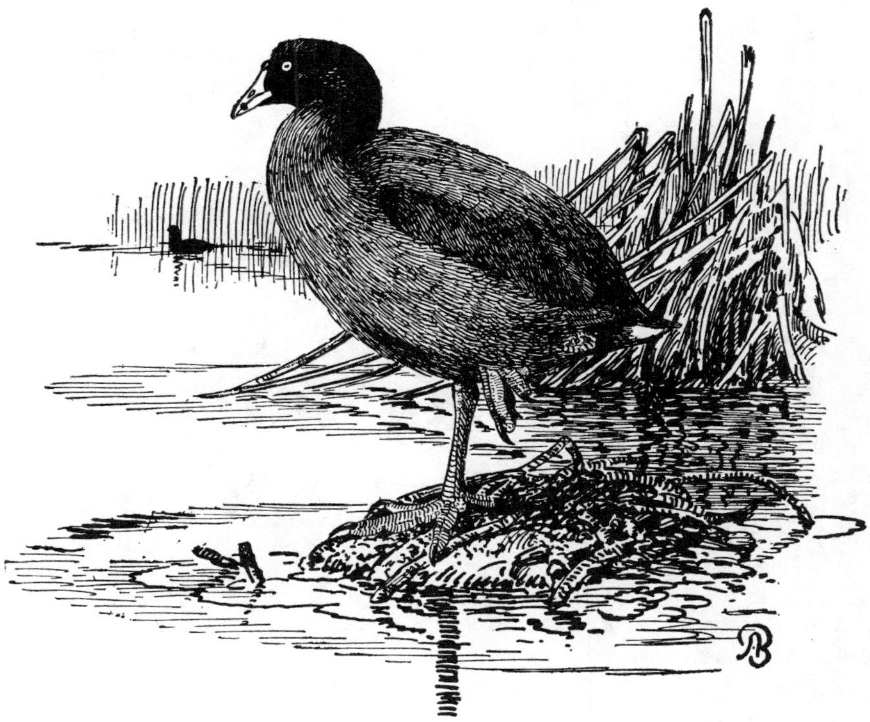

vachpik, vachpikdam, American Coot (Allan Brooks)

Cuckoo, comes to the evening meeting with his estranged wife, who has recently returned. The keeper of the smokehouse suggests that Kadgam bring her to teach the men the rain songs she acquired during her absence. For a woman to attend the men's meeting was not acceptable practice, however:

> She said to her husband, "How shall I go in?" "Step right in." She changed herself into a duck [*sic*] *(vaatac kopik)* and sat on her husband's shoulder. They saw him come in alone. Those who had wanted him to bring his wife to the ceremonial house said, "We wanted you to bring your wife and you have come alone." He answered, "No, she is with me. Put more wood on the fire and you will see her." . . . She began to sing. It was very beautiful. . . . Outside it began to rain. She said to them, "These are the songs to bring rain." They believed her. (Benedict 2001:71–72)

The seemingly random associations in this story share a common metaphor: wetness. While the woman was estranged from her husband, she had lived with Dragonfly, Buzzard, duck (or coot), and turtle. Her original husband, Kadgam, is a summer bird, associated with the monsoons. She also lived with all the different kinds of corn, summer plants requiring wetness. When she at last returns to Yellow-billed Cuckoo, she comes with something useful for the community. She enters the men's communal house as a waterbird. The treasure that she brings are her songs that produce rain.

In a variant of the Killdeer origin story, two brothers gamble, one killing the other, making a utensil from his skin. The dead man's son challenges his uncle to a gambling contest, beats him, and kills him. When this boy tries to resuscitate his own father, the father says he is too far gone, but tells his son to take his skin down to the stream and throw it in, where it becomes a *vachpik*, the coot (Benedict 2001:87n., see also 155).

There is a Piman story of the twin boys and their quest for *vaapk*, Common Reed, to make flutes. When the boys reach the ocean, the Carrizo canes are far out in the water. The younger brother turns himself into a *vachpik*, coot. "The elder brother said, 'I thought you could do better than that. You can't go as far as the reeds under the water. You will lose your breath and die'" (Benedict 2001:107; in Herzog's version [(1929):65a, 65b] the elder brother turns into a gray water snake instead). Although this story is borrowed from Maricopa or Yuman sources, it incorporates an important behavioral characteristic of coots: their ability to submerge and swim under water, as their O'odham name indicates.

As might be expected, the distinction between coots and ducks is less clear among Tohono O'odham, whose country lacks larger rivers or lakes. But at Quitovac oasis, where waterfowl are common, village patriarch Luciano Noriega gave me *vachukuk* for ducks and *vachuk* for coot; he also used *chuk paado*, "black duck" (from Spanish *pato*), as well as *chuk vachuk*. Gary Paul Nabhan was given *wapkik* for coot. I suspect that Luciano's eyes were not all that good and that we did not get an accurate contrast set. Another time I was at Quitovac, Luciano told me he didn't remember the name for coots, which were common on the pond near us, but said that they ate coots "antes—mas bueno como patos," indicating that the breast muscles were good to eat. He mentioned a bird he called *vachmukdam*, which he said is smaller than a duck and dives under the water, most likely a grebe. (The root word is the same as for River Pima coot: *vachum[u]* plus *dam*, to submerge itself and to do it many times on different occasions.) Saxton and Saxton (1969:167) give an additional Tohono O'odham variant: *wakaig* (pl. *wapkia'ig*).

The remaining Pima Bajo speakers, Pedro Estrella and María Córdoba, identified pictures of the American Coot as *s-tuk vaka'ig* (or *vakaig*). Pedro called it in Spanish "pato negro de agua, [con] pico blanco." The Río Yaqui runs quite rapidly at this point, and I never encountered the coot while out with Pedro, so I lack a field verification. It seems he didn't know the *vachpik* (coot) versus *vapkakĭ* (duck) distinction. The Névome vocabulario (Pennington 1979:90) listed *toadakama* as a kind of duck. The name is *toha daak kamă*, "white bill-it-has," most likely the coot. Whether the European compiler or the Pima (or both) considered the coot "a kind of" duck we cannot now tell, but what seems to be a kingfisher word was included in the list as well.

I would have expected more songs or stories about coots among Akimel O'odham, given their long and close association of this confiding and vociferous bird. But I found no more. This lack is probably an artifact of transcribers' and translators' not recognizing the Pima word for coot.

COMPARATIVE LINGUISTICS. In the Yepáchic district of Mountain Pima, I was given by different speakers the words *vaki* and *vakiig* (the latter probably more correct) and even the noncognate *vamulyi.* It seems ducks and coots are not linguistically distinguished here or among Pima Bajo. Culver Cassa notes that the River Pima duck and coot words are permutations of the same root word. The coot's name means it dives under the water or submerges willfully.

The Attí vocabulario, composed by 1774 at this northern Sonoran Pima village, gave several names for sambutiasse, zambullisse, and zambullete, references either to grebe or coot (or perhaps to both). The variants are *vatpinu, vatpin, vatipin, vatpinum,* and *vatpinic*. These terms are contrasted with *vatpi*, pato, duck.

Moisés Valenzuela of Huirivis Yaqui pueblo called the gallineta *ba'a totu'i*. I presume the first part of the name refers to water.

***vakag, vapkagĭ;* var. *vapka'ikĭ* (River Pima); *wapka'ig* (Tohono O'odham)**
***vaka'ig* (Pima Bajo)**

Anatidae, Anseriformes
ducks (all species, possibly including geese and swans); patos

Many species of duck, both pond ducks and sea ducks, even occasionally mergansers, stop at ponds and streams in the desert, usually in transit to larger bodies of water. Some winter, and a few smaller species may even breed if conditions are right (sufficient cover and food). Bones of various species have been recovered from Hohokam archaeological sites (summarized in Rea 1983:155–158).

Because of the ecological abuses that Sonoran Desert wetlands have sustained in little more than a century and a half, it is difficult for today's observer to imagine what conditions were once like. The Gila, for instance, meandered across the desert for several hundred miles, providing oxbows, low islands, and backwaters for waterfowl. Lt. Col. William Emory (1848:92), for instance, writing from camp at Painted Rocks in November 1846, said,

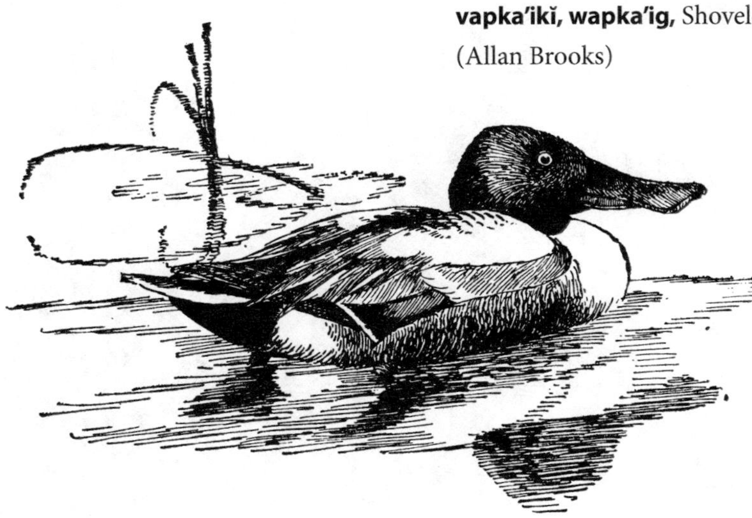

vapka'ikĭ, wapka'ig, Shoveler (Allan Brooks)

"The pools in the old bed of the river were full of ducks, and all night the swan, brant, and geese were passing." John Griffin (1943:35), a physician on the same march, noted: "The River here is some 60 or 80 yard wide—on an average 3 feet deep and rapid. We have seen more water fowl in the last two days, than we have yet met with on the River—ducks, brant geese & swan."

River Pima call all species of ducks *vapkagĭ*, even mergansers. In a group discussion, all agreed that there were no individual kinds named. "I guess they didn't think there were other kinds around," said Joe Giff. "I think they were all just waterfowls. Only *vachpik* is something different—Mudhen." In his vocabulary, Fr. Antonine (1935:8) listed under "domestic animals" *vapkak* as "ducks and geese," giving no plural form. George Kyyitan said goose was called *vapkagĭ*, just like a duck; Leonard Pancott and Irene Hendricks concurred (but see the next account). Theodore Peters said all ducks were *vapkugĭ*, which he contrasted with *vachpikdam*, coots.

Both Sylvester Matthias and George confirmed at the Desert Botanical Garden in Phoenix that the pair of wild Mallards on the small pond along the ethnobotanical trail were *vapkugĭ*, whereas the coot, also on the pond, was *vachpik.*

The Papago word has a long pedigree as well. Explorer-naturalist Carl Lumholtz (1912:373) appended a several-page vocabulary to his accounts of travels in Paguería, noting *vapkaik* as the name for duck (Mallard). (He did not use glottal stops.) Linguist William Kurath (n.d.), in a Papago word list compiled by 1948, wrote *vapka'ik*. In her Papago dictionary, Mathiot (n.d.:235) gives *vapka'ig* as "wild goose, wild duck, crane [*sic*]." Likewise, J. Alden Mason and Juan Dolores (1950:16, 30) used *vapka'ig* for duck. However, Saxton, Saxton, and Enos (1983:49) give *paado* as duck, *paadomaḍ* as duckling. *Paado* is a loan word from Spanish. Saxton and Saxton (1969:166) list *wakaig* (pl. *wapkia'ig*) as the Mudhen or coot. At Akĭ Chiñ, Juanita Norris looked at my elicitation notebook and said all the ducks would be called *paapado*, but that the real name, used a long time ago, was *vapka'igĭ*. She included the coot in this category as well. As nearly as I can tell, the use of the Spanish loan in place of the traditional *wapka'ig* is as a replacement name rather than as an attempt to distinguish the domestic bird from the wild one. So much for honoring the elders! (For a similar situation with the Javelina, see Rea 1998b:233–237.)

Still farther south, Pima Bajo Pedro Estrella called all ducks *vaka'ig* or *vaka'ik*, adding *shuudag vapka'ig*, "water duck." (For the Pima Bajo, Johnson [n.d.] wrote the word without the glottal stop.) Pedro and María Córdoba called the coot *s-tuk vaka'ig*, "black duck." River Pima were well aware of the annual arrival of ducks in their country.

Ducks appear early in the Creation Story (Benedict 2001:18; Lloyd 1911:48). After the Flood, the three Original Beings, now without Ñui, Buzzard, who does not care to participate, sit down, backs to each other, to make dolls from clay. I'itoi asks Tohavs—that is, Ban (Coyote)—why he made dolls with webs between their fingers and toes; they could not point directions. "But

Toehahvs said he had made these dolls for good purpose, for if anybody gave them small seeds they would not slip between their fingers, and they could use the webs for dippers to drink with." His two companions are not convinced by this argument. "And [I'itoi] took Toehahv's dolls and threw them into the water and they became ducks and beavers" (Lloyd 1911:48).

The version recorded by Russell (1908:214), also from Pima narrator Thin Leather, is similar. After the Flood, the three culture heroes, again without Ñui, sit back to back, making new beings. "Coyote had made all the web-footed animals [mammals], snakes, and birds." Then "Elder brother [I'itoi] told Coyote to throw the animals which he had created into the water."

An episode is recorded later wherein two brothers, Yellow-colored Coyote and Sandy-colored Coyote, gamble with each other. They decide to bet on a kickball race. In this game, when the runner encounters his ball, he rolls it up over his toes then thrusts it forward from the top of his foot. One brother has Chukukui, Black Beetle, as his runner, and the other has Vapka'igĭ, Duck. This is the Piman version of the tortoise and the hare tale, but Black Beetle has a dream the night before that he will win, and indeed he does (Benedict 2001:78–79; Russell 1908:232).

Opinions were mixed among River Pimas about whether ducks were eaten. A late version of the Creation Epic recorded by Julian Hayden (1935:8) suggests that they were. In repopulating the earth after the Flood event, Jeweḍ Maakai, Se'ehe (Elder Brother), and Coyote sat down to make creatures. Coyote made some beings that Se'ehe did not like, so he "threw them into the ocean, and those were the fishes and the ducks, and he told Coyote that they were just food for the people." Fish were the main animal food of the River People before their river went dry, so by implication ducks were part of the dietary as well.

Once the Pima acquired shotguns, Joe Giff and Sylvester Matthias said, ducks were eaten at least occasionally. But they were not certain that they were traditional Pima foods or that they were ever taken with bows and arrows. Some years later in a talk, Sylvester enumerated what he considered Piman game birds: White-winged and Mourning Doves, White-crowned Sparrow, and Lark Bunting. Then he added, "The modern generation now eat[s] ducks, but not traditional." Duck eating apparently came as a Euro-American influence. Pimas knew that the *vachpik*, coot, had an unpleasant taste, suggesting that they at least had tried eating these waterbirds as well. I asked Ruth Giff if her family had eaten ducks when she was young. "Not often. Just once in a while, I think, when they find one in the river. Just stew it up, I think." And Leonard Pancott commented, "Pimas are not really crazy about them. Something they start eating later, I guess. Fry them maybe." But *vachpik*, he added, were never eaten.

Their impression about the recency of eating duck is borne out by the remarks of Fr. Pfefferkorn (1949:117), who was stationed among Pimans of southern Arizona in the early 1760s: "At times there are large numbers of wild ducks of all colors and species on the ponds, sloughs, and other bodies of water, where they swim very safely and unconcernedly. They are not at all shy, because no one disturbs their peace or hunts them. The Indians much prefer a roast rat to the best wild duck, and the Spaniard is thoroughly satisfied if he has beef or cow-beef on his table."

Rosita Brown of Santa Cruz discussed hunting taboos and their effects on the hunter's unborn child. When his wife is pregnant, the husband is not supposed to shoot deer (that is, *huai*, the Mule Deer); *chuk chuuvĭ*, Black-tailed Jackrabbit; cow; *cheshoñ*, the Desert Bighorn; fish; or duck. One must even chase away a rattlesnake, not kill it, she said, lest the newborn be affected. If the *vapka'ig* is violated, a duck-child will be born with webs between its fingers. Someone in Santa Cruz born like that died when just fourteen years old.

I have heard no songs about ducks among the Akimel O'odham, where I would have expected them, but Underhill (1946:256, 274–275) recorded two in her studies among the Tohono O'odham. In neither is the O'odham bird name given for verification of the folk taxon. The first song (number 5) is for the girl's puberty ceremony:

> Lo, it is the duck children!
> In the water they swim.
> Thither I run and hear them talking.
> The water is murmuring.

The other is a shaman's divining song (number 22):

> There will be a flood.
> The water is roaring.
> Then a duck girl came out.
> She talked; she wept.
> While around and around the white
> mountain
> She went circling.

In describing contemporary (ca. 1932) ceremonies for the dead, Underhill (1946:323) noted, "Some families keep dishes at a grave from year to year, and one dish seen by the writer was in the ancient ceremonial duck form." I have no clarification of this report.

A few species, such as the Cinnamon Teal and the Ruddy Duck, may breed in Pimería Alta when conditions are right (Witzeman, Demaree, and Radke 1997). Most are migrants. Sylvester said, "As soon as the ducks and geese come to the river, they say it's wintertime. Later on I found out that the lakes in the north are frozen."

COMPARATIVE LINGUISTICS. The colonial Névome vocabulario (Pennington 1979:90) gave pato as *vakaika*, nearly the same as its modern equivalent here. Several other duck names are listed: *toadakama* and *baitituri*. The former, *toha daakkamă*, means "white nose (bill)–it-has," surely in reference to the coot. The second, notes Cassa, may be the kingfisher (now *ba'ivchul* in the north). The word for patos peruleros (Peruvian ducks) was given as *sudagui kokoni*, "water raven/crow." I suspect that it was the name for cormorants. These birds occur along the Río Yaqui in the vicinity of Ónavas, where Pedro Estrella identified them as *s-tuk vaka'ig*, the same name he used for coots. In Spanish, he called the cormorant *corvejón* (*Corvus* is Latin for crow and raven). Linguist J. Alden Mason (1950:16, 30) wrote *vapkaig*, without a glottal stop, in Papago for "ducks."

The Attí Pima vocabulary from northern Sonora, circa 1774, listed pato as *vatpi* after a series of related words that label either grebe or coot (see American Coot account).

Several duck words appear in the Eudeve vocabularios: *babítsi, bavítzi*, and *humuviri* (Lionnet 1986:121; Pennington 1981:143). Among the various Cahita pato names given by Eustaquio Buelna (1890), only *bapo moatela* may be at least partially cognate with Piman names. For modern Yoeme (Yaqui), Molina, Valenzuela, and Shaul (1999:215) list the Spanish loan *paato* and *vetayeka* (*vakoni* is given as well, but this is obviously a heron/egret word). The Rarámuri (Tarahumara) equivalent is *basoná* or *wasoná*. The aquatic signature *va-* is prominent here.

In his Northern Tepehuán vocabulary, Rinaldini (1743) listed ganzo, goose, as *a'a*, a word I have not encountered elsewhere. Also given was "pato o ganzo," duck or goose, *bajaigue (vahaik* or *vakaiki).*

hialak

Anatidae, Anseriformes

Anser caerulescens

Snow Goose (all geese?); Ganso

As any mystery enthusiast knows, evidence sometimes comes from unexpected sources. For some years, the Pima folk taxon *hialak* rested in this book's manuscript in the appendix of unidentified Piman bird names with the following conclusion: "The present evidence is contradictory. It is unlikely the biological referent of the name will be resolved. In the 1870s, the Gila Pima began losing their life-giving river, fewer waterbirds were encountered, and much of the associated folk taxonomy collapsed."

Sylvester Matthias was the first to call this name to my attention. He said *hialak* came in flocks, was a waterbird, white, a fisher, perhaps a heron or crane. This description was quite vague. The Wood Stork; White Pelican, *Pelecanus erythrorhynchos*; Black-crowned Night-Heron, *Nycticorax nycticorax*; and various egrets might fit that description. Later, in a telephone conversation, he gave me this additional information: "Don't see them anymore; black wings; black underneath and tail, 'wearing' a white collar on back, similar to duck family; always in group formation or rank; always together. Some behind. Fishing birds." He thought "*maybe* yellow bill. Not too long legs, not as long as *vakoañ* or heron; stand on land, but not a stork; make noise almost like a crow"; he thought they were year-round residents.

This description was somewhat more definite. Some of what he said suggested the Canada Goose *(Branta canadensis)*, and the calls of some Canada Geese could be interpreted as *hialak*. But the white-and-black pattern he emphasized fit better a Wood Stork or a Snow Goose. The voice, however, definitely ruled out the stork: storks and their kin the vultures are silent. Besides, Sylvester usually had very good intuition about assigning organisms, whether plants or animals, to what he termed families. The White Pelicans are white-bodied with black primaries and outer secondaries and with a yellow bill; they are fishing birds, but they, too, are essentially silent. The word *hialak* itself also suggests the trumpeting sound made by cranes, but that bird is well attested as *kookoḍ* in colonial vocabularies and modern discussions. And the Sandhill Crane lacks black in the wings.

hialak, Snow Goose (Amadeo M. Rea)

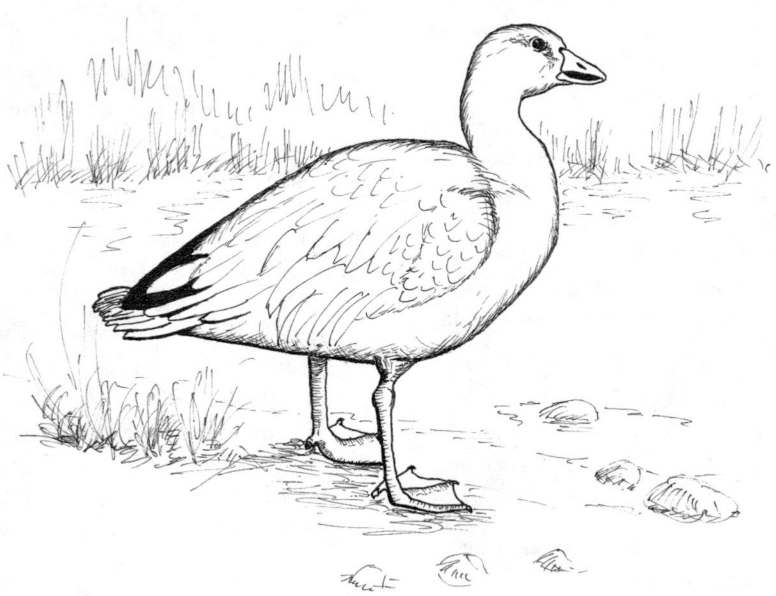

When Sylvester saw a *National Geographic* article on Snow Geese (Bartlett and Bartlett 1973) at the school, he was convinced that this bird indeed was his *hialak*. Here is the wording he worked out with me for the Snow Goose entry in *Once a River* (Rea 1983:156): "a flocking white water bird with black wings that flies in formation. It once occurred on the Gila and lower Salt. The name refers to the sound these birds make when flying over."

Scientists are sometimes too skeptical for their own good, but this attitude goes with the turf. I wondered, Does anyone else know the *hialak?* Leonard Pancott of Bapchule had heard the name, but said, "Not here—hangs around the coast." George Kyyitan, also from Bapchule, gave more information: "A 'stork'—a big bird almost the same as a *vakoañ* —they have very sharp beaks, bills; spear the fish." But Bapchule was an upstream village, and the river here had essentially died before either of these men was born.

In a mythological narrative (Herzog [1929]:63–64), the twin boys approach the ocean. They see some *u'uhig vapkoañĭ*, "heron birds." The boys are supposed to be frightened, but say, "You are just crane birds; cranes [*u'uhig híarak vapkoañĭ*]. They went on. The birds just scattered in all directions." The word *vapkoañĭ* is the plural "herons." In the Pima version, the two bird names seem to stand in apposition, so the *hialak* should be some kind of heronlike bird, as George and even Sylvester once said. In the English version, a semicolon separates the two bird names. This is ambiguous. But this mythological account, too, was taken down at an upriver village many decades after the Gila went dry.

The several colonial Jesuit vocabularios gave nothing I could associate with this word. However, Frederick Grossmann compiled a Gila Pima dictionary in 1877 in which he listed *híarlak* as "goose" (whether domestic or wild is not stated) and *hίalak a'an* as "goose quill for writing." He gave *vapokick* for duck. But for gander, the bird, he wrote *vákuan*, which is heron—thus initiating a question regarding his reliability.

While proofreading the manuscript to this book, I came across a note by Russell (1908:269) listing some unpublished River Pima vocabularies. Two of them I had never seen, one by Ammi M. White, a trader who had a flour mill at Casa Blanca from 1860 to 1870, and another by John D. Walker, who lived in that neighborhood from 1863, had a Pima common-law wife, and free-wheeled in relating Piman mythology (see chapter 3). Each vocabulary gave just less than two hundred words. I fired off a note to the National Archives at the Smithsonian, where the documents are housed, asking if there might be any bird names on either of these lists.

A quick response from Daisy Njoku brought news that there were only two bird names in the Walker list (MS 623-b): *he-a-lak* for wild goose, contrasted with *vap-o-kik*, duck (Mallard). The White document (MS 625a–c) gave just four bird words, including, fortuitously, *he-a-lak* for wild goose and *vap-e-kik* similarly for duck (Mallard). Here were now three nineteenth-century documentary sources confirming Sylvester's duck versus goose distinction. (Another mystery remains, however: at the Smithsonian, someone other than the original authors wrote in the word *wild* before *goose* in both the Walker and White manuscripts.)

Some months later Dr. Andrew Darling phoned saying that the Gila River tribe had acquired a copy of a Pima word list compiled by Rev. Charles H. Cook in the late 1800s (Cook [1943]). I asked him to check one word for me, *goose*. "Humm, here it is, *hí-arl-ák*. And goose quill is *hí-arl-ák-á-an*." The duck word he gave, *va-kdkj*, is actually the coot. Again, this result was fortuitous: Cook recorded only fourteen bird names among the thirty-six hundred entries of his vocabulary.

By 2004, most of my River Pima teachers had passed on. On a visit to Leslie Pasqual, whose ranch is near Sacaton, I asked if he had ever heard of *hialak*. He had heard that it was some kind of bird.

"Do you know what kind?"

"I think it's some kind of waterfowl." So the folk taxon has not vanished from the Pima lexicon yet! (Leslie, a mere seventy-seven at the time, also knew the original Piman name for Javelina, *tasi'ikol*, although it is now labeled by most speakers with two Spanish loan words, *miischiñ koogi*.)

There is still another mystery that is not likely to be resolved. In Piman thought, what is the relationship between *vapka'ikĭ*, duck, and *hialak?* Is the duck domain partitioned, or are these folk taxa mutually exclusive, as is the coot? The generation with whom I worked, born in the early 1900s, said all Anseriformes were included in their term *vapka'ikĭ*, hence *hialak* would have to be a special kind within this domain. But how did their parents cut the cake?

The Snow Goose is a medium-size goose, all white except for its black primaries and pink feet and bill. It once visited the Gila and lower Colorado River in great numbers. In mid-November 1846, Lt. Col. William H. Emory (1848:91) marched down the Gila on military reconnaissance to San Diego. At camp below Gila Bend a few days after leaving the Pima villages, he wrote, "The bottoms of the river are wide, rich, and thickly overgrown with willow and a tall aromatic weed [Arrow-weed?], and alive with flights of white brant, (wing tipped with black,) geese, and ducks, with many signs of deer and beaver." *Brant* is the word an American would have used for a smaller goose.

The Snow Goose's status in Arizona is given as "[f]ormerly common winter visitant to Colorado and Gila Valleys, now reduced in numbers" (Phillips, Marshall, and Monson 1964:11; see also Monson and Phillips 1981). The few that appear in Pima country today, mostly on impounded water, are seen from November through mid-April (Witzeman, Demaree, and Radke 1997). Bones of the Snow Goose as well as of the Canada and White-fronted Geese *(Anser albifrons)* have been recovered from Snaketown, a Hohokam archaeological site on the Gila Reservation (McKusick 1976). Whether all geese are included in the Piman concept of *hialak* (plural *hihialak*) or just the Snow Goose (the most common species) is we now have no way of knowing.

COMPARATIVE LINGUISTICS. The Névome vocabulario gave the Pima equivalent of ánade and ansar—that is, goose—as *sabarhicuha ondama* (Pennington 1979:8, 9). This word does not fit any of the modern Piman names. Culver Cassa interprets the name as *savatk hondam*, "that which has a thick body." This is an apt descriptive name, particularly for domestic birds. None of the several words given for pato (duck) is cognate with *hialak* (see previous species account).

In the Yuman Origin Story, John Peabody Harrington (1908:336) translated *kelák* as "goose." For the Cocopa just to the south, it is *yaalák*. This name seems cognate with the Pima one, though both are doubtless onomatopoeic. I have found no other animal names that Pimans held in common with the Maricopa or other Yuman-speaking neighbors.

chivi-chúuch; sivi-chúuch; piwi-chúuch

Charadriidae, Charadriiformes

Charadrius vociferus

Killdeer; Tildío

There is one plover, the Killdeer, that is ubiquitous throughout Northern Piman country year round. This plump, robin-size but rather long-legged waterbird is terrestrial, running around stream sides, charcos, irrigated fields, and even lawns close to dwellings. It may also be found in quite dry places such as fallow fields. It is so strikingly marked that even the more casual person is likely to notice it. The tan and white body tends to conceal the bird, but the boldly marked head, double black-banded chest, and rusty-colored rump attract attention as the bird moves about. Its eyes are large and surrounded with a red ring. If these markings don't attract one's attention, then the Killdeer's loud voice will. It calls even while flying about on moonlight nights. Its English, Spanish, and Piman names all are attempts to mimic this bird's calls.

Typical of plovers, the Killdeer has no nest, just a scrape where the four eggs are laid on the ground. The heavily spotted eggs are almost impossible to see, even when one knows where they are. So are the incubating adults and the precocial chicks. Any predator or person approaching the nest is led away with a display by the parent, dragging a faked broken wing to one side while calling vociferously. Black and white wing bars effectively add to the distraction display.

River Pima Francis Vavages noted, "They [the Pima] don't eat it. It eats the bugs. They know they eat worms. Anything that eats grain, wheat, like that, is good for the People [to eat]."

Juan Dolores (1936), writing on Papago nicknames, said of the *chivi-chúuch*, "A kind of bird which runs around in the water. A girl who has the bad habit of wearing short dresses."

chivi-chúuch, sivi-chúuch, Killdeer (Allan Brooks)

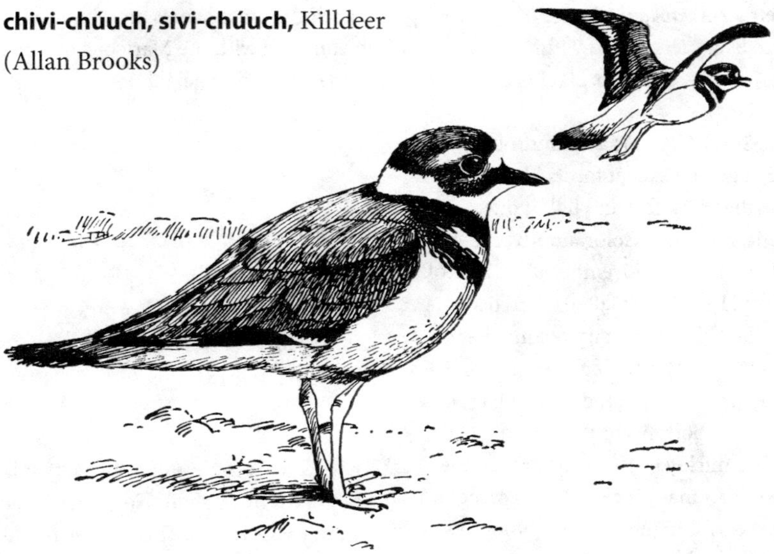

Delores Lewis of Ge Oidag said, "The *piwi-chuuch* runs along the edge of the water. Used to be around." At Aangam, Leon Juan called the bird *chiwi-chuuch.*

Bahr and associates (1997:50, 99; slightly amended by Culver Cassa) recorded this song among a collection of ant songs from Salt River Pima:

> Komagi chivichuuch u'uhig
> gan hu wa jeweḍ hogid an med,
> shaa wa s-ñe'edam kaidam kuihu
> jeweḍ keihi
> a'an chevaj.

The literal translation Bahr provides is:

> Gray Killdeer bird
> Away land edge-at runs.
> Very songlike loudly crying
> Land strides
> Wing lengthens.

And a freer English rendition by Culver Cassa is:

> Gray Killdeer bird
> Away world's edge runs
> So songlike loudly crying
> Land striding,
> Wings spreading.

Anyone who has been led away by Killdeer feigning one broken wing will appreciate the image captured in this sung poem.

As noted earlier, one River Pima story, apparently of Maricopa origin, tells of two brothers who gamble, then one killing the other (see Benedict 2001:80–87, 155). The dead man's son vindicates his father in a gambling contest with his uncle and kills him: "He took the seed bag that his uncle had made from his father's skin and soaked it till it was pliable again. He took ribs of the giant cactus and stretched it to shape, and his [dead] father spoke to him. 'My poor boy, you have worked hard to bring me back to life. It is not possible; my bones are rotted away and I could not walk. There is nothing you can do. But take me down to the stream and throw me in and I will become *tsimitutch* [*sivi-chúuch*, Killdeer].'" Today you can still find this rawhide-colored bird, with its Saguaro rib structure showing below, running along the streamsides, calling plaintively.

COMPARATIVE LINGUISTICS. How River Pima pronounce this bird's name marks their origin. Kuiva O'odham (Downstream People) and Salt River People, both originating from the more western historic villages around 1874, say *chivi-chúuch*, whereas those Ta'i O'odham from Bapchule east will usually say *sivi-chúuch*. *Piwi-chuuch* is a variant heard among the Tohono O'odham. For instance, Delores Lewis of Ge Oidag used this form exclusively and had not heard of the others. Saxton and Saxton (1969:166) give *chiwi-chuuch*, and Mathiot (1973:212) has *chivi-chúuch*.

I have not found cognates farther south. Don Pedro Estrella called the bird *duuk-kakial*, meaning "rain [something]." The second word may refer to *kakiodĭ*, "legs," hence "rain running" or "rain's legs." I heard *telyish* and *telshi* among Mountain Pima (Isidro Cruz, who added that the bird was not eaten). This is also an onomatopoeic rendition of the two-syllable call that English speakers hear as "kill-deer" and Spanish speakers as "tildío."

A few other names have been recorded for Sonoran-branch Uto-Aztecans. The Yoeme (Yaqui) call it *tuivit* (Molina, Valenzuela, and Shaul 1999:233). Three names have been noted among the widespread Rarámuri (Tarahumara): *títuri, togérichi,* and *tucha* (Brambila 1983:547).

shuudagĭ maḍ,* pl. *shuudagĭ maamaḍ

Scolopacidae, Charadriiformes
All small species
sandpipers; Playero, Playerito, Chichicuilote

Tiny "peeps" in winter plumage (mottled gray and tan upper parts, white bellies) appear, sometimes in numbers, throughout desert Piman country. Virtually any spot of water affording them shallow muddy margins to run about and feed upon attracts them: the edges of rivers, creeks, charcos, irrigation tailings, and sumps. They are most common during spring and fall migrations, but a few can usually be found during the winter months as well. These sparrow- or finch-size birds are inconspicuous but active as they probe the mud for food with their short dark bills.

The most numerous species in Pimería are the Least Sandpiper, *Calidris minutilla*, and the slightly larger Western Sandpiper, *Calidris mauri*. More solitary is the Spotted Sandpiper, *Actitis macularia*, seen here in its unspotted plumage, teetering at the edges of even the smallest ponds. Several other species migrate or winter here but are much less common.

The Pima name for all these small shorebirds is *shuudagĭ maḍ*, "water baby," or, as more frequently heard, *shuudagĭ maamaḍ*, "water children." I find it one of the most endearing and fitting bird names coined in the Piman language. The name seems to be unknown away from the Gila River, and even there few now remember it.

There are at least three Piman sandpiper stories, and each involves transformations. The first was recorded by Lloyd (1911:41–42). Elder Brother, or I'itoi, creates Handsome Man, who goes about the country from one maiden to the next, bringing forth children from them overnight. At last he is tricked and sodomized by the daughter of a powerful shaman. He himself conceives and delivers a baby in the morning. The complex linguistic laws of Piman kinship classification are convulsed. The girl's parents are unable to refer to their grandchild by the kinship term normally used because their daughter is the father and their son-in-law the mother.

Handsome Man returns to I'itoi, but, being embarrassed, he first leaves his child under a bush. I'itoi tells him to go get his grandchild, *ba'amaḍ*. "And the young man-woman found his baby crying, and the tears from it were all over the ground, around. And when he stooped over to pick up his child he turned into a sand-snipe, and the baby turned into a little teeter-snipe." Thus began the Flood. We don't learn what the Pima names for these two shorebirds are. A sandsnipe could be any species of sandpiper. A teeter snipe is specifically a Spotted Sandpiper. However, these two designations suggest that the Akimel O'odham once distinguished at least two species of small sandpiper, perhaps singling out the distinctive little Spotted Sandpiper because of its habit of "teetering" while walking alone at the edge of streams. Most other small peeps passing through Pima country are gregarious. (Herzog's [(1929)] version, with Pima and English transcriptions, says at this point only that both individuals died.) Culver Cassa considers this transformation irrelevant, merely a deus ex machina to get these two characters out of the way so we can get on with the Flood Story.

Densmore (1929:77) found another transformation story among the San Xavier Papago. It was told by Sivariano Garcia, whose stories have definite River Pima affinities. In the story, a boy decides to follow his grandmother, who has crossed the ocean. Not finding a passage, he uses his bow magically as a bridge, crossing over. When his mother and aunt try to follow him, crossing over the bow, he twists the bow so they fall into the water. "He said they would always travel along the shore of the ocean and never be with him. They turned into two kinds of birds that run along the sand at the edge of the ocean." Unfortunately once again, what the names of these two birds are in O'odham is not stated. Here also is a suggestion that Pimans once distinguished two named shorebirds.

(Incidentally, John Bourke [1889:188] recorded this same story among the Mohave along the lower Colorado River, translating the metamorphosed birds, *okog*, as curlews [*Numenius americanus*], a likely possibility for the O'odham story as well.)

The third story was related by Sylvester Matthias as an explanation for this song:

Shuudagĭ maamaḍ
lyat i vo'i waha
ha'ichu n aagida
hab ali kaijena a
chepavi yuhuvi
hiosika momotaha
jujuli ihihi.

In English, worked out jointly by Fr. Richard Purcell and Sylvester Matthias, the song says:

Water Children
Came and told me something
They told me that a loose woman
With a burden of flowers
Winded her way away from me.

Culver Cassa's rendering is:

Sandpipers
Here came running
And told me something
Thus they said,
"Coquettish women
carrying flowers upon their heads
zigzaggedly are going [dancing]."

Sylvester told me, "My father had many songs, but when he died they went with him. He would sing this Sandpiper Ñe'i especially in the morning." With most of these pieces of sung poetry, we do not know the contexts in which they were originally received. This one is an exception. Sylvester said that around 1930 his father had a dream where he got this song. "Four little girls were lying along the shore, talking to him, in the water up to their breasts. When they get through talking, they just dive right into the water and came out *shuudagĭ maamaḍ*, running along the shores. [They were] naked, hiding their breasts—just sank into the water and came up [sandpipers]."

Sylvester's song is a so-called prostitute's song, a Piman genre with numerous examples. "Silly woman" or "loose woman" would be a better translation for *cheepa'avĭ* because the woman was not engaged in a commercial endeavor. She was lighthearted and not yet tied down to family obligations. The atmosphere of frivolousness is created between the little birds, the harlot, and the singer. Action is everywhere in this song: the Water Children rushing about carrying gossip, and the more sinuous and subtle loose woman carrying her flowers to attract the singer in the crowd.

Culver Cassa says that the man singing this song has somehow been freed of something; he has undergone a psychic liberation. That's what the women mean. The problem is gone, and he can deal with it now.

At the eastern Pima village of Haashañ Keek (Sacaton Flats), Leonard Pancott said, "I've heard of *shuudagĭ maamaḍ* in the song. It's some kind of a bird, that stays in the water. The river birds are gone."

shuudagĭ maamaḍ, Least Sandpipers (Amadeo M. Rea)

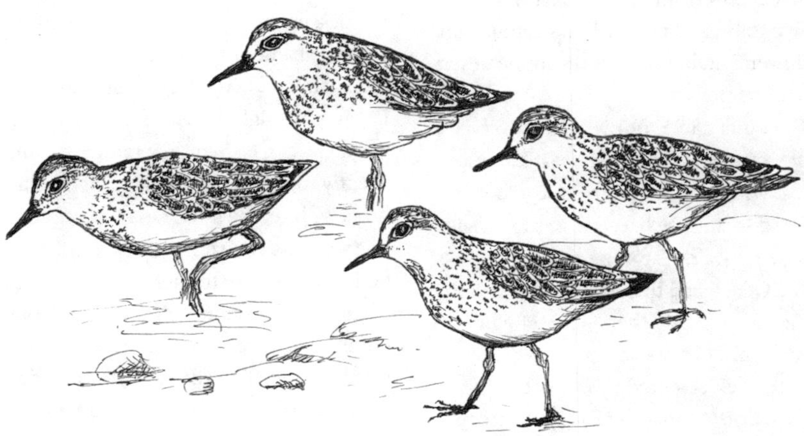

chevel u'uhig, Black-necked Stilt (Takashi Ijichi)

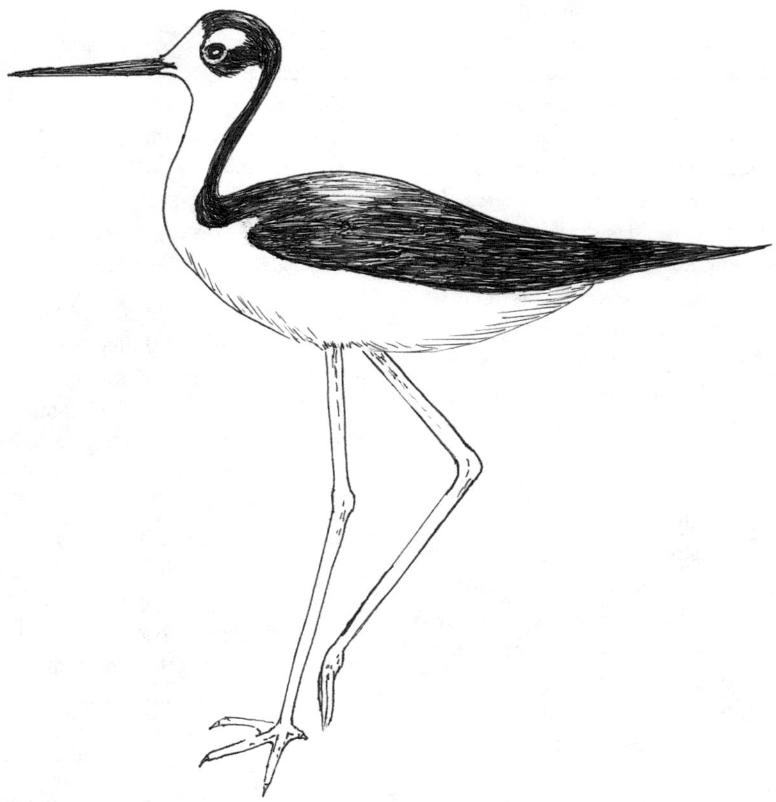

chevel u'uhig

Recurvirostridae, Charadriiformes
Himantopus mexicanus
Black-necked Stilt; Monjita, Candelero, Avoceta Piquirrecta

In the almost unbearable heat and humidity of some desert mudflats, bordered perhaps by Quail Brush, mesquite, or exotic Salt-cedar, a vociferous bird harasses the human visitor. The bird is black and white, with trailing deep-pink legs, and very long black wings. Soon it is joined by others of its kind, perhaps even by a few Killdeers, and the overhead alarm continues until the visitor retreats. Whether the birds are nesting somewhere on the salty flats or just thinking about it or have young hiding somewhere, one seldom learns.

On calmer encounters at an irrigated field or sewage settling pond, this delicate bird is seen to have an all-black nearly straight bill, a large rose-colored eye enhanced by a white eye spot, with black beginning on the crown, running down the neck to the mantle and wings; the rump and short tail are snow white. What confirms the identification is the pair of almost absurdly long rose pink legs. The birds are sociable, often in pairs, where the female may be seen as slightly grayish or brownish black.

In Pimería Alta, stilts are often accompanied by the similar sized but slightly more heavily built American Avocets *(Recurvirostra americana)*. These birds have blue-gray legs and a distinctly upturned fine bill, and they display an intricate black-and-white pattern across the wings and back in flight. If it is the breeding season, a cinnamon wash covers the foreparts of the avocet. Both stilt and avocet have been known to breed in River Pima country on the lower Salt, but most birds in the desert are transients.

The Pima name for the stilt, *chevel u'uhig*, means "tall bird," at least among Kuiva O'odham. A man from Komaḍk village had that name because he sang *chevel u'uhig* songs. Fr. Antonine (1935), working on the Gila, listed this bird's Pima name along with several others whose English name he did not know.

In the field, George Kyyitan of Bapchule suggested that the *kulisgam*, more often mentioned in the plural, *kukulisgam*, might be either the avocet or the stilt. Both species were on the pond we were watching that day. He said he knew of *che'echev u'uhig*, but had never seen them. Possibly his *kukulisgam* are avocets. But with the drying of their country for more than a century, the People had forgotten the names to many of their waterbirds. We are left with tantalizing names and fragmentary descriptions that can no longer be mapped confidently to biological species (see appendix F). Culver Cassa believes the word *kulis* comes from a word for "beak," perhaps in reference to its shape. The avocet's beak is recurved and needle sharp.

kaachki-tab u'uhig
kaachkĭ hugidag
kaachkĭ baasho u'uhig; shuudagĭ baasho u'uhig (Tohono O'odham)

Laridae, Charadriiformes
Larus spp.
gull, "sea" gull; gaviota

In the expansion from their ancestral Tepiman homelands, the O'odham Vupshkam encountered a number of new phenomena that they held in utter awe and that became sources of individual power: Apaches (Oobĭ), Golden Eagle, Prairie Falcon, and the ocean. But these things were so dangerous that they could also be sources of *kaachim mumkidag*, staying sickness, arising from their violations.

For these Upper Pimans, the ocean became a place of pilgrimage. Groups of men, rigorously observing ritual restrictions, went to a certain place at the head of the Gulf of California, carried out certain ceremonies, entered the water, offered ground cornmeal and prayer sticks to the ocean, then gathered crystallized salt nearby. During the night here, some might have private visions. Then they returned home silently in single file. On approaching their home village, the salt bearers were received like warriors returning from a raid. The salt and anything else brought from the ocean, such as shells, were purified in an elaborate communal ceremony; then the individual salt gatherers, like enemy-slayers, underwent a purification lasting from four to sixteen days or more, depending on the village (Underhill 1938:111–133; 1946:211–242; Underhill and associates 1979:37–69).

Salt pilgrimages waned early in the twentieth century. And so did most direct knowledge of the ocean biota. The number of gull names suggests considerable improvising for a group of birds not frequently encountered any longer. Sylvester Matthias provided the first three given in the heading. His first name, *kaachki-tab u'uhig*, means "at the ocean bird," meaning it is the attribute of the bird to be at the ocean. The word *kaachk*, ocean, literally means "where it [water] is spreading" (Mathiot 1973:439). His second name, *kaachkĭ hugidag*, may be more properly a gull name. The second word in this name means "edge" or "shoreline." His third name, *kaachkĭ baasho u'uhig*, means "beach bird"—that is "ocean chest [shore] bird."

All these names were elicited on a trip to the beach while Sylvester was visiting San Diego. "They are brave. When people are fishing [in California] they come right up to get the bait." He had never seen or heard of them on the Gila River. He said Pimas who came over to the coast or to the lower Colorado River to work on the dams saw these birds. And, of course, at an earlier time, those who went on the salt pilgrimage encountered them on the Gulf of California.

Leonard Pancott, also of the Gila, noted, "*Kaachki-tab u'uhig* is something that hangs around the edge of the ocean. They [the old people] didn't specify which [kind of bird]." Juanita Norris of Akĭ Chiñ community used *kaachkĭ baasho u'uhig*, one of Sylvester's names. On the salt gatherers' return from the ocean after the salt pilgrimage, she recalled, the village would *wusot* for a day to purify them and the salt.

On a visit to Laura Kermen at Topawa village, Gary Nabhan and I were admiring a set of Papago dancer figures she had fashioned from her pottery clay. Some carried rainbow arcs, others cloud effigies, and each of the two dancers in front of the procession bore aloft a white bird at the top of a stick. The birds were clearly gulls, and all the objects carried were water symbols. We asked about the gulls. She called them *shuudagĭ baasho u'uhig*, literally "water chest [shore] birds." She said she couldn't remember the real name, but used this phrase for her figurines. The parallel between her name, Juanita's, and one of Sylvester's, *kaachkĭ baasho u'uhig*—from people of different reservations—suggests an underlying communality rather than mere improvisation.

Densmore (1929:136) recorded a song that seems much like a *wiigida* song but is rather from the *chelkona*, Skipping Dance, petitioning fertility for the fields:

The great white birds that fly over the
ocean
came to our singing place,
And over the ground they strung about.

kachki-tab u'uhig, kaachkĭ hugidag, Ring-billed Gull (Takashi Ijichi)

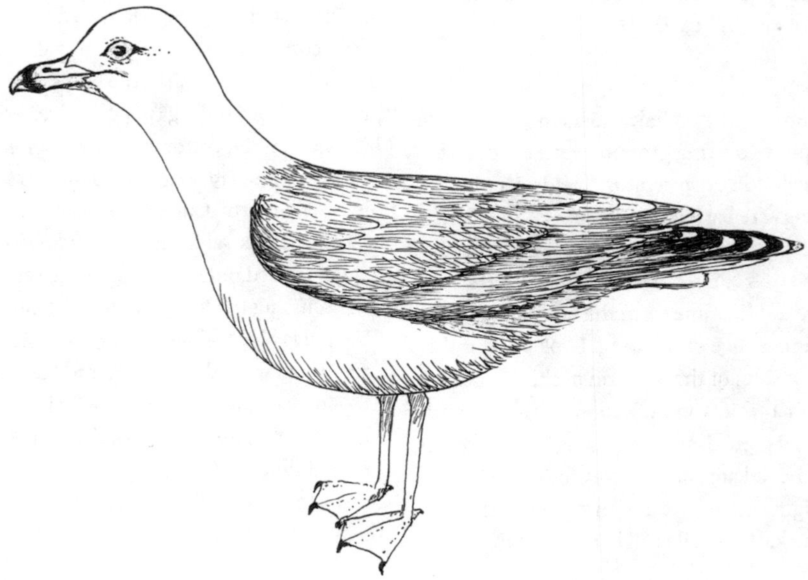

The birds referred to here seem to be gulls, but the original O'odham words are not given. Simon Lopez of Santa Rosa calls the white bird atop the *iagta*, or wand, in his dance group simply *toota u'uhig*, "white bird" (see egret account). They have yellow bills. He said he knew no more specific name for it. Among the other images is also a larger gray bird with a black bill, the *kookod*, and a smaller black bird with a white bill, the *giigdwal*, swallow or martin. Tohono O'odham call the egrets *toha u'uhig*. The other image is a Great Blue Heron.

Wright (1929:32) noted two "sea gull" songs in a story about the "bad people." In the first, the daughter of I'itoi and the children sit at the shore of the Gulf of California and sing:

Oh white birds who cross the water,
Oh white birds who cross the water,
Help us now to cross the water.
We want to go with you across the water.

Hearing their song, the gulls appear and sing:

Take these feathers that we give you—
Take these white feathers that we give
you—
Take the feathers floating round you
And do not fear to cross the water.

Binding these feathers round their heads, the O'odham are able to cross the water. It is said that these white feathers are still carefully kept. The name that Wright uses for the gulls is *kookoḍ*, which is really "crane" (see Sandhill Crane species account).

One of the main symbols of the ocean became these white birds, which in turn are metaphoric clouds. And clouds are precursors of that most precious desert commodity, rain. So the white birds on long sticks carried at the head of O'odham processions even to this day may represent gulls if not egrets.

Before we leave the gulls and powerful ocean experiences, one point needs to be cleared up. Underhill (1946:211, 212, 218) claimed that the River Pima lacked the salt pilgrimage. Luis Velarde (1931:127) suggested the same for the colonial period. However, in describing the Papagos' salt-gathering ceremony, Russell (1908:94) added that after the purification, "they were free to bring it [salt] to the Pimas. The latter sometimes made journeys to the lake for salt, being two days on the way to Quijotoa and two days on the trail beyond." Joseph Giff likewise also told me that the Gila Pima formerly made the pilgrimage.

One day, discussing crop plants, Joe's wife, Ruth Giff, remarked, "And they'd go from here for salt. They'd take an ear of corn [that] grows right to the end [the kernels reach the tip of the ear], grind it. Take that and sprinkle it on the water. (Not many of them [are] like that; I always think of it when I raise corn.) The water would move back so they could get their salt; then it would come back up, so they'd run. They'd have to be fast runners to get salt. That's what I heard. Maybe they go with the Papago, but I know that some *did* go. And I remember when I was a kid, they—Papago—would bring things to peddle here. They'd take anything [in exchange]."

So far as is now known, gulls themselves are not associated with staying sickness, but the ocean is. Ocean sickness might result from violations during the salt gathering itself (there were many restrictions) or from improper purification of the pilgrims on their return. Tohono O'odham Frank Jim noted these infringements, pointing out in his curing kit a *kokoḍk*, shell, and some *toḍshad*, foam, to be used in a *wusota* for healing ocean sickness.

Interesting but revealing fossil concepts are sometimes buried in myth. The songs and stories O'odham told about the Ocean, that powerful being, describe it as located in the West. But as both Akimel O'odham and Tohono O'odham set out to the ocean to get salt and personal power, they traveled south or south of southwest to the Gulf of California, where the sacred salt-bearing lagoons were located. A journey literally west would put them on the Pacific Ocean in the vicinity of San Diego. In the Flute Story, when the grandson of White-clay Eater Woman travels "west" to the ocean, he encounters "all kinds of birds." But it is one bird in particular that allows the crying-singing boy to obtain his goal: the Great Blue Heron. Standing in the middle of the water, growing large, and spreading his wings so that they touch both shores, the heron becomes a bridge for the boy to cross over (Benedict 2001:120–121; Herzog [1929]:96–97).

If the O'odham originated from a southern ancestral home, as all solid evidence seems to indicate, the ocean would always have been to the west of them, with the long Baja California peninsula on the opposite shore. When O'odham pushed northward into Pimería Alta, their traditional cosmology continued to associate west with the ocean, indicating the gulf, not the Pacific. Even when telling a story with Yuman Maricopa origins such as the Flute Story, their fossilized geography was imposed.

TECHNICAL NOTE. The Ring-billed Gull, *Larus delawarensis*, is the only species to be expected inland in southern Arizona and Sonora with any regularity (Monson and Phillips 1981; Russell and Monson 1998; Witzeman, Demaree, and Radke 1997). It is a medium-size gray-and-white gull in adult plumage, with yellow feet and a yellow bill bearing a dark ring near the tip.

On salt-gathering pilgrimages to the Gulf of California, Pimans were likely to encounter additional common species: the small Bonaparte's Gull *(Larus philadelphia)*; the medium-size Heermann's Gull *(Larus heermanni)*; and the larger California Gull *(Larus californicus)*, Herring Gull, *(Larus argentatus)*, and Yellow-footed Gull *(Larus livens)*. All are essentially white birds with gray mantles and black wing tips, except for the exquisite Heermann's Gull, Gaviota Ploma, which is overall sootier below and blacker above, with a white head.

haupal

Accipitridae: Accipitriformes (Falconiformes)

Buteo jamaicensis

Red-tailed Hawk; Aguililla Ratonera, Aguililla Cola Roja

A broad-winged hawk soars in wide circles, rapidly covering ground above the desert vegetation or agricultural fields. Its underwings are barred, but there are three dark areas: the wing tips, a line from the carpals or wrists along the underwing coverts, and another line from the body along the leading edge of the wing. Look also for the dark sides of the face and neck, contrasting with a white throat. Now watch as the bird tilts its broad, short tail, steering in circles. If the tail appears milky rufous below and bright rust above, it is an adult Red-tailed Hawk. If the tail is all barred and there is a heavy concentration of dark spots across the chest, flanks, and leggings, it is an immature Red-tail. Especially when others of its kind are in the air, you may hear its distinctive whistled scream: *keeer*.

This common hawk is often seen perched on utility posts along highways. It is heavy bodied, dark brown above (including the head), mottled and scaled with white. The underparts are variable. In Pimería, adults of the light phase, the most common form, are quite pale below, with only slight markings. But both melanistic (dark phase) and erythristic (red phase) birds occur here, even in the breeding season; these phases are entirely dark below.

The Red-tail is slow and deliberate. It preys on ground squirrels and other rodents, cottontails, jackrabbits, lizards, snakes, as well as many large insects, but almost never birds. Breeding pairs stake out large hunting territories. They build bulky nests of coarse sticks occasionally on cliffs, mesquites, or cottonwoods, but most often in the arms of a Saguaro. In winter, hawks from farther north in the continent greatly augment the local breeding population.

This soaring hawk is still rather well known among O'odham speakers, both Akimel and Tohono. This retention is likely because it is large and commonly seen on the reservations and because its rusty red tail, in adult plumage, makes it easy to identify. Its O'odham name, *haupal*, seems to come from the word *haupk*, meaning "light weight," but also to be associated with words indicating "agile." The root word appears in other biological terminology, as in the Tohono O'odham name for Sweet Bush or Bebbia, *hauk u'us*, meaning "light sticks," and the female fruits of the cottonwood or willow, *haupaldag*, which yield cottony airborne seeds. The plural form, according to both Sylvester Matthias and Fr. Antonine (1935), is *hauhupal*. Mathiot (1973) gives *hahupal* as a distributive form. Culver Cassa says *haupal* is already in plural form and that *hauhupal* would be used only in an aggregate sense.

Discussing the Red-tail, Joseph Giff said, "*Haupal* is bigger than a *vishag* [Prairie Falcon]. That's what they use their tail feathers for—for arrows—as with the *ba'ag* [Golden Eagle]. They raise them until they grow so big, and they pluck the fine feathers. Those fine feathers are called *viigĭ*. Then they let them free, and they can go wherever they

haupal, Red-tailed Hawk (George Miksch Sutton)

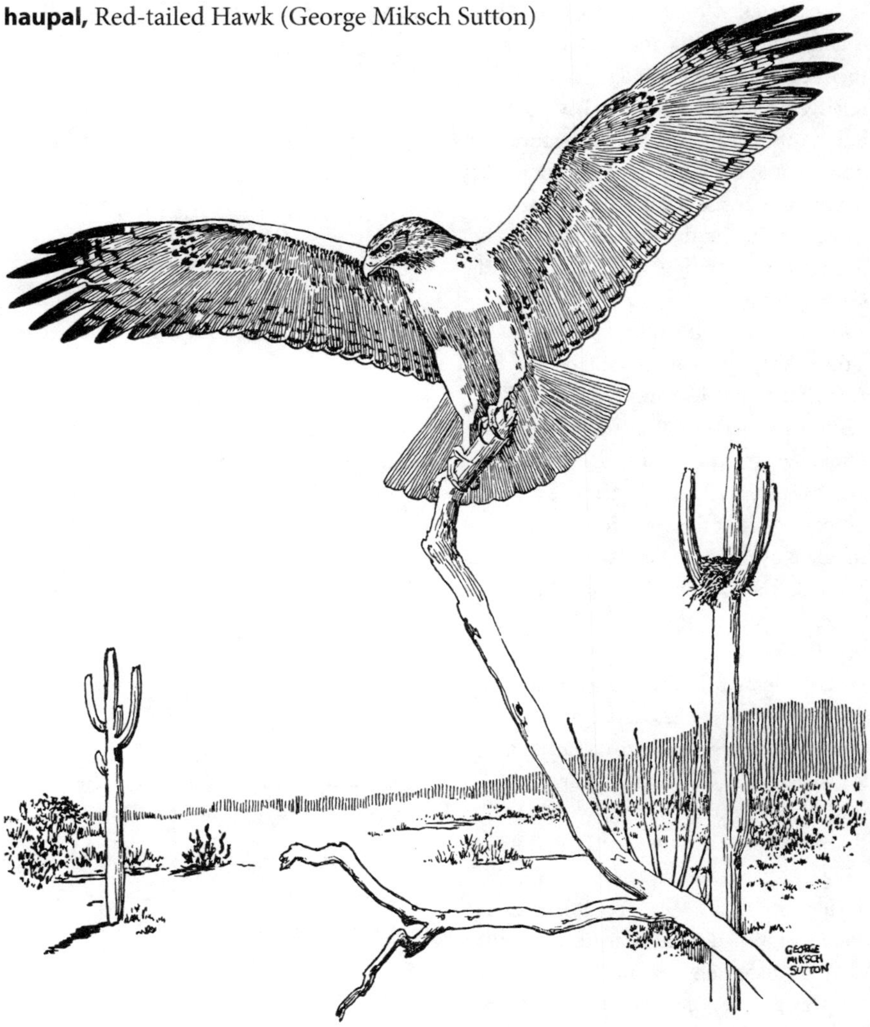

want." He was speaking of both the hawk and eagle being raised for feathers.

Sylvester Matthias spoke of a time when there was subsistence farming and a living Gila River: "There used to be *lots* of *hauhupal* when the cottonwoods [*auppa*] grew along the river. All the way down to the Salt [River]. But the [white] farmers didn't like them [the *auppa*] because they said they take too much water. But the Pima farmers like them because they enjoy the shade. Where there were willows, we got the *kuumul* [edible mushrooms]. Then the river went dry and the cottonwoods died."

He identified Red-tailed Hawks from my study skins and many times later in the field. He pointed out that *hauhupal* nest in the arms of the Saguaro. In early May, those who wanted feathers would lift the nestlings down by means of Saguaro ribs. Either they plucked and replaced the birds or kept them captive, as they did eaglets. The young hawks were released after a year or two, but for a while they continued to return to their former keepers for food (see chapter 6 on bird keeping for details).

He also noted, "The Pimas use that *viigĭ* [soft feathers] for headdresses; both men and women, boys and girls, they use that. It's precious. From *haupal*. They say—S-mohogich [Manuel Peters, the village hunter says]—the best part to make arrows is the tail, *real* red. Looks nice on arrows." Another time he said, "Wing feathers used for arrows, but tail is the best part." Sylvester himself used tail feathers of both adult and immature Red-tails for fletching his arrows.

Sylvester was fond of relating that the *hikvig*, Gila Woodpecker, and *kuḍat*, flicker, excavate nest holes in the living Saguaro, and the *sisikĭ*, kestrel, takes them over. The *hikvig* also makes a hole in a cottonwood and the *kuukvul*, screech-owl, occupies it. "And the *haupal* builds a nest, and the *chukuḍ* [Great Horned Owl], the big owl, takes it."

According to Saxton, Saxton, and Enos (1983:19), sticks gathered for kindling were called *haupal-kosh*, "Red-tail's nest." Juan Dolores (1936) wrote that a woman with the bad habit of scratching others might be nicknamed *haupal* because this hawk fights and kills its game with its claws.

In a set of seven songs that Russell (1908:312–313) recorded for curing roadrunner sickness, the fifth one images, instead, the Red-tail:

Gaḍu cheuv kui keka
kungane kiohosi
haupal maḍ taḍ chui'ika
kungane kioshosi

Somewhere over yonder a tall mesquite stands
The uppermost branches resemble a burden basket
Or perhaps a Red-tailed Hawk chick's talon
The uppermost branches resemble a burden basket.

The image evoked here is the old Piman *giho* or *gioho*, a netted burden basket supported on four poles. Women used it for transporting ollas. It was frequently stacked high with firewood, which in the song is compared to the bulky hawks' nest in the top of an old mesquite (Russell 1908: plate 34).

There is at least one narrative reference to Red-tail (Herzog [1929]). The twin boys, sons of Shining Gopher and White-clay Eater Woman (Toha Kua'adam or Nasia), go on a quest for eaglets. Along the way, they encounter the dead bodies of, first, a *tobavĭ*, Harris' Hawk, then a *haupal*. The younger boy stops to pluck the feathers from each hawk, but the older brother makes him throw them away, saying they will be ridiculed for having gone out for eagles and returning with just these hawk feathers.

After the twin youths are killed, the son of the younger twin goes in search of his grandmother. Along the way, Tamtol, White-crowned Sparrow, befriends him. The two are traveling through hot western desert country, thirsty. Red-tailed Hawks hang high overhead, shading them, singing a song. A cloud forms there. It rains over them. The boy recovers. Tamtol recovers. Because of this incident, the boy becomes known as *haupal eektakam*, Shaded-by-Red-tailed-Hawks (Herzog [1929]; see also Benedict 2001:120).

COMPARATIVE LINGUISTICS. The Red-tailed Hawk is widespread throughout North America and common in Tepiman country, but unfortunately we know little about its folk taxonomy beyond Akimel and Tohono O'odham country. Colonial Névome *au-parhi* and *oparhi*, designating some raptor, surely applies to this species. But I'm not sure I ever clarified the modern Pima Bajo name with Pedro Estrella. Once in Nabogame (Yepáchic district of Mountain Pima), a Red-tail flew over us, and I was told it was *uhíba*.

MISIDENTIFICATIONS. In his English version, Herzog ([1929]) misidentified *haupal* as "nighthawk," while later in the text translating *ñepoḍ*, nighthawk, as "whip-poorwill." Fortunately, because the original Pima text is given in the manuscript, the birds are identifiable.

We are not so lucky in another story (Benedict 2001:92). Here a pregnant widow goes in search of her grieving in-laws. "Her child was born, and she took it up and went on. She had nothing to wrap the child in. When she had gone a little way she saw a hawk dead on the ground, and she stripped off the skin with the soft downy feathers and wrapped the child in the hawk's feathers." There is no Piman word that corresponds to the generalized English word *hawk*; rather, there are a half-dozen individually named biological species that might have been so translated. We can be reasonably sure that whatever species was intended, it carried some symbolic connotation. But that is lost as Benedict recorded the story.

vakav, vakavĭ; wakaw; also *s-heepĭ wuadam*

Accipitridae: Accipitriformes (Falconiformes)

Circus cyaneus

Northern Harrier, Marsh Hawk; Aguililla Rastrera, Gavilán Ratonero

In winter, one is likely to see the Northern Harrier, or Marsh Hawk, over grasslands, marshes, and fields as well as over the saltbush-wolfberry flats along the Gila. The birds arrive in Pimería in the fall just about the time the weather begins to turn seriously cooler. They depart again when spring comes and the mesquites put out their leaves. In some winters, there are relatively few individuals, widely spaced, but in other years they seem to saturate the desert flatlands. While wintering here, they maintain individual diurnal hunting territories, although they gather in the evening to roost communally on the ground (H. Snyder 1998).

There can be little mistaking this harrier, although the sexes are strikingly different. A curiously gull-like appearance marks adult males: light gray above with black-tipped wings, white below, speckled with fine reddish marks. Adult females and immatures of both sexes are warm brown, streaked below, with heavily barred wings and tail. In either plumage, the flashing white rump and base of the tail will identify the bird.

Its foraging behavior is equally distinctive. The large but slim harrier courses slowly over the ground, just clearing the low vegetation. Its wings are long and rounded, held at a dihedral, even more sharply angled than the Turkey Vulture's. Its wing beats are leisurely as it tilts from side to side. The search seems relentless, back and forth. When it sees something of interest, it hovers briefly on fluttering wings. If the prey proves accessible, the bird drops, its quite long legs dangling. If successful, it then flies away directly to some low spot to eat its catch. Marsh Hawks take primarily rodents but add to their diet lizards, frogs, small birds, and large insects as well.

For the River Pima, long agriculturalists of the desert floodplains, the Northern Harrier was a harbinger of cold weather in the fall. Joseph Giff explained the metaphor:

> A bird that's something like a *haupal*, something like a *haupal*, but it's got a white streak across its back. This bird I'm talking about is something similar to that. There's a name for it, but I forgot. [Anyway, we call it] *s-heepĭ wuadam*, "cold dropper." They come right in the fall. And they have a white streak around the back. It comes in the fall and it looks for something—rats or mice—then it dives and comes back up like that—up and down. The Indians believe that's the bird that brings the cold because it comes around the fall of the year. And it goes along and drops [it] down. They say it drops the cold. There's a name for it, but they generally refer to it as *s-heepĭ wuadam*, "the one that drops the cold."

vakav, wakaw, Northern Harrier (Mike Touch)

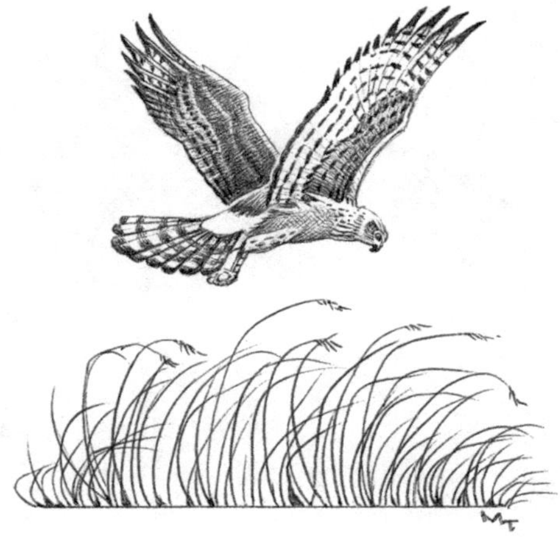

Joe sang a song that describes the hunting behavior of the Northern Harrier:

> Hachune uhing game jumalano himeda
> hachune uhing game jumalano himeda
> doki vakava uhinge game jumalamo himeda
> jevene wesiko komhaiwake.

> Some kind of bird is going low over there
> Some kind of bird is going low over there
> It's a harrier bird that's going low over there
> Everywhere the land is foggy.

Older Pimans knew this bird well. For instance, Francis Vavages said, "*Vakav* flies right over where there's water. Eats fish. They feed on probably frogs or anything like that. They feed on rodents, too. They're big. Gray; tail's got that black [bars?]." Among Tohono O'odham, it seems more are familiar with the wintering *vakav* or *wakaw* than with the resident *tobawĭ* (Harris' Hawk), but the sample size is small.

Sylvester Matthias sometimes switched the names of *tobavĭ*, the Harris' Hawk, with *vakavĭ*, the harrier, the two common hawks in Pimería with conspicuous white rump patches. But he described *vakavĭ* as "color of owl, with white on tail." His comparison is with the Great Horned Owl. Again he said, "It has white spot on the tail. You won't see it when it roosts [perches], but when it flies. Early fall. *S-heepĭ wuadam*—spreading or something like that—the cold. That [name] comes from Santa Cruz people. On the desert or in the fields. Eat? I believe they catch rodents."

In discussing the Harris' Hawk, Red-tailed Hawk, and Northern Harrier, George Kyyitan described the latter's characteristic hunting method, using his hand to show flight pattern. "*Vakav* is white-tail hawk. In wintertime goes low, over fields, all the time, when it's hunting. Then when it finds something, just goes like this—pick them up and go. *Vakav* has white, on tail. *Tobavĭ* always goes fast, not like that [*vakavĭ*]."

The Desert People also know the harrier and its association with cold. Tohono O'odham Jose Enriques noted, "*Wakaw* flies real low in the grasslands catching mice—toward winter time." He did not know the name *tobawĭ* for the Harris' Hawk.

The Marsh Harrier makes a double appearance in the Flute Story (Herzog [1929]). A woman is surreptitiously impregnated by Gopher. She bears twin sons. Various men of the village claim to be the father, bringing her their kill. The first is Vakav, who brings her *totobĭ*, cottontails. The mother rejects his offerings. Other hunters—Prairie Falcon, Golden Eagle, Great Horned Owl, and Coyote—likewise have their efforts spurned.

Later the boys go on a quest for *vaapk*, Carrizo or Common Reed. Their mother makes them love flutes from the reed segments, experimenting with them, until their music floats broadly across the desert landscape. Two sisters, bathing in the morning at Blackwater pond, find the exquisite music irresistible, and they set out in search of its source. Along the way, they encounter six different animals, each one a man who claims to be the source of the enticing sounds.

The fourth creature they meet is *vakav*. He says that he makes the sound at the edge of the brush, *sha'i*. The girls tell him to go ahead and demonstrate. They follow him. "He stood, holding his bow. He made a noise like a *vosho*, 'cotton rat.' And a *vosho* came and sat down. He shot it. Again he cried like a *toobĭ*, 'cottontail.' A rabbit came also; it also sat down. He shot it too." Although *vakav* demonstrates his skill as a provider by obtaining two of the most important game animals in the Piman dietary, the girls hurry on, unimpressed. At any rate, the harrier is not the source of the music they seek.

But the mythological narrative has an interesting biological implication here: that the Northern Harrier, flying low over the open country, makes a sound that entices its prey into the open so it can be taken. This folk science is even more explicit in the information Pimas gave Ruth Benedict in 1927. In her story, "Hawk" offers the girls a rabbit (that is, a cottontail), which they refuse. Then he "whistled, and wood rats came out of their holes and he shot them with his bow and arrow." Benedict's footnote here adds, "Wood rats are always hunted with a whistle that causes them to put their heads out of their holes; and hawks have a call that is like the whistle used" (2001:109). Here is something for an enterprising field ornithologist to look into.

COMPARATIVE LINGUISTICS. Culver Cassa notes that the terminal nonvoiced *ĭ* is unstable in the environment of the breath groups *g*, *v*, or *k*, so O'odham speakers may say either *vakav* or *vakavĭ*. I have not found the name *vakavĭ*, *wakawĭ*, outside the Tohono and Akimel O'odham area. As with many other plants and animals strongly associated with water or damp places, the name of the Marsh Hawk bears the water base morpheme *va-*.

MISIDENTIFICATIONS. The *vakav* song appears to be the same as or similar to one Bahr and associates (1997:127, 155) ascribe to a pelican. Pima coauthor Vincent Thomas translated the song as: "What kind of bird is it that goes low. It must be a pelican bird that goes low. Everywhere the land is made foggy. (Fog exists there)."

What bird goes low?
What bird goes low?
Must be a pelican going low,
Earth fogged.

Bahr (inexplicably) converts the Akimel O'odham spelling into Tohono O'odham, but it is really the *vakav* song according to Culver Cassa.

In addition to the pelican misidentification, Herzog ([1929]) called the bird a crane. Perhaps he had been told that the *vakav* was a gray bird that was associated with open marshy and grassy locations, true enough of cranes. But the *vakav* was said to hunt cottontail rabbits, not a prey item of the crane.

tobavĭ; tobawĭ

Accipitridae: Accipitriformes (Falconiformes)

Parabuteo unicinctus

Harris' Hawk; Aguililla Colorada

Perhaps the easiest raptor in the desert to identify is the Harris' Hawk. Like the much larger Golden Eagle, it has a dark brown body both above and below. It is built like a Red-tailed Hawk, but is somewhat smaller, with longer legs. The Harris' Hawk sports bright rufous underwing linings, shoulders, and leggings. In addition, a large white patch shows both above and below the tail base, and a white bar terminates the black tail. Young birds during their first year have a streaked body and barred wing and tail feathers, but still show the rust and white areas that become more pronounced in adults. Even the dark phases of other broad-winged hawks lack these red and white patches.

Harris' Hawks occupy arid and semiarid subtropical regions of North and South America, reaching their northern breeding limits approximately in Pimería Alta. Their center of abundance here is in the paloverde-Saguaro association, most notably between Florence and Tucson, Arizona. They also use the riparian areas of desert streams. They are increasingly reoccupying more open residential areas that were formerly desert.

The Harris' Hawks' social behavior and their striking plumage set them off from other raptors (Dawson 1998b; Dawson and Mannan 1991a, 1991b; Mader 1975, 1978). Their more musical chattering distinguishes them vocally from the other broad-winged hawks. During the nesting season, a breeding group is made up of four to seven individuals that work cooperatively for the success of the nest. Even larger social groups may hunt communally in the nonbreeding season. Flying low, some birds will flush game while others bring it down. Food is shared according to social dominance. The most important foods are cottontails, packrats, and Gambel's Quail, but Harris' Hawks also take lizards, other birds and rodents, and even jackrabbits.

The Piman name for the Harris' Hawk, *tobavĭ*, is apparently derived from the name for the cottontail, *tobĭ*. We can only conjecture whether this is because both share a white tail patch or because cottontails are

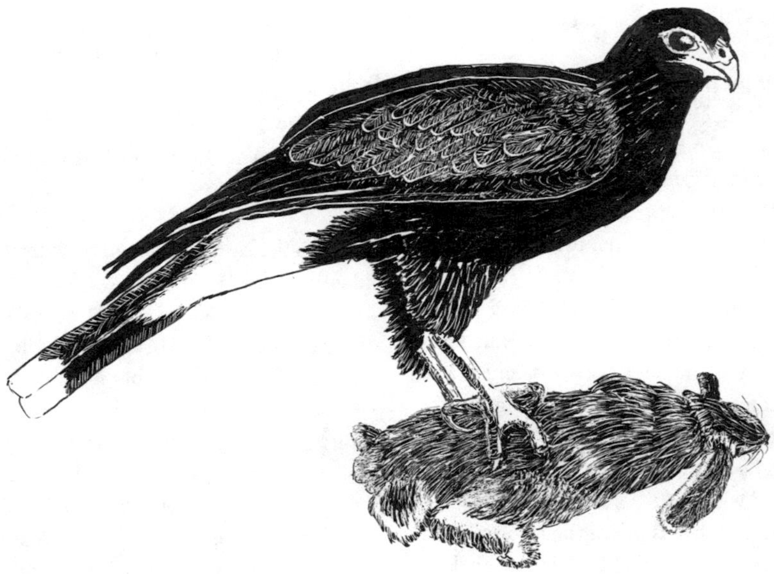

tobavĭ, tobawĭ, Harris' Hawk (Amadeo M. Rea)

such an important prey item to the Harris' Hawk. Delores Lewis switched the Piman names for the Harris' Hawk and the Northern Harrier, the two white-rumped species. Sylvester Matthias occasionally did as well. With the loss of the riparian woodlands along the Gila early in the twentieth century, the Harris' Hawk vacated River Pima country, and the older people with whom I worked no longer had frequent contact with *tobavĭ* as they had in their youth. There is a possible referent shift here with some Papago. Ken Hale (n.d.b) recorded "*s-koomagĭ o heg tobawĭ*, "the *tobawĭ* is gray." This designation needs to be worked out with more Papago cowboys still in tune with their surroundings.

Nevertheless, Akimel O'odham remembered this bird. At Bapchule, George Kyyitan said *tobavĭ* is "white belted" like the *vakav*, Northern Harrier, but smaller. "Does the same thing—flies low over bushes and trees to get the birds. *Tobavĭ* is kind of small, kind of darkish color, smaller than *vakav* and *haupal* [Red-tailed Hawk]. *Tobavĭ* always goes fast, not like that [*vakav*]."

Francis Vavages from Haashañ Keek (Sacaton Flats) said, "*Tobaw* [*sic*] goes right around close to the ground, same as the [*vakav*]. Looks like a *haupal*. Its got that black-and-white tail. They go right around where there's lot of bushes, looking for rodents, too, because sometimes they come out. And that's why they catch them. Yeah. They can see right in the grass there. Come down, catch them, and take them. About the same as a *haupal*. A thunderbird, *visag* [Prairie Falcon], is a little smaller, but it's swift." Another time he described the *tobavĭ* as "almost in the line of *haupal*, only a little smaller. Kind of brown."

Sylvester, from the western village of Komaḍk, said its feathers were not used for anything, not even by the *mamakai*, shamans. He remembered it nesting in the cottonwoods when the river was still running and the people were still farming. Early in the nineteenth century, ornithologist M. French Gilman (n.d.) found it nesting along the river. I never found it during my work here on the reservation until the early 1980s. It is unclear if the cooperative hunting behavior of this species was known, and I did not ask specifically.

There is a reference to this hawk in the bilingual version of the Creation Story that Herzog ([1929]) recorded. The twin boys, sons of Shining Gopher, pass from infancy by killing progressively larger game animals that at first frighten them (quail, packrat, cottontail, jackrabbit, and finally Mule Deer). After these accomplishments, their mother tempts them to undergo the quest for eagle nestlings. Along the way, the younger brother finds first the feathers of a dead *tobavĭ*, then of a dead *haupal*, Red-tailed Hawk. Each time the boy plucks the hawk's feathers, his older brother scolds him that they will be ridiculed on their return: people will say that they went out seeking eagles and returned only with feathers plucked from a dead *tobavĭ* and *haupal*. Both eagle and Red-tail feathers were important for fletching arrows. The juxtaposition of *tobavĭ* in this narrative context would imply that the Harris' Hawk was used similarly. If not, its feathers were of value for some other purpose.

COMPARATIVE LINGUISTICS. The name *tobavĭ* occurs widely throughout Tepiman country, but the biological referent is unclear in the south. On the Gila, Fr. Antonine (1935:9) called *tobafyi* (no plural form) a small "chicken hawk." Saxton and Saxton (1969:167) list *tobaw, totbaw* simply as "hawk." The Névome vocabulario gave *tobabi (tobavi)* as "gavilán," one of several Spanish generics used for hawks (Pennington 1979:56). Pedro Estrella and I never encountered the Harris' Hawk in the field, so I am not certain of its identification in Pima Bajo. Pictures of Harris' Hawk in my elicitation book he referred to as *tuk ba'ag*, "black eagle," and *toha givuḍ*, "white belt." His second name mirrors the bird's specific epithet. I never asked if he knew the name *tobavĭ*, so verification of its contemporary Pima Bajo name must remain uncertain. Mountain Pima use the word *to'oba* for some raptor, probably the Zone-tailed Hawk, *Buteo albonotatus*, a dark-bodied species like the Harris' Hawk. The Harris' does not occur at the elevation where Mountain Pima live.

Pennington (1969:115) notes that among Northern Tepehuán two hawks "are sought because their handsome feathers are desired for use on arrows." One of them, *tobavi*, is identified only as "*gavilán de la tierra*, that preys upon chickens." The same Spanish identification is given for *tobavi* in colonial Northern Tepehuán (Rinaldini 1743). C. Fowler (1983:237) has traced *tobabi* to Southern Tepehuán, again with an unknown referent (Key 1954). Legitimate cognates appear to be lacking in other Sonoran Uto-Aztecan languages.

As noted earlier, this hawk's name (as well as that of the Piman communal hunt leader, *tobdam*) would seem to be derived from *tobĭ* or *toobĭ*, the cottontail rabbit.

ba'ag*, pl. *baba'ag
***ba'aag* (Pima Bajo)**

Accipitridae, Accipitriformes (Falconiformes)
Aquila chrysaetos
Golden Eagle; Águila Real

This giant raptor is still to be seen in Pimería Alta, which is essentially the southern limit of its North American breeding distribution. It has been recorded sparingly in the northern half of Sonora, and Mountain Pima said they still occasionally see it in the sierras. The adult Golden Eagle is dark brown or blackish with a golden wash over the head and nape and irregular whitish gray bars across the tail. Immatures and subadults have extensive white in the tail and a white patch in each underwing. Eagles are often confused with the smaller broad-winged hawks. In flight, the Golden Eagle shows relatively longer wings. Its wingspan is more than two yards (two meters), and the perched bird stands about a yard tall.

The Golden Eagle is really an overgrown *Buteo*, not closely related to the Bald Eagle, *Haliaeetus leucocephalus*, which is in the fish eagle line. Both take four years to acquire their adult plumage. In their younger plumages, both species are blackish with varying amounts of white in the wings and tail. In the young Golden, the basal two-thirds of the tail is pure white, but in the Bald much of the tail is mottled gray. The white patch in the young Golden wing is confined to the bases of the primaries and outer secondaries, not in the wing linings (Clark 1983). Perched, the Golden shows a less massive bill and more feathers on the feet than the Bald. The Bald Eagle is far more vocal at the nest.

The diving speed of the Golden Eagle may be the swiftest of any raptor. William Brewster (1925:337–338) said one that stooped near him "made a sound like that of a strong wing blowing through pine branches." Jackrabbits are one its most important prey, though it is quite capable, in spite of its large size, of capturing much smaller birds, mammals, and reptiles. It can also kill animals larger than it can carry off, such as grown jackrabbits. And it is not above scavenging at carcasses along with vultures, caracaras, ravens, and Bald Eagles.

Eagles breeding in the desert country generally require cliff ledges in the uplifted ranges. Here the birds build huge stick nests, adding to them year after year. Sometimes several nests are maintained, the monogamous birds shifting from one to the other in different years. Courting eagles engage in spectacular aerial displays in the nesting territory. Golden Eagles are silent around nests that hold eggs or young. Unlike the falcons, the Golden parents slip away quietly when someone visits the eyrie.

As with all ethnobiological as well as biological study, there is the primary question of taxonomy: What organism is a name associated with? Piman references to just *ba'ag*, the old people told me, are always to the Golden Eagle.

I asked Francis Vavages, "How about eagle? Was there one kind or two kinds of eagles here?"

"The only thing I know is the eagles around these mountains. They're like one of these hawks but they're bigger. Kind of brown like. They fly low right around the mountains: *ba'ag*. Catch jackrabbits. I've seen them feed on jackrabbits. But they're way up in the mountains. They can catch calves, too. Kill calves. But I've seen them feed on jackrabbits."

George Kyyitan said, "I never saw an 'American Eagle,' just the *ba'ag* [Golden Eagle]." According to George, "The *chuk* [black] *ba'ag* is the only one they use. Other one, with white head [Bald], has no name." Sylvester Matthias knew the *ba'ag* and the *vakoañ ba'ag*, Osprey. He said the Golden Eagle might be called *s-kegchu ba'ag*, "pretty" or "handsome eagle." He thought the Bald might be called *ba'ag* as well. But most denied any knowledge or use of the Bald. This species was known to the Pima only because it was the American emblem.

In northwestern Sonora, Luciano Noriega of Quitovac compared the Golden Eagle to the Red-tailed Hawk, saying, "*Ba'ag* es mas grande [como *haupal*], color como *ñui*; [se llama in Español] 'aguila'" (it's bigger than the Red-tail, colored like the Turkey Vulture, and in Spanish it's called "aguila"). He said it was not common but occurs in the mountains here. Of the Bald, he commented, "No hay." Another Hiá ch-eḍ O'odham, Delfina Cameron, referred to the Bald as *s-toha ba'ag*, "white eagle."

A paired set of songs that Underhill (1938:109) recorded for the eagle purification ceremony suggests awareness, at least, of the Bald Eagle:

A black-headed eagle
On a low rock
Flapped its wings.

A white-headed eagle
On a low rock
Descended.

To understand the pervasive importance of *ba'ag* in Northern Piman culture we must keep in mind that it is a northern species, encountered as Tepimans expanded out of a southern homeland pre-

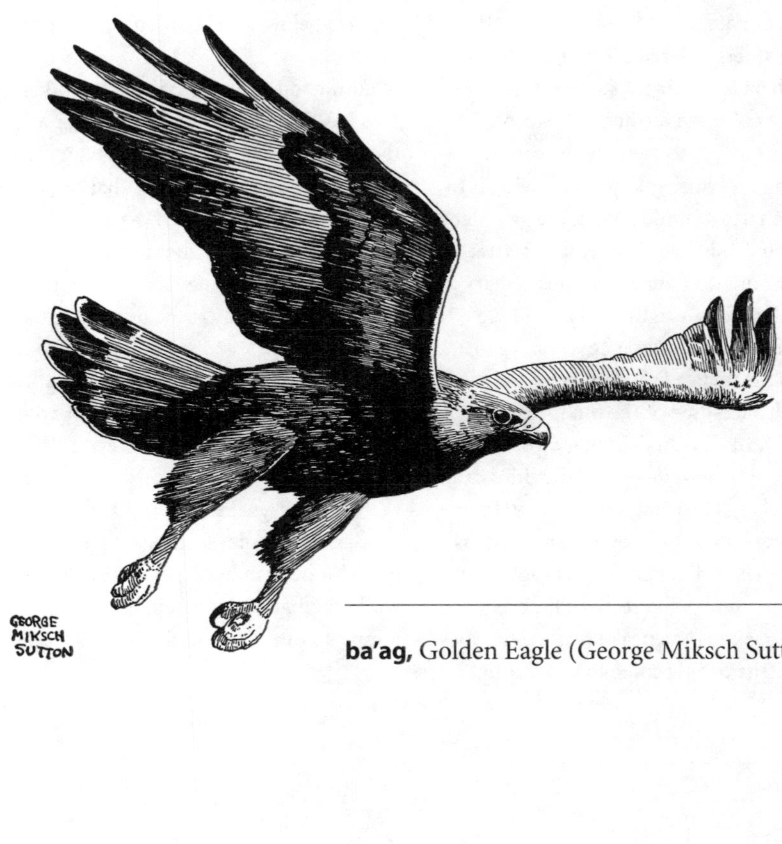

ba'ag, Golden Eagle (George Miksch Sutton)

sumably somewhere in the Sierra Madrean foothill-desert/thornscrub ecotone. Eagles, like the Apache, the ocean, and various other northern animals, were held in awe, being endowed with a special nature or power, violations of which caused illness (see Rea 1998b:28–45).

The Eagle Man episode is one of the most renowned segments in the Northern Piman Creation Narrative (Benedict 2001:29–31, 63–68; Densmore 1929:39–54; Ventura Jose in Evers 1980:110–121; G. Hill 1940a; Lloyd 1911:72–88; Mason 1921:257–261; Russell 1908:218–221; Saxton and Saxton 1973:270–281, 281–285, 299–304; Shaw 1968:20–26; Underhill 1946:243; Webb 1959:99–103). This story also accounts for the origin of lustration (the purification ritual after enemy killing) as well as for certain other ceremonial aspects of Piman culture. It is as pivotal to O'odham thought as the scriptural Exodus is to Judaic thought and ritual.

A handsome young man who is a not-so-successful gambler leads a disreputable life. Lacking a male social partner, he solicits the youth Braided Feather Shaman, who at first ignores him. At length, this other young man agrees to help him in his gambling, providing that he abandon his vices, use the sticks the shaman provides, and split his earnings with him. Subsequently, the handsome man is unbeatable. One of his opponents implores Se'ehe's (Elder Brother's) help, but when Se'ehe arrives to observe the game, he realizes that the gambling sticks are magical.

Se'ehe (I'itoi) instructs someone to have his son or daughter collect the molted flight feathers of large carnivorous birds found beneath their roost trees. He chops the shafts of these feathers into kernel-like sections, roasts them with corn, and grinds the mixture into pinole, giving it to a maiden near a pond. When the gambler comes to drink, he is immediately attracted to the young woman and proposes marriage. She offers him instead the pinole mixture to drink, saying he will have to discuss the matter first with her parents.

He drinks some of the mixture. Initially he feels feverish. On drinking more, he breaks out in gooseflesh. On the third draft, feathers begin growing on his body. Then long feathers grow out on his arms. At last, he becomes such an enormous eagle that the trees cannot support him. He even knocks the top off a peak on which he tries to land. The people come to shoot him, but their arrows are ineffective. He finally makes his home on an inaccessible cliff.

Benedict (2001:66) gave the name of this place as Vav S-Do'iv, Raw Mountain, from its bloody smell, saying it is to the south of the Gila, according to William Blackwater. Thomas Vanyiko's version, page 30, has the mountain in the east. The Santa Rosa version identifies the location as the Sacaton Mountains, although the description fits the Santans better (G. Hill 1940a). Lloyd (1911:79–80) didn't give the direction, saying only that it's a strong cliff. Weaver's Needle in the Superstition Mountains is also commonly held to be this site. Another place so designated is east of Florence, south of the Gila.

From his new home, the Eagle Man kills off all the jackrabbits, deer, and other game, then begins taking people to eat, carrying off one or more a day. He captures the woman who gave him the pinole to drink, taking her to his cliff eyrie as his spouse. They have a son who is an eagle.

Meanwhile, the people who are being ravaged seek the aid of Se'ehe at his home in Greasy Mountain, Muahaḍagĭ (or, for Papago, Belted Mountain, Waw Giwulk). He says he will come in four days, but it is actually four years by the time he responds reluctantly because the gambler is his relative. He asks the people to assemble a number of items for the task of reaching Eagle Man's cliff and slaying him. He makes his approach during the morning hours when the monster bird is away foraging. In some accounts, he ascends by means of special wooden pitons, but in others by planting a gourd below the cliff, then climbing the vine.

At the top is a stack of human bodies the bird has deposited at the nest. Se'ehe asks the wife about Eagle Man's behavior on his return to the cliff and learns that he is very paranoid, even killing flies he finds there. The Eagle Baby, now about three, is just learning to talk. Se'ehe turns into a fly and hides within a body at the bottom of the stack.

On his return, Eagle Man suspects he smells a human intruder, but his wife says no one could reach their eyrie. The young eagle tries to warn his father, but the mother says the baby is just talking nonsense yet. After feeding Eagle Man, she takes his head in her lap to search for lice while singing him to sleep. Several times he jumps up in alarm, but finally is fast asleep. Se'ehe emerges, turns back into his usual form, and cuts off the heads of both the Eagle Man and his child. After plucking the eagles and arranging their feathers in a certain manner, he descends the cliff and initiates the sixteen-day purification ceremony that must follow the killing of an Apache or an eagle, both powerful beings held in awe. Only then can he reenter the human community. The basket with the original feathers taken by Se'ehe was said to have been entrusted to the "ancient" people of Gaka village (Bahr and associates 1994:257; Underhill 1946:243).

The ornithologist in me wonders if these feathers might have been from the California Condor, *Gymnogyps californianus*, a cliff-nesting bird, much larger than the eagle, that once ranged into the Southwest (see Snyder and Rea 1998). In a different story, some great bird flies low over dancing people during the day. The people have never seen a bird like that before (Benedict 2001:94–95). Although the bird is a transformation of a nephew whose father has been killed by his brother, the boy's uncle, one wonders here, again, whether the model for this unidentified bird might have been the condor.

A great deal of Golden Eagle biology is encoded into the Eagle Man story because Pimans once knew this bird so well, in particular its nesting behavior. Still, Eagle Man is really in essence a human. His influence on desert Piman culture is considerable. A number of ceremonies and pervasive metaphors stem directly or indirectly from the Eagle Man episode. They include the lustration or purification after enemy killing, the eagle killing and purification ceremony, and various rainmaking events.

I. Lustration

Even though Elder Brother (called Se'ehe or I'itoi in O'odham) is a powerful culture hero and shaman, the killing of Eagle Man leaves him so filled with power that he has to be "tamed" and "purified" before he can reenter the community of humans. After beheading the monster and his son and throwing their heads east and west, he plucks their feathers. Descending the nesting cliff accompanied by Eagle Man's wife, Se'ehe undergoes sixteen days of rigorous seclusion, moving closer to the human community after each four days, until the power of the event is finally tamed. (Four was the sacred ceremonial number among

Northern Pimans, but the supremely sacred number was four times four.)

Elder Brother decrees that anyone killing an enemy—that is, an Apache—must likewise undergo a similar seclusion until the power received is under control and the person is no longer dangerous. This is the ceremony called the lustration, or in O'odham *sisiakam ha chemo'idag*. Thin Leather described it in detail to Lloyd (1911:90–94), and various other accounts of it exist as well (among them Grossmann 1873:416–417; Russell 1908:204–205; and Underhill 1938:85–92, 1946:192–210). If the killer were married, even his wife became dangerous and had to leave the house and observe certain restrictions. At daybreak of each fourth day, the killer took a cold bath. For the first two days, there was complete fasting, then only a minimum amount of food (weak pinole and atole) brought twice a day by the killer's guardian or "father."

Near the end of this ordeal, the killer would make a fetish from the scalp or at least a lock of the enemy's hair, some long feathers from an owl, and cotton string. This fetish was kept in a special lidded basket, *washa*, surrounded by a bed of *ba'ag viigĭ*, the fluffy feathers of the eagle.

According to Densmore (1929:200), the last of a four-song set concluding the lustration is:

> The eagle is on the highest point of
> rocks
> He is talking
> I was walking below and I heard the
> sound echo among the rocks.

II. Eagle Feathers and Cloud Symbolism

In the arid U.S. Southwest and Northwest Mexico, rain is the essential requirement for the agricultural Puebloan and rancherian peoples. The success of summer crops depends on rain to water the soil directly or to raise desert streams to supply water to adjacent fields. The People, being part of the natural world, did everything in their power to encourage the rains. Rainmaking reached an apex in the kachina cult among the Pueblos. But it is the basis of much of Tepiman ceremonialism as well. In ritual, fluffy eagle feathers, tobacco smoke, and Saguaro wine are symbolically associated with wind, clouds, and rain (see chapter 7). "Papago informants stated that cotton was used ceremonially on their prayer sticks to symbolize clouds, and thereby invite rain. Pimas preferred eagle down for this purpose" (Castetter and Bell 1942:198).

In Piman culture, eagle stands at one pole of a tension of opposites: individual versus community, West versus East, winter versus summer activities. Here is how Culver Cassa explains the overarching "eagleness" metaphor in Piman cosmology:

> Eagle has mostly to do with rain and our mutual desire, as a society, for rain for agriculture. We define ourselves as farmers. That is our obligation towards kin and affines, expressed through exchange and sharing of agricultural produce and labor. Anything that we do that adds to the stability of that system means "eagle."
>
> In other words, it's like a community spirit, for good. What I do as an individual (runner, gambler, *maakai* [shaman], etc.) benefits my kin because it brings support for them. That's why anything done as community brings rain (deer hunt, warfare, agave harvest).
>
> The Rain and Wind of the summer belong to and live in the East. This is the rain of agriculture. Primordial Piman farming is de temporal [summer rain farming]. I suppose you can say that the cosmic essence of Piman humanity is expressed in the summertime.

The roots of this worldview stretch back to an important episode in the Creation Narrative. When Elder Brother kills Eagle Man, he plucks his feathers to make clouds: "At last the eagle was still. Elder Brother went over to him, pulled off his long feathers and downy white tufts of feathers, also those from the baby eagle. The tufts from the baby he threw toward the east and those from the father to the north, west, and south. He did this so that from the east white clouds would rise, and from the north and west black clouds would come, and from the south pinkish clouds would come" (Densmore 1929:52).

From this point onward, the fluffy feathers of the eagle will be used in ceremonies as symbols of rain-bearing clouds: on prayer sticks, on the paired ceremonial wands that dancers carry, and in the dancers' black hair. Rain songs will frequently depict eagle feathers explicitly associated with clouds. For instance, in a shaman's song for sending away evil, we find the relationship between eagle "down" (that is, *viigĭ*) and cloud summoning (Underhill 1946:281):

> Much eagle down spreading out;
> Much eagle down spreading out;
> Let him sing, let him sing, Elder
> Brother!
> Slowly the clouds begin to come.

Or again, in a dance song transcribed by Densmore (1929:136), the same connection is made:

> A powerful eagle left his dwelling place
> and wandered away.
> He found the dwelling place of Elder
> Brother and went inside.
> He got some clouds from Elder Brother's
> house
> And wandered over the village and gave
> rain to the fields.

The two major kinds of feathers connote different storm types, as shown in a set of rain songs recorded by Russell (1908:333, as reworked by Culver Cassa):

> Hi'ihiyanaihu'u bahange machevwina
> yupanha sia'alinge tangio vayulinha
> gengenga chewangimeda'a hijia
> yahina'a.
>
> Bahange vinginga yupanha hunuñnge
> tangio vayulinha
> ya'ala ikumaineka'a hijia yahina'a.
>
> Hamu vaki wecho vanoahima gengenga
> hunuimeda hijia yahina.
> Hamu vaki wecho vañukunha ya'ala
> uñuimeda'a hijia yahina'a.
>
> Hi'ihi yanaihu'u Golden Eagle primaries
> pluck and hold them toward the east,
> Thereupon large clouds appear hijia
> yahina.
>
> Golden Eagle downy feathers pluck and
> hold them toward the west,
> Thereupon little misty clouds appear
> hijia yahina.
>
> There before the sacred house it thun-
> ders
> Large maize appears hijia yahina.
> There before the sacred house it rains
> Small maize appears hijia yahina.

Here the two types of eagle feathers call upon different storm patterns: the massive dark cumulonimbus thunderheads from the east bearing tropical moisture in the summer (the monsoons) represented by the dark primaries and the more gentle winter storms arriving from the Pacific represented by the white fluffy feathers *(viigĭ)*. (This is the more usual directional pattern in Piman thought, although Densmore [1929:52] had it reversed.)

Eagle songs are usually in paired sets, the first about *ba'ag*, the second about *vishag*, the Prairie Falcon (Underhill 1946:248). Underhill (1936:109–110) gave a number of examples. Here are several:

Eagle bird!
When he alights
There is a sound of thunder.

Falcon bird!
When he perches
There is a flash of lightning.

* * *

An eagle is walking,
Toward me it is walking;
Its down feathers blow in the breeze.

A falcon is running,
Toward me it is running;
Its down feathers ruffle in the wind.

For additional songs pairing *ba'ag* and *vishag*, see the Prairie Falcon account as well as Russell (1908:287) and Underhill (1946:248–251).

III. Eagle Keeping

It was customary among the Akimel O'odham to keep both Golden Eagles and Red-tailed Hawks in captivity for a limited time (see chapter 6). Eagles were taken at traditional eyries in the mountains. These eyries were often on inaccessible cliffs, and a person had to be let down on ropes to take the young. The purpose of maintaining eagles and hawks was to provide feathers for war arrows, prayer sticks, the shaman's divining wand, votive offerings, and various other ceremonial uses.

Anyone in the community wanting feathers might help feed the birds by supplying rabbits, rodents, and even fish. In one account (told previously in chapter 6), the person said he was paid two feathers a month for bringing in eagle food, though he was probably one of several providers. After the birds had been plucked one or more times, they were released. Should an eagle in captivity die (which was considered bad luck) or be killed, it was buried away from the village with all the respect given to a person.

This practice is encoded in the Piman story of White-clay Eater Woman and her twins, together with a death and resurrection motif of the birds. (It is a shared Pima-Maricopa story, probably with Yuman roots.) The narrative relates that the twin brothers, in their progression toward manhood, go on an eagle quest. Each acquires a nestling. A cold rainstorm arises, killing one of the eaglets. The boys dig a hole, cover the dead eagle with turquoise beads, place the eagle owner's bow and arrow on top of the bird's body, and bury it. (Herzog's [(1929)] version says that the owner puts down all his clothes as well as his bow.) They then sing a mourning song (keening) without any words (Benedict 2001:104). Surely a high-status burial.

What aspects of the sacred were associated with River Pima eagle keeping can now only be guessed. Was it in the taking of the young birds from the nest? Or was it in plucking them? By the time of Russell's fieldwork in 1903, the reservation had already undergone a quarter century of Presbyterianism, which took a dim view of traditional practices. These practices survived somewhat in a few Catholic enclaves. Eagle keeping, remembered by older Pimas I worked with, faded out in the first third of the twentieth century.

Sylvester thought that eagle keeping was primarily a custom of the Maricopa, who lived adjacent to the Pima. They captured their eaglets in the Sierra Estrella. The two tribes shared a mythological basis for this practice, as just noted. In the Maricopa version (Spier 1933:377–378), twin boys bring two eaglets from a mountain nest. "They went after some willow branches and built a little cage for their pets. (They were the first to build cages for pets, so we always build them the same way.) When it was ready, they went hunting [for meat] for their pets."

IV. Eagle Killing as Individual Power Source

Eagle killing was a source of power to young Tohono O'odham, along with enemy killing and salt gathering. Much of what we know of Papago power seeking comes from the work of Underhill (1938:105–110, 1946:243–260). As she wrote, of the three possibilities for obtaining individual power, scalping became obsolete and salt gathering declined. Eagle killing remained. Some Piman groups kept the birds captive, as already noted. But "[i]t is not 'given' the Papagos to cage eagles. If they want the feathers, they must brave the supernatural dangers of killing the bird" (Underhill 1938:105).

"There was no ownership of nests, and the writer heard of no trapping ritual. The method most often mentioned was to shoot an adult bird when it was heavy after having eaten and therefore unable to fly. The act which gave power was the final plucking of feathers from the dead body" (Underhill 1946:244). The usual hunting time was in summer when eagles were nesting.

The person who sought power through eagle killing was on a quest for direction in life. "Though the eagle he has killed is 'just like a man,' he is not an enemy. He has not really died, but has been released to his original form and that form is human" (Underhill 1938:107). The slain bird became his teacher, coming in his dreams, teaching him songs, giving him the power to be a strong runner or a good hunter or perhaps a shaman. The seeker might find it necessary to kill an eagle during successive years until the answer was given.

Once the killer dispatched the eagle, he was "dangerous" to the community and had to undergo a purification, usually of sixteen days, just as an enemy-killer did. In some communities, only a four-day seclusion was required. During this time, he was not allowed to eat meat, salt, or grease, look at the sun, or go near a fire. He fasted rigorously. He had a guardian, usually an older man, himself an eagle killer or at least one who knew eagle songs. The guardian could not touch the unpurified feathers himself because they too were dangerous, but he would indicate to the killer how to pluck and arrange the eagle's feathers in a ritual manner. The *viigĭ*—the fluffy feathers from the body and underwing, and especially those from the back of the legs and under the tail—were the killer's property. A prayer stick, *omina*, was made by attaching one of the fluffy feathers to a plain short stick with a cotton string. The outer primaries, or *mamchvidaj*, were the guardian's fee.

When the seclusion period was finished, the guardian then directed the all-night purification ceremony. Men who knew eagle songs participated. They sat in a circle with the killer, each singing four songs and blessing the novice with the prayer stick. At dawn, the youth took a cold bath, and the guardian distributed the remaining feathers to the singers. Tail feathers were for war arrows, and remaining body feathers were for ritual use. No mention is made of what became of the dead eagle's body after plucking.

There were local variations in the purification ceremony. Tohono O'odham Frank Jim, from the western part of the reservation, gave me this account from his own experience as an adult:

> [They] kill *ba'ag*; [either] when somebody kills it or somebody finds it dead; go back to the village. They have a meeting in the evening to decide who is going to do it—take the feathers out. They have to pull them out. Can't skin it. Two guys go to do it. Usually get unmarried men for this; in case wife pregnant, no good. [It's] like purifying salt, but more scary [dangerous]. Then they bring those feathers up to the village—bundled up—and give them to the four old men sitting there. They put those feathers down in front. Those two [who handled the feathers] have to dance all night—start about midnight. The singers are the four old men and other singers—sometimes twelve people [in all]. Use the *hiwkuḍ* [notched rasping stick] while singing. Up-down. Up fast toward the body, then slower down. (Ge Wo'o [Big Well] people use the *shaiwkuḍ* [gourd rattle].)
>
> After they dance all night, they have four *mamakai* [shamans]. They take out the four long feathers, the long ones, from each wing, for the *mamakai*. Then other feathers go to other people, according to whoever wants it.
>
> In the morning, after the dance, set a *ha'a* [clay olla], big one, out there full of water. The four old men take that dipper. The two that plucked the feather have to take their shirt off. They pour cold water on them. They can't dry themselves. Have to run, about two miles—mile out, mile back—wearing an *atosha* [breechcloth].
>
> The people who pick those feathers up will have to stay away from their houses for four days. Will have a little shack over there—that's where they stay—an *uksha* [brush enclosure] made of *shegoi* [Creosote Bush], *hukshumkuḍ* [*keskuḍ*, the old term]—about half a mile away from the village. Nobody can go over there. Can leave to go to the toilet. When they pee, have to use a cloth on their hand or a glove. After four days have to bury it. Early in the morning they go out. Only thing they can eat is *chu'i* [pinole] in the morning—nothing at noon. No salt. *Atol* [cooked wheat porridge] in the evening. *Pilkañ chuamai* [whole-wheat loaves baked in the ashes], the one they bury in the ashes. No salt (no *ko'okol* [chile] either).
>
> Hard to see [find an eagle]. But I did twice. I went looking for horses. One that caught a rabbit there. That big [indicating four or three and a half feet tall]. Sound like a wind coming through. Had a rabbit, *toha chuuvi*. I couldn't find my horse. When I came back, it was just sitting there. It ate too much. When I got home, I told my grandfather about it. He said he would tell those people to go kill it. I thought he [the eagle] wouldn't be there. The next day I went back where it was eating that *chuuvi*. That high—one of those brown ones. Other one was in the mountains, just sitting up there.

Not all eagle killing was the deliberately planned event of a young man embarking on his life's career. Frank and I were discussing how, as a youth, he got sinews for making arrows and the bow string and learned archery with a companion under the tutelage of his friend's grandfather. Frank related one story from his childhood.

"One time we went down this wash. We're looking for rabbits. And we get about four rabbits. And there's a bird, a big old bird up there, sit[ting] up there and dark, standing on there. He [Frank's companion] said, 'Let's kill that thing.' So he took his shot, but he got down too low. Then I took mine. I guess I got mine right in here, somewhere [on the chest], and he try to fly right away [with the arrow], then he went down. So we went over there and looked and said, 'What the hell is this?' We couldn't figure out what it is. He said, 'Well, let's take it to Grandpa. He'll tell us.' So we got us some weeds—mesquite tree, one side was long. We cut those. We wrapped his legs up with that. And we got a pole, and run it through there. And we took it home.

"And we got down there, and he's [Grandpa's] sitting there working on the wagon. And we told him, 'We got something here.' 'What is it?' We don't know, we just kill 'em.' And he looked at it and said, 'You guys shouldn't do that.' 'Why?' 'That's an eagle.' It's kinda small and not really big, like a chicken hawk [*visag*]. But he said that was an eagle. Kind of dark. So he got [took] it. 'You guys get away from here; don't get close.'

"So the next day they gather everybody up and start dancing. See, when anybody kills an eagle, they have to dance. But they appointed other guys to [the] part in there; they didn't appoint us to do it. We was too young. We're supposed to be the one to sit over there. They'll dance all night. The next morning they'll tell us what to do and we'll do it. And we have to stay away from the house for four days, and we can't eat salt; we can't eat any sweets. They give us tortillas, but no salt. We was there when they had the dance; they told us to be there, that's all.

"So the next morning they pick [pluck] the feathers off. And they give to the medicine mans first the long ones [outer primaries]. And after that they give to all the singers. Then everybody had some, given way all. It was pretty good. We didn't get any—we're too young to be bothering with that."

"How old do you think you were when you killed that eagle?" I asked.

"We're almost together; we're 'bout twelve, I think. He's about two months older than I am. Maybe about twelve. But we act more like we're grown people going around, doing something. When we was about ten years old we start killing stuff, slingshots and all that—rabbits and doves."

"But you could have gotten *ba'ag mumkidag* from that!" I said.

"Yeah, we could if we touch him. But we didn't touch him. We didn't touch him over there where he fell. We had to use sticks, get him around, see what it is. I don't know, something came up, and I said, 'Maybe it's a *ba'ag*.' 'No,' he said it was one of those buzzards. 'Yeah, but buzzards got red head. So let's just do something, take it home.' So we got that mesquite pole, long one, tie his [feet together]. Don't touch that, just tie it, tie them up. It might be a *ba'ag*. *Then* we don't know what kind of *ba'ag* was it. I thought it was a big one, but that was small. We didn't touch it a bit. So we thought we might be sick with it, but we didn't."

"Now when you said that 'chicken hawk,' did you mean the *haupal* or the *vishag?*"

"Well, *visag* is smaller, yeah. What are those other ones, you know, they're more like eagles? They're kind of dark, but they're not the eagles. We thought it was *haupal*, but. . . ."

"What about *tobavĭ*?" (Being all dark, the Harris' Hawk might be considered a smaller analog of the eagle and given some of the same ritual attention.)

"*Tobavĭ* might be the one. I don't know. Chicken hawk is *visag*, isn't it?"

"So the one that you got was a *ba'ag*, but wasn't as big as a *ba'ag?*"

"No, wasn't as big as a *ba'ag*."

"But you still had to go through the purification dance."

"It's a *ba'ag*, but we didn't know. We didn't know what we got. We never did see one of those [before]. The only thing, we *heard* about it, the old man talks about it. We didn't know about it. 'Let's just take it and see if he'll tell us what it is.' And we got it down there, it's kind of young one; it's not too big. So we got over there, and he said, 'That's a *ba'ag*.'"

As Frank noted, not only the ritually killed eagle required purification, but even one found dead. Vincent Josemaria of Pisin Mo'o related this recent event: "Someone—Edward Lopez—found a *ba'ag* over here that had already died. He had four days of purification. He had to have it. Afraid it would make him sick otherwise. The *wiigĭ* come from under the wing and legs. The *mamchwidag* are used in healing [diagnosing]—two from each side are matched. The ones they use are the first and second fingers from each side."

Another Tohono O'odham, Jose Garcia of Santa Cruz village, said, "When *ba'ag* eats a lot, he can't fly. A man can go after him and tie it up. They sing for four nights. Then take the feathers, *a'an* and *viig*. Then let it go after they get what they need. The feathers are dangerous if not used right. Don't see many *ba'ag* anymore. They order the feathers from another state. I don't know if that's so good."

Here is another conversation I had with a contemporary singer at Santa Rosa, Simon Lopez. His description suggests something more akin to the River Pima procedure.

"I haven't seen any for a long time, the *ba'ag*, but from what I hear, they're around. You can see them, but it's just . . . it's kind of hard to see them now, I guess. I don't know how they used to be around here. I can't recall how long ago it was this guy got one, a *ba'ag*. That's when we have those *ba'ag wiigĭ*, that we [use] during the [singing?]. That's probably the one that we got. What this guy, this medicine man over here, told me, that that's the only way that you can make it strong, and like O'odham *himdag*, they do when they get that *ba'ag*, and pull that feathers, everything out by hand; that's how they use it."

"I've heard there's supposed to be four days of purification with the *ba'ag*," I said.

"Yes. That's what they do. Even the person that gets it, they put him away for four days, by himself. Until after that, then he goes around, the one that gets the bird. So, I don't know now, if we see one we might shoot it, kill it. Can get it that way, that's not *too* good. But, if there's no other way now days, then I guess, that's. . . .

"I don't know if they're going around different than like they used to do before, those *ba'ag*. I guess that's why they're kind of easier to catch or the people *knew* how to catch 'em. And that's why they catch 'em alive and get the feathers and fur [*vopo*, down] and everything. And then let him go again. Then they grow back their feathers.

"There's a lot of talk [speeches], you know, before they start picking the feathers and everything like that. And the same way when they get through with the bird and they talk to the bird and let him go. Then he goes out and lives. Then gets his own, gets his feathers and fur back again. That's the way they do it.

"Now it's going to be kind of hard, you know, for us to [do eagle purification]. Well, I know, I know a little bit, the way, because I hear from my grandfather, who tell us, this is what you do, this is what you say to, to that, those things." Simon was lamenting here that even he, a ceremonialist, didn't know how to conduct the purification correctly.

"I wonder how they trapped them here. *Ba'ag* is pretty strong," I said.

"What *my* dad was telling me, that the way this guy got it—the last one that was taken—this side of Akĭ Chiñ, somewhere around that area, that side. He said that this guy *knew* that there was like a rabbit, and he said that he knew that that *ba'ag* will eat so much of it and get so full that it'll be kind of hard for him to flew, take off. And so he knew that. And so he went and he spotted that *ba'ag*. And the *ba'ag* was sitting there eating. So he stood there and waited and waited. He eats and eats. And I guess he got so full, and then he started kind of walking *away* from that—there was kind of an open spot. There's some greasewoods [Creosote Bushes] around. So he went behind that, real slow, and he said as soon as he turns away, this way, he came from the back and run and by the time he tried to take off, he said it's kind of hard for them to take off, because they're so full. I guess like O'odham, they get so full, which is the same thing. That's what he did. He got so full that he hardly tried to flew away. By that time he [the guy] caught him and brought him way down here, to this side, and tied him up with something, with both feet, tied him up.

"Then [the guy] came and told that old man [the shaman] that he got that *ba'ag* that's over there. So the old man told him to go over there, 'stay over there, and we'll be there.' So he went and told some of the old people: my grandpa, and my other grandpa, and some other older people. And so they went over there and got the feathers and everything. And that's the way he got it, that time. So it must have been like that before. Because he said a lot of them get the eagles before, before him.

"So that's why I say. I wonder if it's different *then* with those eagles, than right now. It's going to be kind of hard [now] to catch [one] like that."

"That eagle was eating a *chuuwĭ? Chuk chuuwĭ* or *toha?*"

"Probably *toha chuuwĭ*. And he said it was a big *ba'ag*. Then he's got some, my dad, those *mamchwidag*, these here [wing-tip feathers]. That's from here. My grandpa, I guess that's what they do. They get all these, then save them, then give it to the *mamakai*, that they use."

"So they let the guy loose, after?"

"Umhum, After that. I don't know. He said he can't fly. He just hops around, gets his food. Live like that until he can be able to grow more feathers. I don't know, they, he said it won't take long to grow more feathers, for them to grow back the feathers. If someone know what they eat, somebody could keep them in a big cage, like now you see them in a museum. And you know."

It would seem that eagle keeping by the environmentally wealthy riverine Pima may have been the norm and that eagle killing by the more impoverished Tohono O'odham was a practice forced by necessity: in some cases it may have been just too difficult to maintain such a large carnivorous bird away from the resource-rich riparian habitats. But was eagle killing the only practice of the Desert People in relation to the eagle? Jose's and Simon's stories suggest that it wasn't. They tell of a middle way: capture, pluck, and release. Underhill said it was not "given" for Papago to cage eagles, and she never heard of trapping, only killing. (The riverine practice was not to trap, but to capture young eaglets at the nest.) However, she noted "some of the songs speak of captive eagles, so trapping may have been practiced also" (1946:244). And Castetter and Underhill (1935:71) had earlier observed that some type of hawk, undoubtedly the Red-tail, was kept caged for its tail feathers to be used on arrows.

One Tohono O'odham "song of a female eagle" implies capture (Underhill 1946:249):

> Did you not treat me ill?
> My beautiful child you took.
> I am anxious,
> All night I scream.

This Female Eagle Song is paired with the another one, ascribed to the falcon, which likewise suggests nest robbing (Underhill 1938:110):

> Did you not treat me ill?
> My beautiful brother you took.
> I am anxious,
> All day I cry.

Even more explicit is a set of songs about caged birds (Underhill 1946:250):

> Eagle, little bird
> In a circular shelter!
> Under it he walks about.
>
> Hawk, little bird
> In a narrow shelter!
> Under it he walks about.

Perhaps Underhill did not have the whole story. I asked Jose Enriques if Tohono O'odham still get the *ba'ag* for their feathers. He described two methods of obtaining feathers, both of them requiring purification.

"I think every one would love to get the feathers, but the main thing is they don't know how to go about processing them. There's a special way of handling [them], even when it's, when they're . . . if they saw it nesting somewhere. I think that's the best way to catch one, get the feathers. *Alive*, not killing it or something like that. That's the *main* thing, you know. And if for some reason, they used to say when the *ba'ag* got through eating his kill, the big one, he won't fly. [laughter] So they'll catch him then, right then they'll do it. They get what they need. But they will let it go, 'cause they grow their hair . . . feathers back again. Those are the more [most] powerful things that you can do, if you succeeding in catching a live one and getting the feathers off him. That's the most powerful [thing]. And I think that's what these medicine men are using, something like that."

"Jose, if they find a nest, like you said, do they get the young ones or do they get the parents?"

"Right. I think that's about the best way they can do it, too, if they can get them where they nest. But usually they'll be on a cliff—a ledge or something; if you can get to it, you will succeed. And I guess that's how somebody will keep one at home, you know, for some time, feeding it, to grow big until they pluck the feathers off and let it grow back again. And they will let it go then. It's another way of doing it. Whoever does that work will have to be a special person, too, will have to go through the initiation [purification rite]. After he'll be doing this, he'll have to stay out of his home until he gets everything out of the way. After purifying himself, he then returns back to his home."

"How many days would that purification be?"

"I forget how much they say it will be. I have no idea, really, the exact days. They will spend some time, isolate themselves, the household or whatever. I guess that's the hard part for anybody nowadays to do that because some young boys just can't stay out there for long, very long. They get bored or whatever."

The Tohono O'odham did not have access to fish, the prized food of the River People. And both *wopsho*, cotton rats, and *kokson*, packrats, were important foods in the human dietary. But possibly before the introduction of Old World ungulates and the subsequent overgrazing of the countryside, at least some Papago had the resources to maintain captive birds even in the desert.

Both the historical and especially the more contemporary accounts just related indicate that capture, bless, pluck, and release formed an important Tohono O'odham practice, in addition to what Underhill wrote of eagle killing.

V. Eagle and Shamanism

As to be expected of such a powerful being, Golden Eagle is intimately associated with shamanism. *Ba'ag* is the one who might come in the dreams of the eagle-killer during the youth's lengthy fasting and seclusion, telling him, perhaps, of a calling to be a shaman. (The shaman held one of the few prominent positions in the egalitarian Piman society, but it was also one of the most risky; the community's recourse with a *maakai* who failed to cure or who was suspected of working evil was to kill him. It sometimes took such a powerful personage as a *ba'ag* visiting in dreams to overcome a youth's understandable reluctance to accept this role.) At this time also, the eagle might teach the young man songs to be used in healing.

In conducting a diagnostic session for sickness, the shaman carries two things: in his right hand he holds the gourd rattle to accompany his singing and in his left the two long divining plumes, usually an eagle's two outer primaries (see chapter 5). He also may consult a quartz crystal during his diagnosis. It is *ba'ag* who tells the young shaman where to find crystals (Underhill 1946:270–271).

Of the three songs for curing *jevho mumkidag*, gopher sickness, recorded by Russell (1908:319), one sings of how these special divining feathers function:

> Bahange macheweina ngingikova
> yewena
> bahange macheweina ngingikova
> yewena
> yakovañ wua koviki ñeina totolinga
> ñimui gayegayomina.
>
> The eagle primaries four times blow.
> The eagle primaries four times blow.
> I turn to look and light shines into and
> crosses my heart.

At least one eagle feather had to accompany warriors on raids to assure victory (Densmore 1929:180). War arrows were fletched with eagle feathers "because the eagle is so brave." These plumes were considered so powerful they could be used for no other purpose. This taboo was part of the violation George Kyyitan and his friend committed: they attached eagle feathers to their hunting arrows (see chapter 5).

VI. Staying Sickness

Any offense to the eagle, including ceremonial impropriety, might put an O'odham at risk of contracting *ba'ag mumkidag*, eagle sickness. Note in Frank Jim's story told earlier that the boys were considered too young even to withstand the seclusion and ceremonial purification and that more mature persons had to stand in their stead. Even the finding of a dead eagle required a rigorous four-day exclusion: the encounter put the finder into a dangerous state. Opportunities for violation of this powerful bird were many. For instance, Frank Jim said that *ba'ag mumkidag* was one of the forms of staying sickness that he himself had contracted. "I didn't know how come I had it. I couldn't figure out." Then he told this story:

> One time I was going along, looking for some horses, and I was going along, way on the other side of the desert. Then, first thing, I heard something just went by like that—*pshew!* And I see smoke [dust] down there. Then the horse started to jump this way, and I looked over there. "What the hell is that?" Then I went around there. It was the eagle. I guess he saw the jackrabbit running, and he came right down there and got that, and [made] the smoke [dust] down there. And I went down there and look at it. It [the jackrabbit] was there and it was dead.
>
> It [the eagle] would fly a little ways and sit there. I went on, went *way* up there. And I couldn't find the horse. And I came back in the afternoon. I came right through there. Then I see that, the *ba'ag*, sitting down under a tree. Then I went over there and went around, went around like this, and tried to chase him. And I see the rabbit wouldn't go. So I don't know how many times I tried to scare him away. He just sit over there, and he just jump up, couldn't go [ate too much]. I just left and went home. Then I got home and told me grandfather about it. And he said, "You shouldn't bother it." I didn't know.
>
> Then for [after] a *long* time, I got sick. Got a headache and I got itches. I got a lot of bugs or something. And I went to the medicine man, and he fix it up. And he said, "I guess you know what you done to the *ba'ag*." And the first thing that came to my head was that one, *ba'ag*. Because he said, "You should just leave him alone, let him go. He's too full, eating, that's the reason he can't fly."
>
> And ever since then, I get it. Once in a while I get it. I had it here a while back. It gets me dizzy—itches like I've got bugs all over me. [laughter] When I get dizzy, boy, it's sure something else. [The rest of Frank's story is in chapter 5.]

Densmore (1929:90) noted that "blue hawk," eagle, and woodpecker sickness have the same symptoms, but she did not say what these symptoms were. Underhill (1946:286, 300), however, wrote, "Head lice came from the eagle, usually infested with them. [Also] [t]he eagle sickness generally showed itself in nosebleed or in lice. It resulted from the breaking of a taboo connected with eagle killing and could be cured only by an eagle killer, using his tuft of down as a fetish."

Russell's (1908:263) findings among the Akimel O'odham are surprisingly similar: "The eagle causes hemorrhage. The remedy is to sing the eagle song and to pass the down of the eagle over the part. The eagle is also blamed for the lice that find refuge in the hair of the Pimas. The remedy is to blow cigarette smoke over the head." With the Desert People, the *wusota*, curing, was done by one who had received power and songs by means of eagle killing, whereas among River People, who kept captive eagles and released them alive, the *wusota* was by a *ba'ag namkam*, eagle meeter, a person to whom the eagle came unsolicited in dream.

George Kyyitan of Bapchule on the Gila contracted eagle staying sickness in his youth. He gave me two accounts several years apart of his experience with this disease: one follows here, and the other can be found in chapter 5.

"You got this eagle sickness, *ba'ag*. Did you have a *wusota* for that one?" I asked.

"Yeah. My dad and one man does that—singing, then do the healing. That one's [*ba'ag mumkidag*] no good. I know, I heard the things what is, there's superstitions, like that. Like an eagle feather. And me and 'nother friend, I think [he] is an uncle to Ted, Caroline Rhoads. He used to go around with me. Way back, when we were young. We grew [up] together. And we worked on them arrows, and his grandfather had some of that eagle feathers a long time. He told him that it was like that, superstitious [dangerous], and you don't have to do this, don't put it in your mouth and smash them, and make arrows with it. Don't put it in your mouth and scrape it. But we did because we were just kids and don't believe things like that, and don't care. And it *really* does."

"You *hivchu?* What do you call that when you take the feather in your mouth?"

"*Hivkua* or *hivñua*, for that feather on that. And it sure did. It sure did! Finally, made sores all over."

"Your mouth, too?"

"No, just our hair. And start with the bugs, too. That's the way *ba'ag* goes that way. And so he got healed up first, *then* they told me to, then I did. He got alright first, and then after that I got alright. So later on I believe *all* that."

"What did he use? Did he use smoke on you?"

"Umhum."

"What kind of smoke?"

"Bull Durham."

"No cigarette?"

"Cigarette has no power in it. They say."

"But did he roll it or did he . . . ?"

"Roll it. And some of them, a while back, [use] what they call Coyote Tobacco."

"Ban vivga?"

"Umhum."

"Did you ever see that?"

"I know one old man used to plant some over here." (The rest of George's description of the tobacco can be found in Rea 1997:320.)

"So they sang and blew the smoke of the Bull Durham on you. Did they use the *a'an* [feathers] from the *ba'ag?*"

"Yeah."

"Who had that—your father or the other person?"

"The other person, because he's the one that sings about that eagle song—he sings some eagle songs."

"So you got cured of that."

"Umhum."

"How long did it take to get you cured after he sang over you?"

"It takes a long time. It takes a long time because it start up, up, up, start going with them bugs, too. All of that."

Eagle feathers were involved with other forms of staying sickness as well. One type, called *kuukpĭ mumkidag*, resulted from careless or disrespectful treatment of an enemy-killer's scalp fetish. The offense might be failure to ritually "feed" the scalp fetish within its basket. The fetish liked especially *viigĭ*, downy eagle feathers, as its "food." In retaliation, the Apache soul might afflict the warrior or his unborn children. "Eagle feathers were commonly used in the treatment of this disease, the doctor stroking the sick man's body with them," noted Densmore (1929:103).

Frank Jim linked the Prickly Poppy (*Argemone* sp.), which he called *kotkok*, and Sacred Datura (Jimsonweed, *Datura* sp.), called *kotḍopĭ*, somehow with the eagle. "We don't look at it; can be affected from your horses getting into it. It causes sickness—stomach; nausea; don't want to eat. It's associated with eagle, *ba'ag*. Need *wusota*. Can use [it?] in place of *ba'ag* feather." In some way, these two plants with large white flowers are linked with the eagle.

VII. Eagle Songs

As we have already seen, *ba'ag* is commonly the subject of Piman songs, often paired structurally with *vishag*, the Prairie Falcon, the next most powerful bird. The poetry is often exquisitely symmetrical. In some of the song sets given earlier, we find such appositions as alights/perches; sound of thunder/flash of lightning; walking/running; blow/ruffle; breeze/wind.

Some songs were specifically for healing *ba'ag mumkidag*, eagle sickness. Although given by Russell (1908:289–291) as circling songs, they are also songs used in the *wusota*, or curing session:

> As the Sun sinks to the westward
> We begin singing the eagle songs.
> The home of the magician rises,
> Standing before me in the land.
> We begin singing the eagle songs.
>
> The evening glows red in the West,
> And the birds here gather about me.
> Now I hear the screams of the Eagle.
> Haiya ha aiya haya haya ha'a!
> Now I meet and see you. Haiya ha!

Eagle appears also in the all-important fermentation songs sung over the Saguaro wine being watched in the feast house. The songs match *ba'ag* flying in a circle with blue falcon flying in a straight line (Densmore 1929:154, 156, see Peregrine account).

One song set was used during the enemy-killer's purification, with weapons hanging on a nearby pole (Underhill 1946:200–201):

> The eagle shield lies yonder far;
> A green rainbow beside it lies,
> A soft rumbling like thunder shakes the mountains.
>
> The falcon club lies yonder near;
> A rosy rainbow beside it lies.
> A soft rumbling like lightning shakes the flat land.

One Circling Song describes the sheer joy of Red Bat's using eagle down (Russell 1908:295, retranscribed and retranslated by Culver Cassa):

> Vengihumi nakamela
> Vengihumi nakamela
> hiovat ñeihenda.
> vokamenanahakime bahanga vivingi hekañi sivonachunaha
> vokamenanahakime bawui yosi'ime.
>
> Red bat
> Red bat
> Here is singing.
> Ecstatically using eagle down I make a headpiece
> Ecstatically the Coral-bean is flowering.

In this song, the downy eagle feathers, representing clouds, are linked with the brilliant red Coral-bean flowers, associated in Piman thought with rain, according to Joe Giff and Sylvester Matthias (see Rea 1997:133–134).

VIII. Eagle Stories and Myth

Although the Creation Story's episode of the killing of Eagle Man is pivotal to understanding so much of Piman culture, it is by no means the only mythical representation of the great bird.

In the Flute Story, Nasia, or White-clay Eater Woman, conceives twin boys. Various characters attempt to claim the children as their own, bringing them food as a father would. Vakav (Northern Harrier) and Visag (Prairie Falcon) bring *totobĭ* (cottontails), which are rejected. Ba'ag brings in his catch of *chuuvĭ* (jackrabbits) and sits down. She sees him and flings away his kill. Next Chukuḍ (Great Horned Owl) brings *nanhagio* (mice), which are likewise scorned. All four large birds of prey fail.

Then Ban (Coyote) brings in *vahamaḍ* (snakes), *vapadjilo* (Western Whiptail Lizards), and *chuchkukui* (Pinacate Beetles), all of which are rejected. Finally, Tondam Jevho (Shining Gopher) brings a stick of reeds *(vaapk)*, which Nasia accepts for her offspring (Herzog [1929]:42–43). Densmore (1929:54) recorded a parallel account wherein the four suitors are Lion (Cougar), Tiger (Jaguar), Blue Hawk (Peregrine Falcon), and Eagle (Golden Eagle), considered to be the four best killers of game in the village.

As the two boys mature and begin to learn hunting skills, they embark on an eagle quest that takes four days' travel (Benedict 2001:101–105; Densmore 1929:56–59; Herzog [1929]). They capture two eaglets of different sizes and squabble over the larger one.

> The mother went out and said that the younger child should take the younger bird as it would grow up to be just like the older bird. Coyote was living there with them and he killed lizards and other small animals for the eagles to eat. He also made a house for the eagles to live in. After the eagles got older their feathers grew very long The boys' mother said, "The feathers are too long. I will pull them but others will grow." The boys pulled out the feathers of the eagles and turned them loose. (Densmore 1929:58–59)

This mythological account is particularly interesting in that it confirms all the aspects of River Pima eagle keeping. Culver Cassa notes that the story has Yuman earmarks, so eagle keeping may be a borrowed trait Pimans acquired from their neighbors along the Gila.

In the Creation Story, the Vupshkam, or emerging Pima, march against the Vipishaḍ, Short-statured People, who occupy the pueblos of the Salt-Gila valley, conquering one settlement after another, often with the aid of animal helpers. They come to one pueblo in the Mesa area ruled by A'an Hitpakĭ Siivañ, Braided Feather Chieftain. They sing a song that describes a small blue eagle alighting: *alĭ cheedagĭ ba'ag*, their animal helper (Russell 1908:229, 282). I do not know what the referent is, but suspect that it is the Peregrine Falcon, usually called *cheedagĭ vishag*. The Prairie Falcon, *vishag*, is considered the younger brother of the eagle, and both falcons are power birds. So calling the Peregrine a small blue eagle would not be inappropriate.

In addition to these major mythic episodes, *ba'ag* is mentioned obliquely in one version of the tale of Coyote and the quail (Benedict 2001:173; see also the Gambel's Quail account). The quail, coming upon sleeping Coyote, their traditional nemesis, open his belly, remove his intestines, replace them with rocks, and sew him back up. When they begin roasting his organs, the fine aroma awakens Coyote, who runs, belly rattling, to the fire and asks the quail where they got such nice fat, expecting some. The birds reply, "An eagle came by and killed a calf, and he gave us the guts." The story was told in 1927 and is one of the few to mention the nonnative animal, calf. But the O'odham knew that the Golden Eagle occasionally killed a newborn.

Finally, let us return once more to the Eagle Man story, mythic basis for much of Piman ceremony and metaphor. The narrative also played a role in the formation of youth. In one of the very few asides, explaining a story, Lloyd (1911:80) was told: "Vandaih [who became Eagle Man] was a handsome man, but he had a bad character, and ever since the beginning parents had warned their children to practice virtue lest they be turned into eagles; because it had been foretold that some good-looking bad person should be thus transformed, and it was to be seen that good-looking people were often bad and homely ones good characters."

The older Akimel O'odham I worked with were familiar with the site Gook Vapchkĭ, Two Ponds, where the handsome man was given the feather pinole to drink and became Eagle Man. Various other narratives are associated with this site as well. George Kyyitan pointed out its precise location. The land has recently been acquired by the Gila River Indian Community, though it is doubtful that anyone engaged in the political negotiations realized they were repossessing an important traditional property.

COMPARATIVE LINGUISTICS. Aguila was recorded as *vaagui* (pl. *vaabaa'igui*) in colonial Névome (Pennington 1979:6) and as *baague* in colonial Northern Tepehuán (Rinaldini 1743). Among modern speakers, Upper Piman *ba'ag* appears as *ba'aag* with Lower Pima, as *ba'aga* among Mountain Pima, as *baagai* in Northern Tepehuán, and as *ba'aa* in Southern Tepehuán. The biological referent is not always known, but the Golden Eagle probably once ranged farther south in the more rugged Sierra Madre, at least in winter.

Among other Sonoran Uto-Aztecans, there is *a'wé, wa'wé* (Tarahumara or Rarámuri), *páve, páwe* (Eudeve), *pake* (Ópata), and *waa'al* (Tubar). For further forms among other Uto-Aztecans, see Fowler 1983. Unlike for the macaw, whose indigenous name graces so many Sonoran locality names, only one toponym, Pague, apparently comes from the Pima eagle name (Molina 1972).

TECHNICAL NOTE. Although eagle taking is no longer of much significance in O'odham culture, the question of the human impact on the local Golden Eagle population needs to be raised. Population densities are necessarily low for such large birds, a breeding pair maintaining a territory of around thirty-five square miles year round. According to Underhill (1938:106), the young man seeking power and direction would stake out a nest and kill one of the adults during the summer—that is, during its breeding season, "when the eagles are feeding their young." Furthermore, Underhill noted that a youth might take a bird on successive summers "until he had the power he wanted." Eagles generally mate for life, and the loss of one of a pair might place a hardship on the remaining bird, possibly affecting that season's breeding success and ultimate recruitment. (There is generally a floating population of nonbreeding birds from which the surviving bird with a territory would soon replace its mate.)

Raptor biologist Rich Glinski (personal communication, 25 February 2000) thinks there would be no net impact on the desert eagle population from these practices. He notes that young fledge during May and June here, so O'odham actually may have been taking the more naive fledglings if the eagle killing occurred in summer as Underhill reported.

Eagle capture, plucking, and release apparently formed an alternative practice. The removal of some body feathers would have little if any effect on an eagle's foraging ability, but pulling out some primaries (outer wing feathers) of such a large bird could be traumatic for the bird, causing temporary or even permanent harm. (So far as is known, they were always pulled out, not cut off.) If the bird was then unable to fly for a period of time, of course, it was at risk of starving or of being taken by another predator. At least one interview, however, indicates that Tohono O'odham may have released birds only after they had regrown plucked feathers.

The Pima (as well as the Maricopa) practice of eagle rearing had a number of biological ramifications. (I suspect that there were never more than half a dozen or so captive eagles at any one time on the Gila, but I never asked.) If both eaglets were taken early in the breeding season, there is the possibility of the adults recycling, making a second breeding attempt. We have ethnographic evidence from both mythology and individual testimony (Sylvester's) that a Pima scaling a nest was free to take all the birds it contained. In the Southwest, Golden Eagles typically hatch two young, but the larger one preempts more food and often kills its younger sibling (Glinski 1998a:113). By taking one bird from the nest, the potential survival of both young is more likely. (In fact, this management strategy is used to enhance the number of young that large birds such as macaws, raptors, and condors can produce in a year.) We know from an account given earlier that just one young might be removed. After all, feeding the captive was a burden that required village collaboration.

We know also from ethnographic accounts that captive Pima eagles were released after several pluckings and that the free-flying birds, while honing their hunting skills, continued to come back to be fed by their former owners. The net effect may have been more eagles in the wild rather than less. Sacrificing captive eagles, known among certain Pueblo groups, appears not to have been practiced on the Gila. There is at least the suggestion here of a sustained-yield harvest. River Pima anticipated returning to a specific eagle eyrie to take new birds in subsequent years.

vakoañ ba'ag
***shuudagit ba'aag* (Pima Bajo)**

Pandionidae: Accipitriformes (Falconiformes)

Pandion haliaetus

Osprey; Gavilán Pescador, Águila Pescadora

The great blackish brown and white Osprey, or Fish Hawk, can be distinguished at a considerable distance by a peculiar bend in the wings accentuated by black carpals. The bird is a little smaller than either species of eagle but still a raptor of impressive size. At a distance, the bird appears snowy white below; on closer view, the underwings and tail are heavily barred and mottled. The Osprey frequents lakes, rivers, and seacoasts, even in the desert.

The wings are quite long for such a sturdily built bird—an adaptation for its spe-

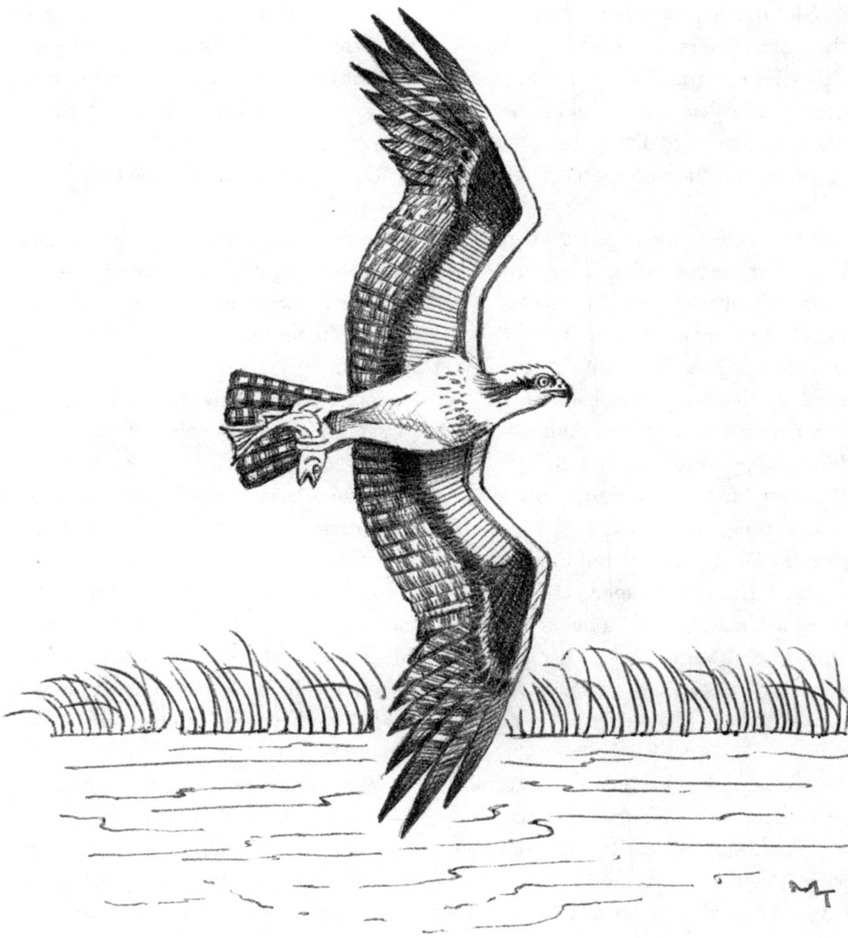

vakoañ ba'ag, Osprey (Mike Touch)

cialization in capturing live fish. The greenish gray feet and claws are stout and rough. Ospreys fly over open water, plunging feet first directly into the water after their prey, with their wings held up. They may submerge momentarily, flap to the surface, and fly off with a fish to some tree to eat it leisurely.

Ospreys may be rather vocal, calling loudly and frequently in wheezy piping or whistling notes, particularly when another Osprey is nearby. Under modern ecological conditions, they are largely spring and fall migrants through Pimería Alta, but pairs may winter along the Río Yaqui in Pima Bajo country.

Sylvester Matthias knew the Osprey as *vakoañ ba'ag*, "heron eagle," hence "fishing eagle," saying he learned the name from his grandparents. "*Vakoañ ba'ag* is just a low-class eagle, heron-eagle. It is a regular fisher. But it isn't a Bald Eagle." Another time he said, "The fish-eater, *vakoañ ba'ag*, goes with the eagle. They don't bother it." Apparently, he was suggesting that the Fish Hawk, just as the eagle, was tabooed to the O'odham. But I am not certain of this association.

Ruth Giff, of the same community as Sylvester, called it "a smaller one—fishing." No one else I worked with on the Gila knew the bird or recognized its name, not even those who still remembered *ba'ivchul*, the kingfisher; the river had been dry too long.

Still, I wonder if the Osprey might have once bred on the Gila among the Pima. Various species of fish were abundant here. Early in the colonial period, Father Kino (1919:197), on his 1699 entrada, wrote while at the middle Gila Pima villages: "In some places they gave us so much and so very good fish that we gave it as a ration to the men, just as beef is given where it is plentiful." The Gileño fish taxonomy distinguished six at least folk taxa of native fish as well as several introduced ones. Joe Giff and Sylvester considered fish the preeminent animal food of the Pima—"*real* Pima food," as one of them commented.

Physical anthropologist Aleš Hrdlička (1908:24), commenting on River Pima foods, added, "They do not eat dogs, cranes [herons], fish-hawks, eagles, buzzards, crows [ravens], or snakes." It is interesting that the Osprey should be singled out here among tabooed birds. Even in the deteriorating riverine conditions of the time, this bird must have been more prominent.

We have altered the desert riparian ecosystem so radically that it is now difficult even to imagine its former condition. So we can only wonder about the earlier status of the Fish Hawk here. Late in the twentieth century, Ospreys bred on the Salt and Verde Rivers, mostly above their confluence (above five thousand feet), and the species winters today from the Salt-Gila confluence downstream to Painted Rocks Reservoir, where there is riparian habitat and open water, mostly sewage effluent from metropolitan Phoenix (Dodd and Vahle 1998).

Tohono O'odham would have far less contact with this great bird. But Luciano Noriega at Quitovac oasis, who must have seen the bird at least during migration, called it *ge'e visag*, which he glossed "gavilán grande," or "great falcon." (He called the Golden Eagle *ba'ag* and said of the Bald, "No hay," there aren't any here.)

On the Río Yaqui, Pima Bajo speaker Pedro Estrella called the Osprey *shuudagit ba'aag*, "water eagle." María Córdova called it simply *ba'aag*. Pedro termed the Bald Eagle that I found on the river near his village (Russell and Monson 1998:62–63) *s-tuk ba'aag*, "black eagle."

***visag, vishag; wisag, wishag*; pl. *vipsag*, etc.**
***visaag* (Pima Bajo)**

Falconidae, Accipitriformes (Falconiformes)

Falco mexicanus

Prairie Falcon; Halcón Pálido, Halcón Mexicano

The heavy-bodied, pointed-wing Prairie Falcon is colored much like another raptor of the arid West, the Burrowing Owl. The falcon is mottled sandy tan above and white below, marked with dark brown spots in adults or larger teardrops in the

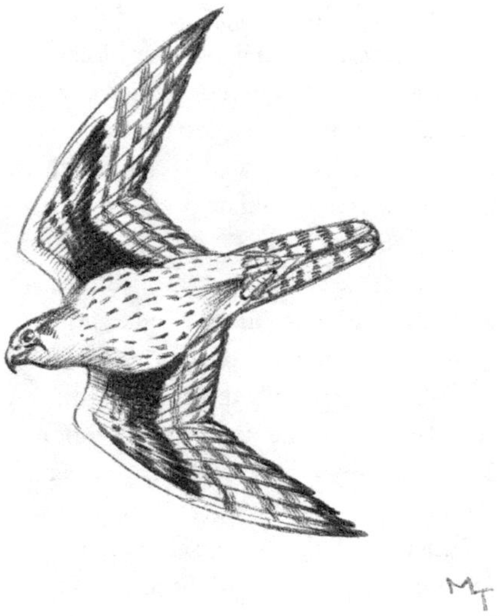

visag, vishag, wisag, wishag,
Prairie Falcon (Mike Touch)

immatures. The tail of the young bird is dark and indistinctly unbarred, becoming paler and completely barred with maturity. The narrow dark brown eye stripe, auricular marks, and moustache identify it at once as a falcon. The black axillaries (the conspicuous "armpits" on the pale underwings) will confirm the species. The birds are fifteen to twenty inches long, with a wingspan of more than three feet.

Prairie Falcons, found throughout the dry open country, are at home in grasslands, desert scrub, and agricultural areas, reaching their southern breeding limit in the northern half of Pimería Alta. In the cold season, individuals from farther north in the continent augment the population here, even spilling over into northern Sonora as uncommon wintering birds, where they do not breed (Russell and Monson 1998:77; van Rossem 1945:62).

This powerful falcon uses several favored hunting techniques. It may circle high, almost out of sight, stooping swiftly on unsuspecting prey. Or it may skim low over the terrain with strong wing strokes, boldly snatching up prey overtaken on the ground. It may even forage on foot, capturing larger insects. It takes many mammals (even cottontails and jackrabbits) as well as birds (meadowlarks, flickers, doves, quail). In winter, it commonly feeds on sparrows and other smaller birds.

The Prairie Falcon is a cliff nester, at home in the desert ranges. It builds no nest of its own, but rather lays its heavily spotted eggs in some bare rocky recess or in a ledge nest appropriated from ravens or hawks. Here it may compete with the similar-size Peregrine Falcon. The Prairie Falcon may be defensive around its nest, screaming loudly and diving ten or fifteen feet above someone who approaches too closely.

Within the Piman folk genus *visag* or *vishag*, there are two folk species: the linguistically unmarked *visag* or "regular" *visag*, the Prairie Falcon, and *cheedagĭ visag*, the Peregrine Falcon. As to be expected, the first is the more common species in Pimería, and most ethnographic references appear to refer to it, though it has been misidentified throughout in Piman writings as a "hawk."

But the O'odham knew this bird well. They (and their ethnographers) often glossed it "chicken hawk," a rather imprecise English vernacular epithet. For instance, Sylvester Matthias said of the "regular" *vishag*, "Gray, like color of a *haupal* [Red-tailed Hawk], only it's smaller." Another time when I asked him what color it was, he responded, "Something like a *kokoho* [Burrowing Owl]." He said it nests on a ledge. Joseph Giff told me, "The ones that I have seen are all gray—their tails are gray. Circles way, way up, then swoops down on its victim." He, too, described it as smaller than a Red-tail. Note from the compared bird species that the Piman concept of "gray" also includes what would be considered in English tan, sandy, or light brown.

In a group discussion, Joe commented, "Usually you find, they fly way up, around in a circle. Then they see the chickens and just dive down. They make a lot of noise, like wind blowing." Sylvester added, "They are fast; they can catch on the wing." Joe continued, "Circle around, when they see something, from maybe way up in the air, maybe a young rabbit or a chicken or something like that, then they just dive straight down." Joe once told me he thought the *vishag* made a clenched "fist" of its talons to strike at great speed any bird prey in flight, killing or seriously injuring it by impact, then flying on to catch the falling bird in its talons.

George Kyyitan also commented on the sound. "You can hear the sound of it coming down real fast." He identified the *vishag* from the exquisite painting by Louis Agassiz Fuertes showing the Prairie Falcon eating a Gambel's Quail. "That chicken hawk, or whatever they call it, that *vishag?*"

"Yeah."

"I've seen that one. Boy, it can sure goes fast. Where was it I just see one? Just goes like this [closes its wings] and go like this—shhh . . . like this. Hit the bird, go right away."

"Did it hit it in the air or on the ground?"

"In the air. Because it will try to fly and hit them. Just go right straight up." He motioned how the bird went.

"Grab them from underneath?"

"Yeah."

Francis Vavages gave this account: "*Visag*, 'thunderbird'—oh, he hides some place—probably between the rocks, or in the thickets or something. Sometimes they would be way up there [in the sky]. When they see a bunch of birds going by, comes like a swift, hit one of them, injure them right away—it can't fly and start to fall down. It [falcon] could come back and catch him—before it gets to the ground. That's how they feed." He, too, described it as a little smaller than a *haupal*.

Papago Frank Jim knew the bird well. "Not all the people see it, just certain people see it. Some are gray—but they're so fast, like one of these jets. If they're after something, all you hear is wind—when they're after doves or something. Probably catch some rabbits, small ones. *Tobĭ* [cottontail] I think. *Ba'ag* [eagle] gets the *chuuvĭ* [jackrabbit], too, big. He's [*vishag*] the fastest bird they see, anybody. When he comes, then get something, then disappear; [it's] just like wind coming somewhere."

Sylvester said, "There are lots of stories of the *visag*, because it's a sacred bird to the Pima." This was definitely an understatement. Prairie Falcon figures in the Creation Story, various tales, orations, songs, and ceremonies that have been documented during the past century. And the irony is that it has never once been correctly identified, so that the vehemence of this biological species as a powerful metaphor has been lost to readers.

In Piman oral literature, the Golden Eagle is often paired with the Prairie Falcon, the smaller bird being considered the eagle's younger brother. We encounter this pairing both in story and in song and even in power-getting ceremony.

This falcon appears rather early in the Origin Story, during the conquest of the Vipishaḍ (the Small People who built the great houses) by the Vupshkam (the Emergenti who invaded the land they now occupy). Densmore (1929:31) related this episode from a San Xavier Tohono O'odham:

> Elder Brother, travelling with his people, drove out the former inhabitants of the country. He tore down their houses and temples and even went to Casa Grande, where the ruins now stand. A man and his daughter lived in that ruin. The man's name was Sivanimaka [Siivañ Maakai].
>
> Among Elder Brother's people there was an old woman and her two grandchildren. She spoke to them and said, "You had better finish them up [off]." She went to her burden basket and took some feathers from a box. She sang a song, and the children turned into birds called Visuk [*vipsag*]. The birds [falcons] flew into the air. The first smote down the man and the second smote down the woman and killed them. Ever since that time the Casa Grande ruins have stood as now.

(Revisionist historians, promulgating the Hohokam-Piman continuum hypothesis, would do well to consider what O'odham themselves have maintained for generations about their relationship with the earlier inhabitants of the country.)

In another important story, a woman conceives twin sons, and various men attempt to win her favor by bringing her their kill. Vakav (Marsh Harrier) and Vishag bring cottontails. Their offerings are thrown away. Vakav and Vishag are rejected as suitors, as are Ba'ag (Golden Eagle), Chukuḍ (Great Horned Owl), and Ban (Coyote) (Herzog [1929]:42–43).

The next time the Prairie Falcon is encountered in Piman mythology is immediately following the conquest of the Vipishaḍ by the Pimans. "Now when the bands were going thru this country they had selected the places for their homes, expecting to return, and each band, as it selected its place, drove down short sticks so as to know it again" (Lloyd 1911:166). The Tohono O'odham did not go to their selected place at once, however, but tarried for a while on the Salt River near present Lehi.

As the story goes, "the *Aw-up*, or Apaches, were a part of the original people of this country." A childless shaman by the name of Sohahnee Maakai (S-huuhañi, signifying something to do with a basket) finds an Apache child who has been left behind, and he adopts and raises him as his own. The shaman is a man of remarkable valor, and when the Apaches return some years later to kill him, he stands them off for a considerable time, single-handedly killing a number of them. When the attackers finally kill him, they say, "'Now let us cut him open and find what it is that made him so brave, and enabled him to kill so many of us.' And they cut him open and found under his heart a feather of the chicken hawk. And the Apaches took that feather, and that is how they are so brave and even if there are only two of them will often attack their enemies and kill some of them" (Lloyd 1911:168; see also Benedict 2001:134, 144–153).

In a parallel Tohono O'odham version, the adopted Apache boy, who is himself subsequently killed by Apaches, is said to be a *wisag namkam*, Falcon meeter (Saxton and Saxton 1973:189–205, 207–209). "One time [the boy] was running again and something came roaring at him and struck him down. He lay there and saw this. There were people warring. And a hawk [*wisag*] was flying along, dropping among the Apache and knocking them flat. The boy was still lying there when the sun rose. Then he came to and got up." The boy learns the Prairie Falcon Song, and the bird becomes his helper, coming to him four times, as does *ba'ag*, the eagle. Through their help, the boy becomes a successful hunter. "No one knew how he killed mule deer [*huawĭ*]. He would go out and right away come back bringing a mule deer. People were talking about him. Some said this young man could become a hawk. Some said he could become an eagle."

In another Tohono O'odham version, Prairie Falcon Meeter (a different person), takes an active role in the skirmish: "and he went, and in this way he hit them [the enemy], and hit them over again; then he would turn over there, and four times he turned over there, and he finished them all off, however many there were" (Frank Lopez, trans. Zepeda, in Evers 1980:147).

Although this event is an aside in a longer story, it establishes the metaphor of what a large falcon means in Piman thought. Still, we may wonder about the identity of Lloyd's vernacular "chicken hawk" both here and elsewhere in the story. In an oration delivered by the "father" or "guardian" of an Enemy (Apache) Killer during his purification, Lloyd's English "chicken hawk" is specifically associated with the Piman term *vishag* (Lloyd 1911:91–92; compare Benedict 2001:203–204 and Zepeda in Evers 1980:146–147).

After the slayer has fasted for two days, his guardian presents him with some pinole to drink, saying that he has run all the way to the ocean to get the water to make it. Four times on his return the guardian encounters a powerful force that enters the dish and throws him down. The first encounter is with S-vegĭ Wihom, Red Thunder Person. The second time is with Tondam Ba'ag, Shining Eagle. According to Lloyd's (1911:91–92) version, "And he turned the water in the dish in a circle, and because of his force I fell down, and when I rose up again and smelled the water in the dish it was stinking." The third encounter is with Prairie Falcon: "and *Vee-sick* the Chicken Hawk, came down from the sky and stood in the dish. And by his force I was thrown down. And when I stood again and smelled the water in the dish, it smelled like fresh blood." The final meeting is with S-koomagĭ T-Veemgal, Coyote ("Gray Paternal Cousin"). Now presenting the dish of pinole to the killer to take a single draft, the guardian says, "[A]nd now I have brought you the water, and so many powerful beings have done wonderful things to it that I want you to drink it all at one time."

Even the little doll that the Enemy-Killer now makes with the lock of hair from his victim must be decorated with feathers from the sacred birds (owl, eagle, and Prairie Falcon) and a string of blue beads (Lloyd 1911:93). The particular associations here are not casual ones. We hear them again from Dr. Hrdlička (1908:243), who was making medical observations among the Greater Southwest tribes about this time. Of Apache sickness, he said: "If a Pima killed several Apache, although the act was lauded, it was believed that some of the progeny of that man would become insane or otherwise injuriously affected. This result could be obviated by use of Apache hair, a tuft of which, tied with a chicken-hawk [Prairie Falcon] feather and an owl feather and burned in a certain way with greasewood [Creosote Bush], would cure any sickness induced by the contact with the Apache." (The story of the monster Ho'ok's soul turning into a falcon after she is cremated refers specifically to the Peregrine Falcon, not the Prairie, according to both Thin Leather's account and that told by Sylvester Matthias, as indicated in the next species account. Curtis [1908:20] gave the bird name only as Víshŭk.) The "owl" feather in each case must be from the *chukuḍ*, Great Horned Owl.

Saxton and Saxton (1973:142–143) recorded another tale in which Wisag, Prairie Falcon, appears. Here the falcon stoops suddenly, striking down some quail. "He would come roaring down from above and flap his wings and strike down a number of them." (Not only the O'odham word but also the behavior marks the raptor as a falcon, not a "hawk," as translated by Saxton and Saxton.) He kills them—all but one, and this one runs home crying out what you can still hear quail saying today (see Gambel's Quail account).

There are several accounts of a young man, usually indicated as the son of an Apache father adopted by an older O'odham couple, who becomes a Prairie Falcon meeter, as we have seen. One account of *vishag namkam* recorded on the Gila is particularly enigmatic (Benedict 2001:144–153). While running, Falcon Meeter has an encounter and receives shamanistic items from both Eagle and Falcon. The people build him a house of *vaapai*, Saguaro ribs. He lives mostly with the great birds, but comes to his own people in the form of a child or a young man or a falcon. As a person, he joins other O'odham on expeditions against the Apache, but he is a loner, holding back and using his special powers when things become hopeless. On the offensive, he goes forth only with a small group, which irritates the larger community.

Falcon Meeter has great powers, sometimes taking the avian form of his mentor. He travels widely and is known everywhere. In the country of the large birds, he lives with the falcons and eagles, who welcome him. But he is not so fortunate with his own people, who try to kill him. He is uninterested in marriage, a characteristic found in O'odham myth most frequently with young women. The People claim he is out for his own glory and still try to kill him. He grows weaker and after a four-year illness dies at Kohadk village.

The Prairie Falcon appears in sung poetry as well. Among the "festal songs," *chechka ñeñe'i*, that Russell (1908:287–288) transcribed, there are several about this bird. They are naming or calling-out songs; the group sings of war exploits in exchange for food, notes Culver Cassa. Underhill (1946:180 fn.) equated them with warpath songs, which seems appropriate, given their content. Song VI pairs *ba'ag* and *vishag* structurally, and Song VIII pairs the feathers of the two raptors used in arrows. (Surely these arrows were for war, not for hunting.) Song XI again pairs the two species, comparing them to two named personages.

Here is Song VI of the set, as reworked and retranslated by Culver Cassa:

Gakodak damai bahange naiwu'usha
yakimeli damai vishange naiwu'usha
hitave Ma'akahi bahange weme ñiohoke
kavahanḍe dange kwushahi
hitave Hivayumi vishange weme ñiohoke soñi'icheki peheka wushame.

Atop Superstition Mountains a Golden Eagle rushes forward
Atop the [Gila] river a Prairie Falcon rushes forward
This Ma'akahi speaks with the eagle and taking his shield rushes forward
This Hivayumi speaks with the falcon and grabbing his club goes forward.

In the two songs used in curing *komkĭchiḍ mumkidag*, Desert Tortoise sickness, the first one pairs Golden Eagle with Prairie Falcon (Russell 1908:306, reworked and retranslated by Culver Cassa):

Sivañi lahai, sivañi lahai,
sivañi lahai, sivañi lahai,
gewesi kehenga siyali wusha
kenḍange bahake wewemia nekai
sivañi lahai, sivañi lahai,
gewesi kehenga hunḍuñ
keḍange vishaka wewemia nekai
sivañi lahai, sivañi lahai.

The shaman is saying, the shaman is saying,
The shaman is saying, the shaman is saying,
The beautiful morning rises
Just then I sing with the Golden Eagle.
The shaman is saying, the shaman is saying,
The beautiful evening sets
Just then I sing with the Prairie Falcon
The shaman is saying, the shaman is saying.

Another Vishag Song that Russell (1908:336) recorded is associated with the enemy-killer purification ceremony. It says:

Mohial novanga keka tamai vishake
vañi vemekai menda chuma petohokime.

The Prairie Falcon on Nightshade Mountain [near Superior, Arizona]
Stood with me beating his wings.

Was there Prairie Falcon killing among the Tohono O'odham as there was Golden Eagle killing as a means of gaining power? Some evidence points in that direction. In discussing the eagle as a source of power, Underhill (1946:243–244) stated, "A hawk would give power also, but in lesser degree. The writer did not succeed in identifying the birds eligible for this ceremonial killing. The Papago said that all eagles gave power but not all hawks. In general the hawk was the younger brother of the eagle, not so important, but able to fly faster. He was the guardian spirit of runners, while the eagle was sought by the would-be shaman." She listed as possible candidates a number of broad-winged hawks, but ironically no falcons. However, in the songs that she collected, her "hawk" is really a falcon, as we know from parallel sources. And, of course, only the Prairie and Peregrine can fly faster than an eagle.

In a novel, Underhill's (1940a:13, 15, 36) main character is a Papago youth named in infancy Hawk Over Whirlpools. But this so-called hawk is said to be the younger brother of the Golden Eagle, hence is really *vishag*, the Prairie Falcon. When his grandfather is instructing the boy in ways of acquiring power, he tells him, "One is killing the eagle. Or his younger brother, the hawk. . . . It is as desperate as killing a man. Yet some day, when you are a youth of mature courage, you will do it. Then you will remain away from people for four times four days, fasting and waiting."

Later in the story, when the protagonist actually does kill the bird for power, it is identified as a Prairie Falcon, accompanied by diagnostic features. The purification ceremony and the ritual removal of the falcon's feathers, similar to those used for eagle killing, are described (Underhill 1940a:126–128). (It is curious that Underhill provided more specific information about falcon killing in this novel than she did in her ethnographic writings.)

It is known that the Prairie Falcon word was used at least once, in the nineteenth century, for a personal name. A man, Vishag Voi'i, Falcon Flying, supplied Russell with a considerable number of songs, including all those that mention this raptor. Perhaps he was a *vishag namkam* and had been taught these songs by his spiritual helper, the falcon, hence his Piman name. None of the songs or orations in Russell's anthology mentions *cheedagĭ vishag*, the Peregrine. Densmore mentioned only "blue hawk," which is the Peregrine Falcon. I suspect that at least some of these references should apply to the Prairie Falcon; it is the more common species in Papaguería.

The Prairie Falcon is among those (generally northern-distribution) organisms associated with staying sickness. George said, "Don't use the feathers."

I asked, "Why, is it dangerous?"

"That's what they say; that's why nobody use it."

Sylvester confirmed this warning, saying that the use of *vishag* feathers was restricted to the medicine man. I asked him how the *mamakai* got this bird's feathers, but he didn't remember how. In discussing arrow making, Frank Jim noted, "You can't use *ba'ag* or *vishag*."

On the Gila, Russell (1908:263) listed "hawk" among the birds causing staying sickness. The symptoms are hemorrhage "in grown persons only," and the *wusota* involved "the hawk song and passing the wing feathers of the bird over the patient." In most instances, Russell's "hawk" is the *vishag*, though the curing song is not given. Like for eagle killing, there was a lustration associated with contact with the falcon: "If one touches a hawk he must be secluded for four days."

As is usually the case, knowledge of *kaachim mumkidag*, staying sickness, has survived far better among the Tohono O'odham than among the Akimel O'odham. Frank Jim explained it this way: "*Vishag* causes sickness—starts in your ear. (Rabbit does that, too.) They call that a boil; when it busts out, and white stuff runs out, it will quit hurting."

"What about *maakai?*"

"Yeah, he can. If you don't see the *maakai* right away, it will bust out. Then sometimes you can't hear too good." He added, "We can use them, but we have to be careful of them. A lot of people don't see it, *vishag*. It don't come around. Used to be, few people, it comes in their dreams. But *haupal* [Red-tailed Hawk] don't got no sickness on it. Just the *vishag* and *ba'ag*."

"But *vishag* has a song?"

"Yeah, it has a song. But I don't know if anybody knows it now. That old guys, used to know it. That song you have to do in the afternoon [*wusota*]."

Through time, as people have had less contact with certain species of birds and mammals, either because of the rarity of the animal itself or because people today spend more time indoors, their knowledge of the biota and its associated folk taxonomy has been collapsing. And the knowledge of how to diagnose and cure certain forms of staying sickness is diminishing as well. After years of inquiry, I began to fear that this might be the case with *vishag mumkidag* and its associated songs.

I visited Simon Lopez of Santa Rosa, son of a *maakai* and a person generally knowledgeable about curing. "Are there any songs about the *wishag?*" I asked during our morning's discussion.

He explained that both a feather from the Prairie Falcon and songs were used in a healing session. Then he added: "I heard some songs from my grandmother before, but I can't really recall how they go. But I know that it's from the *wishag*. Because [*wishag* are] the same way [as *ba'ag*]; they can get you sick. Yeah, they do, *wishag* does. How he does that, too, is the same thing: the eyes and your *iibdag* [heart]. There's another way they say they'll get you, the *wishag* . . . sores? Or something like that. I've kind of forgotten."

"I know that *ba'ag* can do that—sores, or lice." I said.

"Yeah, *ba'ag* does that. But *wishag* is different things. There's an old man over here who does a lot of curing, singing different things. And I tried to ask him, but he said he didn't know about the *wishag* [songs]. But he said if you can get the feather, *a'an*, like the one [I have], you can use that. I know that. Use a feather to try and cure like that, the right feather." Here was a case of using only the fetish in a healing after the curing song itself was forgotten.

Late that afternoon Larry Saraficio took me to see his old cowboy friend Vincent Josemaria of Pisin Mo'o village. Vincent told us *wisag* causes "something in the hair—sores in the hair all over. Almost the same way as *ba'ag mumkidag*; also, increase in heart rate—can make the person dizzy and faint." There are many songs for the *wusota*, but he was sorry that he remembered only two of them. My heart jumped (not, I think, from *wisag mumkidag*). He gave me permission to tape the two he knew and to use them here. The two songs he sang had a similar pattern and melody. He even helped work out the translations, correcting Larry's English glosses.

Nañi wuḍe visange
dañi wuḍe visange
gan hu mek gan hu himenaha
Yahana shoshona gaḍhu
tonlig vaupanaime.

Nañi wuḍe visange
dañi wuḍe vusabge
gan hu meek gan hu himenaha
ñahaha shoshona kuhung an
tonlig vaupanaime.

I am Prairie Falcon
I am Prairie Falcon
Off far I go along
From my wing feather bases light extends.

I am Prairie Falcon
I am Prairie Falcon
Off far I go along
From my feather tips light extends.

In the first song, light is radiating from the base of the wings or at least from the wing feathers, whereas the second says that the light comes from the tips of the wings.

Fetishes used in curing *kaachim mumkidag* and for other shamanistic activities were traditionally kept in long-lidded baskets plaited from *umug*, Sotol. (More recently, because this type of weaving was forgotten, a leather bag, of Plains Indian influence, has been substituted to hold the items.) Kissell (1916:171) recorded that a medicine man's "rain-basket" from Covered Wells contained, among other fetishes, "chicken-hawk feathers, which, when all previous attempts to produce rain have failed, serve as a desperate call for rain for the village, and water and food for the children." (The first plea used turkey feathers and chants, the next "buzzard," and the final "chicken-hawk." Thus, these three birds, along with Golden Eagle, are established as rain birds.)

Another basket from a Topawa *maakai* originally contained, among other fetishes, *visag* feathers said to be for dizziness, headache that goes through the eye to the brain, and nosebleeds (Arizona State Museum specimen). The falcon feathers were among several other items the *maakai* considered too precious (difficult to obtain) or too dangerous for him to part with.

The name *visaag* or *vishaag* was used by Pima Bajo, but I never determined from Pedro Estrella what the biological referent was (see also Pennington 1980:193).

COMPARATIVE LINGUISTICS. The Upper Piman falcon name may be either *visagĭ* or *vishagĭ*, depending on the speaker. Culver Cassa notes that the palatalized form tends to be more northern within this area.

There are various other manifestations of the folk taxon within Tepiman languages: *visagui (visagi)* (seventeenth-century Névome); *vísaag* (modern Pima Bajo); *viishuki* (Nabogame, Yepáchic area of Mountain Pima); *vishig* (Yécora Mountain Pima);and even *pichaag* (pl. *pipchag*) (Tepecano [Mason 1917:329]). Rinaldini (1743) recorded the name *vipicoli* among colonial Northern Tepehuán for gavilán altanero—perhaps also a cognate? In none of these cases is the biological referent known, other than that it is a diurnal raptor. (In Nabogame, I was told that the referent was an otro gavilán, distinct from the Red-tailed Hawk.) Among the Huichol, *wisa (witse)* labels an *Accipiter* species. Catherine Fowler (1983:230, 248–249) has traced the reconstructed **kwisa term broadly throughout Uto-Aztecan languages. Again, what species may be involved is ambiguous.

MISIDENTIFICATIONS. As noted earlier, authors have traditionally rendered the folk taxon *visag* as "chicken hawk," a quite meaningless term applied indiscriminately to diurnal raptors. In an 1877 Pima dictionary, Captain Grossmann ([1877]) came closer, giving *vísak* as "kite," a hawk with pointed wings like a falcon. Russell (1908), throughout his bilingual anthology of River Pima oral literature, translated *virsak (vishag)* as "hawk," as did Curtis (1908:20). Underhill, who gave only English translations, used the word *hawk* throughout her writings, but we know from other parallel sources, such as Russell, that the folk taxon mentioned in song and oration is usually *vishag*, the falcon. Slow-flying hawks were generally not worth singing or talking about; it was the large falcons that captured the O'odham imagination, as they have of peoples throughout the world.

Saxton, Saxton, and Enos (1983:87) identify *wishag, wisag* as both "chicken hawk" and "sharp-skinned [*sic*] hawk." Ruth Benedict (2001:144–153) recorded a lengthy narrative of Falcon Meeter but called him Hawk Man.

cheedagĭ visag, cheedagĭ vishag; cheedagĭ wisag, cheedagĭ wishag; pl. *chehedagĭ vipsag*, etc.

Falconidae, Accipitriformes (Falconiformes)

Falco peregrinus

Peregrine Falcon; Halcón Peregrino

The large, heavy-bodied Peregrine Falcon is similar in size and shape to the locally more common Prairie Falcon, but differently colored. The Peregrine is a dark bird above, the back of the adult slate gray, the head with a black helmet that extends below the eye as a broad black cheek strap. The underparts are light, spotted with dark marks, becoming denser on the thighs. Young birds are very dark brown above, densely marked with brown streaks and blotches below on a buffy background. In overhead flight, the wings and tail appear finely barred, but the wings lack the black axillaries of the Prairie. Whereas the Prairie Falcon has a rather restricted range, the cosmopolitan Peregrine occurs sparsely throughout Tepiman country.

The spirited Peregrine is reportedly one of the fastest-flying birds, its cruising speed around fifty miles an hour. But during stoops or steep dives, the bird attains speeds from two or three to perhaps even five times that amount (Cade 1982:62–63). This dashing raptor takes most of its quarry in flight. It may strike from above or even fly below, flip over and seize a bird from below. The bird has remarkable visual abilities; while it is perched in some open place, it can track birds passing overhead that are completely beyond human eyesight.

Peregrine prey is primarily birds. Shorebirds, seabirds, and ducks are important items. They may be its own weight or even twice that. In the Southwest, favored foods are pigeons, flickers, meadowlarks, jays, and even White-throated Swifts. Desert bats taken on the wing are also a regular part of the diet (Glinski 1998b).

Like the Prairie Falcon, the Peregrine builds no nest of its own, selecting for its eyries ledges on steep cliffs, usually near water or open country best suited for hunting. The three to five eggs are heavily spotted and splotched with rich browns and reds. Peregrines can be highly vocal and aggressive about the nest, driving away and even occasionally striking and injuring intruding humans (Bent 1938:61; Ginzburg 1997).

When Rosita Brown and Joseph Giff were discussing *vishag*, Prairie Falcon, with me, Joe said, "There's another one, they call it the *cheedagĭ vishagĭ*, that nests way on the cliffs, in the mountains." Rosita added, "It's bigger than this [pointing to a Cooper's Hawk specimen]." Joe noted that it was a dark bluish color, and he associated it with the Peregrine. He explained that although the Pima word *cheedagĭ* can mean either "blue" or "green," in this case it means the bird is bluish, in contrast to the pale, sandy-colored Prairie Falcon.

When Francis Vavages of Sacaton was telling me about the Prairie Falcon, I asked him about the other one. He said, "I've heard of *cheedagĭ visag*, but I think it's mostly around the mountains, right around the mountains." He had not seen it himself, just the brown and white Prairie.

Sylvester Matthias supplied additional confirming details. "It's dangerous, especially near its nest and have young ones. Oh, boy! You won't [dare get] close to it."

"Where did it have its nest?" I asked.

"Way up on the Estrellas—way on top; rocky places. And those that go hunt mountain sheeps have to watch out for them. I don't know if it's blue or green."

A few years later I asked Sylvester, "Does the *vishag* nest in the Sierra Estrella too?"

"Yeah, *cheedagĭ vishag*. That's mean."

"What's the difference between the regular *vishag* and *cheedagĭ vishag?*"

"They fly . . . they're *swift*! Just like a ball—hit the ground, pop. But the regular *vishag* is not that fast and doesn't fly right *at* you, but just fly right above you. No harm. No harm for people, the regular *vishag*. But the *cheedagĭ vishag* is harmful, danger[ous]."

"So it would hurt people or hit people, or what would it do?"

"Yeah."

Still another time Sylvester said, "They say the *cheedagĭ vishag* is mean. It'll go after anybody. The *vishag* won't bother anybody."

What he described accurately portrays the differences in nest defense behavior between these two falcons. The key characters distinguishing these two Piman folk species are speed, aggressiveness, and color. Like the Prairie Falcon, the Peregrine plays an extensive role in Piman oral literature, although it has been misidentified as a blue or even green (!) hawk.

There are many accounts of the conquest

cheedagĭ visag, cheedagĭ vishag, Peregrine Falcon (Mike Touch)

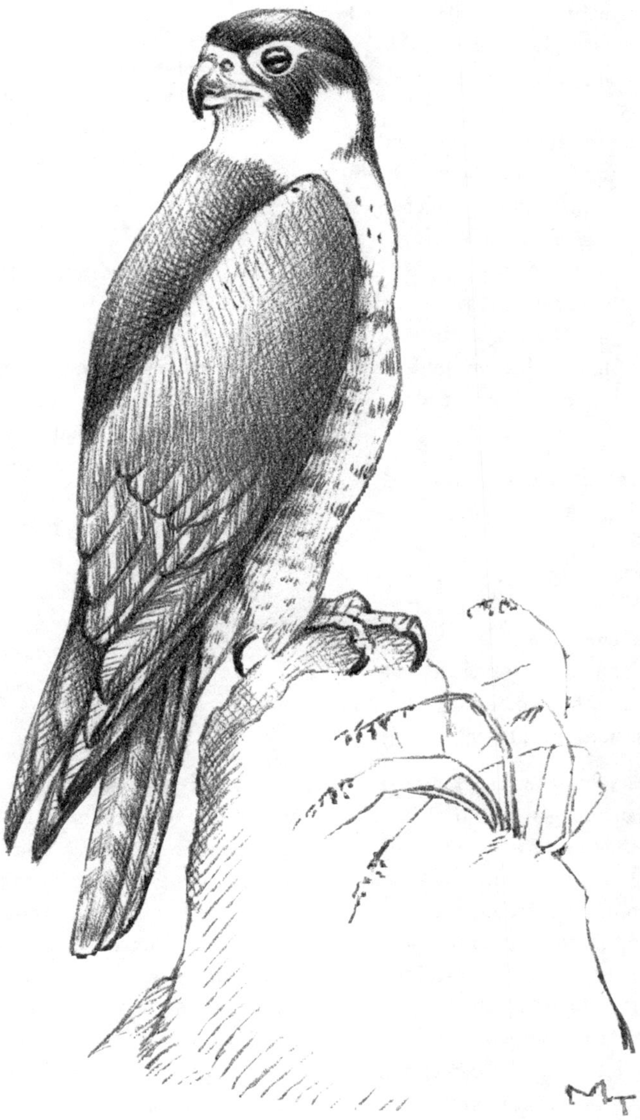

of the middle Gila pueblos by the Pimans, or Emergenti. In a greatly attenuated summary, Pima author George Webb (1959:92–94), raised at the base of the Sierra Estrella, gave an example of one battle to overcome a walled village. The first village the invaders encountered belonged to "Blue Hawk" in the vicinity of Blackwater. (Other versions of the Creation Story assign Casa Grande ruins to a different ruler, named just Cheedagĭ Siivañ, Blue-green Chieftain.)

Early in the Creation Story, the monster Ho'ok is conceived when a boy's wooden kickball rolls to a maiden who is sitting and weaving mats and impregnates her when she hides it under her apron (skirt) (Benedict 2001:27–29, 56–57; Curtis 1908:19–20; Ventura Jose in Evers 1980:110–121; Lloyd 1911:106–117; Russell 1908:222–223; Saxton and Saxton 1973:247–264). The offspring, a part-human witch named Ho'ok, is a cannibal who devours children. Ultimately, the People implore Ho'ok's relative Se'ehe (I'itoi) to kill her, which he does by inviting her to a dance, drugging her with the narcotic smoke of several plants, then cremating her in her cave. But this isn't quite the end of her. Sylvester told this story about Ho'ok's spirit:

> The cave is fifteen miles south of Poso Verde. [My father said] you can still see the ashes in the cave. [It's in Sonora] on the way to Magdalena, [called] Ho'ok Mu'akuḍ [Killing Place]. They took their horses [on pilgrimage to Magdalena]. Se'ehe wasn't fast enough [during the cremation], and the spirit got out through the crack [in the cave top]. And you can see his little footprint. The spirit became a *cheedagĭ vishag*. And they asked Se'ehe again [for help]. And he said, no. This time it's up to you. You have to get a virgin [maiden], one who hasn't had her period yet. And she made an olla. And it was *gai*, firing; it was tilted on the kiln. And Cheedagĭ Vishag swooped down [at her to kill her]. Went into the *ha'a* [olla]. And it burned up. And that's the end of that part. And that start[s] from a woman and end[s] with a woman. (Compare Benedict 2001:29, 61; Curtis 1908:20; and Lloyd 1911:115)

Sylvester had never read Levi-Strauss, of course, but he is pointing out here the structural parallel between the first maiden, who was sitting weaving mats when she conceived Ho'ok by hiding the boy's ball under her apron, and the final maiden, a potter, who captures Ho'ok's still-dangerous spirit in an open olla, a symbolic vagina, baking on the fire. (In three versions, there are two additional manifestations of Ho'ok's spirit: a boiling pot of greens [amaranth] that splashes and kills people, followed by two bears that kill people as well [Lloyd 1911:117; Rea 1998:206–207].)

Incidentally, Sylvester is not the only narrator who specified that Ho'ok's smoke or spirit became the Peregrine rather than the more common Prairie Falcon. In 1927, when William Blackwater told Ruth Benedict this story, she took the type of bird down as "Blue Hawk" specifically, the more aggressive of the two falcons, whereas her other source, Thomas Vanyiko, said only that "[s]he [Ho'ok] became a hawk [*sic*]" (Benedict 2001:29, 61). When Papago Ventura Jose told the story in O'odham, he likewise specified the manifestation as *cheedagĭ wishag* (Evers 1980:117).

Although one would expect Peregrines to have been more common along the Gila than elsewhere in the desert, Russell mentioned only regular *vishag*, the Prairie Falcon, in his extensive anthology of River Pima song and oration. Much of what we know about *cheedagĭ vishag* in Piman oral literature comes instead from Frances Densmore's study of Papago music and legend (but see the technical note in this account).

Densmore (1929:54–77) related the long story of the origin of the flute. At the beginning, Elder Brother's sister lives in the Sacaton Valley. She is called Acorn Eater. (This is an error, notes Culver Cassa. In other versions, she is called White-clay Eater [Woman], Toha Kua'adam; here she is called Acorn Eater, Toa Kua'adam—*toa* being one of several folk generics for oak among Pimans.) The girl has four suitors from the village, each of whom brings her game during his courtship: Cougar, Jaguar, Blue Falcon (Peregrine), and Eagle. They were reportedly the four "meanest" animals in the village; that is, they were the best hunters—two power mammals and two power birds. She rejects them all. Then she is magically impregnated by Gopher and bears twin boys. When they grow up, they marry two daughters of Oam Ñui, Yellow Buzzard, her opponent, who seeks their demise.

Ultimately, the spurned Blue Falcon is implored to attack the house where the twin boys reside, knocking out four of its sides. While doing so, he sings this song describing himself:

> Hanging motionless in the sky
> Yet terrible in its power to destroy.

At the end of his ravaging flights, he kills the two young husbands, occupants of the house (Densmore 1929:70). (In Herzog's [(1929)] version of this story, there are three falcons: *gia'adak visek* [Bowstring Falcon], *gook visek* [Two Falcon], and *koovuḍki wikamĭ visek* [Hollow-eye Falcon]—the first two wreaking havoc, the third killing the boys; see the technical note in this account.)

The specification of the Peregrine in this episode is not unique to Densmore's narrative. The version originating from Thomas Vanyiko and William Blackwater is similar: "[Yellow Vulture] came to the Blue Hawks. 'Come to my home, come to my home. I need you to help kill my enemy.' The Blue Hawks followed all the way to his home, and Bowstring Hawk cut off the top of the [girl's] house with his wings; two Blue Hawks [*sic*] crippled the two brothers; and One-eyed Hawk killed them both" (Benedict 2001:114).

In 1927, Ruth Benedict recorded another telling of this story from William Stevens. "Buzzard said, 'Let us call Blue Hawk. He is left handed' (equivalent to being an excellent shot). Blue Hawk agreed. He flew straight up into the air with his club and dropped upon them and killed them" (Benedict 2001:122 n. 88). This version describes the Peregrine's powerful stoop and the Pima belief that the large falcons kill by striking their prey with their clenched talons.

Pimans, acutely aware of the natural history of these two audacious raptors, appropriated their qualities into their metaphor: *vishag* is an image of power, but *cheedagĭ vishag* is also a dangerous killer.

Similarly, Underhill (1946:179) noted what must be *s-cheedagĭ vishag* in a war oration she transcribed "from south of the [Papago] reservation." Near its end, the speech says,

> Then, after four halts, did I reach
> The very blue hawk, my guardian.
> And I asked for power.
> Then kindly to me he felt, and he drew forth
> His bright blue stone.
> Therewith our enemy waters it made white as with the moonlight,
> And all around them I went striding.

In the heat of midsummer, the Saguaros begin to produce egg-shaped fruits filled with a sweet and flavorful red pulp and oil-rich black seeds. At this time, the People go out to camps to gather and preserve the seeds and boil down the juice into *sitol*, syrup. Some of the cactus fruit will be used in the wine feast to initiate the summer rains, essential for crops, both wild and cultivated. While the Saguaro juice and water mixture is kept warm in the ceremonial house, four prominent older men tend it, singing to assure its proper fermentation. Densmore (1929:154, 156) recorded a song of the fermentation watchers that describes both *ba'ag* and blue falcon:

> The eagle is flying above in a round circle
> and makes a round shadow on the ground.
> I am walking around under that shadow.
>
> The blue falcon is flying in a straight line and makes a straight shadow.
> Under that line I am running.

Four *mamakai* (shamans), meanwhile, attempt to locate and predict the rain through divination. (Bringing or continuing the monsoon precipitation is overtly the primary function of the communal drinking celebration.) While the *mamakai* are divining, the assembled singers assist them with song. Densmore (1929:156) obtained one such song used for this purpose:

> The eagle wing, under that I sat with the tiswin [*navait*],
> I drank that and I am well drunk.
> The water that gathered on the blue falcon feathers at the end of the *wasiw*,
> I drank of that and was staggered.

The word *vasiv* or *wasiw* refers to the rain wand used by a *juukĭ wuadam*, the rain shaman, explains Cassa.

The image of these two powerful birds, Golden Eagle and Peregrine, is invoked during this dangerous time when the wine is in transformation. It might fail and go bad. With song it succeeds. When all is ready, the large assembly from the four invited villages partakes. Soon the desert is drenched and transformed again into a green world.

I wondered if the Peregrine was associated with *kaachim mumkidag*, staying sickness, as was the Prairie Falcon, knowing that even closely related species, such as the Black-tailed and Antelope Jackrabbits, may differ in this regard. Late in his life, when his speech was limited, I asked Sylvester, "Does *cheedagĭ vishag* give you sickness?" He answered, "It will cut you." His response referred to the aggressiveness of this species about its eyries. However, Densmore (1929:90), in her enumeration of animals that cause sickness, listed "blue hawk," together with eagle and woodpecker, noting, "Symptoms not given but were said to be the same in diseases caused by these [three] birds." She listed woodpecker a second time, giving its symptoms as "boils and sores on the head." These symptoms are also characteristic of *ba'ag mumkidag*, eagle sickness, so presumably they would be part of Peregrine sickness as well. She did not obtain songs for curing any bird-caused *mumkidag*.

Simon Lopez, with extensive experience in *kaachim mumkidag*, thought that *cheedagĭ vishag* would pose the same dangers as the regular *vishag*, the Prairie Falcon, and that he would use the same songs and fetish to cure either affliction.

COMPARATIVE LINGUISTICS. Although the name *vishag* (or one of its cognates) occurs throughout Tepiman country, the biological referent is unknown for the most part. In *Birds of Sonora*, Russell and Monson (1998:78) comment, "There is every indication that the great barranca country in Sonora, Chihuahua, Sinaloa, and Durango is an important stronghold of the Peregrine Falcon." Because about nine-tenths of Tepiman country is south of the Prairie Falcon breeding range, the Piman folk generic in its unmarked form most likely refers to the Peregrine throughout this southern region—in all of Tepehuán country, at least. In Pimería Alta, both species breed in the desert, but the Peregrine is far less common (Ellis and Monson 1989).

MISIDENTIFICATIONS. As noted for the Prairie Falcon species account, the two folk species of *vishag* have consistently been misidentified in ethnographic writings as "hawks." Even the "official" American name for the Peregrine Falcon was for generations "Duck Hawk," a misnomer: hawks are in the family Accipitridae, whereas falcons belong to the well-differentiated Falconidae.

Although some of Densmore's (1929) literary references clearly do refer to the Peregrine, the obligatorily marked folk species, I'm not sure that all or even most of them do. It's doubtful she was aware that two folk species were involved here. She made no mention of the unmarked form—that is, just *vishag*, the Prairie Falcon. Yet this is the far more common species in the desert and the one to which nearly all contemporary accounts refer. It is the one most often matched in song with the Golden Eagle and the one whose handling required a four-day seclusion and purification ritual (see previous account).

Bahr (Bahr and associates 1994:312 n. 4) wrote, "Another instance of this [opposition of solar and wetness he discussed on page 139] is a green hawk that is formed from the smoke from the baking witch. I take it that this hawk is solar and yet its color is explicitly green (Saxton and Saxton 1973:295–304)." Here Bahr is reading into the Peregrine's O'odham name the wrong English color of the biological (and folk) species. (The association of *cheedagĭ* with *chuuhuggia*, the cooked amaranth greens, near the end of Lloyd's (as well as Vanyiko's and Blackwater's) story might suggest the color green, but the folk taxonomy here is clear: the Peregrine is slate blue.)

TECHNICAL NOTE. Benedict's accounts from Thomas Vanyiko, William Blackwater, and William Stevens, and Densmore's accounts from Sivariano Garcia, situated in the Sacaton Valley, all specify Blue Falcon—that is, the Peregrine—as killer of the twin boys. Herzog's does not, but his source is Vanyiko also, so we can reasonably presume the narrator intended the Peregrine rather than the Prairie Falcon. What Herzog's ([1929]:85a, 85b) bilingual text actually says clarifies some points: "Again Yellow Buzzard went, this way, toward the East, when nobody could do it [kill the boys]; there in the East dwelled the chicken-hawks [*vipshek*]. One was called Bow-String Chicken-Hawk [*gia'adak vishek*]. Another, again, was called Two Chicken-Hawk [*gook vishek*]. Another, again, was called Hollow-Eye Chicken-Hawk [*koovudki wikamĭ vishek*]. They were three. They did not pity anything."

Herzog's footnote clarifies "hollow-eye" as meaning "one-eyed." Benedict's writing "two Blue Hawks" in the second position (noted earlier) is ambiguous, but Herzog's O'odham version is not: the modified base word is singular; it is someone's name. And his narrator carefully noted, "They were three." Three is atypical in Piman enumerations, but it may signify foreboding from ritual incompleteness. Such texts portend doom.

sisikĭ

Falconidae, Accipitriformes (Falconiformes)

Falco sparverius

American Kestrel; Halconcillo, Gavilancillo

The American Kestrel, long known as the "Sparrow Hawk" (a misnomer in all respects), is common throughout Northern Piman country. This falcon, eight to ten inches long with a wingspan of nineteen to twenty-three inches, is unmistakable, even at a considerable distance. The long tail and pointed wings indicate it is a falcon. When it returns to a perch, it pumps its tail vigorously several times, the kestrel signature.

Its habit of hunting by hovering in the wind, with wing tips fluttering or sometimes even motionless, will draw our attention. So will its shrill, vociferous *kii-kii-kii* or *killi-killi-killi* calls uttered repeatedly and its rather tame behavior. Were that not enough, the dove-size bird will sometimes launch vicious attacks on hawks or other large birds unwelcome in its territory. The sexes, even in first-year plumage, are easily distinguished. The female has a rufous back and tail, extensively barred with dark gray, and white underparts streaked with pale rust. The male has somewhat barred interscapulars contrasting with spotted blue-gray wings. His tail is all rufous except for a broad black subterminal bar and a white tip and outer edge. His underparts, pale or even richly buffy, are spotted. (Young males have streaks on the chest.) Both sexes have a gray crown patch with a rufous cap, broader in females, and two distinct black facial stripes. The cere, eye ring, and feet are bright yellow.

Kestrels are at home throughout North America any place there is open country where they can hunt and holes where they can nest. Because of their small size, desert kestrels can appropriate flicker or even Gila Woodpecker nest holes in columnar cactus or cottonwoods and willows along the rivers. Here they feed on insects (especially grasshoppers), lizards, mice, and occasionally even small birds. In southern Arizona, G. Scott Mills (1998) noted them commonly taking the large cotton rats, once the esteemed game mammal of the O'odham.

This tiny falcon, called *sisikĭ*, was well known to older Pimans. On various occasions it was discussed in the context of small owls rather than hawks and falcons. When I asked Sylvester Matthias to list the kinds of owls he knew, he included *sisikĭ*. This puzzled me. Perhaps it is because the kestrel, like the small owls, is a hole-nester. Both Sylvester and Fr. Antonine (1935:9) said there was no plural of the word *sisikĭ*.

Sylvester noted that the *sisikĭ* nests in the *haashañ shuushud*, "Saguaro's vaginas," as the woodpecker holes commonly seen in the giant cactus are called in Pima. He explained this raptor's natural history with some amusement. "The poor *hikvig* [Gila Woodpecker] or *kuḍat* [Gilded Flicker] makes the hole in the Saguaro and the *sisikĭ* takes over. Just like in the *auppa* [cottonwood]—the *hikvig* makes the hole in the *auppa*, and the *kuukvul* [Western Screech-owl] takes it. The *haupal* [Red-tailed Hawk] builds a nest, and the *chukuḍ* [Great Horned Owl], the big owl, takes it."

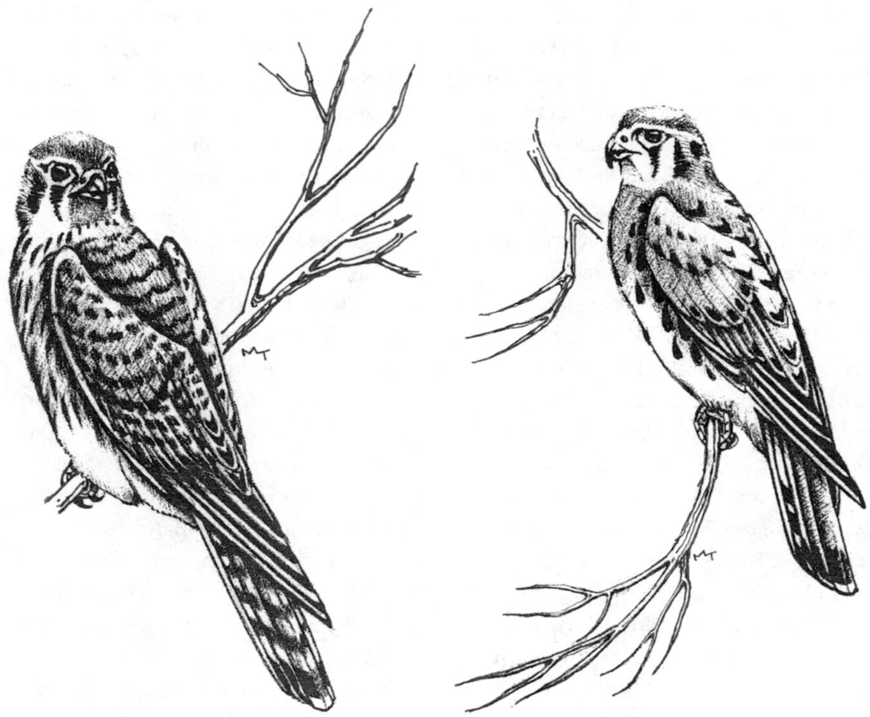

sisikĭ, American Kestrel, female, male (Mike Touch)

In earlier times, the *sisikĭ* was more prominent in Piman consciousness, Sylvester related. They were found in places with thick Saguaros such as the South Mountain area. People said that it would never live if you tried to keep it as a pet. Pima were not supposed to handle or pet them too much because they would die. They won't live in a cage (like eagles and Red-tailed Hawks).

Francis Vavages explained, "*Sisikĭ* is another in that family, too [with the Screech, Great Horned, and Burrowing Owls we were discussing]. Smaller. A little bigger than a quail." But another time he commented, "*Sisikĭ* is almost like that, that hawk family." We had been discussing the Northern Harrier. "It eats birds, too. Rob somebody else's nest. They're a little smaller. They're gray, and you know, the striped tail. Got that hooked beak."

"Where does that *sisikĭ* nest?"

"They got nest in the cottonwood trees, too."

"Have you ever seen them in the *haashañ?*"

"Oh, yeah. The hole—in the hole. That's where they come out. Same with *hikvig, kuḍat, ha'o* [yes]."

In the field, Delores Lewis of Ge Oidag identified a female kestrel that was near us as *sisikĭ*. In this context, he also mentioned the *vishag*, so he associated the largest and smallest falcons with each other.

I mentioned to Jose Enriques, another Tohono O'odham, that I heard little about the *sisikĭ*. "My grandfather used to say it's kind of sacred. It will give a curse [*tto mumkadaj*] or something. We don't bother them. We respect them. They eat small birds, too, like the *go'ogow* [shrike], the masked bandit." He spoke of *sisikĭ* in the context of *wisag*, the Prairie Falcon.

I know of no stories about this bird, but Joseph Giff (1980:137–138) sang this sunrise or "eastward-standing" song:

> Towaiwe'e yokimali
> towaiwe'e yokimali
> da'amaiwe'e kachime wui chume
> himedam
> sisike'e yehi'inge
> heweli kuk abe ya'ahaiwe nangeyaim
> wakwayame yuhinge
> si'iyali kekime
> ke'eḍa mo linekin.

Joe explained to me, "*Sisikĭ* is one of the morning birds—rides on the wind, watching everything below." He said that this particular song came at the very end of a series of swallow songs that tell of the different birds waking up in the night and singing at first light and finally as the sun appeared. This is how he explained it in his 1980 essay:

> That which I have already discussed, which are the eastward standing songs in which are told what the animals [*ha'ichu doakam*] do, and the birds, when the new day's sun shows. Of course, they wake up and get happy, see. And this song says,
>
> Toha hohokimal
> toha hohokimal
> daam kaachim wui chem al hii.
> sisik u'uhig
> hevel kuug ab hahiwa e-nagia.
> Vakoañ u'uhig
> si'al keekkhim
> k eda am hihinnag.
>
> White butterfly,
> White butterfly,
> Sky towards tries to go.
> Kestrel bird
> Wind's end-on hangs.
> Heron bird,
> East standing (dawn).
> Just then honking.

If one were to ask if the People of the desert had an aesthetic appreciation of their environment, this exquisite piece of sung poetry answers that question: in the moments before the sun breaks above the eastern horizon, the sky at this point becomes white, like a white butterfly trying to reach upward; the kestrel hovers with fluttering wing tips in the breeze; and a heron, arising from its roost with a ringing squawk, flies off to the river for its morning fishing.

This song, Joe explained, stands at the end of the series and "is sung when it is fully daylight, when they will close the sing. Then they'll go home when they sing,

> Gidval u'uhig
> gidval u'uhig
> e-ñe'ikaj chuudkhim
> hejel e-ñe'ikaj i an hu chuudkhim
> k am hu wo a'ai upam hihi.
>
> Swallow bird
> Swallow bird
> Dances to his song
> To his own song dances
> And in different directions they go
> home.

The singers, mostly older men, now disperse, going quietly in different directions to their mud houses among the mesquites and paloverdes.

COMPARATIVE LINGUISTICS. The colonial Névome vocabulario gave *sisika* as a gavilán (Pennington 1979:56), but three centuries later Pedro Estrella didn't recognize the word *sisikĭ* when I inquired. I don't think we encountered the kestrel in the field. He seems to have confused the small picture in my elicitation book with the trogon, and the voice he ascribed to it fits that bird, not the kestrel.

Pennington (1969:115) recorded the Northern Tepehuán name *shishikamuli* or *chichikamuli*, which appears to be cognate with *sisikĭ*. The bird was described as having a reddish tail and back, appropriate for both sexes of kestrel. Its feathers were used for fletching arrows. However, this raptor was said to attack chickens, a feat for the tiny kestrel, unless the chicks were meant.

Apparently cognate in Eudeve is a gavilán of unidentified species, *siibí* (in the objective case *siibík*) (Lionnet 1986:117; identified as a large species in Pennington 1981:125).

kusijim

Falconidae, Accipitriformes (Falconiformes)
Caracara cheriway
Crested (Audubon's) Caracara; Quelele, Caracara Común

A raven-size bird flaps powerfully on long wings through Saguaro-paloverde country, flashing a bold black-and-white pattern. On closer view, the windows of the wings and most of the tail show white barred with black, the tail terminated by a broad black bar.

When the caracara perches on a Saguaro or alights on the ground, its impressive features can be appreciated. The bird is large headed and heavy billed, with a black erectile crest and mostly bare reddish orange face. The neck, chin, and throat are white, becoming barred on the breast and back of neck. The rest of the bird is black in adults. On the ground, they move about readily on quite long legs. Young birds for two years are brown and buff, streaked and spotted, but still impressive with an imperial mien. They have class, and they seem to know it.

Caracaras are falcons who have thrown in their lot with the vultures, while keeping their other options open. "When it comes to diet," wrote Seymour Levy (1998:116–117), who has been studying the Papago Reservation population since 1960, "caracaras are the ultimate opportunists; they feed on any animal material, alive or dead, that can be caught or found. Carrion appears to be their main food." The Tohono O'odham are a cattle-raising people. Levy noted, "If unmolested, caracaras do not hesitate to feed on dead dogs and [dead] livestock within Indian villages in close proximity to buildings and traffic." Although powerful flyers, these aggressive falcons are not above hunting on foot as well.

Caracaras must compete at carcasses with Black and Turkey Vultures as well as with fellow opportunistic ravens. They usually forage when the vultures are not feeding, taking advantage of the early hours before Turkey Vultures are abroad or visiting the carcass after the others have left (Wallace and Temple 1987).

Levy (1998) found their nests to be made of fine twigs and stems of semi-woody desert plants, placed in the arms of a Saguaro. Raven *(hawañi)* and broad-winged hawk nests (especially *haupal*, the Red-tail) are made of bulkier sticks. Two heavily marked eggs are the usual clutch.

The Tohono O'odham, living at the Crested Caracara's northwest breeding limit, call this bird by the folk generic *kusijim*. Older people, in particular cattlemen of the southern and central parts of the reservation, know the bird well. It is usually brought up during discussion of Black and Turkey Vultures. For instance, when we were talking about avian scavengers, Delores Lewis of Ge Oidag mentioned *kusijim*, describing its head feathers as "slicked back," with "white on lower back, and eyes red." "What do they eat?" I asked him. "Frogs, butterflies, all that; rats, too." He said he sees one or two at cattle butcherings.

One late afternoon as we drove the several miles from the main highway into the village of Pisin Mo'o, an adult caracara flew with strong wing beats for a considerable distance ahead of the car, just clearing the tops of the mesquites. Vincent Josemaria associated the three scavengers and thought *kusijim* was the right name for the black-and-white one.

Vincent told us that the *kusijim* understands people. He used to live somewhere by himself. Another man in the area would go out to get firewood and was a good shot; that is, he hunted while he was out cutting wood. One time Vincent asked the man to bring him something. He brought a jackrabbit. He fried it. He threw what was left of the carcass outside. A caracara lived nearby. The caracara came and kept inspecting and pecking at the bones. Vincent went out and said to it, "Why don't you take some back with you and feed your babies. Eat with them." So it took it all and flew up to its home. The rabbit was heavy. The *kusijim* was flying low and finally struggled up to its nest.

Hiá ch-eḍ O'odham Luciano Noriega of Quitovac oasis in Sonora used a similar name, *kuhichum (kuhichim?)*. (I suspect he had not used the name in a long while and was a bit rusty.) One July at the oasis on the eve of the Navichu ceremony I questioned a group of men about two Common Black-Hawks, *Buteogallus anthracinus*, that were leaving the pond. A man who spoke only O'odham and Spanish called them *kusijim*—the birds by then having spiraled way high. (Both species share a contrasting black-and-white pattern in flight, although their wing shapes are quite different.) Philip Salcido from the group said there

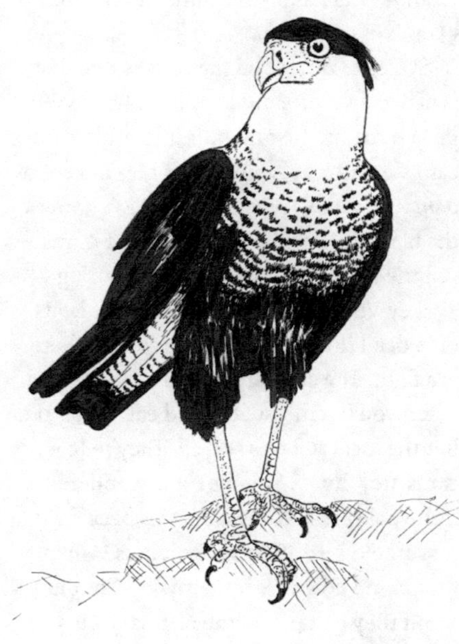

kusijim, Crested (Audubon's) Caracara (Amadeo M. Rea)

was a pair of *kusijim* around his place at Menager's Dam and that they are beautiful birds.

Farther north at Santa Rosa village Simon Lopez noted that the two kinds of *ñuñuwi* (vultures) have always been around, but lately he had been seeing another that hasn't always been there. Someone told him its name, which he thought was the *kusijim* I mentioned.

In my earliest studies with Sylvester Matthias, he told me that the Pima knew the Crested Caracara from their trips to Magdalena, Sonora, for the fall pilgrimage. He noted it was found from Topawa south into Sonora. The Pima didn't know if it was a buzzard or an eagle, Sylvester said, but the Papago told them that it was neither, but an entirely different bird. The Pimas, he said, never saw this bird north of Sells or along the Gila at their own home. Sylvester recalled no name for the bird from the Papago. No one on the Gila seemed to recognize the name *kusijim*.

However, Sylvester regularly referred to *oam ñui* as a species distinct from the Turkey Vulture, presumably the caracara. The Piman word *oam* can indicate yellow, orange, or various shades of brown. Joseph Giff thought the term was used merely as a proper name to a person in legends of the Maricopa, borrowed by the Pima, not to a biological species. The name Oam Ñui is frequently translated "Yellow Buzzard." According to Sylvester, Pimas "believe Pueblo Grande [the Hohokam ruins in Phoenix] is Oam Ñui's mansion."

Sylvester described the bird as one that eats like a vulture but is yellowish. Sylvester picked the brown and buff immature caracara skin as representative of his concept of *oam ñui*, yellow buzzard. He didn't know the black-and-white adult. He said it was not the local bird, presumably meaning Turkey Vulture. "When the local bird eats on a dead horse and this one comes, there is a fight. The caracara would walk right in when you butcher a deer and eat the blood, but the local *ñui* was scared. They're [caracaras are] dumb. Some people wondered if the pale area on the back was because it was an old bird. They make a noise when they come down with their wings, but not when they go up, like the ordinary *ñui* does. After rain, sit in a dead cottonwood, spread wings to dry." It was last seen in the early 1920s (1922, 1923). Before that, it was common.

He never knew where they breed or how they nest. (He noted that Turkey Vultures nest in caves.) There's usually a couple of *oam ñui* around, but there may be five at a dead horse in a field, with about a dozen bald-headed *ñuñui*.

There is some discrepancy between the bird that Sylvester's father's generation saw in southern Papaguería and the birds that Sylvester observed locally on the Gila. Although Sylvester didn't know the "correct" O'odham name *kusijim*, there can be little doubt that the bird he described is the Crested Caracara. He noted its boldness, which the local Pima associated with the bird being unwary or "dumb." The birds appeared not to be local breeders. His *oam ñui* seems to have joined local Turkey Vultures in roosts for spread-wing drying. (Spread-wing behavior in both Black and Turkey Vultures has additional functions as well.) And he contrasted his bird with the featherless head of the vultures. I suspect that Sylvester knew only immature birds at the periphery of the bird's range, north of the normal breeding area, hence his use of *oam ñui*.

Much farther south, Pedro Estrella didn't recognize the term *oam ñui* for the caracara. I doubt that I ever asked him about *kusijim*, a name that was not in my elicitation book at the time. Don Pedro used just the name *kélele*, changing the accent from Spanish *queléle*. Another time he called the caracara *nunui ba'aag*, "vulture eagle," saying, "es como águila, eats dead things."

I am aware of no O'odham legends or songs about the *kusijim*. Because early 1900s naturalists had seemingly not noted caracaras there, Levy (1961, 1998) surmised that the individuals discovered in the 1960s might represent a relatively new northward extension into Papago country as a breeding bird. However, explorer Carl Lumholtz (1912:392) identified the "Queléle" (Spanish for caracara) in Papago as *kúsit*, "a carrion hawk with a yellowish throat," when giving the name of the abandoned ranchería Kusit-mohitatsh, said to mean "Burnt Caracara" (*mehida*, "burned purposefully"). (That Lumholtz got two of the three syllables of this bird's name correct seems close enough to confirm the species; even this approximation was good for Lumholtz.) This village was a little south of the border. On his trip to the Pinacates late in 1909, Lumholtz (1912:192–195) was accompanied by an elderly Sand Papago (Hiá ch-eḍ O'odham) medicine man called "Queléle" ("Carrion Hawk"), then living at Sonoita. Apparently, this name was a Spanish translation of his O'odham name, Kusijim, although this connection was not stated. The nonanalyzable lexeme labeling this bird is presumably ancient, so at least southern Tohono O'odham have had a long association with this interesting bird.

COMPARATIVE LINGUISTICS. The Mexican Spanish common name for the caracara is *quelele*, said to be of Cahita origin, occurring in Ópata as *que'ere*, fide Sobarzo (1966). The Yoeme name I was given in Yaqui country was *keléle*, the same as its Spanish name, but Molina, Valenzuela, and Shaul (1999:34, 200) give *choawe*. For Tohono O'odham, Saxton, Saxton, and Enos (1983:36) list *kuusjim* as "a species of large orange bird (unidentified)."

The caracara should occur in at least the lower elevations of Tepehuán country, but its name does not appear in any of the colonial documents. Pennington (1969:118) notes *kushimuli* as quelele among Northern Tepehuán, evidence for a pan-Tipiman caracara term.

TECHNICAL NOTE. Although the breeding range of the Crested Caracara in Arizona is limited to the Tohono O'odham Reservation (from Menager's Dam to San Miguel, up the western flank of the Baboquivaris, north to Covered Wells and Coyote Pass, fide Levy 1998), nonbreeding individuals range regularly (at least formerly) north to the Gila River. In late winter 1854, ornithologist and physician A. L. Heermann (1859), following the Gila from Yuma to the "Pima Villages," reported seeing one or more caracaras every day. Ornithologist M. French Gilman (1914:261), who taught on the Gila River Reservation for seven years, noted seeing caracaras three times at Sacaton in early spring between 1908 and 1914. Perhaps this number includes the remains of one an Indian had shown him. Edgar A. Mearns collected a November specimen on the Salt River near Phoenix. Another was trapped alive at Casa Grande before 1919 (Phillips 1946). Louis D. Yaeger trapped and released one on or near the reservation in March 1935. I saw an adult at a Turkey Vulture roost in Komaḍk 4 January 1980. Occasional records continue (Monson and Phillips

1981; Witzeman, Demaree, and Radke 1997). All in all, River Pima would seem to have had ample opportunity to encounter these striking vagrants in their own country.

kakaichu, kakachu,* pl. *kakkaichu
***kakatu, kakatuuv* (Pima Bajo)**

Phasianidae, Galliformes
Callipepla gambelii
Gambel's Quail; Codorniz Desértica, Coconita

The plump body of the Gambel's Quail, about ten inches long, should set it off from all other birds in the Sonoran Desert. Where other quail occur, as in the grasslands of the higher desert or in the tropical thornscrub and tropical deciduous forest, the Gambel's is easily identifiable by its black topknot and chestnut flank feathers broadly streaked with white. In addition, the male has a bright rufous crown, an intricate black and white facial pattern, and a bold black patch on a cream-colored belly. Otherwise, the birds are gray above and on the chest and can be quite inconspicuous while foraging in the shadows of mesquites or other trees and shrubs.

Gambel's Quail are sociable birds, spending the nonbreeding season in family groups or small coveys. They communicate by means of an elaborate vocal repertoire. For instance, the low *whut-whut* keeps the group together as it forages across open spaces. A separated bird gives a loud *ka-káa-ka-ka* locational call.

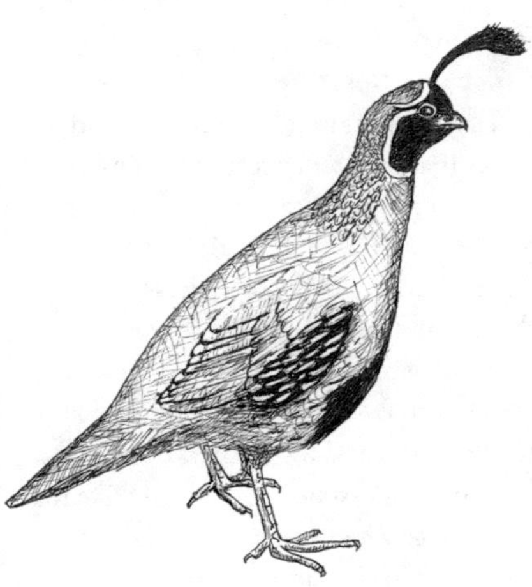

kakaichu, Gambel's Quail, male (Amadeo M. Rea)

At home in both the bottomlands and the bajada uplands, quail take kindly to traditional farming, finding food from disturbance plants, crops, and mesquite and cover in the brush fences that once characterized Pima farming. Here, too, they can find water, although they may live in places where moisture comes only from succulent plants. Grass-lined nests are often on the ground and may contain a dozen or more cream-colored eggs with purplish brown markings. Like other galliform birds, quail dust-bathe, creating depressions in the fine alluvial soils.

The O'odham name for this bird is *kakaichu* or *kakachu*. Fr. Antonine (1935) gave the plural as *kakkaichu*, a form I have heard also from various River Pima speakers. Pima Bajo say *kakatu* (or even *kakatuuv*), as did their more northern relatives at an earlier date. The name seems to come from the loudest call given by the birds.

The mythic origin of quail is specifically noted in one version of the Creation Story (Hayden 1935:8; compare Bahr and associates 1994:78). After the Flood, the three Original Beings—Jeweḍ Maakai (Earth Maker), Se'ehe (Elder Brother), and Ban (Coyote)—first make ants, then "timber," and then two kinds of birds to be sent out to the ends of the earth. Where the birds meet on their return will be the *hik*, the navel or center of the world. They make two quail, sending them west to the ocean, and one roadrunner, who is sent east to the ocean. "Then they got a piece of mesquite tree leaf and placed [it] on top of the quails' heads. That's a sign that at that time was the month in which the mesquite trees bring out the new leaf [March], that's a sign that at that time the water [winter] was gone; it must be spring."

It seems a bit strange that two quail were made, but only one roadrunner was created. But quail are conspicuously sexually dimorphic with strong monogamous pair bonds, whereas in the roadrunner males and females are alike.

Quail are an important Piman game bird. They were once hunted with bows and arrows, swatted with branches during nighttime hunting, and taken in specially constructed traps, *eḍpa*, made of Arrowweed or other small branches. They were considered to be particularly vulnerable when it was rainy or drizzly because then their feathers would be wet and the birds reluctant to fly. Discussing hunting seasons for game animals, George Kyyitan said, "But quails we can get any time."

Nonetheless, there was a danger in handling this bird. Russell (1908:80) commented, "The handsome topknot quail is the most abundant feathered inhabitant of the Gila thickets. It is tabued to the women, though no explanation for this could be discovered." Rosita Brown said that women clean them by dumping them in boiling water before taking off the feathers. "But you're supposed to pull the heads off *first*, or it will make you blind. Throw the heads away."

The head feathers have a specific name. David Brown pointed out, "*Sivoḍa* is the topknot of a quail; [also means] a headdress, something with the feathers standing up. Quail have it. *Hu'u* is a star. Together they mean 'comet,' *sivoḍa[kam] hu'u*."

The Gambel's Quail is among a group of animals associated in Piman thinking with staying sickness. Sylvester Matthias explained, "It isn't really that we honor the animals themselves. We honor I'itoi or Jeweḍ Maakai. We know the Creator. The topknot—can't be abused. Pull the front part off, skin it like a rabbit." He mentioned once about a person's going blind by touching these unusual feathers. "Yeah, there is always something's going to happen." He, too, said they pluck the whole head off and discard it.

People, including women, born early in the twentieth century told me they regularly searched for quail nests to rob. For instance, Sylvester said, "Boil eggs of *kakaichu—nonha*, hard, soft, or scrambled. Or dig a little pit and bury them in the ashes of a small fire. Like soft-boiled eggs. Still eat them even if the chick is started. [But a] young girl in her period, certain things are prohibited." According to Ruth Giff, "Once in a while, if we're lucky, we'd find about a dozen [*kakaichu nonha*]."

"Sometimes they boil their eggs when they got it," noted Irene Hendricks, "but they're too [so] small. The old people are [were] eating them. I just saw them."

Likewise, Francis Vavages of Haasañ Keek (Sacaton Flats) said, "When they were chopping cotton, weeds, sometimes they'll find a nest with a lot of eggs, *kakaichu nonha*. There'd be about fifteen or twenty in one nest. That's why you see *lot* of them when they hatch. And somebody would come along and kind of look over the sun [candle them], see if they're too old. Sometimes they're start[ing] to form, little quail.

"Can you eat them when they start to form?" I asked.

"No. Just let them go."

Theodore Peters recalled that *o'okokoi* and *hoohi* (White-winged and Mourning Dove) eggs might be taken when encountered. "Everybody ate *kakaichu* eggs. They got lots of eggs. When they hatch, sure run fast. These others just got two eggs. If started, don't eat them. After you eat *kakaichu* eggs, you have to wash your hands. I don't know why." Russell (1908:192), discussing hunting by boys, noted that "the warning, 'If you touch quail eggs you will go blind' served most effectually to protect one species at least." Clearly, it didn't: precautions had to be taken with handling the eggs, just as with the whole bird.

Frank Jim of Hodai Shon Wo'o (Charco 27, Western Papago) said, "Same thing [as with hand washing after skinning quail], we used to go out there and hunt their eggs. We get their eggs and take them home and boil them and eat them. But they said we cannot use it, we have to wash our hand before we use it. Sometimes we go out there and take our slingshot and go out for a rabbit, we find some, we pick it up, take them home."

We learn more about staying sickness in relation to the quail from Tohono O'odham, where traditional healing is still preserved. In earlier years, Densmore (1929:88, 90) linked quail sickness with sore eyes. Underhill (1946:294) summarized this sickness for Papago: "Quail [sickness]. *Symptoms*: sore eyes. *Fetish*: the quail's top-knot. *Musical instrument*: scratching stick and rasp." She recorded one Tohono O'odham curing song:

> Quail children under the bushes
> Were chattering.
> The comrade [Coyote] heard them.
> He ran up, pattering softly,
> And stood wiggling his ears in all directions.
>
> Gray quails in a group alight,
> Gray Cousin [Coyote] arrives and glances backward.
>
> Blue quails in a group come running
> Gray Cousin [Coyote] looks around and glances sideways.

Russell (1908:312) obtained two similar medicine songs among Akimel O'odham:

> Kokomangi kakaichuve kakawuḍ miyañida
> vemengali melivak dadai bañeinahime.
>
> Chechenangi kakaichuva kakawuḍe voivekai
> vemengali ñeinoke gakai vañeinahime'e.

Vincent Josemaria of Pisin Mo'o explained that *kakaichu mumkidag*, quail sickness, comes from the *siwoḍa*, "topknot." The symptoms are eye sickness, almost like pink eye, he said. To prevent it, put your foot on the bird's head and pull it off. Don't touch it. The fetish to be used during the *kulañimada* (healing ceremony), together with the singing of the Kakaichu Song, is *wepegi washai*—the seeds, *muḍadaj*, of a red grass. The tassels would be rubbed on the patient.

Simon Lopez of Santa Rosa village told this story:

> They used to tell us, when you eat *kakaichu*, don't touch your eyes or anything. When you get through, wash your hands real good. The same way with the *chuk chuuwĭ* [Black-tailed Jackrabbit], the same thing: when you eat the meat, don't touch your eyes or anything. Wash your hands real good. And then do it, because they're the same thing. But *kakaichu* is the worse ones that'll make you sick. But there's something else that they say [that affects you]. But most I remember is the eyes that they say. That's going to get you blind and from what I hear is from my late grandma from my mom's side, she got blind from that, from the *kakaichu*. Because what they said, that she really loves *kakaichu* meat. And I guess she didn't take care of herself when she eat it or something like that. Yet she eats it a lot. My grandpa used to bring a whole bunch. He shot 'em with a shotgun. He brings it and he cooks them, eats them. And that's when I ate some of that meat, *kakaichu*. But they're good for food for old time, also, like rabbit and *koson* [packrat] and things like that. Like I said, my grandma probably didn't take care of, you know, too carefully with that, and that's how they got the blindness.

I asked him if he would call that *kakaichu mumkidag*. "Yeah," he said, "that's *kakaichu mumkidag*. Yeah, that could be the same thing."

I had a similar conversation with a Tohono O'odham, Jose Enriques of Queen's Well. "What kind of birds' eggs did people from your village eat?" I inquired.

"I think mostly quail."

"Did you have to be careful when you ate *kakaichu* eggs?"

"Of *course!* Yeah. That's what I was going to say. I guess they were real careful, not to, you know, have to wash their hand, handling them, or whatever."

"What happens if you're not careful around the *kakaichu?*"

"Well, to me, I've had that curse on me, one time. There was a black man who really likes *kakaichu*. He come out here and use a shotgun and shoot 'em down. And we just pluck their feathers off, you know. And I think that's the worse thing that we would [could] have done, because to the O'odham it's not good. Like when you pull their feathers out, the dust or whatever is in there will start blowing, get in your eyes, and that's where it really hurts your eyes. It'll go real sore, you know. I had a bad one one time. Really *bad* sore eyes. And I went to the medicine man, and he told me it was that, *kakaichu*. 'Cause at that time, I told you, that colored guy, cook some, and I even eat some. They're real *good!* And I *love* it; but after it happen to me, oh, it was *bad*. Couldn't hardly stand [it]. So I finally went to the medicine man, and he finally told me that it was *that*. The *kakaichu* made you have your sore eyes. And I guess it's about that time they start making eggs and babies, breeding time or something like that. So I went out. And he told me that, 'You go out and you kill one and you eat it, but you have to ask for healing, stuff like that. And I guess you should do that, because you already ate it. You should be, you know, because that's what happened, you know. You went wrong in doing it the wrong way. Next time there's another way you can do it: skin it like a rabbit. And it's easier that way, too, because all the feathers and stuff are intact, and get rid of it, and get the meat out of it.' And that's what I done after that. I kill one. I dress it out there in the fields, not here at home. Left everything out there."

"Do you have to do anything special with the skin and the feathers?"

"No. I understand now why it's like that. Remember when you've seen those little birds, they will go into *anything*, all kinds of brush that grows, yet some of those underbrush got some kind of sickness [?] on there. They rub on there and that's why [it's] in their feathers, the dust in there, mingle with it. [When you] pluck it out, it smokes out, something like that. And the dust [gets] in your eye. That's what causes the blindness, especially that, what you call that brush got little red berries on it?"

"*Kuavul* [wolfberry]?"

"*Kuawul*, yeah. So you usually seen them under those brush. I guess that's why it hurts so much. When you have that contact the eyes. I guess that's the only reason. That's how I figure out after they told me about it. It's not the meat, it's just the feathers that's no good to your eyes. I think it's just the feathers, the dust or whatever, it carries on the feathers. It's mostly the eyes."

"Did anybody sing for you? Did you have a *wusota?* Or did you just do it yourself?"

"I just done it that way, the way I was told to do. Like I said, nobody remembers that song. But it helped, though. I got rid of that sickness. I got over that bad blindness and got better. But I *do respect* that, too [laughter]."

A bit of natural history is embedded in Jose's story. Quail are fond of dust-bathing. They frequent specific dust wallows that are often out in the flats where *kuavul*, Lycium or wolfberry bushes, grow. So the dust in their feathers is real.

When I asked Frank Jim what birds might cause *mumkidag*, he included *kakkachu* among them (he used the shorter version of the name, without the dipthong). One symptom he mentioned was itching. When he had a case in his own family, he went to a female *maakai*, who said to find someone to *wusot*, but nobody in his village knew the requisite singing, so he asked someone to shoot him a quail. He took the feathers of one side and used that in the healing, and the ill family member was then all right. "If you skin a quail and touch the *siwot* (topknot), don't touch your face or you'll go blind. You have to wash your hands. It don't bother us out there—we don't have that sickness—but over here they have." Frank made several points. His district didn't have quail sickness, so no one knew the songs for it. Sicknesses are regional. "We have the *kakkachu*, but we don't have the sickness." If no one was available to sing, then using the fetish alone was the next resort. The person afflicted in this case had not commited the infraction, but her father a long time ago probably had. Frank's use of the feather fetish without the song in this case was effective.

In a long set of songs sung at the curing of *kaav mumkidag*, badger sickness, the last one is about young quail flying to roost at dusk, with darkness settling on their striped crowns (Russell 1908:322). The song may refer to the barred brown and buffy plumage of the downy young.

> The light glow of evening;
> The light glow of evening
> Comes as the quails fly slowly,
> And it settles on the young.

Quail are not major metaphors in longer O'odham narratives but nonetheless play roles in short stories. Many of these stories involve children, in particular boys, and play on the highly social nature of this bird.

In several short tales involving this bird, what the quail reportedly say seems to imitate their actual calls. In one story, a Prairie Falcon attacks a covey of quail. The last one escapes, running home crying, *"Wees t-hugio! Wees t-hugio!" (Ka káa-ka-ha! Ka Káa-ka-ha!)* (Saxton and Saxton 1973:142–143). This is the sound a quail makes when it is separated from its covey. In English, the quail says, "He finished us all off."

In another story, some quail come upon sleeping Coyote, cut out some of his belly fat, and roast it. Coyote is aroused by the smell of the fat cooking and comes to steal it from the quail. They shout after him, in mockery, "Coyote ate his own fat," or, in another version, "Coyote, you ate your own meat." (See Benedict 2001:173; Russell 1908:243; Shaw 1968:90–94.) The Pima-language version, not given in any source, was probably *"Hejel e giig at huu!"* and *"Hejel e chukug at huu!"*—both phrases in imitation of quail calls.

Russell (1908:247) recounted a "nursery tale" of a quail with more than twenty children who were in search of water. They said, "Where can we get some water? Where can we get some water?" The phrase in Pima—*"baatt am t va'igo?"*—is imitative of one of the quail's calls. They found some muddy water unfit to drink, sickened, and died.

Other stories do not replicate the sounds quail make. Densmore (1929:56), Herzog ([1929]), and Benedict (2001:100, 126) have an episode in the story of the twin boys where quail come to eat seeds under their hammock (a swinging cradle), keeping them awake. The boys kill the quail, then cry uncontrollably when their mother returns and throws the dead birds away. She thinks a suitor has deposited the game there out of pity, but the truth is that the boys are beginning to develop their hunting skills. They gain maturity and independence by killing progressively larger prey: next the woodrat, then the cottontail, then the jackrabbit, and finally a Mule Deer. Later in this story, when the boys are themselves killed by Peregrine Falcon, the quail gloat, singing over their dead bodies (Benedict 2001:114):

> The old woman had handsome boys.
> How will she feel now that they are
> dead?

In one version of the mortal competition between Yellow (or Brown) Coyote and Sandy-colored Coyote, the nephew sets out to avenge his father's death by gambling with his uncle for his life. The boy creates a nest of quail eggs in a bush and wagers that he can find such a nest. "This is not the season for young quail," laughs the uncle. They go to the bush, and the boy wins the wager (Benedict 2001:84). The species of bird apparently is not important to the narrative because variants of similar stories involve the eggs of a Curve-billed Thrasher or a dove.

In another frequently told story among O'odham, some quail taunt Coyote, who chases the covey down a hole. The first one in takes some pieces of cholla cactus to the bottom. As Coyote digs out each bird, he asks it if it was the one who made fun of him. Each one denies it and is released. At last reaching the cholla, now covered with quail feathers, and receiving no answer to his question, Coyote decides it is the guilty one, bites into it, and is killed (Benedict 2001:173; Russell 1908:243, 249; Shaw 1968:76–78).

Coyote is not always a failure. In one short story (Saxton and Saxton 1973:90), in remorse for his negligence in supporting his wife and family, Ban decides to go hunting. But chasing game is too much work, so he uses song power. When he spies a covey of quail, he sings:

> Yali kakaichu yuhinge,
> shaam ingi ñielinoke?
> gamu wewesi ñeñehima
> hema pi e memeko
> chum añi ingi yungionga ñi-elihina.

> Little quail birds
> What do you think of me?
> Over there all fly away.
> One does not go very far.
> I wish I could eat it.

Each time he sings the song, one of the quail does not fly well, and Coyote catches it.

The quail's fidelity to the land was noted in another story. There was a severe drought in the land of the Tohono O'odham. People and animals had to leave or go far to find water. "Even kah-kie-choo—the quail—who seldom leaves his own land, was forced to go away" (Wright 1929:225).

In Akimel O'odham country, William Blackwater told another story of a famine that forced the people into vulnerable areas to gather seed (Benedict 2001:155). The Apache kill everyone except a small boy, who crawls under a bush to cry. He tries to hunt, but the Apache had killed the birds as well. There is one remaining quail, and she and the boy form a bond, living together. One day the quail is gone for a long time, and the boy, searching for her, gets lost and dies. The quail is left alone. This deceptively short story of a few lines carries much biological meaning. It tells the listeners that in a food shortage they can eat the seeds of a robust amaranth plant called *giak*. Quail are also fond of these seeds. Both the boy and the bird belong to highly social species that are severely stressed by isolation from their own kind. Both cry when separated from their group. The Pima hunter knew well the quail's loud location call, as noted in one of the stories given earlier. We are left to imagine the quail's cry.

In another narrative, a warrior named Falcon Man sings this song on going out to war against the Apache:

> I drank with the quail
> And with drunkenness
> He painted my face with black paint.

When setting out on the warpath, the Piman man typically blackened his face (see Underhill 1938:71, 1951:94). The male Gambel's Quail has an all-black face, narrowly edged with white. So in this song the bird paints the man's face in imitation of his own. On his return, the nighthawk paints the warrior's face white like his own (Benedict 2001:149).

Like the *hoohi e'es*, "dove's planted crop," there is a *kakaichu e'es*, the "quail's plant," in English coincidentally called the Quail-plant or Alkali Heliotrope. The association with quail is seen in the curved white flower head of this plant. I once heard the plant called *kaakaichu sisivoda*, "[plural] quails' topknots." Sylvester noted that the plant itself is not an *e'es*—that is, something planted or cultivated; the allocation is only allegorical.

Among both Akimel and Tohono O'odham, a constellation consisting of a cluster of small stars was called a flock of quail (Leslie Pasqual and Joe Lopez; also Underhill 1939:126).

Several species of quail occur in Pima Bajo country. In addition to the Gambel's Quail, there is the common Elegant or Douglas' Quail *(Callipepla douglasii)* and at higher elevations the Harlequin, Montezuma, or Mearns' Quail *(Cyrtonyx montezumae)*. Pedro Estrella was familiar with all three. I am not certain of the correct folk-specific Pima Bajo designations for Gambel's and Elegant. (We never encountered Gambel's in the field.) Pedro Estrella noted that the first had *tutk siswodak* (black topknots), whereas the other had *s-vepek siswodak* (reddish topknots). María as well as Pedro called the most common species locally, the Elegant Quail, just *kakatu* or *kaakatu*. Hale (n.d.a) wrote, probably more correctly, *kakatuuv*. Pedro distinguished the Montezuma Quail in the sierra as *hihapsh kakatu*, noting that it flushes at your very feet. (This behavior has earned it its vernacular English name, Fool Quail.) He said that a quail larger than the Gambel's, with a red topknot, is called *kuitu* because of its song, *chitui-tui-tui*. (This sounds quite like the song of the cardinal, which both Pedro and María identified as *sipook*.)

COMPARATIVE LINGUISTICS. Various quail names are used by Uto-Aztecan speakers of the Northwest Mexican Ranchería Complex. Some examples are Eudeve, *kúi*; Tubar, *tsokoará*; Rarámuri, *ŕe'chorí*; and Northern Tepehuán, *váivoli* (Pennington 1969:116). Among Tepecano, Mason (1917) found *súrui, súhrui*. For Gambel's Quail in Yaqui country, I was given *súvai* (listed as *suva'i* in Molina, Valenzuela, and Shaul 1999). Most Mountain Pima gave me *hohog, hohogi*, or *hohok* for Mearns' Quail. The people of Nabogame ranchería, however, called it *tesholi*. Although some of these quail words from the more southern tribes are undoubtedly cognate with each other, none is cognate with desert Piman *kakatu*, in use from the Río Yaqui north to the Gila. In seventeenth-century Névome, it was *cacatuba* (= *kakatuvă*) (Pennington 1979:22). More work is needed to learn if this folk generic extends southward into Tepehuán country. The name may be restricted to locations where people are familiar with the Gambel's (and Elegant?) Quail's distinctive location call.

According to Johnsgard (1973:509), the generic name for the bobwhite quail, *Colinus*, is from Spanish *colín*, which in turn is derived from Nahuatl *zolín*.

s-hadam kakaichu (kakachu)

Phasianidae, Galliformes
Callipepla squamata
Scaled Quail; Codorniz Escamosa

In the semidesert grasslands, sporting interspersed yuccas, mesquites, or prickly-pear clumps, plump bluish gray quail, reluctant to fly, may be seen running along trails or across dirt roads. These birds are slightly larger than the Gambel's Quail of the mesquite desert country. A closer look will show narrow black scale marks on the feathers of the back, neck, and underparts—the source of the bird's name. The olive gray flanks are broadly streaked with white marks. And, most important, the short triangular crest, white or buffy at the tip, distinguishes this "cottontop" from the more elaborately decorated Gambel's Quail.

The Scaled Quail's pale coloration makes the birds hard to see among the shrubs and grasses. Vocalizations give them away. The distinctive *pei-kóos* or *che-kaar* separation calls are heard frequently during the nonbreeding season when large coveys assemble. In summer, other calls are given by these ground-nesting birds (Anderson 1978).

In Arizona, the ranges of the two species may overlap at their peripheries. With overgrazing of grasslands and desertification, Gambel's Quail tend to replace Scaled Quail at this interface. Both species of quail were important food resources to native southwestern peoples, and quail bones generally are recovered among the faunal remains from archaeological sites. Careful identification of these bones can indicate the local habitat at the time of site occupation (see Ferg 1985; Ferg and Rea 1983; Phillips, Marshall, and Monson 1964:29; Rea 1973). The western range of the Scaled Quail generally reaches to the Baboquivari

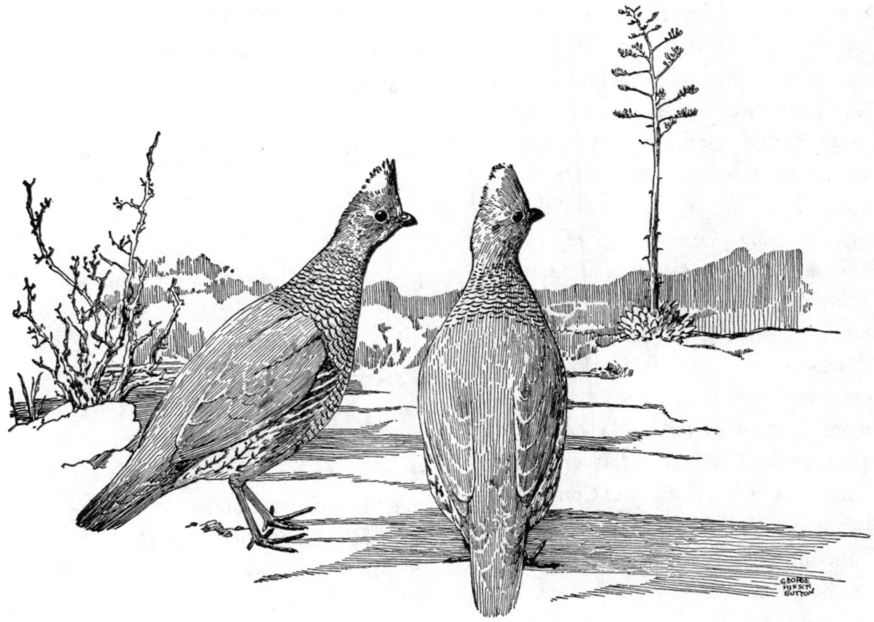

s-hadam kakaichu, Scaled Quail (George Miksch Sutton)

Mountains, but David Brown (1973) has traced the species well into Papago country, a bit beyond Sells, after wet winters.

Near the end of September, when the River Pima made their annual pilgrimage to Magdalena, Sonora, to honor San Francisco on the fourth of October, they encountered along the way two birds that were unknown in their own country. One of these birds was the Black Vulture, for which they used the local Papago name, *ñuipa*. These large, conspicuous birds, usually in flocks, were hard to miss. The other was a much less conspicuous quail.

Sylvester Matthias described the *s-hadam kakaichu* as larger than a regular (Gambel's) quail, but without the topknot and all gray with no distinctive markings, such as on the flanks or belly. He called them "mountain grouse" in English, while noting that they run like a quail. "[We found them] in Baboquivaris, when [we were] making pilgrimage to Magdalena. Also [found them] on other side of the border [in northern Sonora]. Five make a good meal; take a shotgun and get some along the way to Magdalena. *S-hadam* means 'sticky,' like gum or glue; we don't have them round here [Gila River]—need higher places." The species' name he said came from the texture of the meat, considered to be more sticky than Gambel's Quail meat, which the Gileños were used to eating.

The Piman adjective *s-hadam* is an interesting one in that it embraces two quite different kinds of stickiness, much as in English: a gluey type and a spiny type. The Teddy-bear or Jumping Cholla is called *haḍshaḍkam*. Its easily detached branches have the characteristic of adhering, as the O'odham name says.

I long wondered if Sylvester's name for this bird might be idiosyncratic, something invented by transient River Pima. Papago living near Baboquivaris should be familiar with the bird, I thought, and have a name for it. I phoned Jefford Francisco, a Tohono O'odham wildlife specialist from Topawa village at the west base of the Baboquivaris. He said he would check with the old people. A day later he returned the call. "Yes, we have a name, *s-hadam kakaichu*. It's the Scaled Quail." He found various people living in that region of the reservation familiar with the bird and its name. Once again, as with the *ñuipa*, Sylvester's folk taxonomy was vindicated. O'odham, at least in the old days, encountering new species where they traveled through Pimeriá, would adapt local names into their lexicon.

olas kakaichu (kakachu)
o’’ok kakaichu, o’’okĭ u’uhig
(Tohono O'odham)
Phasianidae, Galliformes
Cyrtonyx montezumae
Montezuma, Harlequin, or Mearns' Quail; Codorniz Pinta, Codorniz de Moctezuma

In the larger mountains that rise out of the Sonoran Desert are bands of oak woodlands with perennial grasses. In this habitat, sandwiched between coniferous forests in the higher elevations and desert flatlands in the lower, one might encounter a rather specialized quail, quite unlike the species found at lower elevations. Its intricate dorsal pattern renders the Mearns' Quail practically invisible as it crouches among the grasses and forbes. The encounter might be an explosion of several birds almost at one's feet.

With a good view, however, the bird is unmistakable. The males have a deep chestnut patch on the chest and belly, bordered by gray feathers boldly marked with large white spots. Were this not gaudy enough, the head is white with broad black markings and a short brown crest at the back of the head. This wild Picasso design is the source of the bird's other name, Harlequin Quail. Females have opted for more concealing coloration below.

The bird is shaped unlike any of the several species from the lower country. The body is more compact, the tail and neck short, and the head thick. Sturdy feet, strong claws, and a thick bill enable the bird to dig out food among grass roots.

The prehistoric Mimbres people in their upper-elevation homelands in the Mogollon highlands were well aware of the Montezuma Quail and painted the male unmistakably on their black-and-white ceramics. Pima Bajo and Mountain Pima were also familiar with it, as noted earlier. But did the O'odham know this bird? They might have encountered it on their former counter-raids into Apache country to the east. And the bird survives in the upper parts of the Baboquívari Mountains, which reach higher than 7,700 feet and form the eastern border of Tohono O'odham country.

I went to visit Joe Lopez at Vaya Chiñ. "How many kinds of *kakaichu* do you have here," I asked, "just one or several?"

"Well, we have one that I know, but on

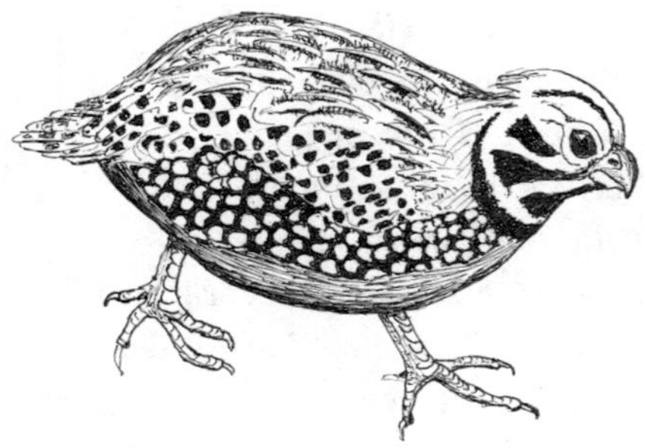

olas kakaichu, o"ok kakaichu

Montezuma (Mearns') Quail, male (Amadeo M. Rea)

the Baboquivari they have that, what they call the *olas kakaichu*."

"Have you seen that one before?"

"Oh, yeah. We used to see quite a few when we go pick acorns, *viyóodi*. We find them in the higher altitude. They fly just [he motioned them flushing at foot] . . . camouflaged."

Joe's name, "round quail," is an apt one. I passed some field guides around to three other O'odham speakers present who had never seen the species before, and they all picked the Montezuma Quail illustration on the basis of the Piman name. All agreed that the local Gambel's Quail they were fond of eating would be called just *kakaichu*.

Joe knew of the Masked Bobwhite, *Colinus virginianus ridgwayi*, an endangered subspecies of quail from northern Sonora that once ranged north into Papago country, but he knew no O'odham name for it, nor did he recognize the word *s-hadam kakaichu* for the Scaled Quail. "The regular *kakaichu* [and] the *olas kakaichu* are the only ones I know."

Later that evening on my return to Sacaton, I visited the Jim family, relatives of the Joe Lopez family, and shared my discovery. Gregory Jim, born in 1950, had hiked in the upper Baboquivaris and had encountered the bird there. "I thought they were just young *kakaichu* up there that had eaten too much!" He, too, found the name perfect.

But the story does not end here. Tohono O'odham wildlife specialist Jefford Francisco from the Topowa area west of the Baboquivaris, asked Joe Enriques (younger brother to Jose Enriques, with whom I worked) if he knew the quail from up on the mountains. Joe did, calling the Montezuma Quail *o"ok kakkaichu*. The word *oks* means "old woman" or "wife," but here it is used in the adjectival form, "old womanish." Jefford was curious about this name. Joe explained that the appelation had several connotations. Mearns' Quail are round, like plump old ladies, and they form tight little coveys like a bunch of old ladies. They are secretive. And they also have a quavering descending call and an alarm note suggestive of old women.

I asked Jose Enriques if he knew any other kind of *kakaichu*. "My mom said there was an *o"okĭ u'uhig*, like a *kakaichu*, kind of round, little round. Its cries are different. On Kitt Peak. And Comobabi [Koom Vavia, Hackberry Well] Mountains. Long time ago. There's different vegetation on the north side, Bear-grass, and oaks, agaves, *howich* [Banana Yucca]." He was unfamiliar with *s-hadam kakaichu*, the Scaled Quail.

Lowland Pima Bajo Pedro Estrella was familiar with this quail, which he called *kakatu hihapsh*, and with its distinctive behavior. "En la sierra" (in the mountains), he said. Pennington (1980:195) recorded the name as *hubigusikat kakatu*, "lost quail." In Mountain Pima country, where this species is common, the bird are called *teshuli* or *teesholi* (Yepáchic) and *hohogi* (lower villages).

COMPARATIVE LINGUISTICS. For the seventeenth-century Névome vocabulario, Pennington (1979) has carefully footnoted annotations that had been made to the lowland Piman lexicon in a different handwriting. These annotations, interestingly, can be traced to Mountain Pima words or to higher-elevation organisms or both. For instance, at the *cacatuba* (= *kakatuvă*) entry, is the addition, "aquí [le] dicen *ioogui* a la codorniz" (here they call the quail *ioogui*). This *io'ogi* is the modern *hohogi*. A synonym given for *kakatuvă* is *hoboca* (= *hovoka?*). Farther south in Northern Tepehuán country, Rinaldini in 1743 recorded the codorniz as *joogue*—that is, *ho'oge*.

TECHNICAL NOTE. Jefford Francisco wrote the word *oo'ok*, and Mathiot used *o"ok*, with a double glottal stop, virtually identical; in either case, the distributive or plural form is intended for this marker.

tova*, pl. *totova*; *towa*, pl. *totowa
***tovo*, pl. *tootovo* (Pima Bajo)**

Phasianidae, Galliformes

Meleagris gallopavo

Wild Turkey; Guajolote, Cócono, Pavo

In spite of its large size, the Common or Wild Turkey of North America, ancestor of the various domesticated strains, can be surprisingly secretive. An entire family can vanish from view in moments. The subspecies found in the Southwest are similar to those of the eastern United States except that the broad tips of the upper tail coverts and of the tail feathers themselves are white or buff rather than russet and much more white shows in the spread wing.

Although the plumage of gobblers and hens is similar in pattern, the males are larger and more intensely colored. The skin of his head is brightly colored red and blue, and a bristlelike "necktie" hangs from his chest.

Today the turkey is resident in most of the larger mountains of the Southwest, where it is at home in the pine-oak woodlands and the higher coniferous forests. In Pimería Alta, it is found in the Santa Catalina Mountains and formerly in the Baboquivaris. During the heyday of mining in the territory, hunters shot out many populations. But the aboriginal distribution must have been more extensive, bringing Pimans into more regular contact with the bird. For instance, coming down the Santa Cruz River in July 1849, Benjamin Harris (1960:77), at the then deserted presidio of Tubac, recorded in his journal: "Thousands of wild turkeys came here to the river to drink. Their fresh tracks were visible everywhere above the river." This is in Sobaipuri Pima country, 35 miles upstream from San Xavier. Turkeys no doubt occupied all of the San Pedro River, home of the Sobaipuri Pima. When Lt. Col. William H. Emory (1848:78), coming down the Gila with the Kearny Expedition, passed the mouth of the San Pedro, he wrote, "Flights of geese, and myriads of blue [Gambel's] quail, and flocks of turkies, from which we got one. The river bed, at the junction of the San Pedro, was seamed with tracks of deer and turkey." This location is forty-two miles east of the current Gila Pima Reservation and within their claimed hunting area (Rea 1998b:62). The expedition passed through here in early November 1846. Hunting pressures must have been at a minimum then because the Sobaipuri had been evacuated from the San Pedro in 1762, and Apache settlement was light.

In the 1880s, ornithologist W. E. D. Scott collected birds on the Gila from Florence to Riverside and all along the San Pedro River. Scott (1886:389) wrote regarding the turkey, "The only records of this species that I have are from the San Pedro River and the oak and pine region of the Catalina Mountains." In the bottoms near Winkelman, not far from the San Pedro–Gila confluence, Alice Carpenter collected a Hohokam shard depicting an exquisite turkey. (The painted turkey was reproduced by Takashi Ijichi in Rea 1983:139.)

Even today turkeys attempt to colonize this area. Robert Schmalzel recovered a road-killed young male Merriam's Wild Turkey in a very dense mesquite bosque near the mouth of the San Pedro (specimen, San Diego Natural History Museum, June 1980). And Mrs. Carpenter reported a family near Oracle about the same year. From these records and observations, I think the turkey inhabited the mesquite bottomlands of the entire San Pedro, once the eastern edge of Pima country. Before the development of cattle ranching and the introduction of domestic predators (dogs and cats), the bird no doubt thrived in bosques, as the more eastern subspecies, the Río Grande Turkey, still does.

The almost complete lack of turkey bones from Hohokam archaeological sites suggests that these people did not have domesticated turkeys, which were so important among the Anasazi and Mimbres cultures of the higher country of the Southwest. But did Pimans who arrived later raise turkeys? In his 1877 vocabulary, Grossmann ([1877]) did not list the bird. Russell (1908) did not include it among the domesticated animals kept by Gileños. Consultants I worked with who were born early in the 1900s thought that it was of minor importance among their barnyard fowl (see chapter 6 on bird keeping).

Joseph Giff said,

> They must know *tova*, that they come from the cold country. Because the *totova* were chopping down the pine [in the story]. See, there's no pines around here. And the *tova* lives in pine country. The Indians talk about it. I don't know where it came from. When I first heard it, I was a small kid. I never know what *tova* is. I thought it was some kind of an animal. Until I grew up. I go to town with my father, and he pointed out the *tova* to me, when we sell wood or wheat. I imagine that they know what *tova* is because they were chopping down those pine trees. (See Rea 1997:175–176.)

Another time he said, "*Tova* is a Pima name, yet the old people never saw it wild and didn't domesticate them."

I asked George Kyyitan if Pima went hunting *totova* a long time ago. "No, I don't think so. 'Cause they don't go for that in those days. But I know the Apaches does, because there's lots of them up there, Wild Turkeys. And here, they don't go for that. They go for jackrabbits, rabbits, and all that."

Sylvester Matthias said, "*Tova* is a Pima name, but they never came around here—just mentioned in the legends, then always as wild. No stories of this as a domesticated bird." He said that the Maricopa, a Yuman people who took refuge with the River Pima, went up the Verde River to hunt turkey. (Spier [1933:65] mentioned turkeys in passing as an occasional food of these people, but does not say where they got them.) Sylvester also said that a man (a Pima?) from the Salt River Reservation hunted *totova* up the Verde Valley using bow and arrow or a .22 rifle or a stick. When Sylvester worked on the White Mountain Reservation, he observed that the Apache would run Wild Turkey up a hill because they could easily escape by flying when chased from above.

Sylvester himself had had experience hunting turkey while working in the White Mountains on the Fort Apache Reservation. "They told us, just run them uphill. We saw some. And we did that, we had just one rifle, a .22. And we got three. Had one for Thanksgiving dinner. And the color is not as bright as the domestic turkey. Kind of faded, like." His plumage description suggests a yearling, perhaps a female.

These three accounts are confirmed by what Castetter and Bell (1942:69) wrote a few decades earlier: "[Turkey] . . . was to be found more plentifully in the mountains formerly than at present, but was hunted [by O'odham] only occasionally and then

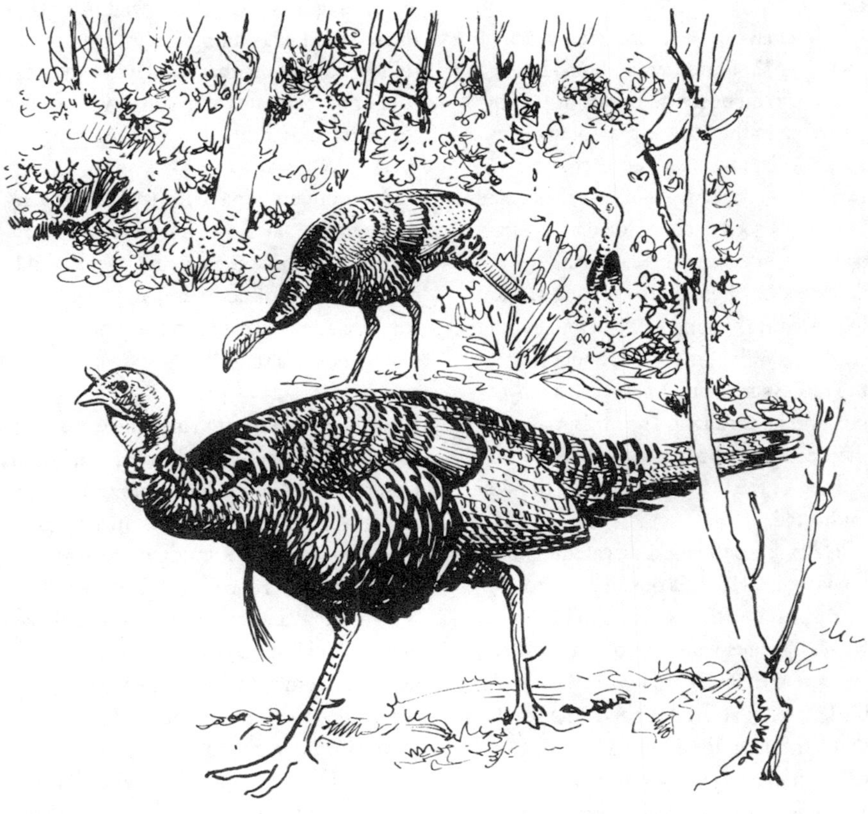

tova, towa, Wild Turkey (Bob Hines)

mostly for feathers rather than food. It was not domesticated."

As with Golden Eagles and formerly with Scarlet Macaw, turkey feathers were important among O'odham (see chapter 6). Hunting arrows could be fletched with them, explained Sylvester, but not war arrows. They were also important in headdresses. Underhill (1938:71) shows several war headdresses made primarily from turkey feathers (see fig. 7.4). The Tohono O'odham formerly had access to turkeys in the Baboquivaris at the eastern edge of their territory, but that turkey population has been extirpated (Phillips, Marshall, and Monson 1964:30). The roosting birds were shot with arrows (Castetter and Underhill 1935:41).

In ceremonial usage, fluffy turkey feathers might replace the soft eagle feathers in certain circumstances, but not those of other species of birds (Underhill 1946:26). These feathers from either bird represented clouds and rain. Perhaps it was because the turkey, like good rainstorms, originated east of Pima country. The type of feathers Underhill was speaking of, called *viigĭ*, are not really down feathers, but body feathers with loose fluffy bases. Among the Tohono O'odham, these feathers are the source of the name for the Navichu ceremony. As Underhill (1946:135) explained, "The [*wiigida*] feast was instituted by I'itoi during his march of conquest. While leading the people through Papagueria, he placed prayer sticks of turkey feathers at two places: Archie (Aji), among the northern Papago, and Quitovac, now in Sonora. This was to signify that the people should settle first of all in these places and that the turkey should be their food [!]."

Neither one of these places (Aji and Quitovac) is habitat for Wild Turkey, and there is no evidence that the Papago were raising the birds, at least at the two places mentioned. The reference might imply that the Emergenti were coming from someplace in the south (such as the Sierra Madrean foothills, as nonanalyzable botanical lexemes suggest) where turkeys were regularly encountered, if not kept.

The association of turkey feathers with rain is explicit in one River Pima tale (Benedict 2001:165): "The turkeys were holding a rain ceremony and they sang, 'Our feathers bring the clouds and it will rain. / Our feathers bring the clouds and it will rain.'"

The kachina-like Navichu, sometimes called a "clown," wears a buckskin or white canvas case mask, in the Santa Rosa region crowned with a mass of turkey tail feathers (Arizona State Museum neg. no. 9132; Hayden 1987:277, plates I, II; Jones 1971; Lumholtz 1912: facing page 92; Underhill 1938:159) (see figs. 7.2 and 7.3). Leon Juan of nearby Aangam called the *tova* headdress a *vonam*. He described its two taller-standing feathers. (According to Sylvester, himself a hereditary Navichu who made masks for this ceremony, the River Pima Navichu lacked this feathered crown, perhaps by default.) At least one ancient turkey feather prayer stick was still being used at the Quitovac ceremony when I last attended. When the Navichu and his attendants go out into the village to beg for supplies, the prayer stick is thrust into a small pile of sand in front of a house where goods are being offered to the dancers.

Such prayer sticks, called *omina*, were used in a number of other ceremonial situations. On the salt pilgrimage, prayer sticks of eagle or turkey feathers were thrown into the ocean before the participants waded in (Underhill 1946:212). River Pima also used turkey feathers in these *omina* (see Russell 1908:plate XIV, b) (see fig. 7.1).

There have been several references to a story about turkeys and pine trees. The story can be found in full, as told by Sylvester, in the Pima ethnobotany (Rea 1997:175–176; see also Benedict 2001:166–167; Neff 1912; Shaw 1968:84–89). A group of turkeys goes to chop down a pine tree, singing while they work. Hearing the song, Coyote joins them on the pretext of wanting to learn the song while helping with the chopping. When the tree is about to fall, all the turkeys run, but it falls across Coyote's tail, and he is caught. The birds make good their escape while Coyote struggles to free himself of the log, finally pulling the skin off his tail.

He pursues the turkeys, catching finally the last one. He pulls out all its feathers, then instructs the bird to go to his house, where Coyote's wife will cook it for their supper. Arriving at the house, the naked bird peeks in at the door and tells Ban Oks that her husband said to boil his sinews (used in making arrows) or his sandals. (Both are starvation foods that can be eaten if no other food is available.) Meanwhile, the featherless turkey climbs a pine snag over a lake or open well, where he will be protected.

On his return home, Coyote and his wife realize they have been duped. Tracking the turkey down to the water, they see its image down in the bottom of the lake or well and try to catch the bird by diving into the water. This doesn't work, so one straps on a heavy metate to reach the bird and is drowned.

Part or all of the Rincon Mountains east of Tucson is called Tova Kusvo, Turkey Neck, in O'odham, a name still remembered. During the emergence of the Pimans and their conquest of the land, Elder Brother taught the people the names of the desert prominences in song. At cardinal points, four particular ranges were named for some event of the journey. "One of the men cut the skin from around the neck of a turkey, turned it inside out and put his lunch in it. He finished his lunch when on this mountain and threw away the skin of the turkey neck, from which the mountain was named" (Densmore 1929:26).

Some people kept turkeys at the Pima Bajo village of Ónavas on the Río Yaqui, undoubtedly domestic breeds. However, Pedro Estrella said that they are (or once were) caught as young from wild parents in the mountains. Pennington (1980:196) was told by older Pima that "turkeys were once sought near streams and 'run down' since the birds cannot fly easily after drinking great amounts of water."

A Mountain Pima in his thirties, Alejandro Sierra Amavizca of Los Pilares, Sonora, said he takes between ten and twenty Wild Turkey a year with bow and arrow or large traps that he constructs in the pine-oak woodlands where he lives.

COMPARATIVE LINGUISTICS. The word *tova* (pl. *totova*) appears in Upper Piman, but in Desert Pima Bajo the form is *tovo*. Even the Névome vocabulario three centuries ago gave *tobo* (= *tovo*) (Pennington 1979:55). However, *tova* is encountered again among both colonial and modern Northern Tepehuán (Rinaldini 1743) and Southern Tepehuán (Bascom 1965 fide Fowler 1983:244). (I recorded *tova* among the Yepáchic Mountain Pima, as did Shaul [1994], who also gives *mumko*, a word for which I cannot account.) More recently I was told *tov, totov* at the Nabogame ranchería of Mountain Pima, where *tova* is also used.

Mason (1917:328) found *pipir* among

Tepecano. I have no idea where this word might come from, but the usual Tepiman turkey cognate had shifted instead to the hen chicken. In Yaqui country, I was given *óowia* for gobbler and *hámut* for hen turkey. Other turkey names among Sonoran Uto-Aztecans include: Tarahumara or Rarámuri, *chiwí*; Eudeve, *kiví* or *zivi*; Ópata, *chiqui (chiki)*; Tubar, *jüwa-lawí*; Yaqui, *kóʼoboʼe* (Johnson n.d.), as well as *chiiwi* and *koovoʼe* (Molina, Valenzuela, and Shaul 1999). In Nahuatl, turkey is *huexólotl*.

MISIDENTIFICATION. In the Neff (1912) collection of stories told by O'odham youngsters, *tova* was translated as "duck" and *ban* (Coyote) as "fox." Otherwise, the tale is remarkably accurate in comparison with versions originating from adult narrators.

TECHNICAL NOTE. During the Late Pleistocene and Early Holocene, the species of turkey occupying most of the Southwest was *Meleagris crassipes*, a species that was contemporaneous with *M. gallopavo* of the east and south and with *M. californica* in southern California. *M. crassipes* was the smallest species in the genus and an evolutionary dead end (Steadman 1980). *M. gallopavo* reached the northern outliers of the Sierra Madre Occidental in northern Sonora and Conkling Cavern in far southern New Mexico in precultural sites (Rea 1980). I have proposed the hypothesis, still not substantively challenged, that the entire range now occupied by the Merriam's Wild Turkey, *M. gallopavo merriami*, was inhabited by this species only after native peoples developed a substantial agricultural base of Mesoamerican cultivars. According to this theory, the turkey was an importation to the area, just as were corn, beans, squash, bottle gourds, and macaws. Either before or after the collapse of the Anasazi and Mogollon cultures, the Merriam's Turkey became feral. I proposed on the basis of plumage that the Southwest turkey was most likely derived from one of the white-rumped subspecies such as *M. g. mexicana* or *M. g. gallopavo*.

However, McKusick (2001:111, 123), on the basis of osteology, believes that Merriam's Turkey is derived from eastern *M. g. sylvestris*, a quite rufous-rumped form.

McKusick's (1980, 1986) studies of archaeological remains have shown that two different breeds were being raised in the Southwest, the Small Indian Domestic (SID) and the Large Indian Domestic (LID), corresponding in size to the modern Merriam's form. The wild subspecies appears in the archaeological record across the Mogollon Plateau between A.D. 400–700 (McKusick 1986:fig. 3), most likely around A.D. 500. Turkey feathers appear before the birds themselves, being used to make cordage for robes and probably other items. She documents the eating of turkey beginning about A.D. 600, the birds being wild-hunted feral Large Indian Domestic or Merriam's Turkeys. It seems that the SID was never eaten.

All of this is outside Pimería. From the core Hohokam area, I have a single substantiated record (Rea 1981a; coracoid, mature female) taken in the Phoenix region. Steven James informs me that the material was excavated from an historic basement. But the bone is too small for any known historic breed. It would seem to be the SID, no doubt a trade item, recovered from a secondary deposition. There are reports of a single bone each of domestic turkey of unknown breed from Casa Grande and Pueblo Grande (McKusick 1986:10–11). McKusick reports other SIDs from the San Pedro area (Reeve Ruin, single specimen) and more tentatively from the University Indian Ruin, Pantano Wash, near Tucson (see Ferg 1985). From the archaeological record so far, it appears that neither Hohokam nor protohistoric Pimans were raising turkeys.

gihodag

Columbidae, Columbiformes
Columba fasciata
Band-tailed Pigeon; Paloma de Collar, Paloma Ocotera

After the Vupshkam, the invading Pimans, had conquered the pueblos of the resident Vipishaḍ along the Gila, they moved downriver to a place along the Sierra Estrella, Komaḍk, where there was a great thicket (Hayden 1935:61; see also Bahr and associates 1994:267). This was near the shrine of the *chumamaik*, the horned lizard, a well-known locality. As told in the Hayden-Smith manuscript, "When they reached this thicket the raven and some bird they call 'gee ho duk,' like a pigeon, had a fight there. So the raven was the Wooshkum and this other one, geehoduk, was the Hohokam. So this bird that was like the pigeon was making a noise like the pigeon does and the people were afraid of it, and it wasn't a very long time since this old woman [buried in the *chumamaik* shrine] died and the people were feeling sad all the time."

This little snippet of narration is inserted into the Creation Story in a section that reads much like history overlaid with magical help from various shamans who were *namkam*, meeters of some animal or natural phenomenon that gave them power. The shrine of the horned lizard and the great thicket are identifiable places in the modern landscape, but the bird has been something of a mystery. The *gihodag* seems to have been forgotten by most modern Pimans. Sylvester and other old people who knew their folk ornithology were gone before I encountered this name.

But Pimans to the south had the same word. The Névome vocabulary compiled centuries earlier gave domestic pigeon as *guiodaca*, the same word as in the River Pima story. Still farther south in the Tepiman corridor among the Northern Tepehuán, Rinaldini (1743) listed *guiodaga* for paloma and *oydiguer guiodaga* for paloma torcaz, a wild gray pigeon. So the word *gihodag* is quite well established both geographically and temporally. But what species of dove is this in Upper Piman country?

The name refers to the Band-tailed Pigeon, which breeds in the oak and pine-oak woodlands in the desert mountains scattered about and around Pimería Alta: the Bradshaw, Pinal, Santa Catalina, Rin-

gihodag, Band-tailed Pigeon (Amadeo M. Rea)

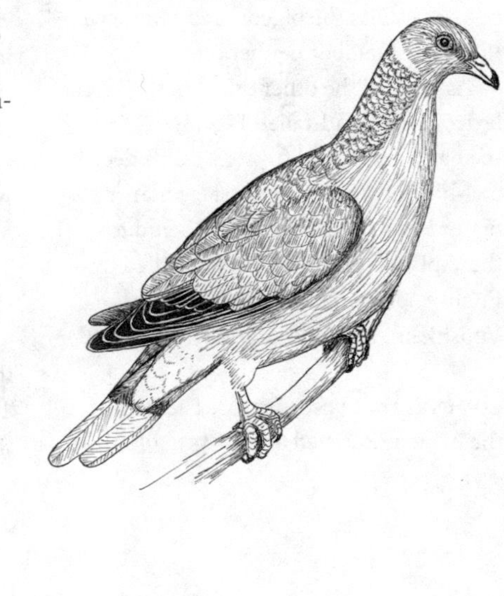

con, Santa Rita, and Baboquívari Mountains. This is a stocky bluish gray bird, larger than a Rock Dove (the common domestic pigeon). It is shaped like an overgrown White-winged Dove. But at closer range the bird shows a vinaceous wash on its crown and breast, bronze-green on the back of the neck, and a white crescent on the nape. It lacks the distinctive white rump of the urbicolous Rock Dove and has a black-tipped yellow bill and yellow (not red) feet. Its English name comes from the broad light gray band on the outer third of the tail.

The *gihodag* has dropped from Piman consciousness, surviving as a relict in the Hayden-Smith 1935 version of Pima origins. Pimans once ventured into the desert mountains in their quest for agaves, acorns, Desert Mulberry for bows, and Wild Turkey for ceremonial feathers. They had names for all these organisms from the uplands, including the oaks and pines where the Band-tails nest. But in the story, the *gihodag* is encountered not in the mountains but at a cottonwood or mesquite thicket above the Salt-Gila confluence. Ornithologist M. French Gilman never came across this bird during his early 1900s tenure in Pima country, nor did I in my middle Gila studies starting in the 1960s. But other ornithologists have found it during migration along the major desert streams (Monson and Phillips 1981:64–65; Phillips, Marshall, and Monson 1964:40–41; Rosenberg and associates 1991:197–198; Witzeman, Demaree, and Radke 1997:86–87). Nearly every natural-history account of the Band-tails mentions their fondness for acorns and elderberries. During migrations, the birds were attracted to the desert riparian oases, often teeming with massive bunches of ripe elderberries, which the Pima called *dahapdam* (Rea 1987:195–197). But with severe desertification, the trees are gone as well as the pigeons, and modern Pimans remember neither one.

As with all the other doves in the Piman lexicon, the Band-tailed Pigeon name seems to be derived from its call, a deep owl-like *hú-whuu*. The mystery that remains is why this bird should stand for the Vipishaḍ (or Hohokam, as called by English speakers) and the raven for the Vupshkam, or modern Pimans, in a battle.

COMPARATIVE LINGUISTICS. The name for the pigeon occurred among the colonial Tepehuán and Névome, as noted earlier. In looking at pictures of the Band-tail, Pedro Estrella, Pima Bajo, told me *dodoak saigam*, with the comment *"s-doadkam*, bronco" (wild). His phrase means that the bird was characteristic (*-gam*) of the woods or forests *(sa'i, sahi)* found in the mountainous country *(dodoaki)*. He may have just been describing the bird rather than naming it. A different word is found among the Mountain Pima. I recorded *makovi* at Nabogame ranchería, and Pennington (n.d.) at Maicoba and Shaul (1994) among Yepáchic were told *makov* and *makowi*. The equivalent in Rarámuri (Tarahumara), the adjacent tribe, is *makawi* (Brambila 1983:414). Pennington (1969:116) says that the abundant, acorn-eating paloma del monte is called *túgu* by contemporary Northern Tepehuán.

Commenting on the Hayden-Smith account, Bahr (in Bahr and associates 1997:267) observes, "Unknown. *Gi'ihodag* means 'thinness,' 'leanness.'" Mathiot (1973:292) lists *"giihoḍag* = (v.) to have a crown; to be crowned; or (n.) marigold." The Band-tail is a heavy-bodied bird, so the first derivation is unlikely, but the bird does have a conspicuous white band on the nape that might suggest a crown. More likely the etiology is onomatopoeic, signifying the bird whose characteristic is to call *giho*.

hoohi*, pl. *hohohi*; var. *hohhoi

Columbidae, Columbiformes

Zenaida macroura

Mourning Dove; Huilota, Cardera

The medium-size, slender Mourning Dove is common year round throughout Northern Piman country, from the riparian vegetation in the north to the thornscrub in the south. Its long, pointed tail is distinctive among the doves; in flight, it shows feathers broadly black and white tipped, except the central pair, which are all brown. At close range, large black spots may be seen on the inner part of the closed wing. The feet are pink, and the eyelids and eye ring pastel green and blue.

In the breeding season, which is long, the pair often feeds and drinks together. The male, somewhat paler, has a pinkish wash across the breast and nape, which shows off a grayish cap. He also has an iridescence on the sides of the neck. Otherwise, the sexes are similar, and both incubate.

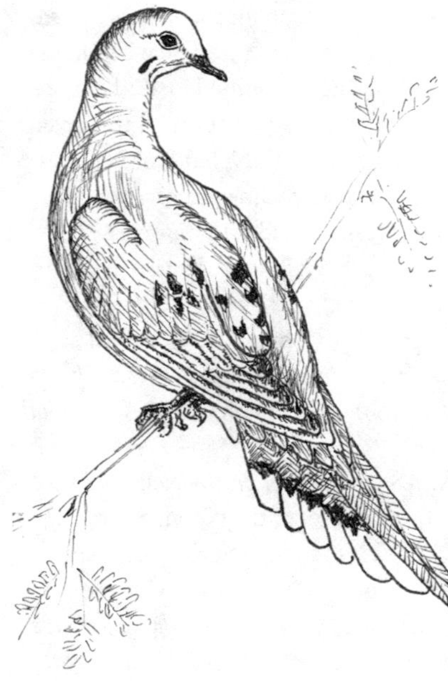

hoohi, Mourning Dove (Amadeo M. Rea)

Although Mourning Doves are ubiquitous, they prefer feeding in fields and open weedy areas. Once on the pale tan ground, they are practically invisible. Traditional Piman agriculture offered an ideal environment for doves. Favored foods are the small but abundant seeds produced by Horse-purslane, amaranth, sunflower, bristlegrass, smartweed, and the smaller cultivated grains.

These doves build such a flimsy nest of twigs, often precariously placed, that it is a wonder the eggs do not fall out more often than is the case. Nests are often relatively low in mesquites, Graythorn, chollas, prickly-pears, embankments, or even on the ground. They may be tucked into protected areas of a building or *vatto* (arbor), where the adults tolerate the daily coming and going of humans. Nests nearly always have two white eggs. Mourning Doves raise successive broods and may reuse the earlier nest.

The incubating bird is a tight sitter, often allowing approach to within a couple feet or less. Then the bird explodes off the nest. The wings make a whistling sound as the bird takes flight. Sometimes the broken-wing ruse is resorted to, the departing bird dragging itself conspicuously and noisily along the ground, distracting the intruder.

The five-noted song of the Mourning Dove, although soft, is distinctive and easily remembered. It is accented on the second, inflected note: *u-húuu huu huu huu*. The Piman name for the Mourning Dove, as for other doves, is onomatopoeic, the name mimicking the bird. The lengthened first syllable of *hoohi* corresponds to the accented note of the bird's call.

As with the White-winged Dove, the Mourning Dove's notes have engendered various renditions of what the bird is saying in O'odham. For instance, Sally Pablo told me a man went under a *kui* (mesquite tree) to defecate. A *hoohi* was sitting in the *kui*. When the man pulled down his pants and squatted, the *hoohi* said, with drawn-out syllables, *"Ge s-chuuk sii-puuḍ."*

Early in the Creation Story, there was a death. As a sign of mourning, in accord with Pima tradition, the hairy animals cut their fur. The birds, however, cut their tail feathers off square. But the Mourning Dove was off somewhere else when the other birds trimmed the tips of their tail feathers, so to this day *hoohi* has a pointed tail, unlike most of the other birds (see also Benedict 2001:98).

At the first wine feast, some of the male participants got into a fight. *Hikvig*, the Gila Woodpecker (some say *chukukmal*, the Western Kingbird), got his head bloodied. Some bird got his head stepped on and flattened. Jose Garcia and Larry Saraficio added that the *hoohi* got his legs kicked and bloodied, and they are still red today.

Among both Akimel and Tohono O'odham, the Mexican (or California) Poppy is called *hoohi e'es*, "the Mourning Dove's crop plant." Mourning Doves are especially fond of its seeds (Gilman 1915a). In the field one day, I asked Sylvester Matthias how the plant got this name. He said *hoohi e'es* was mentioned in Thin Leather's Creation Story. But neither Russell's (1908) nor Lloyd's (1911) version of the Creation Story transcribed from Thin Leather appears to mention this plant. Then Sylvester told me the story.

> It's a grandmother, the Hoohi, it's from the son's side, Hoohi *ka'akmaḍ*. And they were killed by the enemies—the grandsons. The messenger went and told Hoohi, and she said, "Wait till my *e'es* [crops] are ripe. Then I'll cry. That would sound better, when my plants are ripe and harvested. I'll be there working. That would sound way better than right now if I do the singing." So, when the harvest is ready, then she start the song, and cry there, for her grandsons. I heard the song, but I forgot [it]. It's been a *long* time. There's a song she sings with the crying, a mourning song. That's what the Hoohi end up there, with the harvesting of *hoohi e'es*. I don't know what *hoohi e'es* look like, but it's a grain, like *shuu'uvaḍ* [Tansy-mustard] here. That's that story.

Not surprisingly, both the dove's English name and Pima mythology link the bird with mourning. The Pima apparently associated the bird's plaintive call with their own keening, repeating the kinship term of the deceased. She would have cried, *"Heñ ka'akmaḍa."*

Piman children are told the story of *hoohi* and *giisobĭ*, the Verdin (see that species account). There was a storm. The dove, with its open nest on a sturdy mesquite limb, said, "I wish it would blow and not rain." But the Verdin, with its completely enclosed nest out in the swaying branches said, "I wish it would rain and not blow."

Russell (1908:244–245, 250) collected a story of a boy who goes to visit his captive mother among the Apaches, turning first into a dove to escape them, then into a flock of "hawks" that beat the enemy to death. Without the Pima folk names, we have no way of knowing which of the four named doves or which of the various named hawks and falcons is intended in the fable. In another version, the boy turns into a dove of some sort, then "into a great eagle and flew down and killed all the Apaches" (Benedict 2001:153–155).

There is a story where two brothers, Yellow (or Brown) Coyote and Sandy-colored Coyote, are in conflict, and Sandy Coyote kills his brother as the result of a wager game of *kints*. Yellow Coyote's wife is pregnant. A boy is born, and when he is old enough, he engages his uncle in the same game. The boy makes a nest and puts white stones in it as eggs.

"It is about time the birds laid their eggs," says the youth.

"No, it will be two or three months from now before they begin to build their nests," replies his uncle.

"As I came along, I saw that the dove had already laid her eggs."

"No; you are lying to me."

"Well, if I go and bring those eggs to you and show you that I was telling the truth I shall win our wager, if I do not bring them you shall win." By means of this and several other ruses, the boy wins his uncle's life, avenging his own father's death (Russell 1908:236). Without the Pima terms, we do not know which species of dove was intended in this story, but the Mourning Dove is known for its very long breeding season. (See Gambel's Quail and Curve-billed Thrasher accounts for similar stories.)

There is a saying, noted Jose Enriques, that if one imitates the call of the *hohhoi*, one will get a little cut somewhere on the body. "They tell us not to do that." Simon Lopez agreed: "If you copy it, you will have a cut on your finger, thumb, or toe. And it really hurts." Hiá ch-eḍ O'odham Delfina Cameron said, "They told us when we were little if you imitate the *hoohi*, you're going to get a cut under your toe. And it does and really hurts. But we were going barefoot anyway." On the Gila, Culver Cassa says this warning applies to the White-winged Dove, which indeed has a black cut mark on the side of the neck/face.

Looking at a *ku'ukpalk* plant (Verdolaga or Portulaca) one day, George Kyyitan said, "Seeds right in the branch, there. That's where the doves get their seeds, *hohohi*." This plant and other "weeds" such as *chuuhuggia* (Palmer's Carelessweed or Red-stemmed Amaranth) and *kashvañ* (Horse-purslane), supply prodigious quantities of minute seeds, prime dove food. Field edges, canal banks, fencerows, and roadsides provide seeds and shelter for these birds. Here in the fields as well as along the rivers lined with cottonwood and willow trees, Pima boys once learned hunting skills, taking these doves with a slingshot or bow and arrows, contributing at an early age to the family cooking pot. Older Pimans considered *hohohi* a regular game bird, although Gilman (1915a), who taught at the upper Gila River villages, said it was not hunted; perhaps he meant it was not hunted with a shotgun.

In the Creation Story, there is an episode where two small boys, as was the custom, dug a hole in the sand and built a small house over it while waiting inside with their bows and arrows for doves to come. This was probably along the river where numbers of doves accumulated year round. Many elderly Pima men told me that in their youngest hunting days, taking Mourning Doves along the Gila was their chief and daily pursuit.

When I came to the Gila River Reservation in 1963, only one local elderly woman had the edges of her lower eyelids tattooed, a practice once common to young men and women (Russell 1908:161–162). Decades later, one occasionally sees these bluish lines on the lids of an older Tohono O'odham woman. According to Anna Moore Shaw (1974:55), a woman who resisted having her lids tattooed was taunted by being called *hoohi wuupui*, "Mourning Dove's eyes" or "reddish eyes." It seems to me that it would be the other way around: the tattooing renders the eyes more like the doves with their bluish eye rings.

Joe and Ruth Giff explained that a mesquite dish called *hoohi-wichda*, "Mourning Doves made in imitation," was prepared by first cooking *vihog*, mesquite pods, in water until they softened. When the residue cooled, it was mashed and strained. Finally, little whole wheat dumplings—the *hohohi*, "doves"—were placed in an olla and covered with the mesquite pod broth, then cooked again until the dish thickened somewhat.

COMPARATIVE LINGUISTICS. All desert Pimans (River Pima, Papago, and lowland Pima Bajo) call the Mourning Dove *hoohi* (pronounced háw-hi). In the Névome vocabulario, *ohoi* and *ococoi* (White-wing) were listed as perdíz ("partridge") (Pennington 1979:92, 116). Some Tohono O'odham speakers still use *hohhoi* for this folk genus. The vocabulario also listed *utubucama* as tórtola ("turtledove"). Mountain Pima call this dove *vohokoly*, which may also be onomatopoeic; they did not recognize the desert Piman term.

okokoi*, pl. *o'okokoi

Columbidae, Columbiformes

Zenaida asiatica

White-winged Dove; Paloma Pitayera, Paloma Ala Blanca

During hot weather, from about April to early September, a heavily built bird, the White-winged Dove, calls its O'odham name from mesquites, Saguaros, and farmland: *ók-kok-koi*. It will not let you forget it, repeatedly pronouncing the word distinctly. The singer is small pigeon size, with broad white tail tips and white patches on the secondaries, visible on the perched bird, but most noticeable in flight. This dove is gray overall, with a dark malar strip and slight iridescence on the sides of the neck. There is a distinct purple wash from the crown to the upper back, strongest in males.

In the breeding season, White-wings tend to come in numbers to favored drinking places, such as charcos and ponds, and feeding areas, such as Organ-pipe Cactus and Saguaro stands and fields, particularly where grain has been harvested. They even nest in congregations where there are mesquite bosques or Salt-cedar stands. Large flocks are conspicuous as they travel between watering, feeding, and nesting areas, often several miles apart.

White-winged Doves are especially noticeable as they come to the tall Saguaros, probing for nectar (and for insects or pollen?) among the stiff white flowers clustering at the tips of the arms. Later, as the ripe red fruits split open, the doves are back to enjoy the mass of sweet deep-red pulp and rich black seeds.

The White-wing's migratory movements were well known to the O'odham. David Brown and Luke Thomas told me, "O'okokoi Do'ag—the place in Sonora where the White-wings go in winter. They come back here in summer." Sylvester said he had heard of this place but didn't know where it was. Irene Hendricks noted, "*Okokoi* comes around in June and July when it's hot, but now [October] they're gone back to the hot country."

Changes may occur even in bird tradition. Most White-wings leave Arizona abruptly with the opening of dove season in September. But some may stay around in urban areas. At Santa Rosa village, Simon Lopez noticed a change even out in the country. "*O'okokoi*—I see them a lot *now*, like in December. Some are staying now, surprising—that they are starting to do that. I can see them—I don't know how long ago I see them around here—in December."

The last spring frosts and the first fall freezes limited the growing time of the warm-season crops that Pimans largely relied on—namely, various varieties of maize, several species of cucurbits, several species of beans, and some lesser crops. River Pima watched for the mesquite trees to put out their new green leaves in spring. The mesquites know, they said. But there was one even surer sign that killing frosts were over: the return of White-wings from the south. When the birds suddenly began calling in numbers around early April, it would surely be safe to plant.

okokoi, White-winged Dove (Amadeo M. Rea)

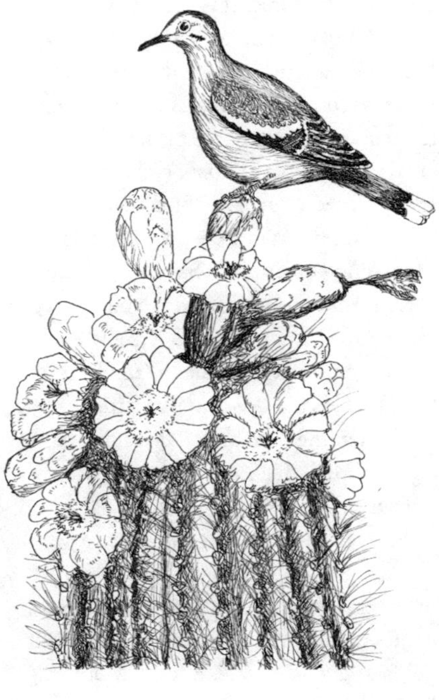

The May morning was getting hot in southern Sonora. Two white-haired Pimas, from speech communities long separated in time and some three hundred miles apart in different states and countries, were comparing notes. A White-wing began calling near us along the Río Yaqui. Don Pedro began explaining in his quiet manner: "*Okokoi kuihim*, 'the White-winged Dove is singing.' Could also say *ban kuihim*, 'Coyote is singing,' or *keli totol kuihim*, 'gallo cantando, the rooster is calling,' or *totol neihim*, 'the chicken sings,'" using a different verb. Sylvester Matthias then added that in his version of Pima he would say *ban hiink* (or in Papago *ban shoak*) and *keli chuchul kok* for the "cock crows."

No matter what language, people have been fond of interpreting what the White-wing is saying in its easily imitated call. Ruth Giff said that the doves are saying *gogoks bo kaij*, "the dogs said it," and Sylvester Matthias rendered it *gogoks hab kaij*, with the same meaning. Clemencia Leos once gave the *okokoi* call as *wepaj ab keek*, "the red fruit is standing" (literally, "its redness is up there"), in reference to ripe Saguaro fruit. Another time she said *vepeg ch-ep keek*, "they're red and they're up there." Delores Lewis hears the White-wings saying *gigik ki kuuk?* "Where are you going to eat?" For Juanita Norris of Akĭ Chiñ, the birds say to the people going out to pick Saguaro fruit,

> Ga hu keek.
> Ga hu keek.
> Hetasp o huu.

> It's standing over there [the Saguaro].
> It's standing over there.
> Eat five [*bahidaj*, Saguaro fruit].

Still other versions are more complicated. George Kyyitan explained that when a man was rounding up his horses, he usually climbed a tall *kui*, mesquite, to look around. One day a *pako'ola*, a pascola dancer, was looking for his burro. An *okokoi* saw the pascola searching and said,

> Pako'ola shoiga,
> Ia o keek!
> Ia o keek!

> Pascola's animal (pet),
> It's over here!
> It's over here!

When the pascola found his burro, the dove told him,

> Cheshajk hek him!
> Cheshajk hek him!
> Cheshajk hek him!

> Get on it and ride,
> Get on it and ride,
> Get on it and ride.

This story was widespread on the Gila. Clemencia Leos of Haashañ Keek, Sacaton Flats, said that the *pako'ola*'s name was Hus Anton (that is, Jose Antone) and that he was looking for his horse, not his burro. Either way, the dove told him the same thing. I have heard this story even southward in Papaguería.

William Blackwater told Ruth Benedict a brief story wherein "Dove" and "Chicken Hawk" challenge each other to a race, and Dove wins. It is unfortunate the English is so careless. We are told that Dove's supporters paint his legs red. All four species of dove that the Pima designate with folk generic names have red or pink legs. Perhaps a clue is provided in the phrases that Dove repeats: "I wish it were morning, I wish it were morning. I would beat him. I will beat him. You shall see." We lack the original O'odham words, but even the English rendering suggests the call of the White-Winged Dove. The raptor, also ambiguously named, "sat in the top of a tree and wept" (Benedict 2001:87). "Chicken Hawk" has usually been traceable to the Prairie Falcon in Piman literature.

In midsummer, many Tohono O'odham abandoned villages temporarily for Saguaro camps where they harvested the year's supply of the delicious, sweet fruit. The White-wings would see the people arriving, said Jose Enriques. "I guess during that time they were asking where they were going to get their Saguaro fruit." The dove would say to the People: *"Shaako apt o oimmed? Shaako apt o oimmed?"* (How far are you roaming [today]?). Then the bird would say to the other doves: *"Has cheegig? Has cheegig?"* (What's their name? What's their name?). Jose even had an English version of the White-wing's call: "Cool Coors for you! Cool Coors for you!" (Coors is a brand of beer once popular with the O'odham.)

Wright (1929:75–86) recorded a story of a young injured girl abducted by Ho'ok Oks, the mythic cannibal woman. The girl holds in her hand a white feather of the White-winged Dove. With the help of Chukuḍ, the Great Horned Owl, the girl escapes as far as a *chepo*, or tinaja, a water-catchment basin in the rocks. Here she still resides happily under the water. The White-wings every summer congregate at the water, attempting to get their white feather back.

When Culver Cassa was little, he was told that if one imitated the call of this dove, it would bite your toe. This seemed to be something said to small children, like the story in Euro-American society that the blackbird would steal a child's nose.

In enumerating River Pima prey items, Pimas I worked with usually included both White-wings and Mourning Doves (see Kyyitan's comments about hunting them, chapter 7). However, Russell (1908:86) mentioned neither of these birds in his list of animal foods, noting later in the text that "Sonora doves were and are yet confined in log-cabin cages built up of arrow rods [*Pluchea sericea*]." Ornithologist M. French Gilman (1911:52), who taught on the Gila River Reservation from 1907 to 1915, said specifically that the Pima do not hunt either of these two larger doves. Perhaps a tradition was in transition, as were many others among the Pima.

In Papago country, Delores Lewis told us he didn't eat *o'okokoi* "because they make the *bahidaj* [Saguaro fruit] go right." Years later I asked Simon Lopez of Santa Rosa village, "You aren't supposed to kill them?"

"Yeah, that's what they say, because they're like—we use that name [*okokoi*] for our tradition, like during the summer for the [Saguaro] cactus. That's when they believe the *okokoi* has got something to do with the *bahidaj* and stuff like that. Those Aangam singers over here, they sing about the *okokoi*. All their singing is about the *o'okokoi*, And that's why they won't [kill them]. Well, my grandparents said, used to tell us, you can't kill them—leave them alone. Only the *hohohi* [Mourning Doves]." Here he sings the one Okokoi Song that he knows. "The White-winged Dove made a wine feast on top of the cactus, *nawait*, and that's when we drank and got drunk, and go back and forth, staggering, *shakal meḍ*."

Other sung poetry preserves the association of White-wings with the columnar cactus. Densmore (1929:217) recorded a song from Jose Hendricks of Vamori village, which she entitled "The Pigeon and His Tiswin Lodge." The bird, of course, is the *okokoi*, and the tiswin is the *haashañ nawait*, Saguaro wine. She gave only an English translation of the song:

> The pigeon pretended that he was setting up a tiswin lodge.
> The frog doctor drank his wine,
> Got drunk and pulled out his cloud.

(Her "frog" is really one of the desert toads that emerge at the summer rains, puffing out its white "cloud" throat to call for a mate.)

Larry Saraficio agreed that they were taught when young that when hunting Mourning Doves always to avoid the White-wing because of its special association with the Saguaro. And, of course, the Tohono O'odham themselves had a special relationship with this giant cactus, camping at big groves during midsummer while they processed the fruit into syrup and jam.

Likewise, Jose Enriques said, "The *kekelibaḍ* ['deceased old men'] told us, don't hunt them. They don't encourage us to shoot them because of the Saguaro. Usually the *o'okokoi* will be singing about the time they're ready [for harvest]." If people raised an injured White-wing, he added, they released it around September when the birds were ready to migrate.

Perhaps this was also once the tradition among the Akimel O'odham that was being lost at the turn of the twentieth century as the Pima became more like non-Indians living around them. Historical River Pima grew wheat extensively, attracting these birds. Ruth Giff said, "*Hoohi* and *o'okokoi*, would be a lot of them around when they are cleaning the wheat, thrashing, just sitting around. They shoot them for their meal. Make a big stew out of them."

Leonard Pancott noted that Gila Pima hunted both kinds of doves extensively when he was young, even taking the birds while they were incubating, but he recalled that there was once some prohibition on hunting White-wings between the time the Saguaros flowered and the time they fruited.

When I asked Leonard if he knew a bird the Papago call *bahidaj u'uhig*, he said, "It could be the [Mourning] Dove and the White-wing. 'Cause, you know, when you hear them doves and White-wings, especially at daylight in the mornings, I used to hear the old people would say, *'Haashañ bahijik jek kaij.'* When they're singing in the mornings, used to be a lot of them. You could hear them all over. And that's what I would hear the old people would say that, *'Haashañ bahijik jek kaij,'* you know, it's helping the cactus to mature, ripen. They're singing the Saguaro fruit ripe. Ripening the Saguaro fruit; I guess that's what it says."

Frank Jim from the westernmost part of the Tohono O'odham Reservation said his people hunted both Mourning Doves and White-wings. And Delfina Cameron, Hiá ch-eḍ O'odham, said they ate *o'okokoi* as well as Mourning Doves and quail. The prohibition may have been relaxed among People with limited animal resources. Among Gileños, many taboos fell with the rise of Protestantism and the decline in shamanism.

For the Pima Bajo, Pennington (1980:196–197) notes that the *o'okokoi*, a term that he ascribes to two different doves, were killed with stones and roasted. They were also sometimes kept as pets in homemade round or square cages. "When it reaches maturity, the bird is eaten." The Pima Bajo of Ónavas have taken on most of the customs of the blancos (non-Indians) with whom they have shared their pueblo for centuries. Few native traditional restrictions have survived. Also, this is not Saguaro country, although several other species of columnar cactus are important components of the local thornscrub and tropical deciduous forest. But these species may not have had such a central cultural significance as did the Saguaro among Pimans to the north, in particular the Tohono O'odham.

COMPARATIVE LINGUISTICS. The name of this dove, as for all other doves, is onomatopoeic and uniform throughout desert Piman country. We find "paloma torcaz" given as *ococoi* in the Névome vocabulario (Pennington 1979:89). Both Hale (n.d.a) and Pennington (1980:196) wrote *okoko'i*. I checked this presentation with both María Córdova and Pedro Estrella, but neither used a glottal stop in the terminal syllable. Incidentally, Pennington here includes both the pitayera and the cardera under White-wing, *okokoi*, but cardera is the Mourning Dove, called in Pima *hoohi*.

All Sonoran groups must have had a name for this large, loud dove. Rarámuri call the paloma pitahayera that eats the fruit of columnar cactus *kóchora* (Brambila 1983:414).

MISIDENTIFICATIONS. Mathiot (1973:384) lists *huchkam* as Sonora Dove, a former English name of the White-wing. *Huchkam* is a variant of one of the Tohono O'odham names for the Common Ground-Dove (see next species). It literally means "nail" or "claw" plus the attributive marker.

In their Yoeme (Yaqui) dictionary, Molina, Valenzuela, and Shaul (1999:214) list *ommo'okoli* as Inca Dove and *kuuku* as White-winged Dove. Surely the English identifications of these onomatopoeic names have been switched here.

***chehopĭ*; also *vuhigam, wuhigam hupigdam, tehop* (Pima Bajo)**

Columbidae, Columbiformes

Columbina passerina

Common Ground-Dove; Tortolita, Tórtola Coquita

In the heat of summer, a sparrow-size dove bursts from the ground with a jerky flight and disappears into the feathery green foliage of a mesquite tree. The rounded wings flash rust color, and the short tail is mostly black with a tan middle. This pattern should identify the Common Ground-Dove, which is seen in Pimería Alta usually during the hot weather. It is most often associated with mesquite bosques along the floodplains of desert rivers or the mesquites that spring up at the edges of charcos. It is also partial to Piman fields with weedy edges and fencerow growth. Even in what appears to be good habitat, the birds are localized at the northern fringe of their range.

These tiny doves are rather private birds. They will often be in pairs or at the most in family groups. If you manage to catch them feeding on the ground, you will notice the black tips of the primaries and the large black spots on the wing coverts. The male has a distinctly vinaceous cast, most noticeable on the chest and sides of neck, setting off his gray cap. The grayer female lacks most of this pinkish wash. Both have red irises and flesh-colored bills with dark tips. The chest is faintly scaled, blotched on the central part of each feather with gray. These field marks distinguish the quiet ground-dove from the also small and scaled Inca Dove (see next species account).

Perhaps its call is more revealing than the markings in distinguishing this bird from the vociferous Inca Dove. M. French Gilman (1911:54), writing from the upper Pima villages, said of it: "He might properly be called the 'woo-woo bird,' as his note is a single 'woo' long drawn out and uttered at short intervals. The sound is very misleading, even to a greater extent than that of the Mourning Dove. The first time I heard it I started to cross a ten-acre field to search for the bird in some trees on the far side. I had gone but a few yards when the dove flew from a fruit tree about three rods [fifty feet] away, where he had been the entire time."

Doves generally make rather flimsy nests, but Gilman noted those of the ground-doves to be rather more elaborate, with a distinct cup. On the Gila, he found the birds mostly summer resident, but later observed some in winter as well, as have I (Rea 1983:161).

This species is one of several birds that have two names in Pima. Once during a discussion, Ruth Giff mentioned the *gugu*, Inca Dove, locally a well-known bird. Rosita Brown brought up the other little dove

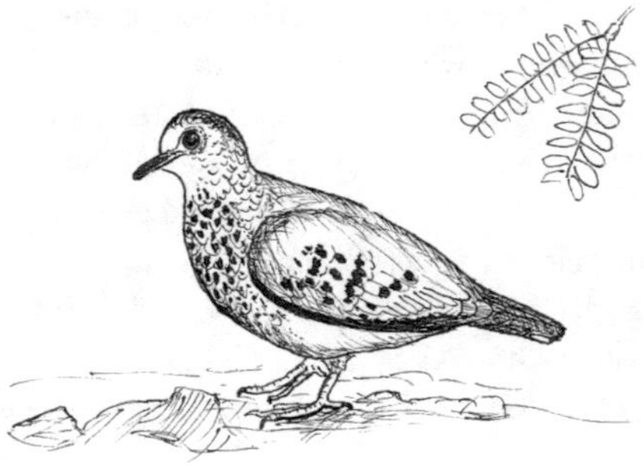

chehopĭ, wuhigam, Common Ground-Dove (Amadeo M. Rea)

and said that *chehopĭ* is the same as *wuhigam*. Joe Giff added, "*Chehopĭ* is something the color of a *hoohi* [Mourning Dove], only their bodies are a little short and round—kind of round."

"Do they both say that *gúgu?*" asked Ruth.

"No," said Sylvester. Then he imitated the ground-dove, "*Kuhúuu*. They used to be common when I was a kid. I would say about middle August [you would hear] *huwúuu! huwúuu!*"

Joe Giff added, "This one—*chehopĭ*—has spots on the wing. The smaller one." On another occasion, Sylvester related, "I don't know any stories about it, but I know the *chehopĭ*. There used to be lot along the rivers, when the creeks were running. You could just hear them: *chuhú, chuhú, chuhú*. Those are *chehop*. It's a family to the *gugú*."

Juanita Norris, Huhu'ula of the Akĭ Chiñ community, called the smallest dove *wuhigam*, but did not know the name *chehopĭ*. The small dove shaped like a *hohi* she called *gugú* (see next species).

Three points are worth noting here. Rosita (and later Theodore Peters) linked the two names, *chehopĭ* and *wuhigam*, each of which has a distinctive Piman pedigree southward (see the comparative linguistics section). So it is a sympatric folk synonym. River Pima recognized the two species of small doves by their voices. And they were aware of the different markings and shapes distinguishing the two.

The ground-dove's name is similar to that of the Antelope (Ground) Squirrel, which is *chuchupĭ* in O'odham (see Rea 1998b:146). I suspect that both names are onomatopoeic. Southward in desert Pima Bajo country, the metered, inflected *huwú* notes of the ground-dove, uttered repetitiously in the heat of the day, are a characteristic of the countryside.

Among Tohono O'odham, there are some variations on the *wuhigam* name. Delores Lewis of Ge Oidag (Big Fields) called the ground-dove *huchigam*, as did Jose Garcia of Santa Cruz. Mathiot (1973:384) wrote the word as *huchkam*. At Santa Rosa, Simon Lopez said, "*Wuhigam*, that's familiar [common] here. That's that little [dove]; that's the one I was trying to think of. Yeah, the ground-dove [looking in my book]—that's what you call it. Okay, that's *wuhigam*. I think you can find them around. I see them; they're around." He also commented on the Inca Dove. The *-gam* part of the name means that what precedes it is characteristic of the organism.

Don Pedro Estrella and I were hiking through la Barranca just off the Río Yaqui. We flushed a tiny ground-dove. Pedro said, "*S-wepeg aangam*, otra paloma" (it has red wings, another dove). I think he was describing rather than naming the bird. When he went through my elicitation book, he gave me *hupigdam* for the ground-dove, calling it Cardera in Spanish, but he called the Inca Dove *tehop* or Tortolita in Spanish. However, María Córdova said both species were called *tehop*.

Both Akimel and Tohono O'odham (with one exception) said that the Common Ground-Dove was never eaten. It was said to be too small, though it is heavier than the White-crowned Sparrow, an important food item. Mountain Pima at Nabogame call this palomito *vuupuí*, also saying it is too small to eat. The bird is not associated with staying sickness or with any legends that I was able to discover. At Santa Cruz Tohono O'odham village, Jose Garcia noted that the little *huchigam* used to be found around the charcos (artificial ponds) but is seldom seen anymore. He was the only one who recalled it being eaten.

COMPARATIVE LINGUISTICS. The *chehopĭ* name is onomatopoeic. Switching the accent to the last syllable and inflecting it makes a good representation of this dove's call. The ground-dove's voice has been described as "a soft, ascending *wah-up*" (National Geographic Society 1983). The word comes from colonial *tehop*, which survives in Pima Bajo. The nearby Eudeve called the tórtola *tehóp*. (This is another example of where the Eudeve, apparently originally from a different environment with a different biota, borrowed folk names for the newly encountered flora and fauna from their Piman neighbors as they moved adjacent to Pimería.)

The first part of the other name for Common Ground-Dove, *wuhigam*, with its cognates *huchigam*, *huchkam*, and probably even *hupigdam* and *hupidgam*, I think is also onomatopoeic. The eighteenth-century Attí vocabulario (1774) gave *archucam* for tórtola, probably cognate with contemporary forms and most likely even with the Névome entry for tórtola, *utubucama (utuvukamă?)*. The final syllable in each case is attributive.

gúgu; also *gugú* *tehop* (Pima Bajo)

Columbidae, Columbiformes
Scardafella inca
Inca Dove; Tórtola Coluda, Tórtola Cola Larga

Like the Common Ground-Dove, the Inca Dove is small, red eyed, with conspicuous rufous wings and a pinkish cast to the body. But the tail is longer, the outer tail feathers white, and the body conspicuously scaled with gray tips to the feathers. The Inca Dove is in many other ways the contrary of the Common Ground-Dove. It has thrown in its lot with humans, at home in cities, parks, villages, and ranches, and is almost never found far from these settled areas. It is heard constantly throughout the year, but its insistent *no-hope* call seems most intense in the heat of a summer day. In winter, it may congregate in flocks of many dozens, feeding in abandoned fields or clearings that supply minute weed seeds, in particular *chuuhuggia*, Carelessweed.

In spite of this sociability, the birds are an irascible lot, frequently scrapping. Ornithologist M. French Gilman (1911:56) wrote from the Pima villages:

> Two of them will face each other with one wing on guard, held straight above the body; then close in and mix it, buffeting with wings till the sound of the blows is audible at a distance of fifty yards. The bill is also used with bloody results about the head. I have been told that one will sometimes kill the other but never saw such an extreme case. When arranging for a fight the com-

gugu, Inca Dove (Amadeo M. Rea)

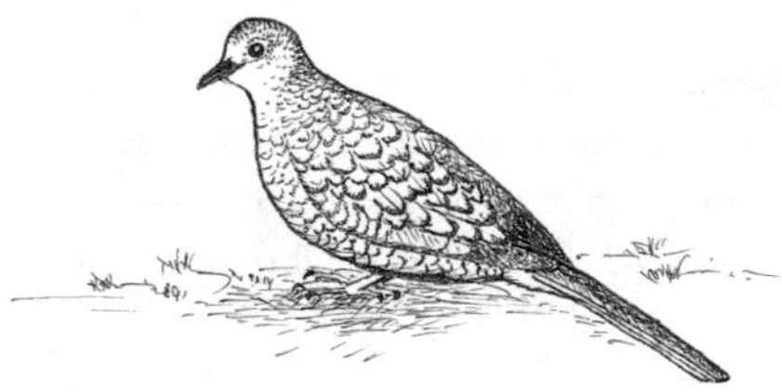

batants utter a sort of growl, if it may be so described: a very guttural, anger-expressing sound. In animated talk, gossip perhaps, they excitedly utter sounds something like "cut-cut-ca-doo-ca-doo." In all, quite a vocabulary is at their command. In motion they are quick and lively, and have the same jerky flight as do the Ground-Doves.

The Inca Dove is called *gúgu* in Akimel and Tohono O'odham, but I have occasionally heard it pronounced *gugú*, which is closer to the birds' monotonous call. Gilman (1911:55) said the Pima of his day called the bird *coo-coo*, without indicating the accent. Eva Marie Lopez at Vaya Chiñ called them *gugudam*.

Joe and Ruth Giff contrasted the *chuhopĭ* or Common Ground-Dove with "*gugú*, the little dove with the longer tail."

Simon Lopez of Santa Rosa didn't recognize the ground-dove name *chuhupĭ*, saying, "*Wuhigam*, that's what's familiar here."

"What other kinds of doves do you have?" I asked him.

"*Hoohi, okokoi*—and there's one other dove that I can't ever know what they call [it]. I've been trying to ask. Those little ones—[say] *kúku*. And somebody was telling me, that is their name—*gúgu*. I heard that here. I know there's a lot in Akimel [O'odham country]. Those are the other kind that look like they're [Mourning] doves. They're small. Then those other two: *hoohi* and the *o'okokoi*. You see them around a lot in Tucson—in parks. There's a lot of them in the park. That's the one that looks like a [Mourning] dove—small, real small."

Sylvester noted, "Some call the *gugú chehopĭ*, too."

"Can you eat those, *chehopĭ?*" I asked.

"No."

"How about *gugú?*"

"Some do. In fact, I do. Fix it with the Mourning Doves." But this seems highly exceptional. Francis Vavages of Sacaton said, "*Gugú* looks like a little [Mourning] dove. I don't think anybody eats that kind."

I asked George Kyyitan of Bapchule, "How about *gúgu?*"

"No. Nobody eats that."

In Pima Bajo country, both Pedro Estrella and María Córdova called the Inca Dove *tehop*, and María said she would call the Common Ground-Dove by this name also. Pedro, however, called the ground-dove *hupigdam* when he saw it in the elicitation book and *s-wepeg aangam* when he saw it once in the field (see previous species).

There is a plant in Pima Bajo country Pedro called *tehop-gaat*, rendering it "rifle de pajaro," otherwise known as yerba de lisa. (Pedro's gloss of *gaat* as "rifle" might more properly be "bow," given the original meaning of the Piman word *gaat.*) This plant, a croton, is known in English as Dove-weed. It is an important seed plant for doves (Martin, Zim, and Nelson 1951:409). I wondered if this species "shoots" its seeds as do certain other euphorbiaceous plants. That would explain its Pima Bajo name.

COMPARATIVE LINGUISTICS. The best evidence available indicates that the Upper Pimans invaded the north country some centuries earlier than the Inca Dove, which arrived in Arizona after 1870 and has continued to spread slowly in the state (Phillips 1968:138–139; Phillips, Marshall, and Monson 1964:43–44). It was found at Florence, a settlement just upriver from the Pima, by 1882, and was common in the Sacaton area at least by 1908.

The dual name for the ground-dove might be explained as a result of the O'odham moving from an ancestral homeland with two very small doves, each of which was named, to a new country with originally just one tiny dove. Here in the north, *both* names were preserved in different speech communities, some calling it by one name, others by another. Then, when the Inca Dove arrived late in the 1800s, this confiding bird was readily named for its most common call note.

Among the Northern Tepehuán, Pennington (1969:116) recorded *túgu* as the Band-tailed Pigeon (in Pima, *gihodag*). I suspect this designation is an independently derived name, imitating the pigeon's low two-note call, rather than being cognate with the little dove's name.

Other Sonoran-branch Uto-Aztecans had contact with this vocal little dove. Those Rarámuri living in the western and lower-elevation Tarahumara area of Cerocahui called the tórtola (variedad pequeña de paloma) *sohó* (Brambila 1983:414). Still farther west on the coast, Yoeme (Yaqui) called a dove *kuuku*, undoubtedly the Inca, not the White-wing, as indicated.

aaḍo
***wúaka* (Pima Bajo)**
Psittacidae, Psittaciformes
macaw; guacamayo, guacamaya (original meaning)
Phasianidae
peacock; pavo real (transferred meaning)

Archaeological remains of the Scarlet Macaw, *Ara macao*, are commonly recovered from southwestern sites. They are represented in all cultural areas: Hohokam, Anasazi, Mogollon, Mimbres, and Sinagua. The entire skeleton is often found. But the natural range of this great parrot comes nowhere near Pimería. It occurs in the humid Caribbean lowlands from southern Tamaulipas, Vera Cruz, northern Oaxaca, and Chiapas southward through much of the tropical zone of South America. The Military Macaw, *Ara militaris*, however, breeds from southern Sonora and Chihuahua south through Oaxaca in more arid pine-oak forests. The two species are readily distinguished by most osteological elements.

In spite of the geographic proximity of its range, there is only one verified recovery of the Military Macaw in the Southwest (Hargrave 1970). The rest are Scarlet Macaws. An elaborate prehistoric trade network linking the Southwest and Mesoamerica supplied this neotropical lowland bird to targets within and even beyond the range of the Military Macaw. There is even evidence that macaws were being raised in numbers at Casas Grandes, Chihuahua, a Medio Period trading center (Di Peso 1974a:267, 269, 272–273, 1974b:182–185).

But why was the southern bird in such great demand when a more local species of about equal size was available? To a large extent, the answer is the plumage:

macaw feathers were (and among some southwestern Native American groups still are) prized for decorating ceremonial items such as kachina dancer masks and prayer sticks (Ladd 1998; Parsons 1939:750, 991–992; Schroeder 1968:1087). In colonial times, missionaries noted Piman macaw headpieces and other personal adornment. Still today red macaw tail feathers are conspicuous in photographs of the daily attire as well as rituals of practically any tribe living in the rain forests of Central and South America. Even high in the Andes, tribes use these feathers ceremonially (Randall 1982:60–61).

The Scarlet Macaw is a quite large red parrot (length thirty-three to thirty-six inches) with blue primaries and secondaries and yellow wing coverts forming a broad band tipped with greenish. Its crissum and rump feathers are pale blue. The bird has very long all-red tail feathers, above and below, the middle ones twenty to twenty-four inches long. Even the undersurfaces of the blue primaries and secondaries are red.

The Military Macaw of western Mexico is a large green-bodied bird (length twenty-eight to thirty inches), with light-blue primaries and outer wing coverts, a red forehead, pale blue crissum and rump feathers. Its tail, about sixteen inches, is red with a distal third blue on the upper surface, all golden olive below. (For an excellent comparison of the two species showing feather colors above and below, see Guy Tudor's plate 8 in de Schauensee and Phelps 1978). Although impressive, the local green bird held little of the esteem accorded its scarlet relative from the south.

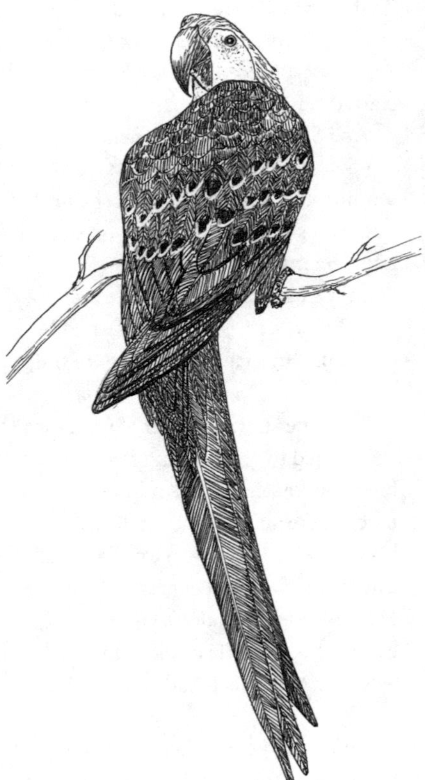

aaḍo, Scarlet Macaw (Takashi Ijichi)

A third species of parrot, the Thick-billed Parrot, *Rhynchopsitta pachyrhyncha*, smaller than a macaw (about sixteen inches) but shaped somewhat like one, ranged throughout the Sierra Madre of western Mexico, formerly spilling into the borderland mountains of southeastern Arizona and southwestern New Mexico in considerable numbers, at least in the nonbreeding season.

There is an archaeological record of the Thick-billed Parrot from Snaketown, a Hohokam site on the Gila River Reservation, dating from A.D. 100–900, and of a Scarlet Macaw dated between A.D. 1 and A.D. 700, both the earliest occurrences for these parrots in the Southwest (McKusick 1976). These occurrences are quite in advance of the macaw trade into other parts of the Southwest that subsequently developed around A.D. 1100 (Hargrave 1970:53–54).

Trade or at least successful rearing of macaws among O'odham did not stop with the disruption of trade routes caused by the Conquest. The 1660s Névome vocabulary from the middle Río Yaqui (Ónavas) preserved some interesting information from the colonial period (Pennington 1979:57). Here the Jesuit identified the bird as the guacamayo, a mexicanismo specifically for the macaw as distinct from other parrots. The spelling of its Piman name is *arho (aaḍo)* or *varo*. The guacamayo was of special economic importance. Its small feathers were called *arho vopo*. In contemporary Piman, *vopo* indicates "down"—that is, small body feathers—as well as "fur." The *capacete*, "headdress" or "helmet" made from the macaw plumes, was called *arho vonoma* (pl. *arho voponama*). Presumably this was a headpiece made from the long, brilliant red tail feathers. (As an aside, the present Piman word for "hat," *wonamĭ*, appears to be a legitimate aboriginal term, not one torturously derived from the Spanish word *sombrero*.)

Half a century later and 275 miles to the northwest, macaws continued to be noted in at least one Piman (Sobaipuri) settlement. In April 1701, Fr. Kino (1919:291–292) wrote from Pimería Alta, "On the tenth we rested here at San Xavier [near Tucson]. . . . Much kindness was shown us by these very excellent natives. They gave us many of their provisions, many of their good fabrics and blankets of cotton, numerous baskets, buck-skins, and red feathers of the many macaws which are raised here."

In 1716, Fr. Luís Velarde (González 1977:53) wrote,

> Y en San Xavier del Bac y rancherías comarcanas hay muchas guacamayas, que crían los pimas por sus hermosas rubicundas plumas, y de otros colores, y cuasi como las del pavo real, que las pelan por la primavera para su adorno.

> At San Javier del Bac and neighboring *rancherías*, there are many macaws, which the Pimas raise because of their beautiful feathers of red and other colors, almost like those of the peacock, which they strip from these birds in the spring for their adornment. (Velarde 1931:129)

Both accounts establish that the local Pimans (the Sobaipuri) were actually raising the birds, that they were macaws, and that the macaws were predominantly red—that is, Scarlet not Military Macaws. The birds were reared for their feathers. Even the Papago apparently kept macaws. In 1701, Capt. Manje (1954:167–168) and his companion Fr. Kino were given red macaw feathers at the village of Guactum eighteen leagues (forty-five miles) west of Mission San Xavier.

Another Jesuit, Fr. Philipp Segesser (1945:258), writing in 1737 from the Pima Bajo settlement of Tecoripa somewhat northwest of Ónavas, said, "The *quagamais* [guacamaya] bird, large as an ostrich, as well as red and yellow parrots, I could do without. Both kinds of birds are very ill tempered and talkative." The wording suggests that two species were being raised here, the Military and the Scarlet Macaw. Segesser made a passing reference to "little parrots" as well, without implying that they were kept. Their Piman names were not given.

Still a fourth Jesuit, Fr. Pfefferkorn (1949:124), stationed in Pimería Alta between 1756 and 1767, wrote of various psittacine birds found in the warmer country south of his mission district. He described the guacamayo as being essentially red with some yellow and light and dark blue in the feathers, clearly the Scarlet Macaw. "Many Indians living in the interior of Sonora raise this species [kind] of bird, especially the guacamayo, in their houses. The pretty feathers of these birds are sometimes plucked for use in ornamentation."

His companion, Fr. Joseph Och (1965:163), stationed at San Ignacio on the Río Magdalena in 1756, did not mention macaw rearing and left us to guess the source of the "beautiful varicolored feathers" his Pima flock were using in their ceremonial dances. Fewkes (1900:692) observed that "Parrot-feathers are highly prized by the Hopi for the ornamentation of their masks, and in former times were brought from the settlements on the Rio Gila and from the northern states of Mexico, where they were obtained by barter." Pimas were macaw feather suppliers until the early 1700s according to Parsons (1939:991–992n.)

Beyond its use for feathers, the *aaḍo* figures as one of the primary birds in Piman oral literature. Throughout myth, oration, and song, the macaw is the metaphor of blood—brilliant, scarlet red blood, usually shed by the enemy.

There are two different mythic narratives accounting for the origin of the *aaḍo*. (There are similar duplications in the origin of tobacco, maize, and especially Saguaros, all species of great economic importance in O'odham life.) Origin I is found in several places (Benedict 2001:27–28, 56–57; Lloyd 1911:99–101; Sylvester Matthias, notes; Russell:221–222). Lloyd entitled it "The Story of the Turquoises and the Red Bird."

Si'al Cheedagĭ Siivañ, Morning Green Chief, is ruler of the pueblo now known as Casa Grande ruins. Someone there sinks into the ground, becoming a blue stone, turquoise. After these pieces are distributed, Tash Siivañ, Sun Chief, of a neighboring *va'aki* (pueblo), "thought he would do something wonderful, too, being envious, and he opened one of his veins and from the blood made a large, beautiful bird, colored red." Sun Chief instructs the bird to go to the city of Morning Green Chief and eat nothing but turquoises. Morning Green Chief's daughter finds the bird and reports it. Her father says it is a very rare bird, but that she should not try to pick it up. "'Offer it a stick, and it will take hold of it, with its bill, and you can lead it here.' And Nawitch [the daughter] offered the bird a stick, and it caught hold of the end by its bill, which was like a parrot's bill, and she led it to her father" (Lloyd 1911:100).

They try feeding it first pumpkin seed, then melon seed (this is an anachronism), then devil's claw seed, but it will not eat them or broth of corn. Then it starts eating turquoises, so they feed it these beautiful blue stones. When full, it flies back to Sun Chief's "city." He gives it water twice, and each time it vomits turquoises, amounting to four basketsfull.

Origin II links the macaw with the blood of Ho'ok, the cannibal witch. The story accounts for the origin of tobacco as well. Various published versions are available (Benedict 2001:61–62, 62–63; Densmore 1929:17; Hill 1940b; Russell 1908:223–224; Saxton and Saxton 1973:265–270; Underhill 1946:83–84; Wright 1929:96–103; this Ho'ok's blood episode is absent from Lloyd's account [1911:106–117] as well as from Hayden's [1935]). I give here an unpublished version narrated by Joseph Giff in 1973:

> There's another legend on how the *aaḍo* was made. It's a long story [about] Ho'ok *ee'eḍ* [Ho'ok's blood]. There was a Ho'ok one time. There was a woman sitting here weaving a basket or something [a sleeping mat]. A boy was playing the kick-ball. The kick-ball rolled right by that woman. And she got it and stuck it under her clothes, to fool that boy. And he came around her looking for it. And she looked up and said, "What are you looking for?" "I'm looking for my throwing ball." "I don't know [where it is]." Then this boy started going off, and she said, "Here, I've got your throwing ball over here." And when he came back, she couldn't find it. She couldn't find the throwing ball. It just disappeared.
>
> And a few days later she found herself pregnant. The ball went in there and she got pregnant. And when the baby was born, it had claws like an animal. It looked awful. It had an ugly-looking face, ugly-looking form. It had claws like an animal. Somebody said, "Get rid of it! Kill it! We can't bear this all the time to see this baby grow up and look like that and be among us and look pitiful." And another said, "No! It's a human being. You don't want to kill it. It's a part of us."
>
> And they tried to find out what it is, what they going to call it. And they couldn't find a name for it. So they called Ban. Ban was a smart man. He just went and looked at it, and that's what he said, "It's Ho'ok," he said.
>
> And she grew up. Every time she's playing with the kids, every time she touched one, she'd leave claw marks on it. So all the kids were afraid of her. And the people around there wanted to kill Ho'ok. And she had an aunt who was taking care of Ho'ok. And Ho'ok decided to go, to leave, up to the mountains. And she left. She was a young girl then. She left, and there's a song that goes with it. This aunt was crying, begging Ho'ok to come back. And she keep on going and climbed up on a mountain and lived on a cliff.
>
> And after that she got to have revenge on the people who were mean to her, treat her like that. She comes around and eats their children. Grabs them and eats them. So everybody's afraid of Ho'ok. And they couldn't find a way to kill her. They couldn't kill her. The medicine man tried and all this and that. Everybody tried.
>
> So this time they finally got I'itoi. And they made brew out of *kotoḍopĭ* [Sacred Datura] we were talking about. And they started singing. And they started singing and dancing.
>
> And Se'ehe—I'itoi's—dancing with Ho'ok. And once in a while, he'd give her a drink, give her a drink until she passed out. And still the people kept on singing and dancing. They had to carry Ho'ok. So they carried her up into the mountain cliff, in a cliff, and they burned her in a cliff. ["In a cave," interjected Ruth Giff.]
>
> Then she started to tremble, because she wanted to get out. And that rock broke on top, cracked. And I'itoi jumped up on there and stomped it down with his foot. And they say to this day you can see his footprint somewhere in Papago country—I don't know where exactly. ["Across the border," inserted Sylvester Matthias.] And so they killed her.
>
> And then the people started eating her. Eating Ho'ok. And this woman had two grandsons, paternal grandsons. And they were left out. They didn't take any part of it. And they were hungry. And then they scraped up the blood, from

the rock, inside the cave. And made it into little balls and carried them home. They came to a nice soft sandy place. And the grandmother told the boys to bury those in the sand.

So they buried them, and she said, "Four days from now you come and look at them." The four days passed, and she said, "Now go ahead and look at those." So they went and found them, and they were eggs, some eggs. Then they left them and came back in another four days, they were hatched, they were little birds. There were little goose pimples coming out. And they left them. And she told them not to take them until their feathers started coming out, like a young *hoohi* [Mourning Dove]. So they went and got them and brought them home. And raised them.

And they had the prettiest feathers. No one had ever seen a feather like that. So they picked the feathers. And decorated their bodies, and they go to dances and whatever. And the people there got envious of them. And they wanted to know where they got the feathers, those pretty feathers. And the boys wouldn't tell them. Then they wanted to buy the feathers from them. But the grandmother told them, "Don't you sell it, don't you give any one to anybody. They're yours. They didn't give you what they had over there. See, this is all you got. Why should you give them anything?"

So they [the people] started plotting against the old lady's life. And they had these birds. And the grandmother told the boys to take those birds, and you go so many days, over the high mountain, and from that high mountain there's another high mountain you can see the caves in there. And you point your birds that way and you let them go. And they let them fly, and they [the birds] went to live in those caves. And whenever you need pretty feathers here, you go in the caves and you pick them up, because they shed them for you. You might never see the birds again but they'll be there. They'll be there.

So that's the origin of the peacock [*sic*, actually *aaḍo*, the macaw].

And they did kill the grandmother. The boys were coming back. (I don't know how many days' journey they walked.) And they were getting close to home. The older one kept awake and heard their grandmother singing, coming closer. And he woke up the brother and said, "You hear our grandmother? She's singing. She's singing." Then they know that she was dead. So they came home. And they [the people] had already killed her. She had already told them where to bury her. So they took her to a nice soft sandy spot there, and they buried her [in an arroyo]. She had said, "You come around in so many days and you'll find something growing in my grave. You leave it. Till it's ready to pick. And I'm going to give this to the people. That's what they will grow for the medicine man only, to smoke. And they can see things *way* around, and they can predict what's coming, and they can cure illnesses." And that's tobacco.

So they went and they found it. And it grew up. And this Ban was fooling around there—he's a mischievous person. And he knew what it was. So he went and picked some already, before it's ready. And the people get around in there [in the communal round house], once in a while. I don't know what they smoke, but they do smoke once in a while. And Ban [was] asking them, "Who wants to smoke?" Nobody, nobody. Pretty soon he said, "Well, I've got tobacco." So he smoked. So that's why they call that *ban vivga*.

Those boys got that tobacco, and they gave it to the medicine man. And ever since then it's the custom for the medicine man to smoke. It was an honor for a medicine man, *then*. See, that's the origin of the tobacco and the peacock [*sic*].

Both versions place the origin of the macaw in the cultural context of the people whom the River Pima call the Vipishaḍ, the Small People who built the pueblos in the vicinity, and both have the bird coming from someone's blood. One prehistoric *va'aki*, pueblo, on the Gila is even described as having many macaws. Origin I associates macaws with turquoise. In the mythic context, the stones are obtained by stealing. In prehistory, the beautiful blue stones from northern mines were major exchange items for the brilliant red birds from the tropics (Di Peso 1974b). In Origin II, the birds are always associated with the origin of tobacco, sometimes specified as *ban viv*, Coyote's tobacco, which at times refers to a local wild species found in arroyos, but at other times to a cultivar (see Rea 1997:236–238, 316–320, for the ethnobotany). At least one cultivated species, *Nicotiana tabacum*, eventually filtered into the Southwest from South America and the Caribbean region. So the association of macaws and tobacco in this story may be more than coincidence: both come from distant southern lands.

In a curious twist, two narrative versions of the Origin I category associate the conception of Ho'ok herself with the macaw (Benedict 2001:27–28, 56–58). After Tash Siivañ, Sun Chief, has stolen the precious turquoise from Si'al Cheedagĭ Siivañ's (Morning Green Chief's) *va'aki*, Morning Green sends cold and rain on Sun Chief's *va'aki*. Fires are quenched, and the macaw, a tropical bird, is frozen to death. Sun Chief cremates the bird, forming a kickball from its ashes. A young man kicks his ball near a maiden (sometimes specified as Morning Green's daughter) who sits weaving *main*, sleeping mats. She playfully hides the ball beneath her skirt, and when the youth asks about his lost ball, the girl goes to retrieve it, only to find that the ball has entered her. The offspring of this odd impregnation is Ho'ok, the cannibal woman.

As mentioned earlier, one Vipishaḍ *va'aki*, located somewhere along the middle Gila in the western part of what is Pima country today, was noted in myth for its abundance of pet macaws, and another had many captive eagles (Benedict 2001:50). The invading Vupshkam, the modern Pimans, sang a song as they approached this pueblo:

> Mountain with many macaws in the west,
> I am approaching.
> My heart grows red [with macaw feathers].

"They fought and overcame the *va'aki* and took the macaws." Here is something for zooarchaeologists to watch for: a Huhugam (Hohokam) ruin with specializations in aviculture for trade.

Macaw feathers, in particular Scarlet Macaw tail feathers, have continued in use among some southwestern groups (Hopi, Zuni, some Río Grande pueblos), into the twenty-first century. But Akimel and Tohono O'odham have long ago forgotten the original referent of the term *aaḍo*, substituting for it instead the one long-tailed bird that they saw among the Euro-American farmers who settled around the reservations, the peafowl. This shift may have begun around the turn of the twentieth century. Although Thin Leather supplied the Creation Epic to both Lloyd (1911) and Russell (1908), the referent is clearly a red bird with a parrotlike beak in Lloyd's careful transcription. However, it appears as *s-cheedagĭ u'uhig*, a green bird, in Russell's version. It may be that this version was translated by Jose Louis Brennan, who no longer knew the original biological species as Thin Leather clearly did.

The referent switch is not unique to Russell. Anthropologist Ruth Benedict, transcribing Pima mythology at Sacaton in 1927, remarked that interpreters always translated the word *aaḍo* as "peacock" (Benedict 2001:27 n. 39). Densmore (1929:17), recounting Origin II, said the peacock came from Ho'ok's blood. Wright (1929:96–103) told the parallel story. Unlike in the others, in Wright's version the boys go to the west to release their birds. "[T]he beautiful birds were free. They spread their wings and rose in the air, and the sun, which was just setting, shone on their bright feathers and all the sky was full of their colors—gold and blue and green. The colors of the two birds were so strong that even to this day, as everyone knows, the sunsets are often that way." However, sunsets, as we know, are basically *red*. Wright did not conceive of a tropical macaw being the bird in his story. Likewise, Sylvester Matthias, telling Origin I, and Joe Giff, narrating Origin II (just given), both called the *aaḍo* "peacock," as did Saxton and Saxton (1978:265). (In this latter version and in Gertrude Hill's [1940b], the grandsons return with some of Ho'ok's bones rather than with her blood, thus moving away from the red imagery.)

"The Papago myths," noted Castetter and Underhill (1935:51), "tell of Elder Brother wearing bright feathers, probably macaw." Before the Flood, Elder Brother creates Handsome Man, decking him out in the finery of the day: a Cougar-skin quiver *(mavid eldagĭ vogsha)*, a macaw-tail-feather headdress *(aaḍo bahivua siivda)*, his light spread out around him (Herzog [1929]). Thus arrayed, Handsome Man's amorous conquests are irresistible and his progeny bountiful, until he is sodomized by a maiden. The tears from the child he bears and abandons the next morning initiate the Flood.

Until the species became identified from the analyses of archaeological bones by Hargrave (1970) and by subsequent zooarchaeologists, writers had trouble conceiving of *aaḍo* as a macaw. Some opted for green parrots and others for peacocks, the current but totally illogical referent; Pimans had no contact with these latter Old World barnyard birds until after American settlement of Arizona Territory.

Throughout her work, Underhill (1946) glossed the word *aaḍo* as "macaw." Though she does not say which species, the blood metaphor soaks through every piece of oral literature she so carefully transcribed. (Unfortunately, she gave only English translations, not the original O'odham words.)

For instance, Underhill (1946:208–209) recorded an oration that was included in the lustration, or ceremony of purification, required after an O'odham warrior killed an Apache. One part describes an Apache house, apparently the scene of mourning following a raid by the Pimans:

> Far yonder, beside the rising sun,
> A house stood—
> Red, with a domed roof;
> There young women sat,
> With headdresses of macaw feathers,
> With headdresses of hawk [*sic*] feathers.
> That I much liked to see.

The location is the high country to the east, inhabited by the Apache. The imagery is obscure, and the two birds are atypically paired; the falcon ("hawk") is usually matched with the eagle. But evidently red was needed here, to go with sunrise, the red house or wickiup, and blood of the slain enemy. The site of the mourning widows pleases the Pima warrior.

Additional blood imagery appears in the same oration:

> Yonder three poles
> Stood interlocked.
> From it a fine male hawk
> Hung downward.
> From the end of his tail
> Blood was dripping
> That I liked to see.

Underhill's (1946:181) warpath songs continue the metaphor of the macaws as blood, here again of slain Apaches:

> The land was covered with macaws.
> The land was covered with macaws.
> I came there.
> I came there,
> And saw my enemy
> In the likeness of a macaw.
> The mountains were covered with macaws.
> The mountains were covered with macaws.
> I came there.
> I came there
> And saw my enemy
> In the likeness of a macaw.

The two songs tell what the singer hopes to see in the upcoming raids: land and mountains saturated with the blood of the enemy. Note that each of these four pieces ends with a coda mentioning *seeing*—that is, envisioning—the bloodiness (macawness). The bloodiness pictured is the hoped-for result of the war expedition.

Occasionally, this large parrot does not stand for blood in Piman oral literature, but the association is still with the color red. For instance, in the version of Elder Brother's (I'itoi's) resurrection speech recorded by Benedict (2001:37), there are springs of different colors at the cardinal points, each associated with a different bird:

> I made the red water-hole of red water in the north
> And caused bulrushes to grow out of it
> And heaped the surface of the water with macaw feathers.

Aaḍo also feature in the O'odham version of the worldwide theme of ill-starred lovers. Wright (1929:209–214) told the Tohono O'odham story of a boy and girl in love about to be separated. Her childhood lover is big and strong, but slow and poor. She is about to be married to another boy her parents have chosen who lives in a distant village behind the mountains.

The unfortunate boy seeks help from a *kulañ oks*, medicine woman, in the south. She gives him a small basket with beautiful feathers—green, blue, red, and yellow. "They were so bright they looked as if they were burning." Although the bird's name is never mentioned, the feathers come "from a bird that was lost. The bird was sometimes called 'Sunset Bird.'"

In a big storm, the lovers think of each other. Afterward, they take out the feathers and wave them in the sunshine. "The bright colors would shine from one to another" even though there were mountains between them. Everyone could then see an arc of brilliant colors in the sky between the lovers, the symbol of their longing. If the girl waved the feathers in both hands, there would appear a double arc, one above the other.

Emergence from an underworld seems to be a prominent metaphor in the mythology of southwestern peoples who undertook pre-Contact migrations. In Pima tradition, the Vupshkam together with Elder Brother emerge from the underworld, through the help of Jevho, Gopher, to wrest the world from the Vipishaḍ, or the Puebloan peoples who occupy the *vapaki*, pueblos, now found as ruins in the Salt and Gila River valley. The emergence song of Elder Brother and the Vupshkam is given twice in Russell (1908:226–227, 280–282). The first verse conjures an unmistakable military air even in English:

> Together we emerge with our rattles;
> Together we emerge with our rattles,
> Bright-hued feathers in our headdresses.

In reading this verse, it would be nice if we knew *what* these brilliant feathers were that decorated the invading warriors. We find out in Russell's native-language text of the Pima song, where the words are *yaaḍo mamai iyingingvoike*. *Yaaḍo* is song language for *aaḍo*, which Russell gives in interlinear translation as "parrot." The next word in spoken Pima is *babhai*, "tail feathers." The headdress word is *giigkoa*.

Two additional words are used for the headdresses, *voponama* (sing. *wonamĭ*) and *sisivoda* (sing. *siivoda*), the more usual word for a feathered crest headpiece. Hayden (1935:44; see also Bahr and associates 1994:218) used still a different word that he wrote *buch po ka*. According to Joe Giff and Sylvester Matthias, this word is either *bichpoga* or *bichpodam*, depending on one's dialect. Hayden noted that a medicine man made these headdresses "out of some kind of feathers [that] meant that whoever wore it would die." The song says "peacock," but the metaphor refers to the tail feathers of a macaw. But in Hayden's version the waving scarlet feathers in the headgear are an ominous sign.

In addition to the Scarlet Macaw tail feathers waving in the headpiece, the next part of the song tells of *iuke* tail feathers waving. This word may refer to a bird from Pima Bajo or Mountain Pima country. Such a bird might have important ramifications for the source of Pima emergence into the north country. Culver Cassa traces the name to the Military Macaw (see the comparative linguistics section).

North of Pima country, in the vicinity of Fort McDowell near the Salt-Verde confluence, is Red Mountain, where Sylvester obtained a brilliant red ochre for painting arrows, bows, and masks. In a legend, this place is called "Land of the Macaws" (Benedict 2001:144–145). A man makes a racetrack here where he runs to obtain power. He is met by Eagle and Falcon, who give him items used in curing staying sickness. (Gifford [1933:391–392] recorded a Northeastern Yavapai story that describes the origin of the Scarlet Macaw somewhere in Black Canyon country just north of here.)

The O'odham acquired various new plants from European colonizers. Some were of economic significance, such as wheat. Others were purely ornamental. For instance, a leguminous shrub, the Yellow Bird-of-paradise, was taken up by the River Pima. Its name among these new cultivators became *aaḍo o'ohoḍag*, meaning "macaw color, design, (or) sign." Sylvester Matthias explained its origin. "My grandmother named it. It's not a native here. When [Pima] ladies went to Phoenix to peddle their baskets they saw it. Brought back its seeds to grow in Komaḍk. Called it something to do with 'peacock,' *aaḍo*."

Unlike Sylvester, his elderly grandmother, born before the settlement of whites around Pima country, clearly still had the macaw referent in mind. This bush bears large flowers with yellow petals, a greenish calyx, and long exserted red stamens reminiscent of the long tail feathers of the Scarlet Macaw. It has no resemblance to a peacock (see Rea 1997:293).

Interesting bits and pieces relating to the biology of the *aaḍo* are woven through the various O'odham narratives. Joe Giff's story compares the young birds that emerge from Ho'ok's blood to those of the *hoohi*, Mourning Dove. Anyone who has seen the naked nestlings of psittacids (parrots) and columbids (pigeons) will appreciate the comparison. (In contrast, the precocial peacock, a galliform bird, hatches covered with a fine plumage, ready to be about.)

The stories also document the sensitivity to cold of these magnificent tropical birds. Lloyd's (1911:103) account notes that Morning Green Siivañ, after the theft of his turquoise, sends a heavy cold rain upon his rival's pueblo, almost killing Tash Siivañ. The macaw's fate is not specified in this version, but two accounts by William Blackwater say that as a result of the cold wind and rain, the macaw froze to death (Benedict 2001:28, 57).

Embedded in this story is information on macaw breeding habitat. Russell's version of Origin II says the boys released their birds in the "forest." Joe said the birds were taken into the high mountains, where there were caves in which they would live. He used the word *s-u'usikam doag*, "forested mountain," for their habitat, adding that *s-u'usik* means "a place of many trees, forest," whereas *s-u'usgam jeweḍ*, in contrast, means "wooded flatlands." In Wright's version (1929:100), the birds are released in "a very high mountain with many high steep bluffs." In the bilingual version recorded by Saxton and Saxton (1973:268), we hear the grandmother telling her grandsons: "At dawn, take your pets [*shoshoiga*] and run toward the east. There's a mountain there called Many Holes Peak [Waw S-Jeejeg]. Leave them there and scatter this seed below the mountain. It will grow there and they'll eat it and always stay around there. When she said this she gave them devil's claw seed [*ihug kai*]."

Ruth Benedict's collection of Pima narratives (2001:62, 63) includes four variants of the macaw origin story. Two of these (Origin II, told by William Blackwater) specify release sites. In one, the grandmother instructs her grandchildren to take the birds to Steep Pitted Mountain, Vav (Waw) S-Jeejeg. "They opened the cages of the macaws and the birds flew up to the holes on the face of the mountain. So they escaped." In the other, she says to "take the birds to Mount Kiwatkam and hide them." (Neither Underhill nor Densmore specified where the birds are released.)

The Saxtons' account implies and Joe Giff states that *aaḍo* feathers can also be gathered in the wild, where they are molted at roost sites. The mythic association of *aaḍo* with bluffs and caves preserves aspects of traditional folk science. Macaws as a group congregate and roost in such places, at times eating earth or clay found there.

Lloyd's version of Origin I mentions four foods for these birds: pumpkin or squash seed, melon seed, devil's claw seed, and a broth made of partially roasted and ground corn that O'odham call *ga'ivsa*. The modern aviculturist raising parrots couldn't go wrong with these foods.

Ornithologists know surprisingly little about the nesting habits of either the Military or the Scarlet Macaw, but nesting in tree cavities has been reported for both (Forshaw 1973; Munn 1994). The two large parrots that Tepimans would have encountered in the wild, the Military Macaw and the Thick-billed Parrot, are birds of rugged mountainous country, known to congregate at cliff sites, if not actually to nest there (Russell and Monson 1998:128). (In northeastern Mexico, Maroon-fronted Parrots, *Rhynchopsitta terrisi*, nest colonially in cliff excavations.)

Well, so much for what ornithologists have recorded. During the rainy season of July 2001, a family of Mountain Pima in Mexico took Robert Nagell and me across a green mesa to a steep-banked barranca called in Pima a name that means "Arroyo Where the Macaws Sing." We stood in a wild place, with the tops of pine trees at our feet. High on a cliff rising above our heads was a round hole with telltale whitewash, entrance to a nesting cavity where the Pima said the *var*, or Military Macaws, had nested every summer for as long as they could remember, returning each year with the rains. They asked if I would like to visit a spring at the far edge of the cliff below the nest. We scrambled down along the slippery cliff base to the dripping spring, flushing the pair of macaws, who loudly protested our invasion. A few bright feathers were dropped where they drank. For me, the boundary between myth and reality temporarily blurred.

COMPARATIVE LINGUISTICS. Saxton and Saxton (1969:166) originally gave *aaḍho* or *a'aḍho*, but later (1973) switched to *aaḍo*, as used here. Culver Cassa considers the two to be legitimate variants.

As noted, the Névome (colonial Pima Bajo) for guacamayo is *arho (aarho)*, where *rh* is the orthographic precursor of *ḍ*. (Long versus short vowels were not indicated in the vocabulary.) Another entry, *baro* (= *varo*), for papagayo grande, is the same word (Pennington 1979:57,89). In contemporary Upper Pima, the word used is *aaḍo*. Tepecano, at the opposite end of the Tepiman continuum, had *vaar* for "parrot" (Mason 1917:328). I have nothing for other Southern Tepehuán or Northern Tepehuán speech communities. Cognate words appear in other Sonoran Uto-Aztecan languages: *háro* (Eudeve); *waró (walá)* (Tarahumara or Rarámuri). Local Tarahumara in Urique Canyon called the Thick-billed Parrots that passed over our camp each day *álo*. The same word is used in places for the guacamayo. Molina (1972:16) recorded some macaw (or at least psittacine) Sonoran place-names as: Barobabi (Varo Vavia, in Pima, "Macaw Well"), Barobampo, Barocahui, Baromico, Baroyeca, Haro, Haropa (Jaropa).

The *ara/aro* macaw word spans two continents and many language families. In the Amazon basin, Jensen (1988:49–50, 72) found *arara* among two of three Tupi-speaking groups and one Carib tribe. Even the generic name in Western taxonomy is *Ara* Lacépède. (The word also appears in the Linnaean *Ara ararauna* and the diminutive *Aratinga* species.) This span speaks to the enormous and widespread salience of the Scarlet Macaw.

Linguist Cecil Brown (personal communication, n.d.) writes:

> The Portuguese word for macaw is *arara*. This probably entered Portuguese from the General Language (Tupi) that was a widespread lingua franca in Brazil and abutting countries. According to Santamaria, its ultimate origin is Guarani *ara*, "loro, periquito." Santamaria also cites *arara* as a widespread South American Spanish word for macaw. Because the word is common to at least three widely spoken lingua francas, it is not surprising to find it in New World areas outside of South America, especially given the fact that macaws were widely traded. The Piman words almost certainly are loans and *not* cognates. Also, *ara* is clearly onomatopoetic, which would contribute to it being found widely.

Forshaw (1973:360, 363) gives the call as a loud, harsh *rraa——aar* in contrast to the Military Macaw's raucous, drawn-out *kraa——aak*. Various Native American groups raising Scarlet Macaws had ample opportunity to hear them. Differentiating between cognates, loans, and independent invention in imitation of the bird's call is a problem, particularly because the trade in macaws and their feathers spanned some five centuries or more in the Southwest.

Pima Bajo Pedro Estrella called the Military Macaw *wúaka*, a name he said did not apply to other parrots. Culver Cassa notes that this is the same word *(iuke)* that appears in song language at the emergence on the Gila River (Russell 1908:280). Accordingly, the Vupshkam arrived with the tail feathers of both species of macaw waving in their headdresses.

I have found no contemporary Upper Pimans who were familiar the macaw except for Delores Lewis, who called the macaw illustration in my elicitation book *o'odham u'uhig*, "human bird." For most, the word *aaḍo* now signifies the domestic peacock or Common Peafowl, *Pavo cristatus*, and the metaphoric connotations are lost.

TECHNICAL NOTE. In his interpretative construct of red things for "sunness" and green things for "wetness," Bahr (in Bahr and associates 1994:138–140) was puzzled that a green bird should be associated with the so-called Hohokam in mythology. His thesis would have been strengthened if he had correctly identified *aaḍo* as the Scarlet Macaw, as long ago pointed out (Rea 1983:162–163), rather than as a green parrot, and the following conundrum would have been solved: "That this parrot is green does not cancel its solarity, I submit. There could be a green sun as a kind of limiting or bridging condition between hot light and rain, and indeed I suspect that this is the meaning of Morning Green's name. He approaches sunness from the side of wetness, and the parrot approaches wetness from the side of the sun."

Bahr was not alone in having problems with macaw stories. The Yavapai told E. W. Gifford (1933:391–392) their version of the origin of the Scarlet Macaw. A naked hunter jumps into his wives' fire, is badly burned, so plunges himself into a pond and on emerging transforms into an all-red parrot. His wives do likewise, one becoming a macaw, the other a Wild Turkey. The macaws fly away, but the turkey remains in the vicinity. Gifford was loath to believe his storyteller, recording the account as the origin of red "parrots."

tavir **(Pima Bajo)** ***cheedagĭ u'uhig*** ***s-ñeokdam u'uhig*** **(Akimel O'odham)** ***pilgíido, pilíiko*** **(Tohono O'odham)**

Psittacidae, Psittaciformes

Amazona spp. (esp. *A. albifrons*)

Amazon parrots; perico, cotorra, loro

Writing at the Pima Bajo settlement of Tecoripa in 1737, Jesuit father Philipp

tavir, cheedagĭ u'uhig, White-fronted Amazon Parrot (Amadeo M. Rea)

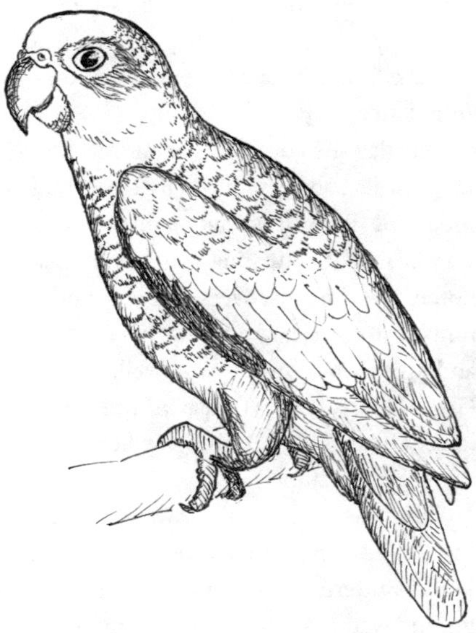

Segesser commented on the macaws, evidently being raised there, then made a passing reference to "little parrots" as well, without implying that they, too, were kept. Unfortunately, he did not give their Piman names.

Another Jesuit, Ignaz Pfefferkorn (1949:123–124), working in Pimería Alta from 1756 to 1767, described various parrots and macaws from the warmer country south of where he was stationed. Those types he called "cotorras" and "cotorritas" were clearly parrots of the genus *Amazona*. They had short tails and "a mixture of iridescent blue, red, and yellow feathers augments noticeably the predominantly green color of these birds." They also had loud voices. "Many Indians living in the interior of Sonora raise this species [kind?] of bird, especially the guacamayo, in their houses. The pretty feathers of these birds are sometimes plucked for use in ornamentation." In retrospect, one wishes he had been more precise as to which ethnic group he was referring. Pfefferkorn worked seven years with the Upper Pima and four years with the Eudeve and Ópata, but he still often lumped descriptions of them under simply "Indians" in contrast to "Spaniards." The passage just quoted seems to imply that some indigenes were keeping Amazon parrots in addition to the usual macaws.

Pimans have used or coined various names for parrots. *Cheedagĭ u'uhig*, "green bird," goes back at least to Grossmann in 1877. However, a strange metamorphosis appears in Russell (1908:222): the Scarlet Macaw of legend is given as *s-cheedagĭ u'uhig*, literally a green bird. It may be that the legend was translated by Tohono O'odham Jose Louis Brennan, who no longer knew the original biological species that Thin Leather, Russell's source, clearly did. (Lloyd [1911:99–101], taking down the same story from this same Pima narrator, wrote "red bird.") More recently, Sylvester Matthias used *s-cheedagĭ u'uhig* for "parrot."

However, Joseph Giff and various other Pimas called it *s-ñeokdam u'uhig*, "talking bird." Joe explained that this name originated "when white people brought the parrot here." The biological referent is usually some Amazon species.

At Ónavas, Pedro Estrella and María Córdova used the name *tavir* for the White-fronted Parrot, *Amazona albifrons*, which occasionally wanders up the river. Here Hale (n.d.a) recorded *tavir* (= *taviḍ*) for papagayo, the local name for a parrot. The word has changed little in three centuries. The Névome vocabulario recorded in this settlement translated *tabirho (taviḍo* or *taviro)* as "papagayo," in contrast with *baro* (= *varo*), papagayo grande—that is, the macaw, and *arho* (*aḍo*), the guacamayo.

The Amazon parrots are short-tailed, rather short-winged, heavy-bodied birds that fly with rapid wing beats through the tropical and subtropical forests. They flash patches of red and blue, sometimes yellow, in the bright green plumage. When they alight in some green tree top, they stop their raucous screeching and become almost invisible. Several species occur in Tepiman country, the northernmost being the White-fronted Parrot. Amazon parrots have been popular cage birds, but there is less evidence that Pimans were raising them aboriginally, as they were macaws (see previous account).

Saxton, Saxton, and Enos (1983:90) recorded the loan word *pilgíido* (Sp. *periquito*) for a parrot among Tohono O'odham. The referent would seem to be one of the smaller species of parrot or parakeet that are common as cage birds. Similarly, Hale (1964) recorded *períik* (Sp. *perico*) as "parrot" among Mountain Pima. I asked Joe Lopez of Vaya Chiñ, originally from Pozo Verde, Sonora, what he would call a green parrot. *"Pilíiko,"* he said, "they talk too much." He did not know the word *aaḍo* for either macaw or peacock.

At Urique, the local canyon Tarahumara and mestizo name for Thick-billed Parrot is *álo*, pronounced with a Spanish *l*, but the word is cognate with the O'odham macaw word.

COMPARATIVE LINGUISTICS. Yoeme (Yaqui) *tavelo* is cognate with Pima Bajo *tavir*. They are adjacent language communities. Among other Sonoran-branch speakers, for perico or loro we find *zíra, círa* (Eudeve); *yoali* (Huichol); and *kaknali* (Mountain Pima).

kadgam

Cuculidae, Cuculiformes

Coccyzus americanus

Yellow-billed Cuckoo; Cuclillo Pico Amarillo

The Yellow-billed Cuckoo is one of the latest birds to arrive from the south in hot weather and one of the first to leave at the end of its breeding duties, being in Pimería Alta scarcely more than three months of the year. Here it is partial to dense cottonwood and willow trees along rivers. In Pimería Baja, it is both more common and more widespread, occupying mesquite and other legume bosques as well as riparian communities.

The cuckoo is almost unbirdlike. It sits solitary and still below treetops, watching a person tranquilly before it is spotted itself. When it does change position, its movements are deliberate, almost reptilian. It seems to crawl among the branches of the canopy, searching out its favorite food, caterpillars. One is more likely to hear than to see it. Its most common call is in a long series, first of faster *kuk-kuk-kuk* notes, then slowing into *koow-koow-koow*, with more deliberate intervals. Even these wooden notes, which may be given in flight, seem nonavian.

The Yellow-billed Cuckoo is the size of a Mourning Dove, but its head and neck seem almost as thick as its body. It is grayish brown above, white below from chin to vent, with quite a long and mostly black tail with white tips to the feathers. In flight, the primaries or outer wing feathers flash rusty brown. At close range, its eye seems large, enhanced perhaps by the ring of yellow

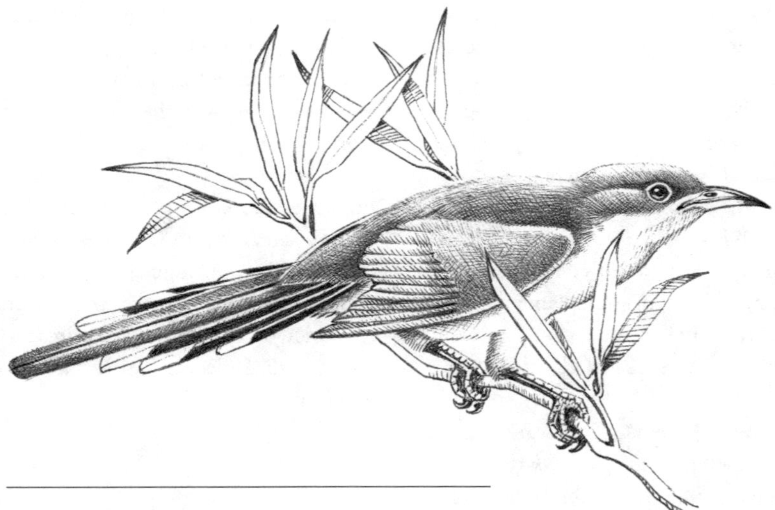

kadgam, Yellow-billed Cuckoo (Mike Touch)

skin. Its decurved bill is yellow at the base, black along its ridge and tip.

In Pima country, cuckoos still breed in the cottonwoods of the lower Salt River. But I once found them (1970) as far east as Sacaton, in the dense growth of small canals and living fencerows with a mature cottonwood canopy, where they were no doubt breeding (Rea 1983:164). Such habitats have now been cleared for "clean" agribusiness or else have died for lack of water.

But the Pima once knew their cuckoos. Sylvester Matthias was the first to mention the *kadgam* to me in our discussions of birds. He said that it was not common but could be found in the willow and cottonwood thickets formerly found along the Gila, though he had never seen the bird itself, only heard it. "A river bird usually found in cottonwood trees. It may nest there. Same size and has a loud noise like a *chukukmal* [Western Kingbird]. It's in a story." Later he told me, "A common bird in my grandmother's time. Common around Bapchule. There is a story of *kadgam* where New York Thicket is now." He disassociated this bird from another bird known as *kahagam*.

Joseph Giff knew the bird better. "*Kadgam* is in thick willows and cottonwoods along the river. It's rare. About the size of a *hoohi* [Mourning Dove]. It makes an odd sound. It's the size of a *kuḍat* [flicker]. Tail like a *haupal* [Red-tailed Hawk], sort of brownish. When they fly, like you can see through their wings. They're hard to see, but you hear them most often. Not a sociable bird." He then imitated the sound they make by clapping his slightly cupped hands, describing their call as "sort of a wooden sound, not like a bird."

Like Sylvester, George Kyyitan disassociated the *kahagam*, or "bluebird," of his songs from the *kadgam*. He said that the *kahagam* is a small bird. "*Kadgam*, I think, it's kind of a big bird. I think it's almost like these blackbirds [grackles in his yard], but it's a different color."

All three Pimas are clearly describing the Yellow-billed Cuckoo, largely gone from their country with the destruction of the riparian woodlands. Although Joe had not seen the bird in decades, he described it excellently except for one point: the rufous color, which he compared to the tail of the Red-tailed Hawk, is in the primaries or outer wing feathers rather than in the tail.

The Yellow-billed Cuckoo does not nest in Tohono O'odham country. Although it is common in Pima Bajo country, and no doubt we heard it while afield, I failed to get a name for this bird from Pedro Estrella. No word in the Névome vocabulary seems to fit cuckoos.

By good fortune, the story that Sylvester alluded to has been preserved in the version of the Creation Epic that Julian Hayden (1935:17–18) recorded at Bapchule from storyteller Juan Smith (see also Bahr and associates 1997:113–120; Benedict 2001:69–72). A girl from Gook Vapchkĭ, Two Tanks (Ponds) village, is called Ba'ivchul Uvĭ—that is, "Kingfisher Young Woman."[2] A boy is in love with her, so her parents and the boy's mother arrange a marriage. Although Ba'ivchul Uvĭ is in love with another youth in a distant village, she acquiesces. On her marriage night, having lain with her new husband, Kingfisher Woman gets up and goes downriver to Maas Akimel, a large lake in the Gila around New York Thicket, then west of here to Sand Hills, south of Santa Cruz village, where she finds Kadgam, Cuckoo, whose son she likes and subsequently marries. She is happy here, where her new father-in-law lives in agricultural abundance.

One day she goes to Maas Akimel for water, where she is seduced by Hevachuḍ, Blue Bird (Blue Grosbeak), a man who is very beautiful, returning with him to his home. Here she discovers that husband number three is lazy and that his corn and pumpkins don't grow so well. (Blue Grosbeak lives east of Maas Akimel on the lower Queen Creek drainage, where the soil is highly saline.)

Kingfisher Woman runs away from Blue Bird, taking up residence next with Hohokimal, Butterfly, who farms with underground water. But his type of farming is too heavy of work for a woman. She eventually leaves him as well.

Although this story is superficially about marital fidelity and is associated with several people with bird names, at a deeper level it tells about the Pimas' attempts to farm under varying desert conditions. Each husband lives in a different ecological setting, at real places that can be identified today. Just as the woman abandons her husbands, the agricultural Pima had sometimes to abandon their chosen field settlements, trying alternative strategies. The people living at Gock Vapchkĭ on the upper Queen Creek drainage actually did abandon their settlement, according to Pima oral tradition, settling just north of Santa Rosa village, where they are found today as the Aangam (Desert-willow) People. We are left with the impression that the best type of farming, though, is in the fields adjacent to the riparian woodlands, home of the cuckoo.

Culver Cassa suggests that the story also relates to Kingfisher Woman's sense of isolation and need for female companionship, which she had earlier. With dry farming or floodwater farming, just one or two men could manage the system. In contrast, the streamside irrigation system required more manpower, and consequently there were more wives for company.

Culver conflates this account with the Corn Man and Tobacco Woman story in Lloyd (1911:218–229), noting how the Piman narrator incorporated more details of cuckoo biology into the imagery. In a pique, Tobacco Woman leaves the People. Without tobacco to make smoke to bring clouds and rain, there is a long drought. As a result, the People are abandoned by Corn Man and Pumpkin Man, who go east to better-watered country. Giisobĭ (Verdin) is sent to entice Tobacco Woman back with offerings. Still in a huff, she sends Verdin back with tobacco seeds and some hornworm "dirt" (feces) to be smoked. These items are enough to get the rain cycle and consequently the growing cycle started once again.

The association of the hornworm here seems obscure. Several species of this sphinx moth larva can be found in Pima country. All are large, plump, and green and are called *ta'apa* in O'odham. Some eat tobacco leaves, others willow and cottonwood. In turn, the hornworms are important food for Kadgam, Cuckoo. Both the larva and the bird mark this tale as a summer story. Neither can be found at other seasons.

Culver pondered why a social theme should be introduced into the story at this point. Kadgam, with the People only briefly, represents humans reassembling for agricultural activities, which require communal efforts to bring rain. Some ritual is needed to reacquaint the People with one another and to reestablish their communal bonds. This is accomplished through the smoking ceremony in the big house, as taught by Verdin. Here the village men assemble at night, light a cane tube of tobacco, and pass it to one another in a circle. As they offer and receive the pipe, they address one another by some kinship term, a custom initiated by Cuckoo.

One Pima song about the Yellow-billed Cuckoo has been preserved. It says simply, "*Kadgam* bird / far off is singing." In the two-part song, the cuckoo is paired with the hummingbird (see that species account for the complete set).

MISIDENTIFICATIONS. While transcribing the Corn Man and Tobacco Woman story, Lloyd (1911) took down the name *"kah-tee-gum"* for some edible greens (presumably an *iivagĭ*). In discussing this word, Bahr (in Bahr and associates 1994:99 n. *l*) says: "Sounds like what I would spell as *kadigam*. In story 6, 'The Whore,' Hayden spells what is undoubtedly the same word as 'Cadigum.' Here in this text the word is apparently taken as the name of a plant with fragrant leaves, but Hayden was told it is the name of a kind of bird unfamiliar to Smith and Allison. I have not found anyone who knows the word, so I cannot confirm either interpretation. The word does not sound particularly Pima-Papago and may be adapted from another Indian language. In any case, it seems to be obscure."

But in discussing the name of a person that Hayden wrote as Cadigum (Hayden 1935:18), Bahr (in Bahr and associates 1994:114 n. *e*) further muddies the water. "This might be Kuhigam, identified by Saxton and Saxton (1969:27) as a 'black bird species' and on p. 167 of the same dictionary as 'unidentified.' The name would seem to mean 'neigher,' 'one who neighs (like a horse),' or 'crows,' 'caws,' or 'screeches.' (All those are meanings of the stem *kuhu*). Alternatively, as discussed . . . the word may not be a bird name at all but the name of a leafy plant."

Almost certainly the dictionary entry *kuhigam* referenced here is a mistranscription (or a faulty recalling) of *kuigam*, the Phainopepla, common in the desert and indeed a black bird (see that account). Saxton, Saxton, and Enos (1983:35, 36) report both *kuhigam*, "a species of a black bird (unidentified)," as well as *kuigam*, the Phainopepla, even giving the latter's correct scientific name. Bahr (in Bahr and associates 1997:28–29) repeats the error from his 1994 work. Yet again in a footnote (in Benedict 2001:69 n. 24), he confounds the bird with a plant, although storyteller William Blackwater's Pima translator told Benedict, "A bird about the size of a mockingbird but unidentifiable by the interpreter."

Similarly, Hayden's text says Cadigum is the name of a bird, and Joseph Giff, who lived in the immediate vicinity of Kadgam's village and, incidentally, was one of Bahr's own major consultants, had long ago identified *kadgam* as the Yellow-billed Cuckoo (see Rea 1983:163). Furthermore, Joe Giff, Sylvester Matthias, and George Kyyitan, all elderly Pima singers and storytellers, were of one mind that the *kahagam* and the *kadgam* were two different species of birds. There were once songs about the *kahagam*, but these men did not know what that bird looked like, other than that it was blue and small (see appendix F, unidentified birds). Also, on two occasions, Sylvester Matthias told me that *kadgam*, the cuckoo, occurred in legend and that the story was associated with New York Thicket, as indeed the Hayden-Smith story is.

If there is a plant by this name also, its identification remains obscure (see Rea 1997:387, *katikum* in "Unidentified Gila Pima Folk Plant Taxa"). Cassa suspects that *kadgam* is the name of the bird only and that we have misinterpreted the sentence in Lloyd, which he would render, "the man, Katigam, who had eaten the greens" rather than "the man who had eaten the greens, *katigam*." The word order could have easily been switched during translation and transcribing. (The antecedent would be unambiguous to a native speaker.) This interpretation would reflect the real-life situation where the cuckoos in summer eat so many green caterpillars that the feathers of their faces may become soiled green from larval body fluids, like someone who has eaten too much *iivagĭ*, cooked greens.

Furthermore, in their transcription of Pima words, both Lloyd and Hayden seem to have written the schwa or vowel-like interlude between two strong consonants (as did others, including Russell), resulting here in a three-syllable word. Living speakers pronounced the bird's name as *kadgam*.

Contra Bahr, all Pimas I talked to identified *kadgam* as a bird, not a plant. But as the water table and riparian woodlands declined in the Upper Piman villages, the People became less familiar with the cuckoo. Ornithologist M. French Gilman, who lived in upstream Gila Reservation settlements from 1907 to 1915, noted this cuckoo only as "[s]een occasionally in thickets along the river in June and July" (quoted in Rea 1983:163). But Downstream People, Kuiva O'odham, could still identify the cuckoo even in the mid–twentieth century. (And in mid-June 2004, I watched one passing through residential Sacaton, pausing leisurely at a utility post.)

taḍai, pl. *tataḍai*
taḍa (Pima Bajo)

Cuculidae, Cuculiformes

Geococcyx californianus

Greater Roadrunner; Churea, Correcaminos Norteño

In the popular imagination, the Greater Roadrunner has become a hallmark of the desert and chaparral country, spanning

taḍai, Greater Roadrunner (Mike Touch)

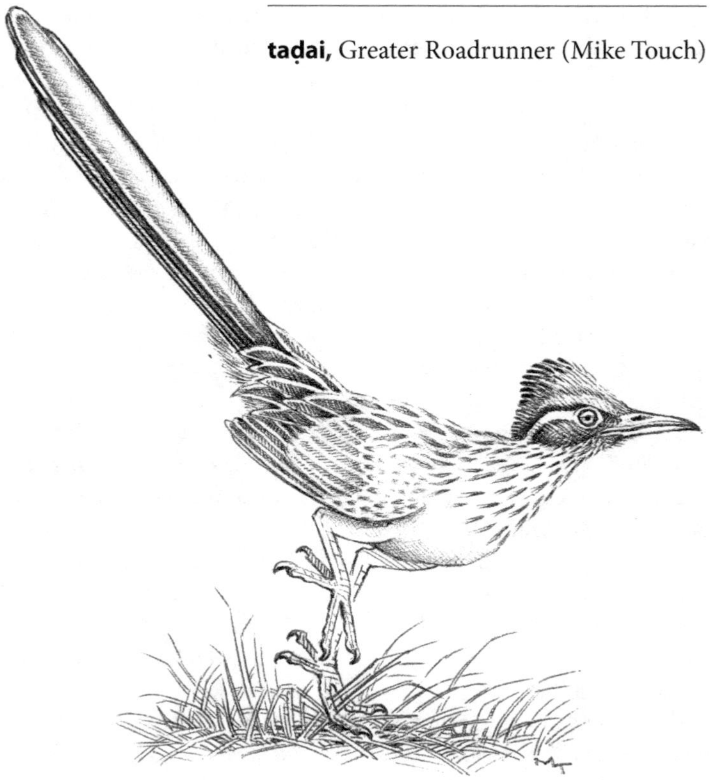

both the Upper and Lower Sonoran Life-Zones. Its sleek form, blackish crest, and very long and mostly black tail can hardly be missed. Throughout Pimería, this terrestrial cuckoo is most often seen as it sprints diagonally across the road, spreads its rounded, blackish, white-flecked wings and tail like a parachute, and sails impudently from the road edge into the desert scrub, where it vanishes. Some of us with paleontological proclivities suspect it is the closest we will get to knowing Archaeopteryx from the Mesozoic, some 150 million years ago.

Roadrunners' plumage is rather coarse. The feathers of the back are black with a metallic purple and greenish sheen, edged with white or buff. The buff-colored underparts are heavily streaked with dark brown and black on the neck, throat, and chest. The brilliant hues of the orbital skin and sides of head are usually concealed with feathers.

Unlike its local relative, the Yellow-billed Cuckoo, the roadrunner is more often seen than heard. In the spring, the male courts with a descending series of dovelike coos. The birds may coo at other times of the year as well. More frequently one hears a series of *hrrrt-hrrrt-hrrrt* sounds made by a rattling of the bill. The bird is often loitering behind some bush, nowhere in sight.

The roadrunner actively hunts on foot for small animals, leaving in the sands of arroyos its distinctive X-shaped footprints (two toes forward, two back). Lizards and other reptiles are important foods. At night, the birds drop their body temperatures, conserving energy. Then in early morning they sit with their backs to the rising sun, absorbing heat energy in the black areas (bare skin and soft black feathers), an economy that helps tide them over when reptiles and other prey may be scarce (Ohmart and Lasiewski 1971).

Roadrunners take up in Piman villages where they run over the tops of houses, perch on cars, check out the pet dish, visit corrals, drink from faucets, and sometimes tease the dog or cat into a playful chase. Their antics are accentuated by the bushy crest, which is expressively erected, and the long tail that may be pumped up and down. When it does catch a lizard, snake, or mouse, it grasps it in its long beak and pounds it to a pulp on something hard before swallowing it whole.

The Old World compiler of the Névome vocabulario didn't quite know what to make of the roadrunners he encountered in Pima Bajo country along the Río Yaqui. Under *faisán*, "pheasant," he said, "Así llaman a unos pajarillos que no son faisanes *tarha*, plural *tatarha*" (So they call some smaller birds that are not pheasants) (Pennington 1979:49). The Jesuit was correct insofar as he went, never guessing that the ground cuckoo was related to the well-known cuckoo of his home country.

Centuries later in this same community, Pedro Estrella and I frequently came upon the Greater Roadrunner as we traveled about the countryside collecting plants. Don Pedro always pointed out the *taḍa* gleefully, but I never learned any lore that may have been attached to this bird. The Lesser Roadrunner, *Geococcyx velox*, from somewhat higher elevations in the Tropical Deciduous Forest, was probably not distinguished. This bird is smaller and lacks the streaking on the neck front and chest. Both species have broad white tips to the tail feathers.

The roadrunner is generally associated with preying on reptiles, smaller mammals, and invertebrates. Sylvester noted that they bring snakes and mammals to their nests. The *taḍai* kills a snake by repeated attacks until the prey is tired out. I have seen roadrunners also taking carrion from road-killed mammals. They may be implicated in more than that. Some O'odham farmers complain that animals make holes in melons and eat the insides out. *Vavuk*, the Raccoon, is probably the main culprit. But when I asked Frank Jim about this problem, he responded, "*Havañ* [raven] and *taḍai* in the daytime. They'll do that—make a hole and just suck all the juice out and just leave it there. One time up here where they were growing watermelons, my grandfather didn't know what it was. I didn't either. I was young then. One time we went over there, we was doing something, kind of rest a bit under a big tree there, sitting there. And we saw that roadrunner went by and watermelons out that way. He was sitting there, chipping [pecking] those things. He got a hole in there, stuck his bill in there, and I guess he suck all the juice out, and took off. No wonder we couldn't figure [it] out."

According to Juan Dolores (1936), *taḍai* is the Papago nickname for a person who carries the news back and forth. A Pima man was called Taḍaikaj Shalivim, "Roadrunner Pants." In earlier days, when they depended on gathering, hunting, and subsistence agriculture, both Akimel and Tohono O'odham, experienced a low ebb in the food supply each spring. The roadrunner once served as a well-known marker for this season. Joseph Giff related: "In those days the Pimas have no calendars. They have nothing. And they go by nature. Like, for instance, take the sun.

Where it rises and where it goes down, this and this. And about the *taḍai*, *taḍai kuk* [calls]. In the spring, spring around April, they start running around. When it gets out, when the weather gets warm, they get up and *kuk*. They [the Pima] store things away for that season, because at that time everything was gone. When the People were hungry, they say *taḍai kuk* [meaning the lean season]. They go by the birds and by anything like that."

Joe said that when the roadrunner started rattling its bill in the spring, it was the time to bring out and cook the least desirable foods that remained, such as dried lungs. Likewise, Tohono O'odham Frank Jim recalled, "When they put food away [in the fall], they say, *'To'akch ko huu wenog kuhu g taḍai,'* 'till the roadrunner starts calling' in spring. Then they get [out] the dried *haal, bavĭ, ga'ivsa* [squash, teparies, ground roasted corn], and dried meats—the lungs and hooves."

Ruth Giff one day was discussing this spring stress period when the Pima made *s-cheedagĭ hidoḍ* and *sitḍoi* from the almost ripe but not yet dry wheat, roasted in a *bid ha'a*, clay vessel (see Rea 1997:330–341). "Doesn't even taste good, but they [we] were so starving at that time of year. [The Pima] would be starving, but would never eat snake, *ñui, taḍai, hikvig, ban*, or *jevho* [vulture, roadrunner, woodpecker, Coyote, or pocket gopher]. Don't see why not *jevho*—they were everywhere and clean animals." All the animals on her list, however, are animals associated with staying sickness that must never be offended by O'odham, not even by handling.

For the most part, Pimans seem to have been on good terms with the familiar roadrunner. Francis Vavages said, "Never chase a *taḍai*. You'll never catch it, and you're going to get disease: gonorrhea or claps—*chiomĭ*."

When we were talking about animals that cause staying sickness, I asked Sylvester Matthias, "How about *taḍai?* Is it special?" He responded, "In a way it's a sacred bird. They say don't eat them, but other People do say they can be eaten. [Gives] VD. That's what they say. It's the feathers that has the curse [*taach*] on it."

Leonard Pancott, a younger Pima (born 1919), said, "Kill a roadrunner, get more meat, white meat. Taste like chicken."

I said, "I thought you weren't supposed to chase a roadrunner or you would get some kind of sickness?"

He responded, "Yeah, they tell us don't chase the roadrunner, but they don't tell us we're not supposed to eat them!" He added they were taken with slingshot.

During a discussion with Leonard Soke and his wife of Santa Cruz village, I made this ambiguous note, "Two Akimel O'odham ate *taḍai*." I didn't learn what the consequences were, if any, but this probably happened at the time that converted Gileños were repudiating traditional beliefs.

Frank Jim from the western part of Tohono O'odham country said that they would run the young roadrunners down to catch them. "*Don't* eat adults or eggs." On another occasion, when we were discussing specifically birds with the power of causing staying sickness, Frank explained, "Once in a while. When they lay their eggs, if you go over there and try to bother them, that's when you get sick with it. There's only one-two-three persons used to know how to sing that [curing] song, but they're old people now. Nobody knows how. It makes you dizzy. Kind of walk around like you're drunk or something like that, but you don't feel good. If you happen to bother it. They say, 'Don't bother their eggs. Leave them alone. Or anything. What they're going to do, you're going to get sick, you're going to get the claps.' [laughter] One time down there we were bothering it. You get sick if you bother the eggs or the little bird[s]. And they tell [us], 'Don't bother it.' We used to chase them, catch them. 'Make you dizzy or give you the claps.' That's what they say. I don't know. I never did get it, because we don't bother it."

At Ónavas, Pima Bajo María Córdova said the *taḍa* was not eaten. Among Southern Pimans, Pennington (1969:114) reported that the Northern Tepehuán ate the fast-running bird that local mestizos called *faisán*, for its long tail. "After being plucked and cleaned, the *tarágai* are boiled in an olla or roasted on a spit."

Among a set of diagnosing songs used by the shaman on the Gila, Russell (1908:305) recorded one that deals with *taḍai*:

Alwese kokomangi taḍai wuḍe makai
gamhe mamadaj biuhugime chum shuakimunha
hia todaiwuñk jejewena chechemoka memehelima
al vopoyuli memuahak kevashaj yuhi yan motoi chunga

Gray Roadrunner is a shaman
Over there her children are hungry and crying.
Here she'll rush out and cover the country running
Little fuzzy caterpillars she kills and yonder across carries them (on her head).

Russell's translation has the bird capturing millipedes. Although roadrunners are notoriously catholic in their tastes, the creature they eat here isn't millipede, which is *kommo'ol*, nor centipede, which is *maihogĭ*. In song language, it is *al vopoyulĭ*, small something in the plural. Sylvester Matthias identified *vopoliu* as "woollybears," the larva of tiger moths, Arctiinae. The verb here, *mooto'o*, is interesting also, meaning "to carry something on one's head," there being no simple English equivalent. Culver Cassa sees this word as comparing the roadrunner's erectile bushy crest and this caterpillar.

Underhill (1946:294) noted the symptoms of roadrunner sickness as "sharp pains in the stomach. *Fetish*: a sharp ironwood stick (because the pains are sharp), No song recorded." Although Russell (1908:263, 312–313) overlooked the roadrunner in his list of birds that cause staying sickness among the River Pima, he did record a set of seven roadrunner medicine songs. Songs I–IV describe the behavior of the roadrunner, in particular its feeding habits. Song V is about a *haupal kosh*, Redtailed Hawk nest (see that species account). Song IV describes the roadrunner hunting the *koomagĭ jusukal*, "gray Desert Spiny Lizard," a heavy-bodied lizard with strong phallic connotations among O'odham. It introduces Songs VI and VII, which are overtly sexual. Russell left them untranslated. Culver Cassa has reworked all but the first two here:

Taḍai, Taḍai sivolik mo'oka
Taḍai, Taḍai sivolik mo'oka
kita vengañ sheshele pui, pui, pui.

Roadrunner, Roadrunner, spikeyheaded
Roadrunner, Roadrunner, spikeyheaded
Around the house always [calling]: "Pui, pui, pui."

Tatenḍai, tatenḍai piwupeso
tatenḍai, tatenḍai piwupeso
masikahuhujeḍ echiawa.

Here is the lonely roadrunner
Here is the lonely roadrunner
He eats whiptail lizards in the morning.

Alisi wepengi wupeikam etanḍai
alisi wepengi wupeikam etanḍai
hinange yakawonḍa chenga chuviñ chuviñ
alisi wepengi wupeikam etanḍai.

Here is the red-eyed roadrunner
Here is the red-eyed roadrunner
Who runs about the mistletoe
This is the red-eyed roadrunner.

Kuñi gamu yangitoka memelihime
Kuñi gamu yangitoka memelihime
Komangi jusukali muakai vokaj ki'ima hu'u

Over there I stealthily run
Over there I stealthily run
A gray Desert Spiny Lizard I killed and opening its belly I eat.

Kuñi ngamu shohiga memelihime
ama aichuvasi ñhoigeida ama aichuvasi ñhoigeida
hiyo vañ musi yenḍa siviahawa.

Over there I miserably run and run
Then something pities me, then something pities me
Here in my vagina a penis surely strikes.

Mui yo'otama, mui yo'otamaka
sivengivia, sivengivia
mo'odama ikitañi
hiyami ñemḍa shudagi vichiwuañ.

Many people, many people
A red penis, a red penis
Atop its head, a crack
Here from within me water rushes forward.

As to be expected from a bird with such a strong persona, the roadrunner finds its way into various O'odham oral narratives. In a war oration, a roadrunner's call is the agreed signal to begin the attack on an Apache encampment (Russell 1908:372). There is a reference to Green (or Blue) Roadrunner in I'itoi's resurrection speech, although this section naming the four birds is more characteristic of a war oration (Lloyd 1911:141; see also Benedict 2001:36, 189).

In a late version of the Pima Creation Story (Hayden 1935:8; see also Bahr and associates 1994:78), Jeweḍ Maakai (Earth Doctor), Coyote, and Elder Brother must determine the center of the earth where real people, that is O'odham, are to live after the Flood. They make two quail and one roadrunner. "Then they sent those two quails toward the west, to find out how far the water had gone west, and they sent the roadrunner to the east, to find out how far the water had gone that way, and coming back they would find out where the middle of the earth was. After these quail came back from the ocean, and the roadrunner came back from the ocean, they came and met right in the center of the earth."

In the narrative related by storyteller Thin Leather two decades earlier, Hummingbird and Woodpecker are the ones sent out to the edges of the earth four times (Lloyd 1911:45–46). By 1935, when Hayden took down Juan Smith's telling, the Pima had acquired a more sophisticated concept of North American geography. Now the slow-walking *kakaichu* (quail) are sent to the Pacific while the rapidly running *taḍai* is sent to the Atlantic, and they return to Pimería, the *hik*, navel, of the world. The outcome is rigged.

In another story (Lloyd 1911:231–233; Russell 1908:239–240), twin boys of a Hohokam mat weaver who conceived by Cloud, go to visit their father. On their return, they are told not to speak to anyone along the way. They meet Raven, Roadrunner, Hawk (Falcon?), and Eagle. The boys are firm in holding their tongue while passing the birds. But then Coyote meets them, and they inquire of their mother. At that moment, they are turned into great mescal (agave) plants near Tucson.

I do not know the significance of the four birds encountered en route home. However, those familiar with Arizona plant distributions will recall that the Desert Agave that grows in the vicinity of the Pima villages is rather small, but two larger species, the Golden-flowered Agave, *Agave chrysantha*, and Palmer's Agave, *A. palmeri*, are found in the Santa Catalina Mountains, where the twins were transformed. Ethnohistorical accounts tell of River Pima going this far to gather mescal.

Yet another story (Lloyd 1911:103–104; Shaw 1968:95–96) explains the origin of a number of things. Following the stealing of turquoise from Morning Green/Blue Siivañ by Tash Siivañ (Sun Chief), Morning Green sends three days and nights of rain on Sun Chief's city and puts out all the fires. Tash Siivañ is dying of cold. First Ban, Coyote, is sent to the home of Wihom, Lightning, keeper of fire, to get fire for the people.

When Ban fails, Taḍai is sent. "[S]o the first chance he had, he snatched one of the fire-sticks from the blazing fire. Hastily placing it across his back and curling his tail over it, he scampered away." Roadrunner once had beautiful long plumage. But his head plumes are burned to a tuft in this act, and his "back was singed so that it became a brownish color, and his eyes turned red from the smoke he endured." Taking up his *wihom gaat*, lightning bow, Lightning shoots flaming arrows at Taḍai, grazing his head first on one side, then on the other. To this day the wounds can be seen as the bare spots on the sides of the roadrunner's head where the skin is vividly colored purple, blue, red, and orange.

The wounded "Tottai whirled around then so that the sparks flew every way, and got into all kinds of wood, and that is why there is fire in all kinds of sticks even now, and the Indians can get it out by rubbing them together to this day" (Lloyd 1911:104). This explains why Pimans are able to make fire with their two-piece fire drill, called the *iivdakuḍ*.

On Roadrunner's successful return to the pueblo, wrote Shaw (1968:96), the women say, *"Sho'ig, sho'ig, sho'ig"* (Poor thing!) And Roadrunner calls, *"Poi, poi, poi."* Sometimes he rattles *"Thra, thra, thra!"* These words are just imitative.

In still another version of this tale in the Creation Story (Hayden 1935), one that seems quite Yuman, at the first death, Coyote is sent off for fire so that he won't steal the corpse from the pyre. Roadrunner follows him, ultimately getting the firebrand, but being struck on each side of the head by Lightning's arrows. Because of his absence at the funeral, the tips of his tail feathers are not trimmed square in mourning, as are those of other birds. (The Mourning Dove is also away at the time and likewise sports a pointed tail.) Roadrunner plays the role of fire seeker in various other tales as well (compare Benedict 2001:171–172).

I long puzzled over the association that older Piman men regularly made between *taḍai* and venereal disease. The other connections from stories made sense to me, but not this one. Then one day I was sitting in the yard at Sacaton when the local pair of roadrunners came through. One ran down the walk between the house and the *vatto* (arbor). Surprised at seeing me, it stopped suddenly, cocked its tail and flashed wide the normally concealed red skin of its face. The image of a woman's pudenda was inescapable. Georgia O'Keeffe would have grasped the metaphor instantly.

COMPARATIVE LINGUISTICS. The Piman word *taḍ*, foot, is the root for the name *taḍa, taḍai*. The Névome *tarha* and modern Pima Bajo *taḍa* amount to little more than orthographic difference. Yoeme (Yaqui) *taruk* is likely cognate (Molina, Valenzuela, and Shaul 1999); I recorded *taarup chu-rea*. Among the Mountain Pima (Oob), I was given *puulyi* and *talpuí* at Nabogame (Yepáchic area). This name is likely onomatopoeic (note Shaw's call of the roadrunner given earlier). Shaul (1994) recorded *pu'i, tarpui*, and *u'oki*, also from the Yepáchic area. Only *tarpui* seems cognate with lowland Piman names. The Oob word there for foot is *tar*. Northern Tepehuán call the bird *tarágai* (Pennington 1969:114).

With Rarámuri (Tarahumara), the Correcaminos (faisán) is *pú, púari*, or *ipú*, again most likely based on the roadrunner's cooing call.

ee'eḍ vahuḍam
***ee'ed vahuḍ* (Pima Bajo)**

Tytonidae: Strigiformes
Tyto alba
Barn Owl; Lechuza

Oftentimes we know the Barn Owl only from a quick passing of white wings overhead at night, sometimes accompanied by an explosive, blood-curdling *reeeek* scream, often followed by a long series of rapid cluckings as the bird flies off gull-like in the sea of blackness. Unless disturbed from some dense tamarisk grove or deserted building or cave during the day, the bird is strictly nocturnal. Most people are unaware of these rat catchers until they see one as a roadkill.

When cornered during the day in a roosting or nesting site, the Barn Owl rocks its head and body from side to side eerily like a reptile, while perhaps hissing. It is quite different from other owls. The white face is heart shaped and the small bill appears at the bottom of a medial ruff. The eyes are dark brown, sometimes enhanced by a brown spot on the facial ruff. The bird's upper parts, including the long wings, are honey colored, overlaid in places with varying amounts of gray spotted with white. The black-specked underparts range from silky white in males through rich buff in females.

Long legs with bare toes equip this raptor for taking its major prey, rodents. As with other owls, the soft plumage muffles the slightest sound that might warn mice at the last moment. "For nesting sites [in the desert], it often chooses holes in ledges and stream banks and corners in dark interiors of caves and man-made structures," write Russell and Monson (1998:135). When food is abundant, a pair may raise up to six young.

A bit of mystery surrounds the O'odham name for this unusual owl. Does its name mean "blood *sucker*," from the Piman word *va'u* or *va'o*, or does it mean "blood *sweater*," from the word *vahud*, "to sweat"?

M. French Gilman (1909) studied the owls of the Gila River Reservation and tried to untangle their Piman names. For the Barn Owl, he recorded *er-er-tvá-ho-tum*, translating it "a blood sucker or vampire." Russell (1908:388) gave the name in a song as *e'et vauhutam*, which he glossed "white blood sucker." Sylvester Matthias said *ee'ed vahuḍam* means "it squeezes out the blood." George Kyyitan pronounced the name *ee'eḍ va'odam*, "blood sucker." Fr. Antonine (1935:9) listed *oe-oet vahudam* as "small owl (lives in wells)."

Three hundred miles to the south, in another dialect, Pima Bajo María Córdova called the Lechuza (the specific name for Barn Owl in Spanish) *ee'ed vahuḍ*, which she translated "sangre sudora" or "something that sweats blood." (That a derived and compounded name has such an extensive geographical distribution came as a surprise to me. It is usually the shorter, nonanalyzable folk taxa that have such wide distributions.) In this area, Pennington (1980:242) recorded the name for Lechuza as *u'ubahut* (that is, *e'e vahuḍ*), confirming doña María's name. (I think he missed part of the word.)

Culver Cassa observed that the northern form "*ee'ed vahuḍam* means 'blood drinking,' that is, drinking water or juice that 'bleeds' from a pulverized, squeezed, pounded mass—probably best rendered as 'blood extractor.'"

The bird also has several alternate names, perhaps more in the line of poetic nicknames. Joseph Giff explained that *ee'eḍ vahuḍam* could be referred to as *vainam mumkidam* because of its piercing scream. *Vainam* means "metal" or "knife" (as well as "money") and *mumkidam* means "sharpener." The name describes the sound of a tool being sharpened with a file or a mechanical grindstone. Sylvester said this owl could also be called *kokokokdam chukuḍ*, "clucking owl," based on its other commonly heard chickenlike call as it plies the night sky. Irene Hendricks gave its alternate name as *olas mo'okam*, "round head–it-has." Tohono O'odham seem not to know much about the name *ee'eḍ vahuḍam*. For instance, Delores Lewis of Ge O'idag called the bird *chaango chukuḍ*. *Chaango* is a Tohono O'odham loan word from Spanish *chango*, monkey. (Akimel O'odham, by the way, use *goks o'odham*, "dog person" for monkey, though they understand their neighbors' use of *chaango*.) The base word is *chukuḍ*, Great Horned Owl, the most culturally salient of the owls.

At any rate, the bird was well known among River Pima. Sylvester reported it nesting in holes in the cliffs and banks of

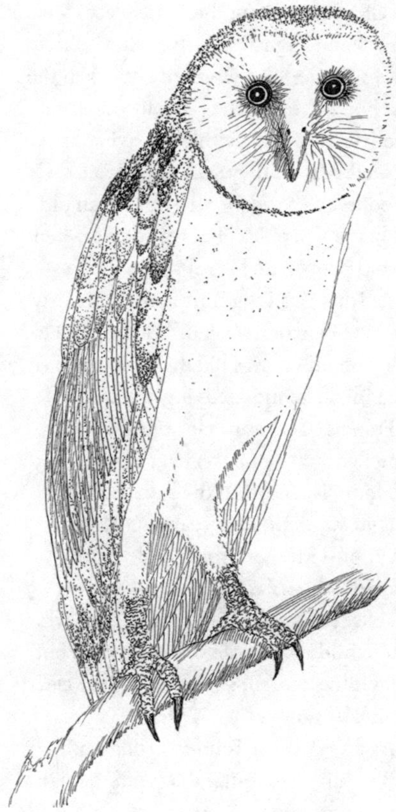

ee'eḍ vahuḍam, Barn Owl (Takashi Ijichi)

the Gila River, about the same places where the Belted Kingfisher nested. He said that as a youngster he would poke sticks in the nest holes to make the birds fly out. Irene Hendricks said, "They always living in the well. Maybe they suck their [victim's] blood, *ee'eḍ vahuḍam*. In Pima it's an old, old name—people call it way down when the old people are [were] around for a while [that is, a long time ago]." Similarly, Leonard Pancott recalled, "There was a lot of it when there was old dug-out wells. They lived in old abandoned dug-out wells."

Francis Vavages made an important comparison when observing that "*ee'eḍ vahuḍam* is almost in the same class [as *chukuḍ ka'amaḍ*, Long-eared Owl], too, but almost kind of brown and white; but *chukuḍ ka'amaḍ* is different, like." This establishes that the "Great Horned Owl's grandchild" is the size of a Barn Owl but differently colored, so the names are clearly not folk synonyms.

As noted, most Tohono O'odham I spoke with were unfamiliar with the name *ee'eḍ vahuḍam* and seemed unaware of the Barn Owl itself. The reason may be biological: the species becomes rare or absent in the open deserts of southwestern Arizona (Monson and Phillips 1981:69; Phillips, Marshall, and Monson 1964:46).

The Barn Owl is not a cause of staying sickness, nor is it in any way associated with souls of the dead, as is the Great Horned Owl. So far as I know, there are no Piman songs or stories about this unusual creature. But it does appear in a warpath speech recorded by Russell (1908:388–389) in both Pima and English:

> I called on [Barn] Owl, above for assistance. He was friendly to me and came, sailing four times on the way. He looked about him and saw my plan. In accordance with this he slowly reached the enemy after sailing four times. He cut the power of the enemy, their springs, their trees, and their dreams. He grasped their bows and arrows and bit them in twain. He bit off their flesh and sinews, and in their bones made holes. From the things collected he made a belt from a bowstring and turned homeward. He came through his gray mist in four flights. Leaving all that he had secured in my behalf, he reached his home in four flights. At early dawn he rejoiced in his success and the darkness rattled with him.

Although Great Horned Owl is usually associated with the east and souls and is the "meeter" who helps the shaman locate the enemy, in this particular passage the helper is specifically named *ee'eḍ vauhuḍam* in O'odham, although the English translation gives just "owl."

Orations are characteristically rich poetry. The original language of one of them, "Tonḍam Ee'eḍ Vahuḍam," contains complex references to the orator, the owl, the enemy, and a war shaman. The antecedents are sometimes confusing, but Culver Cassa points out that this ambiguity is part of the beauty of the piece; the style presupposes some understanding without actually saying the names themselves. Culver has retranscribed the entire native-language version and provided this new translation:

Tonḍam Ee'eḍ Vahuḍam

O va i tdaam ge s-chuhukumum sikolkch kaach.
O vak eḍa s-io'om eulinch ge tonḍam ee'eḍ vahuḍam.
Kuntak hab chui'igam mamchek ab iimk shoshoa.
T ab s-ñhoi'i gei'idam etaatk
Amjeḍ gi'ikho i komalwuak maakai naada cheemo'ok t am naggia.
Am a'ai ñenhogĭ chum ñeik.
O va maakai voogta tonoḍch kaach.
T heg oidk am gi'ikho i komalwuak ab s-baabagĭ cheemo'ok heg ñ-oobda(g).
O vak jeweḍ kaach.
Doa'akaj chum s-naggi'ich keek eelid.
Tak am si sho'oshpol ki'ikumia.
Shoshonkam shushugĭ toto'ach uus keesaj chuuch chum si s-naggi'ich eelid.
Tak am si sho'oshpol ki'ikumia.
Hemajkam heg daam ab oimmeḍdam va chum si s-keeg cheechkadk eelid.
Tak am si sho'oshpol ki'ikumia.
Va chum s-ap'ekam gaat gi'aḍkam shel da'adam vaapk chui'idag shaakchidk eelid.
Tak am si sho'oshpol ki'ikumia.
Am va k heg chuuhugaj hahava ki'ikumia.
Gaataj i si gi'ahuntk oo'oj (kahai)jeejegashk.
Ab huabudk ab giisha'adk.
Ab uupam ge tonḍam e chuhugig wuak.
Heg oidk ab gi'ikho i komalwuak.
S-va'usim sijkĭ.
Maakai naada cheemo'ok hugid am dagĭto.
S-maakaidkam o'odham i si vakusidk t am dashwua.
Vi'ikam s-baabagĭ gawulkadk bek.
Am uupam ge tonḍam chuhugig wuak.
Oidk am gi'ikho i komalwuak.
S-va'usim sijkim.
Evo'ikuḍ cheemo'o.
Hekaj s-hai juukam etaatkch.
Am a'ai dai'iñuñk ova hekaj s-chuhukumum am eḍa si sijkch kaach.

Shining Barn Owl

Above us the darkness swirling exists.
Rightly therein existing is the shining Barn Owl.
I figure his relationship to me and greet him and cry.
He feels mercifully toward me.
Therefore four times he sweeps low and reaches the shaman's fire and hangs.
He looks all about and sees.
It's the shaman path brilliantly showing.
There along he four times sweeps low and cautiously reaches my enemy.
There the country is.
There the mountains are, firmly hanging, themselves thus believe.
He then surely bluntly chews off. [The land seems to be hanging, and the owl bites the suspensions.]
The springs forming ponds here and there, trees being placed firmly hanging, themselves thus believe.
He surely then chews off.
The person in that country believes he'll beautifully be dreaming.
He [the owl] then surely bluntly chews off.
He [the enemy] believes he'll be grasping his perfectly strung bow, his straight-flying reed arrow.
He surely chews off.
He [owl] enters his [the enemy's] body and then chews.
His bowstrings [meaning his tendons] really loosen; his bones perforates.
With that [which he has bitten] he [the owl] makes a headband; with that he makes a stripped skirt.
Homeward he throws the shining darkness.
And through there he four times sweeps low.
It moistly rattles.
He reaches the shaman's fire and leaves it [enemy's power just taken].
The shaman spreads the mat and seats him [the owl].
The remainders he [the shaman] carefully bunches and takes.

Homeward he throws the shining darkness.
And through there he four times sweeps low.
It moistly rattles.
It reaches his bed [this experience of the shaman].
Therefore he believes he is a powerful man.
Now he rushes out and quickly the night rattles in itself.

So much of southwestern ceremonialism centers on the one essential for life in the desert: rain. Culver points out that the rattling in the final line refers to rain: as soon as the Barn Owl has overcome the enemy with his power, the rain wants to become manifest. Even war and the killing of an enemy are associated with rain, as we see in the Scarlet Macaw metaphor (see that account).

chukuḍ, pl. *chuuchkuḍ*
tukuḍ (Pima Bajo)

Strigidae, Strigiformes
Bubo virginianus
Great Horned Owl; Tecolote, Búho Cornudo

The Great Horned Owl is one of our most widespread raptors, and being largely sedentary, has evolved many subspecies. Large pale forms breed up to the Arctic, darker forms across the forested continent, and somewhat smaller and again pale forms occur in the arid and semiarid West. While canoeing on tiny Amazon tributaries, I was surprised to flush again and again a Great Horned Owl of a race so dark that it seemed a gigantic silent moth in the rain forest, a feathered noctivid.

In the desert thornscrub and tropical deciduous forest of Pimería, it is the largest, most powerful nocturnal avian predator. It has become a cliché to refer to it as the "tiger of the night." Perhaps Cougar or Jaguar of the night would be more appropriate because, like these two mammalian predators, the Great Horned Owl is at home on both continents of the New World.

Sheer mass will identify this tomcat-size bird. Its upper parts are finely mottled dark brown and white, often with buffy ochre mixed in. The white underparts are heavily barred, the powerful feet often pure white. Its head is broad, adorned with two widely spaced "ear" tufts that give the bird its English name. Western birds lack the rust in the facial disks usually shown in field guides.

At dusk, Great Horned Owls begin hooting with a deep, hollow, resonating voice. At close range, the timbre has an eerie, nonavian quality. Pairs may duet or call antiphonally, the male usually with four or five notes, the deeper-voiced female with six or seven notes (Peterson 1961:159). When calling, the bird flashes out its white throat patch and may stoop forward, raising its tail. The birds use other conversational notes that we less often hear.

Breeding territories in the desert may be surprisingly dense (Dawson 1998a). As with other owls, Great Horned Owls build no nest of their own. They preempt those of other birds, perhaps ravens or eagles, most often Red-tailed Hawks. Where no ready-made stick nest is to be appropriated, a hollow tree, pothole, or cliff ledge will suffice for the several pure white eggs. Two or three young are generally reared.

In common with other nocturnal and diurnal raptors, the females are conspicuously larger than their mates. She spends more time incubating, being fed by the male. Although capable of taking cottontails and the largest rodents, Great Horned Owls are flexible in their diet. Stomachs I have examined from desert birds often contained large scorpions, beetles, and centipedes. One was packed with caterpillars, which may have been crossing the road, where the bird was feasting when hit by a vehicle.

The folk taxonomist's first job is to determine the biological domain of any taxon or name for an organism or group of organisms. Throughout Northern Piman country, a nocturnal raptor is called *tukuḍ* or *chukuḍ*. Either word has often been glossed simply as "owl," which isn't really correct. This is what I call the fallacy of the faulty domain (see chapter 4). English *owl* encompasses two bird families in an order, the first subdivision below the class Aves. The Piman referent does not. For older O'odham, the word *chukuḍ* always means Great Horned Owl. No other owl is included except the folk species *chukuḍ ka'amaḍ*, the "Great Horned Owl's grandchild," apparently the Long-eared Owl—a

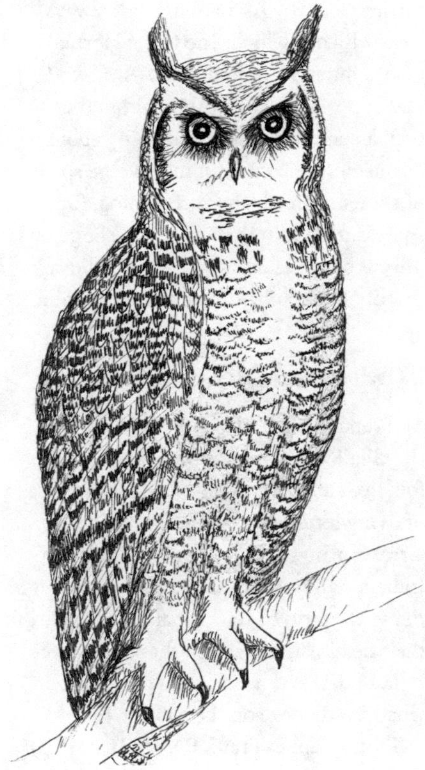

chukuḍ, Great Horned Owl (Takashi Ijichi)

smaller, similarly patterned version of the regal predator.

There are at least three other named Piman folk taxa encompassed in the English word *owl*. I mention this because younger Piman speakers, who now know fewer and fewer species in the wild and know less of their own names to label them, tend to collapse their taxonomy. Sometimes the lesser-known organisms become telescoped into some better-known, more conspicuous organism. But the danger of this linguistic erosion is that certain folk taxa carry considerable cultural baggage, as we will see in the case of *chukuḍ*, whereas others, even though biologically related, are endowed with little or none. English writers recording Piman oral literature who translate any one of these five folk names merely as "owl" leave us in the dark as to what species the native speaker intended.

Having cleared up this bit of folk taxonomy (which seems to be a pan-Tepiman phenomenon), we are left with another, greater difficulty. Although *chukuḍ* means a biological species, the word can also mean "ghost." And to complicate the matter further, for the O'odham it is not always clear which is which. The Great Horned Owl is inherently ambiguous. A *chukuḍ* that appears may be nothing more than that: a common desert bird going about its business. However, it may be the spirit of someone dead who is returning, for some purpose, in the form of the large owl. Throughout the discussions that follow, this dual possibility must be kept in mind.

I. Owls, Death, Reincarnation

M. French Gilman (1909:145), teaching on the Gila River Reservation, attempted a folk taxonomy of owls based on examples of living birds and specimens. For the Great Horned Owl, he wrote, "The Pima Indians call this bird Chú koot, and say it is the soul, spirit, or reincarnation of some of their dead. Their ideas on the subject seem rather hazy, and it is hard to get at just what they believe on the subject."

Ethnographer Frank Russell (1908:252) fleshed out these Piman thoughts in more detail:

> Some [Pimas] declared that at death the soul passed into the body of an owl. Should an owl happen to be hooting at the time of a death, it was believed that it was waiting for the soul. Referring to the diet of the owl, dying persons sometimes said, "I am going to eat rats." Owl feathers were always given to a dying person. They were kept in a long rectangular box or basket of maguey [*sic*, Sotol or yucca] leaf. If the family had no owl feathers at hand, they sent to the medicine-man, who always kept them. If possible, the feathers were taken from a living bird when collected; the owl might then be set free or killed. If the short downy feathers of the owl fell upon a person, he would go blind.

Russell added, "Even to-day [1903] the educated young people are very chary about entering an abandoned building tenanted by an owl." This comment is curious because the owl most likely to be in an old building is a Barn Owl, and I have found no negative associations among Pima with this or other owl species; the Piman apprehension is exclusively reserved for the Great Horned Owl.

One day, discussing birds with Leonard Pancott at Haashañ Keek (Sacaton Flats), I said, "I know the *chukuḍ* can be the bird or the person who has died."

"The dead person's spirit," Leonard clarified.

"That doesn't apply to any other owl, like *e'eḍ vahuḍam?*"

"No, just . . . all I heard is the regular *chukuḍ*. They say that *chukuḍ* could represent a dead person's soul. That's the best I could try to explain it."

When I asked Joseph Giff, "Do the Pima have the idea that the owl could be *somebody*, like the soul of somebody who died?" he responded, "That's the reason why they say, *'Nt o ep uupam kuud'* [I'll return glowing]. If you ask a man for something, I'll tell him, *'Nt o meek dagito g kuud(j)'* [I'll far away leave the glowing]." The second phrase is a way of saying thank you, notes Culver Cassa, "meaning the benefactor won't hang around as a reminder of unpaid obligations. He'll go and be an owl elsewhere."

Similarly, Rosita Brown of Santa Cruz village explained to me that someone who is obligated to a deceased person will be bothered by a Great Horned Owl coming to him. Therefore, the O'odham sought never to remain indebted to another for long lest that person die. Then she promised, "When I die, I will go sit on a perch a long way away and not bother anyone." She has kept her word.

Ruth Underhill (1939:191–192) left the most precise statement of O'odham belief in this matter based on her work with Tohono O'odham:

> The souls of the dead went to the east, escorted by their ancestors, who came for the dying man in the shape of owls. This was true whether they had died of illness or of enemy wounds. The soul on its journey must pass through a gap, which was very dangerous. His owl companions went through it easily but the newly dead had trouble. Some turned back at this point and found themselves alive again. Beyond the gap is the town where the dead live, just as people on earth do. If the newly dead mingles with them and speaks to them, he cannot come back. If he does not speak, he may still regain life. In the town he will find his ancestors who "bring him up like a child," teaching him all their ways. The life is like the life on earth, except that there is plenty of rain, for this is in the east, whence the rain comes.
>
> The dead can leave this land at any time and fly back in a moment in the form of owls and they come thus to call the relatives whom they wish to have as companions. Warriors killed in the Apache country meet war parties and guide them through the enemy's land. Shamans come back and instruct living shamans.
>
> If a person is being called by the dead, he may dream of them or he may fall unconscious, his soul gone. He then applies to the shaman, but the shaman does not send his own soul on search. He sharpens four sticks, paints them and puts them outside the grave to "shut the door." Or he smokes a ceremonial cigarette and asks the dead to come out. An owl may appear and then the shaman urges him to go away and leave the living alone. Meanwhile a group of helpers, some distance off, sings the song cycle which removes an evil charm.

The ambiguity I mentioned about whether an owl is a biological creature or an appearance of a dead person is not a new problem. It is embedded in myth. The twin boys of White-clay Eater Woman embark on a journey to see their wives. On the third night out, the boys are visited by a big owl at their campfire, "talking and trying to tell them something. He awakens his elder brother and says, 'You had better wake up and listen. Perhaps we are already dead'" (Densmore 1929:68).

The Piman language has accommodated this phenomenon. The verb *oo'id* means to forewarn of death by the appearance of an animal (Saxton, Saxton, and Enos 1983:47). Anytime the *komalwupdam chukuḍ* is seen, Culver notes, it also is an omen of death.

Arthur Woodward (1933:167) interviewed River Pima Evanico, apparently Thomas Vanyiko, who also helped Ruth Benedict with myths. "[Great Horned] owls were of ill omen; they were of the land of the dead, and it was believed that spirits of the departed returned in the form of owls and bothered people. Some medicine-men, it is said, could converse with owls and interpret their speech." Then the seventy-some-year-old Evanico related this story:

> Near the Pima town of Santan lived a man who thought some people were stealing his melons. Determined to put a stop to the thievery, he lay in wait by his field one night with a gun. He heard soft voices and faint laughter of people in the melon patch; but the sounds came from

> above, not from the ground. He raised his gun and fired. The next morning he searched the field but saw no tracks; but in a dry wash a short distance away he found a dead owl, riddled with shot. Then he was convinced that the people who had been stealing his melons were ghosts, and he was afraid.

One song from the Gila incorporates this spirit/bird duality. It is inherently ambiguous. It is ostensibly a ghost song, but Culver Cassa notes that it may be about two entities: the bird flying toward the sunset and a spirit, the narrator who hoots.

Hunuñim wengiumhime
hunuñim wengiumhime maku
im wui u'uhig chem daiwuñime
ioñi nama yahaichuli kuñ heg
wui chem daiwunime
kama gi'iko kuhunama.

Westly reddening
toward a bird tries to rush.
Here above me there is something;
toward it I try to rush.
There four times hoot.

In the late 1800s, River Pima began to abandon many of their traditional beliefs, as a 1989 conversation with Sylvester Matthias shows. I asked him, "What do people think about the *chukuḍ?*"

"Years ago they used to be scared of them. They said if an owl hoots or cries, it means someone is going to die. We used to hear a lot of owls, when we lived on our farm—lot of willow trees. You could hear the owls. Way down there, where cowboy Felix Enos [lives], right west of there. Now it's all grown into mesquite thickets."

"S-kuik?"

"Umhum. Everything went to brush. The owl comes around our house after the chickens."

"Which owl ate the chickens?"

"The *chukuḍ*, the big owl."

"Would the people ever bother the *chukuḍ?*"

"Yeah."

"They'd shoot it or get it away?"

"Yeah. My father done that. He'd shoot them down."

"But how do you know that isn't somebody? How do you know that *chukuḍ* isn't one of your relatives?"

"The *chukuḍ* usually come to the chicken house, and he [his father] would come up because he's always on the lookout for Bobcats or Coyotes or owls—one of them should come for the chickens. He had his gun ready. Oh, boy, he would really shoot at them."

"But isn't he afraid that *chukuḍ* is a dead relative that's coming back to talk to him?"

"Some people believe that."

"But your father didn't think so?"

"My father didn't believe it. Nor the ghost—that's the ghost. Some people believe the dead people come out from the grave. He said, no, it's not true. A dead person won't come near a living person because we have, the living have, flesh and blood. And they're spirits and they have no blood, no nothing. They're spirits, but they won't come to the living. They're scared of the living, or whether [either] they don't like to be near the living because they're blood and flesh and it smells to them."

"So the living smell to the dead?"

"Umhum."

"So the *chukuḍ* could be a ghost? Only the big one?"

"Yeah, only the big *chukud* is the one that they are afraid of. [They are] sent, carries the message that someone is going to die, and they could come around, around, try to talk to [them] and all that. They say sometimes it's true, somebody is going to die."

"Do you think that hooting, when they hoot, is the message, or do they talk in some other way to the people who are alive? What is the idea?"

"Yeah. Sometimes they sound [like] the owl is crying. And they say a Coyote, too, he'll sound crying just like a human when someone's going to die. The Maricopas have a strong belief in that: the owl and the Coyote. They keep that tradition."

"Is it the same species of owl, the *chukuḍ*, that they're worried about?"

"Yeah, not the *kuukvul* [Western Screech-Owl], but the owl."

The compromised belief system Sylvester indicated here seems to be characteristic of the Akimel O'odham (his father died in the early 1940s), but we find a much more traditional approach among many Tohono O'odham at present. For instance, Vincent Josemaria, discussing Great Horned Owls, said, "When we die—at your first death—become *o'odham chukuḍ* [meaning 'human ghost']. At second death, then you become an *u'uhig chukuḍ* [the owl, the bird]. Never shoot any *chukuḍ*." As I understand it, the Great Horned Owl is always a soul in this second stage.

Traces of this Upper Piman tradition remain among the Pima Bajo. For instance, Pedro Estrella said that the *tukuḍ* announces something bad, not necessarily death. Another time he said that when this owl sings at night, someone is dying. María Córdova called the bird by this same name, adding, with her dry humor, "They say they may announce death, but when you hear them, they want to eat the chickens, I think." Both of them said that *tukuḍ s-kuihim* means "owl calling."

II. Staying Sickness

The Great Horned Owl is one of those special animals that has the strength or capacity to inflict *mumkidag*, a form of illness, on humans, at least on O'odham-speaking ones. There are numerous contemporary occurrences of owl sickness. According to Russell (1908:263), "The owl throws people into trances and fits." Dr. Aleš Hrdlička (1908:244) was a bit more explicit about *chukuḍ mumkidag*: "Owl's feather is used in curing a person who steadily loses flesh and feels ill. The cause of such illness is supposed to be a dream about long-dead relatives. A medicine-man is called, who generally perceives at once that the patient dreamed of dead people." What he described here seems to be wasting such as that preceding death from old age. If the patient is younger, help is sought, but if it seems to be that the proper time for death has come, the sickness is considered just a matter of course and accepted as a sure sign of approaching death.

Underhill (1946:293) gave the symptoms of owl sickness as sleepiness and dizziness or as "heart-shaking," evidently heart palpitations. That sickness occurs when the person has contact with *chukuḍ* as just a biological manifestation. But in a footnote on the following page she added, "One who 'meets' the owl may also commune with the dead who take owl form. This permits of curing sicknesses caused by the dead." Densmore (1929:90, 102) likewise noted sleepiness or association with the dead as symptoms.

After discussing buzzard sickness with Sylvester, I asked, "What about any other birds, like *chukuḍ?* Could any of those give you trouble?" He responded, "Ah, *chukuḍ* can put a curse on you, *taach* [*to mtaach*], like dizziness, again, or bad dreams." Note once more the duality: the owl may cause a sickness either as an animal or as a dead person who comes to the patient in a dream, attempting to call him or her to the next world.

Theodore Peters of the next village upriver explained more about the sense of obligation that Rosita Brown spoke of: "In his, her dream, [*chukuḍ* says], 'I won't come and bother you [in your dreams]. I'll sit far away, *muuk dak chukuḍ*. I'm never going to bother you in your dreams.' My daughters have dreams of their mother. She asks for eggs fried on one side. But me, I haven't seen nothing like that. Like we've been saying, if someone is dreaming about someone all the time—*chukuḍ*—they have to get someone who knows the *chukuḍ* songs, *chukuḍ ñeñe'i*, to cure him."

Songs and owl feathers are used in the *wusota*, curing session, after owl or ghost sickness is diagnosed. Russell (1908:263) said that six owl feathers mounted on a stick were swung over the patient during the singing. Joseph Giff, discussing fetishes, said, "Same way with the owl feathers the medicine man uses for curing. See, that's not to be handled just by anyone. I think they use mostly the wing feathers." Underhill said that "tufts of the down from under the owl's wing" were used. Frank Jim's medicine bundle had a Great Horned Owl's flight feather, a secondary, as I remember. Theodore Peters recalled that when the wand of two joined owl feathers was being swung over a patient during the *wusota*, "sometimes it sparks like a cat." Sylvester reiterated, "The only four birds the *maakai* uses are *ba'ag, ñui, vishag*, and *chukuḍ* [Golden Eagle, Turkey Vulture, Prairie Falcon, and Great Horned Owl]."

Russell (1908:311) recorded four songs used in curing *chukuḍ mumkidag*. Culver Cassa has reworked three of them as follows:

Yahali Komenḍak nonovange
muiva hunḍuninga kungana wepengiuhumi
kuñ henga wewei ñeunendek imekai
mumui ñeñei wewesi namekime.

Little Sierra Estrella mountain
Many evening tips begin reddening
I go toward there singing
Many songs, all of them (I'm) encountering.

Chukuḍ yo'otama meemeka ñehena
kuñ kaichungaji weweiñ haimiehena
mumui yu'uvī kmomotama
chepavīinga'a amivoiveke
kayehinei kainakime'e.

Great Horned Owl man far off is singing
I am beckoned toward there moving to and fro
Many burdened woman; frivolous women arrive running
Giggling sounds.

Vengihumi shunangi chumengosi
mamangia ñ-i'ichuhu'una
piañingi babagi navahamu
kam siyalika wuhuishakal
himemelhime'e.

Reddish water [Saguaro wine] you try to slowly make me drink
Yet I am not slowly drunken
There eastward I stagger.

The fourth song in the set compares the Great Horned Owl with the screech-owl (see that species). Another song, collected by Underhill (1946:294), says:

Just before me,
The evening is growing red.
I fly out, and four times
I hoot.

Densmore (1929:117–134) recorded in English a number of songs used by a Papago called Owl Woman in her healing sessions. Some of these songs capture words provided by the dead, but others are about the Great Horned Owl. Here are several of these owl songs:

Gray owls came here in the blue evening,
They are hooting about,
They are shaking their wings and hooting.

The owl feather is rolling in this direction and beginning to sing.
The people listen and come to hear the owl feather,
Rolling in this direction and beginning to sing.

Early in the evening they come hooting about,
Some have small voices and some have large voices,
Some have voices of medium strength, hooting about.

Ahead of me some owl feathers are lying,
I hear something running toward me,
They pass by me, and farther ahead
I see spirit-tufts of downy white feathers.

The owl feather is likely to find the daylight.
He is looking for it.
He is looking to see the dawn shine red in the east.

At Santa Rosa village, Simon Lopez said that there were many *chukuḍ* healing songs that could be used in a *wusota* for Great Horned Owl sickness. Asked if he would like to contribute any to this work, he said he would give as an example this one, which is the Going in Song:

Huḍuñinge wengiumhime
huḍuñinge wengiumhima kuhu
i añe wuwui chume dahiwuni'ime
io añe namai yahaichulihi
kune nge wuwui chume dahiwuñe
kame gihiko kunaam.

As the evening reddens,
as the evening reddens, hooting;
toward me it tries to rush out;
here above me something goes along,
toward that I try to rush out
and there I hoot four times.

"That's the very start from the *chukuḍ wusosig*," explained Simon, "that's the very start of it. It just tells that the evening are getting done [beginning], like the color red or somethng like that. And it's towards that I'm going and *kuhu*, 'hooting,' you know."

"And so the songs could take clear to midnight or more than midnight?"

"It can go *more*, all night, if they really sing. Like I said, there's a lot of them. It's just like the *kehina ñe'i* and things like that. There's a lot of them. If somebody knows a lot of these, he could sing to the sick person for a long time, over . . . after midnight. Same way with the *ñeñawul* [devil (songs)]—they [are] the same thing [in duration]. Like I said, it can go up to the morning. Yeah."

Apache spirits, associated with the scalp basket, might also cause sickness (Densmore 1929:102–107). Eagle feathers were commonly used as the fetish in the *wusota* for this sickness, but the accompanying songs that were sung might have been taught to the ceremonial healer by "singing owl feathers." One of these songs, reworked by Culver Cassa, says:

Maakai kii wui ñvañim eḍa ñwua
kii ñsikol chukuñi a'an
biñiminak ñeichuna.

Shaman house toward leads me, inside
places me
my younger brother. Great Horned Owl
feathers
circle around [the shaman house] singing.

Owl-ghost may be involved in sorcery as well as the more conventional forms of *kiikam* or *kaachim mumkidag*, staying sickness. This sickness may be caused when a bad shaman *(ko'ok maakai)* supposedly plants a disease-causing agent (such as a human bone or owl feather) around the person's house. The *maakai* consulted in the case discovers this object during a divining session.

Great Horned Owl may figure in the curing songs for other types of sickness as well. For example, a song for healing Black-tailed Jackrabbit sickness mentions this bird (Russell 1908:315):

A Great Horned Owl feather headpiece
he makes for him.
A Great Horned Owl feather headpiece
he makes for him.
He went far away in the land searching
and met him.

Russell (1908:321) collected a set of eight songs for curing *kaav mumkidag*, badger sickness. They are beautiful pieces of poetry that have been reworked by Culver Cassa, rendering their literal meaning more accessible in English (see Rea 1998b:218–220). One of these songs is about Great Horned Owl:

Tashai yunḍuñikuḍa ñeñwenḍa
tashai yunḍuñikuḍa ñeñwenḍa
am miyenḍanga gesekomangi chukunḍ
ñiñivime
hano vañ wengañi ñungusha'ali kuhuta
idala ñimunanga gayuvaiya heihime.

The sunsetting-place land,
The sunsetting-place land,
Over there then a gray Great Horned
Owl arrives and greets me.
There around me he softly hoots.
This my heart he's taking away.

III. Great Horned Owl's Guardian/*Namkam* Role

Someone may have a dream helper, usually an animal, who supplies him with power. The person is a *namkam*, or "meeter." The *chukuḍ namkam*, Great Horned Owl meeter, was a particularly powerful animal meeter, as were eagle and falcon meeters. For example, Underhill (1939:132) explained the *chukuḍ namkam*: "Each [Papago] village if possible brought a shaman [on Apache raids] to 'see' the enemy and to recite spells to disable him. These shamans were of a particular class, the 'owl meeters,' who had owls for their guardian spirits. The dead are supposed to fly back from the east, where they live, in the form of owls. Dead Papago warriors, in owl form, haunted the enemy country where they died and with these the shaman communed." Again she said: "Owls are the bodies taken by dead Papagos when they wish to visit their home country, and a man who can meet the owl and talk with him can get news from almost anywhere, especially if he can call an owl who was once a Papago warrior killed in the Apache country and familiar with all its happenings" (1938:76). A song from Densmore (1929:182) illustrates the cooperation between the owl and the shaman in this endeavor:

Komang chukuna maakai tvevemi'a
himda
gaduli ñomaida shuligake si'om kokosi

Gray Great Horned Owl shaman with us
goes yonder
my enemies upon knock down, deftly
puts to sleep.

Discussing this owl and spirits, Leonard Pancott volunteered, "And the *chukuḍ namkam* is supposed to know what is going to be happening. Or what it's going to be. Like the story I've heard—the last battle with the [allied] Maricopas. There was other tribes come along with the Yuma warriors. And the *chukuḍ* showed up at their camp. And they *koo*. And when they make that different *kui* sound, it's telling you something's bad going to happen. And it happened that night. And the other warriors from other tribes turned around and went home. 'The bird has told us something bad is going to happen.' So the old people believed that, when they hear or see that *chukuḍ* hanging around, they say he's crying, make that funny sound."

"It's a good thing those ones who went back went back," I commented.

"They stayed alive!"

The episode to which Leonard referred was the 1857 attack by Quechan, Colorado River Yuma, on their former neighbors, the Maricopa, then living on the middle Gila. The Pima came to the rescue, and in less than an hour about a hundred husky warriors, nearly the entire contingent of aggressors, lay dead on the battlefield. It was a happy day for scavengers as the victors left the bodies unburied and, worse, uncremated (Kroeber and Fontana 1986). Various Mohaves as well as Yavapai and Tonto Apache allies had already turned back. According to Pima tradition, they did so because they had been warned of the impending disaster by their own *chukuḍ namkam*.

Not all *namkam* were men (males). Densmore (1929:114–134) collected songs at San Xavier from Owl Woman (presumably Chukuḍ Oks), renowned for curing ghost sickness. Over a period of thirty or forty years, she received hundreds of songs from the spirits of fellow villagers who had died. After a diagnosis of *chukuḍ mumkidag* by a shaman, she was frequently summoned. During an all-night *wusota*, Owl Woman would sing, accompanied by perhaps fifteen or twenty others who had learned her owl songs. She would use owl feathers sprinkled with ashes from the patient's fire to stroke his body, brushing away the disease. Underhill (1946:293–294) was told by some that owl songs were the only ones a woman could sing in a *wusota* because they are simple and other songs would make her insane.

Both owl feathers and the spirits of dead people might teach someone curing songs, noted Densmore (1929:101–102, 105–107). She recorded a case in which owl feathers circled around the apprentice, teaching him a song (cited earlier).

Thin Leather gave Lloyd (1911:92–93) a detailed account of the purification ceremony that followed the killing of an enemy. The enemy-killer had to undergo a sixteen-day seclusion. "The slayer finds an owl and without killing him pulls long feathers out of his wings and takes them home." The species of owl is not indicated. Although the Great Horned Owl is most often associated with mystical power, tackling this powerful owl barehanded would not be recommended. More likely the *kuukvul*, Western Screech-Owl, was the source of the feathers. Near the end of the purification, the feathers of the owl and falcon and the *viigĭ* of the eagle were tied to the buckskin bag housing the lock of hair from the killed Apache. A ball of lac from the Creosote Bush was stuck on one end of the lock.

If the scalp bundle is not properly cared for, the effigy it contains may cause its owner to have Oob *mumkidag*, Enemy sickness. There are various manifestations of this disease, the worst leading to insanity. For the River Pima, Hrdlička (1908:243) recorded that a bit of Apache hair tied with Prairie Falcon and Great Horned Owl feathers was burned with Creosote Bush. (This sounds much like the makings of the scalp fetish, except for the burning with Creosote Bush; perhaps the cure paralleled this action from the earlier lustration.)

IV. Mythology

There seem to be no stories specifically with the Great Horned Owl as a major character. This is because it's his own thing, notes Culver Cassa: *chukuḍ* either as a bird or as a manifestation of the dead was so pervasive in Piman culture that there was no further need to tell stories about it. But this large bird makes a number of lesser appearances in Piman mythology.

Various birds were used by the Vupshkam, or Emergenti, to help them overcome the Puebloan dwellers who occupied the Salt River valley where the River Pima now live. In one case, a gray owl is sent by the conquerors into a "Hohokam" settlement to frighten the inhabitants by its calls at night, "a sign that these people were all going to be destroyed pretty soon" (Hayden 1935:50, also in Bahr and associates 1994:240). The O'odham bird name is not given at this point, but it is *chukuḍ* by context. The symbolism here hinges on the owl/ghost duality in Piman thinking. The original inhabitants were demoralized by the appearance of these manifestations of the dead among them, thinking they themselves were being summoned to the next world. In the Flute Story, a Yuman story taken up by a Piman storyteller, White-clay Eater Woman (Nasia), conceives and bears twins. Various suitors try to claim them, capturing game to give to her. Chukuḍ brings her *nanhagio* (mice), Ba'ag (Eagle) brings her *chuuvĭ* (jackrabbits), and Vakav (Northern Harrier) and Visag (Prairie Falcon) bring her *totobĭ*, cottontails. Ban (Coyote) tries to get into the act as well, dragging in *vahamaḍ* (nonpoisonous snakes), *vapadjilo* (whiptail lizards), and *chuchkuko* (black stink beetles). In each case, she scolds the suitor, flinging out his offering (Herzog [1929]:42–43).

This owl appears again later in this story. One morning while bathing, the two daughters of Oam Ñui, Yellow Buzzard, the impersonation of melancholy, hear some musical notes, the love charm of White-clay Eater Woman's twin sons. The girls go off to find where this enticing music comes from, encountering along the way various individuals who claim to be the source. One of them is Chukuḍ, the Great Horned Owl. He offers them mice to cook, but the girls want only to hear what music he has for them. He flies high into a tree and begins hooting. "They said, 'And is that it?' 'Yes. This was indeed it! It is very hot just now. That's why it doesn't sound right. At dawn I sound it like this and all these trees *(uus keekam)* here would echo, so it sounds nice'" (Herzog [1929]:75; see also Benedict 2001:109–110).

From the southern Papago, Wright (1929:265–273) recorded a story that accounts for the origin of the *chuchukuḍ* (Wright's spelling) from a great grandmother, two grandmothers, and six aunts. As they became old and white headed, they turned into owls. (These stories don't seem to me to be authentically Piman, having the odor of Mexican tales, as might be expected from their provenience. Wright's plural is not conventional.)

V. Folk Science

Many aspects of the biology of the Great Horned Owl are incorporated into Piman folk science. In addition to the bird's strength or way to cause staying sickness, its physical power was commented on by Sylvester: "My dad said a *chukuḍ* can really hurt you, too, if you come close. It will poke [puncture you]—just anywhere—with the claws. They're not fast, but they really get you if you bother them, make the *chukuḍ* mad. And they scratch you anywhere. And that sore will last."

Francis Vavages of Sacaton said, "*Chukuḍ* always hunts at night with its mate; they go out and hunt. They don't really have a nest. They just have a place to stay up in a tree, and they raise their young like that. They stay in touch with each other [by hooting]." Francis then imitated the birds on his flute, correctly differentiating the male from the female calls. Theodore Peters likewise commented, "You know, one has a man's voice, the other a woman's voice, going around [in a pair]."

Sylvester said the *chukuḍ* nests in the arms of Saguaros. He noted with mirth, "And the *haupal* [Red-tailed Hawk] builds a nest, and the *chukuḍ*, the big owl, takes it." This bit of folk science was so well known it was embedded in at least one version of the Creation Story. The Hayden (1935:60) manuscript, with the Creation Story as told by Juan Smith, says:

> Then they [the Invaders or Vupshkam] said that they are going to do something so that it will show that the owl, "chukut," does not belong here [on the Gila]. So they made a rule, when a big bird builds a nest someplace, and this owl was to come and fight this bird and chase him away and lay its [own] eggs in that nest. And so it was true. When the owl found the nest of another bird, he fought that bird and took the nest away from it and laid its eggs in that nest [aside]. So it is true today that the owls still do that. There is nobody who know[s] or saw an owl building its own nest anywhere.

George Kyyitan of Bapchule was talking about how an old man who lived at Snaketown used to call in jackrabbits so he could kill them. "Like owl, *chukuḍ*; if you hear an owl hooting somewhere, *way* out there and you hoot right there, and he'll come. That's what I done one time, right there where we used to live."

"Weren't you afraid that it might be *somebody?*" I asked.

"Naw. Because I saw that one, got there. And the man that live there said, 'There's a *chukuḍ* out there,' he said. 'I hear it there.' He came out, and he flew away. Owl can say anything, too. He can sound like a cat, when they were [are] fighting or something like that."

There remains the mystery of the *tondam chukuḍ*, "shining owl," also known by several additional terms. This phenomenon is described in Pima folk knowledge as an owl that makes sparks or a small light in the night, much like the glow of a firefly. It was well known to older people, in particular cattlemen who spent a great deal of time outside at night. In each case, the reference seems to apply specifically to the Great Horned Owl.

Sylvester Matthias called it *komalwupdam chukuḍ*. He said, "I never get close to them, but I've seen them. But it's hard to see. It must be gray. It will make light; when it flow [flies] over you, you can see like lightning, that quick. Then it drops right in front of you, then start running. It won't fly." He referred to this owl also as *jumal da'a*, "lowly fly," as well as *memhedam chukuḍ*.

"Where would you find it?" I asked him.

"Anywhere."

"How big is it?"

"I don't know, because I never get close to it. I chased one, but I couldn't catch it. *Komalwup* means 'slides' [glides?]. And they say it's more like a ghost—it'll just disappear. Any kind of owl [!] makes light."

Another time he said, "Any kind of owl. [When we are] in the field irrigating, you can see them making light. In the cottonwood, you can see them making light. The regular *chukuḍ*." His implication seems to be that any Great Horned Owl is capable of this behavior, not just some special category of them. Still another time Sylvester said, "Little owl, runs; not too little. *Komalwapudam chukuḍ*. Not really a bird; halfway a ghost."

Joseph Giff (born 1907) was familiar with the *tondam chukuḍ* or *memhedam chukuḍ*, "shining owl" or "glowing owl." He said, "A reddish glow, *memhedam chukuḍ*, 'burning owl.' *Kuuj oimmeḍdam* means 'going with a torch,' torch bearer, such as a person going to a neighbor to 'get fire.' *Kuuchkĭ* is the torch used to keep fire going at night. Sometimes the light gets very bright, illuminates the bushes around. Then the light gets very small, about the size of a cigarette." He suspected that the birds must ruffle their feathers to make static electricity so they can see their prey.

Culver Cassa writes that the word *komalwua* means to glide closely against something with a flat surface and *komalwup* means to sustain a series of gliding flights. Joe Giff used this term also to describe the nuptial flight of dove: *komalwupdam hoohi*, "gliding Mourning Dove."

Leonard Pancott is somewhat younger (born 1919) than Sylvester and Joe, but an experienced cattleman who has spent much of his life outdoors. One day he told me that Great Horned Owl feathers are never used for arrow fletching because they have "ashes" that might get into the shooter's eyes. He continued, "Whatever that is, them owls could spark with their feathers."

"What do you call that when the owl sparks with its feathers?" I asked him.

"*Nanda*. Just like building a fire. You know, you'd see them alive, like a little fire, way the hell out in the middle of nowhere, and you know that's what it is. And they say, *nanda chukuḍ*. When you're traveling at night from any distance, like Santa Cruz, someplace at night, way late at night, you'd see that, down the road. And you know that's what it is. You'd think someone's there, had a fire going. Looks like a little fire."

"What color?"

"Same as a little fire."

"Would this be at any time of the year that you'd see those?"

"Yeah, any time of the year, late at night. And it usually happens when you're alone. I don't know what it is—trying to scare you, or what."

"Does it move, or does it stay still?"

"It moves. You'd see it here, then the next time you'd see it some place else. And, you know, that don't happen every night. Off and on you'd see that. You'd see that and then you'd see it another place down the road—somewhere further off down the road. Maybe twice then, then it'd be gone. It disappear[s]. Nobody seems to know what kind of positions he gets into to do that."

When I asked Theodore Peters, who is a bit younger still than Leonard, about *memhedam chukuḍ*, he responded, "'Lightning ghost'—I just heard of it—a light. That's when they're hunting—using that light—like a flashlight. That's what he [his father or grandfather] told me. I don't know if it's true or not."

The idea has persisted. Culver Cassa (born 1968) said, "In my younger days I was told not to wave glowing firewood; it attracts ghosts. I also heard that such wondering lights are ghosts. I suppose like attracts like. I can now see the connection with Great Horned Owl." We thus have two Piman explanations for the glowing owl, one that the owl makes sparks to find its prey, another that the owl doing so is a ghost.

I am not familiar with any technical report of bioluminescence with regard to owls so am unable to explain this phenomenon of "shining owls." The Pima along the river knew of fireflies, calling them *memhedam muuval*. Saxton, Saxton, and Enos (1983:44) give *naaj oimeḍdam*, a noun they define as "a phosphorescent light seen at night over marshy ground, a will-o'-the-wisp." But the accounts of shining owls do not seem to be associated with marshes, so the mystery remains.

VI. Plant Names

A story is told of a literal-minded orphan who is instructed by his grandmother to go out and collect an *iivagĭ*, greens for cooking, that was called "Owl's-feathers," perhaps *chukuḍ a'an*. Only the English is given (Lloyd 1911:62; see also Rea 1997:388), so we have no way of knowing which owl folk generic is associated with the plant. "And he went and found an owl and pulled its feathers out and brought them to the woman, and she said, 'This is not what I want! It is a vegetable that I mean.'" He is also told to get a plant called "blackbird's eyes" and another called "crow's feet." Of the three plants, only the last has been determined. It is *havañ taataḍ*, or wild-heliotrope, still remembered as an *iivagĭ*, though it is no longer eaten.

In Pima Bajo country, Pedro Estrella identified a wild tomatillo, or ground-cherry, as *tukuḍ wupui*, "Great Horned Owl's eyes." There was no doubt about species as he added, "Muy grande Tecolote, muy bravo." The "owl's eyes" of Pima myth possibly refer to the local species of ground-cherry, now called *kekel viipiḍ*.

The date palm is known to Upper Pimans as *chukuḍ shosha* because the dates bear a resemblance to the preying mantis's egg case, also called *chukuḍ shosha*, "owl's mucus." In addition, there was formerly among the Akimel O'odham a small musk melon called *chuuchkuḍ ha-miilinga* (plural) and described as a little larger than a big grapefruit. People said they were unsure if this name means the "Great Horned Owls' melon" or "ghosts' melon." On this point, Culver Cassa observes, "This means that it smells 'heady,' a smell associated with ghosts."

When the pigmentally challenged brought doughnuts to the region, Pimans devised a number of names for them. The one I've heard most commonly among River Pima is *siipuḍ paan*, "anus bread." But once at Sacaton, June 1995, someone told me it is also called *chukuḍ daikuḍ*, "Great Horned Owl's chair."

VII. Place-Names

A small village in far southern Papago country has the name Chukuḍ Kuk, "Owl Hooted," popularly known to outsiders as Tecolote, the name for "large owl" in Mexican Spanish. Chukuḍ Kuk gives its name to the entire district, which runs along the southern border of Arizona. This village is apparently where the tale of the nine elderly women, keening at the death of a young relative, turned into Great Horned Owls. Jose Enriques said the formation east of Redrock is called Chuuchkuḍ Mo'o (Pima, Moomĭ), "Owls' Heads." It might be either the Black or the Tortolita Mountains.

There are also two place-names in Sonora based on the O'odham name for this owl. One is Tucuru, El Tecolote; the other Tucurubabi, Agua del Tecolote (Molina 1972:74). Piman *vavia*, "well," often became *babi* when taken into Spanish place-names. And the retroflex *d* was often transliterated in colonial writing as *r* or *rh*.

I would be less than honest if I didn't relate one particular personal incident related to this species. I had been working with George Kyyitan quite regularly for a number of years, learning about ethnobiology and sung Pima poetry and many other things. Then for two years we worked together on a special project of inventorying the tribe's traditional cultural properties. George knew the sites of Pima country the way some of us know the warts on our face. In his eighties, he was indefatigable. It was a delight to work with him.

Shortly after the project finished, George was diagnosed with cancer. Later, in October, a few minutes after he died, his family called me in San Diego to tell me. They asked if I would come over to give the eulogy at his funeral.

Three nights later and back at home in San Diego, I had been asleep for some time when some sounds outside the studio where I slept awakened me. They seemed to be some of those low conversational notes that Great Horned Owls sometimes make. It was 11:30 P.M. I got up and opened the door. Right overhead came the three fast and two slow hoots of a Great Horned Owl. I waved a greeting to the cold night air and heard nothing more.

I mention this event with proper scientific skepticism, but at no time, before or since, in the twenty-five years I have lived in this house in the middle of metropolitan San Diego, has a Great Horned Owl visited me. With *chukuḍ*, it's a question of polysemy, level of meaning. You can decide what level.

COMPARATIVE LINGUISTICS. The name *tukuḍ*, later *chukuḍ*, seems to be related to the root word *tuk (chuk)*, signifying "black," "darkness," or "nighttime." We can trace the Piman name for this owl back to the colonial period. The Névome vocabulario gave *tukurhu* (= *tukuḍu?*), and the Attí manuscript *tucurch* (pl. *tuturch*) (where *rch* is or approximates modern *ḍ*).

The Great Horned Owl name is probably pan-Piman. Recorded names for "large owl" include: *túkuur* (Tepecano); *tukur* (Southern Tepehuán); *tukurai* (Northern Tepehuán); *tukur, tuukura* (Mountain Pima). Pennington (1980:198) wrote the modern Pima Bajo name as *tukutch*, and Ken Hale (n.d.a) wrote *tukur*. I recorded it from Pedro Estrella as *tukut* or more likely *tukuḍ*. (Mason [1917:328] listed two Tepecano owl names, *tukuur* and *tekkur*, without specifying the biological referents; I'm assuming the first is the Great Horned Owl and the other a screech-owl.)

The Proto-Uto-Aztecan reconstructed form is **tukurĭ*. However, apparent cognates occur well outside this language family, as discussed by Gursky (1967) and Hunn (1975b). For instance, Proto-California-Penutian reconstructs the term as **tukun* ~ **tukul*. This might represent cognates, ancient borrowings, or independent onomatopoeic coinages.

An aside for word lovers. There is a dialect of Tohono O'odham called the Kokololoti. The principal village where it was spoken, Tecolote, sat on the international border, with fields on both sides. Underhill (1939:67) explained how this name came about. "Its old Papago name tchukuht kuk, (owl cry) became translated into Mexican [Spanish] Tecalote (Nahua *tecalotl* [*tecolotl*], owl) and its people, by a doubly reduplicated plural[,] are called kokololoti. Papago words are all accented on the first syllable and since the language could not accept a word accented in the middle, it split it into two, *teca loti* and reduplicated both. The *te* got lost in the process."

Another matter concerns the Piman names for the different owls. Ornithologist and Indian Service schoolteacher M. French Gilman resided on the eastern section of the Gila River Reservation from 1907 to 1915, actively collecting and publishing on the local avifauna. Gilman (1909) found six species of owls along the middle Gila. He showed skins and live specimens to the Pima, asking for their indigenous names. He was given *chú koot* for Great Horned Owl, *er-er-tvá-ho-tum* for Barn Owl, *kau'kau-hä'* for Burrowing Owl, and *koó-ah-kohld* for both Elf and Ferruginous Owls, but no name for the screech-owl. I've never been able to confirm the latter name (it sounds very much like the Piman word for chile, *ko'okol*), nor has anyone ever volunteered to me a name for these two smaller species of owl.

The second puzzle is the screech-owl. "The Indians would give me no name for this owl," wrote Gilman (1909:50); "one man said it had a name but he had forgotten it; another lookt puzzled and said he thought it had a name but he had never heard it. They all knew the bird however." In my experience, this species and the Great Horned are the best-known owls among O'odham, and just imitating their call is sufficient for older people to identify *kuukvul* and *chukuḍ* at once.

One of Gilman's perplexities was the apparent failure of the River Pima to distinguish between the Ferruginous Pygmy-Owl, *Glaucidium brasilianum*, and the Elf Owl, *Micrathene whitneyi*, seemingly calling them both *koó-ah-kohld*. "I showed both species to them, and pointed out the difference, but it was all one to them." The modern ethnobiologist, however, would be concerned with working out the domain of a folk taxon rather than seeking one-to-one correspondence between Western and folk systems (see chapter 4).

At Quitovac oasis, Luciano Noriega reportedly called the Elf Owl *chem vahum*, "small owl" (Nabhan and associates 1982). I hesitate to include this folk taxon without further work verifying it with at least one other speaker. The first word in the compound means "small"; the other may be related to the verb that means "to sweat," in reference to the Barn Owl—hence, "little Barn Owl." But this is conjecture. That such a locally common, confiding, and highly vocal species would go without a name among O'odham would surprise me.

chukuḍ ka'amaḍ

Strigidae, Strigiformes

?*Asio otus*

?Long-eared Owl; Tecolote Orejón, Búho Cara Café

An owl called *chukuḍ ka'amaḍ* was known among the oldest Gileño Pima. Francis Vavages said, "*chukuḍ ka'amaḍ* is in the class with the *chukuḍ* [Great Horned Owl], but doesn't have the horns. Head round. Stays in the trees. Round like. It doesn't have that. It kind of just whistles—whistle sound. *Chukuḍ ka'amaḍ* always stays in the trees. *Ee'ed vahudam* is brown and white. But *chukuḍ ka'amaḍ* [is colored like regular *chukuḍ*]." Another time he reiterated that this bird was smaller than the Great Horned Owl and whistles rather than hoots. He again implied that it was the size of a Barn Owl.

Sylvester Matthias said, "[Great Horned] Owl's grandson or granddaughter from the *father's* side. Just makes a little sound." The same relationship from the mother's side would be *moos*. He, too, said it whistles rather than hoots and lacked ear tufts.

George Kyyitan knew it also. "*Chukuḍ ka'amaḍ*—I *heard* of it. Don't know what it might be. *Chuuchkuḍ ha ka'akmaḍ*. One old man used to sing with them."

Leslie Pasqual said, "Yeah, I know that *chukuḍ ka'amaḍ*."

"Do you ever see that anymore?

"Not anymore. Since there's nothing here for it to sit [in], nest [in], raise little ones. I don't know where they're at now."

Another evening we sat on the tailgate of Leslie's pickup as the nighthawks sliced the air around us. I asked him again about *chukuḍ ka'amaḍ*. "It likes to hang around in the cottonwood trees."

"Did it nest here?"

"I think they did."

"Is it like the *chukuḍ* or the *go'ogov* [shrike], about news, about death?"

"No. It's probably a good bird. Not warning of death. I'm not sure. But I think so."

I was never able to identify the bird *chukuḍ ka'amaḍ* with certainty, but the best evidence suggests that it is the Long-eared Owl, a species that I encountered a number of times during my fieldwork in Gileño country and that might have been more common there when the riparian habitat, in particular willows, was healthy. It has been known to breed in the Sonoran Desert and conceivably once even bred locally. (I did not have an illustration of this owl in my elicitation notebook, so during most years I worked, I never asked people about it specifically.) Pima folk science suggests that it bred in trees.

The Long-eared Owl is a slimmer edition of the Great Horned Owl but it is smaller (about fifteen inches rather than more than twenty). The two are similar in having finely mottled blackish brown, tan-buff, and gray upper parts, but streaks through the bars on the lower parts (as does the screech-owl) rather than just the bold scales of the large owl. Also, the facial disk of the Long-eared is distinctly rufous, unlike the western and southwestern races of the Great Horned Owl.

The Long-eared Owl is strictly nocturnal, concealing itself during the daytime in leafy trees. It has a considerable repertory of calls, including whistles, and is most vociferous when its nest is disturbed. As with other night birds, vocalizations are the best means of identifying the Long-eared Owl.

Francis contrasted *chukuḍ ka'amaḍ* with the Barn Owl, which is the size of the Long-eared, but quite differently colored. The only character he gave that doesn't fit is the lack of "ear" tufts. This alone might suggest the Short-eared Owl, *Asio flammeus*, a more diurnal species that visits Gila country uncommonly in winter. But it is not patterned like the Great Horned Owl, as its Piman name implies it should be, and it is seldom heard in winter. Also, it is a bird of open grasslands and marshes. The once great mesquite bosques and gallery woodlands of the middle Gila adjacent to Pima fields and open grasslands maintained by frequent fire drives would have provided excellent Long-eared Owl habitat.

The Long-eared Owl is a northern bird that reaches its southern limits in Pimería Alta. It is unknown through the rest of the Tepiman corridor. It would have been a species new to northward-invading Pimans, but whether it was associated with staying sickness or shared any of the other attributes of its larger relative, the Great Horned Owl, I cannot say. Most individuals who knew *chukuḍ ka'amaḍ* were long gone by the time I wrote this account.

With the recognition of the taxon Great Horned Owl's grandson or granddaughter, there become two folk species: unmarked, type-specific *chukuḍ*, which for Pimans means the Great Horned Owl, and *chukuḍ ka'amaḍ*, the smaller species. At another level of meaning, *chukuḍ* can designate "soul" or "ghost," but there are no field marks to distinguish them. Culver Cassa explains the implications of Piman relationship naming in chapter 4.

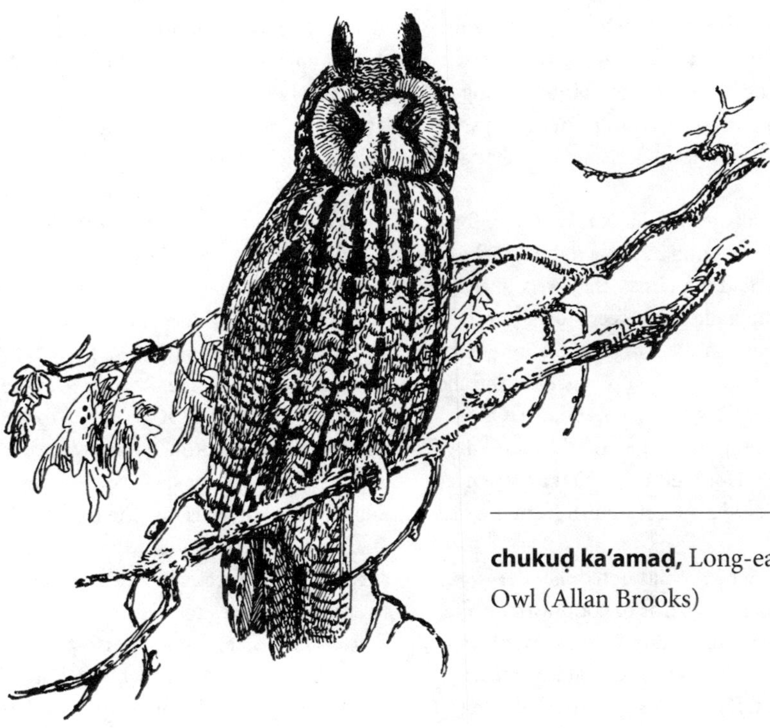

chukuḍ ka'amaḍ, Long-eared Owl (Allan Brooks)

kuukvul, kuukul, kuukwul,* pl. *kukuvul

***tuukul* (Pima Bajo)**

Strigidae: Strigiformes

Otus kennicottii (asio)

Western Screech-Owl; Tecolote Occidental, Tecolotito

The cottonwood-willow forests bordering streams, the mesquite-ironwood fringes of arroyos, and the stands of Saguaros

kuukvul, Western Screech-Owl (Amadeo M. Rea)

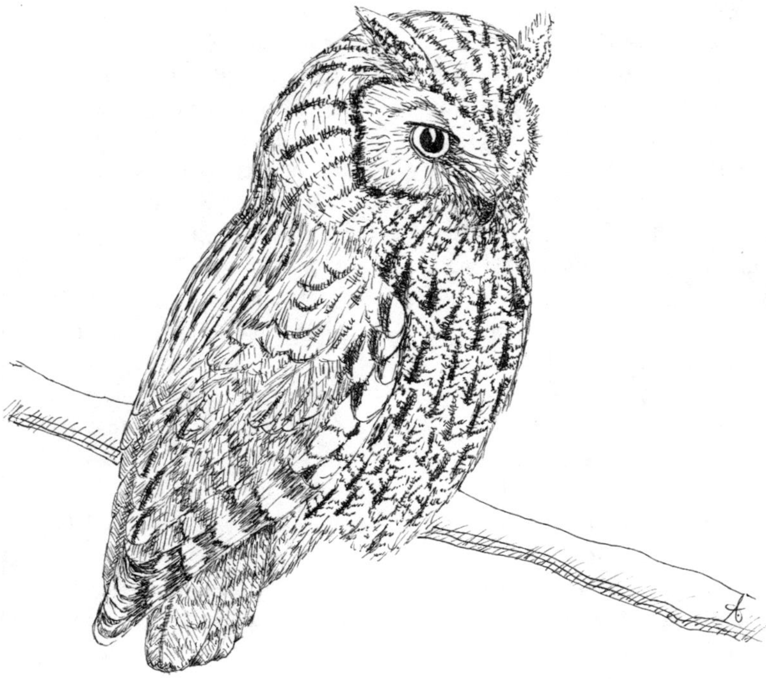

mixed with leguminous trees spreading across desert bajadas are favored home to the Western Screech-Owl. This little owl is quite common in the desert where Gilded Flicker (and Gila Woodpecker?) nest holes are available. The eight- to nine-inch bird is finely marked gray on the back, with bold "black ant" marks on the pale underparts. The facial disk is rimmed with black, and an incomplete V of white spots marks the back.

In the daytime, when perched in the pale gray underbranches of a Desert Ironwood, this little owl may be easily overlooked, even as one passes nearby. It draws its body into a narrow cylinder, erects its ear tufts, and narrows its eyes to mere slits, appearing for all the world like just another dead branch. At other times, it is a feisty ball of feathers harboring piercing yellow eyes. Anonymity ceases at night as pairs duet and territories are reaffirmed. Western Screech-Owls have quite a vocal repertory. The most easily remembered vocalization is the common "bouncing ball" call, a series of easily imitated *puup-puup-puup* notes, beginning slowly and picking up speed, like a tennis ball dropped on a hard surface.

The screech-owl needs open spaces where it can hunt. Its prey is largely scorpions, centipedes, and other terrestrial arthropods, supplemented with small snakes, lizards, mice, and smaller birds, pounced on from above.

Among Northern Pimans, this bird is one of the better-known owls. Its imitated call was readily associated with *kuukvul*, this small gray owl. (None of the other three small owls here is gray.) Many Tohono and Akimel O'odham, at the mention of this owl, related its behavior of turning its head to watch you as you walk around it. Rosita and David Brown raised several nestlings as caged birds, feeding them raw meat. Irene Hendricks said she once had one nesting in a hole at her back door.

Joseph Giff said that *kuukvul* were formerly very abundant. When the Apaches came to raid the Pimas, they would whistle in the night like the screech-owl. But the old Pimas could distinguish the imitation and alert the other people that the Apaches were preparing for attack at dawn. Or sometimes the invaders would sound like the Great Horned Owl, which experienced Pimas could also distinguish from the real bird.

Sylvester Matthias noted that the *kuukvul* nests in Saguaros or cottonwood trees, using the woodpecker holes, sometimes with the *sisikĭ*, the American Kestrel, as neighbor. He said it sleeps much during the day. Youngsters used to pick them up by hand from their roosts in the mesquites, taking hold of their heads. The birds, he said, would not be especially disturbed or alarmed at being handled and when released would fly a short distance and immediately resume "sleep." (I might add here that on moonless nights, with the aid of a flashlight, I have caught adult screech-owls by hand in ironwoods, but never succeeded when there was moonlight, nor did I try by day.)

Delores Lewis said, "*Kuukwul* and *koolo'ogam* [Common Poorwill], they go together." Juanita Norris of Akĭ Chiñ contrasted the *kuukvul* of the night with the *koka* or *kokohol*, the Burrowing Owl, of the day, saying that they were very different birds.

When the twin sons of White-clay Eater Woman, Nasia, play their amorous flutes in the morning, the daughters of Oam Ñui, Yellow or Brown Buzzard, hear the sounds from their bathing place and go in search. They encounter various males along the way who claim to be the source of the enticing music (Desert Spiny Lizard, Great Horned Owl, Antelope Squirrel, Ladder-backed Woodpecker, Northern Harrier), then finally a group of *kukuvul*, Western Screech-Owls. We are not told what food these owls have to offer. The song they sing to the girls is a repetition of the phrase "*Vaashan g i dahiwa vuhio keekam*"—that is, "Move over, cross-cousin" or "Sit over, cousin," in imitation of these birds' call (Benedict 2001:110; Herzog [1929]:75). (This description doesn't seem to fit the "bouncing-ball" call of the Western Screech-Owl, though.)

Another segment of the Flute Story purportedly pertains to this bird. The twin boys embark on a quest for their separated wives (Densmore 1929:68). On the fourth night, a screech-owl visits the campfire of the youths but is silent. Two earlier visitors were a headless Wildcat and the big owl associated with death (that is, the Great Horned Owl). Only the big owl speaks to the boys, as a manifestation of an O'odham ghost. Without the Piman name, we cannot be sure of the smaller species: the irony of English bird names is that the Barn Owl regularly gives a blood-curdling screech, but the several species of North American screech-owls never screech.

Russell (1908:279, 311) recorded two songs about the screech-owl. The first is listed among "archaic songs" that accompany the telling of the Creation Story. It is the last of a set of five songs that Elder Brother and the people sing as they prepare to cremate Ho'ok, the witch.

> Vav wetama vupahimu
> chechenuñi kukuvuli sisvunuka'a
> vav wetama vupahimu.

Russell's free translation of this enigmatic song-poem is:

> Throw a great stone.
> The Blue owl is brightest [flaming].
> Throw a great stone.

The other song occurs in a set of curing songs for Great Horned Owl sickness, where the two species of owls are contrasted:

> Chukuḍ yuvĭnge
> kukuwul(d) yo'otama chuma siñi wepo-
> vingi chuhikame ya'ahana
> kuñ yedañi amevelhongi taime
> hanu kukungane kokomaiwuakimu.

> Screech-owl resembles Great Horned Owl.
> The winds rise from [Great Horned] owl feathers
> With their ashy tips he starts them.
> Screech-owl is like the Great Horned Owl.

Culver Cassa's retranslation of this song is:

> Great Horned Owl bird
> Screech-owl Man perhaps has similar wings
> Just then I cause a gust of wind.
> Along my wing-tips fog appears.

When the River Pima were practicing canal irrigation from weirs on the river, their earth ditches used to get clogged with a water plant, Common Poolmat or Horned Pondweed (Rea 1997:100). Its Pima name is *kuukvul huch*, "screech-owl's nail." The comparison is to the cluster of minute green fruit or seeds, which resemble the plump, fully feathered toes of the Western Screech-Owl, each with a small claw. Not only were the Pimas who named the plant on familiar terms with the owl, but they had very good eyes to see the little *huch* at the seed's apex.

The Gileño Pima distinguished at least six named species of native fish in addition to the life form *vatop*, fish. One of them, a small but edible form, was called *kuukvul*, just like the screech-owl, Sylvester said, but he didn't know why.

COMPARATIVE LINGUISTICS. The screech-owl name appears to be pan-Piman, with *kukvulyi* (Hale 1964: *kuukvily*) among Mountain Pima; *tukul, tuukul* among lowland Pima Bajo; and *tekkur* used by Tepecano. In colonial Northern Tepehuán, Rinaldini wrote *cucubuli*—that is, *kukuvuli* for "lechuza," as the screech-owl sometimes is called in the sierra above Barn Owl range. Culver Cassa considers *kuukvul* the base variant in the north. Note that all these words are contrasted with a *tukuḍ/chukuḍ* word designating the Great Horned Owl.

MISIDENTIFICATION. Both Herzog and Benedict identified the *kukuvul* from the Flute Story as Elf Owls. In an oration with three species of owls, Benedict called *"kukowol"* the Dwarf Owl, which editor Bahr "corrects" to *kokoho*, which is the Piman name for the Burrowing Owl, appearing a few lines later (Benedict 2001:189 n. 18).

TECHNICAL NOTE. Eastern and Western Screech-Owls were long considered conspecific, with the scientific name *Otus asio*. However, eastern and western birds have different calls (critical in species recognition among night birds) as well as different plumage patterns. Hybridization is so restricted that they are now considered separate species (Marshall 1967:3).

Incidentally, the subspecies of screech-owl occurring throughout both Pimerias is *O. k. gilmani*, named in honor of the Gila Reservation ornithologist M. French Gilman (see Marshall 1981:69; Rea 1983:168).

kokoho

Strigidae: Strigiformes
Athene cunicularia

Burrowing Owl; Tecolote Llanero, Lechuza de la Tierra

The Burrowing Owl defies much of what we think we know about owls. It is active primarily during the day. It perches conspicuously in the open and is more often seen than heard. It eats more arthropods than rodents. For its nest, it excavates or enlarges a burrow in the ground if a natural one is unavailable or unsatisfactory, rather than appropriating a tree cavity, abandoned nest, or some other elevated place. It even adds some dried dung to the brood cavity by way of nest lining. In suitable locations, the birds are colonial. And the sexes are readily distinguishable, at least in the breeding season, when males are bleached pale from guarding territories in the bright desert sun while their dark-colored mates are attending household duties underground. (Whether there is also a genetic component in this color difference I have not been able to tell: molted, anatomically sexed, unfaded fall specimens are scarce in collections.)

Furthermore, the Burrowing Owl is easy to identify. Its legs are long and the toes bare, much like the Barn Owl's, but the bird is smaller (about ten inches). The sandy tan colors, white-spotted brown above and brown-barred white below, match the pale colors of desert and prairie soils. When a human approaches their burrows, the birds are likely to fly a way off and chatter in protest. Underground, the young make a very convincing rendition of a rattlesnake.

Before the advent of mechanized farming, Burrowing Owls were widespread and common on the Gila River Reservation, taking advantage of traditional farming and irrigating methods. Kangaroo rat and Round-tailed Ground Squirrel burrows, in particular those enlarged by badgers, are appropriated for nesting. Most older cemeteries harbor a breeding pair. These owls are supposedly year-round residents, but while surveying traditional cultural sites for the tribe, I had occasion to revisit all known cemeteries in midwinter and found not a single Burrowing Owl at home at that time (see also Rea 1983:171). The entrances to the summer-occupied nesting burrows were now screened with cobwebs.

The Piman name for the Burrowing Owl, *kokoho*, is onomatopoeic, mimicking the common cackling call given when the bird is disturbed. Once, as evening fell at Blackwater, I heard the bird out across the Desert Saltbush flats call out its name precisely: *kokoho*. Various of its other common calls are similar.

I have heard variations on the name, including *kokoha, kokova, kokawa*, and *kakuhua* (see also Mason 1950:30). Some of these names may be mistranscriptions or lapses of memory. Juanita Norris of the Akĭ Chiñ community called the bird *kuka*, saying it is the same thing as *kokohol*, the owl that flies about in the day. She contrasted

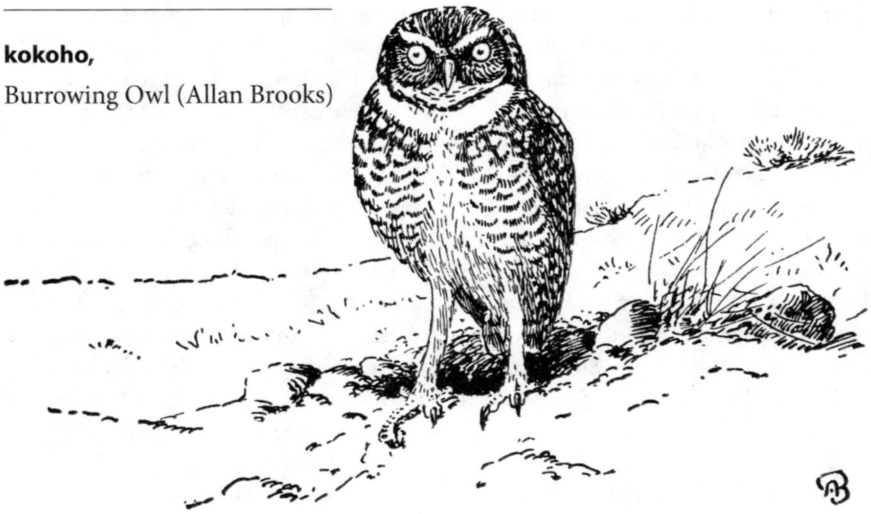

kokoho,
Burrowing Owl (Allan Brooks)

the bird with *kuukvul*, the screech-owl. Downstream at Santa Cruz village Rosita Brown also said *kokohol.*

Joseph Giff said, "When you pass them, they keep looking at you and they twist their head." Sylvester Matthias added mirthfully, "Twist off its neck!" This latter was intended as a joke, not a fact. He said Burrowing Owls were formerly common about west end villages and would nest close to Pima houses.

Luciano Noriega of Quitovac oasis called the bird *koko'i*, ghosts, saying this owl is the one related to souls of the dead, rather than the *chukuḍ*, Great Horned Owl. This association is so much at variance with well-established Piman thought that it requires verification as a local tradition.

In Tohono O'odham country, someone told Culver Casssa and me that the Burrowing Owl might also be called *go'ol hikkam*, "differently positioned navel." The word *go'ol* means "in the wrong place" or "not in the normal place," a humorous reference to this bird's habit of turning its head around or even upside down while watching you. The synecdoche is even funnier because birds don't even have a *hik*, navel.

The owl occurs throughout Tepiman country, so it will be interesting to learn what it is called by southern speakers. It is not associated with staying sickness, nor have I found it mentioned in song or story, although Burrowing Owl occurs in various Puebloan tales. (Russell [1908:279, 311] used the symbol *o* for the sound *u*, so the two songs naming *kókovoli* are about the screech-owl, not about this species.)

In William Blackwater's version of Elder Brother's resurrection speech (also a war oration), four birds are teamed (Benedict 2001:189). First is the Great Horned Owl—or, as often termed in Piman literature, Gray Owl, *koomagĭ chukuḍ*. Next is *kuukvul*, the Western Screech-Owl, then the Burrowing Owl. One might logically expect Barn Owl to be the fourth (it appears in another war oration), but the translation has Roadrunner. Over each, Se'ehe blows smoke, then the bird goes to the enemies to weaken them. In Thomas Vanyiko's version, only Blue Roadrunner and Gray Owl are paired (Benedict 2001:36). In Thin Leather's earlier (1903) telling, which is the most detailed, we find Gray Owl, Little Blue Owl, Green Roadrunner, and Gray Nighthawk paired in the parallel segment (Lloyd 1911:141). The birds are not associated with any geographical directions.

In Pima Bajo country, we never encountered the Burrowing Owl in the field (they are spottily distributed in the interior of Sonora), but Pedro Estrella called pictures of this species *duis kupadak*. The first word probably should be *devel*, earth.

koologam; var. *koolo'ogam*

Caprimulgidae, Caprimulgiformes
Phalaenoptilus nuttallii
Common (Nuttall's) Poorwill;
Tapacamino Tevíi, Jarapena

The Common Poorwill seems almost more like a giant noctuid moth than a bird. Its plumage is perhaps the most exquisite of any bird in the world, velvety and finely marked. The ground color is brown, tan, or gray, or even pinkish gray, depending on subspecies. The upper parts (crown, back, and upper surface of the folded wings and tail) are marked with black chevrons and lines. The underparts are finely etched with black bars, coalescing into a black chest patch. The flight feathers are buffy to ochraceous, broadly barred with dusky tones. Adults have three white patches, one across the throat, two on the tips of the outer tail feathers. Pure white in males, these patches tend to be less extensive and partially invaded with buff in females.

These compact birds are difficult to see during the day, so well are they camouflaged. When flushed from the ground or some low perch among rocks, their broad wings and tail seem owl-like. In the hand, their soft dense feathers and light weight also suggest an owl. Like owls, poorwills forage at night, seeking various insects, especially moths, near the ground. However, they are goatsuckers or nightjars, relatives of the nighthawks, as shown by their minute bill, tiny feet, and gaping mouth that opens wide past their ears.

Poorwills are more likely to be seen at night, perched along unpaved roads. In the headlights, they have a pinkish eyeshine, like two small embers. But they are far more often heard than seen. In the desert ranges, they are surprisingly abundant. As soon as darkness seriously sets in, they begin calling their name from the rocks: *poor-wíll, poor-wíll.* If the bird is very close, one hears a third note, *poor-wíll-up* or *poor Philip*. The cadence may be casual, or it may be more intense, as when several birds sing antiphonally across the expanse of rocks and paloverdes. As dawn approaches, one may hear all the birds from the entire range erupt in a simultaneous signing-off for the night.

Unlike most other ground-nesting birds, including their nighthawk relatives, poorwills lay pure white eggs. Sonoran Desert naturalist John Alcock (1990:47–50) surmises this is because the exquisitely patterned Common Poorwills sit tight throughout the daylight hours, unlike the nighthawks, which take advantage of dawn or late-afternoon flights of insects.

Supposedly migratory here in the north country (desert winters get cold, and frost is not rare), poorwills would occasionally be found at odd times and places in midwinter. It remained for desert naturalist Edmund C. Jaeger (1948, 1949, 1953) to discover that Common Poorwills actually hibernate, or at least some individuals do. Feeding heavily and storing up a supply of fat, the birds find a sheltered spot—a rock hole or crevice—and drop their metabolic functions (body temperatures, breathing) to a minimum for extended periods, all winter if necessary (Austin and Bradley 1969; Bartholomew, Howell, and Cade 1957; Brauner 1952; Marshall 1955; Thorburg 1953). Some individuals have been found even in exposed situations. Whether whole populations or just certain individuals resort to torpor or hibernation is not yet known.

It is surprising that this phenomenon should have gone so long undetected by ornithologists. But, then, no one ever asked the Indians. In my earliest interviews with Sylvester Matthias at Komaḍk, he identified this bird from skins and pictures, calling it *koologam*. He said his grandmother claimed that this bird hibernates in winter, though Sylvester added that he himself never believed that any bird actually hibernated. The hibernation idea he said was based on the fact that "in the olden days" birds were found apparently dead, but when handled proved to be still alive. "That's what my grandmother told me: it was laying right on the flats and weeds grow over it. Seems it's dead. And it's alive! That's that *koologam*." Several other Pima still knew about the poorwill's behavior. (Interestingly, none of the Papago I worked with mentioned this phenomenon.)

Poorwill hibernation was not some obscurely known item of folk science. It was once such common knowledge that it was incorporated into the Creation Narrative as a major metaphor. This is most graphically explained in an episode Julian Hayden (1935:45; see also Bahr and associates 1994:220–221) recorded from Juan Smith at Bapchule on the Gila.

During the conquest of the Gila valley, the Vupshkam (conquering Pima) noted that some of *va'aki* (big house, pueblo) dwellers were escaping. "So they worked and made Kal da kam [Koologam], a bird, something like a nighthawk, and sent him over there. They did this because it was this bird's habit upon landing in winter to lay there all year without anything to eat or drink. So this bird came here among the people and they heard him singing this song at night. And when winter came he laid down. So it happened to these people—laying down seemed a great pleasure and all wanted to rest because the bird held down their strength." The poorwill's unique physiological characteristic becomes pervasive. Another bird, *giisobĭ*, Verdin, which retires early to its domed nest, was also sent in to help put the people to sleep (see that species account). By this ploy, the invading Pima conquered the pueblo.

The Common Poorwill figures in another Piman story that does not involve the sleep metaphor (Saxton and Saxton 1973:222–224). A man, sad that his wife has run away at the puberty rite dances, makes a reed flute to imitate a poorwill so that he can cry without people knowing who it is. He goes into a cave to play his plaintive song. But it turns out to be such beautiful music that it arouses the curiosity of women all around, and he forgets that he is sad. Four sisters, attracted to his songs, go in search of him, bringing him home with them. He marries the youngest. At first, the *koologam*'s song may seem sad, but it is also seductive.

Other aspects of the poorwill's behavior and characteristics were incorporated into Piman folk science and lore. Sylvester said that there used to be more poorwills around the village, closer to Pima homes than now. He described their nesting behavior as similar to the nighthawks'. At dusk at Quitovac oasis, we heard two poorwills coming in to drink. Luciano Noriega commented, "Muy flojo *koologam, ñepuḍ, chivi-chúch*" (Very lazy the poorwill, nighthawk, and Killdeer). This was in reference to their nesting behavior. When I asked Frank Jim about the *koologam*, he responded, "Those that lay on the ground? Those gray ones?"

Delores Lewis said people were afraid to mimic the *koologam* because they might burn your blankets. This warning might be in reference to the bird's eyeshine at night, resembling a glowing ember. Jose Enriques related, "Don't imitate the *koologam*; it will burn your house down or something." Culver Cassa points out that both claims are references to the sleep metaphor: one sleeps on blankets or in a house. "If the bird is mimicked, you'll sleep so heavily, you won't put out the fire."

I questioned George Kyyitan, "Do you know the *koologam?*"

"Yeah. It's something like that *ñepuḍ* [nighthawk], but it's kind of gray, that *koologam*." He offered this humorous interpretation of its call in Pima:

Dooñ hug wuḍ va'i?
Muuñ hidoḍ!

Who could I be?
Boiled beans!

Juan Thomas gave Russell (1908:302–306) a set of seventeen medicine songs. The subject of the fourth was said to be "bluebird," but it is actually the poorwill (retranslated by Culver Cassa):

Alisi chechenongi kologame
jewenḍ lungidane chedongi yewelhela voishpaiime
toto yewelhela vipiahime
vatoi hevenḍak jeñewena kubañima.

Little blue poorwill
At the earth's edge lies down upon the blue wind;
The white winds remain uncovered [?];
They will blow; the lands will become dusty.

Anyone who has handled these bundles of feathers will appreciate the image here of the buoyant poorwill floating on the wind.

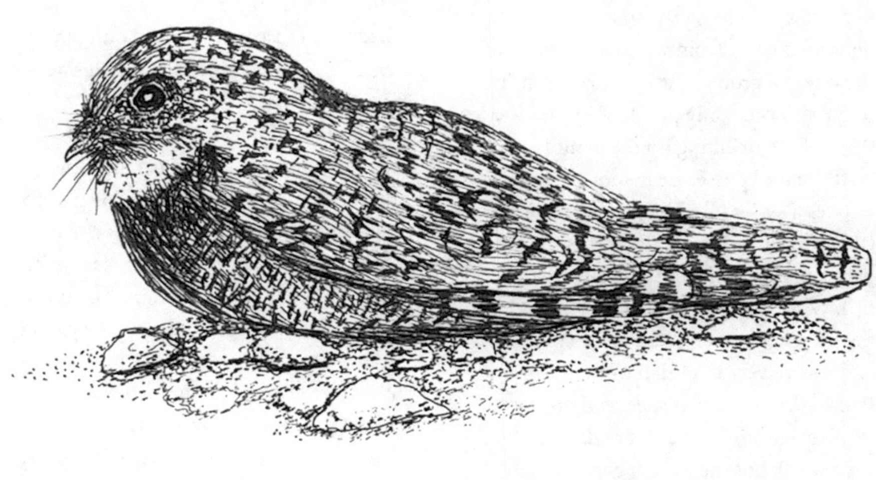

koologam, Common Poorwill (Takashi Ijichi)

COMPARATIVE LINGUISTICS. Little information is available on goatsuckers and other night birds outside Pima and Papago country. In Desert Pima Bajo country, either Pedro or María called the tapacamino *koha*, but whether this name refers to the poorwill or to the nighthawk I never found out.

With three nightjars common in Mountain Pima country, it's difficult to tell which name applies to which species there. For the Common Poorwill in the Yepáchic area, Isidro Cruz gave *kukulyi*, whereas Lorenzo Cruz said *kokoloki*. While we were interviewing, a Ridgway's Whip-poor-will or Préstame-tu-cuchillo *(Caprimulgus ridgwayi)* started calling. Donasieno Cruz called it *kokoloki*, saying that it has the same name as the Common Poorwill. Possibly this word is related to Desert Piman *koologam*, but I suspect it is rather an independently derived onomatopoeic name.

MISIDENTIFICATIONS. Hayden (1935) and his translator were unsure of the exact identification of the Common Poorwill, but knew it was similar to the nighthawk. Bahr (Bahr and associates 1994:321 n. 18) correctly guessed the word in the Hayden manuscript as *koolo'ogam* but followed Saxton, Saxton, and Enos (1983:33) in misidentifying the bird as the Whip-poor-will, a species that does not occur in Pimería Alta. Consequently, Bahr missed the hibernation metaphor entirely. In their transcription of the Flute Story love song, Saxton and Saxton (1973:221–225) again misidentifed the bird in question as a Whip-poor-will.

Mathiot (1973:486, n.d.:15) confused the Burrowing Owl folk taxon with the Common Poorwill. She equated the terms *kuuku'ul* and *kuukulul* with *kologam* and *hol[o]gam*. The last name is a variant I have never encountered before, and I suspect here that Mathiot's native consultants had gotten rusty in their nightbird names and identifications.

TECHNICAL NOTE. According to avian physiologist William A. Calder (personal communication, 2000), *dormancy* describes the state of suspended function of plants and cold-blooded animals. *Hibernation* is a lowering of body temperature in homeotherms or warm-blooded animals in winter, with reduced response to external stimuli. *Torpor* is similar, but for shorter time periods, as for hummingbirds at night. In Piman metaphor, hibernation is intended, whereby the enemy becomes unresponsive and vulnerable like a poorwill in winter.

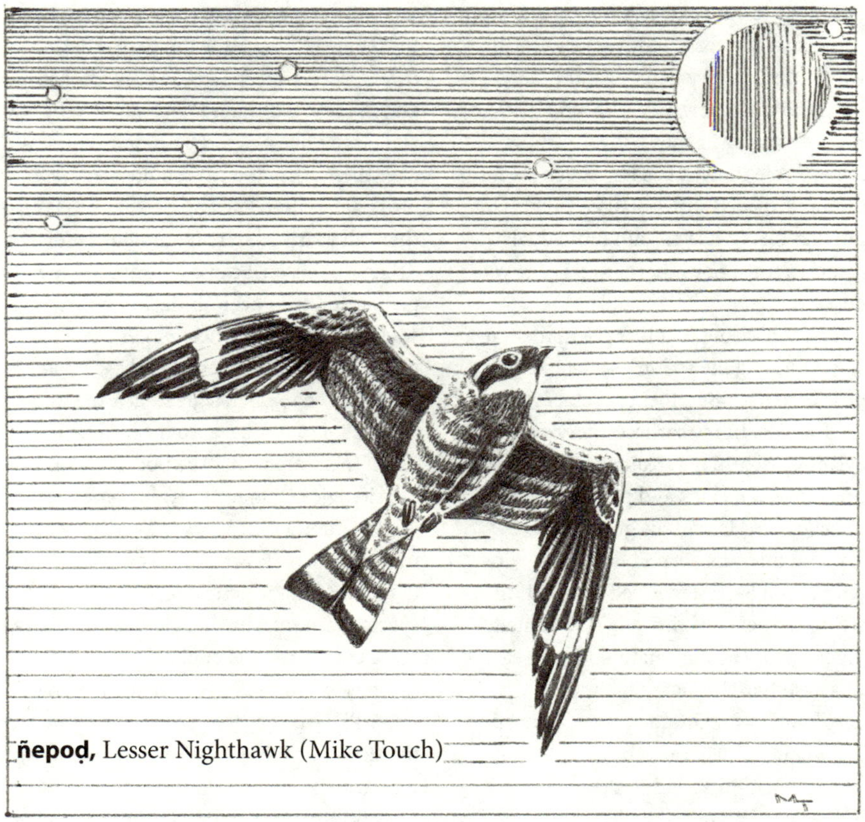

ñepoḍ, Lesser Nighthawk (Mike Touch)

ñepoḍ, ñepuḍ*, pl. *ñeñepoḍ
(?)*nepoḍ* (Pima Bajo)
Caprimulgidae, Caprimulgiformes
Chordeiles acutipennis
Lesser Nighthawk; Jarapena, Chotacabras Menor, Tapacamino Zumbón

The long-tailed, narrow-winged Lesser Nighthawk suggests a small falcon in outline, but its flight is buoyant and fluttery, more like a giant slow swallow. When the sizzling summer heat of the desert begins to diminish in the late afternoon, a few nighthawks may venture out in broad daylight, impatient from waiting out most of the day at a ground nest or low perch. They often seek out some pond, where they skim the surface, drinking on the wing.

If the bird passes near enough, one notices a series of white patches: one fore (the throat) and three aft (across the outer primaries of each wing and near the tail tip). When the bird is held in the hand, the plumage is seen to be more compact and less velvety than that of the Common Poorwill, but the two species share some similar markings. The feet and bill are very small, but the mouth gapes widely, from ear to ear. The birds literally scoop mouthfuls of small flying insects from the air. They may gorge on insects attracted to street lights. In flight, particularly when several are together, nighthawks utter a short nasal twang, *bzzzt*. On the ground, the birds may make a subdued trilling noise that sounds quite like a distant motor. This sound is often the only indication that a nighthawk is nearby, until the bird is flushed at foot.

Nighthawks lay their several speckled eggs directly on the ground without any pretense of a nest. Short-distance movement of eggs and longer-distance movement of young (probably under their own power when called by the parent) have been reported (Bent 1940:246–248). Egg movement may be to take advantage of shade, and discovery of the young by potential predators probably initiates nestling movements.

A Lesser Nighthawk flew over us at Quitovac oasis in Sonora. Luciano Noriega commented, "*Ñepoḍ* pone su nido en puro tierra. Muy flojo pájaro; cria en la tierra también." The village patriarch was observing that the nighthawk is so lazy that it puts its nest right on the ground, and the young as well.

When we were afield in northern Sonora, I was puzzled by a bird that Delores Lewis had been calling in English "nightingale." Finally at dusk at Mission San Ignacio, we saw the mystery bird, a nighthawk. "*Ñepoḍ* lay eggs just in [on] the rocks [gravel] over there. Maybe [someone] step on them." Vincent Josemaria of another Papago village said the People would eat nighthawk eggs when found, just as they do quail eggs.

According to Sylvester Matthias, these birds were once of considerable interest to the Pima youngsters who found them nesting always in the salt flats with *kauk u'us*, Greasewood; *itañ*, Pickleweed; and saltbushes. They have no nest. Pimas maintain that this bird moves its eggs around the bush during the day in order to keep them always in the shade. The boys who were told this would locate nests so they could observe this behavior. Sylvester said that when his grandfather told them about the nighthawk, he and his cousins had to find out for themselves. The youngsters would never see the nighthawks actually moving the eggs but suspected that they must push them with their breast. They would watch the eggs morning, noon, and afternoon, and always the eggs would be in the shade around the same bush, even after the site had been discovered. The birds made no attempt to hide the eggs or move them to an entirely new place. Eggs were difficult to find, he said, because they were speckled and colored like the bare alkaline soil. Also of fascination to the young Pimas was the parents' defensive behavior once the young had hatched. Sylvester added that some Pimas thought this bird hibernated like a poorwill. However, only the latter serves as the metaphor of sleep in Piman mythology, not the *ñepoḍ*.

Not only were nighthawks common nesters in places where the River Pima had their villages, but they were attracted to houses by the O'odham practice of storing great quantities of mesquite pods in Arrow-weed bins *(hohomta)* on the flat roofs of arbors *(vapto)* and storage houses *(kosin)* as well as on the ground in yards. Sylvester recalled, "*Nanakmel* [bats] live on some kind of little insects, little bugs. That's why they always hang around those *hohomta*, that *homta*—that storage basket—when those *vihog maamad* [emerge]. They come and eat those, just like the *ñeñepoḍ*." The *vihog maamaḍ*, "mesquite pod babies," are the bruchid beetles that infest the pods in great numbers, emerging from the granaries at dusk, attracting both bats and nighthawks.

As to be expected of a bird with such prominence in Piman consciousness, nighthawk has an important place in legend. Various versions of one story have been recorded (e.g., Curtin 1949:136; Saxton and Saxton 1973:218–219; Wright 1929:119-120; also told by Sylvester Matthias, Leon Juan, Jose Enriques). At the first wine feast, Chuuk Ñepoḍ, Black Nighthawk, came dressed in fine gray and yellow, bringing a *vaapk*, Carrizo (cane), straw with which to drink the wine. The girls, who much admired Ñepoḍ's fine attire, were further impressed with his fineness and tiny bill.

Shoo'o, Grasshopper, was envious of all this attention and set out to make nighthawk laugh by means of various antics. Finally, he took off his hind leg and danced with it on top of his head. Nighthawk thought this was so funny that he laughed until he split his mouth open to the sides of his head as far as his ears. "And that is why the night-hawk never flies in the daytime. His mouth is so big and white and ugly that he has to fly at night so people will not see him. And that is why he darts past you so quickly in the evening" (Wright 1929:120). *Ñeepoḍ* cries at night, added Leon Juan, because ladies no longer like him on account of his large ugly mouth. And if you look closely, you are likely to see grasshopper still missing one of his large legs. A key to appreciating this story is that one of the Piman canons of beauty was a small mouth, such as Nighthawk originally had (Curtin 1949:136).

Among the River Pima, there are *ñepoḍ* songs, George Kyyitan told me, that had been taught to the original singer by the nighthawk. Not many are now known, but he sang this one:

Haiya ñi ñeikun
haiya ñi ñeikun
voho si shohigo va chuikahachi kahaj
inoñ oimelkuḍ i oina geñ hiosig veechim
haiya, haiya, haiya
heñ hiosiga mumukim.

In spoken Pima, this song would be:

Ai ñ-ñe'ikuḍ!
Ai ñ-ñe'ikuḍ!
Voho 'at si sho'ig o va chu'ig ch kaach
Id ñ-oimelkuḍ i oid heg ñ-hiosig'o weechim
Ai, ai, ai
Ñ-hiosig 'o va mumku.

Alas my singing place!
Alas my singing place!
Truly it will poorly appear lying
Along this my wandering my flowers are lying
Alas my flower is sick.

Sylvester also noted a nighthawk constellation. It includes the three stars we call the belt of Orion, representing the white wing and throat (and tail?) patches of the bird in flight. Culver Cassa interprets the constellation as two nighthawks (with three stars each) facing the three-star belt of Orion. George sang a Swallow Song referring to the nighthawk constellation:

Shashakal himena
shashakal himena
jing amhu chuma himeda
ab tonolikahachi kukua i tabi nanangia
doki ñeñepoḍ uhuhing huhukung tabi nanangia.

Stars following in a line [Orion]
Stars following in a line [Orion]
Should be trying to go along
Up against its light they hang
They're nighthawks that at the stars' edges hang.

George explained his song this way: "Three stars, *shashkal himdam*, still going. At the end of that *shashkal himdam*, there's a bunch that was hanging on there. But it's a *ñepoḍ* that was hanging there, at the end, at the end of that *shashkal himdam* when it shines, there. See, it hangs on there when it shines. At the end, it's a *ñepoḍ u'uhig* that was hanging on there. They didn't know that was *ñepoḍ* hanging there, but at the end of the song you can find out. They don't know that was *ñepoḍ* that was hanging there, shining there."

At the first wine feast, one bird participating had his head stepped on and has since been known as *komalk mo'okam*, "it has a flat head." Different storytellers associate various birds with this designation. For Simon Lopez of Santa Rosa, it is *ñepoḍ*.

On setting out on a war expedition, it was customary for the Northern Piman to paint his face black (Underhill 1938:71). In a narrative, Falcon Man sings this song:

I drank with the quail
And with drunkenness
He painted my face with black paint.

When the warrior returns, successful, he paints his face white. Then he sings this song:

I drank with the nighthawk
And with drunkenness
He painted my face with white paint.

The imagery of the two songs is taken directly from nature. The male Gambel's Quail wears an all-black mask, whereas the lower half of the nighthawk's face is white.

O'odham war orations depicted with vivid often metaphorical imagery what the People hoped would happen to the enemy Apache. These images frequently used various kinds of birds to indicate what the People might do to the Oobĭ. For instance, one of Elder Brother's resurrection speeches, after invoking Gray Owl, little Blue Owl, and Green Roadrunner, concludes:

And they selected the small Gray Nighthawk;
And he blew a gray dust all through the enemy's houses and swept their ground.
And their springs of water were left dry, choked with driftwood and covered with cobwebs.
And their *kiik* (houses) were full of soot, and their trails like old trails;
And after that the fresh foot-tracks could be seen—
And they went out and found the enemy by his fresh tracks and captured him, for he had no weapons.
And from the sending out of the birds, even to the end, all this is a prophecy. (Lloyd 1911:141–142)

Among the Sonoran Papagos, there was a village called S-Ñeñepoḍ(k), "Lots of Nighthawks," known as Represa de Enrique in Spanish. It was between Taak (La Lesna) and before Quitovac, roughly (Jones 1971). Explorer Carl Lumholtz (1912:393) gave the village name in the singular form. The large pond here apparently attracted the nighthawks from the surrounding desert. The birds skim the water's surface to drink like swallows.

COMPARATIVE LINGUISTICS. Both the forms *ñepoḍ* and *ñeepoḍ* have been recorded by professional linguists. The plural is *ñeñepoḍ*. Culver Cassa said the name comes from *ñep*, signifying something that is round, narrow, and tapered at the ends—that is, cigar shaped; *ñepoḍk* means "shaped like that." Saxton, Saxton, and Enos (1983:46) define the word as "to be loaf-shaped."

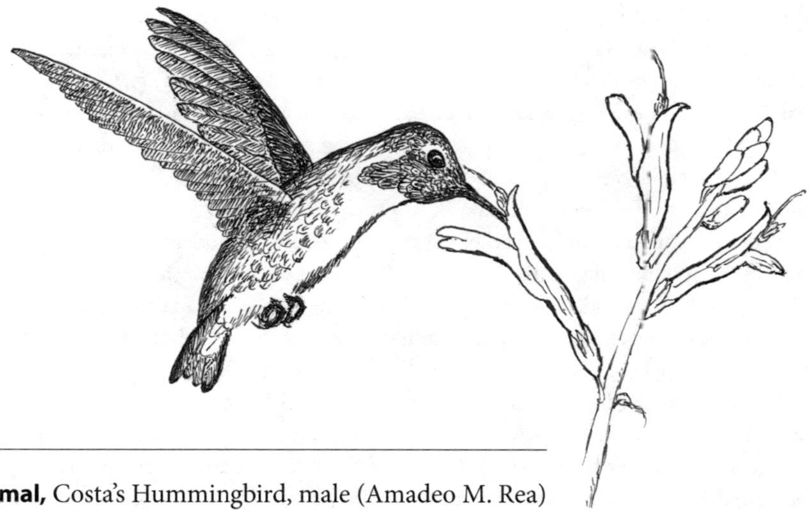

vipismal, Costa's Hummingbird, male (Amadeo M. Rea)

Night birds have been poorly explored among Tepiman speakers. Pima Bajo Pedro Estrella used the name *nepoḍ*, noting that it was the same as the *ñepoḍ* about which Upper Pima Sally Pablo asked him. We did not confirm the species. He also called one night bird *duudual*, but I am not certain of the biological referent because several other nightjars occur in the area.

vipismal, wipismal
***vipsumaal* (Pima Bajo)**

Trochilidae (all species), Apodiformes
hummingbirds; chupaflor chuparrosa, colibrí

Hardly anyone fails to notice the metallic green avian jewels, darting from flower to flower on wings that move so fast we can make out no more than a blur. These tiny birds can hover as well as fly backward. This flying ability, slender bills, and exsertable capillary tongues enable them to extract nectar from tubiflorous plants.

Hummingbirds and certain plants, especially those with tubular red flowers, have coevolved. The plant rewards the bird with nectar in exchange for cross-pollination. These tiny birds have developed highly pugnacious behaviors to defend territories that are particularly rich with nectar-producing flowers. Intruders, in particular other hummingbirds, are driven off and scolded.

Each species of hummingbird has a unique aerial courtship display punctuated with sounds. In some species, the male performs deep dives, genetically choreographed, at speeds almost too fast for the human eye to follow. After this show of agility and copulation, the female assumes all the remaining domestic duties: building the tiny, deeply cupped nest, incubating the two white eggs, and feeding the young.

Three species, all with green-and-white females, breed in Northern Piman country. The most common in true desert situations is the Costa's Hummingbird, *Archilochus costae*. Males are dark green above, with a metallic purple crown, throat, and gorget extending to each side of the neck. It breeds from late winter to spring along arroyos in the desert mountains, then departs for its "wintering" grounds. The Black-chinned Hummingbird, *A. alexandri*, prefers the cooler and leafier habitats, as along desert streams. The males have a black chin bordered broadly with a purple throat, but an all-green crown. Starting in the mountains of Tohono O'odham country and becoming the most common species of Sonora is the Broad-billed Hummingbird, *Cyanthus latirostris*. Unlike the other two, this one has dark blue and green underparts and a solidly black tail. The coral red bill of the male makes it unmistakable. Various other species, mostly showing bright rufous in the plumage, are abundant migrants through Pimería.

All hummingbirds regardless of species, are called *vipismal* or *wipismal* among Upper Pimans and *vipsumaal* among desert Pima Bajo. Hummingbird appears prominently and early in the Piman Genesis. In addition to the four primordial beings, four birds are saved from the Flood: a thrasher, Gila Woodpecker, Verdin, and

Hummingbird. They fly up to the sky and hold on with their bills. "Finally, as they were threatened with destruction, the god Vikärskam [Viigĭ Koshkam, 'Fluffy-feather Nest–It-Has'] took pity on them and gave them power to make 'nests of down' from their own breasts which floated on the surface of the waters and so enabled them to survive the flood" (Russell 1908:211). Of the four birds saved, Hummingbird makes the finest nest—a deep cup of plant down and other fine materials held together with cobwebs, the inside often lined with fine feathers.

At the height of the Flood, Jeweḍ Maakai (Earth Doctor) gives Hummingbird the power to make the waters go down by singing (Benedict 2001:15-16). Hummingbird's song is: "Hummingbird is trying to make the waters go down." When Verdin is asked to help, he joins Hummingbird with the same song. Then the two are assisted by Earth Doctor himself, who sings:

Hummingbird, hanging above me,
Has power to make the waters go down.
He is causing the waters to fall.

At this point in the story, Russell added, "If anyone harms the little Vipismal to this day the flood may come again. Accidental injuries to the bird must be atoned for; if it be killed, its tail feathers must be kept for a time to avert disaster; if it is found lying dead, it must be buried and appropriate gifts must be placed upon its grave." (I have not heard of this practice among contemporary River Pima, and as far as I know the bird is not associated with staying sickness, although Russell's comments might suggest otherwise.)

As the floodwaters recede, the first task on the wet and muddy land is to locate the center of the earth, its navel. In one version of this story, Hummingbird and Gila Woodpecker are chosen for this task, using a bit of geometry: bisecting a line. Woodpecker is initially sent east, the Hummingbird west, until they reach the margins of the earth. Then the birds are to fly back until they meet. Then they switch directions and try again. First one then the other flies faster. But on the fourth trip, averaging the results, the culture heroes standing as judges decide that the *hik*, the center, has been determined (Benedict 2001:17; Lloyd 1911:45–46). There is another version with different flyers (Wright 1929:20–21). Russell (1908) had the culture heroes themselves performing this task.

The Vipishaḍ, or Small People, who inhabited the ruins now found in Pima country, were thought by the O'odham to have not only miraculous powers, but fantastic material possessions as well. They had Scarlet Macaws, turquoise, and radiant clothing, in addition to substantial architecture. The leader of each pueblo, the *siivañ*, held a position somewhere between that of shaman and chief. One episode from the Creation Story (Lloyd 1911:235–239) describes a rather fabulous article of apparel.

A *siivañ* from Blackwater area invites a *siivañ* from the Gila Crossing area to bring his wives and gamble. At first, all the wives go to dance with the host, but when the guest takes out his rattle and begins to sing, all the wives, his own and that of the host, begin to dance with him. The next morning he returns downriver with all the wives, his own as well as the host's, who have chosen to go with him.

The losing *siivañ* sends a messenger with a beautiful cloak covered with live butterflies and hummingbirds, requesting that his most beautiful wife and her daughter return to him. The request is spurned. "Tell him I do not want his cloak. I have one just like it, and I have all I want, and I will not send back any of his wives. It was his wish that we should gamble, and if he had been the better singer and had won my wives I would not have asked for any of them back."

Fewkes (1912:47–48) recorded from storyteller Thin Leather an episode that is missing from both Lloyd's and Russell's accounts from the same narrator. Morning Green Siivañ of Casa Grande pueblo banishes Wind Man and Rain Man from his pueblo because of an indiscretion on their part and causes Rain to become blind. (For this reason, he must be led by Wind, who precedes him.) A four-year drought follows. Hummingbird is sent to try to entice the two back. "Humming-bird carried with him a white feather, which he held aloft to detect the presence of the wind. Three times he thus tried to discover Wind-man by the movement of this feather, but was not successful. When at last Hummingbird came to a place where there was much green grass he again held up the feather to see whether it showed any movement of the air. It responded by indicating a slight wind, and later he came to the spot where Wind-man and Rain-man were, but found them asleep." When Hummingbird awakens them, the three negotiate conditions for the return to the Casa Grande *vahki (va'aki)*, initiating a four-day downpour causing every roof to leak.

This story is more likely Yuman than Piman in origin. The two gods depart for the west, the important Maricopa direction. They end up at a mountain in southern California, in Yuman country. In a Piman story, Wind and Cloud (who bring rain) are associated with the east, which is also the final destination of human souls. In a different story, when Tobacco Woman goes west, then Corn Man leaves and the People go east, it is another small bird, the Verdin, who is sent after Tobacco Woman with offerings to stay the drought. (One may question whether this tale is really a hummingbird story rather than a Verdin story; the Piman bird name is not given, and Fewkes depended on a dubious interpreter. A parallel Tohono O'odham version calls the bird Giisubĭ [Saxton and Saxton 1973:321–333]; see Verdin account.)

A popular tale involves a number of animals participating in the first cactus wine feast after the Saguaro is discovered. Depending on the teller, the story often involves the kingbird, Canyon Wren, Gila Woodpecker, nighthawk, and grasshopper (see Lloyd 1911:229; Rea 1998b:222–223; Saxton and Saxton 1973:79–84; Wright 1929:120–122). A wise person decides that it will be in the community's best interests to dispose of the seeds of this new plant. In some versions, *kaav*, Badger, is sent off to get rid of the seeds. As Jose Enriques tells the story, *wipismal* is the courier. Coyote, however, circles around and meets Hummingbird on the path, encourages him to open a finger so he can see what he is carrying, then hits his hand so that the winds scatter the Saguaro seeds all over.

Vipismal occurs in both sung poetry and oration. Sylvester Matthias said there was a Blue Hummingbird Song among the *shashañ ñeñe'i*, the blackbird songs, from Santa Cruz village. The two-part song pairs *vipismal*, the hummingbird, with *kadgam*, Yellow-billed Cuckoo:

Chenangia vipisimali
hengenamo ñiñivia
gamhuya nangia
gamhu ñeheda

Kodakang u'uhinge
gamhu mek ñeheda
gamhu mek ñe'enam
kadangam vema sho'o
gamhu mek ñe'enam

Blue Hummingbird
just then arrives
over there hangs
over there sings.

Kadgam [Yellow-billed Cuckoo] bird
far off is singing
over there far off is singing
[Kadgam ?cries]
over there far off is singing.

There is also a Hummingbird Song series said to have been given to someone by this animal. One day I asked Clemencia Leos of Haashañ Keek (Sacaton Flats) if she knew any of these songs. Her maternal grandfather, Jiosh Keli, originated the *vipismal ñeñe'i* sequence. In a beautiful strong voice, she at once sang this one:

Vivingi chupulima keka
oinaju vi nanaiwua
tana hihivakunava shavakuhuk nanaiwua
vipismali ñeñei
oinaju (hu) vi nanaiwua.

Clemencia's free rendition is:

There's a square *viigĭ* standing [*viigĭ* squarely stands]
Coming through.
Behind that the rasping stick and the rattle come through,
Then the hummingbird songs come through.

Culver Cassa proposes a more literal translation:

Eagle down in a square stands
One after another we sit
Our scraping sticks echo and sound as we sit
Hummingbird songs!
One after another we sit.

The *viigĭ* is a fluffy feather used on ceremonial objects such as the *omina*, prayer stick. The song describes in the abstract the assembling of the items needed for a ceremonial sing: the prayer sticks stuck in the ground, the scraping or rasping stick, the rattle, and finally the songs themselves. (The wry part is that the humans are omitted, and the ceremonial items assemble themselves as if in an animated cartoon, at least in Clemencia's version.)

When we were discussing the song about the Rock Wren's making wine from cholla cactus flowers, Simon Lopez said that he knew two songs that told in a similar fashion of Hummingbird's making wine from the *hasahañĭ*, the Saguaro, and from *chuchuis*, the Organ-pipe Cactus. Then he sang this song:

Wipismal u'uhinge
wipismal u'uninge
haashañi hiosike nawaitok
kame ñehedam.

Hummingbird bird
Hummingbird bird
Saguaro flower wine brewed
And there sings.

He said that the second song was the same except that name *chuchuis* was substituted for *hashañĭ*.

Underhill (1946:181) took down a warpath speech in which two birds accompany the warrior and help overpower the enemy:

Gray owl shaman
Comes to me.
Yonder he finds my enemy
And makes him helpless.

Blue humming bird shaman
Comes with me
Yonder he finds my enemy
And makes him helpless.

Without the original O'odham, we are left in the dark as to the specific identification of the "gray owl," but in keeping with other usage in similar contexts, it should be *chukuḍ*, the Great Horned Owl. We have earlier seen the role of Great Horned Owl when the bird is the spirit of a deceased O'odham in avian form, locating and helping to overcome the enemy to the east. Although the word describing the color of the hummingbird is here translated as "blue," the original must have been *cheedagĭ wipismal*, which could equally indicate "green hummingbird," a more likely possibility. In this odd paring with the powerful owl, the hummingbird metaphor is quite different, embodying the bold and pugnacious nature of biological hummingbirds.

In Mesoamerican contexts, hummingbirds are generally a metaphor for warriors, appropriately enough. In Puebloan mythology, hummingbirds are associated with rain or at least moisture, particularly as contrasted with drought (summarized in Tyler 1979:98–104). There seem to be two facets of this metaphor. Hummingbirds are able to locate moisture-laden tubular flowers anywhere in the desert. Important among these plants are tobacco flowers. Tobacco is smoked to produce, by mimetic magic, the clouds that bring rain. Similarly, in most Piman mythological contexts, *vipismal* figures in rain and flood situations. But in one episode of the Piman Creation Narrative the birds are instead compared to *chechpa'avĭ*, promiscuous women (Benedict 2001:102; Herzog [1929]). (Although the term *prostitute* has been widely used in translations of this word in Piman narratives, it is rather imprecise because these women's role was not economic; "loose women" would be a better translation.) In the story, the twin boys, sons of White-clay Eater Woman, are on a quest for eaglets. They had been warned of supposed difficulties along the way.

And there were things in their time of blooming, and in the midst of them hummingbirds busied themselves. They stopped and looked at them. "Perhaps this is what the old lady [*oks*, their mother] meant when she spoke, and perhaps now we'll fall into this?" Then they spoke again: "What will you do to us indeed; you are just hummingbirds, and you are always acting like this: wherever it is lively on the earth, there you turn into loose women and run around. If somewhere there is a lame person, you let yourselves be had. If somewhere there is a blind person, you let yourself be had." (Herzog [1929]:52, 52b)

Songs were sung during the planting and growth of traditional crops. Hummingbird appears in the Tobacco Planting Song. Here again Underhill (1946:84) translated the word modifying "hummingbird" as "blue," where "green" would seem more appropriate:

Am I not the tobacco shaman?
Here I come forth and grow tall.
Am I not the tobacco shaman?
The blue humming bird finds my flowers.
Above them softly he is humming.
Hilo á'alo.

The tubular flowers of tobacco (yellow in three of the species known to the O'odham, pink in one of their cultivars) attract hummingbirds, an important pollinator.

Hummingbird has lent its name, in both O'odham and in English, to a bush that grows in desert mountain arroyos, *Justicia californica*. In English, it is known as Chuparosa, a Spanish name for the hummingbird. In Upper Piman, it is *vipismal jeej*, "hummingbird's mother" (Rea

1997:137–138). The bush blooms in winter when the Costa's Hummingbirds are nesting in its range. The lower lip of the red tubular flowers is turned downward and is just the right length to allow the Costa's bill to reach the nectar reward at the base of the corolla while the bird's crown is dusted with pollen from the anthers above. The bird then flies off to another bush, cross-pollinating it. This symbiotic relationship is captured in the plant's Piman name.

COMPARATIVE LINGUISTICS. In addition to Upper Piman *vipismal* and Lower Piman *vipsumaal*, and *viipsimal*, we have *vípish* and *vípishi* among Mountain Pima and *vipíshi* among Northern Tepehuán (Pennington 1969:117). Rarámuri (Tarahumara) *semuchí* is not cognate. The chuparrosa is *séma lúlukut* or *séwa tulukút* (*sewa*, "flower") among Yaqui.

The hummingbird name is at least partially analyzable. *Viip* means "to suck [an] object" (hence *vippi*, breast, and *viib*, milk). I suspect that the last syllable, *-mal*, comes from an earlier form of what is now *-maḍ*, the "offspring or young of" whatever noun it is attached to. In this sense, the name means "sucker baby." The permutation of the connecting syllable between modern Upper and Lower Piman is probably of no significance.

The suggested etymology of the last syllable may be simplistic. Among the Luiseño, a Uto-Aztecan group of southern California, animal names frequently end with *-mal* (see kingbird account). Here hummingbird is *túshmal*.

MISIDENTIFICATION. Although the hummingbird itself apparently has not been incorrectly identified in the literature, linguists and ethnographers sometimes confuse the Verdin, *Auriparus flaviceps*, with it (see the Verdin species account).

TECHNICAL NOTE. There is an inbuilt ambiguity in translating the Piman word *cheedagĭ* because the word encompasses that segment of the color spectrum included in English as blue and green. As we have seen, mistranslating the name of the Peregrine Falcon produced a "green" raptor. Green is a pigment lacking in the order Accipitriformes. Most hummingbirds in Pimería Alta are predominantly green. However, native speakers have usually translated songs as being about the "blue" hummingbird. For Tohono O'odham, this would be appropriate for the Broad-billed Hummingbird, which has a conspicuous blue throat and a blackish green body. As far as I have been able to determine, no folk species of hummingbirds are distinguished in O'odham. Culver Cassa notes that *cheedagĭ*, blue/green, is often added to the name of animals in song and oration to indicate some ethereal or spiritual aspect of the organism. So the term may be more metaphorical than morphological.

***duuk vaidam* (Pima Bajo)**

Trogonidae, Trogoniformes
Trogon elegans
Coppery-tailed (Elegant) Trogon; Coa, Trogón

In summer, this exotic bird comes to nest in the Sierra Madrean outliers of southeast Arizona. Here it is at home in pine-oak country with sycamores, Arizona Cypress, ash, Arizona Madrone, and walnut trees growing in the barranca bottoms. Farther south in Pima Bajo country, as along the Rio Yaqui, it occupies much lower elevations in tropical deciduous forest and tropical thornscrub as well (Russell and Monson 1998:163–164). In both settings, it eats small wild fruits and large insects captured on the wing.

The trogons are a pantropical family of birds with very short legs with two toes forward, thick and short bills, and often the brilliant colors one expects of tropical birds. Coppery-tailed Trogons will not disappoint in this regard. They seem like something a mad artist, perhaps a Picasso, created. The stocky birds are about a foot long. The males are bright bronze green above and on the chest, with a white collar separating the red underparts. The tail is white below, with fine barring except at the tips, and coppery above, hence their name. Each set of tail feathers is a different length, and these feathers are squared off at the tips. The central pair is broadly edged black. The bill is yellow. The females and young are similar but duller, with the green replaced with brown.

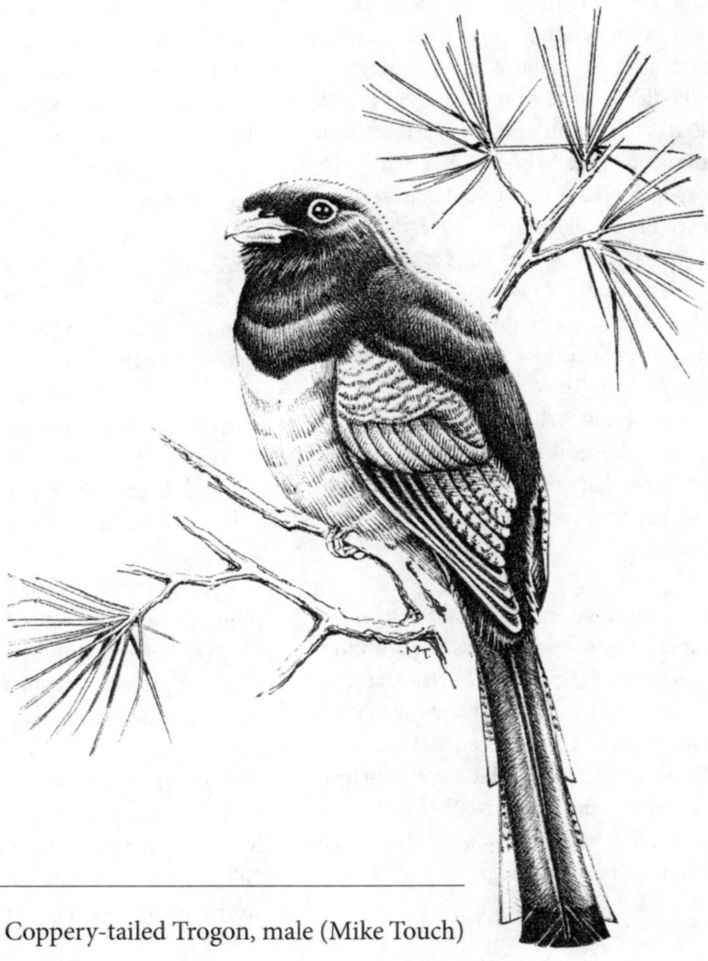

duuk vaidam, Coppery-tailed Trogon, male (Mike Touch)

The sedate trogon is easily called up by imitating its repetitious unbirdlike call: *kum! kum! kum!* This call has often been compared to that of a hen turkey. Another repeated call, *kóa, kóa, kóa*, is the source of the trogon's name in Sonora. The buoyant fluff of feathers may come close to an intruder, landing low in the canopy to peer at him or her. It sits erect, tail vertical. For such a strikingly colored bird, it is curious and confiding.

In Pima Bajo country, this trogon is common and don Pedro Estrella called the bird *duuk vaidam*; doña María Códoba concurred. We encountered it often, and once at Rancho Palmar one of his sons and his son-in-law showed us a nestling they had captured and were raising. I never discovered the meaning of the second word, but the first signifies "rain," which is *juukĭ* among Pima Alto speakers. He said that they sing during the tiempo de las aguas—that is, during the summer rainy season—which is also when the lowland trogons begin to nest. Hence its name.

Pedro also noted its song, *tum! tum! tum!* in the night! (This song is also suggestive of that of the Yellow-billed Cuckoo, which sings throughout the night). He compared it to the sound made by one who is "sharpening" a metate or grindstone with a smaller stone. Although Joseph Giff didn't know the bird (it doesn't reach Pima Alto country proper), he knew well the sound that Pedro was describing. He explained that when a woman has ground corn for a while on her metate, the grinding surface wears down and is too smooth to be efficient. Someone then takes a hard stone, perhaps an old stone ax found in a ruin, and pecks the surface to make it rough again. The rhythmic pounding of stone on stone was well known to indigenous maize-growing peoples. Joe said that the masa that came from the newly sharpened metate would be sort of gritty for a while.

COMPARATIVE LINGUISTICS. Mountain Pima know the trogon well. I was given two names for this bird. At Nabogame, Yepáchic district, Isidro Cruz called it *mai kusalyi*, but both Lorenzo Cruz and Donasieno Cruz called it *kurta*, which is more likely the flicker. Neither is cognate with the desert Pima Bajo name. *Mai* is the name for maguey, or agave.

ba'ivchul*, pl. *baaba'ivchul

Alcedinidae, Coraciiformes

Megaceryle alcyon

Belted Kingfisher; Martín Pescador Norteño

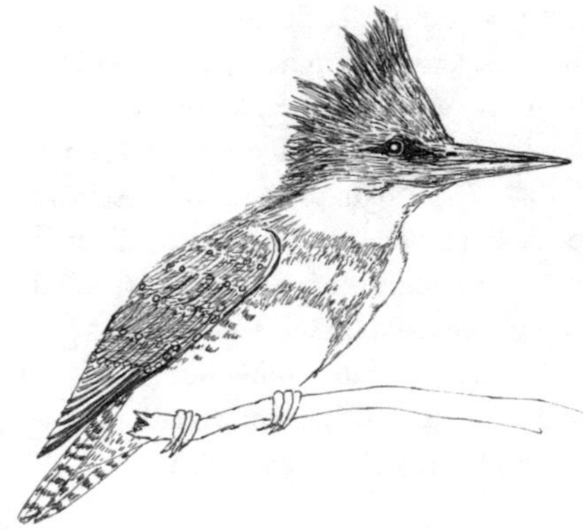

ba'ivchul, Belted Kingfisher (Takashi Ijichi)

The loud rattling cry of the Belted Kingfisher as it hunts up and down streams and rivers is in itself unmistakable. But so is the bird. Its foot-long body is compact, mostly slate blue above, silky white below. The head and bill seem disproportionately large, the feet and legs tiny. The shaggy blue crest accentuates the large head. A broad white band encircles the neck, and a broad blue chest band sets off the belly. In the female, there is a rusty band below the blue one, separated by white, making her the more colorful of the pair.

Kingfishers are primarily fishing birds, flying over the water, sometimes hovering in place, then diving for small fish that they capture in their bills. At other times, they sit motionless on an exposed perch over the water, waiting for prey. As a spring and fall transient through Pimería today, the Belted Kingfisher may visit ponds and irrigation sumps seldom harboring fish. Here it must hunt insects, lizards, small snakes, and small rodents, supplemented at times with fruit (Bent 1940:120; Scott 1886). Along those few rivers that still flow through the desert, the kingfisher may occasionally winter, establishing well-spaced linear territories where the water is clear and fishing is good.

Although Belted Kingfishers seem incapable of such a task, they excavate their own nest burrows in banks, usually near water. Using their strong bills to loosen the sand or soil and their small feet to push it out the entrance, a pair of these birds alternates at the chore until they have made a tunnel three to six feet or more into the embankment. The entrance is only about three-by-four inches.

Belted Kingfishers breed throughout most of temperate North America to the north of Pimería. But River Pima knew this bird well, and there is ethnographic as well as other evidence that it once bred even in the desert.

My initial Piman information about these kingfishers came from early sessions with Sylvester Matthias of Komaḍk on the Gila. He said that *ba'ivchul* nested in the steep banks (cliffs) of the river where the water ran fast, about where the Sundust and Shelde families now live. He noted the kingfisher's distinctive head shape and crest. Nesting holes were about six feet or higher on the bank. Sylvester said that he would poke a stick at the nest entrance so the parent would fly out. The river at the time contained abundant fish on which the kingfishers fed. Several pairs nested in a cliff. (Barn Owls also nested in these bluffs along the river.) The last fish to survive the dying river were carp. Kingfishers were last seen about their nesting places approximately in 1928.

Some years later, in a conversation with Joseph Giff and Sylvester, I asked, "Where does the *ba'ivchul* nest?"

"Up in the cottonwoods," responded Joe.

"Where did you know their nests, Sylvester?" I asked, remembering what he had told me years before, but also knowing that Pimas are not likely to contradict each other.

"On the side of a cliff, on the river. There could be two different kinds—that nest on the cottonwoods or either the side of a cliff, on the bank," was his properly diplomatic response. (It should be pointed out here that the species has been reported as nesting at times in the hollow limbs of trees or even on an open stump where banks are unavailable [Bent 1940:114].)

"Do they make the hole themselves, or what do you think, Sylvester?"

"Like [the] *ee'eḍ vahuḍam* [Barn Owl]. There's a cliff there, where there's a cliff." He parried my question, perhaps because he did not know, implying that whatever the owls nesting here did, the kingfisher did also.

"What do those things eat?" I asked.

"Usually live on mostly fish," said Sylvester. He had no knowledge of the Green Kingfisher, *Chloroceryle americana*, nor did anyone else I asked on the Gila. One of Joseph Giff's male relatives was named Ba'ivchul.

The song group from Bapchule is called the Kingfisher Singers (Baaba'ivchul). George Kyyitan said the Hevel Kii Song, "The Wind's House," is one of the *ba'ivchul* songs.

Yevela kiiki keekang
tam anoñ uapa
tonad.omiñ imuidag
ia aiv yevelonga tang
a no ñi vevegañi
heveli kukuhim ge(ñ) aihima a.

As rendered in spoken Pima by Culver Cassa, the song says:

Hevel kii o keek
t am ñ u'apa
tonod. o ñ-iibdag
a'ai hewelogi
n o ñ vegaj
Hevel o kuukhim (k ñ aihim).

Wind's house stands
There it brings me
My heart shines
To and fro it blows (me)
Around me it roars
(It reaches me).

This song, originating from the kingfisher, is about a shrine in the Sierra Estrella, but tells us nothing about the bird itself. George mentioned once, in our discussion of the unidentified bird *kahagam*, that there is a song about the blue *ba'ivchul*, but he must not have known the song itself because he did not sing it. However, he clearly described the antics of this species on the basis of traditional songs, though he himself had never seen the bird; the river went dry here sooner than in the Komaḍk area.

For a bird that is almost never seen today in Pimería Alta, the Belted Kingfisher figures prominently in Piman oral literature. In the Corn Man and Tobacco Woman story, Corn makes a speech (Lloyd 1911:225–226). In it, the great doctor, *siivañĭ*, who lives in the east in the Tondam Va'aki, Shining Ancient House, is speaking:

> And I came to Bives-chool, the kingfisher, and asked him for power.
>
> And Biveschool asked the Bluebird to do some great thing to show his power, and Bluebird took the blue grains of corn from his breast and then planted them, and they grew up into beautiful tall corn, so tall its tops touched the sky and its leaves bowed over and scratched the ground in the wind.
>
> And Biveschool took white seeds from his heart, and planted them, and they came up, and were beautiful to be seen, and came to bear fruit that lay one after another on the vine—these were pumpkins.

From this oration, it would seem that the Pima once knew the bird intimately: the dense white feathers of the breast, helping to cushion the kingfisher's impact on diving, have a texture like the white pithy coating of Pima pumpkin seeds.

The story of Ba'ivchul Uvĭ, Kingfisher Young Woman, is told in the Hayden-Smith version of the Creation Story (Hayden 1935:17–20; see also Bahr and associates 1994:112–122 and Benedict 2001:69–72). Ba'ivchul Uvĭ lives at a Queen Creek settlement nine miles east of Chandler, Arizona. Here her family is associated with deer hunting and wood gathering rather than with a plant-gathering and agricultural subsistence. She is given by her parents to a young man. Although the girl loves another, she obeys her parents. That night, after she and her new husband had lain together, she gets up and comes west to Maas Akimel, to the place of Kadgam, Yellow-billed Cuckoo, whose son she likes. There she marries this one she loves. Then another man called Hevachuḍ, Blue Grosbeak, seduces her, but she soon sees that he is very lazy. Next she goes to a man called Hohokimal, Butterfly. She leaves him, too, and returns eastward up the Gila.

This story superficially seems to be about marital infidelity (indeed, Hayden's deacon translator here heaped on connotations of sin and wickedness, and Bahr titled the story "The Whore"). However, each family that she marries into has a different subsistence ecology in a different environmental setting. The story, it seems to me, is more about the need for plasticity in making a living as marginal agriculturalists in the desert (see Yellow-billed Cuckoo account). Both in the oration cited earlier and in this story, the kingfisher is associated with seeds, crops, and fields. In addition, Culver Cassa points out that the *female* kingfisher, specifically, shows all the colors of O'odham maize: white, blue, and red (rust).

Before we leave the Gila and the Gileños' knowledge of this bird, let me comment on some folk science. Sylvester and probably other older Pimas as well knew the kingfisher as a locally breeding species, not only as a winter visitor. However, conventional western ornithological wisdom holds that the Belted Kingfisher does not breed south as far as Arizona or probably in most of lowland New Mexico (AOU 1957; Bailey 1928; Hubbard 1978; Phillips, Marshall, and Monson 1964:67). The River Pima information helps clarify this bird's clouded former summer status on the middle Gila (see technical note).

I am not certain of the identification of kingfishers among Pima Bajo. The Belted Kingfisher is common along the Río Yaqui throughout the nonbreeding season, and the much smaller (seven- to eight-inch) Green Kingfisher occurs the year round. Pedro Estrella called the Belted Kingfisher *komalk mo'okoi*, apparently "flat head–it-has," and said the Green Kingfisher was the same. The next year he called the kingfisher *vatop muuval*, "fish fly/bee," describing its call and feeding correctly, but saying it was *azulito*, not *verde* (bluish, not green).

COMPARATIVE LINGUISTICS. Mountain Pima Isidro Cruz of the Yepáchic area gave me *baivkuli* or *baykulyi* for kingfisher. (The elevation here is too high for the Green Kingfisher.) Speakers from the Yécora-Maicoba district knew the bird and its name well. The word appears among Northern Tepehuán as *baivukali* (Pennington 1969:114). The description suggests this name refers to the Belted Kingfisher. The word, with a *k/ch* substitution, is cognate with desert Piman. The Névome vocabulary listed *baitituri* as a kind of duck (Pennington 1979:90). If the first *t* is actually an *f*, the word is a prepalatalized form of Upper Piman *ba'ivchul*.

MISIDENTIFICATION. Bahr (in Bahr and associates 1994:113 n. *a*) confuses the name of the girl in the story told in this account. The name Ba'ivchul Uvĭ means simply "Kingfisher Young Woman." Bahr tries to make this name into Ba'i-guklim, "throat-black," the black-throated desert sparrow (*Zonotrichia querula* [*sic!*]). "Smith does *not* seem to have the same bird in mind, and his description baffles me because I think the name literally means 'throat-black.'" A simple solution would have been to ask any elderly Pima singer what the bird is called in English. For more on this confusion, see the Yellow-billed Cuckoo *(kadgam)* account.

TECHNICAL NOTE. Phillips, Marshall, and Monson (1964:67) note about the kingfisher that "there is no good evidence of nesting in the state in the present century. The past status in Arizona is much clouded in doubt."

Notwithstanding this statement, W. E. D. Scott (1886:425), who collected on the Gila from Florence, just east of the Gila River Reservation, to Riverside, near Kelvin, and south along the San Pedro, considered this kingfisher "[a] resident species. Met with at the several points where I collected, but it retires from the mountains in winter." And Bendire (1895:2:35) wrote, "in southern Arizona . . . I have found Kingfishers breeding in localities where fish must have formed but a small percentage of their daily fare." Finally, in 1980 Steve Emslie discovered a nesting attempt on a Verde River tributary and salvaged the young that had died (Monson and Phillips 1981:88–89).

The Pima information helps uncloud the local picture a bit. The anthropogenic changes that have occurred in the Southwest, particularly since the advent of Europeans with their animals, have been extensive and far reaching, the ecological connections sometimes subtle. It is often difficult for us even to imagine former conditions in the desert on the basis of its present state.

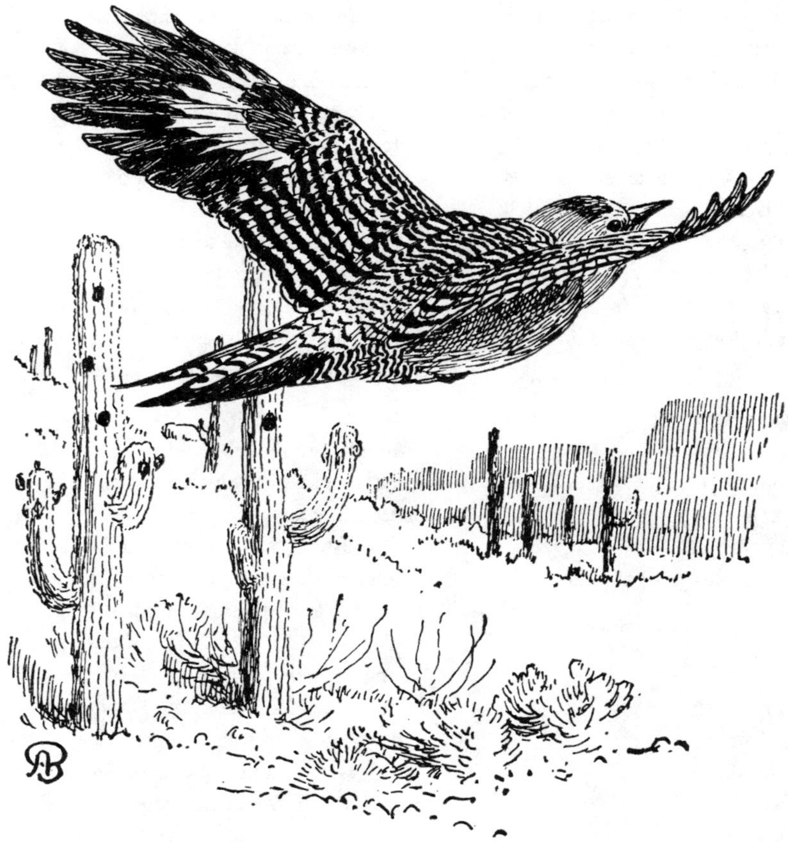

hikvig, hikwig, Gila Woodpecker (Allan Brooks)

hikvig, hikwig
***hiikvig* (Pima Bajo)**

Picidae; Piciformes
Melanerpes uropygialis
Gila Woodpecker; Carpintero Desértico

The antithesis of the retiring Ladder-back, this woodpecker is not shy. Rather, the Gila Woodpecker is one of the most conspicuous birds of the Sonoran Desert, taking to villages, towns, and traditional Indian fields, as well as to natural habitats. It is perhaps most frequent in the Saguaro stands, where it excavates its nest in the succulent flesh of the cactus between the tough epidermis and the ring of tough supporting ribs inside. The cactus responds by sealing the walls of the excavation with woody scar tissue.

This medium-size woodpecker (about nine inches long) has a tan head and body contrasting with black-and-white-barred wings, back, and tail. White wing patches show in flight. Its species name comes from the yellow patch on the lower abdomen. Males sport a distinct red cap.

Even if the bird is not seen, it will be heard. One of its common calls is a *churr-churr-churr*, issued in forced spurts, ringing through the Saguaro forests or cottonwood and mesquite woodlands. The bird drums as well, using a mesquite or ironwood branch for a sounding board.

Like the Gilded Flicker, the Gila Woodpecker contributes to the avian biodiversity of the desert by creating available nest holes for secondary cavity nesters. In the Sacaton area, M. French Gilman (1915b, 1930) found Gila Woodpecker holes hosting Elf Owls, Ferruginous Pygmy-Owls, Ash-throated Flycatchers, Wied's (Brown-crested) Flycatchers *(Myiarchus tyrannulus)*, and Lucy's Warblers *(Vermivora luciae)*. Farther south in Tohono O'odham country one finds the Purple Martin occupying woodpecker holes. Today we can add to this guest list the all-too-abundant European Starling, an aggressive species that crowds out native birds.

The *hikvig* figures importantly in Northern Piman culture. Early in the Creation Story, the four primordial beings (including Ñui, Vulture or Buzzard) make plans to escape the coming Flood. Four small birds are also saved by clinging to the sky. One of them is *hikvig*. The birds are almost drowned by the time the waters stop rising. Delores Lewis explained, "The shape of the [*hikwig* tail] feather is stiff and sharp because during the Flood the water touched it; came up that high." Juan Smith's version (Hayden 1935:6) told at Bapchule says, "Now today the signs you see on the tail of the woodpecker show that it was saved from the Flood. You see the water rose up to a certain height, almost to where the woodpecker was hanging, and it touched his tail and made signs on it." (The Piman word *o'ohaḍag* means "signs," "designs," "markings," or even "writing.")

The song accompanying this event, as recorded by Danny Lopez (1980:125), calls the woodpecker a *maakai*, shaman:

Al wa s-koomagi hikiwigi maakai
suunañi s-eemunia
kia t-nanamai hihiihime
do k al totosagi kia ñ-ab wiwihime.

Little gray Gila Woodpecker medicine man
The water he feared
To the heavens he flew
"This is the water-foam that is on my tail," he said.

Simon Lopez tells this story: "When the Flood came, Hikwig hung on to the sky with some other birds. He hung feet up [horizontally]. The waters started going down. The *toḍshagĭ*, foam, left marks on his back and across his wings. Then he got down. He got cold. He tried to build a fire, but everything was wet. So he borrowed a *kuudagĭ* [firebrand] from someone [one of the other seven saved]. He took it, put it on his head. And now he has red on his head."

After the Flood, the survivors had the task of locating the *hik*, the navel of the world—that is, the center where the People were to live. In most versions, Hummingbird and Gila Woodpecker are assigned to fly to the edges of the world, east and west, and to return. Then each bird switches directions and repeats the flight. After four flights, the *hik* is located (Benedict 2001:17; Herzog [1929]:23; Lloyd 1911:45–46; Saxton and Saxton 1973:57–58).

At the first wine feast, an important rainmaking celebration for the monsoon season, there were fights, and some birds participating in the feast ended up bloodied or stomped on. Delores Lewis explained, "Also, old people say that when they first made *nawait*, these *hikwig* didn't have any red on their heads. But these got into a fight and someone hit him on the head." (See also Wright 1929:120.) Several species of birds, preeminently the *hikvig*, reportedly became *komalk mo'okam*, "flat headed," because of these fights. Leon Juan of Aangam said that at this first wine feast *hikiwik* dressed up for the occasion. He poured water or something on himself and let it trickle down to form the designs now seen on his plumage.

In the Eagle Man story, a disreputable young gambler is given special pinole to drink that makes him break out with feathers and become a giant eagle. This Eagle Man begins to prey on people after he eats all the local game (see Golden Eagle account). In several versions of the story, the gambler is called Vanḍai. In Shaw's (1968:21), he is named Hikvik, but the reason for this name is unclear.

In another episode of the Creation Story, when Corn Man abandons the People, he takes with him his pets, the blackbirds and Gila Woodpeckers. "That's why these two birds are always in our cornfields."

The ubiquitous Gila Woodpecker can cause *kaachim mumkidag*, staying sickness, in O'odham. Densmore (1929:90) listed simply "woodpecker" twice as a cause of staying sickness. In the first listing, woodpecker appears with "Blue Hawk"—that is, falcon, *visag*, and eagle, *ba'ag*, with the notation "symptoms not given . . . but the same" for the three birds. Then "woodpecker" is listed by itself as causing "boils and sores on the head." In the latter case, "Gila Woodpecker" is implied because it has a round crown patch. But how can one abuse this bird, which seems on such familiar terms with the O'odham? Like other animal sicknesses, *hikvig mumkidag* can be caused either by offending the animal itself directly or by bothering something that it has claimed as its own, much as one must avoid tubers stored by the gopher or berries gathered by a bear.

George Kyyitan explained that there were two categories of wood that were tabooed: that struck by lightning and that with a woodpecker nest. If a *hikvig* hole is found in wood, that wood must be avoided: "That's what I heard. Because we're not allowed to take that wood to burn it or anything. Just put it away [aside]. Not to build for a fire. Like, if you're looking for a wood, fire wood, and you find that one on there, you have to cut, and throw it away. Not to use it. Then you can cut the rest of it."

"Is that just if the woodpecker makes little holes or big nest hole?" I asked.

"Same way, anything, because it's been built that way, by that *hikvig*."

I wondered about other woodpeckers that are named in the Piman lexicon. "Just the *hikvig* or would that include the *kuḍat* [flicker]?"

"Yeah, I think it is, because it goes with that, together with that." He associated the two woodpeckers even though they have different Piman names.

"How about that little *chehegam* [Ladder-backed Woodpecker]?"

"Oh, yeah. I don't know. I don't know that. I haven't seen it [its nest]. But the *kuḍat* and the *hikvig* make holes in there, goes in there, sleeps in there."

"The *chehegam* makes holes like that, too."

"Yeah, I think it does."

"Sometimes the *kuḍat* or the *hikvig* has that nest in the *haashañ* [Saguaro]. When you're getting *vaapai* [Saguaro ribs], do you have to do anything with that or not?"

"Yeah, got to cut it off. If you can use that rib, you can cut that off and leave the nest there, *kosh*, nest." So any wood that had been claimed by the Gila Woodpecker, either by pecking into it or by making a nest hole in it, must be avoided, just the same as a tree that has been struck by lightning, although the consequences are not the same. Lightning-struck wood produces blisters on the hands where touched and might burn down one's house if used for fuel, George explained; woodpecker-claimed wood produces just staying sickness.

The same restrictions apply among Tohono O'odham. Frank Jim was speaking of trees that had been hit by lightning. "No matter how old, how big you are, they still get sick with it. That's one thing that you [always] have to be careful of."

"Like cutting wood, if that tree got hit by lighting?" I asked.

"Can't take it. Can't build a fire with it. Never do. It's just the same with . . . if you go out there and get some wood, if you see the woodpecker being in there, you cannot use that. Don't touch it. People getting wood aren't supposed to burn it. Or the smoke will get into them and get you sick. So they have to cut that piece of it off and take the rest, throw the other one away. You cannot burn it. Anything. Like *vaapai*, if you're going to use it on your *kii* [house] or *vatto* [arbor], you've got to cut that off where you could see the hole, in the *vaapai*. You've got to cut that off and just throw it away. You cannot use it."

"But don't you have some *haashañ shuud* ['Saguaro's vagina,' woodpecker nest] out in your shed, Frank?"

"It doesn't matter because we don't burn it, just hang them up. They use that to get a *juuñ* [dried Saguaro fruit], put it in there; so it don't matter about that, because the *hikvik* don't bother it, inside. It's just that if you burn it." He was not sure what the symptoms of the sickness were.

At Santa Rosa village, I asked Tohono O'odham Simon Lopez how the *hikwig* would affect someone.

"It's the same thing. Like they say, if you kill *hikwig* or treat him mean and he'll get to you. And it's the same way, like you said, if you see a wood that the *hikwig* did it and you cut it for fire wood or something like that, it'll get you, too. There's [one] right there, we hear it all the time from the sides [of the house], on the attic. And then I go out and then he'd take off. Then I saw he started to make holes on it. And my dad was alive. And I told him, 'That *hikwig* is starting to make holes here and there.' And he said, "Don't bother it, 'cause you know, he's going to get you, *tto mumkadaj*." And I told him, 'Well, if I just get a [piece of] metal thing and put it there?' 'Yeah, you can do that. Just leave it alone. Just do that. Or just try to shoo the *hikwig* away and maybe he won't come back.' So I tried to do that. Well, I know if you kill him or hit him with a rock, I know that things like that will happen [to you]."

"I heard that even that nest in a tree, if you're cutting it down, you have to leave it alone."

"Yeah."

"What about if it's in the *haashañ?* Because they make that *haashañ shuud*. If you're getting *vaapai*, and you find one of those, is it okay to pick that up?"

"I don't think so. I think it will be the same as—just leave it alone. Just forget it. Don't, because it's from that, *hikwig*."

"And if it's from the *kuḍat* [flicker], is that okay? The *kuḍat* is the bigger one."

"Yeah, that's okay. I think that's okay, *kuḍat*."

In the adjacent village Aangam, Leon Juan said that the nest was used as a fetish in the curing of *hikwig mumkidag*.

Farther south in Santa Cruz village Jose Garcia looked at a picture of the Gila Woodpecker in the field guide. "Don't bother the *hikwig*'s nest. Don't cut it, don't burn it, don't use it. It will make a sore that doesn't heal. Never heal unless you know the song and *wuso* and use [as a fetish] anything from the *hikwig*, part of the nest or whatever."

The prohibition extended even to the Hiá ch-eḍ O'odham (Sand Papago). Delfina Cameron said, "Don't bother the *hikvik* or you'll get sick."

Densmore (1929:199) obtained three songs sung at the Wind Dance concluding the lustration (purification ceremony for enemy-killers). One of these songs is about a "woodpecker," but there is no way of knowing which of the three named folk generics is intended because only the English is given. The song says:

Who is that bird flying over there?
Oh, it is the woodpecker,
He does not seem to want to go anywhere,
He cries and pecks at a cactus.

Don Thomas of Komaḍk had a group of singers and dancers called Hikvig Eḍa Mel, Woodpecker Inside Run. The *eḍa mel*, "inside run" or "end run," is a ceremonial dance. The songs used by this group were supposed to have originated from this bird, who taught them to the singer.

Clemencia Leos of Haashañ Keek (Sacaton Flats) related that a cloth of different colors was called *hikvik so'oi*, "woodpecker plaid." The word *so'oi* can mean anything with markings, even writing. This particular usage derives from the intricate black-and-white pattern of this bird's back and wings.

In the early twentieth century, Gilman (1915b) noted an interesting commensal relationship between the Gila Pima and the Gila Woodpecker: "The Indians store corn in the ear on the flat tops of their houses and sheds, and each home has one or more of woodpecker retainers or pensioners hanging about most of the time. This corn provides an abundant and sure source of food, and the birds make the most of it. They seem to be allotted on the ratio of a pair of birds to a home, and it is but rarely that more than two are seen at the same corncrib." There is no indication that the Pima took any measures to deter the birds. At first, this seemed puzzling to me. But when seen in the light of Piman cosmology and beliefs about sickness, it is understandable. As Simon's father told him, "Just leave it alone."

Far to the south on the Río Yaqui, Pedro Estrella said that the *hikvig* doesn't do much damage to the corn crop, just picks a little. This seems to reflect the attitude of his more northerly relatives.

COMPARATIVE LINGUISTICS. On the Gila River Reservation, Fr. Antonine (1935:9) wrote the name *hikwikhi* (pl. *hi'ikwikhi*). I have heard both *hikvig* and *hikwig* here. At Vaya Chiñ, Joe Lopez, originally of Poso Verde, Sonora, used the variant *hikwich*. For desert Pima Bajo, Ken Hale (n.d.a) wrote *hikvig*. María Córdova said *hiikvig*, and I recorded both *hiikvik* and *hiikwik* from Pedro Estrella. I did not find this word among Mountain Pima. For the latter, the common Acorn Woodpecker (*Melanerpes formicivorus*) is *chiakalyi* or *shiakli* or even *pishkalyi*.

MISIDENTIFICATIONS. Russell (1908:211) called the *hikvig* a flicker. Saxton and Saxton (1969:166) identify the *hikiwij* (pl. *hikikiwij*) as a Hairy Woodpecker (*Dendrocopus villosus*), a species of the coniferous forests high above O'odham country. In a subsequent lexicon, *hikiwigi* is correctly identified as the Gila Woodpecker (Saxton, Saxton, and Enos 1983:21).

chehegam
***tehegam* (Pima Bajo)**

Picidae; Piciformes
Dendrocopos scalaris
Ladder-backed Woodpecker; Carpinterito

A tiny woodpecker, scarcely seven inches long, works its way slowly through a mesquite bosque or along the paloverdes and ironwoods lining an arroyo, occasionally uttering a short subdued *wik* or *pik* note. Its back and wings are black, boldly striped and spotted with white. The pale underparts, mostly black speckled, are suffused with a dirty smoke color, as if soiled from the surfaces it works over so intently; the color is genetic.

Males have a considerable amount of red from the crown to the nape, this area being black in the females. Paired individuals are usually not far apart, even in the nonbreeding season, keeping track of each other with an occasional whinny note when one moves to a new location. The birds are confiding, coming in close to O'odham houses, peeking cautiously from behind the small branch it is scrutinizing.

Nest holes may be excavated in such hard woods as mesquites or the exotic Athel Tamarisks so common in Piman country. In the Sierra Estrella foothills above Santa Cruz village, I found one

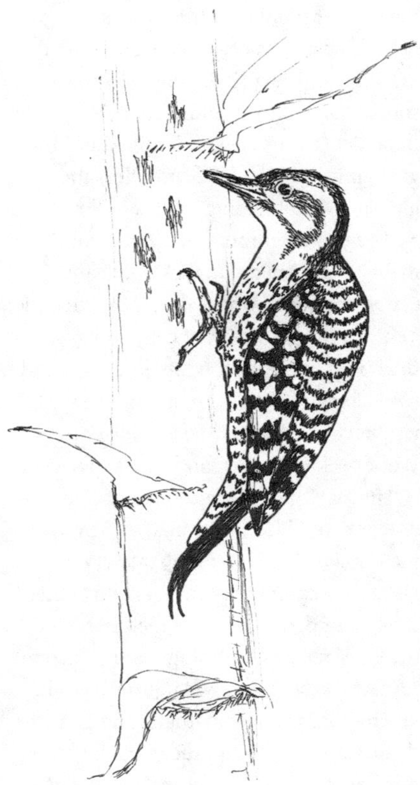

chehegam, Ladder-backed Woodpecker (Amadeo M. Rea)

chiseled into a Desert Ironwood branch, an astonishing feat. Earlier in Gila Pima country, the indefatigable M. French Gilman (1915b) found nest holes in mesquites, Screwbean, ironwood, cottonwood, willow, paloverde, and Chain-fruit Cholla. Openings averaged about an inch and a half. Had he searched the Sierra Estrella, he probably would have found them in Desert Agave stalks as well.

Many older people throughout Northern Piman country knew this diminutive bird and identified it for me in the field. David Brown of Santa Cruz said that a pair nested every year in the churchyard where he and his wife, Rosita, were caretakers. "It's like a *hikvig* [Gila Woodpecker] only smaller—just *little*." He showed me the small nest hole excavated in an Athel Tamarisk stump.

Irene Hendricks pointed out that the *chehegam* "is not too big, red head," but the *hikvigĭ* has "just a little red on the head."

Juan Dolores (1936:45), explaining Papago nicknames, noted that the name *chehegam* was given to someone who is a gossip. He wrote, "A kind of woodpecker which, when it strikes a tree once, makes a note or two and then flits around to the other side. This bird is said to have no especial purpose in pecking at the tree, but it pretends to be doing something while it watches and hides behind the tree. The name is applied to a person who tells everything he or she hears."

At Rancho Palmar in Pima Bajo country, a Ladder-back flew into the mesquite don Pedro and I were sitting near. I asked about it and was not sure he saw the little bird until he answered, *"Tehegam, u'us tupuidam."* I think he was describing how it taps on trees or wood *(u'us)*. This is the same name as that used farther north, where *te* has been palatalized to *che* in post-Contact times. Pedro said the *tehegam* was not a kind of *hiikvig*, a name he used exclusively for the Gila Woodpecker.

At Quitovac oasis, also in Sonora, Hiá ch-eḍ O'odham Luciano Noriega said he did not consider *chehegam* to be a type of *hikvig*, a name he, too, reserved for the Gila Woodpecker. However, speakers tended to associate in discussions these three birds: *hikvig, chehegam*, and *kuḍat* (flicker).

Underhill (1938:46, 1946:81) was given a corn-growing song (number 24) in which Corn speaks:

> The little striped woodpecker
> Descends right down into my heart.

Two types of striped woodpeckers frequent O'odham cornfields, eating worms from the ends of the ears: the Gila and the Ladder-back. Without the original words, we cannot be sure of which bird this song is about, but the stipulation that it is little suggests *chehegam*, one of the smallest woodpeckers known.

Some episodes in the Piman Creation Story also include an account of White-clay Eater Woman and her twin sons (Benedict 2001:97–122; Densmore 1929:54–77; Herzog [1929]). This story, notes Culver Cassa, seems to be of Yuman origin. After the twins' quest for reeds, the boys' mother makes flutes, which they use to seduce girls. While out bathing, the two daughters of Yellow Buzzard hear the beautiful music and begin searching for its source, encountering various animal-men along the way. Each one first offers the girls his characteristic food to cook, which they refuse, then auditions his own music, which he assures the girls is what they had heard.

In Herzog's version of this account, one of the contenders is Chehegam, Ladder-backed Woodpecker. His food is not specified, but he drums on a tree: "I don't sound it here, I sound it right over there, on the tree that stands there. And he jumped on it and clung up high, he lay against the tree. He sang indeed; singing he circled around the tree. He sounded a little like a flute" (Herzog [1929]:73). Upstaging the queen mother, the two girls said in essence, We are not amused.

Incomplete Identification. In one version of the Coyote brothers story, the younger brother, who is always down in his luck, meets a girl with a pet woodpecker who awakens her in the morning with his drumming (Benedict 2001:79–80; see Blackbird account for details). We are not told which of the three woodpeckers that the Pima named is intended here, but all three species drum to advertise their territory.

kuḍat, kuḍad
dudgat (Pima Bajo)

Picidae; Piciformes

Colaptes auratus

Northern Flicker (Red-shafted Flicker, Gilded Flicker); Carpentero Collarejo Alirrojo y Desértico

The Northern Flicker is the avian ecological equivalent of the horned lizard: both are adapted primarily for eating terrestrial ants. Unlike most others of its tribe, this large woodpecker is predominantly brown above, barred with black, making it difficult to see as it feeds on the ground. Here a flicker may probe into the soil, causing its prey to swarm out. Further adaptations include a long but relatively weak bill and a long exsertable tongue aided by large salivary glands that produce a viscid secretion—all aids in ant eating.

Once the bird takes wing, it is far more conspicuous, flashing a white rump and bright underwing linings. In the desert, two forms (subspecies groups) are easily distinguished. The form breeding throughout the cactus forests of Northern Piman country is the Mearns' or Gilded Flicker. It is paler brown, with deep yellow under the wings and basal two-thirds of the tail, and a thick black patch on the chest. The Red-shafted Flicker is variably darker brown, with salmon wing and tail linings, black

kuḍat, Northern Flicker (Allan Brooks)

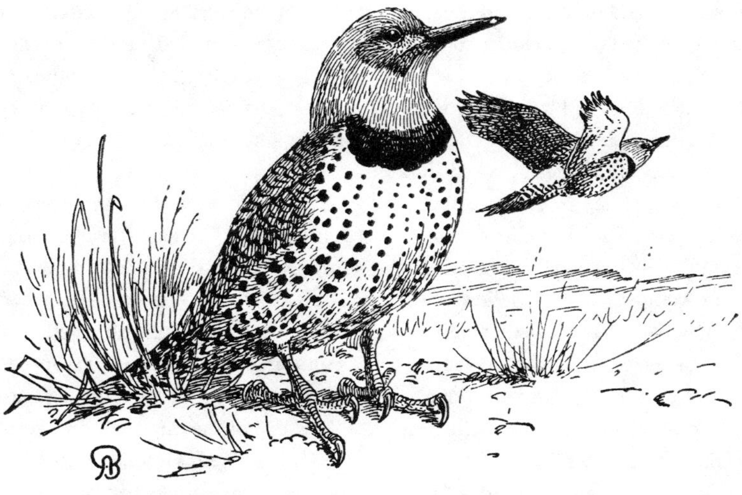

tips to the tail feathers, and a black crescent on the chest. It breeds throughout the open pine country of the West, descending to the desert during winter, when it far outnumbers the smaller Gilded Flicker.

The voice of the Northern Flicker is a loud ringing *kiiiiu* and a somewhat more subdued *wika-wika-wika* call, supposedly the source of the bird's English name. The Gilded Flicker excavates nest holes in living Saguaros or in dead branches of willows and cottonwoods that once lined the desert rivers. Nest openings range from two and three-quarters to four and three-quarters inches across, whereas those of the Gila Woodpecker (which may be in the same Saguaro with the flicker) average around two inches (Gilman 1915b). Flicker nest holes in cactus may be appropriated by kestrels, screech-owls, and several smaller birds.

Although the status of the Red-shafted Flicker as a winter visitor, arriving in the fall and departing in the spring, must have been known to the O'odham, everyone told me the same thing: *kuḍat* refers to all flickers, no linguistic distinction being made between this larger, darker form and the resident Gilded Flickers.

Discussing this bird, Joe Giff sang this children's song:

> Kuḍat ab pi oi jivia
> oi pi emiishmad
> oi pi ejioshmad.

Here's how he explained the song: "*Ejioshmad* [means] when you go to church, you become sanctified, I guess, or something like that—a little godliness [literally]. *Kuḍat* came late to *miishmad* [mass] and didn't get sanctified."

Sylvester noted that the *kuḍat* makes a nest hole in the *auppa*, cottonwood, and it is used by the *kuukvul*, Western Screech-Owl.

In most renditions of the Creation Story, four birds are saved from the Flood by flying up to the sky and holding on to it with their beaks. But in the version recorded by Herzog ([1929]:14, 19), Kuḍat is saved, along with Gila Woodpecker, Hummingbird, Verdin, and Curve-billed Thrasher. The white on the tail of the woodpecker shows that the foam of the floodwaters reached that high on the clinging birds. But this would explain the white at the base of the flicker's tail as well. Later in Herzog's story, as the floodwaters recede, it is *hikvig* not *kuḍat* who is paired with *vipismal* (hummingbird) to measure the earth to locate the *hik*, navel, of the world. And still later in the story, Herzog ([1929]:86) confuses the flicker with the Cuve-billed Thrasher, *kuḍvich*.

At Sacaton, Leslie Pasqual also knew the *kuḍat* as the bird that was saved from the Flood. The tips of its tail feathers were in the water as it clung to the heavens, and that's why they are curved that way today. He called the bird "flicker" in English and knew it as the one with red underwings.

In the story of Nasia, or White-clay Eater Woman, and her twin sons, the boys seek reeds that their mother makes into flutes, which for the O'odham are powerful love charms. Two girls bathing at Blackwater pond hear the sounds and go off to discover its origin. Various animal-men waylay the girls, claiming to be the source. One of these individuals, whom Benedict (2001:97–122) identified in English only as "Woodpecker," offers the girls ants to eat, then goes to a hollow tree and begins drumming. The Pima name must have been *kuḍat*, flicker, the ant-eating specialist, whose dorsal brown and black coloration so cryptically conceals the bird while it is foraging on the ground. (In Herzog's [(1929)] rendering, the bird is given as *chehegam*, the Ladder-backed Woodpecker, but his food is not specified.)

Ultimately, the twin boys are killed by one or several falcons. Kuḍat is the boys' maternal uncle (Herzog [1929]:86, 86b). He comes upon the slain boys and bathes himself in their blood. He then goes to the boys' mother, claiming that he tried to protect the boys and was injured as well. The mother is unsympathetic and in anger throws dirt on fleeing Kuḍat, cursing him by saying that he will eat *che'ul vatopaḍk*, willow caterpillars. Flicker then leaves, afterward being seen eating ants at the base of a willow. He is still covered above with dirt, and he still forages for ants. Blood still covers the undersides of his wings and tail.

In yet another version, after White-clay Eater Woman throws the dirt on him and curses him to wander, "The bird flew a little way and sat down on a tree to rest. The ants came and crawled over his feathers and one crawled into his mouth. He swallowed it and it was sweet. He ate all the ants he could find and they bit him inside and he died" (Benedict 2001:115).

One might easily overlook an important bit of folk science worked into the legend at this point (Herzog [1929]:86, 86b). In addition to the dietary condemnation, White-clay Eater Woman says here: "You shall go and you shall roam about, somewhere, crazed!" The bird in this episode is specifically the Red-shafted Flicker, which has been dyed red in blood. Unlike the yellow Gilded Flicker, a year-round resident in the desert country, the Red-shafted bird is migratory, departing for the breeding season, returning in autumn to the lowlands. Because of his failure to protect his younger nephews, then lying about it, he is condemned to this annual wandering.

A character called Kuḍat makes a cameo appearance in one version of the Ho'ok episode (Curtis 1908:18). Here Kuḍat is the name of the boy who kicks the wooden ball near a maiden, who teasingly hides it under her dress. When she tries to retrieve the ball, she cannot find it; it has impregnated her. She becomes the mother of Ho'ok, the cannibal witch. In another version, the ball is made from the ashes of a cremated *aaḍo*, Scarlet Macaw. If Kuḍat is a wintering flicker, the color red would seem to be significant, as in the parrot case. As the story is told in Cheedagĭ Wahia (Poso Verde, Sonora), by Ascencio Antone, the ball is a ball of red dust (Fontana 1981:26–29). The red ball belongs to Sun. Bright red may be associated with rain (something good) or with blood and doom. The conception of Ho'ok brought bloody devestation. If there is a significance here, Culver Cassa believes it is an archaic one. In other versions, it is just a boy who kicks the wooden ball.

As noted in the Gila Woodpecker account, use of any wood appropriated by that bird, either by pecking or for nest excavation, is strictly tabooed. But I wondered if this taboo also applied to the *kuḍat*. Both species excavate similar nest holes in the Saguaros. Gilded Flickers nest higher, in the central part of the cactus. Because this part tends to be thinner, the flicker cuts through the Saguaro ribs into the center (McAuliffe and Hendricks 1988). The cactus forms a protective callous on the exposed soft tissue, producing a bootlike structure that survives after the Saguaro dies. There are references to O'odham using these "desert moccasins" for storing items, such as tobacco. Simon Lopez said he thought it was okay to get the *haashañ shuud*, "Saguaro's vagina," from the *kuḍat*, "but if it's from the *hikwig*, just leave it alone."

Both size and shape differences would have allowed O'odham to distinguish the nests of these two birds. In the story of Corn Man and Tobacco Woman, after Giisobĭ (Verdin) Shaman reacquires tobacco, he "gathered a lot of the leaves and filled with them one of the gourd-like nests which the woodpecker, *koh-daht*, makes in the *har-san* or giant-cactus" (Lloyd 1911:220). Thin Leather, the Pima storyteller, leaves no room for ambiguity here.

COMPARATIVE LINGUISTICS. Pima Bajo Pedro Estrella called the flicker *dudgat*, cognate with the Pima-Papago name, though it does not follow any predictable pattern. Mountain Pima and Northern and Southern Tepehuanes live in country where Red-shafted Flickers are common. Mountain Pima told me both *kurta* and *pishkali*. Pennington's (1969:117) *kúratu* and *vepégi kúratu* woodpeckers among Northern Tepehuán are likely flickers. One of the carpentero names that Brambila (1983:113) recorded for Tarahumara, *koratza*, may also be the flicker.

MISIDENTIFICATIONS. In quoting from the Corn Man and Tobacco Woman story as told in Lloyd (1911)—"When the tobacco was ripe, Geeheesop [Giisobĭ] gathered a lot of the leaves and filled with them one of the gourd-like nests which the woodpecker, *Koh-daht*, makes in the *har-san* or giant cactus"—Bahr (in Bahr and associates 1994:98n.) says, "Might be *hokkaḍ*, 'cactus-wren' *(Heleodytes brunneicapillus)*." *Kuḍat*, the flicker, is a kind of woodpecker, as Lloyd originally had written it, not a Cactus Wren. There were cultural reasons for specifying which woodpecker.

Herzog ([1929]) misidentified the *kuḍat*, in this context the Red-shafted Flicker, as a thrasher. Benedict (2001:114n.) wrote "Kotat (a woodpecker)," and editor Bahr added "not further identified." (See Rea 1983:178.)

TECHNICAL NOTE. The various subspecies of Red-shafted Flickers, long known as the species *Colaptes cafer*, interbreed freely with the Yellow-shafted Flicker, *C. auratus*, wherever their ranges meet. Intergrades of the two, with various combinations of characters, can be found on migration throughout the West. Although opportunities for interbreeding are less extensive between Red-shafted Flickers and the various subspecies of Gilded Flicker, long known as *C. chrysoides*, they, too, interbreed freely wherever their ranges meet, as in riparian woodlands of the Southwest that link the cactus stands of the desert with the pine-oak woodlands of the uplands. Intergrades of various combinations can be found here. With the riparian destruction throughout the Southwest in the past century, opportunities for gene flow are becoming more restricted. After many decades of hesitation, the AOU (AOU Check-list Committee 1973) at last recognized what the birds themselves already knew: that they are all the same species, much as humans from different continents, in spite of observable physical differences, are the same species. All Northern Flickers take the oldest name—that is, *C. auratus*—with the various subspecies within the groups given a trinomial. The AOU (AOU Check-list Committee 1995) again waffles, but the birds themselves ignore pronouncements and interbreed wherever and whenever they possibly can.

hevel moos

Tyrannidae, Passeriformes

Sayornis saya

Say's Phoebe; Mosquero Llanero, Papamoscas Boyero

A medium-size (seven-and-one-half-inch) flycatcher flies buoyantly up to a perch—a post, dead tree, or rock—affording it good visibility from which to forage. On landing, it pumps and slightly and quickly spreads its dark tail. The bird's upper parts are soft gray-tan, its chest paler, and the rest of the underparts pale rust. There are no conspicuous markings. The bird is a Say's Phoebe, a flycatcher inhabiting the open arid areas throughout western North America.

Say's Phoebes maneuver in the wind as if weightless, capturing insects on the wing. Overall, they are rather quiet. They occasionally whistle a plaintive *pheee-uuur* call, accented like a Pima word on the first syllable. The male's courtship song is a fast *pit-tsée-ar*, often given in fluttering flight (Hoffmann 1927:202).

Say's Phoebes will throw in their lot with humans, nesting in wells, caves, mine shafts, deserted buildings, outbuildings, and even occupied Pima houses where afforded a minimum of protection. For years, a pair occupied Rosita and David Brown's yard in Santa Cruz village, placing their nest in the low church tower nearby. When I began following the little flycatcher's breeding patterns here, the Browns told me its Pima name, *hevel moos*, "Wind's grandchild." One could hardly dream up a more appropriate epithet.

In this open country of sparse but violent storms, it is not surprising that the elements are personified in O'odham narrations. Both Thunder and Lightning are identified in myth as masculine figures. So is Cloud. But the name for this flycatcher tells us that Wind—that all-important figure in the desert—is feminine. The kinship term *moos* indicates a granddaughter or grandson from a maternal grand-

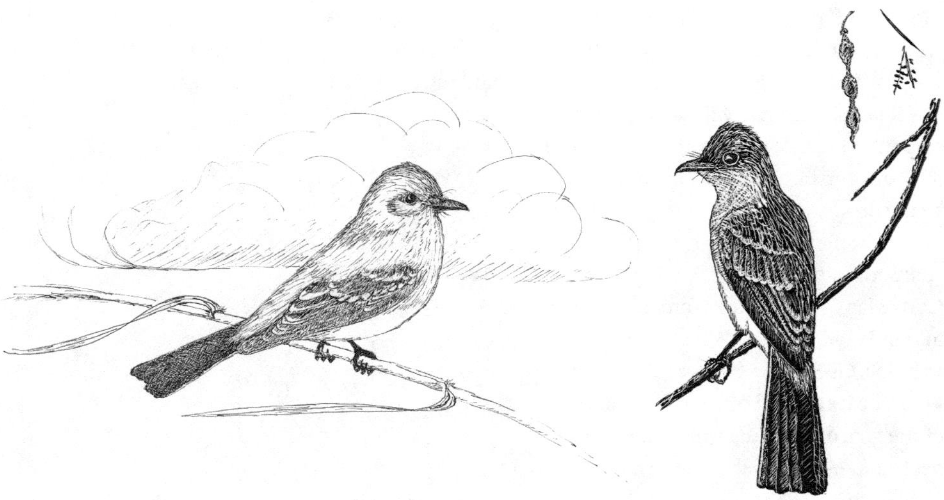

hevel moos, Say's Phoebe (Amadeo M. Rea) and Ash-throated Flycatcher (Amadeo M. Rea)

mother—that is, a grandmother's daughter's child.

I was surprised at how well known such a small and relatively quiet and inconspicuous bird was among both Akimel and Tohono O'odham, but nearly all the older people, in particular cowboys, recognized and named it when I pointed it out in the field. Far to the south, at Quitovac oasis, Luciano Noriega called the Say's Phoebe *hewel moos*, which he translated (not quite precisely) as "abuela de viento," wind's grandmother.

At Sacaton, I mentioned a bird that nests in a well to Francis Vavages. "They call it *hevel moos*," he responded. "It's a gray bird, sort of gray brown. He has a nest in the well. When they come over to get the water, he will fly out." As we sat outside talking at Leslie Pasqual's one evening, a Say's Phoebe came in to roost under the eves. Although Leslie didn't know the bird itself, he did know the name *hevel moos*, and its penchant for nesting in the old wells. He pointed to the location of a former well not far from the Gila where the water level reached to fifteen feet. Today the water table here, close to the Gila, starts at six hundred feet.

In her study of Papago music, Densmore (1929:198) collected some songs that were used during the lustration, purification ceremony, after an Apache-killer returned to his home village. Near the end of this sixteen-day exclusion, the Wind Dance was performed. She noted, "This song was said to be concerning the 'wind-grandson-bird.'" She labeled it, "A whirlwind is singing":

> Where is that bird going?
> Where is that bird going?
> With that bird I whirl around,
> whirl around higher and higher,
> Just before we touch the sky,
> the bird faints and falls to earth.

There is a hint here that the song might be describing the nuptial flight of a male Vermilion Flycatcher *(Pyrocephalus rubinus)* or a Horned Lark, but little seems to have been written on Say's Phoebe courtship.

Among O'odham, there are apparently several different birds referred to as the *komalk mo'okam*, "flat-headed (bird)." The ascription is usually to some bird that got drunk during the first summer wine feast and was not careful, getting his head stepped on. I have heard this said of a woodpecker and a wren as well as of a kingbird. For Rosita, this flat-headed bird is the *hevel moos (maḍ)*, or Say's Phoebe. "Knows good Saguaro fruit harvest coming—got drunk ahead of time." David also used this designation for the phoebe.

For at least some Tohono O'odham, *hewel moos* is the Ash-throated Flycatcher, Copetón Cenizo, *Myiarchus cinerascens*. This very widespread western bird is somewhat larger than the Say's Phoebe, gray-brown above, with pale yellow underparts except for its gray throat and chest. It shows bright rufous in the spread wings and tail. Like the phoebe, the Ash-throat has a buoyant flight and a confiding nature that often brings it close to human habitations. Tohono O'odham Jose Garcia of Santa Cruz said that *hevel moos* has a yellow belly, and he picked out the Ash-throat in the North American field guide, not the rusty-bellied Say's Phoebe. He said *hevel moos* is never to be molested for fear that violent desert winds will blow away your property, even your house. Larry Saraficio agreed that you should never mess with this bird, or you will suffer the consequences from Hevel, a powerful force.

Unfortunately, I never discovered how extensively the Ash-throat is the referent for *hevel moos* away from Akimel O'odham country. (Gila River people seem to have no name for *Myiarchus*.) However, far to the south on the Río Yaqui, don Pedro Estrella identified my specimens of a similar-size tyrannid, the Nutting's Flycatcher *(Myiarchus nuttingi)*, as *komosh mo'okam*, meaning "flat head–it-has," lending credence to Rosita and David's ascription of the similar name to the Say's Phoebe. In addition to the phoebe, there are three *Myiarchus* flycatchers in the local avifauna around Ónavas, but I obtained no contrast set of these four to work out the proper domain of Pedro's name.

MISIDENTIFICATION. Saxton, Saxton, and Enos (1983:21) give *hewel moos* as the name for a much larger flycatcher, the Western Kingbird. I don't know whether this is a legitimate local usage in some Papago dialect or just another misidentification. The kingbird has a different name among people I worked with (see the next species account).

chukukmal

Tyrannidae, Passeriformes
Tyrannus verticalis
Western Kingbird; Madrugador Avispero, Tirano Pálido

Kingbirds are large flycatchers (about eight inches long) that perch conspicuously in open country, calling in loud voices. The Western Kingbird, found throughout the warm season in Pima country, is partial to riparian woodlands, agricultural lands with utility poles or cottonwoods, mesquite steppe with open grasslands, and villages where taller trees such as planted eucalyptus provide foraging perches. It seldom is found in Papago country.

Western Kingbirds are large-headed birds with conspicuously yellow underparts that fade into a light gray chest and white throat. Above the head is gray, the back olive, and the tail black with narrow white outer margins. Their rather heavy bills aid their taking flying insects. In O'odham fields, kingbirds are particularly fond of yellow and white butterflies that they capture with skillful maneuvers. Here they may be abundant if there are nest sites.

Kingbirds are aptly named. Nothing shy or retiring about kingbirds. They are the avian tyrants of their territory, boldly and noisily chasing hawks, ravens, cats, and other members of their own kind from the vicinity of their nests. I have, however, seen them set up housekeeping in the same tree with a pair of Red-tailed Hawks, apparently having worked out the domestic details. Another reason for the name is the red crowns of adult males, but this color is usually concealed except when the birds are excited.

Pimans could scarcely overlook this bird, and it is one of two flycatchers formally named in their lexicon, so far as we now know. A quarter century ago I recorded the following conversation with a group of older River Pimas as we looked at a collection of local bird skins.

David Brown: "What about a *chukukmal?* Do you have one here?"

Sylvester Matthias: "It's in the same family [as the orioles we were talking about]. It nests in the fields; just like a meadowlark."

Ruth Giff: "Aren't they gray with a lot of white?"

Rosita Brown: "No, they've got a yellow part."

David: "Yellow breast."

Rosita: "And a black head. It's yellowish, yellowish gray."

Ruth: "And they're on the wires?"

Rosita: "They nest up on the wires?"

Sylvester: "It mostly nests, too, up in a cottonwood."

Amadeo: "On the transformers. [They all say the bird's name in Pima.] They're real lively, and they chase *every*thing away."

Ruth: "They chase cats!"

Rosita: "They go after little birds."

Sylvester: "They go after butterflies. They go after it."

Amadeo: "They come back in late March or April."

Sylvester: "*About*, yeah. It's common about that time."

Amadeo: "They leave . . ."

Sylvester: ". . . in late fall."

Ruth: "I didn't know they ever leave. I thought they were here all the time."

Sylvester: "Naw. Hu'uh."

David: "Yeah, they leave. They go someplace."

On another occasion, as we sat outside at Bapchule on a hot June afternoon, a Western Kingbird was eating George Kyyitan's mulberries near us. He knew the name *chukukmal*. "Did you ever hunt that bird?" I asked. "No," he said.

Later that summer Sylvester told me, "*Chukukmal* is the size of a *tosiv* [meadowlark] but hangs around cottonwoods. It fights buzzards!" The disparity in size and the smaller bird's audacity he found most amusing.

Upriver, Leslie Pasqual used the name *chuchukmal* to refer to some kingbirds calling from a dead Athel Tamarisk over his corral. "They have a song—long time ago. I don't know how it goes, but there's a guy by that name, who sings over at Salt River. I guess the white man calls that the Arkansas Kingbird." This is one of the old names for the Western Kingbird.

The association of *chukukmal* with *hohokimalĭ*, butterflies, was well known. Wright (1929:217–222) recorded an O'odham story explaining the origin of the butterflies. One autumn I'itoi, watching children at play and wishing to make something that never grew old, gathers into his bag yellow leaves of the cottonwood, the gray shadows under the leaves, some sunbeams that filter through to the ground, some brown leaves that had fallen, even some tiny white flowers. Needing more yellow, he reaches for more cottonwood leaves above, grasping some bits of blue sky as well. All these he puts into his bag and with the children goes to sleep under the tree, with his bag as a pillow.

When he awakes, he shakes the bag, and out flies the yellow and gray and brown *hohokimalĭ*, with some small blue and white ones as well. The children, singing a song, chase after these forever young creatures. The birds sitting in the tree are jealous of the bright colors of the butterflies, but then everyone notices that these newly made creatures have no song: when he put the leaves in his bag, I'itoi had gone to sleep with the children and hadn't given them a song.

This origin tale has several interesting implications. The story as recorded here doesn't tell what kind of birds are in the *auppa*, cottonwood, but given the association of kingbirds with butterflies, their major food, it would seem that the birds are *chukukmal*. (Kingbirds are yellow, gray, and olive, the colors I'itoi selected.) The narration also says that I'itoi always carries a bag, with which he makes things. This detail suggests a possible association with the Kokopelli figure, the hunchbacked flute player of the Southwest, often depicted with a bag. And finally, the insect metamorphosis theme is rendered in mythic imagery: caterpillars consume the leaves and flowers of plants, go into a bag, the chrysalis, and after a sleep emerge as butterflies.

Jose Enriques flipped a page of my field elicitation book, coming to the kingbird. "*Komalk mo'okam* ['flat head–it-has']. Kingbird. He fought at the first wine feast. Also the *hikwig* [Gila Woodpecker]." Both birds still wear bloody heads as proof.

COMPARATIVE LINGUISTICS. The name *chukukmal* might refer to "becoming black, *chuk*," notes Culver Cassa, but the reasons are no longer evident. Fr. Antonine (1935)

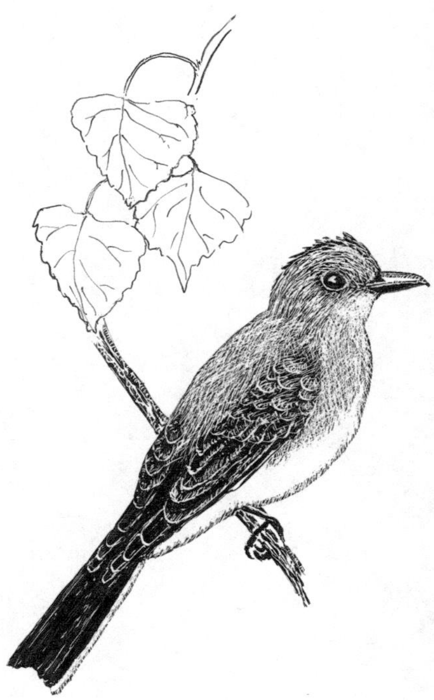

chukukmal, Western Kingbird (Amadeo M. Rea)

listed *chuekukumal* along with several other Pima birds whose English name he did not know. Although the name was once well known to Akimel O'odham, no one I asked among Tohono or Hiá ch-eḍ O'odham seemed to recognize it. Kingbirds are more abundant where there are riparian habitats and open places such as fields for foraging. Few Tohono O'odham would encounter kingbirds except in the Baboquívari Mountains and along the Santa Cruz River at the eastern periphery of their country. The kingbird name has apparently not been checked among other Tepiman groups. The name appears again in coastal southern California. Among the Luiseño, a Uto-Aztecan group that shares few avian cognates with Pima, kingbird appears as *cheshéemal*. And one of *its* food items is *chukwíimal*, a species of fly (Bright 1968). Culver Cassa thinks this Takic language to be too distantly related to Pimic for this word to be cognate. The Yaqui name for the Tropical Kingbird, *Tyrannus melancholicus*, is *chirk* fide Moisés Valenzuela.

The flycatchers in the desert as a group seem to be underdistinguished, only the kingbird and the phoebe being named. Perhaps those entrusted with recording the Piman lexicon have not asked sufficient questions. The *Myiarchus* flycatchers are common throughout the desert. For some, the Ash-throat replaces the Say's Phoebe as the referent. And the Vermilion Flycatcher is confiding and conspicuous, but so far I have found no name for it.

ba'i chukul

Alaudidae, Passeriformes
Eremophila alpestris
Horned Lark; Llanero, Alondra Cornuda

In winter, loose flocks of up to several dozen sparrow-size birds may be seen flying over saltbush flats, playas, or fallow fields. Their soft tinkling calls, white underparts, and contrasting black tails narrowly edged with white will identify them in flight as Horned Larks. When the birds drop to the ground, they seem to disappear. Their mouselike behavior of running and freezing and their tan upper parts make them almost invisible as they search for weed seeds and insects on the pale desert soils. Unlike sparrows, larks walk rather than hop, and they avoid landing in trees or bushes.

With binoculars, one may catch a glimpse of the black, yellow, and white facial pattern. A black streak runs from the bill to the eye and down. The black bib on the lower throat shows on most birds, more distinctly on adult males, suggesting a miniature meadowlark. The tiny "horns" are not easily seen. Most of these birds coursing through Pimería at this season have abandoned the cold, snowy country to the north.

In summer, Horned Larks may be found in spots of Piman country with short grass or even in alfalfa fields. This is one bird that has increased with the overgrazing so rampant in the arid Southwest. By this season, the pale feather tips have worn away, and the facial pattern has become more obvious, but the black "horns" are still inconspicuous. Some birds are very pale above, whereas others are somewhat rusty colored (see technical note).

As with several other birds that have evolved in open, essentially treeless regions, male larks deliver their courtship songs while hovering hundreds of feet in the air over their territory. Somewhere below this display area, the female scrapes a depression in the soil, building in it a nest of rootlets and fine grass.

In February 1983, I turned in page proof to my book *Once a River* and by way of celebration picked up Sylvester Matthias to spend the day afield. I stopped at Hidden Valley, and I was watching some Horned Larks with my binoculars when Sylvester asked if he had ever told me about a bird called *ba'i chukul*, "black neck" (actually "black throat"). "No, but it's too late now; I just sent the book back to the press for publishing, Sylvester." I was not sure if he had seen the birds I was looking at, but I was interested in this new bird, of course.

He went on to say that it has a black throat, lives in fields, eats worms, has a brownish back and a bluish white breast. Maybe a little yellow on the tail. No black on head, just gray. It puts its nest in a bush.

ba'i chukul, Horned Lark (Mike Touch)

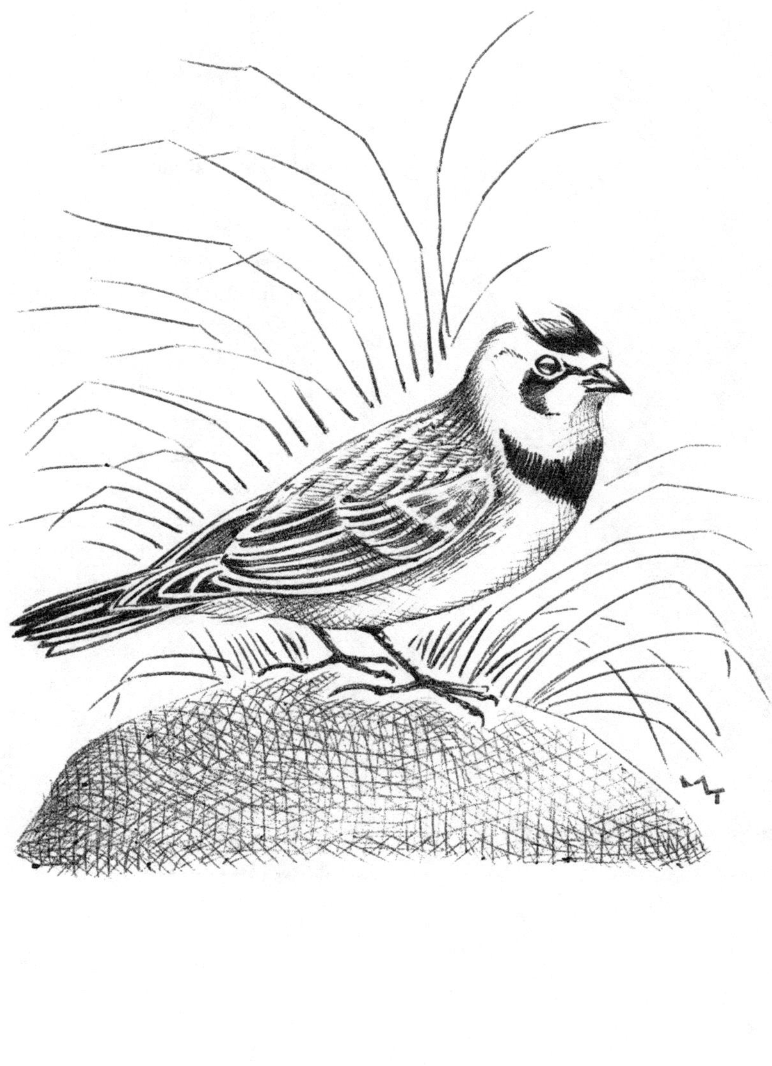

The characters he described were contradictory. But he went on, "In plowed fields, along with meadowlarks. Feeds in open fields." This last detail was more suggestive, and by good fortune somewhat later that day we found a road-killed Horned Lark, which he promptly identified as *ba'i chukul*. "Belongs in the fields, just like the meadowlark."

Over the years, he mentioned the bird at various times as we saw it in the field or looked at specimens.

"What do they sing like?" I asked him.

"Oh, I forgot! It's been so *long*. This is the first time in I don't know how many years . . . fifty years or so ago. They used to have a lot in the fields. *Ba'i chukul* disappeared for a long time. That['s] first I've seen [that] you picked up. It belongs in the fields, just like the meadowlark," he reiterated.

People, including many Pimas, often tell me that an animal is not around anymore, be it a bird, lizard, or insect. Sometimes it isn't. But more often, it's the person who isn't outdoors any longer to encounter wildlife of any sort. Few now hunt, farm, or herd cattle. People have become housebound television watchers. But in this case, I think Sylvester was right. I had made ornithological surveys of the Gila River Reservation for nine years before encountering my first Horned Larks here. For some reason, the birds were making a comeback, two subspecies invading from two historic breeding ranges (see Rea 1983:191–194). What I didn't realize at the time was that Horned Larks were *re*invading.

MISIDENTIFICATIONS. In their Tohono O'odham dictionary, Saxton, Saxton, and Enos (1983:90) call the Horned Lark *chukul-ba'i chuk*, probably a legitimate variant of the Pima name given here. However, in the same work (1983:5), they identify *ba'i chukulim* as "the black-throated desert sparrow, *Zonotrichia querula*." This scientific name does not apply to this bird. The Black-throated Sparrow (also known as the Desert Sparrow) is *Amphispiza bilineata*. However, it is prominent throughout the Sonoran Desert and should have been named by Piman speakers. *Zonotrichia querula* is actually the Harris' Sparrow, a species so rare in the desert that few people will ever encounter it here in a lifetime of bird-watching. Jose Enriques of Queens Well used the name *ba'i chukuliw* for the introduced Old World House Sparrow, *Passer domesticus*, which builds its nests in people's houses and arbors, but he thought that the Desert Sparrow had a different name, which he did not know.

For Bahr's confusion regarding the name *ba'i chukul*, see the Belted Kingfisher account. In Bahr and associates 1997:28, he further confuses it with the Harris' Sparrow.

TECHNICAL NOTE. Selective pressures on ground-dwelling Horned Larks have produced numerous subspecies across North America. Two forms may be found breeding in Pimería Alta: the extremely pale, rather ghostly Sonora Horned Lark, *E. a. leucansiptila*, and the also rather pale but more richly marked Scorched Horned Lark, *E. a. adusta*. The Sonoran race generally breeds in the lower, more western desert. During winter, the sparse local populations are augmented by great numbers of northern birds from the prairie country, which are also pale (see Rea 1983:191–194).

giidval*, pl. *gigidval*; *giidwal*, pl. *gigidwal
***gidoval*, pl. *giigdoval* (Pima Bajo)**

Hirundinidae, Passeriformes

swallows (all species including Purple Martin); golondrinas, Martín Azul

Various species of swallows are spring and fall transients through Northern Piman country. Four swallows may breed in Pimería. The Barn Swallow, *Hirundo rustica*, of higher elevations, throws in its lot with humans and is most at home on ranches, nesting under the eaves of buildings and in wells—places that give protection for its open mud nest. Its deeply forked tail, metallic blue back, and rusty orange underparts make the bird unmistakable. It is found throughout the northern hemisphere.

The Cliff Swallow, *Hirundo albifrons*, is a chunkier bird with no tail indentation, a broad pale forehead, and a more intricate body pattern of rust, blue-black, and gray, with white streaks on the back. It builds jug-shaped mud nests on cliff faces and beneath bridges, overpasses, and the eaves of buildings.

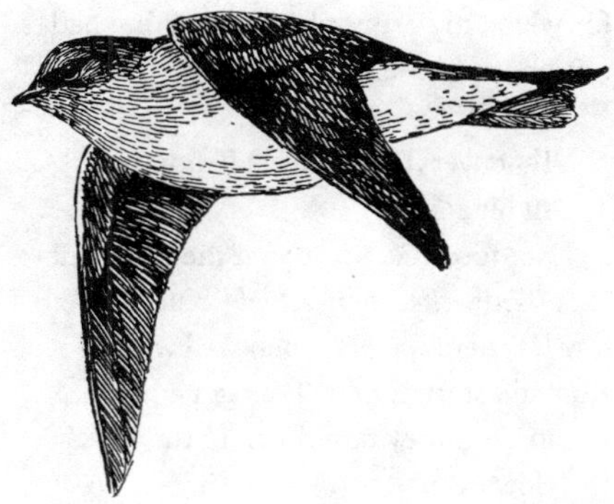

giidval, Rough-winged Swallow (Allan Brooks)

The most plainly colored is the Northern Rough-winged Swallow, *Stelgidopteryx serripennis*—uniformly dull brown above, tan and white below. Its tail is distinctly indented. Rough-wings nest in burrows in earth banks often near water. On the middle Gila where the Pima live, it was formerly the most common breeding swallow (see Rea 1983:194–195).

Much larger than these three as well as the several transient species is the Purple Martin, *Progne subis*. A sexually dimorphic swallow, males are all glossy blackish blue, whereas females and young are gray-brown and white, with some metallic blue on the upper parts. Its tail is conspicuously V shaped. In Pimería, it nests in woodpecker holes in Saguaro and Cardón cactus.

Migrants include the Tree Swallow, *Tachycineta bicolor*; Violet-green Swallow, *Tachycineta thalassina*; Bank Swallow, *Riparia riparia*; and perhaps in Pima Bajo country the Mangrove Swallow, *Tachycineta albilinea*. Barn, Cliff, and Rough-winged Swallows breeding farther north also pass through the desert in numbers. Although the songs and calls of each species are distinctive, they are generally twitterings with limited melody.

Gigidval (the word is most often used in the plural) are well known to Northern Pimans, figuring prominently in song and occasionally oration. Most speakers seemed to agree that the name covers all members of the family, though the focal species varied among different consultants. For instance, I asked Sylvester Matthias, "What color are *gigidval?*"

"Different colors. The popular ones are blue. The common ones are gray." This response introduced a taxonomic character for which my ornithological training had not prepared me.

"Popular?" I asked.

"That everybody knows [from songs]."

"All blue?"

"Not too blue. Mostly on the back, tail . . . called *s-cheedagĭ giidval* ['blue swallow']. [There was] an open well where Komaḍk store is now. They got a hand pump. And they nest down in the well."

"How deep?"

"Between fifteen and twenty feet," he responded. (I'm not sure now whether he meant the height of the water or the placement of the nest within the well.)

When he looked at skins, he selected specimens of the Rough-winged Swallow as representing the common *gigidval*. He picked specimens of Barn Swallow as his *s-chedagĭ gigidval*, reiterating that they are blue and fork-tailed, and that they nest in wells. Another time he said, "*Gigidval* are different colors. Most are blue, solid, not light. Some are gray. Fly over water. When wings are open, slightly yellowish or reddish." He was describing the light coming through the flight feathers.

Joseph Giff of the same village used the name *gigidval* for swallows but called the Barn Swallow specifically *hevaichuḍ*, usually glossed "bluebird": "*Hevaichuḍ*, swallow, fork-tailed blue swallow. Blue, real pretty blue. *Very* blue! You know, it shines. The only place I used to see them nest was down in wells. The Jacksons had a well. I used to come from school. And I used to go by these. And I'd see them fly out. They kind of glide. They nest down there under the surface [of the ground]."

"How long ago?"

"It must have been 1919 when I last saw them—somewhere around there—could be '20, could be '18, but somewhere around there. They also used to be at Santa Cruz, also in [a] well."

The well he spoke of in Komaḍk is the same one Sylvester had described. I could remember my own days as a youngster on Sierra Nevada ranches, flushing Barn Swallows from shallow rock-lined wells. Joe's assigning the name *hevaichuḍ* to this swallow is probably idiosyncratic; it properly refers to the Blue Grosbeak. What is most remarkable is that, except for these reports from the two Pima villages, the Barn Swallow is not known to breed in the desert, being found today at much higher elevations in eastern Arizona.

Herb Narcia, also a Kuiva O'odham but living at Blackwater, said, "*Gigidval*, their tail kind of V shaped. They go into washes where there's caves because it's hot. Come out only when it's cool, kind of cloudiness. When it gets hot, they go back into the caves. Kind of gray and black, forked tail." His description best fits the bank-breeding Rough-winged Swallow or the Barn Swallow. Martins are not found here.

I asked George Kyyitan, a Ta'i O'odham, about *giidval*. "It's kind of dark. It's really a peppy bird. When it's kind of windy, or in the evening, or . . . come out and flew all over. You know, their tails go like this."

"So they have the really forked tail. So they're shining on the back?"

"Umhum."

Another time as we sat outside talking, some swallows, probably Cliff Swallows, flew over. George, devoicing, called them *kikidval*. "*Kidval* is swallow. The Papago sing about them also. They're blue and black. That's why they're mentioned that way, those in Papago [songs]. That's a swallow."

Farther south, Tohono O'odham Delores Lewis of Ge O'idag called the Purple Martins that we saw "*gigidowal*, comes out only in summer." Another time he said that the name includes all swallows, even the Purple Martin. Joe Lopez knew the *gigidowal* as black. Luciano Noriega of Quitovac oasis picked the Purple Martin as his *gidowal*, noting that it is "negrito." This species doesn't reach as far north as River Pima country as a breeding bird, even though there are apparently suitable columnar cactus stands here. The martin is the bird that is compared to a black summer rain cloud in the Papago song given in chapter 3.

For Jose Enriques of Queen's Well, *gigidwal* included all the different swallows. "They're common in the summertime. When we see them in the clouds, making noise, singing, it's going to rain. We know it's going to rain." I asked him what kinds he was familiar with. "Mostly those, dark brown. Some are black with light breast and have swallow tails. Mostly in the mountains, we see them in the mountains."

Not everyone knew the bird as a breeding species. Leon Juan of Aangam said he knew the *gigidval* as part of Rain's entourage. But since it's a bird, it must lay eggs and have a nest. Culver Cassa said he would equate this statement with saying that a dove is a sign of the Holy Spirit; a dove as an animal must have a nest and lay eggs, but I know it only as a sign.

In Pima Bajo country, Pedro Estrella called the Rough-winged Swallow (and perhaps other species as well) *gidoval, giigdoval*. María Córdova couldn't remember the name. Three centuries earlier here the Névome vocabulario gave golondrina as *huiguidovari*, in the more usually encountered plural form (Pennington 1979:56).

These buoyant, aerial feeders, seldom resting, seem to have captured the Piman imagination and are associated either directly or indirectly with a great number of songs. Sylvester said, "There are Swallow Singers also from Sacate. The singer went from Komaḍk to Kyrene—Chuk Kavij (Lone Butte)—to Sacate, and died [out] at Sacate. And Don Thomas took over the *hikvig ñeñe'i* [Gila Woodpecker songs] and brought the songs back to Komaḍk [with him]. The Blue Hummingbird songs are included with the swallow songs of the Salt River." Although we often associate songs with a particular village, we see from these comments that a song series is instead associated with a particular singer, often the one who dreams the songs with the help of a particular animal. If the singer moves, so do the songs.

Others may learn and sing the songs and even become experts in them. Joe Giff (1980:138) was especially knowledgeable about the Blue Swallow songs. At a conference, he first explained the distinguishing features of this particular group. Then he gave a Swallow Song in both sung and spoken Pima, and finally in English:

Gitovaliyu hiinge
gitowaliy hi ge
ñeika ñe chunekim
hejele ñeika ji
yaanho wa chunekim
gamhu we ya ahai we
yupa me himeda.

Gidval u'uhig
gidwal u'uhig
e-ñe'ikaj chuudkhim
hejel e-ñe'ikaj i an hu chuudkhim
k am hu wo a'ai upam hihi.

Swallow bird
Swallow bird
Dances to his songs
To his own song dances
And in different directions they go home.

On another occasion, Joe offered this poetry for another Swallow Song:

Give me my scraper
I am over here
I have reached the mountain and I am hanging on there
Scraping my wing feathers.

Then he sang:

Yoikeñe maakeñe hivikune
Yoikeñe maakeñe hivikune
Yahali komanakehe nonovang oho iaihime
bashoñ ena nangiya
yahana hihiva.

In spoken Pima, this is:

Oig ñmaak ñhivkuḍ
Al Komaḍk do'ag aihim
baasho oñ nagia
ña'an hiva.

Give me my rasp
Little Sierra Estrella at I arrive
Against there I hang [hover]
My wing [feathers] I scrape.

Joe explained, "Blue Swallow didn't have his *hivkuḍ* [rasping stick], so he used his wings, his wing feathers." The association here with the notched scraping stick, a musical instrument used to accompany singing, is not accidental, I think, but rather is based on a surprisingly intimate knowledge of avian anatomy. The outer edge of the outer primaries (wing feathers) of one swallow, the Rough-winged Swallow, is roughened and in the male is strongly serrate, hence both its English and scientific names. At some time, the Pima must have handled this bird that once nested commonly along their rivers, hence the association with the notched musical instrument.

Sylvester said that both the Sandpiper Song and the Three-star Song that he sings are swallow songs (see sandpiper and Lesser Nighthawk accounts).

George sang this South Mountain Swallow Song, given first here in song language, then in spoken Pima, and finally in English, as rendered by Culver Cassa:

Gam hu huñuimada
muahanangi dohonovange
ga hu keekang
am hega yeḍang gigidoval ñeheda
wesameyangena si [vach] ñeheda
hejel yahane gigikoa[ch]k ñeheda.

Gam hu o huḍuñim
muhaḍg do'ag ga hu keek
am heg eḍa g giidval ñe'e
wes o am angevk si ñe'ed
hejel a'an giigidk o ñe'edam.

Over there evening is falling
Over there stands South Mountain
Just then swallows are singing
All flapping their wings, really singing,
Their own feathers ruffling and singing.

At Santa Rosa village, Tohono O'odham Simon Lopez sang a song about the swallow and Gakoḍkĭ, Superstition Mountains:

Jupiñi tañio keḍake gakoḍake nowange kekaa
Jupiñi tañio keḍake gakoḍake nowange keheka
kuñ hege uhunghi da'a hii mek ap ñeidachu
wa sape mamasime tonashhana giuhuḍe jewena wausimee.

"So he's flying by, and the *giidwal* sees this mountain."

"Yeah."

Culver Cassa noted that the swallow is carried north to Superstition Mountains, which still bear a pale band from the Flood:

In the north looms Crooked Mountain.
I'm translated far and behold it.
"Truly a fine foam belt could be made."
The earth drips.

Russell (1908:292–295) collected a set of ten swallow songs, *gigidval ñeñe'i*. They are circling songs accompanying social dances, where singers sit in the middle keeping time by beating on baskets (or, later, on cardboard boxes) with sticks. These songs are a marvelous series with vivid desert imagery. The opening three songs are specifically about swallows. The words suggest at least two different species. Song IX pictures the Turkey Vulture circling in the blue sky. Song X features Reddish Bat, eagle down, and Coral-bean flowers. Here are the first three songs as reworked by Culver Cassa:

Gigitovali uvhinge hiyañ wewema ñeñei ivamo shoshonechu
hiyañ wewema ñehed a'a
shoingama su'uvi yemapim ulina'a
hiyañ wewema ñeheda'a.

Swallow birds here with me begin the songs
Here with me they sing
Poor women, here they gather and loiter;
They sing with us.

Wavaisa ngingitovali nam keka'a
keḍamo ñuvapa'a, keḍ ñuvapa'a
hanoñ wewegañ chedongia kikihioḍe wuvakime.

Atop the cliff the swallows stand
Then they brought me there, then they brought me there
There all about me blue rainbows emerge, [that is, the birds are flying above the singer in an arc].

Chuchuka gingitoval hiatovoiveka
hiatovoiveka muñi vañimeḍa
kame koi-iñ vañemeḍa.

Black swallows here came running
Here came running and somewhere lead me
Over there far they lead me.

In a series of seventeen songs used by the *maakai*, shaman, in making a diagnosis, Song VII features white-headed swallows, *tota moomim gigidval* (in song language, *tota mohomama gingitovalĭ*) (Russell 1908:304). The reference would seem to be to the Cliff Swallow, which has a conspicuous white forehead patch, the only species of swallow in the region with any white on its head. Again, as reworked by Culver Cassa:

Kuch yemuiva luḍuñ daime
kuch yemuiva luḍuñ daime
ha sesepiñime gese tota mohomama gingitovali machewuina ngingi kaya betoñinaha
kiya luduñ daiime.

We now set the sun
We now set the sun
Their younger brothers, the white-headed swallows, cause their primaries to four times rumble
And here set the sun.

The idea of this song, Culver says, is that we are starting the dancing, dancing down the sun.

There remains the question of whether the White-throated Swift, *Aeronautes saxatilis*, common in Pimería Alta, is included in the Piman concept of *gigidval*. I never had occasion to point one out to consultants in the field. These narrow-winged aerial feeders are even more zippy than swallows, appearing and disappearing quickly. They nest in cliffs in the more rugged desert sierras. The white-headed *gigidoval* song might be about this swift.

comparative linguistics. The presumed more primitive condition of *gigidval* is *gigidoval*, with the medial *o* syllable, and is found in colonial and modern southern Piman dialects, as well as in contemporary song farther north.

Mountain Pima of the Yepáchic area gave me *gigikolyi* (Lorenzo Cruz) and *gigijulyi* (Isidro Cruz) for the Barn Swallow, apparently the only species they recognized. It breeds locally at these higher elevations. Shaul (1994) recorded the name as *gigokuli*. Pennington (1969:117) found *pipídamuli* among Northern Tepehuán for golondrina. Although more distant, it would seem to be cognate. Among the Tepecano, the southernmost Southern Tepehuán dialect, Mason (1917:327) recorded *gídaawal* as "sascuines." I have been unable to identify the Spanish word, but the Pima word must mean "swallow," not much changed a thousand miles from the Gila.

Among other Sonoran Uto-Aztecans, we find noncognate terms for swallows: Rarámuri (Tarahumara), *sowé, sowéperi (sowépari)*; Eudeve, *váidarus* (Brambila 1983; Lionnet 1986); and Yoeme (Yaqui), *kutapapache'a* (fide Molina, Valenzuela, and Shaul 1999).

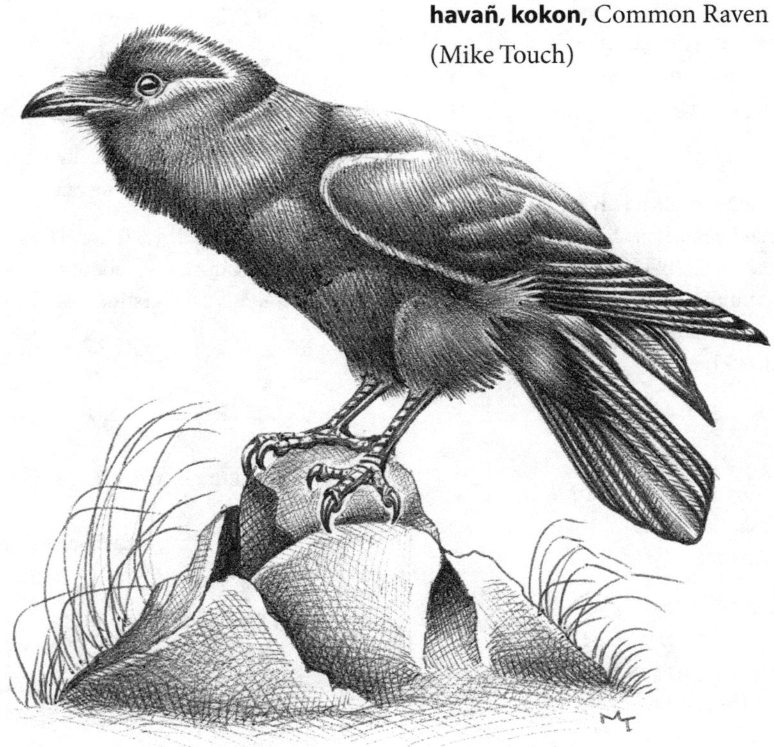

havañ, kokon, Common Raven (Mike Touch)

***havañ*, pl. *hahavañ* (Akimel O'odham)**
***havañi, hawañi*, pl. *hahavañi, hahawañi* (Tohono O'odham)**
***kokon, koko'on* (Pima Bajo)**

Corvidae, Passeriformes

Corvus corax

Common Raven; Cuervo Grande

Corvus cryptoleucus

White-necked (Chihuahuan) Raven; Cuervo Llanero

?*Corvus brachyrhynchos*

?Common Crow; Cuervo Norteamericano

At a distance, an all-black bird with slotted wing tips and heavy body suggests some dusky broad-winged hawk. But its prominent bill, often turned down as the bird flies strongly over open country, searching the ground, gives the bird away as a raven. It is an aerial master, flying in formation with a partner or two, dodging, even turning somersaults, often seemingly for no other purpose than the sheer kinetic joy of it all. It may circle effortlessly high over the desert or fly low with strong, determined wing beats, sometimes with an audible *wushhh* as it suddenly changes course.

Spying something of interest in a village, it lands with a bounce and begins walking. Its hackled throat flares out in the breeze or when the bird voices a deep croak. Omnivorous, it takes both plant and animal material, often searching the highways early in the morning for roadkills, competing with late Coyotes or early vultures and caracaras.

The Common Raven is the largest living passerine, songbird, ranging across much of the Northern Hemisphere. In the New World, it is at home from the polar ice south to the Mesoamerican tropics. It is reputed to be the most intelligent of birds. It also has the largest vocabulary, communicating with from eighteen to sixty-four distinct sounds, a feat rivaled in the animal world only by us humans (Savage 1995:87).

Most of the ravens in O'odham country are probably Common Ravens. Some, particularly in winter, may be White-necked (Chihuahuan) Ravens *(Corvus cryptoleucus)*, a somewhat smaller species, partial to mesquite and yucca grasslands. This species is distinguished in the hand by the snowy white rather than dark gray bases to the body feathers. Its status across Papaguería, where it may breed, needs to be verified with specimens. Outside the breeding season, it often roves in large flocks.

Much easier to identify is the Common Crow, a smaller bird with a straight tail, rounded at the end, rather than diamond shaped as in ravens. It tells you who it is with a tenor *caaw*, so different from the base croaks of both ravens. There are suggestions that it once visited River Pima country, but don't look for it in today's ecologically depauperate environments. Even the raven is seldom seen here any longer, being more at home in Tohono O'odham rancherias. There are no crows throughout the rest of Tepiman country, all the black corvids being ravens of one species or the other.

As farming people, Northern Pimans were familiar with these large corvids, which are associated with agriculturalists throughout much of the world. Joseph Giff said, "[*Havañ*] is like a packrat, *koson*. They do that, the raven family." "Big flocks," noted Francis Vavages. "Eat mice, rats, worms, whatever." Delores Lewis said, "Real black one. It shines when it gets in the sun. Eats anything—corn, melon, all kinds. Will steal your melons, corn, sugar cane [sweet sorghum] seeds."

Once, discussing the raven with a group of River Pima, I asked, "Did *havañ* ever bother your fields?"

"Yeah, Joe's corn," responded Sylvester.

"Yeah, they like corn," Joe confirmed.

"If these bother your field, what would you do? Would it be a problem, or were there that many?" I asked.

"Not that many," said Joe.

"There was more corn then," added his wife, Ruth.

"Everybody had corn—corn everywhere," agreed Joe.

"I think they're about gone now," said Ruth.

Scarecrows were sometimes set up in Pima fields. They might be called *ebchud akuḍ* or *hu o'odham s-du tosdakuḍ* or *ha-o'odham chuda* (Sylvester) or *maas o'odham* (Culver Cassa). They are probably of no great antiquity. Traditionally, the man working in his field wore only his *atosha*, loincloth. But once white man's clothing became the norm and there were some items to spare, scarecrows were elaborated. Of course, they were intended as deterrents to problem mammals as much as to birds. Young hunters in fields were probably the first line of defense.

Although Frank Jim said of *hawañĭ*, "[We have] nothing to do with them!" implying avoidance, both Tohono and Akimel O'odham said the bird was not associated in any way with *kaachim mumkidag*, staying sickness.

In parts of North America, particularly in the Pacific Northwest, Raven is a major character in the legends, often a trickster on a par with Coyote or even surpassing his importance (Bierhorst 1985:25–34; Nelson 1983:79–84). We find none of this in Piman culture, perhaps because ravens are usually not all that common in this desert country. But even rarer animals such as Cougar, Jaguar, and Gray Wolf wander in and out of important mythic episodes. Raven enters rather obliquely when it does.

In the account of Children of Cloud (Russell 1908:239–240, 249), on their return from visiting their father in the east, the twin boys, admonished not to speak to anyone along the way, first encounter Raven, then Roadrunner, Hawk (probably Falcon), and Eagle. All are power birds. The boys pass the test. But when confronted by Coyote, they fail and are turned into agaves. They can no longer return to their community in person, but only as a useful food plant.

Plants or animals of high cultural salience such as macaws, corn, and tobacco may have two or even more different stories of their origin encoded in Piman oral literature. This is true of the Saguaro, the giant cactus that for most of us typifies the Sonoran Desert. One story in particular is associated with the origin of both the Saguaro as a plant and of the wine feast as an annual festival celebrated by O'odham to begin or continue the monsoon rains in midsummer, and this story features Raven (Bahr and associates 1994:125–129; Rea 1997:253–260; Underhill and associates 1979:17–35). There are at least five published accounts of this dual origin (Benedict 2001:88–89; Densmore 1929:149–150, 161–162 [said to be a Pima version]; Saxton and Saxton 1973:211–219; Underhill 1946:42; Wright 1929:109–121). These accounts are quite similar in essential elements.

In all versions, a young mother is so inordinately fond of playing *tokaḍa*, a woman's game of field hockey, that she neglects her child (usually a boy), leaving it with a gourd of milk or other nourishment. The child seeks its mother, who travels from village to village playing, betting, and winning. At last finding her at some distant village, the child sends a message to her to come, while awaiting her in an arroyo. The mother is too busy to stop playing, and by the time she does come, the child has sunk into the sand (usually into a tarantula hole). It turns into a Saguaro. In two versions (Wright's and the Saxtons'), the village children mistreat the vulnerable Saguaro, and in all versions it leaves, coming up somewhere else (usually Quijotoa Mountain, but once in Superstition Mountains) so that it has to be searched for.

Havañ, Raven (glossed "crow" in four versions), is ultimately the one who is successful in locating the cactus. (Densmore's [1929:161] Pima version says just "a bird.") In three accounts, Havañ then gathers the *haashañ bahidag*, ripe Saguaro fruit, from the top of this first cactus, returning with it to the People (who then included the animals) and vomiting it into baskets or ollas. This fruit is fermented, becoming the first *haashañ navait*, Saguaro wine.

Certain animals come to the resulting ceremony dressed very festively, just as they are today. The episode of Grasshopper and Nighthawk outdoing each other in entertaining occurs here (see Nighthawk account). During the ensuing drinking, a number of birds fight and get their heads bloodied, as you can still see today. (The narrators apparently didn't point out that only the *males* of these species have bloodied heads; perhaps this was common knowledge.)

At the termination of the celebration, the *kaij*, Saguaro seeds, are gathered up and sent away by means of either Badger or Jackrabbit or Hummingbird or a messenger. In transit, however, the seeds are somehow tossed into the wind and spread across the south sides of mountains and bajadas, where Saguaros still grow today.

Raven is important here in locating the Saguaro fruit and returning with it, but one has to wonder, Why this bird? Why not *hikvig* (Gila Woodpecker) or *okokoi* (White-winged Dove), two birds traditionally associated with this cactus? "Ravens go way up there and can see far, always circling around," offers Culver Cassa. "And they like mountains. Maybe they eat Saguaro fruit." He considers the raven a discoverer of plenty and perhaps a harbinger of it. "It has to do with the escape of plenty." Indeed, Raven may come to a man in his dreams and give him power to become a cunning scout (Underhill 1939:170). Culver also notes that in a parallel Paiute story, the Common Raven is the bird who discovers piñon nuts, a staple for these more northern people as the Saguaro is for desert people.

This story has some interesting structural characteristics. The neglected child is deprived of the liquid that it needs—that is, its mother's milk. When it metamorphoses into the Saguaro, it provides the liquid, Saguaro wine, that the people need in order to bring rain (another liquid) to their fields of summer crops. After finding the cactus fruit, Raven returns to the People and vomits it. When the wine is distributed and the People celebrate, vomiting again ensues, one of the several metaphors of rain in the *haashañ navait ii'e*, drinking ritual. Of the several Saguaro origin stories, Culver Cassa believes this is the primary one.

In 1935, Julian Hayden took down a version of the Creation Narrative from storyteller Juan Smith on the Gila. During the Vupshkam (Emergenti) invasion of the territory occupied by the Vipishaḍ (pueblo builders), a *maakai* (shaman, presumably a *havañĭ namkam*, Raven meeter) sends a raven into a *va'aki*, pueblo. He sings this song:

You have made a raven out of me
I am now hanging above the va'aki
And I am making my enemies' heart
To go to sleep.

So this raven came down to where Siivan was, and it happened that this Siivan had some sons, and these boys ran out of the house, and they were trying to shoot this raven down, and yet this raven was not a real bird.

In another version of the emergence, Elder Brother first sends out a *maakai* with the power of Coyote to search for the Vipishaḍ. He finds no one. He digs a hole to show how far he had been. "After four days Elder Brother said to one of the medicine men, who had the power of raven, 'Go out and see what you can see.' He put on his raven's skin and went as far as he could fly. He could not see any of the people of the earth. He piled up rocks to show how far he had been" (Benedict 2001:42). Again, the underlying metaphor is of raven the searcher.

In the conquest of Vulture's pueblo, the invaders sing a song that calls Vulture "Child of the Raven" (Russell 1908:228)—the only reference to this relationship I have seen. Both birds scour the country searching for food. And ravens feed at carcasses with vultures, so this kinship follows from their ecological niche.

Later in the Creation Story, there is an episode where a boy called Paḍ Aangam ("Ugly Desert-willow"), whose father has been killed by Apaches, incites the men of his settlement to depart on a retaliatory raid. The story has many of the earmarks of an historical rather than a mythic episode, and it is no doubt a late addition to the canon.

In the vicinity of Gakoḍkĭ, Superstition Mountains, the warriors discover sign of the enemy. The boy and Coyote scout ahead of the main group, going up an arroyo. They accidentally stumble upon some Apache children playing in the sand, who run home to give the alarm. Knowing they have bungled things, they look at each other and say, "What shall we do now! for if we go back the people will blame us for letting the Apaches see us first." Paḍ Aangam tells Coyote to return to the encampment, stepping in the boy's tracks, while he himself turns into a "crow" and flies up on a rock. The ruse works, for when the adult Apaches arrive, they say, "There are no human tracks here. It must have been a coyote the children saw," and return home. The boy flies back to Coyote and resumes his human shape (Lloyd 1911:173–187; see also Benedict 2001:141). The raid is ultimately successful.

Lloyd (1911:236) recorded another story that also belongs with the so-called Hohokam inhabitants, though it is misplaced in the published sequence. A *siivañ* (leader) from the Blackwater area invited another *siivañ* from the Gila Crossing area (Shell Field) to bring his wives and come gamble. Along the way, the invited chief, Cheenashaḍ Siivañ, stops to bathe and decorate himself for the occasion. He takes a stick of *uupaḍ* (Catclaw) wood, sharpens it, splits the other end into four pieces, bends them over, and ties "crow" feathers to the ends of them, then puts the whole thing in his hair.

The usual story of a bird changing its color to bright blue is about *hevachuḍ*, the Blue Grosbeak (see that account). But there is a similar story told by William Blackwater (who also supplied the conventional one) about the "crows" who did not like the color of their feathers. They plucked their old feathers, put them in hot water, then bathed in cold water for a day while singing:

We are making our feathers all colors.
We are making our feathers all colors.

New beautiful black feathers grow out of their bare skin. Coyote envies the birds and asks them if he can try changing his color too. The "crows" fly away laughing in the evening when no feathers appear on Coyote, and he ultimately dies of cold (Benedict 2001:170–171).

Havañ appears more frequently in oration and sung poetry, where almost invariably the Piman name has been translated "crow" rather than "raven." The association is often with war exploits. Here raven is a bird of power who can "see" the enemy. For instance, we find this imagery in a Gileño warpath speech (Russell 1908:366, retranslated by Culver Cassa):

Ova id daam ge s-oam havanta
kuntak chu'igam mamchek ab iimk shoshoa
kut ab s-ñho'ige'idam etaatk
gi'ikho am eḍa si kokovk
i huḍuñk g jeweḍ an kaachim gaghai si gevshp.
Hemajkam ab oimeḍdam
heg i pi maasko ee'eḍ i cheshajidk hohodai daḍhadam s-hahamasma i siibañmad.

It was above us, the Yellow Raven.
I figured his relationship and greeted him and cried.
And he mercifully felt toward me.
Four times within himself he popped.
He descended and the country over there on each side crashed down upon.
(They didn't see that they were doomed.)
The inhabitants' blood he took up and upon the rocky hills it was brightly sprinkled.

(Russell's briefer translation omits an important Piman cultural trait: in order to interact with someone, in this case the speaker with Yellow Raven, it is necessary to know one's relationship with the person addressed and greet the person with appropriate kinship terminology.)

In another version of this war oration (Lloyd 1911:202), Sun (or perhaps Blue/Green Fog) "took me toward the enemy. But instead of taking me to the enemy it took me up into the sky, to the Yellow Crow [*sic*]. And the Yellow Crow, as a powerful mahkai, went down to the enemy and divided their land four times, and slew the human beings, and painted the rocks over beautifully with their blood." (Although the macaw is not mentioned, the situation is one where we might expect this bird of blood to appear.)

Underhill (1946:180–181) found three warpath songs that feature crow/raven:

There sits a raven
And sees my enemy
There stands a raven
And sees my enemy. (Song IX)

Here I stand and look
What is the mountain
Far off, blue?

Here I stand and look
It is Raven Mountain
Far off, with a cloud on it. (Song X)

Raven! Little shaman!
He lights up the flat land
Raven! Little shaman!
He lights up the mountains. (Song XII)

Densmore (1929:26) recorded what is no doubt the same song as Underhill's number 10:

Here we are on our way and see the distant mountain.

See, the mountain far from us that has
the cloud is Raven Mountain.

Densmore provided the musical annotation, but unfortunately neither author gave the original O'odham. Herzog (1936:225–226) recorded a series of Pima songs used in the rainmaking ritual that picture storms atop Raven Mountain and Frog Mountain (Mt. Lemmon near Tucson).

In much of Piman oral literature, the raven embodies the role of searcher and discoverer, an image built on its foraging behavior. In addition, the black feathers of this glossy bird were an analogue for the slain enemy's shining black scalp. Upon hearing the return of Pima warriors from a successful raid on the Apache, the Pima women sang a victory song (Russell 1908:338). In it, the enemy scalp, soon to be hung upon a pole and danced around, is called a *havañ*, raven. In the song, the scalp speaks: *"Name toñ yavañ hichunaki yu'usi kugabe vañ nagihiya?"* (Will you all upon making me a raven at the tip of a staff hang me?).

This scalp/raven metaphor was widespread among Pimans (see also Herzog 1936:332n., 395). The elderly Papago Maria Chona gave Underhill (1936:15) this song:

Poor raven.
There it hangs.
Poor raven.

And she explained, "It is a good thing to have the enemy's hair raised on a pole above our houses. It calls the clouds to it and brings rain on our land, and then the corn comes up."

During the conquest, the Vupshkam, or Emergenti, sang the following song as they approached Vulture's pueblo (Russell 1908:282, reworked by Culver Cassa):

Havañi yali ñanga
Havañi yali ñanga
tonolike nanavai'i
hiyañi yameko neida tonolike kuñuñui.

Raven child tells me
Raven child tells me
Light glistens
Here I watch the light form heat waves.

Not all references regarding the raven are to war. Paired hunting songs among Underhill's (1946:93) anthology, for example, do not make this connection:

The black ravens go flying,
Having seen the snapping bow;
Toward it in a narrow line
They go flying.

The yellow comrade [Coyote],
Having seen the humming arrow,
Toward it crouching
Goes running.

The first song is odd in that ravens appear rather than Vulture, the moiety mascot, normally paired with Coyote. And usually the word *oam*—designating brown, orange, or yellow—is applied to Buzzard, whereas *koomagĭ*, gray, marks Coyote. At any rate, the two songs are paired antiphonally, as in many Piman songs. The birds fly and the mammal runs, learning of the killing from a snapping bow and a humming arrow. The ravens move in a narrow line, the Coyote crouching. There is often a color contrast as well, here black and yellow (or brown).

One version of Elder Brother's resurrection speech (which seems to be a war oration) features four birds and their respectively colored springs: the black waterhole in the west is strewn with raven feathers; the blue one in the south with "soft feathers of the bluebird"—that is, Blue Grosbeak; the red one of the north with macaw feathers; and the white one in the east with eagle fluffies (Benedict 2001:37). Elsewhere in a myth, Benedict (2001:44) talked of a mountain where there were many crows. Whether a different corvid species is intended here or this is just a variation from the narrators (William Blackwater and Thomas Vanyiko) and their English translators we do not know. No Pima is given.

There is a lengthy and apparently late narrative about a half Apache boy called Prairie Falcon Meeter (Benedict 2001:144–153; Saxton and Saxton 1973:169–209). While still a boy, he incites the people to raid the Apache, who had killed his father. On going forth, Falcon Meeter is described as wearing a headdress of raven feathers. When the O'odham see him, they sing (Benedict 2001:146):

From a great distance
You brought these feathers
And came to us.

I asked Jose Enriques of Queens Well about his knowledge of *hawañ*. He described both a negative and a positive attribute of this great bird.

I guess the reason the *hawañ*—it's always a tattletale—whenever something bad will happen to a person, he will come around and call around. And sure enough, later something will happen. Somebody will die or whatever.

And I guess my uncle always say that when you are a hunter—like to hunt a lot—you always have him [*hawañ*] as a companion. *Hawañ* is pretty helpful to the hunter in a lot of ways. He can tell where you can find game or whatever. If there is game nearby, you can see him come to the area or even flying close by so you'll know where it's at. And start looking for, down that place [where the raven flies], then you'll find it—deer, Javelina, or whatever.

And I notice one of my grandfathers singing about it. It's just kind of like telling a story about it. He was out all that time trying to look for game and having a hard time finding game. He'd sit up there on top of a peak looking around. He couldn't find anything. He'll point his arrow here and there, but there's nothing to shoot at. Then all of a sudden he heard a crow [*sic*] cawing, cawing. And that's what I guess he was talking about, the *hawañ* being a hunter's companion. He was telling him something, that relates to his life, spending his life in the desert, trying to find game And what do you say when you feel sorry for yourself, doing all the hard work [self-pity]?

And all of a sudden I guess that *hawañ* finally came and heard him saying it, crying up there. And now I guess he knows that he's going to get his game. He dreamed this, in his own way; he end up there on top of that little hill. And he probably dozed off and he dreamed that song. It came to him. It's a pretty good song. I like it. When I used to go out and I sing it. That's the only song I remember, because I like to hunt. And that's why they taught me that song.

Shoiga añi wua'a[da]
dovange kug ab hu wa chum ñ ulina
ida ñ gagaat babaij i ulina
ke ñeinachu
hawañi u'uwhigi uwua ñ damai himena
kongolongdange.

I'm sad
On top of the mountain I try to hang
around
This my bow I hold out in front of me
and sing to it
Raven bird is up on top going along,
doing something

And he's on top of me going along
Making a sound *(kolog)*.

The O'odham word in this song for the particular sound that the raven makes, *kolog*, describes a deep hollow wooden tone or rattle, a characteristic but less often noticed note in the raven's vast repertoire. The raven's more frequent call is termed *kaakag*.

Jose's reference to the raven's being a harbinger of bad news is the first I've heard among O'odham, although it fits a broad pattern found among both Old and New World peoples. Raven as helper to the hunter is also widespread (see Nelson 1983:83–84 for northern Athabascan), probably based on associations this highly intelligent bird is capable of making between the hunter and the acquisition of leftovers after the kill.

His associating raven with hunters helps explain an enigmatic passage in one of the Piman versions of the story of the confinement of the game animals by some evil person (Saxton and Saxton 1973:232; see also Russell 1908:217–218 and Shaw 1968:43–55). Two hunters, *moombdam*, bearers of a head disguise, are unable to find game. They camp overnight. The next morning one says, *"Kutt am wo hii'hii abai k wo ñei mas haschu aag k hab e juu. Ab at g hawañi hud ga hu m ab do'ag baasho."* That is, "Something happened there at the foot of the mountain. Let's go and see why the raven went down there." They discover a man who has killed and butchered a Mule Deer.

The raven also enters O'odham culture as a figure of speech, *hahavañmad*. Sylvester Matthias said it's something to do with crows. Culver Cassa explained the meaning of the phrase as: "[It] really pours on you; something gets on you [bothers you]; 'swamped'; 'deluged.' Literally as a field, figuratively as a human condition or state."

The Pima word for raven is included in an O'odham plant name: *havañ taatad̩*, "raven feet" (Hrdlička 1908:183, 245; Lloyd 1911:63; Rea 1997:205). This is the wild-heliotrope, *Phacelia* spp. The reason for this name is not clear, but the blue or purple flowers are arranged in scorpioid heads, perhaps suggestive of the bird's foot. In former times, the new growth was picked in early spring and eaten as cooked greens, *iivagĭ*.

Several O'odham place-names include the raven. Hawañĭ Do'ag, "Raven Mountain," is mentioned as part of the emergence and in war songs (Densmore 1929:26; Underhill 1946:183, 191). When the Tohono O'odham looked to the east and contemplated a raid on the Apache, they saw Hawañĭ Do'ag. It is presumably somewhere in the neighborhood of Tucson (or at least on the horizon in that direction), but I haven't learned its exact location. There is a ranchería in Tohono O'odham country called Hawañĭ Naggiak, usually called "Crow Hang" in English. It should be called "Raven Hanging." Whether this phrase refers to the actual bird or to what it stands for as a metaphor, an Apache scalp hung on a pole during the victory dance, is unknown to me.

Another place name, Havañ Kosh, "Raven's Nest," is in the vicinity of the Salt-Gila confluence northwest of the Gila River Reservation, near the present speedway. It was well known to the older Pimas I worked with as the location where there was a settlement of Apache during times of peace. The Gila Crossing calendar stick for the year 1876–1877, narrated by Ko'oi Mo'o, Rattlesnake Head, mentioned this ranchería: "There was an Apache village called Havany Kas at the junction of the Gila and Salt rivers while a truce existed between the Pimas and Apaches. During this year an epidemic of smallpox prevailed in that village, as well as in all those of the Pimas and Maricopas."

But the Havañ Kosh settlement was not completely wiped out by the epidemic, noted Sylvester Matthias. He related with some humor that when the San Carlos Reservation was established, the government rounded these Apaches up and moved them upriver to the new location. Pretty soon the Pima looked out and saw them coming back, some on foot, others with their wagons, carrying their belongings. He placed the village of Havañ Kosh somewhere between Arlington and the hill by that name. He related that he went to Phoenix Indian High School with a girl from these people, then living near Arlington, but the settlement has since dissipated.

Joseph Giff explained this place-name: "And that little peak—the *last* before the river—that's what they call *kots keek*, 'cross standing' [Monument Peak, the point from which Arizona is surveyed]. And that ridge—from there to the mountain—is Havañ Kosh. There must [have been] a lot of *hahavañ* nests there. They nest [roost] like the *ñuñui* [vultures], like that on dead trees."

Ruth Giff added, "A long time ago, when there's water in the river."

Sylvester elaborated: "The Maricopas come to this area to hunt. Creeks that run into the Gila, farther down, that's where we used to come to fish—just narrow creeks with cattails. And [the creeks] enter the Gila River. Then the Gila turns and enters the Salt River. So that must have been a pretty good sized river *here*. Two creeks enter and the Santa Cruz [River]."

The biological referent for this folk taxon is always some member of the genus *Corvus*, but the species is not always clear. Most of the time it refers to the Common Raven, usually a rather sparse inhabitant of desert country. But some (or most) of the ravens nesting in Papago country may be the somewhat smaller White-necked or Chihuahuan Raven (Monson and Phillips 1981). I have seen no specimens from there, but the comment by Delores Lewis of Ge Oidag that the bases of the *hahawañ* feathers are white, not gray, clearly points to this latter species.

Sylvester Matthias always maintained that the word *havañ* properly referred to the "crow," a still smaller species of corvid, whereas the large bird he called *s-chuk u'uhig*, literally "black bird." For instance, one day as we were traveling east of Florence to Cochran, we spotted a raven in the distance. *"Havañ?"* I asked. "No. *Chuk u'uhig*—out here on the desert. *Havañ* is crow. Usually in the field where there is corn. Or in mesquite forests. They say they hide corn in the mesquite mistletoe. Dozens of them usually."

Another time he said, "*Chuk u'uhig* has a large beak. My great-granduncle used to tease [me], because [my nose is] big like that. *'Hegio chuk u'uhig shakachk a daak.'* Those are the ones that stole their corn. They take whole ear of corn and hide it in mistletoe of mesquite." He characterized *hahavañ* as coming in the fall, always in flocks, about corn harvest time, unlike the raven, *chuk u'uhig*, known as a solitary bird.

I might be inclined to consider Sylvester's folk taxonomy idiosyncratic, but there are several suggestions that it might not be. There is, for instance, the statement by ethnographer Russell (1908:92) that "[n]ot with the withering drought alone has the Gileño planter to contend, but also with the myriads of crows that are extravagantly fond of a corn diet." In all other contexts, Russell referred to ravens. Sally Pablo, a generation younger than Sylvester, told me she remembers flocks of them coming (to Santa Cruz?) when she was a child. Also, Pima cattleman Leonard Pancott said that *havañ* was the size of a *haupal*, Red-tailed Hawk, and a lone bird. "There is a bird, smaller than *havañ*, bigger than *shashañ* [Red-winged Blackbird], but like them, travels in a bunch." But he couldn't remember the Pima name. This suggests an additional category in the folk lexicon. At Quitovac oasis in Sonora, Luciano Noriega called the Common Raven *ge'e havañ*, "large corvid," implying a distinction with some other species. I didn't ask with what it might contrast.

A sentence in Wright (1929:129) uses the name Sylvester gave, *chuk u'uhig*: "But Neu-neu-fee—the buzzards, and Schook Oo-oo-fick—the blackbirds—and all the larger birds were on guard." The context implies a large bird. The gloss may be just a literal translation of the O'odham words. If an icterid had been intended, the O'odham storyteller would have used its own name, *shashañ* (Red-Winged Blackbird). (There are additional references on pages 164 and 166 in Wright.)

At Contact, there was a Sobaipuri settlement on the middle Río San Pedro recorded as Gaybanipitea or Jaibanipita. This name could be Haivan Biit, "Cow Dung," but it is doubtful that a native village would have been named after a cow before Kino and Manje's early entradas. More likely the name is Havanĭ Biit, "Raven Feces." The San Pedro today abounds with ravens, both Common and, along its upper reach at least, White-necked.

Was the Salt-Gila confluence, when it was luxuriant with the two major rivers and with the additions of several rivers from the north and a number of major washes from the south, once the location of colonial nesting corvids such as the Common Crow, or was it just the congregating place for roosting corvids such as the Common and perhaps White-necked Ravens? We can never know now in its present degraded condition.

COMPARATIVE LINGUISTICS. The unanalyzable lexeme *havañ (hawañ)* is found throughout Upper Piman. But the word seems absent in the rest of Tepiman country. Both seventeenth-century Névome and eighteenth-century Northern Tepehuán had *coconi*. At Ónavas, both Pedro Estrella and Maria Córdova identified ravens as *kokon*. Hale (n.d.a), probably more correctly, recorded *koko'on* from the same speakers, as did Johnson (n.d.). Here the Purple Nutsedge, an important weed in summer rain milpas, is called *kokon tatat*, "raven's feet." Among Mountain Pima, I have heard the bird called both *kokon* and *kokoñi*. The word continues into Northern Tepehuán, where Pennington (1969:115) found *kokoñi* for some species of *Corvus*. The palatalized *n* in these two groups is curious. But Mason (1917) recorded *kókon* among the Tepecano (southernmost Southern Tepehuán).

No cognate of *kokon* has been found so far in Upper Piman folk taxonomy, but Culver Cassa suspects that a residual form may survive in a mythic narrative where a character appears as Kokoñip (Bahr and associates 1994:258–260; Hayden 1935; Lloyd 1911:166–167). Culver suggests the boy's name means Raven. He is an abandoned Apache who grows up to become the father of Paḍ Aangam, "Ugly Desert-willow," in the story related earlier. The particular episode in the Creation Story seems rather late, perhaps sixteenth century, and has an historical ambience.

The more southern word for cuervo is well attested among other Sonoran Uto-Aztecans: *koni* (Cahita); *komát* and *kárach* (?) (Eudeve); *gorachi, korachi* (Tarahumara or Rarámuri). Except for the Yaqui-Mayo name, these names are more likely onomatopoeic than cognate.

For modern Yoeme (Yaqui), Molina, Valenzuela, and Shaul (1999:79, 120) give *kooni* for "crow" and *sanku'ukuchi* for "raven." The crow referent must be the Sinaloan Crow, *Corvus sinaloae*, a social species here at the northern edge of its range long the coast. They do not reach as far as Pima Bajo country on the Río Yaqui.

TECHNICAL NOTE. Although Saxton, Saxton, and Enos (1983:19) correctly identify *hawañĭ* as Common Raven *(Corvus corax)*, they misidentify the plant *hawañĭ taatad* as filaree, *Erodium cicutarium* (which they give elsewhere [their page 33] correctly as *kookoḍ-oipij/hoho'ibad*, "goose's [= crane's] awls/needles"); as noted, *havañ taataḍ* is the wild-heliotrope, *Phacelia* spp.

***shadav* (or *shalav?*) (Pima Bajo)**

Corvidae, Passeriformes

Calocitta colliei

Black-throated Magpie-Jay; Urraca Hermosa Cara Negra (Carinegra?)

This enormous jay, almost a yard long, reaches the northern end of its range in Pima Bajo country. Its body is bright blue above, white below except for a broad black bib from chin to chest. Black cheeks and a tall black crest ornament the head. The bird's tail accounts for much of its length, sixteen to twenty inches. The feathers are strongly graduated, black below, broadly tipped with white, and blue above.

Such a large and showy bird can be notoriously secretive as it forages through the tropical deciduous forest, usually in small flocks. When they vocalize, their various calls ring through the treetops.

Pedro Estrella and I are working our way down an arroyo through a barranca north of Ónavas called El Cajon. It is midday in late May. The sun is merciless. We are neither mad dogs nor Englishmen. The thornscrub provides little shade. We pause to admire a leafless tescalama, a giant fig whose intricate network of roots clings to the barranca wall, a living filigree in white. We are discovered by two great magpie-jays, who begin scolding us from the top of the fig, their calls echoing through the canyon. Don Pedro calls them *shalav*. Later in my elicitation book he calls them *s-shalav*, in Spanish *urraca*. He adds, *"tete babahi*, colas largas" (long tails) (I don't know if this is an alternate name or just additional information.

In October several years later, Pedro and I wander onto a small flock or family group of magpie-jays. We're in tropical deciduous forest. This time I write *shadavĭ* for urraca. I tape the birds communicating with each other in tropical tones, then Pedro saying their name. He also calls the jay *s-tuudagĭ bahikam*, "blue tail–it-has" [attributive].

For urraca, the Névome vocabulario gave *saiabi* (= *sai'avĭ* [?]) (Pennington 1979:118). This 1660s entry added: "su copete [se llama] *sisibota*"—that is, its crown is called *sisivotă*. This word is still in use in the north (for example, *siavogĭ*, meteor). But the jay name differs from Pedro's contemporary word. One might suspect a mistranscription except that the Attí vocabulario gives *saiabi* for urraca pajaro. This is

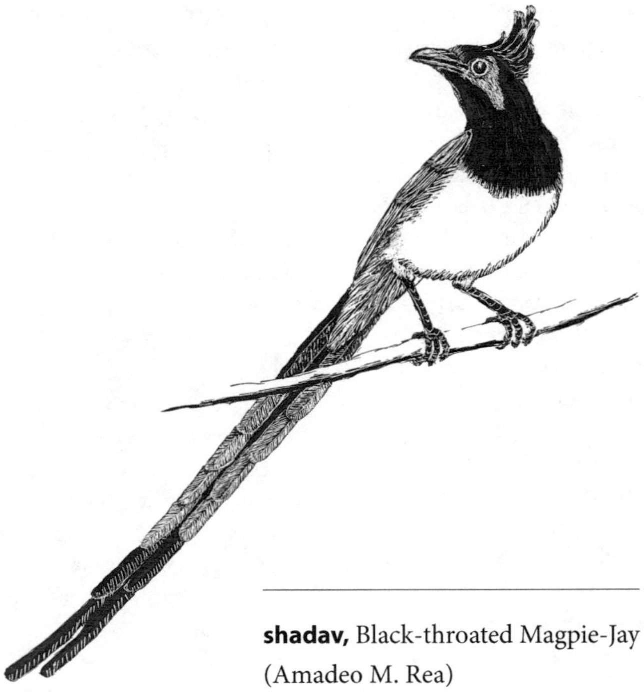

shadav, Black-throated Magpie-Jay (Amadeo M. Rea)

Upper Pima, near the Arizona border in desert scrub, far north (by two hundred kilometers) of the magpie-jay's range. The most likely referent here would be the Mexican Jay *(Aphelocoma ultramarina)*, a social species from the higher oak country to the north of this mission.

COMPARATIVE LINGUISTICS. The names for this bird appear to be lacking among Upper Pimans, who live below the wooded and forested mountains occupied by jays. The smooth-headed Mexican (Gray-breasted) Jay and the crested Steller's Jay *(Cyanocitta stelleri)* occur in Mountain Pima country. For the Mexican Jay, I was given *shishokama* and *wishokama* by different speakers. For Steller's, I heard both *si'alkam* and *shi'elkama*. But these names need to be worked out more carefully here and also with the Tepehuán, where additional species occur.

TECHNICAL NOTE. The Black-throated form of magpie-jay to the north and the White-throated form to the south meet in Jalisco. Their taxonomic relationship is uncertain. They have been treated as full species or as intergrading subspecies. For a discussion of the problem, see Phillips 1986:58.

giisobĭ*, pl. *giigisobĭ* (all groups); var. *giishobĭ

Remizidae: Passeriformes
Auriparus flaviceps
Verdin; Verdín, Valoncito, Baloncillo

In the Piman genesis, there are four original Beings: Jeweḍ Maakai (Earth Shaman), Ñui (Vulture), Ban (Coyote), and Se'ehe or I'itoi (Elder Brother). After the peopling of the earth, Elder Brother makes the most handsome man yet seen, who does a bang-up job of producing children with maidens in rapid succession, until one, with some magical help, sodomizes him, and he himself bears a child. He leaves the child under a bush, and its tears begin the Flood. Four small birds escape the Flood: *giisobĭ* (Verdin), *vipismal* (hummingbird), *hikvig* (Gila Woodpecker), and *kuotvich* (= *kuḍvich*) (thrasher) (Benedict 2001:11, 15; Lloyd 1911:43–45; Russell 1908:211). These four fly to the sky and hold on to it with their bills.

In another version of this story, only two birds are holding on to the sky. One is the thrasher, and the other "the tiny bird that builds its nest like an olla, with only a very small opening to get in. The larger one cried and cried, but the other just held on tight and said, 'Don't cry. You see that I'm littler than you, but I'm very brave'" (Russell 1908:245; see also Neff 1912:52–53).

What is this tiny bird, *giisobĭ*, that has endured so much confusion on the part of ethnologists, folklorists, and linguists (see "Misidentification" at the end of this account), yet is so important in Piman thought?

A compact tan bird with a relatively short tail foraging among the paloverdes, mesquites, saltbushes or even Creosote Bushes utters a whistled *cheet-cheet* or *cheet-cheet-cheet*, disproportionately loud for its small size. This is the Verdin. When this bird is seen at closer range, the pale gray underparts, an all-yellow head, face, and throat, and a spot of rufous in the bend of the wing will distinguish it from all other desert birds. The Verdin is usually solitary, but family groups forage together for a while after young have fledged. The young are all tan and gray. The sharp, cone-shaped bill will separate young Verdins from similar-colored Bushtits, *Psaltriparus minimus*.

The nest, like the voice, seems outsized. It is a sturdy globe of usually spiny twigs, held together inconspicuously with spider webs. The nest is almost as large as that of the Cactus Wren, but a bit more tidy. The entrance is at one side, facing downward. Nests are constructed for winter dormitories as well as for breeding. Breeding nests have more of a lip between the nesting cavity and the opening than do roosting nests. Relatively early in the afternoon the Verdin retires to its fortified roosting place, secure from predators except snakes.

"The *giisobĭ* we have around here," explained Joseph Giff, "builds their nest mostly in *u'us chevaḍbaḍ* [Graythorn], something like *hegai a'agovĭ*, that Crucifixion-thorn. Their nest is like this, with the opening in the bottom. They crawl in from the bottom." He then continued, "There's a story about the *giisobĭ* and the *hoohi* [Mourning Dove]. The birds [are] in their nests. The *hoohi* builds a nest on the fork of a mesquite, up there and lays its eggs. The *giisobĭ* builds its nest on something that swings with the wind. When the wind was blowing hard, rocking its nest, the *giisobĭ* said, 'Why doesn't it rain instead of blowing?' And the *hoohi* [on a sturdy mesquite branch] said, 'Why doesn't it blow instead of rain?' The *hoohi* was out in the open."

Sylvester Matthias told a somewhat different version. "There is that bedtime story about the tarantula and the *giisobĭ* complaining about the storm. The tarantula has its hole in the *akĭ* [arroyo]; it wants the wind but not the rain to come and flood

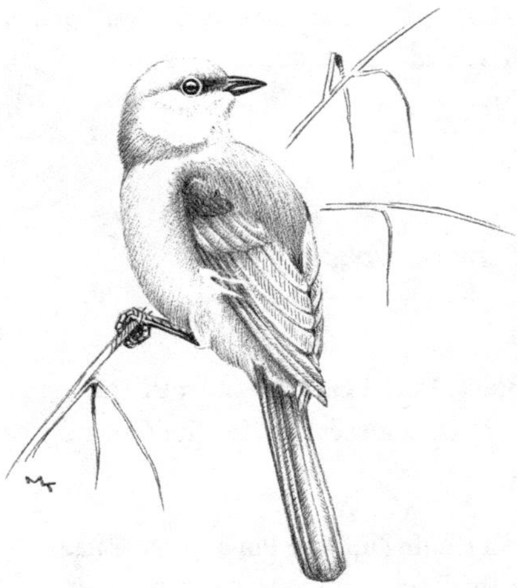

giisobĭ, Verdin (Mike Touch)

it. And the *giisobĭ* has a good nest in the branches. It doesn't mind the rain. But it doesn't like the wind to shake up its nest."

The Verdin is also a power bird. Tohono O'odham Jose Garcia explained that the *giisobĭ* is a bird that should not be molested because it could cause a violent and destructive rainstorm. Larry Saraficio agreed, adding, "Don't mess around with them, or they will mess up your house, bad rain."

The Verdin in Piman thought is the paradigm of one who retires early. For instance, Tohono O'odham Juan Dolores (1936), discussing Papago nicknames, said of *kihSuhP*, "A very small bird which goes to sleep at sundown. A person who goes to sleep early in the evening." Clemencia Leos of Sacaton Flats characterized the Verdin as "the bird that puts us to sleep." Theodore Peters said, "That guy would be in his nest *early!* When we were kids, we would go catch it." Apparently here on the Gila they were no longer warned of the dangers of molesting the Verdin.

Tohono O'odham Jose Enriques called this bird by a variant name *giishobĭ*. "They make the door downwards in their nest. In my younger days, we would plug the hole and catch them before it gets dark."

In the Pima Creation Epic recorded by Julian Hayden (1935:45), two birds are used by the invading Vupshkam as agents of sleep, the *koologam*, or poorwill, and the *giisobĭ*. After sending in the poorwill to put the people of one pueblo to sleep, one of the Emergenti also makes the Verdin.

> Another Wooshkam made another bird called "gi i sop," a small bird that built [a] small nest like a basket with a hole in the side, and sent him here.
>
> And he sings a song:
>
> Blue oriole [*sic*]
> You are going
> You have found the enemies
> And are putting them to sleep.
>
> When this bird got here he made a nest and before the sun went down he went in the nest and went to sleep. And so it was with these people, everyone went to sleep.

In this manner, the inhabitants are overcome. The bird that the Verdin is paired with, the poorwill, was well known to the Pima as the bird that sleeps in the winter and is now known even to Western scientists as a hibernator (see Common Poorwill species account).

In reference to the same song Hayden recorded, Herzog (1936:334) said it could be used either as warpath magic or as a lullaby to put a child to sleep. He used the word *green* in the translation of the song, but the same word in O'odham, *cheedagĭ*, is used for both blue and green. Why this color should be used in reference to the Verdin escapes me. But this little bird is respectfully called a *maakai*, magician, in Piman stories. Here is Herzog's version:

> Little green *giisobĭ* Magician
> Going with us
> He saw my enemies yonder and
> Put them well to sleep.

Densmore (1929:182) noted a related warpath song in which the Verdin assists the *maakai* in rendering the enemy vulnerable: "The little green bird [*sichunang gisok (sic)*] and the medicine man together were running about over there, locating the enemy and causing their minds to have nothing in them." In a parallel song, she gave *koomagĭ chukuḍ maakai*, "Gray Great Horned Owl Shaman," as the agent in making the enemy sleepy.

One of Thin Leather's most eloquent stories transcribed by Lloyd (1911:217–229) is the account of Corn Man and Tobacco Woman. The two quarrel over gambling losses, telling each other to leave the community. Tobacco ultimately departs to the west, at which point a drought ensues because the People can no longer entice rain with tobacco smoke. Corn and Pumpkin abandon the People, going eastward. In the second year of drought, "a powerful doctor, *Gee-hee-sop*," goes to Tobacco Woman with four votive offerings: a shaman's crystal, a shaman's slate pallet such as those discovered in the ruins, some *viigĭ*, or soft feathers, and some eagle tail feathers. Although Tobacco is still in a huff, she gives Giisobĭ four balls of tobacco seeds and the droppings of the Tobacco Horned Worm to smoke. Just enough rain is produced to get the seedlings started, then some of their leaves are smoked to begin more rain and revive the wilting tobacco plants. With these leaves, Verdin initiates the communal smoking ceremony. After many years, Corn and Pumpkin return to the People.

Discussing the Akimel O'odham calendar, Russell (1908:36) noted, "It is said that when Elder Brother was leaving Pimería for the last time he told the people to count the tail feathers of the little bird, Gisap, which are twelve in number, and that they should divide the year into that number of parts." Another version says, "[Jeweḍ Maakai] also did this. He spread out the tail feathers of the *giisobĭ* in the sight of the people. 'The tail feathers of the *giisobĭ* are twelve; that many will be the months. That will be the year. On this side [of the winter solstice] there will be six months. And on this side there will be six months'" (Herzog [1929]:31). (See also Benedict 2001:24, 97–98, where this calendric lesson is associated with learning how to farm.)

Nearly all song birds (passerines) in the world have twelve rectrices, tail feathers. Why this particular species was chosen to represent the calendar is not clear. The Piman month was a lunar month (*mashad*, moon, twenty-nine and a half days), with roughly twelve moons in an annual cycle. We know that the winter solstice was carefully noted by O'odham, so perhaps the discrepancy in days was "corrected" at that time. The new year began, however, at the summer solstice, about the time of the Saguaro harvest moon, *haashañ bahidag mashad*, and the anticipated monsoons. Perhaps *giisobĭ* was chosen because he represents a powerful mythological shaman.

There are many versions of the story of how a neglected child becomes the first Saguaro. In one related by Susie Ignacio Enos (Evers 1980:176–180), a little gray bird befriends the child during her search for her mother. Later it is a little gray bird with one eye that discovers the child transformed into the Giant Cactus. Saguaro says, "Now I am a *hash'an*, I can help the birds. They will be the first to eat of my fruit when it is ripe. When I stood near the village the Aw'awtam children used to shoot and throw rocks at the birds when they tried to eat my fruit, but I didn't like that. If the birds couldn't eat of my fruit then no one else could." The O'odham name for the little gray bird is not given, but was likely *giisobĭ*, the Verdin, which figures so importantly in Piman story.

COMPARATIVE LINGUISTICS. The name *giisobĭ* appears to be widespread among desert Pimans. Pima Bajo María Córdova on the Río Yaqui used the name. (Strangely, Pedro Estrella didn't know the nest or the specimens we collected in the field.) Luciano Noriega at Quitovac oasis switched the names for hummingbird and Verdin, I think probably in error.

With the folk genus *giisobĭ* being used by most Piman speakers for both *Auriparus* and *Polioptila*, the word becomes polysemous and thus unmarked as the folk species name for the Verdin. See the next species account.

MISIDENTIFICATIONS. The Verdin, *giisobĭ*—such a common and conspicuous element in the Sonoran Desert and southward throughout the tropical thornscrub, a bird so well known to O'odham—seems to have befuddled ethnographers. Russell (1908:211) as well as Benedict (2001:11) gave the bird's name only in Pima. Lloyd likewise gave only *gee-ee-sop*, supposing this bird to be a gnatcatcher (which he described, but did not identify). Densmore (1929:182) translated *gisok* as "hummingbird." But the modifier in the song she recorded is supposedly *green*. Saxton and Saxton (1969:167) called *giisoki* "large hummingbird," in contrast to *wipismal*, "small hummingbird." Verdins (passerines) are unrelated to hummingbirds, which belong to a different order.

Bahr (in Bahr and associates 1994:221 n. *n*) misidentifies the bird in the Hayden-Smith manuscript and thereby misses the metaphor entirely. "Gi:supi, 'black-phoebe-bird.' These birds are like orioles in that they build mud nests [*sic*], but in my experience Pima-Papago have a different word for orioles, *wajukuk*. The latter are yellow and black, while the phoebes are gray and black. Smith's song, however, is about a blue 'oriole/phoebe,' which, I take it, is a mythical magical bird that we cannot see." And in an earlier footnote (Bahr and associates 1994:97), Bahr explains his (mis)identification of *giisobĭ*: "Gi:supi, 'black phoebe bird,' *Sayornis nigricans* (Saxton, Saxton, and Enos 1983:167). Saxton and Saxton (1969:167) spell this bird name as *gi:soki*, and at that time they identified it as a large hummingbird. No doubt, the later 1983 spelling and identification are corrections. These 'flycatcher'-type (at least to Americans) birds always live near water. They make their nests from leaves mixed with mud (Peterson 1961:182)."

Unfortunately, the original O'odham words of this particular conquest song appear to be now lost (at least, the Verdin Song is not among the emergence songs recorded by Russell [1908:280–282]). However, I think that the reference to "blue oriole" in the song was simply Hayden's (1935) best attempt at translating the Pima word *giisobĭ*, a species that he apparently did not know. After all, the translator had described its nest (somewhat inaccurately) as like a basket, and orioles weave basket-like nests. Actually, the Verdin belongs to an Old World family, the penduline tits (Remizidae), and is not closely related to either the tyrant flycatchers (Tyrannidae) or the oriole-blackbirds (Icteridae), both being exclusively New World families. And not only are they nonmythical and quite visible, but they have been depicted on a series of U.S. postage stamps honoring the Sonoran Desert (in 1999).

Bahr (in Bahr and associates 1997:129 n. 112) subsequently repeats this error. "Different versions of the story give lists of the birds, but they commonly include hummingbird, woodpecker, buzzard (I didn't expect this), and black phoebe." I disagree with this list and the claim that "five birds in all were saved," as given by Russell (1908:211); Buzzard was saved as one of the pantheon of four Original Beings or culture heroes, not as a bird. Four *birds* as such were ultimately saved.

In editing Benedict's anthology of Pima myths, Bahr thrice more misidentifies the Verdin, called "Kiisop" and *giisop*, as the Black Phoebe (in Benedict 2001:11n., 15, 76, 98). It is doubtful that any O'odham will ever see a Black Phoebe, but all who are familiar with their Sonoran Desert biota will know the ubiquitous Verdin, the powerful shaman.

Mathiot (1973:293), to her credit, got the identification at least partially correct, calling *giisobĭ* the Verdin, but also the oriole (which it isn't). She added that the word is "also used as a pet name for babies," as in *ñgisobiga*, "my Verdin."

s-chuk mo'okam giisobĭ

Sylviidae, Passeriformes

Polioptila melanura

Black-tailed Gnatcatcher; Pisita, Perlita Desértica, Perlita Colinegra

I sit reading under a Pima *vatto*, ramada. A small sound attracts my attention. It is easily imitated by drawing air slowly between the upper teeth pressed against the lower lip. A tiny gray-and-white bird with needle thin legs and an expressive black tail pops into view on the Ocotillo wall. It forages between the supporting beams and the Arrow-weed roof, searching for creatures too small for me to see. I mimic the call and shortly the gray mite is joined by its mate. They bob about the branches peering at me.

I imitate their call note again, and the two approach confidently, tipping and cocking their heads to discover who the interloper might be in their territory. Now I can see the white tips on their tail feathers that identify this local species. Satisfied that I am not another gnatcatcher, they depart, one trailing low after the other, for some nearby scrubby mesquites.

This bird, usually in pairs, is found commonly in the Sonoran Desert in the small-leafed vegetation: mesquite, paloverde,

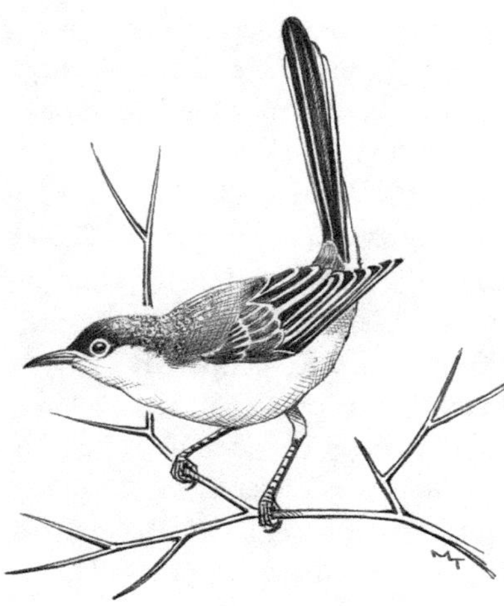

s-chuk mo'okam giisobĭ, Black-tailed Gnatcatcher (Mike Touch)

acacia, Creosote Bush, Graythorn. During the breeding season, the male sports a black crown, hence its folk species name. But in winter he is indistinguishable in the field from the gray-headed female and young. The two often forage in the somewhat taller and thicker vegetation along arroyos. They keep in touch by call note and are seldom more than hearing distance apart.

The gnatcatcher nest almost rivals that of the hummingbird. It is a finely constructed deep cup of grayish fibers, often close to the ground. The adults alternate nest duties. The black eyes, bill, and upright black tail of the incubating bird are all that show above the nest rim. The gnatcatcher set is usually four eggs, faintly bluish green and finely speckled, whereas hummingbirds lay two immaculate white eggs.

The Black-tailed Gnatcatcher is distinguished in Akimel and Tohono O'odham at the folk-species level from the similarly small but very different and unrelated Verdin. The Verdin is called just *giisobĭ*, but the gnatcatcher is *s-chuk mo'okam giisobĭ*, the "black-headed Verdin."

In her search for Tohono O'odham songs, Densmore (1929:212–213) recorded two lullabies, both widely known to the River Pima as well. One of them is the song about the Round-tailed Ground-squirrel and the green mesquite pods (see Rea 1998b:153–154). The other is the Black-headed Verdin Song. Densmore provided only the English words:

> Gisuk [*sic*], black-headed gisuk.
> Run and come so the baby's eyes will go to sleep.

She was told that the *gisuk*, which she didn't identify, was "a funny little gray bird that runs on the ground." (Small wonder she couldn't identify it!)

The lullaby sang to Pima children compares the black-headed baby with the black-headed bird asleep in its soft, finely constructed nest. (Only the male has the black head, but both sexes incubate and are very tight sitters on their eggs.) The song was well known to the older people with whom I worked. Joe Giff of Komaḍk provided this version of the lullaby:

> Chuk mo'okam giisobi
> ab g hu i meḍk shulij wuupui
> kokosim.

George Kyyitan of Bapchule sang for me this slightly different version of the lullaby:

> Chuku mo'okam a giisobi
> apakwi muda Amadeo wupoi
> kokosim a wa(n)
> piu! piu! pium!
>
> Black-haired [-headed] Verdin
> go back and make
> baby's eyes tired
> piu! piu! pium!

In Joe's version, a little boy is named in the second line, *shulij*, as the one being put to sleep. If it were a girl, the word *chehia* would be used in this place. In George's version, the actual name of the child is inserted; in this case, George used my name. The piman word *mo'okam* can mean "black-haired" as in the case of the O'odham baby being sung to, or "black-headed," as in the case of the gnatcatcher.

Anna Moore Shaw (1974:111) provided still another version:

> chuk mo'okam giisob
> ku'imediga chulachga
> wuupuik kokosima
> biu, biu, biu.

Lloyd (1911:53) left us two entries from his Gila River journal, both clearly applying to the gnatcatcher, which he supposed to be his "gee-gee-sop." But there is no indication that anyone actually pointed out for him the Black-tailed Gnatcatcher or the Verdin itself as being *giisobĭ*.

hokkaḍ*; var. *hogkaḍ*, pl. *hohokaḍ

Troglodytidae, Passeriformes
Campylorhynchus brunneicapillus
Cactus Wren; Saltapared, Matraca del Desierto, Matraca Grande

These clowns of the desert seem the antithesis of wrendom (typically small, furtive, dull-colored birds with elaborate melodious songs). Cactus Wrens do everything wrong. These heavy-bodied eight-inch birds, almost as large as a Bendire's Thrasher, go boldly about scouring the environment with rather long curved bills and insatiable curiosity. Big feet equip them for climbing as well as getting along on the ground. The brown upper parts and white underparts are boldly spotted, streaked, and barred, seemingly negating protective coloration. When wings and tail

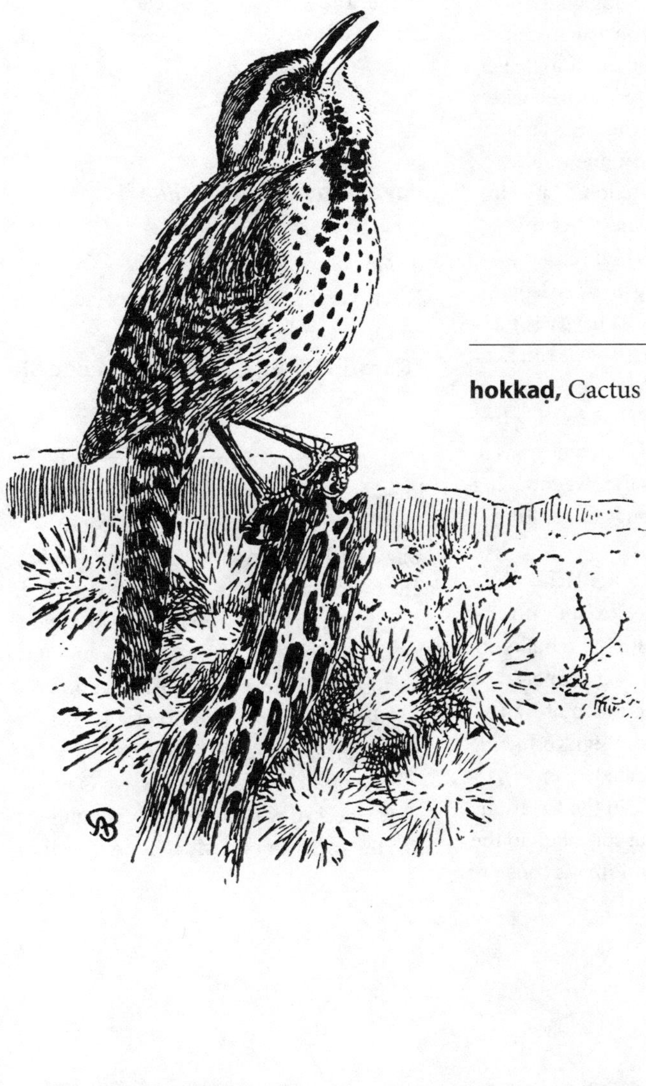

hokkaḍ, Cactus Wren (Allan Brooks)

are spread, the harlequin effect is still more noticeable.

Were all this not enough, when a Cactus Wren opens its bill to sing—and they are not at all silent birds—all resemblance to expected wren melody vanishes. The results, although variable, are loud, harsh, and rather grating. These sounds seem not to discourage them one bit: Cactus Wren calls are as defining a characteristic of the Sonoran Desert as are Saguaros and paloverdes.

Then there are its decorating proclivities. The football-size nests, with a small opening at one end, are built largely of grass or other fine vegetation, lined with whatever is available: chicken feathers, toilet paper, rags. If there are chollas, they are convenient places for nests. If not, some spiny shrub, such as Graythorn, gets decorated. About Piman villages, nests may be tucked into the roofs of the *vapto*, arbors, that stand near the adobe houses. In these quarters, the birds—pairs and families—scour through dog dishes, bean pots, outdoor cooking areas, saddles, corrals, baled hay, clotheslines, and even parked cars with open windows.

These bulky nests are used for incubating eggs and rearing young. But nest building does not stop there. Additional nests in the territory serve as sleeping quarters for the family, keeping the construction abilities honed. (This building of multiple nests is at least one orthodox trait shared with some other members of the wren family.)

With all this familiarity, one would expect the Cactus Wren to loom large in Piman literary consciousness, but this seems not the case. Ruth Giff related a conversation with her husband Joseph:

Joe: "The *hokkaḍ* doesn't have a song."

Ruth: "Yes it does. That's what you hear now, it trying to sing."

Joe: "I mean, it doesn't have a song [or group] named for it." His reference was to groups of songs, such as the swallow, kingfisher, or woodpecker songs originating from that animal.

Herb Narcia at Uus Keek, Blackwater, compared the nest of the *hokkaḍ* with that of the *giisobĭ*, Verdin. Both build bulky enclosed nests with an opening at one side.

In a curious twist, Leon Juan of Aangam equated *hogkaḍ* with Viigĭ Koshkam Maakai (Soft-Feather-Nested Shaman), the bird of the Flood episode in the Creation Story. This is the bird that stayed up in the sky in its safe nest. (Most Pimans consider this bird to be the Curve-billed Thrasher.) "During the Flood, Hogkaḍ made a *kosh*, nest, in the sky out of its *wopo*, short feathers, and was living there. The Flood reached and dislodged it. It descended with the waters. *Hokkaḍ* laid eggs. These hatched and became various peoples who dispersed."

This is the only mythological reference to this conspicuous Sonoran Desert bird that I have come across. But the bird does lend its name to the term *hogkmaḍma*, meaning "to be dotted with black" (Mathiot 1973:374). This usage parallels another Piman term, *ooshaḍ*, "to be spotted like a Jaguar," *ooshaḍ*.

COMPARATIVE LINGUISTICS. Pedro Estrella on the Río Yaqui said of the Cactus Wren, "*Séo kua'adam*, come sebo." I don't know if he intended this as the bird's name or was just observing its behavior (*sebo*, tallow, suet; *kua'adam*, "eater").

There is some variation in published forms of the bird's name. Mathiot (1973:374) wrote it *hagkaḍ*, whereas Saxton, Saxton, and Enos (1983:23) give *hokkaḍ*. Fr. Antonine (1935:9) listed the plural of *hawkat* as *hawhawkat*, glossing it "bee martin," an old name for the kingbirds. Sylvester Matthias likewise used *hohogaḍ* for the plural. The word seems unrelated to *ogaḍ*, a species of lizard.

vavas, vavuk (wawusijim?)

Troglodytidae, Passeriformes

Salpinctes obsoletus

Rock Wren; Chivirín Saltarroca

?*Catherpes mexicanus*

?Canyon Wren; Chivirín Barranqueño

Two small wrens, about five to six inches long, are at home in the desert sierras. Both bob like Water Ouzels that have lost their way and shrunk in this rugged dry country. They often carry their tails cocked upward, wren style, as they creep under rocks and boulders and scale canyon walls foraging for insects. Their colors aren't what one usually expects for wrens, typically dark furtive birds of the understory.

One of the two is mostly warm gray above, flecked with white, colored quite like the rocks among which it forages. Its rump is rusty, and its spread tail bears a broad black subterminal band ending with buffy feather tips. The underparts are whitish, faintly flecked with gray streaks. This is the Rock Wren. Its ethereal call, *ki-trrree*, ringing out in the silence, is characteristic of these wild, solitary places. In winter, Rock Wrens may appear in Piman yards, foraging through woodpiles, climbing adobe house walls, bringing wildness with them. Whether they have descended from the local desert sierras or are migrants from the cold north country is still their secret.

The other bird is equally unorthodox for a wren. It is mostly strikingly rufous (deep rusty brown), darker above, speckled with dark and white. Even the tail is all rusty, narrowly barred with black. It sports a silky white bib from chin to chest. Its head, more grayish than the back, is flattened and the bill quite long. This is the Canyon Wren, at home on the steepest cliffs and canyon walls of the desert sierras. Few will fail to notice its easily imitated piercing cascade of whistled notes, dropping and slowing, ending in a rebounding flourish, *tsii-tsii*. In winter, it, too, may descend into the villages, climbing on mud walls rather than on steep rock faces, but far less frequently than the Rock Wren.

The crowds of Pimas assemble slowly on this cold November day as the village of Santa Cruz celebrates its patron, St. Catherine. Sylvester Matthias sits out in the sun by himself. Although only in his early eighties (as are some of the women working here to get the feast ready), he is beginning to slow down, and it takes a bit of priming to get a conversation going. A Rock Wren comes right next to us, bobbing first on the top of a fence post, then down on the ground, ignoring the dozens of people around. It calls *ki-trrree*. I ask Sylvester about it.

"*Vavas!* It's Rock Wren or Canyon Wren—it's the same [in Pima]." But it has taken some three decades of equivocation to reach this conclusion. *Vavas* means specifically Rock Wren, the most frequently encountered species in the lowlands and bajadas where most Piman activity takes place. But are there two folk species?

Years ago Joseph Giff had said that the *vavas*'s song is about *komalk vavas*, meaning "flat or thin wren," not *koomagĭ vavas*, "gray wren." Sylvester said he agreed with this interpretation. In one discussion, Joe

vavas, vavuk, Rock Wren (Allan Brooks) and Canyon Wren (Allan Brooks)

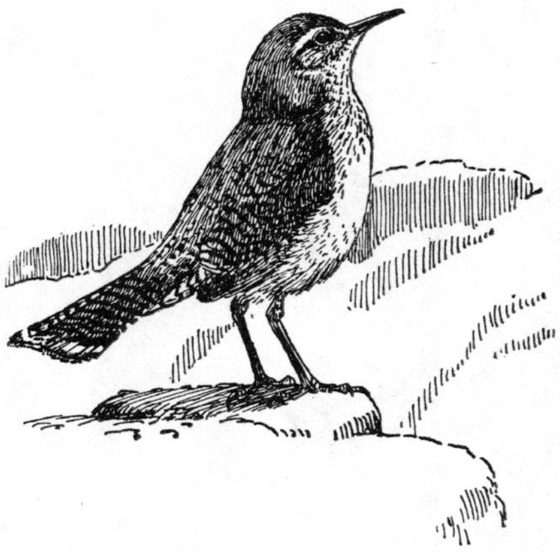

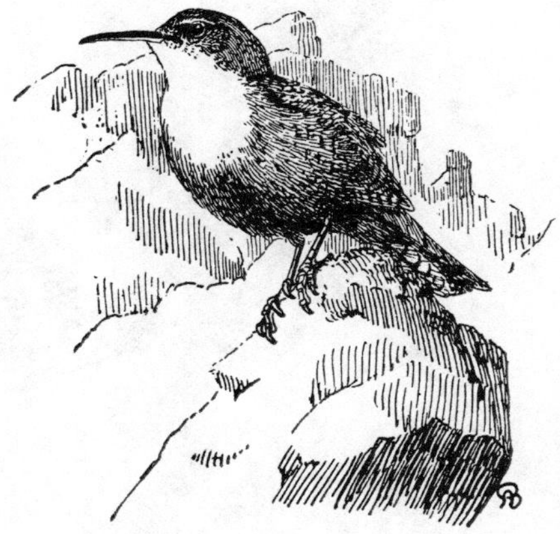

first thought the bird was the Rock Wren, then changed his opinion to the Canyon Wren. "You can't see them, but you hear them." But once when David Brown of Santa Cruz village was telling me about *vavas*, he said, "You know, the little gray bird that comes down [presumably from the adjacent Estrellas] in the winter."

The song that Joe and Sylvester mentioned is about a Rock Wren, *vavas*, who makes wine from the *hannam hiosig*, Buckhorn Cholla blossoms. It is sung by an oriole who gets drunk. Here is Blaine Pablo's version first in song language:

Koomagĭ Vavas Ñe'i
Kokomagi vavchus
hannami yosike navaitoke
iyañe melvita.

kuñama wemaji yo ke navamo
hiaya vañe piñe mamache
gamu shakalano meneta.

Then in spoken Pima:

Koomagi vavas
hannam hiosig navaitk
iiya at o ñ-meliv.

kuñ am wemaj ii k ñ-navamch
vañ pi ñ-maach
shakal meḍ.

And in English:

Rock Wren Song
Gray wren
Cholla flower makes wine and
Here runs to me
[to invite me to drink].

And I with [him]
Drink and get drunk
I don't know,
Diagonally run.

Francis Vavages at Sacaton also sang the song. His version is slightly different:

Hanama hiosikai'iyo ke'e navama
vañ pi ñmamache he
shakali memeḍa
kuñ ama wemaj haiyo ke'e navama.

Here is how Francis rendered his song in English:

[I] got drunk with cholla flowers
And I don't know what I [am] doing
Running side by side [or crooked]
With [them] there I get drunk.

Francis said that the *vavas* was a little gray bird that sleeps in the rafters of people's houses. This song is clearly about the Rock Wren, a gray bird, seen more frequently in the villages than is the cliff-loving Canyon Wren.

However, Russell (1908:303, 305), in a set of seventeen diagnosing medicine songs, included two (numbers 3 and 13) that are about a "kind of yellow wren." Culver Cassa has reworked these two songs, the first one as:

Alisi yua'ame vavajusi
hejel ya'ana yupanaka
yekaj vañi chepavchuna hiyañi momodame
maveshapaiime jewenḍa wesiko memehelime.

Little yellow wren
Using his own feather pluckings
He makes me a harlot placing them atop my head.
Holding hands, all around the country I run.

Culver notes that what is understood is that the wren makes a feathered crown and puts it on the dreamer-singer's head. To hold hands means to dance; he's going to dance. The second song says:

Alwes yua'am vavchusi
gahamuva mehemeko chechokame dodovange
wuika beñi beichukime
gamudañi mamaichumhi pokiñim chekima.

Little Yellow Wren
Over there far off Cheecho [Chuichu] Mountain
Toward there he's taking me
He's following someone's tracks.

Both songs are about *oam vavas*, supposedly "yellow wren," but the word *oam* can equally mean orange or brown. These songs depict the rusty chestnut Canyon Wren, not the pale and grayish Rock Wren. Also, as the second song tells, the Canyon Wren nests in caves in the desert sierras, and the bird forages here in rock crevices.

For some Pimans, *vavas* is the *komalk mo'okam*, or flat-headed bird, whose head was stepped on at the first wine feast. Those who hold this view associate *komalk mo'okam* with the Canyon Wren, which indeed has a very flat head, an adaptation for foraging in rock crevices. Others ascribe this incident to various other birds (see chapter 4).

One time Sylvester spoke of a *cheedagĭ vavas*, literally a blue or green wren. *Cheedagĭ* is the same modifying term used by Pima to distinguish the Peregrine from the Prairie Falcon. The use of these two different modifiers suggests that there are two folk species of small wrens: *koomagĭ* or *cheedagĭ vavas* (or just *vavas*), the Rock Wren, and *oam vavas*, the Canyon Wren. Appropriately, the root of the generic name is *vav*: cliff, boulder, rock. The primary habitat for both species is the rocky country of desert prominences.

Delores Lewis used a somewhat different name, *wawusijim*, still incorporating the root word *vav (waw)*. On the basis of pictures, he thought the name probably included both Rock and Canyon Wrens. When I asked Joe Lopez, a relative of Delores from another village, about *vavas*, he called it *wawuk*, "from among the rocks," but also knew the name *wawusijim*, saying, "It lives among the rocks, too." Ruth Giff on the Gila also used the name *vavuk*, saying, "It is real tiny." Saxton and Saxton (1969:167) list an unidentified bird as *wawasjel*, undoubtedly a wren. All four of these variants include the *vav/waw* word that characterizes the wren's habitat.

Jose Enriques of Queens Well called the Canyon Wren *wawaga* or *wawga*. He said they hear it in the mountains when they go hunting, but were warned not to whistle its easily imitated cascading song or the bird will confuse the hunter. He will become lost, not know whether he is going up or down, and be unable to get out of the mountains. Of course, the lost hunter becomes a reflection of the acrobatic wren that scales up and down the steepest cliffs while foraging.

Still farther south, in Pima Bajo country, I whistled the Canyon Wren song to Pedro Estrella. He said it was *saltapared*, using the Mexican vernacular name for "wren," then described them flying up and down steep canyon walls feeding. He called it in Pima *bidovak dada*, explaining that *bid va'aki* meant "casa de pared" (mud house wall) and *dada* meant "brinca, jumper." I think his name was a back translation of its Spanish name, saltapared, "wall jumper," into Pima. But his use of the word *va'aki* is interesting. To the north, the Pimans apply the word to the puddled adobe ruins found in their country. *Va'aki* is the old (archaic) word for "house."

Although I have identified Puente Babasic near Ímuris, Sonora, as the place where the Yerba Mansa *(vavish)* grows, based on the locally marshy ecology, it might also mean the place of the Rock Wrens, *vavasik*.

COMPARATIVE LINGUISTICS. Perhaps the Yaqui name *bai wakas* for all wrens, including the Cactus Wren, is cognate with the Piman name. The Rarámuri (Tarahumara) of the Sierra Madre distinguish the white-chested Canyon Wren, *ripíkuri* or *ripírowi*, from other wrens, called *chuépari* (Brambila 1983:504).

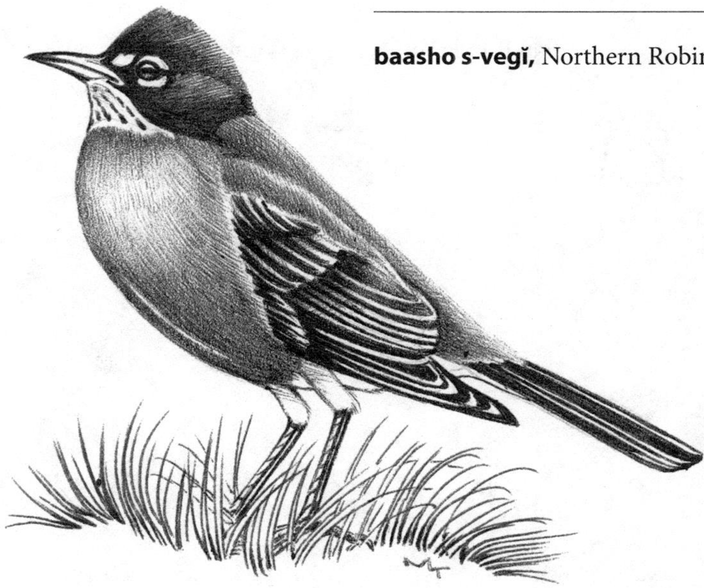

baasho s-vegĭ, Northern Robin (Mike Touch)

baasho s-vegĭ, pl. *bapsho s-vepegĭ*

Muscicapidae, Passeriformes
Turdus migratorius
Northern Robin; Primavera Real

The golden light of dawn hits the top branches of a bare tree. Several plump-bellied birds greet *tash*, the sun, with loud, explosive, single syllable call notes. In seconds, the growing light reveals the gray tops, rust breasts, and white bellies of overgrown thrushes. It is mid-November in Sacaton. Summer and winter are in tug-of-war. For the O'odham living in this desert country, the appearance of the robin meant that colder weather was soon to follow. Its spring song, known to most people of temperate North America, is unheard here in Pimería.

The birds call like businessmen in suits, loud and assertive. Their expanded waistlines are beyond salvage. There is nothing of the furtive behavior exhibited by the Hermit Thrushes or other thrushes that winter more commonly in damp, overgrown spots of the desert.

The Piman name *baasho s-vegĭ* means "chest red." One might be tempted to think this is just a back translation of the common English epithet, "robin red-breast." But far down in the Sierra Madre of Mexico I recorded the name as *oam baasham*, "orange chest," from Mountain Pima Lorenzo Cruz of Yepáchic. He said the birds were eaten. In the Yécora area, I heard both *oam baash* and *oam apach*. The first means "orange, brown, or yellow chest," the second apparently "orange Apache." Another Mountain Pima called the robin *tolonchi*.

Densmore (1929:214) recorded a Tohono O'odham song, "The Robin Brings the Cold Wind."

> The robin spreads the green wind on us.
> He spreads the coldest wind on us.
> Now knock him down with a stick.

Unfortunately, she did not give us the O'odham name for "robin" but identified the bird as such in English, with information about the bird's behavior that is appropriate: "It should be remembered that the robin goes to southern Arizona at the beginning of the cold weather, being a harbinger of winter instead of a sign of spring, as in northern climes."

Joseph Giff explained that the "green wind" was a metaphor in Pima for sadness or melancholy, or what we call in English "the blues." The last line of the song suggests that it is a hunting song, although I was never told this bird was a prey item among desert Pimans.

I asked Joe Lopez at Vaya Chiñ if he knew *s-vegĭ baasho*. "That's a robin, isn't it?" Although I inadvertantly switched the word order, he knew the bird.

Pima Bajo don Pedro Estrella used the name *s-vepek baasho* for the Vermilion Flycatcher, *Pyrocephalus rubinus*, an appropriate name for these strikingly black-and-red little birds of mesquite country. Oddly, no one among the Tohono O'odham could recall the name for this conspicuous flycatcher found about their charcos, though many knew the bird itself.

go'ogov; go'ogow

Laniidae, Passeriformes
Lanius ludovicianus
Loggerhead Shrike ("Butcher Bird"); Chonte Cabezón, Verdugo Americano

The Northern Mockingbird and the Loggerhead Shrike are both common Sonoran Desert birds, about the same size, and even similarly colored with gray-and-white bodies and showy white patches in the wings and tail. Both may perch conspicuously atop taller mesquites and paloverdes. The shrike has black (not dark gray) wings and tail. Its white wing patch is more narrowly confined, and there is a large white rump patch. But, most important, the shrike's head gives it away. It is large, supporting a heavy, more raptorial bill. A black eye patch extends forward across the bill.

The shrike flies heavily with rapid wing beats, close to the ground, in an undulating path, arcing upward just before it reaches its perch. It has none of the buoyancy of the acrobatic mockingbird. A songbird (passerine) turned raptor, it has earned one of its vernacular names, Butcher Bird. As a predator, it seeks large insects, lizards (including horned lizards), smaller rodents, and other birds. What it doesn't eat, it impales on thorns of desert trees and bushes, or even barbed-wire fences. My first year in Pima country, I watched a shrike attack, kill, and attempt to carry off an Inca Dove, a bird almost as large and heavy as itself.

The common call of the Loggerhead Shrike is loud, harsh, and discordant, seemingly appropriate for its behavior. But at times the most soft and melodious song will waft out over the saltbush flats or paloverde uplands of Piman country. When the sweet warble is traced to a perched shrike, one can only wonder if it is taking lessons from its look-alike, the mockingbird. Or practicing delusion?

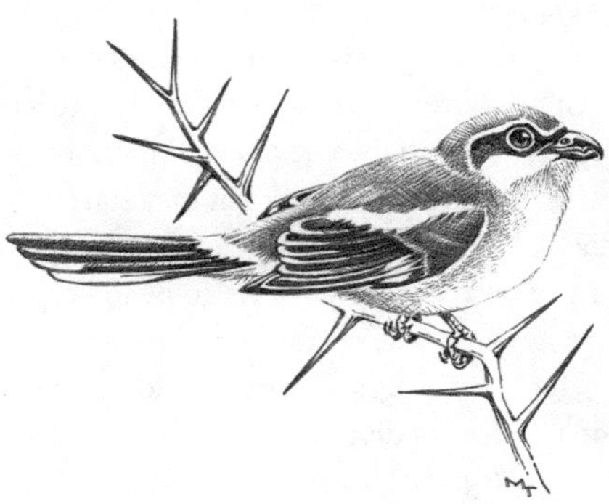

go'ogov, Loggerhead Shrike (Mike Touch)

Sylvester Matthias and other older Pimas were familiar with the shrike's food-storing tendencies. Sylvester noted that it catches lizards and eats their eyes. He related this story: "Once a Red Racer, *s-vegĭ vamaḍ*, went up a mesquite after a shrike. Chased it around. Finally, when the snake got tired following it, the shrike pecked its eyes out and the *s-vegĭ vamaḍ* fell to the ground dead." Similarly, Joe Giff remarked, "They're a cannibal bird. They go after other bird, eat the eyes out of it." Theodore Peters said, "*Go'ogov* don't like anything, just put it through a thorn." I asked what kinds of animals he had seen impaled. Grasshoppers, birds, lizards. Delores Lewis called this bird *go'ogow*. "Sure mean. Stick little animals in a tree—even horny-toad. Put the grasshopper on the thorn."

Leslie Paqual observed, "*Go'ogov*'s got something like the Lone Ranger," motioning over his eyes. "That's a bad bird. When somebody's dying, they will hang around there, you know. It's about as bad as the owl, *chukuḍ* [Great Horned Owl]. That's what I heard." This is the only association of the shrike's announcing death that I heard.

The shrike appears in a war oration that Lloyd (1911:203–204) took down in the Creation Story:

> And I asked the Butcher-bird for power, and he followed his Road of Light, and touched the ground four times with his tail, and came to me. And he went on the road that is lighted by a mahkai, and following that reached my enemy.
>
> And my enemy thought himself a good dreamer, and that his dreams were fulfilled for good, and that he had a good bow with a good string, and good cane arrows, but the Butcher-bird had already punched his eyes out without his knowing it. And all the animals and birds of the Apaches think they have good eyes to see with, but the Butcher-bird has punched their eyes out without their knowing it. And the winds of the Apaches think they have sharp eyes, and the clouds of the Apaches think themselves sharp-eyed, but the Butcher-bird has punched their eyes out without their knowing it.
>
> So he treated the enemy like that, and left him there as a woman, and then pushed me toward him, and I went and captured them.

This oration tells what the Pima hopes magically will happen when he encounters his enemy, the Apache: that he will be rendered helpless. The shrike's behavior is encoded in the oration, even to its undulating flight pattern. In the oration previous to this one, the warrior is helped by Yellow Spider and "Yellow Crow" (see Common Raven account).

shuug, pl. *shushug* (all groups)

Mimidae, Passeriformes
Mimus polyglottos
Northern Mockingbird; Cenzontle Norteño, Chonte

Through much of the year, the most varied singer in Pimería is the Northern Mockingbird. Its song may be heard periodically throughout the day and on moonlit nights. At the height of the breeding season, males will sing at various times throughout the night even without the moon. While the nesting territory is being defended, the sheer volume of the song seems completely out of proportion to the size of the singer.

The songs are a hodgepodge of bits and pieces, strung together in short repetitive series, followed by a variable period of silence. Certain songs are typical of mockingbirds in general, such as *pretty-boy! pretty-boy! pretty-boy!* Others are mimicked from other sounds in the vicinity: other animals, even mechanical sounds. Sometimes a "new" bird species heard in the neighborhood can be traced to a mockingbird that has been living in a different habitat where it learned its neighbor's calls and brought them along to its new home.

The mockingbird is rather drab, as are many talented vocalists, and makes little effort at concealment. It often sings from the highest, most exposed perches in its territory. Sometimes in a burst of exuberance, the singing male leaps up from his perch and flutters back down. It is the size

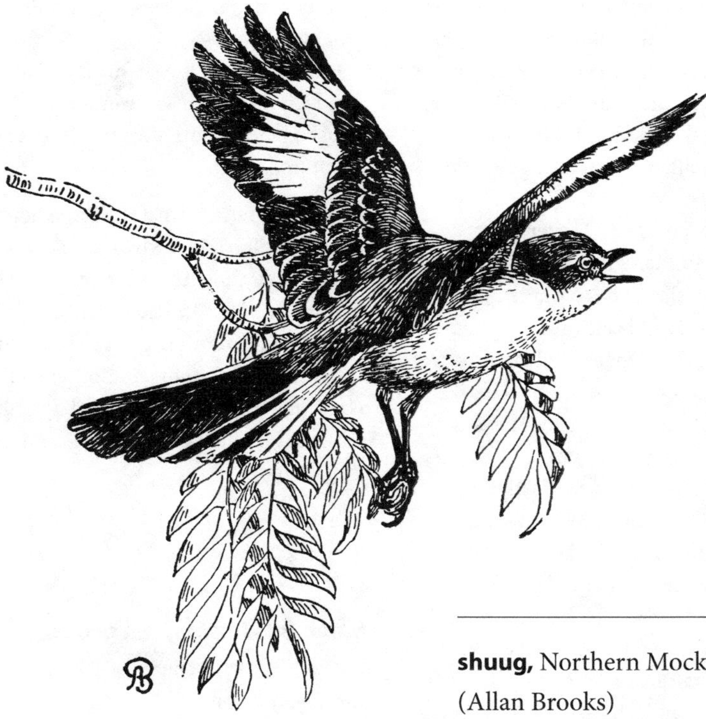

shuug, Northern Mockingbird (Allan Brooks)

of a small thrasher, about ten inches long, medium gray above, light gray to whitish below. The dark gray or blackish tail and wings have white patches, and the outer tail feather is white.

Like its other thrasher kin, the Northern Mockingbird is omnivorous, taking wolfberries and other small fruits and small invertebrates, mostly insects. Its method of foraging on the ground and at times even on branches will identify the species. It walks or hops to a spot, extends its wings horizontally while opening them in three successive jerks, flashing its white wing patches, then closing them. This calisthenic presumably startles prey into moving, allowing the peering hunter to see and take it. The white wing and tail patches are also displayed in acrobatic maneuvers to ward off intruders or to attract mates. The birds are definitely feisty, aggressively mobbing falcons, hawks, ravens, or owls that happen to perch in their territory (or even cats in urban areas). The mockingbird will continue to fly over and harass a predator for a long period, uttering a growling *churrr* call, finally chasing the offender away.

The mockingbird, *shuug*, is very well known to the O'odham. One day Sylvester Matthias told me, "They nest about anywhere, in the mesquites, in the fields [fencerows]. And years ago, people get them for pets."

"How do you keep them?" I asked.

"In a cage."

"What do you feed a *shuug?*"

"Anything. Grasshoppers. Any kind of meat. The thing I know when they water—they'll drink the water out of your mouth."

"Why did people keep them?"

"Just to hear them sing."

Sylvester then related a major medicinal use: "That Valentine Soke [a Pima baby of Santa Cruz village] should be talking. But he didn't talk. Leonard Soke took that *shuug* and roast it on the coals [and fed it to him]. He started talking. Then he talks a lot!"

Another time several Gila Pima and I were talking. I asked, "If somebody doesn't learn how to talk, what do you do?"

Ruth Giff responded, "Feed *shuug* to him."

"Any part of it?" I asked.

Everyone responded, "The whole bird."

"How do you fix it . . . in a soup?"

"Any way, I guess," said Ruth.

"Roast it, fry it, on the coals," added David Brown.

"Yes, get one and we'll see if it will work," said Ruth.

Sylvester added, "You'll be talking, talking, talking. You'll be talking in your sleep. Or you'll be mocking somebody. They always say, if somebody mocks somebody, *'Shuug! Shuug!'*" (See also Dolores 1936.)

This custom is not restricted to River Pima. Leon Juan of Aangam said *shuug* is roasted and fed to a youngster to aid speaking. I asked Tohono O'odham Frank Jim, "Have you heard of using *shuug* to get a baby to talk when it doesn't develop speech?"

"Yep. They [we] have the same thing. After he's born. They get a *shuug* and cook it, when it's dead. More like you pray to it first before you cook it. After you doing, your praying quit, then you cook it. You have to pray to give to him. Then he eats it.

"Then later on, if somebody's around, if he happens to say two or three words, talk, if that guy starts laughing at him, those people start laughing at him, well, he's lost it. It won't come back up. I don't care what they do with him, try to make him talk. You shouldn't do anything like that. You shouldn't say anything about it. Just act like you didn't hear him or act like you didn't see him or what he does. Give him about four days, then. Four days, after that he going to start talking. (In those days everything's four days.)

"There's a man, old man, down there. He died a long time ago. I guess his brother was like that. Guess he was half *maakai*, too. What they call, *tosiv* [meadowlark]? They talk a lot. So he got some of those and he let him eat that. And he got alright. He talked. It's good, too, but he said *shuug* is better. *Shuug* is more like a human. He learned to talk. You got to train him that way, to be good."

I asked Frank, "What do you call that when the baby isn't learning to talk—is that also *kaachim mumkidag?* When he doesn't talk at the right time or he doesn't crawl at the right time? Or is that something else?"

"No, they call [it] that, I guess, because it'll be the same thing because the thing is, it happens when he comes out [is born]. Everything happens when he comes out. You just have to wait, to see when he grows up a little bit. It'll come out, and it'll show what's wrong with him. Then they start to figure out what they [are] going to do. But if you don't do anything, it's going to keep on: he won't talk, or walk, or whatever."

Francis Vavages of Sacaton told a story of a healing of defective speech that implies *mumkidag*:

> The mockingbird, *shuug*. There was an old man here. He's just by himself. He's from Bapchule. They call him Mogiiya. I heard that when he was small, I guess he

couldn't walk when he was just a baby or something like that. He just crawls, sit here and there like some kids do, just crawl and sit over there. And he couldn't talk. [Francis explained here how his walking was taken care of.] Here's his talking: he just mumbles. So then another man said, "You know, they say that, if you sing a song about the mockingbird and doctor him by putting that feather around his neck, around his voice [larynx]." [So they did.] And he started talking. And he talked smart, too, after that. They used the feathers, *shuug*, to *wusot*, you know, sing a song; a person who knows how to sing about Mockingbird Song. That's what they use. That's a true story, too.

This is my only reference to the mockingbird and *mumkidag*, but the association of the specific songs and feathers used as a fetish in a *wusota*, or curing session, are clear from Francis's account. What is not given in either Frank's or Francis's story is what the offense might have been to cause the postnatal deficiency. Russell (1908:116) collected a Pima fetish composed of a set of mockingbird tail feathers, but he didn't elaborate on it (see fig. 5.2).

The Pima were familiar with the vocalizations that belong to the mockingbird itself. One of them is a sharp double-noted phrase, repeated in rapid succession. One spring day, a story goes, some Pima women were out in the flats with their baskets, picking the plump orange berries of the *kuavul*, or wolfberry bushes. A territorial mockingbird was singing, and the women interpreted its song as *"Keliv! Keliv! Keliv! Keliv!"*—that is, "Pick! Pick! Pick! Pick!" encouraging them along in their work. But among the women was one with a husband much older than she was, and she thought that the *shuug* was making fun of her, saying, *"Keli, Keli, Keli, Keli"*—that is, "Old Man, Old Man, Old Man, Old Man. Your husband is an old man."

Mockingbird figures importantly in O'odham mythology. In the Pima Creation Story, Lloyd (1911:214–215) recorded several speeches attributed to Navichu, a borrowed kachina figure. These speeches, encapsulating vivid images, are archaic pieces of literature, recited verbatim, like poetry. In Navichu's second speech, describing things right after the creation of the world, there is the following cosmological setting: a black mockingbird in the west, a blue mockingbird in the south, a mockingbird of light in the east, Soft-Feather-Nested Doctor (Curve-billed Thrasher) in the zenith, and Yellow Spider in the north. (Note the peculiarly Puebloan reference to the fifth direction here, seldom mentioned in Piman poetry or speech; it recurs among Huicholes and Tepehuanes at the southern end of the Ranchería Complex.) The four peripheral creatures stabilize the earth around its circumference by means of webs, fringes, and arrows. "And after this was done, the earth stood still."

Another version of the creation is equally poetic (Herzog [1929]). The newly formed earth is quivering. So Jewed̥ Maakai, Earth Doctor, makes *shuug*. "Passing around all the trees, it talked; in front of the sky it turned back, tied its talk onto the tops of the mountains, tied again its talk to the sky, went all over the Earth and the mountains, tied again its talk to the sky, went all over the earth and the mountains, and by this the earth was quieted."

If you watch Mockingbird even today, notes Culver Cassa, you can still see it going about the business for which it was created, stabilizing the shaky earth. It flies jerkily from one mesquite to the next, white wing and tail patches flashing, singing all the while, never still for long, uniting the sky, earth, mountains, and trees together with its song, weighing everything down with song.

In the wild imagery of a war oration, after the psychological disruption of enemy killing, not only must the warrior be purified, but the world must be put back into order. *Hevachud̥*, Blue Grosbeak, sings and spreads his blue feathers over the land. Then *vajukukĭ*, oriole, sings and shakes his yellow feathers (spring flowers) over the mountains. Finally *shuug*, mockingbird, sings and spreads his white feathers of sunshine over the land (Russell 1908:367–368).

One day at Bapchule I was sitting outside with George Kyyitan under his *vatto* (arbor). In the distance, a bird softly whistled a song.

"There's our *biit keishnam* [Bendire's Thrasher] singing," I commented.

"No, it's a mockingbird," corrected George. At that point, the bird switched into a regular mockingbird motif, louder and harsher than a Bendire's.

"Yeah, you're right. Yep."

"See?" said George.

"Yeah. *Shuug*," I said, correcting my identification.

"*Shuug*," he confirmed.

"Are there mockingbird songs? I know there's a Mockingbird Speech."

"Yeah. There's a song. In that *vajukuk* [oriole] song[s]. They mention about the Flood. When they had the Flood. And it says that *every*thing is quiet. *All* the birds are quiet. Still. They don't go anywhere. They don't fly. And they don't know how to fly after that. And *no*body sings. *Every*body's quiet. Only the mockingbird that is singing there."

"That's after the Flood?"

"That's *when* the Flood is. During the Flood. Because he always sings, always cheerful. That's why he says that." George then sings this song:

Tashai hahava mumukim
tashai hahava mumukim
je vene we si ko chuka ke
ke da ge we sai chu yu hinge ku kuive
da ngi to
je vene kachim ga mu pi sha mu ñe
shuhuge yu u hi nge
shoi ngeve kai je
hejeli ganu
ve ñe ño ke.

"The Sun then dies [an eclipse],
The Sun then dies.
Everywhere the world is dark.
Just then all the birds leave off singing;
The world is silent.
A mockingbird bird
Pitifully speaks,
Alone, over there
He talks.

"It's the only one that is talking in there. Nobody is . . . everybody quiet except *shuug*, that is only *shuug* that is cheerful, talking" (see also Bahr and associates 1997:131, 158).

Half a dozen years later, George related a similar (or perhaps the same) incident about mockingbird, originating from a song about a solar eclipse. We were discussing the sounds that animals make. George said:

Baaban hinkam, Coyotes singing, from *hink*, yell, something like that. *Tod̥k* [is] barking, [like a] dog. *Kuuk, kuhuk*, horse whinny. [The word] *kuuk* means a bird sings. Or a *shuug* could be *ñiok*, talking, just the same [as people]. [There is a] song about when the Sun died, whatever it is—eclipse—in Blaine Pablo's song. And it's dark, *everywhere, all* over.

And everything, all the birds and everything stopped their singing. Everything's quiet. And later, in the dark, only the Mockingbird was talking, and nobody was singing anything. But the mockingbird was—keep on singing, keep on talking, like that. Yeah. I think it's that way they say it, that mockingbird is talking, still talking; he's the only one that is talking. But everything's quiet.

Yeah, *shuug* never stops. That's why anybody, you know—someone will be talking all the time—then they'll call him, her, *shuug*, just like a *shuug*, talking all the time. I think that's what it meant, like that. It's the only one that can talk all the time—*any* time.

Among a long series of songs that Ruth Underhill (1938:139, 1946:257–258, 259) obtained from the girls' puberty dance are two are about *shuug* that incorporate bits of the bird's distinctive behavior. The first alludes to its nocturnal and matinal bursts of singing, the second to its wing-flexing habit.

The shining mockingbird
At evening could not sleep.
When the moon was in mid-sky
He ran to the puberty dance.
At early dawn
High did he raise his song.

Above Quijotoa Mountain
Stand terraced clouds.
Above them the mocking bird sings,
Flapping his wings.

In the wine feast or rainmaking ceremony of midsummer, Saguaro juice or syrup is diluted, mixed with water, then fermented into a kind of cider or wine, to be drunk by participants from four villages. Its purpose is to initiate or continue the monsoon or summer rains on which the People depend for their crops, whether they be dry farmed, as with the Tohono O'odham, or irrigated from the river, as with the Akimel O'odham. (I have argued elsewhere [Rea 1997:253–260] that the River Pima, like the Papago, once had a wine feast.)

The complex liturgical celebration, called *dahiwuak ii'e*, "sit and drink," begins with an invitation speech delivered by a hereditary orator from the host village, which is responded to in like manner by the orator from the guest village (Underhill 1946:52–55; Underhill and associates 1979:32–35). Culver Cassa offers this interpretation of the *dahiwuak ii'e*.

To understand the elements of the *dahiwuak ii'e*, "sit down and drink," we must first have a clear concept of a rudimentary Piman ceremony. (Its other components can thereafter be seen.) Such a nocturnal ceremony includes a shaman, orator, chorus, and observers. The shaman is a specialist in the desired end of the ritual and acts as its intermediary. (Some fields of specialty are deer hunting, enemy tribes, and rain.) The orator may be a specialist in a particular field but is usually acquainted with a village's entire ceremonial repertoire. There may be more than one orator, each speaking at his turn. The chorus consists of the village men whose prayerful singing lends support to the shaman's divination and intercession. Their songs are pertinent to each rite. At times, the chorus includes dancers, who together with the singers, constitute the prayerful action of the ceremony. The observers are just that and include those resting from their participation in the rite. Their interactions are often imbued with the emotional character of the ceremony, and many songs of the ceremony recount the deeds of the crowd.

The prelude to the "sit down and drink" is the brewing of the Saguaro beer. The nights of these brewing days contain some of the elements already stated. The shaman (or shamans) is called *juukĭ wuadam*, "rainmaker." Throughout the nights, he divines the rain toward its symbolic home in the east. He uses Golden Eagle feathers. His chorus includes singers and dancers. Their prayerful actions aid the beer's fermentation, their steps resembling a drunken stagger.

Anytime after the "sit down and drink" when the people desire rainfall, they will use their standard format. The rain diviners will be acting in their shamanic roles. This performance will include orators reciting rain speeches and the chorus accompanied by rasps. Basic rites of this type are situated in the way shown in the Rain House drawing I have done here [see the Rain House figure]. A rite is oriented toward the perceived home of the desired end, in this case, rain from the east.

The actions of the nights of the brewing are based on this format just stated. Thus, the night ritual can stand by itself as a complete ritual expression. The historic ritual is the joining of complete night and day performances.

The focus of the daytime ceremonial action is the strengthening of the reciprocal bonds of the regrouped community through the drinking of the cactus beer. Calling each other by their kinship terms, they pass around and serve each other the beer.

The time of the Saguaro harvest is considered the beginning of the new year, and so it follows that it was summertime at the world's primordial completion and subsequent initiation of its cycles. (The people's past is reckoned by its present references.) Prior to this point in the cosmic order, the world was unstable and quivering. The Creator summoned mockingbird beings from the four directions to stabilize the newly created earth. They did this by means of their vocalizations. These were placed all over upon the various landforms to weigh them down and hold them fast and thus stabilize the earth. Thereupon, the universe's support and fabric could be completed on a firm foundation. It is said that these birds returned to their homes and would help the people when they were called upon. One oration states this creative process in ceremonial language [see Russell 1908:347–352]. It is the featured oration at a nighttime summer rain ceremony. It is also used to set the mood at the "sit down and drink" and when a village establishes its meetinghouse.

There is a re-creative and renewal element in the "sit-down and drink." The ceremonial drinking is cast in a kind of re-creation of the world.

In the beginning, four houses of the rain beings were established at the four corners of the world. The invited and home drinkers so seated represent these beings. They also symbolize the created world. The ritual orations allude to the drinkers as equivalent to these beings who, upon being invited to the ritual space, bring their beautiful summer storms with them. The ending speeches are the Mockingbird Speeches. They tell of the primordial mockingbirds who are anxious and restless in their houses. Now has come the time for the People, their relatives, to call upon them.

> They've divined and communicated with the rain. They've cemented their bonds with drink and prepared for farming, thereby re-creating and renewing their world. The mockingbirds communicate to the cosmos that the People have prepared, just as they had in the beginning of creation, for their reinitiation of its cycle (which, of course, begins with the thunderstorms of summer). And so the rains follow.
>
> It must be understood that the past creation is viewed with a modern perception of the universe and its cycles. The past is fitted with the present conjectures.
>
> The mockingbird is busy doing its thing just prior to the summer rains (symbolic of starting of the world) and Saguaro harvest. This is especially noticeable where an area is favorable to support a large population of these birds. In which case, it can be seen how appropriate a symbol the bird is for the preparation and onset of the Piman new year. Many times the Piman knowledge of the natural world is essential to understanding both myth and ritual.
>
> Summer is the time of O'odham culture in its agricultural expression. O'odham culture is truly alive. In the winter, we can contemplate, speculate, and pronounce upon it, but in the summer we live it.

At the end of the drinking ceremony, there is a final round of orations. The name for these formal speeches varied by village. They were called *shuugaj ñi'ok*, Mockingbird Speeches, at Santa Rosa and Bapchule, the most common designation, but other names were used as well: *s-hevelim ñi'okĭ*, "wind speeches," at Pisin Mo'o; "brave speeches" at Gaka; and *cheḍum ñi'okĭ*, "cotton blanket speeches," at Sacaton. "They were delivered not in the abrupt, staccato style of the war and salt speeches, but were smooth and flowing, with heavy emphasis on the last word in each sentence," noted Underhill (1946:52). "[I]t is the mockingbird who 'talks' and is responsible for the release of winds, clouds, and rains."

It is said that at the original wine festival mockingbird was the first one to get drunk, as one song records (Underhill 1946:64, song no. 44):

> Mocking bird, mocking bird!
> First drunk!
> He talked much;

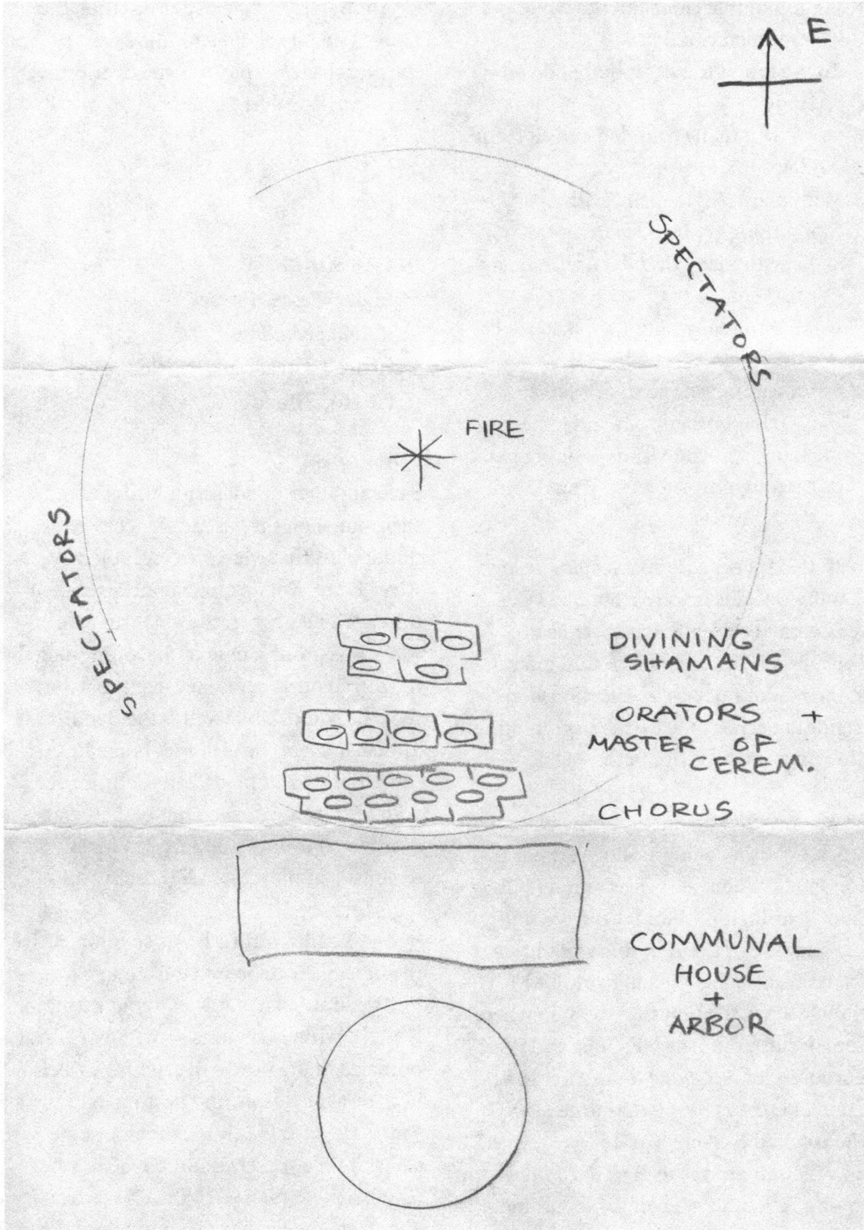

Figure A.1. *Culver Cassa's ceremonial seating. Squares represent mats.*

> Loudly he talked,
> Going around drunkenly.
> Not knowing [what he was doing].

Of the ending speech at the wine feast, Bahr wrote (in Underhill and associates 1979:35):

> Although the speech doesn't say so in so many words, a plausible reading of the story is: "The people are now drunk and celebrating. The mockingbird hears them and passes the sounds on. Through him the wind hears of the event, and it "understands and runs." It is ambiguous whether the bird understood and meant what he said, or simply copied the sounds. He is described as if drunk in his rainhouse, with his head bobbing this way and that, as if he copied not only the sounds of the ceremony but also its mood.

As with many formal Piman orations, the Mockingbird Speeches describe hoped-for events. We hear vivid poetic descriptions of the violent summer storms that are expected as a result of the wine ceremony: first come the dry winds, shaking the trees, sweeping everything clean, readying the earth for the downpour, as in this speech transcribed by Underhill and associates (1979:33):

On top of that came out the white-
bottomed cloud
To the top of the sky it bumped and
stood,
with many rounded ends it stood
And it went.
With many dust storms, with many
lightnings,
with many thunders, with many
clouds,
with many rainbows, it started
to go.
From within wet mountains, more
[clouds] came out and joined it
Plucking out white from their breasts,
coughing out white from their
breasts.

With this speech, the cosmos now knows that the world is finished, observes Culver Cassa. It can now start over again. The renewal includes the clouds and rain in the new annual cycle. The monsoon's beginning echoes the earth's beginning. Mockingbird stabilizes both.

COMPARATIVE LINGUISTICS. The Northern Mockingbird is *shuug* among Akimel and Tohono O'odham as well as among Pima Bajo. Pennington (1980:199) wrote *shu'ug* for Ónavas Pima, but in this work he characteristically splits the long vowel with a glottal stop while omitting the stop where it does occur; both Pedro Estrella and María Córdova told me *shuug* for censontle. The name in Ónavas pueblo has time depth. The colonial Névome vocabulario has *suca, suga* (Pennington 1979:21). Far to the south among Tepecano, Mason (1917:329) found the word *suug*, with the plural *susug*, nonpalatalized as in the colonial Ónavas document. I do not have the word used among other Tepimans.

I did not get a certain identification among Mountain Pima, though the bird is surely common there. Shaul (1994:346) recorded an unidentified bird from the Yepáchic region as *so'ok*, which would seem to be cognate with the lowland word. Pennington (n.d.) was told *masad* for censontle at Maicoba. This name is not cognate with the name used by desert O'odham. People in this upland region told me *sinsont*, a derivative of Spanish, which in turn is from the Nahuatl word *centzontli*. Several thrushes and solitaires at these higher elevations are also powerful singers, so the biological contrast set needs to be explored among knowledgeable speakers.

Although Sylvester Matthias thought there is no plural, Fr. Antonine (1935) gave the plural as *shushug*. It is the distributive form, notes Culver Cassa.

biit keishnam

Mimidae, Passeriformes
Toxostoma bendirei
Bendire's Thrasher; Cuitlacoche Pico Corto, Pitacoche

Several rather nondescript birds make their home on the vast adobe-colored floodplains that were once occupied by River Pima, Kohadk, Sobaipuri, and their unrelated neighbors, the amalgamated Maricopa groups. One of these long-tailed terrestrial birds whose plumage matches the pale soil color so well is the Bendire's Thrasher. Here the bird is at home in Piman yards and fields, especially those with fencerows, or even out on bare flats dominated by low wolfberry and Desert Saltbush, with occasional scrubby mesquites.

The Bendire's Thrasher is so unobtrusive that it is likely to be overlooked in River Pima country. Here it goes about gathering seeds in winter (swiping some Pima wheat or maize in the process), catching insects in the warm season, and nesting in Pima yards. The bird seems to combine some traits of three relatives: the Sage Thrasher *(Oreoscoptes montanus)*, the Curve-billed Thrasher (see the next account), and the Northern Mockingbird (see the previous account). Its body markings are like a subdued or faded Sage Thrasher, but its bill is longer and just perceptibly more curved. The Bendire's sings a lovely warbling song much like the mockingbird's, but not as loud or as overbearing. You have to listen for it. The medley may continue uninterrupted for half a minute or more. And, finally, it is a smaller version of the Curve-billed Thrasher. Both have yellow eyes. But the Bendire's is a smaller bird (about nine and one-half inches rather than eleven), with more distinct triangular breast marks and a pale base to the mandible, even in young birds.

M. French Gilman (n.d.), teaching school in various Pima villages in the early 1900s, wrote: "The Bendire Thrasher is one bird that from all indications takes kindly to settlement. These birds nest near houses, on which they perch to sing, come into the yards, and seem fearless if not molested. If their natural shelter is cleared up they take kindly to artificial or planted growth and I believe will persist in the face of civilization. All of this, of course, provided they receive some measure of protection and encouragement."

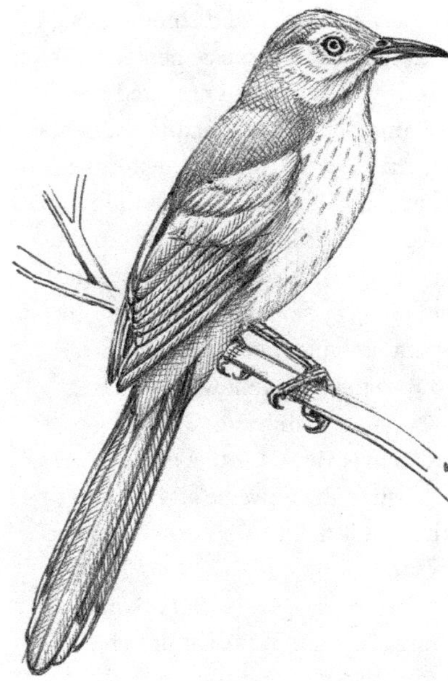

biit keishnam, Bendire's Thrasher (Mike Touch)

The Bendire's Thrasher was well known to older Gila Pimas, who called it *biit keishnam*, "excrement-it-stepped-on." I also heard the variants *biit keishapdam* (Santa Cruz) and *biit keishud* (Sacaton). Sylvester Matthias said, "It's something like gray—in the fields." Various people noted that it can cause damage in the gardens, particularly in digging up germinating seed. Whether it eats the seedlings as Ruth Giff thought or primarily is thrashing the soft soil for grubs I am not sure.

Looking at a series of different thrasher skins, David Brown called one, the Crissal Thrasher, *chev chiñkam*, "long bill–it-has." Rosita added, "They call one 'he stepped on kaka [feces],' *biit keishnam*." This was the common spotted Bendire's. "Its beak is sort of straight," noted David. I asked, "What do you call the other thrasher with spots on the breast?" David responded, *"Kulvichkam."* This is the Curve-billed Thrasher. We had just worked out a biological contrast set of the three most common thrashers found about Pima yards, fields, and woodlots.

Late one May in Santa Cruz village a Bendire's Thrasher was catching insects in the churchyard. The spots were rather worn on its breast, but the Browns identified the bird as *biit keishnam*. In addition to its rather straight bill, Rosita noted its similarity to the *bichpoḍ*, Abert's Towhee, with which David agreed.

Another day, as George Kyyitan and I sat outside in Bapchule talking about birds, a Bendire's Thrasher came into the yard from the adjacent field. *"Biit keishnam,"* he noted. He was familiar also with *chev chiñkam*, with its much longer bill, a bird that he would encounter while out cutting mesquite wood.

Unlike the Curve-billed Thrasher, the prosaic Bendire's seems not to be memorialized in any Piman song or legend. It may be another of those folk taxa known only (or primarily) to the Akimel O'odham. At least from my small sample, I have found no Tohono O'odham who recognized the name.

But the Tepecano name for huitacochi, *kurupiit*, may actually be *kuru-biit*. Also, the Nahuatl thrasher name, *cuitlacochtótotl*, said to be based on the birds' whistled call, comes from *cuitl(a)-tl*, excrement (see Karttunen 1983:73–74; Macazaga Ordoño 1985:47). The seeming localized Gileño Pima name may be a manifestation of a broad Uto-Aztecan thrasher-naming phenomenon.

kuḍvich, kuḍvik, kulvichkam; kul-wichgam
***tuv kuagam* (Pima Bajo)**

Mimidae, Passeriformes

Toxostoma curvirostre

Curve-billed Thrasher; Pitacoche, Cuitlacoche Pico Curvo, Cuitlacoche Común

On the more elevated lands of the Sonoran Desert, in particular the bajadas, a piercing two-noted whistle *whit-wheat!* rings through the Saguaros and Little-leaved Paloverdes. The call is urgent, almost frenetic, the second note higher, sometimes with a third added. Those new to the desert might suspect a human source until the singer is located: a strong-bodied thrasher, adobe brown above with indistinct gray spots on its pale underparts. Its all-dark bill is distinctly curved, its eye deep yellow or sometimes light orange. There are indistinct wing bars and pale tips to the tail. The Curve-billed Thrasher can be distinguished from its relative the Bendire's Thrasher in being larger (about eleven inches rather than nine and one-half) and having more diffuse breast marks.

The birds can be extremely wary, but around some villages and even urban areas set in native vegetation they become more familiar. Here they hunt for small animal life, thrashing vigorously through the litter, tossing leaves aside with their powerful beaks, and probing into the soil. Their bulky nests are most often found in the protecting arms of some taller cholla. The inside cup of fine materials is quite deep for a thrasher.

Well known to the O'odham, the Curve-bill has two basic variations on its Piman name: *kuḍvich* (or *kuḍvik*) and *kul-vichkam* (or *kul-wichgam*). I have found the two variants among both Tohono and Akimel O'odham, so these names do not sort out by dialect groups. (In Pima Bajo country, where the bird is quite common, a different name is used.)

It should come as no surprise that a bird so vocal and so characteristic of the Sonoran Desert, tropical thornscrub, and tropical deciduous forest appears prominently and early in the Creation Story. In the Flood episode, four small birds plan to save themselves by flying up and holding onto the sky with their bills: *hikvig*, Gila Woodpecker; *vipismal*, hummingbird; *giisobĭ*, Verdin; and *kuḍvich*, Curve-billed Thrasher (Benedict 2001:11, 15; Lloyd 1911:43–45; Russell 1908:211). "And Quotaveech [Kuḍvich] was cold and cried so loud that the other birds pulled off their feathers and built him a nest up there so he could keep warm. And when Quotaveech was warm he quit crying. And then the little birds sang, for they had power to make the water go down by singing, and as they sang the waters gradually receded" (Lloyd 1911:43).

Russell (1908:245) added a detail: "When the waters covered all the earth two birds were hanging onto the sky with their beaks. The larger was gray with a long tail and beak; the smaller was the tiny bird that builds its nest like an olla, with only a very small opening to get in [Verdin]. The larger one [thrasher] cried and cried, but the other just held on tight and said, 'Don't cry. You see that I'm littler than you, but I'm very brave.'"

After the floodwaters recede, important events transpire. Three of the primordial beings—Jeweḍ Maakai (Earth Maker), Ban (Coyote), and Se'ehe (Elder Brother)—find one another by their tracks in the mud. The fourth being, Ñui (Buzzard), is not mentioned; presumably, he is flying instead of walking, unlike his three companions, leaving no tracks. At any rate, Ñui does not join in the repeopling of the earth. His work was to create the geomorphology and a functional system of waterways in the desert.

"And the [small birds] came down from the sky. . . . But Quotaveech staid up in the sky because he had a comfortable nest there, and they called him Vee-ick-koss-kum Mahkai [Viigĭ Koshkam Maakai], the Feather-Nest Doctor" (Lloyd 1911:45). In a cosmological readjustment that follows, Vipismal (Hummingbird) and Hikvik (Gila Woodpecker) locate the *hik*, navel, of the world by east-west measurements. But Curve-billed Thrasher, with his new title, Viigĭ Koshkam Maakai, Soft-Feather-Nested Shaman, is at the zenith, the ceremonial fifth direction. Culver Cassa sees

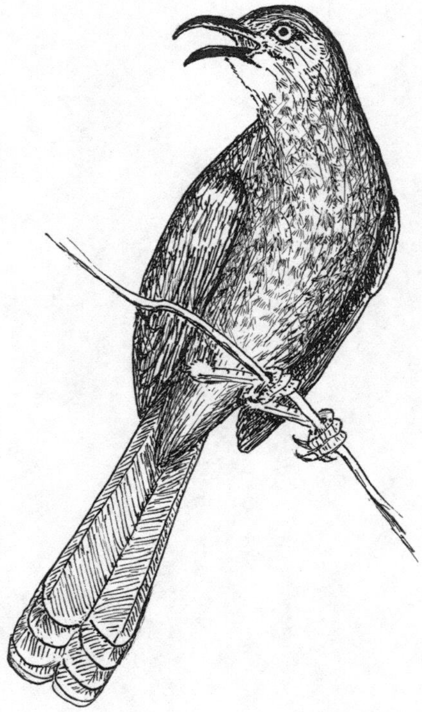

kuḍvich, kul-wichgam, Curve-billed Thrasher (Amadeo M. Rea)

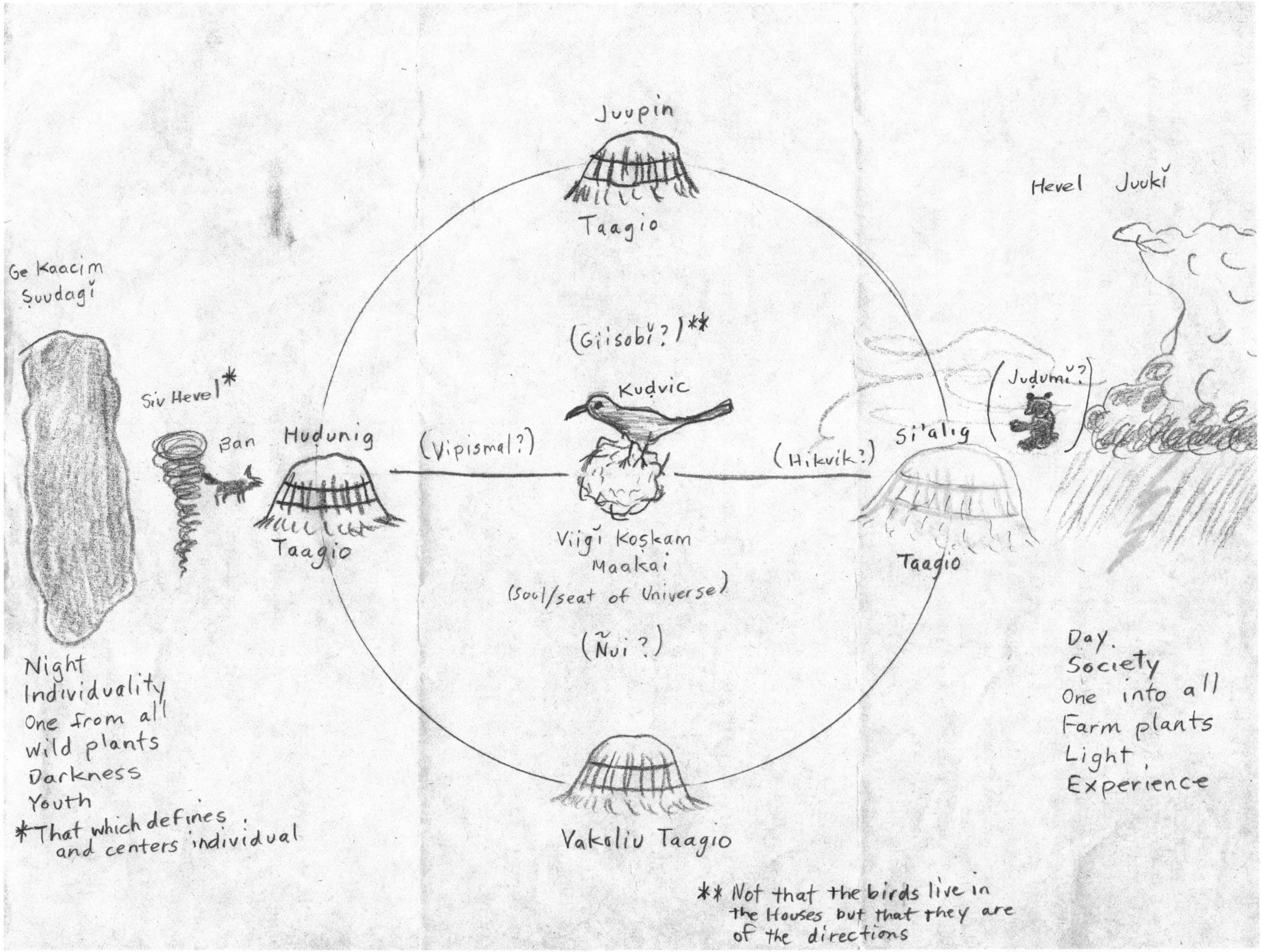

Figure A.2 *Culver Cassa's Pima cosmology drawing:* ge kaachim shuudagĭ, *ocean; Siv Hevel (Bitter Wind); Hevel Juukĭ (Wind Rain); Kuḍvich (Curve-billed Thrasher); Viigĭ Koshkam Maakai (Soft-Feather-Nested Shaman); Ban (Coyote); Juḍumĭ (Bear); Vipismal (Hummingbird); Ñui (Turkey Vulture); Giisobĭ (Verdin); Hikvik (Gila Woodpecker); Si'alig Taagio (east, sunrise direction, yellow); Vakoliv Taagio (south direction, green); Huḍuñig Taagio (west, evening direction, black); Juupin Taagio (north direction, red).*

this zenith as the seat of the universe (see Culver's cosmological drawing, fig. A.2.). Note that these species are common to a southern Piman homeland; they are here in the story for some reason. They have a special relationship to the ground; that's why they came back to earth. Culver sees the cosmological centrality of the Curve-billed Thrasher or Feather-Nested Doctor in O'odham mythology as a parallel to the Crested Caracara eating a snake in the Nahua or Aztec migration story.

There are other references to this thrasher as being at the fifth direction. For instance, in a speech by Navichu (Lloyd 1911:214), the Feather-Nested Doctor is said to be "famous for his power." He lends his aid in attempting to stabilize a shaky earth. In another oration, he helps a boy with powers against the Apache (Benedict 2001:140). In a rain ceremony speech recorded on the Gila River, the four colored mockingbirds of the four directions are unable to stabilize the earth by flying to each tree and mountain. Then Soft-Feather-Nested Shaman of the zenith "smoothed the earth and the hills and overturned the mountains and the trees. He finished his work, and the earth stood firm." Likewise, Gray Spider Shaman of the nadir spins the trees together around the edges of the world and the cloud encircling the edges of the earth, and the earth stands firm (Benedict 2001:179–180; compare Lloyd 1911:214–215). (In Piman events, the personification of the zenith is infrequent, but mention of the nadir is even rarer; all six ceremonial directions, in contrast, are common among Pueblo ceremonials.)

Another oration found in one version of the Creation Story (Benedict 2001:21–22) tells of Kuḍvich Medicine Man, who paints a black horizontal line across his face and weakens the Apache cane arrows. The reference is puzzling because this thrasher does not have a facial line, contrary to what the footnote in this source says. The bird's behavior of biting the "eyes" or joints of the arrows are much more suggestive of a Loggerhead Shrike, as is the black facial mark. In fact, there is a war oration where *go'ogov*, the shrike, does that to Apache arrows (see that species account). Perhaps the narrator, Thomas Vanyiko, slipped.

Nevertheless, the *kuḍvich* is still very much a part of the everyday world. Sylvester Matthias identified it as a gray bird with black bill and feet, little spots on the wings, between white and gray below. "Nests in the *u'us chevaḍbaḍ* [Graythorn] so nobody can get to them." David and Rosita Brown also knew it as being the other spotted thrasher by its whistle.

Delores Lewis of Ge Oidag called the bird *kul-wichgam*, identifying it in the field. "Eats everything we plant; gives us lots of trouble." He said its name means "beak something has" "because they can sing any kind of notes." He added that this bird is fed to youngsters when they don't learn to talk. He said one was roasted and fed to one of his grandchildren who had this problem, and after a while he started talking. (The mockingbird is used by others as the remedy for this problem; see that account.)

Far to the south in the thornscrub, Pedro Estrella called the Guitlacoche, as the bird is named locally, *tuv kuaga*, mentioning "liña larga" (long firewood) in reference to the long sticks used in nest construction. His identification was based on a specimen as well as on our encountering this thrasher in the field. Earlier Jean Johnson (n.d.) wrote *tu kúagam*.

Later in the Creation Story, the Curve-billed Thrasher appears again in a puzzling reference in a narrative that seems much like oral history (Lloyd 1911:173–189; Saxton and Saxton 1973:191–192; see also Benedict 2001:136, 137). There is a boy with a Pima mother and an adopted Apache father who has been killed. The young widow and her child are poor without a game provider. The boy hangs around the young hunters. "And they found a quotaveech's nest near them, with young ones in it, and one of the men shot into it and killed one of the young ones, and then the children ran up and join[ed] in the killing. And when [the boy] Pahtahnkum came up, one of the men threw him one of the young birds, and said: 'Here, take it, even if your mother does not wish to marry me.'" The mother, affronted by this gesture, begins to cry, saying that the *koson*, packrat, and *kakaichu*, quail, are good to eat, but "[t]his bird is not fit to eat" (Lloyd 1911:174). The passage seems to reinforce the idea that the rodent and bird are legitimate game, but the thrasher is not.

I have never heard of the Curve-billed Thrasher's causing staying sickness. However, as one of the four surviving small birds from the Flood, it should have been endowed with that status. Note that in the various versions of the hunters' giving the little orphan a *kudvich* to take home to his mother, along with their comments about being rejected as suitors, the mother's response is always strong. Piman listeners must have well understood the symbolic connotations of the gesture: the suitors are sending her the carcass of a tabooed bird. More specifically, it is a *nestling* from Soft-Feather-Nested Shaman.

But there is more to the story than that, notes Culver Cassa. The thrasher, because of its role in the Flood narrative, is part of the stabilization of the world. The boy's reckoning of the world is now destabilized, gone. The device indicates that he must now make a new world for himself.

White-clay Eater Woman's grandson, whose father and uncle (the twins) have been killed, seeks revenge against his maternal grandfather, Oam Ñui, Yellow Buzzard. He challenges the old man to a game of *gins*. The boy makes a nest with *kuḍvich* eggs. As they are making their wagers, the boy says, "As I was coming this way I saw a *kootovitc* nest with eggs in it." "It doesn't lay eggs at this time at all. A long time off, and after that, it will lay eggs," replies the elder. "All right we'll go and see, and if there are no thrasher eggs, I will give you all the stakes." The boy wins (Benedict 2001:121–122; Herzog [1929]:99–100). (In other versions of this story, the reference is to dove or quail eggs; see the *kakaichu*, Gambel's Quail, account.)

COMPARATIVE LINGUISTICS. The Curve-billed Thrasher is a widespread species ranging throughout Tepiman country. Its name needs to be worked out among other Piman-speaking groups. Mason (1917:331) recorded "huitacochi" (presumably this thrasher) as *kurupiit* among the Tepecano—not cognate with the usual Curve-bill name in the north, unless it is the equivalent of the Bendire's Thrasher name (see previous account). Yaqui Moisés Valenzuela of Huirivis pueblo said the Curve-bill is called *wivis* and in the plural *wivisim*, hence the name of his town.

Although Ruth Benedict wrote *kootovitc* (= *kuḍvich*) as her two Gila Pima narrators gave the name and as Lloyd took it down as well (1911:174), Bahr, in footnotes, "corrects" this spelling to *kul-wichgam*, a Papago variant of the name (Benedict 2001:122 n. 87, 136 n. 11).

MISIDENTIFICATION. Fr. Antonine (1935) called *kuḍvich* a "small woodpecker." For more misleading and erroneous information, see Bahr 1997:129 n. 112.

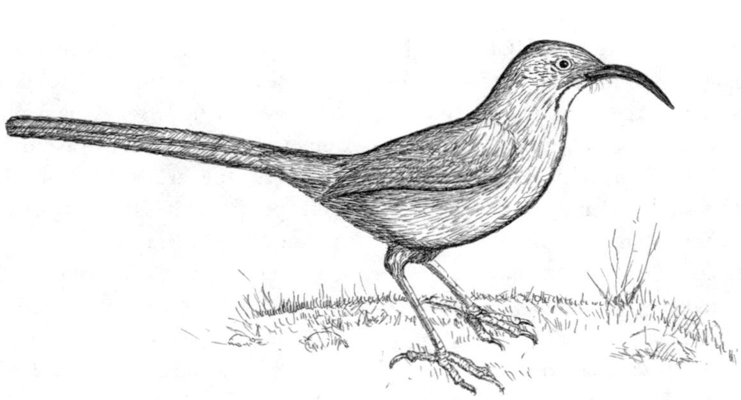

chev chiñkam, Crissal Thrasher (Amadeo M. Rea)

chev chiñkam*, pl. *che'echev chichiñkam
keli huch

Mimidae, Passeriformes

Toxostoma crissale

Crissal Thrasher; Cuitlacoche Crisal

The secretive Crissal Thrasher stays mostly in dense mesquite thickets or the underbrush along desert streams. It is more likely to be heard than seen. Its dawn song emerging from bosques shows the bird to be much more common than expected, but an hour or so after daylight the bird becomes silent again, except for its occasional call note, *queety-queety*.

When finally seen foraging under some brush, the Crissal Thrasher shows a tail and bill that seem disproportionately long for its body. They give it a very slender look, although it is our largest local thrasher (eleven and one-half inches). Quite like the Abert's Towhee, it is rather uniformly colored dark gray-brown, both birds matching the desert shadows rather than the sun-bleached open areas. The bright rufous crissum (the undertail coverts) gives the bird its common name. A dark whisker stripe on each side of the pale throat and the straw-colored or pale grayish eye also are good field marks.

Ironically, we owe to Fr. Antonine (1935) the confirmation of the Pima name *chev chiñkam* as well as its plural form. He misidentified the folk taxon as the Gray Catbird *(Dumetella carolinensis)*, a different mimid that doesn't occur in the desert. But, like the Catbird, the Crissal Thrasher is dark gray-brown with the exception of the bright rufous crissum. Both are mimids. The Pima name means "long bill–it-has" (with *-kam*/*-gam* an attributive marker).

But some Pimas still remembered this secretive thrasher. George Kyyitan knew the bird well and associated it with the mesquite thickets where he would cut wood. "*Chev chiñkam*, 'long bill.' One time I had some squash plants in there, and they ate them up, their leaves, when they first start up. It sort of sings like a *shuug* [mockingbird]. Almost. But the *shuug* uses *his* [the *chev chiñkam*'s] song."

Leonard Pancott said, "Two birds almost look alike; *chev chiñkam* almost like *bichpoḍ* [Abert's Towhee]. It's mentioned in the legend, during the Flood; it hung on to the sky. Got scared and started crying. Long bill with kind of a hook. Similar but not the same as *biit keishnam* [Bendire's Thrasher]—maybe difference in the bill." The bird in the Flood episode is usually said to be *kuḍvich*, the Curve-billed Thrasher. Another time Leonard contrasted the *chev chiñkam*, having no spots and a longer bill, with *biit keishnam*, having spots and a shorter bill. Another Ta'i O'odham, Leslie Pasqual, associated four similar-size birds: *bichpoḍ, biit keishnam, kuḍvich,* and *koomagĭ u'uhig* (Le Conte's Thrasher, see next account). But he didn't know the name *chev chiñkam*.

But they also used another name, *keli huch*, for the Crissal Thrasher (a name that Leslie didn't recognize either). "Has long beak, bent down," noted David and Rosita Brown of Santa Cruz village. They disagreed on whether the name means "old man's *finger*nail" or "old man's *toe*nail," because the same word in Pima, *huch*, is used for both. They finally decided that it referred to the old man's big toenail, which gets long like the thrasher's bill, particularly if he has used the kickball a lot in his life (see also Russell 1908:160). They described the bird as being colored like the Abert's Towhee.

The name *keli huuch* is also known to some in Papago country. Leon Juan of Aangam said that *keli huuch* has a *chev chiñ*, "long bill," and runs on the ground. Of course, the Aangam people originated from the Queen Creek area north and east of the modern Gila Pima villages, which may explain this bit of folk taxonomy.

koomagĭ u'uhig

Mimidae, Passeriformes

Toxostoma lecontei

Le Conte's Thrasher; Cuitlacoche Pálido, Cuitlacoche del Desierto

An elusive thrasher makes its home on the vast alluvial flats dominated by Desert Saltbush interspersed with wolfberry bushes and occasional stunted mesquites. As the country becomes lower and dryer westward, this mimid lives among scattered Creosote Bushes and bursages. This is the Le Conte's Thrasher, a bird that few will ever encounter.

In these desolate areas, the thrasher prefers to run rather than to fly, dodging swiftly behind the dispersed bushes, often cocking its long tail upward, roadrunner fashion. If you're lucky enough to catch a glimpse of this will-o'-the-wisp, notice how perfectly its plumage matches the very pale tan surface of the desert flats. It is reputed to be the palest land bird of North America.

A slightly smaller, bleached-out version of the Crissal Thrasher, the Le Conte's Thrasher has a whitish throat, faint whisker stripe, and buff undertail coverts. It is the only thrasher in Piman country with a dark iris (Crissal's is grayish white, Bendire's is yellow, and Curve-bill's is yellow-orange).

The Le Conte's is not only elusive but also rare where its habitat hasn't fallen victim to development. Where it survives, as it does at the northeastern edge of its range in Gila Pima country, one is more likely to hear than see it. Like other thrashers' songs, its song is loud, most often heard at dawn and again at dusk.

The Kuiva O'odham, or Downriver Pima, among whom I worked the longest, had five names for four species of thrashers resident in their country. There was never any question about *biit keishnam* (Bendire's), the small spotted species that ran around in everyone's yard, or about *kuḍvich* or *kuḍvik* (Curve-bill), the large spotted one on the bajadas with its distinctive, powerful whistle. The ambiguity came with the two plain-breasted and far more secretive species. About these birds, my teachers were understandably unsure. It was agreed that *chev chiñkam* and *keli huch* were the same darker bird of the mesquite thickets around their ranchos. But they would also bring up another name, *koomagĭ u'uhig*. Could the Crissal Thrasher really be endowed with three sympatric folk synonyms, almost unheard of in one Piman community? From my extensive series of study skins, Sylvester Matthias said in the 1960s that *koomagĭ u'uhig* was either Crissal or Le Conte's.

More than three decades later, I visited Leslie Pasqual, a Ta'i O'odham who still runs about a hundred head of cattle. It was dusk, and he had just come in from the range. He gave his horse to a grandson to unsaddle, and we sat down to talk about birds. He started with *bichpoḍ*, the Abert's Towhee, then mentioned *biit keishkud*, his name for the Bendire's Thrasher. He knew the loud whistles of the *kuḍvich*. "Yes, there was one out here singing this morning." He spoke of another bird, *koomagĭ u'uhig*. "The one that's got the curved beak. Different from *biit keishud* and *bichpoḍ*. Just solid brown. Sings real loud, just like the *shuug* [mockingbird], sitting up there, just singing away." Curiously, he did not recognize the names *chev chiñkam* or *keli huch*. Surely this cattleman would have been familiar with both Le Conte's and the more common Crissal Thrashers. The folk names must have become telescoped. But *koomagĭ u'uhig* was clearly a long-billed and unmarked thrasher.

The word *koomagĭ* may mean "pale" as well as "gray." That the Pima would name Le Conte's simply "gray bird" is singularly appropriate: it is the palest, grayest bird to be found in the desert. Crissal is dark gray-brown.

I haven't found anyone among Tohono O'odham who recognized the name *koomagĭ u'uhig* for a thrasher (one person used it for Great Blue Heron), and I am unsure of the Le Conte's distribution in Papaguería.

koomagĭ u'uhig, Le Conte's Thrasher (Larry Toschik)

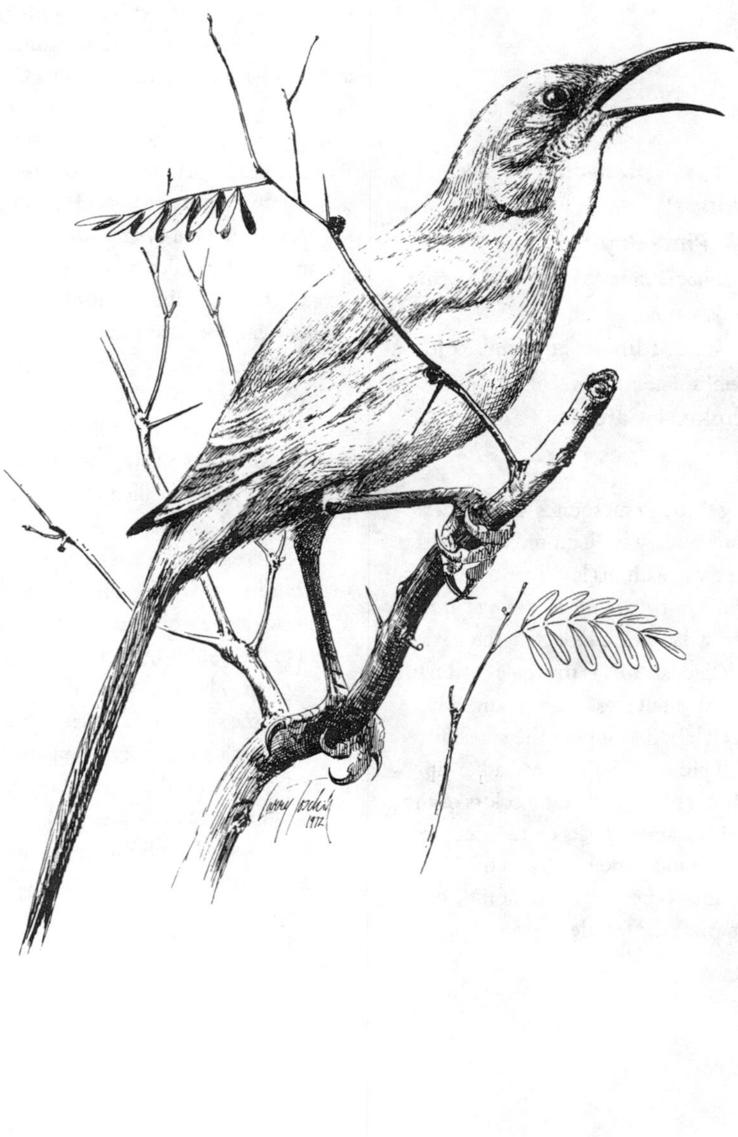

kuigam

Ptilogonatidae; Passeriformes
Phainopepla nitens
Phainopepla; Chivito, Capulinero Negro

As I step out my back door, I am surprised by a single short *wurp* note. I soon trace the sound to its source: a slim, glossy black bird with an elegant crest and red eyes. My pomegranates must be getting ripe and splitting open, exposing the red berries within. The bird flies buoyantly to an elevated perch, briefly exposing broad white wing patches. Here it raises its long tail periodically. After several more casually whistled soft *wurps*, my visitor departs with an almost butterfly-like flight. The pomegranates have not opened yet.

Phainopeplas are most likely to be found wherever Desert Mistletoe infests the mesquites and ironwoods. They are inordinately fond of the sticky white berries. In the desert ranges, they may take the bright orange and red fruits of wolfberries. The birds also congregate in numbers along desert streams, particularly where elderberries are available. And everywhere it sallies out from a perch after insects, earning its family name of silky-flycatcher.

But it is primarily in the mesquite bosques that this bird is to be found, staking out territories wherever mistletoe berries grow. It even puts its nest within the defended clumps. Here the gray female and the black male share incubation duties. Once the young are reared, the mistletoes are stripped, and the desert begins to heat up, the delicate birds congregate and depart, presumably for cooler locales. They turn up in the higher chaparral country and along the coast, even in my backyard.

kuigam, Phainopepla (Allan Brooks)

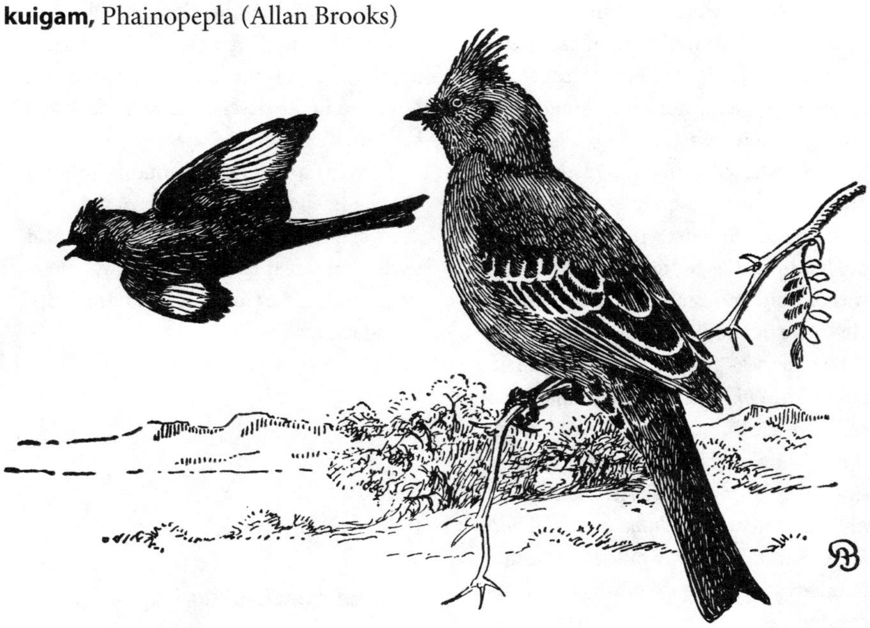

There are suspicions that they raise second broods in these areas, but, so far as I know, no one has traced individually marked birds to prove this claim.

Joe Giff explained this bird's Piman name, elegant in its simplicity: *kuigam* means the bird that owns the mesquite, *kui*. The attachment *-gam* means something that has an association with or is attributable to something. (In biological lingo, using a genetic approach, we might say "coevolved.") Probably no wild resource was so important to the River Pima as the biannual crop of *vihog*, the mesquite pods. Mesquite products, I am told by Pimas born near the turn of the twentieth century, found their way daily into the diet. Families gathered the sweet pods literally by the tons. And the people knew well the *kuigam*, guardian of the mesquites.

Sylvester Matthias noted that the *kuigam* lives in the *haakvoḍ*, mistletoe, growing on paloverde, ironwood, mesquite, cottonwood, or willow. (The leafy Yellow Mistletoe grows on these latter two broad-leafed trees.) Francis Vavages likewise knew the *kuigam*. "They're black. *Siivda* [motioning, topknot] they call that, anything like that is *siivda*. *Sipok* [cardinal] has a *siivda*, too."

In Pima Bajo country, María Córdova said there is a name for this bird, but she couldn't remember it. (The name presumably was not *kuigam*, which I must have mentioned to her.) One hot October day, dozens upon dozens of the buoyant birds congregated in several ornamental trees planted on the plaza near her house, eating the small fruits.

Saxton, Saxton, and Enos (1983) list *kuhigam* on page 35 as "a species of a black bird (unidentified)" and *kuigam* on page 36 as Phainopepla. Surely these birds are one and the same.

sipuk, sipok, pl. *sispuk, sispok* (all groups)
sipook (Pima Bajo)

Cardinalinae, Emberizidae, Passeriformes
Cardinalis cardinalis
Northern Cardinal; Cardenal Rojo
?*Cardinalis sinuatus*
?Pyrrhuloxia; Cardenal Pardo

Slinking through mesquites, paloverdes, and ironwoods, a brilliant red bird gives himself away with his loudly whistled song: *huuu-iit, huuu-iit, huuu-iit, chur, chur chur.* The first syllables are inflected, the next series deflected. This is the male cardinal, all red with a tall crest, deep orange bill, and black bib that stops at the nares. In fresh fall plumage, light gray feather tipping obscures somewhat the colors of the back, wings, and tail. (Its eastern relatives are smaller and darker red, and the black bib continues above the base of the bill.) Not far away, the female answers. She shows somewhat duller red in the crest, wings, and tail, but the back is gray and the underparts rich clay color (buffy ochre). At any time of the year, the two sing antiphonally, occasionally duetting.

Leguminous seeds are an important part of the diet of these powerfully billed birds. Where there are fields with fencerows and large grains such as sorghum, cardinals become common. They frequent Piman yards, in particular those with spilled seeds and plantings such as pomegranates. Here they feed on the ground familiarly with doves and sparrows, jerking their tails and raising the crests expressively while chipping.

In both Pimerias, there is a related bird, the Pyrrhuloxia. If anything, it is more elegant than the cardinal, being smaller and more slender, with a narrower crest. The males are mostly gray with deep pink on the face, throat, and chest. There is red in the crest, wings, and tail. The similar female is more buffy, less gray, with reduced pink areas. Both have a short, strongly angled, parrotlike bill that is yellowish in summer, horn colored in winter. (Young cardinals have a black bill that changes to orange during their first winter; young Pyrrhuloxias also begin life with dark bills.)

Given its loud song, bright plumage, and familiar habits, the Northern Cardinal is well known to the O'odham. There is some question as to whether the similar Pyrrhuloxia is included in this folk taxon and whether it was perceived as a different species. In Gila Pima country, where the smaller Pyrrhuloxia is only a recent invader, local O'odham considered it to be a female cardinal (Monson and Phillips 1981:178; Rea 1983:230–231). Farther south Delores Lewis included both species in his concept of *sipok*. In Pima Bajo country, Pedro Estrella as well as María Córdova said both would be called *sipook*. Pedro Estrella called a female specimen *sipuk*. At least here people must have been aware that there were two breeding kinds. Among all speakers, the word *sipok* specifically refers to the red bird, whether or not the pink and gray bird is included.

The name *sipok* or *sipuk* seems to refer to the cardinal's crest. *Sipulk* means "to be in a pile." *Siivda* is the topknot or crest of the cardinal or of any other bird, such as the Phainopepla. It also designates a headpiece someone wears, with feathers sticking up prominently. Perhaps *siivda* is even related to *siivañ*, the name given by the

Pima to the chief or leader of the Puebloan *vapaki*, ruins, of the Salt-Gila valley, whom the Pima claim to have conquered. These leaders reportedly wore long plumes of macaw or other birds in their hair.

Sylvester Matthias of Komaḍk remembered that sometimes the *sipok* entered his father's quail traps. Ruth Giff, raised in the next village up the Santa Cruz, similarly recalled, "We used to trap the *taatamtol* [White-crowned Sparrows] and sometimes those red ones would get in there—*sipok*." Anyone who has had experience trying to remove these powerful-billed birds from a trap or mist net will appreciate why these incidents were considered so noteworthy even half a century later.

In the Creation Story, Lloyd (1911:210, 213) accounts for the origin of Navichu, a kachina-like character found among the northern O'odham. When Navichu appears (from the north?), the People entice him to stay with them. They say he will be pleased with two *vapaki* (pueblo-style ruins). One is called *sipok va'aki*, "cardinal ruins," and is covered with red flowers. The other is called *vajukukī va'aki*, "oriole ruins," and is covered with yellow flowers. "And where they came together the mingling of red and yellow was very pretty." Navichu at this point is something of a captive, and the People do not allow him to visit these mysterious structures. "One day he slipped away, when they were not looking, and opened one, and saw in it many wonderful things, clouds forming and sprinkling all the time; and the other was the same."

The color red is associated in Piman thought with lightning, the precursor of storms, hence with rain. We see this association in the red flowers and red seeds of the Coral-bean, a rain bush. The rinds of the Saguaro fruits, when emptied of their juicy contents, must be thrown on the ground with their brilliant red insides facing skyward. The cactus wine brewed for the summer rain festival is blood red. When summer rains begin, thousands of Giant Red Velvet Mites may emerge from the earth and ply its moist surface. Even war, the bloody killing of the enemy, and the associated Scarlet Macaw are ultimately harbingers of rain. So it is not a surprise that the cardinal should be a rain bird.

Anna Shaw (1968:82–83) recorded a tale in which Redbird (presumably the cardinal but possibly the Summer Tanager), wanting a new spring coat, sings from a willow tree. (The words apparently have no meaning, just imitating the cardinal's song.)

> Chee chee paw nee
> Chee chee paw nee
> Chee gall lee
> Chee gall lee.

The bird plucks out its feathers while singing, then flies away with a new red coat. Expecting the same results, Coyote climbs into the willow, tries pulling out his fur, singing the same song. It seems to work, but the red that he sees is his own blood. During the night, he dies of cold.

It is a puzzle that such a well-known, common, vocal, and showy bird does not figure more prominently in Piman oral literature. But we have seen other similar cases of prominent birds being slighted in song and story.

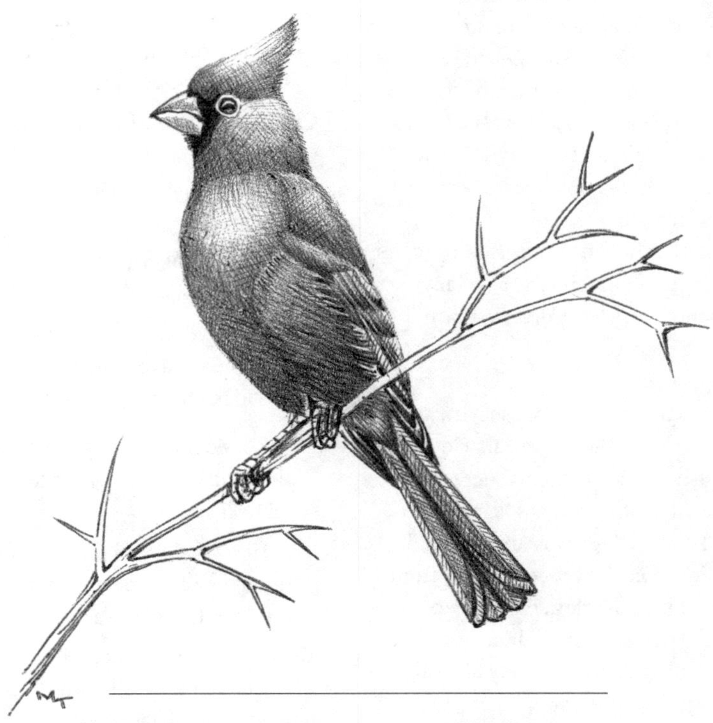

sipuk, sipok, Northern Cardinal (Mike Touch)

hevachuḍ, pl. *hehevachuḍ*; ?var. *hevaichuḍ*

Cardinalinae, Emberizidae, Passeriformes
Passerina caerulea
Blue Grosbeak; Gorrión Azul, Picogordo Azul

It is midsummer 1985. I am censusing birds that are nesting along the lower Salt River in the cottonwood and willow woodland that forms the northern boundary of the Gila River Reservation between 107th and 115th Avenue, where the Gila and Salt join. Even though the day is warming and the nearby Sierra Estrella peaks shimmer in heat waves, the bird life here is active, with many species now at the height of breeding. From where I rest in the sand, I try to capture in my notes the scene before me. Song Sparrows *(Melospiza melodia)* are busy feeding their begging young in a Seep-willow. Bright and noisy male Red-winged Blackbirds are flying back and forth between cattail patches on a sandy island, trying to maintain territories. A slinking Yellow-billed Cuckoo makes a mechanical wooden sound from a nearby cottonwood. The continuous question-and-answer song of a Bell's Vireo *(Vireo bellii)*, the *spreeee* call of the Red-wings, and the full melodious notes of a Blue Grosbeak in the willow are making sure that there is no silence, even in the late morning heat. The cottonwood and willow groves are proof that ecological degradation is not complete—at least not quite yet.

The Blue Grosbeak was an enigma to me during my first two decades of studying the middle Gila avifauna, as it was for M. French Gilman nearly eighty years before me. I had found occasional individuals in summer, as had he, at wet field edges, in sunflower patches and sorghum stands,

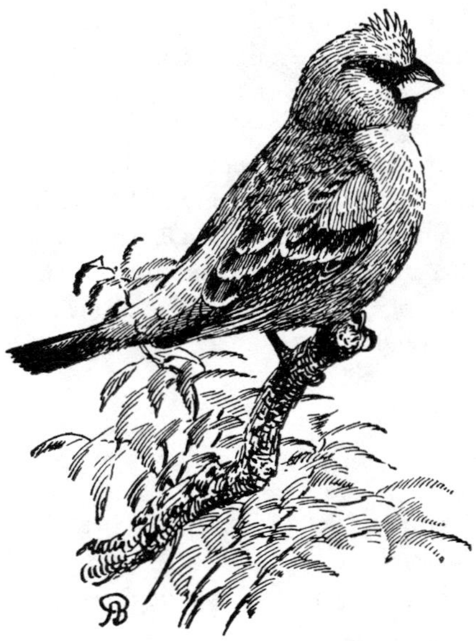

hevachuḍ, Blue Grosbeak (Allan Brooks)

and in marshes, but never nesting—just wandering individuals. Finally, this summer (1985) I have found a pair rearing a family of fledglings, along with other interesting riparian birds that have recolonized this riparian area as the trees have matured (Rea 1988).

The adult male grosbeak is hard to miss. It is largely deep blue with two broad cinnamon wing bars. The thick finchlike bill, black above, pale gray below, has a base of black feathers, extending through the lores. Much of the time at a distance he appears all black, quite like a Brown-headed Cowbird in size and shape. The female, less distinctive, is gray brown, thick billed, with double brown wing bars. Young males start off like their mothers but are richer brown. They return the next summer with the body a mixture of brown and blue blotches. They sing in this patchwork attire and try to attract mates. It is not until their third year that they appear as all-blue adults.

In Pimería Baja and Alta, the Blue Grosbeak arrives in mid- to late May, nests in midsummer, and leaves in September once the young are fledged (Phillips, Marshall, and Monson 1964:179; Russell and Monson 1998:279). In the arid West, at least, Blue Grosbeaks nest with water close at hand somewhere, particularly where willows and mesquites are near farmlands, pastures, or marshes, where the birds can forage for insects and seeds. The nest is made in a bush or low tree, sometimes with a base of cast snake skin. Fine rootlets and often horsehair line its inner cup.

Although practically extirpated from Piman country today with the drying of desert streams, the *hevachuḍ* looms large in Piman song and story. It has generally been glossed just "bluebird." (It is a blue bird, but not a bluebird.) Indeed, this Piman taxon had long eluded identification until close attention was paid to its role in legend and the folk science surrounding it. Most O'odham said they had forgotten what the bird looked like, and some confused it with the Barn Swallow, which once nested in their shallow wells. But this swallow is properly called *s-cheedagĭ giidval* in Pima.

Sylvester Matthias said, "*Hevachuḍ* is here all year. Hangs around cottonwood trees or willows where it's damp. Heavy bill." Twenty years earlier, however, he had said that it was migratory, coming back each summer, which is correct. He mentioned a place called *hevaichuḍ ha ñe'ikuḍ*, "the place where the blue birds have been singing." He explained further, "On the Gila, down beyond, tail end of, Co-op Village."

Some years later we were discussing the *eḍa mel*, "end run" ceremony. Sylvester noted, "Santa Cruz [village] way back [had the] *hevaichuḍ eḍa mel*—that 'blue bird.' My aunt was a dancer in there. [It's] more a river bird—the color is blue. Eat along the river. That's where they sing over there—they sing down below here somewhere along the river. They call it *hehevaichuḍ ha ñe'ikuḍ*. West of Co-op somewhere. I guess they rehearsed there. My father helped them sing, too, [although he was] a Navichu [by heredity]. But his sister is in there, so he has to help them. Songs were about the *hehevaichuḍ*."

Still another time he said, "*Hevachuḍ*, it's a bluebird. It's . . . the color is blue [anyway]. It's more a waterbird, a river bird. Right along the river."

These characteristics, as well those encoded in myth, as we will see shortly, apply specifically to the Blue Grosbeak. However, some of Sylvester's comments suggest the possibility of another species being (occasionally?) included. For instance, he said, "I really don't know. I've seen [them] from a distance. Like a swallow, only larger. Around the fields. Flying in bunches. Sign of rain." And again, "Somebody plow a field, they just *swarm* around there, in a plowed field, around people working." This suggests one of the true bluebirds, either the Mountain Bluebird, *Sialia currucoides*, or the Western Bluebird, *S. mexicana*, both of which, sometimes in numbers, visit Pima country in winter when much of the plowing takes place. (Fr. Antonine [1935] lists a *cheedagĭ u'uhig*, literally "blue bird"; the referent might be these winter visitors.)

George Kyyitan and others upriver said they knew the name *hevachuḍ* and knew its story and song, but didn't know what the bird itself looked like. They lived where the river had dried much earlier, and riparian birds were few or none.

In O'odham oral literature, the blue *hevachuḍ* is usually paired with the yellow oriole, *vajukukĭ*, another bird of mesquite thickets and cottonwood-willow woodlands, both summer residents. For instance, Russell (1908:333–334) recorded two series of rain songs. Three of them are referred to as corn songs. The middle song deals with Taatkam, Picacho Mountain, the shrine of Corn Man. The song describes the storm clouds rising and thundering over this rugged desert outcrop. The blue *hevachuḍ* holds thundering clouds in his talons. The yellow bird holds clouds that are raining in his talons. Culver Cassa has reworked this song:

> Hi'ilu'u ya'a'a! Ta'atekam he donovange dadamana chevwangi shaika'a amonges sikai'idame toahime. Hi'ilu'u ya'a'a!
>
> Ta'atekam he donovange dadamana hikum he shaika'a amonges sikai'idame jukuna. Hi'ilu'u ya'a'a!
>
> Che'enangi hevwachuna chewagi yukañi sanavwu shaika'a amonges sikaidame toahime. Hi'ilu'u ya'a'a!
>
> So'oham he vañkunga hikum ha yukañi sanavwu shaika'a amonges sikaidama junkuna. Hi'ilu'u ya'a'a!
>
> Hi'ilu'u ya'a'a! Picacho Mountain clouds atop there adhere and seem to loudly thunder. Hi'ilu'u ya'a'a!
>
> Picacho Mountain misty clouds atop there adhere and seem to loudly rain. Hi'ilu'u ya'a'a!
>
> Blue Hevachuḍ raises the clouds thereupon they adhere and seem to loudly thunder. Hi'ilu'u ya'a'a!
>
> Yellow Oriole raises the misty clouds thereupon they adhere and seem to loudly rain. Hi'ilu'u ya'a'a!

Several warpath speeches were preserved by Russell. In one (1908:367–368), the orator uses birds as images of renewing the world after the chaos of killing:

> And the Blue Grosbeak Shaman descended; he was ecstatic with happiness and sang.
> His blue feathers he ruffled, and thereby the land boldly adorned with algae presented itself.
> And the Oriole Shaman descended; he was ecstatic with happiness and sang.
> He ruffled his yellow feathers, and thereby the mountains boldly yellowly adorned at their bases stood.
> And the Mockingbird Shaman descended; he was ecstatic with happines and sang.
> His white feathers he ruffled, and thereby the land everywhere shiningly presented itself.

Russell (1908:344–345; see also Bahr 1975:34–50) also recorded a call-to-arms oration in which Elder Brother sends up Gray Gopher to scout the upper world inhabited by the Vipishaḍ. He makes four entries. At the first entry, he discovers a shining water, and it is covered with eagle tail feathers. On his second emergence, he finds a black water, and it is covered with raven feathers. On his third emergence, he finds a blue water, and it is covered by *hevachuḍ* feathers. And the fourth time he finds a yellow water, and it is covered with oriole feathers.

As is usually the case with Piman orations, the significance of this oration is obscure. Culver Cassa suggests that the black water means that the enemy who are discovered occupying the upper world are not going to make it. The white one means the victory is surely "ours"—that is, the Emergenti's. The *hevachuḍ* signifies there is probably a lot of rain up here. And the yellow means that it is a beautiful land here. All of these elements are leading to the enemy's doom. The covering of the waters means that the emerging People are covering the enemy's waters with their skills and expectations in the battle that will ensue.

Among several sets of "festal songs" in his anthology, Russell (1908:283–284) recorded six that are assigned to the *eḍa mel* ceremony. One of these songs is about *hevachuḍ*.

> Hevachunḍ voivimena
> mumui yuhuvi voivimena
> chevangi momotok voivimena
> wesat momenama chevahangi yungeñi
> nak vat vonge ñihina'a.

> Blue Grosbeaks came running
> Many women came running
> With cloud burdens they came running
> All shook the clouds above their heads
> Thereupon shake bodies.

Culver Cassa, who reworked this song, notes that the verses are instructions to the dancers. The rubrics suggest that men, probably, come forth with sticks bearing at their tops images of the grosbeaks. They are followed by women who come forth with images of clouds supported on two sticks held over their heads. This is the ceremony that Sylvester's aunt participated in and for which his father helped sing. Ultimately, it's a prayer for rain.

The place where this bird lives seems to be anthropogenic, Culver notes. "People shaped the environment, and the birds liked it, like the Red-wings and grackles do today when we're plowing with the tractor. They are wild things that bring their wildness to our tameness."

This grosbeak ceremonial, *hevaichuḍ eḍa mel*, seems to have once been of some importance, as Sylvester noted. The River Pima calendarist for the year 1888–1889 observed, "A Papago who knew the bluebird series of songs sang for the Santan people during the festival held by them" (Russell 1908:61).

These references to the Blue Grosbeak in sung poetry and oration are characteristically sketchy, depending on common knowledge to carry the impact of the metaphor. However, it is in the more narrative types of O'odham oral literature that the biology of the Blue Grosbeak is most carefully encoded.

During the invasion of the Gila valley by the Vupshkam (emerging Pimans), various creatures, in particular birds or their powers, are called upon to assist in the conquest (Hayden 1935). One of them is a blue bird, earlier in the text identified as *hü wut jut*. More precisely, it is a *maakai*, a shaman, with the power of the Blue Grosbeak. The invaders take captives from one settlement. They decide not to sacrifice their victims, so they must somehow immobilize them to keep them from giving aid to their neighboring pueblos. So the *hevachuḍ maakai* sings:

> Blue Grosbeak Shaman
> You called me
> To the land that lies before me
> I am breathing.
> I am tying the hands
> Of my enemies.

Grosbeak's work is not finished. The story of conquest continues. In what is now Casa Grande ruins there lived two brothers. Hevachuḍ uses his power to entangle the hair of each brother with that of his wife in their sleep so that when they arise, they are unable to get loose and can be taken. (Both the narrator and his listeners were familiar with the nest-building behavior of the Blue Grosbeak.)

There is also the story of young Ba'ivchul Uvĭ, Kingfisher Woman, who, abandoning her husband on their marriage night, wanders down the Gila to the Santa Cruz dunes and takes up with the son of Kadgam, Yellow-billed Cuckoo, who is a good farmer; then with Hevachuḍ, who is a lazy farmer; then with Hohokimal, Butterfly, who farms where water moistens the land from below (you can still see butterflies at such seeps); then finally goes back upriver to the east to Gook Vapchkĭ, Two Ponds (Bahr and associates 1994:112–122; Hayden 1935:17–20; see also Benedict 2001:69–72). This story—usually called a story about a "prostitute," but really more a story about adaptation to the environment—is discussed in more detail in the Yellow-billed Cuckoo and Belted Kingfisher accounts.

"Bluebird" appears in one version of the story about the boy whose father had been killed by Apaches and his mother abducted. Bluebird comes to the house. "Let's go to the ruin and get nuts." Bluebird is lazy and doesn't gather nuts, but he eats the nuts the boy gathers. The boy sees this, and Bluebird laughs. He steals nuts again from the boy's pile. The boy is angry, so he takes a stick and throws it at the bird so that it breaks both his legs (Benedict 2001:154).

The references here are puzzling. The character is presumably Blue Grosbeak, and again he is portrayed as being lazy. But what is the food? There are no nuts growing in Pima country. They are said to come from the ruins, presumably a *va'aki*. The Pima knew the black walnut, *uupio*, that grew upriver, but there is no Piman name for nuts as such. Once again, the O'odham-language version would have clarified the reference. Perhaps it is to seeds of Desert Ironwood or one of the paloverdes.

There is an apparent reference to this grosbeak in the Flute Story. When the grandson of White-clay Eater Woman travels west in search of his paternal grandmother, he is said to come upon the feather of a bluebird, a sign he is approaching the ocean (Benedict 2001:120). But in Herzog's bilingual manuscript ([1929]:96) we read, "He went on again and the bird followed him, toward the West. He came to some place, and there lay bird feathers *(u'uhig a'an)*, blue, beautiful *(s-chehedakĭ, s-kegach)*. He picked them up and said: 'Truly the feathers are beautiful; I will make a plume for myself.'" What he makes is an *a'an siivdach*, a feather headgear.

The story of *hevachuḍ* and Coyote was popular among the O'odham (Benedict 2001:171–172; Hayes 1957:9–16; Russell 1908:245–246; Shaw 1968:70–72). There is a bird that is colored brown like the dirt. One day Coyote sees it go up a mountain to "a lake where no river flowed in or out, and the bird bathed in this lake four times every morning for four mornings." Each time it sings this song:

> Gaḍhu siche anunima shungga
> genangesha
> wes siko sivañ chechenunga.
>
> In the bluish pond
> In it I fall
> All over I become blue.

Each time it comes out of the water somewhat bluer. On the fourth (or fifth) morning, the bird emerges all blue. Coyote is envious of the beautiful color and asks Blue Grosbeak how he did it. So the bird teaches Coyote the song, giving him instructions, warning him to avoid vanity when he returns to his friends in the lowlands. In four days, Coyote is also all bright blue. "He was so proud that as he walked along he looked about on every side to see if anyone was noticing how fine and blue he was. He looked to see if his shadow was blue, too, and so he was not watching the road, and presently he ran into a stump so hard that it threw him down in the dirt and he became dust-colored all over. And to this day all coyotes are the color of dirt" (Russell 1908:246).

The published versions do not name the bird involved in this metamorphosis, but Ruth Giff, Sylvester Matthias, George Kyyitan, and others knew the story and said the blue bird was *hevachuḍ*, who was originally dull brown. Ruth heard that the magic lake was somewhere on top of the Sierra Estrella. George told a somewhat different version of the story:

> And these bluebirds, they were by the sea. Bluebird, I don't know what kind of blue bird, [they] call them *hehevachuḍ*. They are the bluebird and they were by the sea, there.
>
> Coyote came there [and said], "How did you get that . . . [color]? You look so beautiful, that blue color," he said.
>
> "Well, we go in there, in the ocean, and take a bath in there, then sing a song, then go in there take a bath, come right out, see how it looks. If it's not too blue, have to sing another time and go in there. Keep going, going. Now we got the blue."
>
> "Well, let me do it. I'm going to turn blue."
>
> "Alright, but we'll tell you what you're going to do. You have to go in there and sing first. Then go in there. See how you can look. Then, if you got that blue, don't say anything; don't go anywhere; don't say anything to anybody."
>
> So he went in there. Sing. I forgot the song that they sang. Anyway, it's blue. He went in there and jumped in the sea. When he come out, not too good. So he got to sing another song. He went in there. And not too good. Just the same way. Not too blue. Third time he went. And he got blue.
>
> And he said, "Now, how do I look?"
>
> "You look pretty good now," they said. "But we tell you, don't say anything to anybody. Even if they say something about you, don't say anything. Even if they call you that you are a fox, Coyote, but don't say anything because they are going to make fun of you."
>
> He went away. Them birds flew away and left him there, thinking what he's going to do. Some people came and asked him, "Why? Why did you make yourself blue? Where did you get that blue color?" they said. "I thought you were gray."
>
> "No, it's like that. I was born like that."
>
> "Aaaaah. Don't tell us stories like that."
>
> "Yeah, it's true," he said.
>
> So they start throwing dirt on him. Because they was making fun of him, like that. He said [to] them, "Don't do that! Don't do that!"
>
> "Well, tell us how you [turned blue]." He never did say anything. [They] keep on throwing dirt on him. That's why he got his gray color. Finally them birds caught [found] him again. Ask him why did he get like that.
>
> "Somebody came and throw dirt on me when I was still wet." The end of story.

George's version, narrated in 1994, differs somewhat from how the story is usually given. In three of the published stories (Benedict's, Hayes's, and Russell's), the blue is derived from a local lake, and Coyote stumbles from pride. George adds the device of lying; his setting is the ocean not a local body of water. All five variants emphasize the importance of following directions explicitly.

In mythic time, it takes *hevachuḍ* three or four days to turn from dull brown to bright blue. In biological time, it takes Blue Grosbeak three years (and three molts along with two migrations) to become deep blue. But he must sing all the while to do so.

Although the name *hewachuḍ* (pl. *hehewachuḍ*) is in the Tohono O'odham lexicon, no Papago I interviewed had any information about the bird. Saxton and Saxton (1969:167) list an unidentified bird as *heewasjel*, a variant of the Blue Grosbeak name in some Papago dialect. Most likely the bird was found primarily in Gileño country before that area's disastrous desertification.

MISIDENTIFICATION. Bahr (in Bahr and associates 1994:117) misidentifies Hayden's "Her-va-chut" (1935:19) as *he:wacud*, "bluejay," based on an incorrect dictionary entry (Saxton, Saxton, and Enos 1983:20). Again (Bahr and associates 1994:251), "He: wacud Namkam" appears erroneously as "Bluejay Meeter." The Blue Jay *(Cyanocitta cristata)* is an eastern North American species. Other species of jays with blue in the plumage may descend into the Sonoran Desert on extremely rare occasions—not the stuff of folk taxonomy. Mathiot (1973:342) identifies the bird as the Red-backed Junco *(Junco hyemalis dorsalis, J.h. caniceps)!* Cassa notes that *hevachuḍ* is a summer bird, based on Piman literature, where it is matched with oriole, hence eliminating the bluebirds (*Sialia* spp.), jays, and juncos, all mid-winter birds in Pimería Alta.

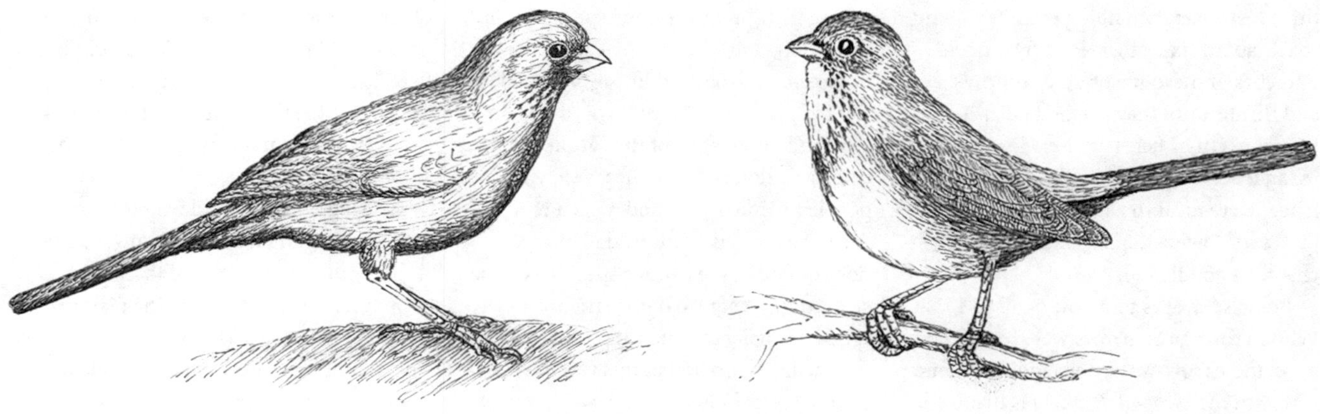

bichpoḍ, Abert's Towhee (Amadeo M. Rea) and Canyon Towhee (Amadeo M. Rea)

bichpoḍ, bichpuḍ,* pl. *biibichpoḍ
***?bitpoḍ* (Pima Bajo)**

Emberizinae, Emberizidae, Passeriformes
Pipilo aberti
Abert's Towhee; Toquí Enmascarado, Rascador Desértico
Pipilo fuscus
Canyon Towhee; Toquí Pardo, Pájaro Viejo, Rascador Pardo

The river bottoms of southern Arizona host to an oversize sparrow, the Abert's Towhee. Here the bird is at home in the underbrush of the cottonwood-willow gallery forest, older mesquite bosques, and Arrow-weed thickets, as well as around old-style farmlands with dense fencerows and huge mounds of Quail Brush. In these dense, shady retreats, the towhee becomes abundant, sometimes even familiar.

These plump terrestrial birds, about nine inches long, superficially suggest a Crissal Thrasher or one of the other local thrashers. They are brownish gray above, pinkish or cinnamon brown below, with a conical pinkish bill set off by a charcoal facial patch and some ashy flecks on the throat. (All the thrashers have long dark bills.)

When feeding in open areas, usually in the shade and not far from cover, an Abert's keeps up a *chink, chink* note, communicating with its lifelong mate. When disturbed, it takes wing into the underbrush with a flurry of rapid *chink* notes, patently complaining.

In the higher country of Papaguería and in the mountains along the Gila with Upper Sonoran shrubs and grasses, the Abert's Towhee is replaced by the Canyon Towhee. This chunky sparrow is slightly smaller but more distinctively marked. The underparts are whiter, the crissum (undertail coverts) more obviously cinnamon, and pale gray chest streaks converge into a pectoral spot. The bird's pale eye ring gives the face a soft appearance. The crown patch, raised when the bird is excited, is rather rufous.

Canyon Towhees are partial to broken, rocky country but are at home in Tohono O'odham rancherias, foraging familiarly in yards. Paired for life, individuals keep within earshot of their mates by means of a low "bark" note. Both species of towhees are primarily seed eaters in winter, switching to an insect diet when breeding.

River Pima knew the Abert's Towhee well, it once being such a familiar bird when traditional agriculture was practiced, houses were interspersed in fields, and the riparian community was intact. Rosita Brown compared it to *biit keishnam*, the Bendire's Thrasher, another dooryard bird. Leonard Pancott said, "*Bichpoḍ* is almost the same color as *chev chiñkam* [Crissal Thrasher]. The two birds almost look alike." Sylvester Matthias said he didn't consider *bichpuḍ* to be a game bird for the Pima. However, Francis Vavages upriver commented, "Yeah, they're good [to eat]. And that *tamtol* [White-crowned Sparrow]; they are a winter bird." (Even though the two thrashers are similar to the towhee in size, color, and habits, I'm told the thrasher would never be eaten, whereas for some, at least, the towhee was fair game.)

Joseph Giff explained the significance of *bichpoḍ* in more detail:

> *Bichpoḍ* is something the size of a *hoohi* [Mourning Dove]. They're like a sparrow. Dark around eyes. They used to make fences of poles and branches, cut the tree limbs and put them in there. And *bichpoḍ* used to nest [roost] in there in the wintertime. We used to hunt them. They're better eating than a *tamtol*. They're kind of a bright pinkish color. They have a *short* beak, anyway; kind of a brownish color—light brown and gray. Kind of a light orange color. They nest in the brush, in the fences like I told you; they nest in there. We used to charge them out and kill them. They don't fly very far, or very high—they're like a quail: they fly a little ways, and then they hide. And I think they have a little bit of something on the head—a plume, maybe just the male. Not anything outstanding—just a bit taller feathers there. That's why they call them *bichpoḍ*. You see, a long time ago they [the Pima] used to decorate themselves. And they say they have a *bichpoga*—that means something, a little something [stuck in hair] right here [back of head, high]. *Bichpoḍ* is built something like a Plain Titmouse [he was thumbing through my field guide], but is the size of a quail, the size of a *hoohi*.

So the Piman name for this bird derives from the small feathered headpiece that the O'odham once wore, suggesting the towhee's erectile crown. (A different name was used for the larger headdress made of macaw or other long feathers.) In earlier days, the Pima enclosed their fields with brush fences made by double rows of mesquite stakes thrust into the ground, about three feet apart then filling the intervening space, about six feet wide, with trimmed branches of mesquite or other thorny trees and shrubs until they reached taller than a person. Other fields were enclosed with living fencerows of spiny plants. These fences were ideal roosting and nesting places for towhees, quail, and doves (see chapter 8 and Rea 1978).

Francis Vavages told how quail and White-crowned Sparrows were trapped in *eḍpa*, the Arrow-weed trap, adding: "Same way with the *bichpoḍ*. *Bichpoḍ* is that bird, too, that likes to eat [whatever] you plant, melons or squash; he'll go in there. It's got that wide beak, too. They'll crack those, like those birds, those talking birds [parrots]. They'll crack those. Lot of times we'll go around there, you know, and try to plant melons. They start digging. They seem to know where the seeds are, too. All you'd see is the shells around there. Yeah."

The name *bitpoḍ* is found southward among the Pima Bajo. Although Canyon Towhees are common here in tropical thornscrub and tropical deciduous forest, Pedro Estrella described his *bitpoḍ* as having a "capetito," little crest, and being completely green, including the topknot; same size as the cardinal. I don't know what bird this might be. It is not the Canyon Towhee, though it might be the Green-tailed Towhee, *Pipilo chlorurus*, a wintering species that, though common, is quiet and retiring.

There is one reference to the towhee in Piman legend. The son of one of the twin boys, journeying to find his grandmother, White-clay Eater Woman, stops for a while to visit his maternal uncle, Tamtol, White-crowned Sparrow. Wishing to cover his tracks when he leaves, he spreads the sparrow's seeds over his footprints. The quail and *biibichpoḍ* eat the seeds, obliterating the boy's trail. (More detail is given in the White-crowned Sparrow account. All three birds in this narrative segment are seed eaters that were trapped by the River Pima.)

Jose Garcia related that there is a place called Bichpoḍ north of Pisin Mo'o village, where someone used to farm, but nobody lives there now.

MISIDENTIFICATION. Herzog ([1929]), to whom we are indebted for the plural form of the word and the narrative reference, did not know the English name for the *bichpoḍ*, calling it "[a]n unidentified species of bird, with a long tail."

Saxton, Saxton, and Enos (1983:6) identify *bichpoḍ* as the Pyrrhuloxia, a slender pink and tan relative of the cardinal. This determination is at variance with what people I worked with said, which was usually backed up with field identifications of one of the towhee species. No one seemed to recognize the Pyrrhuloxia as a species, usually commenting at pictures in my elicitation book that this was the female cardinal. Pyrrhuloxias are new to Akimel O'odham country, but someone among the Tohono O'odham should have recognized that the tan birds form nesting pairs and are sexually quite dimorphic. Most likely we have not asked enough people or the right questions.

tamtol*, pl. *taatamtol

Emberizinae, Emberizidae, Passeriformes
Zonotrichia leucophrys
White-crowned Sparrow; Gorrión Corona Blanca, Zacatero Mixto

Anyone who has trapped birds for banding in the western lowlands knows that the most easily and abundantly taken species in winter is the White-crowned Sparrow. These medium-size, plump, grayish sparrows have brown wings, tail, and back streaks. But their most distinguishing marks are the three white and four black head stripes in adults (buffy and brown in immatures).

During a sharply cold winter morning or in the waning late afternoon sun, one is likely to hear a chorus of these birds from their brushy retreats. The plaintive, leisurely whistled series of ascending notes in a minor scale are as characteristic of the desert as are the gray saltbushes and bare mesquites.

Throughout Pimería Alta, White-crowned Sparrows are common or even abundant from October to mid-May. They congregate in any brushy country, frequenting farmlands, fencerows, ditch banks, roadsides, house yards, and even undisturbed desert country. They are less common in tropical thornscrub and tropical deciduous forest of Pimería Baja (Russell and Monson 1998:300). Their food at this season is weed seeds and waste grain, so they become particularly abundant in agricultural country, feeding on the ground. The trapping and banding of individual birds shows that wave after wave of these sparrows pass through an area on their way south, then again on their return north to their high mountain and subarctic breeding grounds.

These little birds, weighing just under an

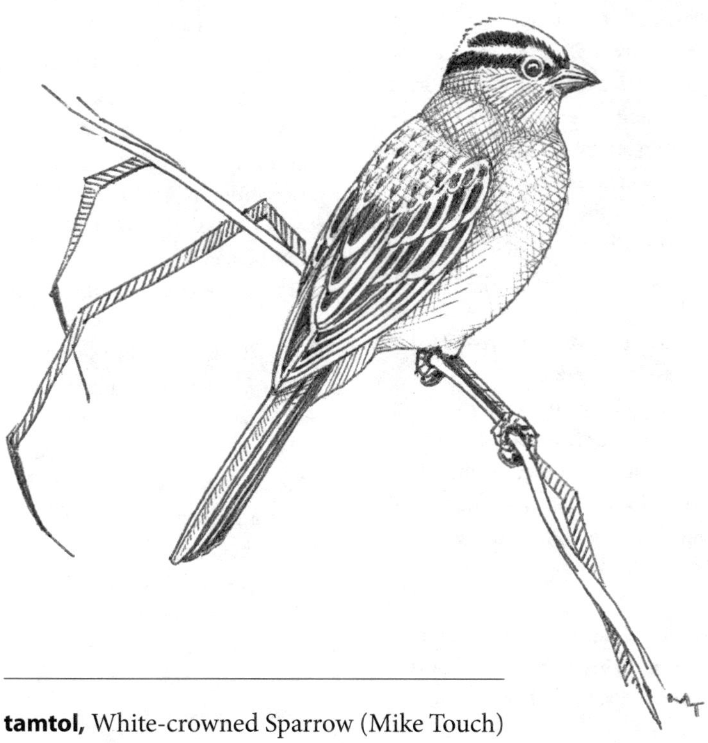

tamtol, White-crowned Sparrow (Mike Touch)

ounce, seem an unlikely food for humans, but, in fact, White-crowned Sparrows were a major protein resource for the River Pima. What the birds lacked in size they made up for in numbers. Perhaps they contributed more to the diet in winter than any other species of bird. In the early 1900s, Frank Russell (1908:81) noted, "During the winter months these birds [*taatamtol*] are caught at nearly every house by means of traps." These traps, made of Arrow-weed rods, were of two types: a flat, shelf-type trap and a cage-shaped trap. The Pima, too, had discovered how easily trapped these sparrows are. In addition, White-crowns were taken by several other means of hunting. Older Pimas all remembered them.

Joseph Giff observed, "*Tamtol* is a little sparrow. There are two kinds, maybe two different species. Sort of brown and white. White streaks above the eyes—brownish color." He identified them from pictures. "The pretty one is the male, just like anything else!" he said, teasing his wife, who was listening. "We used to eat them. One bite and they're gone. They 're common around here. We see them every winter. There's nothing outstanding about a *tamtol*."

Ruth Giff added, "We used to trap them [*taatamtol*] and sometimes those red ones would get in there—*sipok* [cardinal]." Another time she said, "*Eḍpa*, the trap; catch *kakachu* [quail]. *Taatamtol* always get in the trap."

In the next village, Leonard Soke recalled, "Hunt *taatamtol* with a slingshot. Takes two to make a bite."

One early April I pointed out to George Kyyitan an adult White-crowned Sparrow foraging near us. "What do you call that in O'odham?" I asked.

"*Tamtol*. Supposed to be here in the *winter*time. The other day I heard them. *Taatamtol*—can eat the bones with it, whole thing."

"How do you fix them?"

"The way doves are."

"How is that? Cook the whole thing or take the feathers off?"

"Feathers off. Take everything inside."

"Pull the insides out?"

"Umhum."

"There's not much left."

"That's why just going to eat the whole thing. [Not?] much bones in there, [just] little ones."

"How about the head?"

"No, not the head."

At Sacaton, Francis Vavages described the function of the quail trap, *eḍpa*, adding, "Oh they [also] catch *tamtol*, winter birds. Little birds. All you got to do put one in your mouth. They're little birds. It's got little brown and white on the head. They're winter birds. I don't know why people eat them when they're just that small. After the feathers are all off, well looks like there's nothing left."

Trapping and using a slingshot were not the only ways of capturing White-crowns. Two Pimas mentioned how the birds were taken in the communal night hunt called *wepegi*. Leonard Pancott explained the hunt while he and George were talking of sparrows. "*Tamtol* has black stripes on head," he said, noting that it was a bit smaller than *o'oḍopiwua*, the Lark Bunting. "Hunt *tamtol* in winter at night. Fix a five-gallon can for light—wire for handle. Punch holes in it. Cut mesquite limb with branches. Set light under tree. They start flying. Get them [by swatting them down with the mesquite limb]. Throw them on the fire. Clean them up—pull feathers and gut them. Lucky if you get two bites. Spit them [the bones] out."

Francis Vavages also commented on this *wepegi* hunting. "A lot of times they're doves in there, too. And those little winter birds."

"What kind?" I asked.

"*Tamtol*. They come out in the winter. They're so small. I don't know what you could get out of them."

Tamtol appears as a character in a complex subtext of the Flute Story. The grandson of the younger slain twin boy journeys to find his paternal grandmother, White-clay Eater Woman (Nasia) in the West. En route, he meets Tamtol, a maternal uncle, hence a relative of Yellow Buzzard, the sad and depressed family lineage. Tamtol, recognizing the boy because of his likeness to his father, embraces him and sings a crying song. Although we don't know the song, we can have little difficulty in imagining what it sounded like from the rising minor thirds of the White-crown's three- or four-noted winter song. The bird's Maricopa name, Shinawawi, occurring in a keening song, appears onomatopoeic (Benedict 2001:119). In Pima, a mother's younger brother, uncle, is *tatal*.

In the morning, Tamtol tells the child to wait at home while he goes out to gamble with neighbors, who always beat him. When the boy insists on coming along, the sparrow puts him on his back. Tamtol loses, as usual. The boy tells him to challenge his opponents again by wagering his nephew. Tamtol does this four times, taking all wagers. White-crown "bundled up his winnings, put them on his back; he set his nephew on top. They went back [home]" (Herzog [1929]:93–94).

This story was told in midwinter. By then, the People had been trapping White-crowned Sparrows for several months already. The listeners, intimately familiar with this bird's plumage, recognized the complex markings of Tamtol's interscapulars as the winnings he was carrying back home and the four black lines on Tamtol's head as the arms and legs of the boy piled on top and holding on tightly.

The next morning the boy asks again to be taken along to gamble, but is refused. Later the boy departs, taking Tamtol's stored seeds *(kaichka)*, scattering them along his path. The quail and Abert's Towhees find the seeds and eat them, thereby obliterating the boy's footprints. (Another version says that blackbirds and quail ate the seeds.) When Tamtol returns, he tries to follow his nephew, eventually discovering where the seeds ran out and the footprints resumed. He follows the boy from hillock to hillock through the desert until they both fall from thirst.

Pitying the boy, the Red-tail Hawks hang over him, shading him, making a cloud and then rain so the boy recovers, as does the sparrow. The pursuit continues to the edge of the ocean. Here the boy encounters Koomagĭ Vakoañ, Great Blue Heron, and asks for his help. The heron arches his wings so that the boy can pass across to land on the western side. When Tamtol, following along behind, comes to the shore, he tries to cross on the bridge made by the heron's outstretched wings, but he falls into the water. As a member of Yellow Buzzard's lineage, the depressed people, he cannot enter Nasia's land of happiness.

Along the flank of the Sierra Estrella, near the confluence of the Santa Cruz and Gila Rivers, is a prominence called Tamtol Do'ak, "White-crowned Sparrow Hill." The reason for its naming is no longer remembered, but the immediate area was the original site of the Komaḍk ranchería, discovered by Fr. Kino and Lt. Manje in

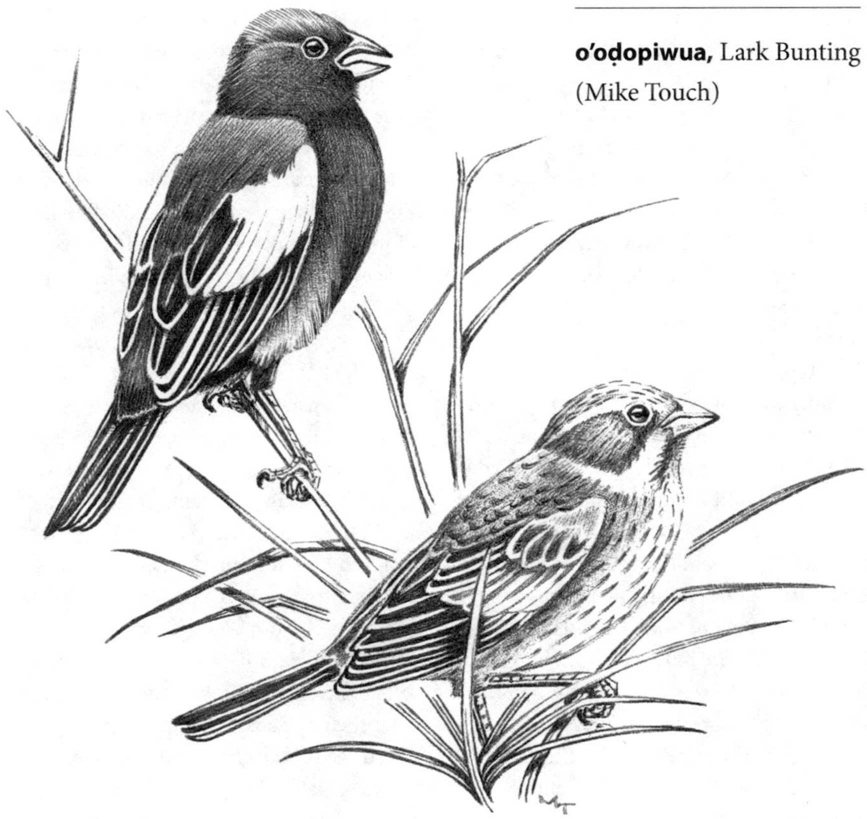

o'oḍopiwua, Lark Bunting (Mike Touch)

March 1699. Once well watered by the merger of several streams, the adjacent fields must have hosted abundant White-crowns.

o'oḍopiwua
s-banmakam, pl. *s-baabanmakam*
kauk (kavk) oo'okam
chechetopiwua;
o'odopiwis, o'odopiwua (Papago)

Emberizinae, Emberizidae, Passeriformes
Calamospiza melanocorys
Lark Bunting; Gorrión Ala Blanca

From August through April, sometimes even well into May, small to quite large flocks of sparrows wander about the open desert flats and fallow fields on the floodplains. They may number several hundred individuals. The birds are gregarious, feeding as a group on the ground, flying up and off as a unit when finished or disturbed. They are noticeable not only by their numbers, but also by their continuous call notes, which increase in vigor as spring approaches. They are Lark Buntings, a species that breeds on the central plains of North America but winters from the deserts southward over northern Mexico.

When the birds arrive in Pimería, they are mostly in winter plumage, looking more like finches than sparrows. They are plump, short-tailed, conspicuously streaked below, dusky streaked on tan or grayish above. Broad buffy wing patches, showing in flight, distinguish them from the finches. At this season, individuals in the flock seem rather uniform, but a close watch will reveal differences. Adult males have darker wings and tails; the blackish streaks below coalesce into a dark patch on the chest. With the progression of winter, the sexes become more dissimilar as the males slowly molt into their breeding plumage, patchily acquiring their black body plumage and broad white wing areas.

During their winter sojourn in the desert, Lark Buntings feed on seeds, largely weed seeds. They become especially common in areas where grains are grown, energetically gleaning spilled wheat, oats, or barley after harvests. The flocks are vagrants, however, wandering about the countryside rather than being more uniformly dispersed as are White-crowned Sparrows.

As we traveled about Pima country, Sylvester Matthias was the first to identify these birds for me, using their several names. "These are the ones, flocks. They even come close to the house. *Ooḍopiwua, o'oḍopiwua.* But [Ta'i O'odham, Upriver People] would say *s-baabanmakam.* Same thing." Another time he said, "*Kauk oo'okam,* 'hard bones, hard back.' Striped back, black and white. In the fields. Same as *s-baabanmakam;* [same as] *chechetopiwua.* It's gray, sort of like the doves."

"What do they eat?"

"Grain. *Chechetopiwua.*"

Ruth Giff, contributed, "That's what we used to eat."

"They're in the fields, in a bunch," added Sylvester.

There was no doubting that here were at least four different names—sympatric folk synonyms—for the same species. Synonyms of any kind, for either plants or animals, are unusual in Pima. However, while discussing this bird, People typically knew and used at least two different names for them. I think a clue comes from the species' high utility.

But what do all these different names mean? *O'oḍopiwua* seems to be the basic name, a nonanalyzable lexeme or word. With slight variations, it is both well known and widespread among both Akimel and Tohono O'odham. Most frequently, this bird was contrasted with *tamtol,* the White-crowned Sparrow.

Another widely used name, *s-baabanmakam,* seems more of a nickname, or perhaps an indirect reference, to avoid giving offence to a prey item. (The *tobdam,* village hunt leader, typically referred obliquely to the animal to be hunted that day, not naming the animal directly.) Here is how Joseph Giff, in his inimitable way, explained the name:

> *S-baabanmakam.* That must be from the word *ban,* "Coyote." They say *ban* is into everything and eats anything. And eats all he can. And a person, who is like that, is an eater—you know, eats a lot—they say he's *s-banma, s-banmakam.* That's singular; and the plural for that is *s-baabanmakam.* That's what they call the birds [Lark Buntings] because they say those birds eat like a Coyote. Because in the old days, *ban,* I guess, is about the only thing that they would compare anything that would eat so much. Because the *kooji* [pig]—the *tasi'ikol* [Javelina] wasn't domesticated—

they never know how much they eat. But the *ban*, they know about him.

And they know that the *ban* is into everything—that is, I think, is taken from the fact that *ban* is in all the legends—he's in *everything*. He has a part in *everything*, you see, and if a person is like that he wants to get in[to] everything, well, they call him [?] *s-dapchum ban*. See, they compare him with the *ban*. And for the eating, why I suppose it's the *ban* that eats anything, will steal and eat, or anything like that. So they call a person *s-banma* when he eats a lot.

And these birds, kind of like Sylvester said, that these birds will eat a lot at one time, you see. And so one bird would be *s-banmakam*. But they're talking about a flock of those birds, [so] they call them *s-baabanmakam*, see, which is plural for *s-banmakam*, [the] *o'oḍopiwua*.

So the Piman equivalent of "piggish" is a Coyote-like eater. The roving flocks of intensely feeding Lark Buntings suggested this behavior to the O'odham.

One day I asked George Kyyitan, "Do you know *s-baabanmakam u'uhig?*"

"Yeah. Got black on the chest. Those are the bulls [males], got black. Females just kind of gray, no black. If you will talk about one, *s-banmakam*. There's one old man, old man by name [of] *pi-banmakam*, just like me, not eat [too] much; not *s-banmakam*."

An alternate name, *kauk* (or *kavk*) *o'okam* means "hard bones." Sylvester said *o'okam*, but George said it's just *oo*, "back," not *oo'o*, "bones." His name refers to the broad avian pelvic bone. Either way, this name no doubt derives from the Piman method of eating the bird, as we shall see.

Sylvester used still another synonym, *chechetopiwua*. Mathiot (1973:172) gives the word *cheche* as meaning "to make, emit, a certain sound, noise, utterance; to make strange repeated unintelligible noises." This name refers to the calls that these highly gregarious birds use to maintain flock cohesiveness while traveling and feeding in winter. It is a rising three note call, given in chorus as the flock takes to the wing or alights. Culver Cassa says *topiwua* means "whatever it does, it's doing it."

On an early January day, George Kyyitan and I are watching a large flock of Lark Buntings feeding in a fallow field—two, three hundred strong.

"How do you call them in O'odham?" I ask him.

"*Ooḍpwa*." (Another time he used the name *oodakua, o'odokkua*.)

"What do those words mean? Is 'bone' part of that name?"

"I don't know. [It's] not *o'o* [bone], just *oo* [back]."

"Did you ever eat these?"

"Yeah."

"Did you like them?"

"Yeah, taste good."

"And how do you fix them?"

"Just the same way like a dove. Yeah. Put it in tortilla and eat it."

"Just a *chemait?* Do you have to take the gut out? Clean them out?"

"Yeah. Some way with a dove, like a dove. Clean them out, put them on the pan. Boy, it takes just one at a time. The whole thing will be bones. Because it's not hard. Not hard bones."

"So you eat the bones too?"

"Yeah, just chew them up. Just like that sparrows."

"*Tamtol*."

"Those are good, too."

"Which one has more meat on it?"

"This one here, because these are bigger than that, bigger than the *tamtol*."

"And how did you used to get these?"

"They shoot them. When they're in a bunch, you can shoot them. One time we were plowing, planting down there, on this side of that canal there, from church on down that way. My brother had been planting here. We were over there, and there was *lots* of these. Yeah. Tried to broadcast the wheat there. They just all go over [eating the seed]. So he come over and get his shotgun, killed them. Fixed them up right away, and fry 'em."

"How many does it take to make a meal?"

"Ah, it takes a *lot!* [laughter]. Pull the feathers off, clean them up. Put them in a fry pan, fry them. They're good!"

"How did they get them before the shotgun?"

"I don't know. I don't know how they get them."

"You didn't see those traps, those *eḍpa* they used to make?"

"Yeah, I know what you mean—*eḍpa?* Yeah, they used to, that one there."

Leonard Pancott, also from the Gila, said, "*O'oḍopwa* has a little white on the wing. A little bit bigger than the *taatamtol*. We ate them." A Lark Bunting weighs about an ounce and a third, the White-crowned Sparrow less than an ounce.

But I wondered if the Tohono O'odham, living out on the desert away from the great alluvial floodplains, away from extensive agriculture, knew much about these birds or even ate them. Delores Lewis of Ge Oidag, Big Fields, told me, "*O'oḍopĭ*, little one. They eat any kind of things—seeds, grain. Clean them up, pull out feathers, gut them. Lucky if you get two bites [from one]!" The Papago form of the word used in the species heading comes from Saxton, Saxton, and Enos (1983:48, 103).

At Vaya Chiñ, Joe Lopez, Delores's first cousin, did not recognize any of the Lark Bunting synonyms until I came to *s-baabanmakam*. "My father used to call those—what do you call that, stick together all the time?—birds, hanging around the house. When they feed his chickens, they all come around. He called them *s-baabanmakam*." Like various other Tohono O'odham, he did not know the name *tamtol*. When I asked him, "Did you people make *eḍpa* to catch birds in?" he retorted, "Twelve gauge!"

The importance of Lark Buntings in the Piman dietary should not be underestimated. What they lacked in size, they made up for in numbers, much as did the somewhat smaller White-crowned Sparrows. Wandering flocks of these birds were attracted to Piman fields, particularly after the introduction of wheat by Fr. Kino in the late seventeenth century. They congregated around villages where various seeds were processed and where they could be readily trapped. But, most important, flock after flock of Lark Buntings funneled through Pimería Alta in the spring, leisurely returning to their northern breeding range during the critical months when Pimans were most stressed for food and most animal populations were at low ebb.

In February 1854, Dr. A. L. Heermann (1859), U.S. Army physician and naturalist coming up the Gila with the railroad survey, wrote, "[F]irst observed [the Lark Bunting] on approaching the Pimos villages, associated with large flocks of sparrows, gleaning grain and grass seed from the ground. When started it would fly but a short distance before again resuming its occupation." Piman fields have long been attracting these symbiotic sparrows and buntings, fit game for Piman hunters. Perhaps, given Piman garden hunting practices, the relationship might best be termed *mutualism*.

COMPARATIVE LINGUISTICS. Both Tohono and Akimel O'odham have multiple names for this bird (see discussion of folk synonyms in chapter 4). Of the *chechetopiwua* form, Cassa notes that it "refers to how the whole flock descends—sort of hovering descent, like a dragonfly comes up to and lands on a bulrush." The dragonfly is *vaak chechetopĭ*, the first word meaning "bulrush." One of these two folk taxa, he thinks probably the dragonfly is the original *chechetopĭ*. *Topiwua* confirms the action of the word to which it is attached.

MISIDENTIFICATION. On the basis of study skins, David Brown segregated the adult (black-and-white-headed) White-crown as *tamtol* and the immature (brown-and-buff-headed) White-crown as *kauk oo'okam*. This, I think, was a mistake (and we were not looking at the living birds); the winter Lark Bunting, in particular the female, is similar to the immature White-crown, but is streaked below. I mention this incident as a cautionary note: even with skins, a species may be misidentified. The two species' behavior is quite different, as are their vocalizations.

tosiv, pl. *totosiv*; *tosiw*, pl. *totosiw*

Icterinae, Emberizidae, Passeriformes
Sturnella neglecta

Western Meadowlark; Pradero Occidental, Pradero Gorjeador

Meadowlarks are icterids (blackbirds) that have invaded the quail niche. Like quail, they inhabit low growth: grasslands, weedy fields, and saltbush flats—shrubby country with open spaces for foraging. Also like quail, they are short and compact, with strong feet. Their dorsal coloration of dark brown or even black with buffy tan and grayish brown makes a fine pattern that serves to conceal the bird as it forages on the bare ground or in dry grass and weeds. When it is seen in flight, its wings and tail are short and broad. Even the flight pattern—alternate rapid wing beats and gliding—suggests a quail. The outer tail feathers are white, a signal to help these social birds keep track of one another.

Once the meadowlark mounts a post or hillock to sing, much of this quail-likeness vanishes. The now visible underparts are mostly bright yellow with a bold black V splashed across the chest. The bill is long and rather flattened for working the soil. And the rich notes that bubble out are anything but quail-like. The loud whistles are worthy of the bird's icterid heritage. The common song, a flutelike series ending in a trill, has been rendered many different ways in English—for instance, *Oh, yés I ám a pretty-little-bird* (Pierce 1921).

Western Meadowlarks nest on the ground, often in a depression with dried grass stems arched over the nest. In Pima country, they prefer to nest in fields such as winter wheat or alfalfa—artificial grasslands. At one time on the middle Gila, there were actual grasslands, some several miles long, interspersed with mesquite woodlands and alkali flats. Pima fire drives formerly helped maintain these grasslands by burning out woody shrubs. Eastern Meadowlarks, *Sturnella magna*, winter in Pimería, but the folk biology applies primarily to the Western.

Tosiv was a well-known bird among Pimans. Sylvester Matthias said, "Meadowlarks can follow right behind the plow—get all the worms they could, called *taatam kua'adam*, 'teeth-eaters.'" He described this creature as a white insect, either a grub or a pupa, turned up in the soil, probably a cicada or beetle. "Oh, I could hear them singing in the cottonwood trees around three o'clock in the morning. That's the time the birds start singing." (At that time, kingbirds would most likely be heard.) He associated *tosiv* with *ba'i chukul*, the Horned Lark, a smaller bird of similar open habitats.

Joseph Giff said the meadowlarks nest right in the fields. There was a Pima rendition of what the meadowlark sings: *baat o chipia sikul keli*. The saying is old, and he could not translate it except for a few words. But the cadence clearly mimics the bird's song. George Kyyitan and Culver Cassa worked out this literal meaning:

tosiv, Western Meadowlark (Mike Touch)

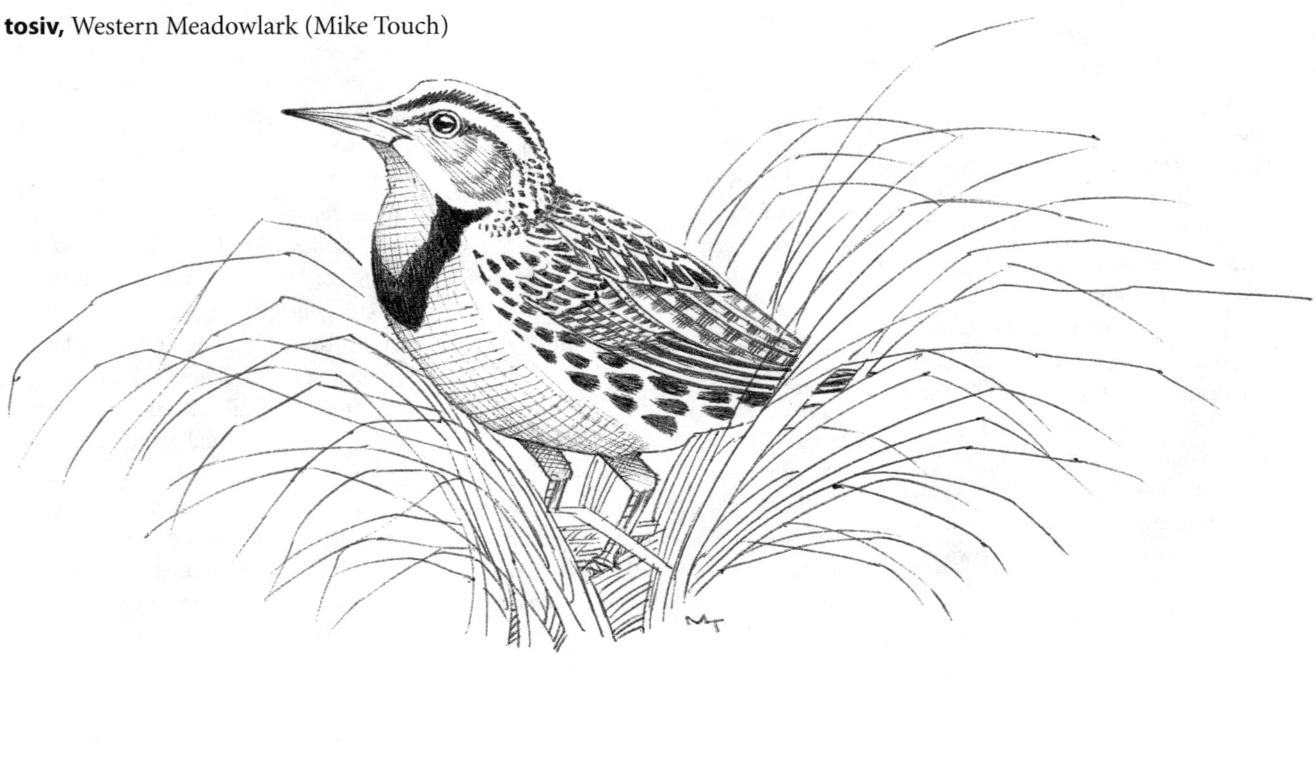

"Where is cousin's uncle moving?" (*Chipia* means to "move, go someplace else"; and *sikul keli* is a relationship term.) The bird is saying that he, *sikul keli*, has closed up his house door, so he'll be away for a while. When someone was to be away for some time, he blocked the door (*kii jug*, "house opening") of his *olas kii* (round house). The closure was a lattice work, *eḏpa*, of Arrow-weeds or Saguaro ribs or a pile of cotton-wood logs stacked against the opening. For the meadowlark, it is a screen of grass. The Pima had watched the nesting behavior of the meadowlark closely!

Whenever we would hear a meadowlark singing in the field, George Kyyitan would say *miligaan ha vijina*, meaning "American's or white man's rope".

At Santa Cruz village, Theodore Peters observed, "*Tosiv* usually comes when everything is green, when the fields get green. I never heard of anybody eating it."

When Francis Vavages noted that nobody eats the Inca Dove, he also said, "Same way with the meadowlark, *tosiv*. They're supposed to be growing the plants. Eats the bugs that spoil the plants. They say that. Don't kill them." Although the meadowlark was not a source of staying sickness, molesting it was prohibited for practical reasons.

Perhaps a teaching that Leonard Pancott mentioned is related. "They say *tosiv* got something on one of his legs. One of his legs is sour, *siv um*. But I never tasted it." Then he added with laughter, "And nobody told me whether it was the right or the left." The play on words between this bird's name and the Pima word for bitter, *siv*, was probably addressed to young hunters, discouraging them from hunting this plump quail-like bird.

Leon Juan of Aangam village said that there were once social dance songs called *tosiv ñeñe'i*. I've not heard any or found anyone who still knows them.

Naming of new items is sometimes whimsical, as shown in this account by Francis. "Bull Durham, [tobacco in] bags, came when the white man start to come in. There's another tobacco they call *tosiv viv*. That's 'meadowlark [tobacco].' That's kind of a brown package, wrap around. *Tosiv viv* they call it, [from] the cover. A little bigger than the Bull Durham [bags]." The label showed a bright yellow meadowlark.

The meadowlark's exquisite vocal abilities did not go unnoticed by the O'odham. Western Tohono O'odham Frank Jim, speaking about the mockingbird being able to cure speechlessness in children, recalled that someone had used a cooked *tosiv* for this same purpose. "They talk a lot. So he got some of those and he let him eat that. And he got alright. He talked. It's good, too, but he said *shuug* [mockingbird] is better."

COMPARATIVE LINGUISTICS. Mountain Pima (Nabogame) *tochi* is cognate. The people here also said the bird would not be eaten.

shashañ, shashañi
***shashan* (Pima Bajo)**

Icterinae, Emberizidae, Passeriformes

Agelaius phoeniceus

Red-winged Blackbird; Magalón, Tordo Sargento

Almost no one can miss the Red-winged Blackbird. Adult males are all glossy black except for a brilliant red shoulder patch bordered behind with yellow buff. Female Red-wings are dark brown and blackish above, with bold black streaks on all the white underparts except the throat. They have a pale eye stripe, a median crown streak, and often, most noticeable in fall, a pinkish or apricot suffusion over all the anterior parts. First-year males are intermediate, somewhat streaked like females, with the red and buff of the epaulets invaded with black. In fall, the feathers of both sexes are variously tipped with tan, brown, or cinnamon, according to subspecies. Males are about nine inches long, females somewhat smaller.

Even if one should overlook the Red-wing's bright plumage, its song and behavior can hardly fail to attract attention. In winter, same-sex and even same-age flocks, numbering in the hundreds or thousands, wander about feeding in plowed fields, charcos, corrals, and feedlots, all singing simultaneously. At night, they retire to cattail or bulrush marshes or even to thickets of Salt-cedars. At this season, many of these birds are from more northern areas.

As spring approaches, local males set up territories in marshes or even in very wet fields that are irrigated by flooding. Here they sing, on the wing or perched, display-

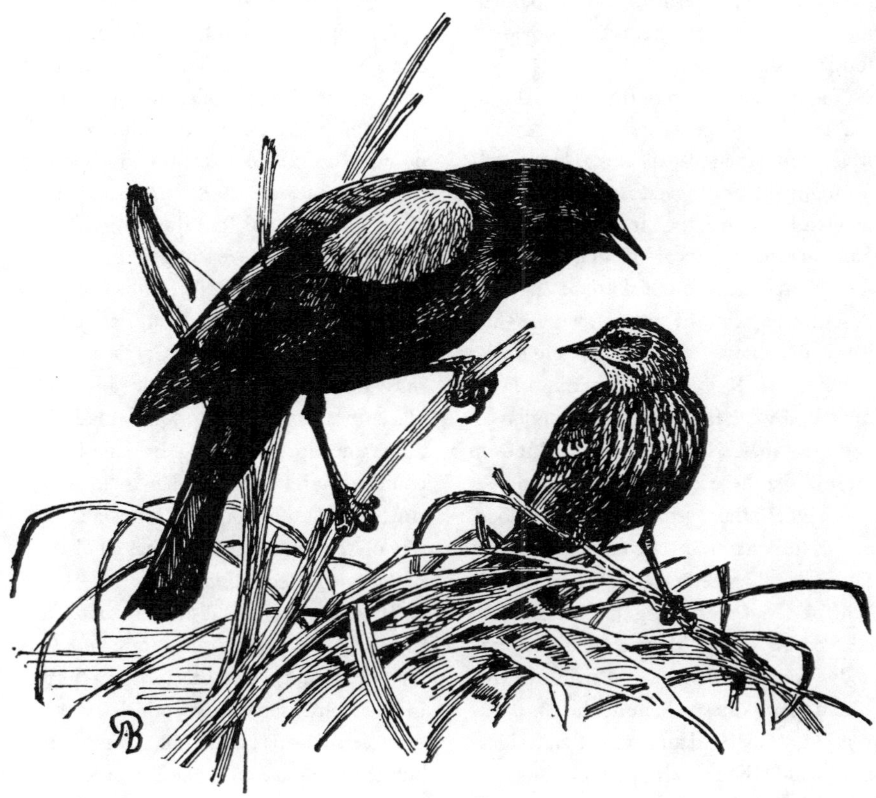

shashañ, Red-winged Blackbird (Allan Brooks)

ing their scarlet and yellow epaulets. The song is a loud *kong-kla-riii*, the first note low, the last a high trill. When a number of males hold adjacent territories, as is often the case, these squeaky gurglings virtually define a marsh. Meanwhile, the females go about their task of building nests of wet strips of vegetation sometimes patched with mud, incubating the eggs, then catching insects for the young, now assisted by the male.

Particularly in winter, Red-wings may be accompanied by various other icterids such as the purple-and-green-glossed Brewer's Blackbird *(Euphagus cyanocephalus)*, the Yellow-headed Blackbird *(Xanthocephalus xanthocephalus)* with a white wing patch, and the smaller finchlike Brown-headed Cowbird *(Molothrus ater)*. The Bronzed or Red-eyed Cowbird *(Molothrus aeneus)*, with its curious puffed-neck display, visits Piman country only in summer. During the past half century, Pimería Alta has been invaded by the very large Great-tailed Grackle *(Quiscalus mexicanus)*, with glossy purple and blue-black males and much smaller brown females. Its folded tail is unmistakable, as is its loud and varied vocalizations. The grackle has long been among the Pima Bajo. These slender but almost crow-length blackbirds have now colonized many O'odham settlements that provide food and nesting trees.

There is a question of the proper domain of the folk generic *shashañi*. Among River Pima, the focal species at least, and for many the only species, is the Red-winged Blackbird, the bird they are most familiar with as a local breeder. If asked for a characterization of *shashañ*, they describe a black bird with red wings. This is the bird that is pointed out in the field. For instance, Ruth Giff picked out the male Red-wing from various specimens and the female's skin as his mate. But even on the Gila there is some equivocation about what other birds might be included in this folk taxon. In a group discussion at Komaḍk, all agreed on the Red-wing, but Ruth added, "Everything that is black is all *shashañ*," including both Brewer's and Yellow-headed Blackbirds. David Brown added that it doesn't include the *bapsho s-vepegĭ*, robins. And all agreed that it does not include the grackle, a puzzling new bird on the Gila.

Leslie Pasqual of the Sacaton area, who has spent a lifetime around corrals, said *shashañ* was the Red-wing. I asked if he had seen any other kinds. "There's another one that's got yellow; that's the only two that I know." He did not know a specific name for the Yellow-head, but said, "If I'm talking to somebody, I'd just call it *s-oam mo'okam shashañ* or just *s-oam shashañ*. It's my own, how I'd call it." He had not heard anyone else use this name, however. Leslie's name means "yellow-headed blackbird." Farther down the Gila, another cowboy, Theodore Peters, said "Once in a while they come around," but he didn't know a special name for them either. Western Papago Frank Jim called them *ge shashañ*, "large blackbird."

However, Tohono O'odham Simon Lopez considered all the smaller blackbirds to be *shashañ*, but excluded the grackle, which he said was new about the villages. He was not sure whether grackles were related to ravens or blackbirds. Another time he said he had been told that real *shashañ* were the smaller ones (cowbirds) that followed the livestock and often rode on the backs of his horses. I asked him what their nests were like. "You know, that's strange. I've wondered about that. I've never seen their nests—I don't know whether they all nest together somewhere or separately." Brood parasitism of cowbirds seems not to have been part of folk knowledge.

Farther south Eleanor Lewis-Valenzuela of S-Gogosik, Delores Lewis of Ge Oidag, and Luciano Noriega of Quitovac included all the blackbirds, even cowbirds, in the domain of *shashañ*. Luciano's concept included grackles as well, but he added, "Pero no hay antes" (but there weren't any before). When we stopped to watch a puffed-up male Bronzed Cowbird courting a female, Delores called them *shashañĭ*. At Santa Cruz village, Jose Garcia said of *shashañ* that there are gray ones and very black ones. Some are big, some small. He didn't know the Red-wing. In Pima Bajo country, María Córdova likewise included cowbirds. At the Mountain Pima community of Yepáchic, I was told that all blackbirds were called *shashani*.

Then there is the question of *koomagi shashañĭ*. In his 1935 Pima vocabulary, Fr. Antonine (1935) listed this bird as a "grayish blackbird" separately from *shashañ*. Sylvester Matthias said of it, "A *shashañ*, but it's gray; *shashañ* is black, that Red-wing. But *koomagi shashañ* is gray. Same size. But it disappeared long time ago. I don't know whether they are natives or what. You could find them a lot around Sacate, but mostly we would find the black ones. When someone was irrigating, they'd be a whole swarm of them. They bred in the cottonwoods. There used to be a lot of cottonwoods there along Maricopa Highway and on down this way [toward Santa Cruz]. They used to nest there." But Ruth Giff, Leslie Pasqual, and George Kyyitan thought *koomagi shashañ* was just the Red-wing female. If so, this case is one of the few (perhaps the only one) where Pimans have overclassified the avufauna (here giving a special name to one of the sexes).

Thus, among some Pima language groups, this taxon might include, aside from the Red-winged Blackbird: Brewer's Blackbird (Tordo Ojo Amarillo); Yellow-headed Blackbird (Tordo Cabeza Amarilla); Brown-headed Cowbird (Tordo Cabesa Café); Bronzed Cowbird (Tordo Ojo Rojo); and even Great-tailed Grackle (Shanate Mexicano) for some southern Pima speakers.

The biology of the Red-wing was well known to River Pima. Joseph Giff said, "*Shashañ* make a nest and line it with mud." Francis Vavages observed, "*Shashañ* follow the plow and eat the worms." Sylvester noted of the peaks to the north of Pima country, "The White Tank Mountains [have] springs all over for drinking water. Blackbirds, *shashañ*, always singing in that area."

Blackbirds apparently were never eaten, perhaps for the same reason that meadowlarks were protected: they were associated with agriculture and were considered coworkers with O'odham, not to be molested. In the great story of Corn Man and Tobacco Woman, Corn Man and his sidekick Pumpkin desert the People during a drought (which has been brought on by the abandonment of the flighty Tobacco Woman after she argues with Corn Man). Later they return and take up residence again among the People. When Corn leaves the People once more, he takes along this time his pets, the blackbirds, sprinkling corn seeds along the way for them to follow (Lloyd 1911:218–229; Rea 1998b).

A similar story of birds associated with Corn Man was told by William Blackwater in 1927 (Benedict 2001:72). "In the beginning Earth Doctor called the living creatures. He called wind and clouds, the blackbirds, the woodpeckers, and all the birds that like corn, and he said to them, 'Whenever you find people, use your power.' They started on their journey [east]." Corn Man and his associates eventually come to the People. "That is why we have rain and corn and all these vegetables. The birds who had brought Corn Man (all the birds who like to eat corn) lived there with them for four years. That is why we have all these birds in the cornfields." What a radical departure from Western thought. For the O'odham, the blackbirds and woodpeckers found in every milpa are the ones, along with wind and rain, who *bring* Corn Man; rather than being varmints and pests, they are his friends.

In one version of the story of the two Coyote brothers, the youths meet a woman in the north who declines marriage, her sole companion being her pet blackbird (Benedict 2001:79–80). The elder brother shoots the blackbird with his arrow. "Why did you kill my pet? It was always with me, and every morning it woke me with its pretty song." The woman, Corn Girl, becomes the elder youth's wife.

Next the brothers go south and meet a young woman who also remains unwed, her sole companion being a woodpecker. The younger brother kills it with an arrow. "Why did you kill my pet? Every morning it woke me with its tapping." This girl becomes Iitañĭ, Pickleweed Woman, who marries the younger youth.

This deceptively simple subplot is a study in contrasts. Both birds are male, and they make some sort of noise in the morning, the blackbird with its exquisite territorial vocalizations, the woodpecker (we don't know which species) with its unmelodious drumming also identifying its territory and attracting mates. In each case, the girl is bonded to the bird, and this bond must be broken before she can accept the boy as a spouse. The boy kills his competitor. The boy's erotic arrow penetrates the pet; the girl pulls out the arrow, runs after the boy, and offers it back to him. He ignores her, but she runs ahead, presenting the arrow again when the boy arrives along the path. This time he accepts the arrow and the girl as well for his spouse.

The contrasts continue. The first girl is Huuñĭ, Corn or Maize, the staple cultivated grain crop of these agricultural and wild-gathering people. It grows in their fields of fertile alluvial soils. The second girl is Iitañĭ, Pickleweed, a famine food for the Pima. It grows in the most heavily saline flats along the Gila (Rea 1997:122–123).

Among the Tohono O'odham songs Underhill (1946:257) obtained for the *wuagida*, girl's puberty ceremony, one seems to refer specifically to the male Red-wing's lifting its red epaulets in courtship display:

> The black crow [raven]
> Knows nothing.
> The little blackbirds
> Made everything.
> There they erected a mountain;
> Yonder, a red mountain.

Several people I talked to mentioned a Gileño man who was named Red-wing. Francis Vavages said, "One man at Gila Crossing [was] called Koomagĭ Shashañ, the mother bird." Sylvester recalled, "The singer, Shashañ Ñe'edam, from Santa Cruz [village] predicted already when the river was running full and all the creeks—that the river would dry up and burn us. And they called him a liar. Shashañ was not a very old man when he died—maybe sixties. Maybe in 1923, if I'm not mistaken, *before* Santa Cruz had its church [St. Catherine's, built in 1924]. He had cattle and would bring us homemade cheese, just like you buy in Mexico." His songs are apparently now gone, but one was about the blue/green hummingbird.

Pima author George Webb (1959:124–125), also of Santa Cruz, left us this poignant account of the demise of the Gila wetlands of his homeland:

> In the old days, on hot summer nights, a low mist would spread over the river and the sloughs. Then the sun would come up and the mist would disappear. The red-winged blackbirds would sing in the trees and fly down to look for bugs along the ditches. Their song always means that there is water close by as they will not sing if there is not water sloshing somewhere.
>
> The green of those Pima fields spread along the river for many miles in the old days when there was plenty of water.
>
> Now the river is an empty bed full of sand.
>
> Now you can stand in that same place and see the wind tearing pieces of bark off the cottonwood trees along the dry ditches.
>
> The dead trees stand there like white bones. The red-wing blackbirds have gone somewhere else.

An edible plant, referred to only as "blackbird's eyes," occurs in several versions of a story about a grandmother raising her orphan grandson. The plant presumably would be *shashañ wuupui* in O'odham, but no one I worked with recognized that name (Rea 1997:388).

A Rhynchosia vine (Fabaceae) in lowland Pima Bajo country is called *shashan m-upui*, "blackbird's eye," or Ojo de Chanate on the Río Yaqui. Its striking seeds, resembling Japanese lacquer ware, are half black and half red.

COMPARATIVE LINGUISTICS. George Herzog told J. Alden Mason (1950) that in Gila Pima the plural of *shashañĭ* is *shash'añĭ*. Though this plural form appears in no lexicon, Culver Cassa considers it legitimate. Mathiot (n.d.) regards *shashañĭ* as an aggregate noun.

The colonial Attí vocabulario (ca. 1774) lists *san* as tordo, "blackbird," with its plural as *sansan*. This is interesting because the present form of the word *shashañ* seems to be a fixed plural (aggregate noun); there is no single syllable form in modern Pima. The document adds, "chanate lo mismo." This phrase refers specifically to the grackle, from Nahuatl *tzanatl* (Karttunen 1983:311), and is particularly significant because it fixes a northern point in the range of a species that several centuries later was to expand dramatically northward. The Névome vocabulario, compiled almost a century earlier, listed tordos, in the plural, as *sasani*. No singular is given. We see similar patterns in things typically encountered in aggregate: teeth, leaves, ants. Rinaldini's 1743 vocabulario, compiled among the Northern Tepehuán, gives tordo (in the singular) as *sasane* (aggregate form).

Shashañ appears to be a pan-Piman word. In addition to this form in Akimel and Tohono O'odham, *shashan* (Pima Bajo) and *shashani* (Mountain Pima) have been recorded. The Red-wing is found throughout Tepiman country, as are grackles, both species of cowbirds, and (at least in winter) the Yellow-headed Brewer's Blackbird.

Yaqui Moisés Valenzuela said the Red-wing is *chana* and the Yellow-head is *vai chana*, adding that they don't sing very well!

MISIDENTIFICATION. Saxton, Saxton, and Enos (1983:52) list *shashañĭ* as "blackbird"; *s-chuk shashañĭ* as "any black bird"; *s-oam shashañĭ* as an "oriole"; and *s-wegi shashañĭ* as Red-winged Blackbird. I have never heard any of the last three names (folk specifics) being used. The oriole already has a Piman name, *vajukukĭ*, and its behavior

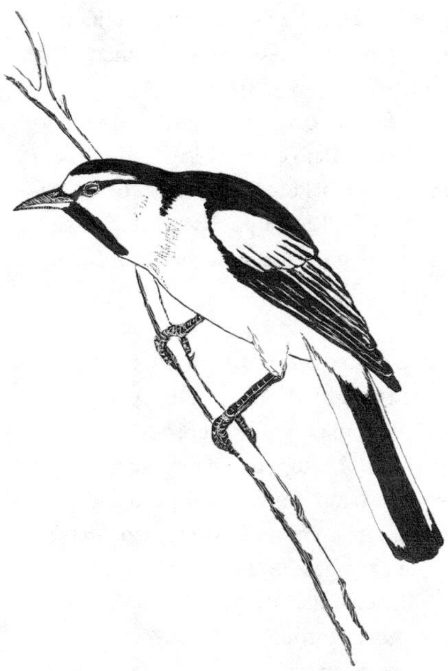

vajukukĭ, Bullock's Oriole, male (Amadeo M. Rea)

is quite different from that of blackbirds. Almost certainly the authors' reference is a misidentification of the Yellow-headed Blackbird, which sounds, acts, and looks much like other blackbirds, its closest relatives. As for *s-chuk shashañĭ* being "any black bird," we must remember that both *kuigam*, the Phainopepla, and *havañĭ*, the raven, for instance, are black birds, but they are not *shashañĭ*.

vajukukĭ, wajukukĭ

Icterinae, Emberizidae, Passeriformes
Icterus galbula bullockii
Bullock's Oriole, Northern Oriole; Bolsero Calandria, Bolsero Norteño, Calandria Norteño

A flash of orange, black, and white along with a loud chatter in deciduous broad-leafed trees or in mesquite bosques or even out in paloverdes should identify the Bullock's Oriole. This blackbird-size bird is here in Pimería only in the warm season, sojourning with its tropical relatives in winter. Males are orange bodied with black on the chin, crown, back, and part of the tail. There is also a black eye stripe and a broad white wing patch. Females are similar to other orioles in being less brightly marked. They are washed with yellow on the head and throat, and lack the black entirely, having instead an olive gray wash on the back, with whitish underparts.

With the decline in riparian habitats in Pima country, Bullock's Orioles have shifted to nesting in Athel Tamarisks, long the favored shade tree across the reservation once cottonwoods could no longer reach water. Here they build their pendant nests, most often woven of black horsehair with additions of string and other fibers. The pair begins nesting early in spring, working together on a nest, passing strands to each other through the walls of the nest.

Except for some melodious notes when these birds seem to be talking to themselves, the Bullock's Oriole's calls seem rather harsh. Dawson (1923:104) described one call as *cut cut cudut whee up chooup*. Hoffmann (1927:293) rendered it *kip, kit-tíck, kit-tíck, whew, wheet*. Both sexes give a rattling call as they search through the foliage for insects and other arthropods. Orioles also have a sweet tooth, harvesting nectar from flowers or taking soft fruit and berries, much like their tropical relatives.

Its brilliant colors, the loud carrying calls, and familiarity about gardens and villages made the Bullock's Oriole a prominent bird in Piman consciousness. Sylvester frequently remarked on their black horse-hair nests. Once, while trimming palm fronds about the school where I was teaching in the 1960s, he brought me one with a pale fiber nest of a Hooded Oriole suspended from it. He was unfamiliar with this species, but thought the bird would be called *vajukukĭ*, just like the Bullock's. Joe Giff said, "I've heard them. They make their music early in the morning. I used to hear them, because my wife never gets up early enough." He considered the oriole one of the four birds saved in the Flood.

Ruth Giff, herself an accomplished basket weaver, told me the story of how one day Se'ehe, Elder Brother, says he is going to teach the birds how to build their nests. All the different kinds of birds come together. The *ñepoḍ* (nighthawk) and the *chivi-chúuch* (Killdeer) say, "Oh, there's nothing to it," and they get up and leave. They build no nests at all. After listening for a while, the doves say, "Oh, there's nothing to it." They too get up and leave, building very flimsy nests. Little by little various other birds leave. The *giisobĭ* (Verdin) and the *hokkaḍ* (Cactus Wren) stay for a very long time. They learn how to make enclosed nests with a little opening at the side. But the nests look rather rough. However, *vajukukĭ* stays to the very end and learns to weave a nice tight nest, suspended from some branches, like a finely woven Pima basket.

Leslie Pasqual asked if I had seen the socklike horse-hair nest of the *vajukukĭ*. "One time that bird got caught in there and was hanging down like that. Died, yeah. Now you don't see them anymore. Used to be a big cottonwood tree, right there. Used to be *a'uppa* all along the canal. They took them out because they say they get fifty gallons of water every day. That's what they say, but you know how the white man is, they can lie, too." More than just vegetation is lost with environmental degradation. Important cultural metaphors, adding to the quality of life, vanish as well.

As with many other tribes, the O'odham had a mythical fire quest in order that they might preserve some of the sun's heat for the night. At first an old man tries to get fire from the sun. He fails but reports, "Sun, at the end of his journey, jumped into a big hole and the Indians would have to send some of the flying people to bring the fire" (Wright 1929:39).

Oriole is the first animal to make a stab at getting fire, starting off early in the morning to the east. "Oriole did not return home until it was very late. And he was all changed. Some of his feathers had turned to the color of the sun, and some were black. Oriole told the people that he had gone so close to the Sun that his feathers began to burn and so he turned back and flew until he saw some water and dropped in. And now you know why, even in these days, some of Oriole's feathers are yellow and some are black" (Wright 1929:39).

In oratory, this bird is often chosen to represent the color yellow. For instance, in a war-expedition speech, after the enemy is overcome, the bird *mamakai*, or shamans, sing, filled with gladness, and rebeautify the earth. *Hevachuḍ*, Blue Grosbeak, sprinkles its blue feathers over the land, making it smooth as algae. *Vajukukĭ*, oriole, sprinkles its yellow feathers over the mountains, seen as the yellow flowers that color them in spring. *Shuug*, mockingbird, sprinkles its feathers over the land, bathing it white with sunshine (Russell 1908:367–368). (The fourth bird was omitted in this telling; presumably it was the cardinal.)

In one version of Elder Brother's resurrection speech, there is reference to "everywhere oriole skins," *wesko s-waakchu hogidag*, which is paired in the imagery with swallow skins (Bahr 1975:62–63). Here whole skins seem to be used. In other versions, the feathers are attached to a string (or to several strings), a common Puebloan form of votive offering.

In another oration, Navichu, after he is created, is being induced to stay with the People and to settle down to the farming life. The People tell him of two *vapaki*, "mythical structures, rain houses," where he will find something that will please him. One is the cardinal house, *sipok va'aki*, the other the oriole house, *wajukukĭ va'aki*. One is covered with red flowers, the other with yellow, and inside clouds are forming, and it is raining (Lloyd 1911:210, 213). As Culver Cassa interprets this imagery, Navichu is taking the rain from its wild setting and bringing it back to farming.

Oriole, Blue Grosbeak, and Northern Cardinal are rain birds, as are eagle and turkey. The first two are summer residents only, present during the rainy season, whereas the others are resident.

Because the word *vajukukĭ* begins with *va-*, usually the water signature, I wondered if it might somehow be associated with water. "The name has something to do with a color, like rock, mineral, ochre powder, something like *vajuk*. It's an old word," notes Culver Cassa. "The name might be short for *vajuk u'uhig*." He adds that the dreamer of the oriole songs may have connected oriole with the sun because it is yellow or, rather, orange. At the beginning of the world, Sun surveys everything from above. Events unfold under him. The yellow oriole-spirit may similarly be reporting events from a high vantage point, giving them to a song dreamer.

The *vajukukĭ ñeñe'i*, oriole songs, form a prominent group of River Pima sung poetry (see Bahr and associates 1997:111, 129, 157). These songs originated from Roy B. Hill of Haashañ Keek (Sacaton Flats) on the Gila, who taught them to his stepsons, the brothers Zachary and Blaine Pablo, Clemencia Leos's father and uncle. Nearly seventy oriole songs have been recorded from Blaine. Only one of these supposedly has the bird oriole as its subject:

Wahame vañupiñi
damai kachime wui koñe beichukim
gumu wiwini kosekame maakai
kumeseke wuwi wo ñuaahapa
ge weesa momoika wiwini ginito
heko ñeyona gewechuke
yupame ñuñifta.

Yellow Oriole [?]
take me to the sky
soft feather shaman
bring me to him
very soft feather will clasp
will cover me and
lower me home.

At Quitovac oasis, Luciano Noriega called the Hooded Oriole's nest *wakokam kosh*, "bottle gourd–it-has nest." He said that all the different kinds of orioles pictured (Hooded, Bullock's, and Scott's [*Icterus parisorum*]) would be called *wakokam*, the one who possesses a bottle gourd, *wakoa*. Surely Luciano had ample opportunity to see two oriole species building gourd-shaped nests in trees about the oasis. Luciano's name is both poetic and appropriate, but I never learned if it was a replacement name for *vajukukĭ* or a local (allopatric) synonym. This name needs to be checked among other Tohono and Hiá ch-eḍ O'odham. They seemed not to recognize the name *vajukukĭ* and seemed generally unfamiliar with orioles, even though Bullock's Orioles must nest in mesquite at larger charcos of Papaguería.

In the tropical deciduous forest of Pima Bajo country, Streak-backed Orioles, Calandria de Fuego or Bolsero Pustulato, *Icterus pustulatus*, hang their enormous slender nests in conspicuous places. So far as I could tell, Pedro Estrella called this common bird just *s-oam u'uvig*, "yellow bird," or *chichar*, the latter name apparently local Spanish. (In his *Vocabulario sonorense*, Sobarzo [1991:80] gives this oriole's name as *chilic*, derived from Cáhita.)

MISIDENTIFICATIONS. Russell (1908:345), normally careful with biological identifications, called *vajukukĭ* a "hawk" in Elder Brother's speech, an error repeated by Bahr (1975:49). The imagery in this oration matches oriole's yellow feathers with *hevachuḍ*'s blue feathers (see Blue Grosbeak account).

Saxton, Saxton, and Enos (1983:61) give *wajuk* as goldfinch. Surely this is a misidentification of an oriole. They don't give a *wajukukĭ* form here, but do list *o'am shashañĭ* as an oriole. The words mean "yellow blackbirds," almost certainly in reference to the Yellow-headed Blackbird. The folk generic *shashañĭ* refers specifically to Red-winged Blackbird, at least in River Pima country. In winter, flocks of blackbirds in Pimería may contain numbers of brilliant, noisy Yellow-heads. Although both blackbirds and orioles are in the family Icteridae, behaviorally they are not particularly close.

heet kawuḍk

Thraupinae, Emberizidae, Passeriformes
Piranga rubra
Summer Tanager; Tángara Roja

After six months in the tropics, the exotic Summer Tanager arrives in the Southwest to breed in mature stands of cottonwoods

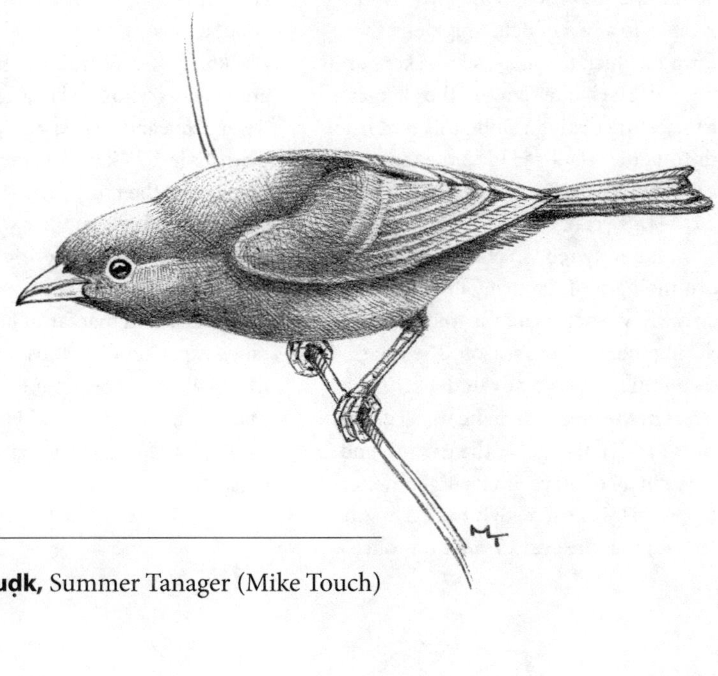

heet kawuḍk, Summer Tanager (Mike Touch)

and willows or even old mesquites along the desert rivers. The brilliant, all-red male is secretive, heard more often than seen in the tree canopy. The female is olive-gray above, dull yellow below, lacking contrasting marks. The bill is rather heavy, with a toothed maxilla (upper part of bill), pale in summer, dark horn in winter. The head is rounded, in contrast to the crested cardinal, the only desert species likely to be confused with the tanager.

Young males in their first-year plumage may have a motley mixture of colors—yellow and buff with admixtures of green, pink, and even bright red—producing odd patterns that suggest some extraneous species. They may breed in this intermediate plumage.

Summer Tanagers are rather sedate and deliberate birds. Their loudly whistled song resembles the Black-headed Grosbeak's, *Pheucticus melanocephalus*, or even the Northern Robin's. The call note is like the *petrrít* of the Western Tanager *(P. ludoviciana)*, but descending. They are adept flycatchers, taking bees, wasps, beetles, and other insects on the wing. Tanagers glean some small fruits as well. Once found throughout River Pima country, they now breed only in riparian remnants, as along the lower Salt River.

Sylvester Matthias was the only one to remember the Piman name for Summer Tanager still, *heet kawuḍk*. *Heet* is the red ochre once prized for body decorations and for painting items such as arrows. *Kawuḍk* usually indicates a rounded knoll or hill, separated from a main range. (This is the source of the Spanish name "Caborca," the former Piman settlement in northern Sonora.) But as used in the bird's name, it refers to a ball of ochre pigment, the form in which the material was kept until needed. The name signifies the characteristic feature distinguishing this bird from the other local all-red bird: the tanager has a smooth round head, whereas the cardinal has a large crest.

In the early 1900s, schoolteacher and ornithologist M. French Gilman found only a few tanagers on the middle Gila during the breeding season. The riparian community was already under serious stress in the areas where he worked. The trees survived longer at the western end of Gila Pima country with its higher water table. This was where Sylvester grew up. He said that the *heet kawuḍk* was once a common bird about the villages "as long as the cottonwoods lasted." One doesn't need a degree in ecology to appreciate the relationships between habitat and biodiversity.

It is not certain whether Anna Shaw's (1968:82–83) story about the "Redbird" applies to the resident and better-known Northern Cardinal or to the tanager: the males of both species have all-red plumages. A better storyteller than naturalist, Shaw used the pronoun *she* in her tale. (Piman pronouns are nongendered.) Spier (1933:405–409) recorded a Maricopa story about a tanager, his friend the Red-winged Blackbird, and a cardinal. The Maricopa are Yumans who lived adjacent to the River Pima and shared many legends with them. Spier's folk taxa have never been verified for the most part. The Maricopa should have been familiar with the Summer Tanager from their original home on the lower Colorado, but the cardinal had not yet reached that far in its northwestern range expansion (Phillips 1968:151; Phillips, Marshall, and Monson 1964:176–177). The Maricopa, however, would have known the cardinal in their new home on the middle Gila.

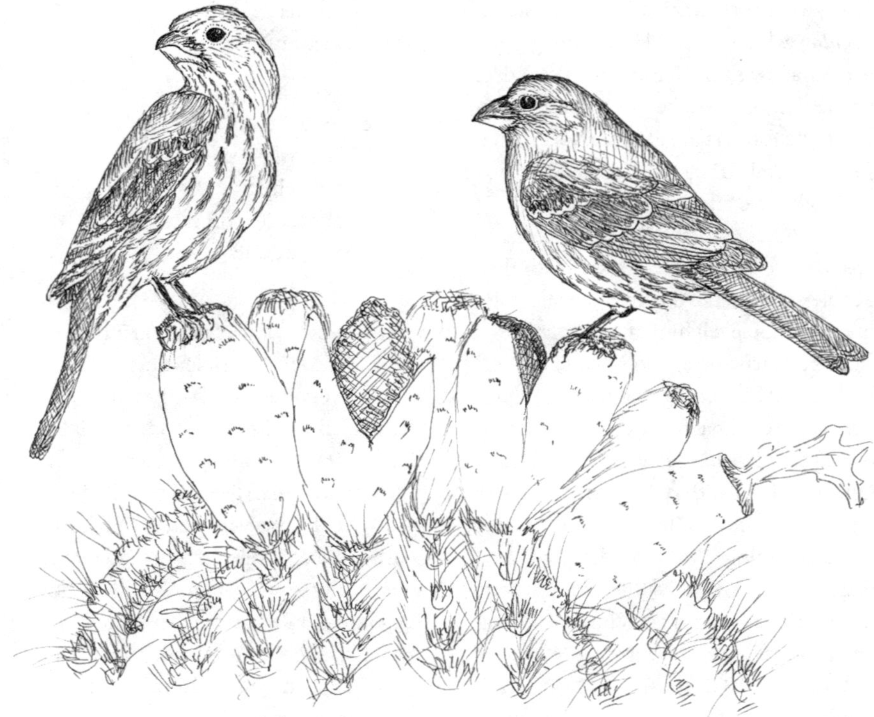

bahidaj u'uhig, House Finch (Amadeo M. Rea)

bahidaj u'uhig

Carduelidae, Passeriformes

Carpodacus mexicanus

House Finch; Pinzón Mexicano, Carpodaco Doméstico (?), Gorrión Común

A loud, cheery warble may draw our attention to House Finches coming to drink at a tinaja, desert spring, cattle trough, or ranch yard. As they take wing, they give a *cheep*. The birds have no objection to taking advantage of the water and nesting niches that humans may provide. And House Finches are inordinately fond of windfalls from the fruit trees people raise. Urban dwellers familiar with these birds in more domestic settings may be surprised to see them among paloverdes, acacias, and Saguaros.

These small stocky birds have small eyes and stubby finch bills. They are gray-brown above with indistinct tan streaks over their whitish underparts. In addition, the males are splashed with bright red beginning at their foreheads and washing down over the face and chest. Even the rump sports a splash of red.

Delores Lewis of Ge O'idag called these birds *bahidaj u'uhig*, "Saguaro fruit bird," certainly a good name. The Saguaro cactus, *haashañ*, along with the mesquite, *kui*, provided the Piman Desert People with their two major wild crops. So important were these two plants that there are separate names for their fruits. Mesquite pods are *vihog*. Saguaro fruit get two names. When the fruit, *haashañ iibdag*, ripens and splits open on top of the cactus, it is called *bahidaj*. When it falls to the ground in a dark dried mass, it is then *juñ* or *gakidaj*. (The fruit of plants in general is called *iibdag*, their "heart," or *bahidag*.)

In midsummer, when the Saguaro fruit splits open to reveal a juicy core of brilliant red pulp and rich black seeds, the People went out into the cactus stands, made camp, and knocked down the delicious fruit with long poles, *ku'ipaḍ* (Crosswhite 1980; Densmore 1929:148–162; Rea 1997:253–260; Underhill 1938; Underhill and associates 1979). Vying with the O'odham for the fruit would be a number of birds, including the *okokoi* (White-winged Dove), *hikwig* (Gila Woodpecker), *kuḍvich* (Curve-billed Thrasher), and the most voracious frugivore of all, the House Finch.

Most people I asked about this bird didn't know it. But at Aangam Leon Juan said he knew the name *bahidaj u'uhig*. We asked what color it was. "Red," he replied. I was really expecting him to associate this name with *o'okokoi*, the White-winged Doves, which are so important ceremonially in his community (see species account). Jose Enriques, who had heard the name *bahidaj u'uhig*, said it was not an *okokoi*.

No River Pima I asked recognized the name *bahidaj u'uhig*. House Finches have been eating their apricots, peaches, and in particular their pomegranates for several centuries, but Akimel O'odham abandoned Saguaro camps many generations ago.

Appendices

Appendix A

Alphabetical Listing of O'odham Bird Names with English Equivalents

aaḍo (macaw, Scarlet, Military) N

ba'ag (Golden Eagle) N

baasho s-vegĭ (Northern Robin) D

bahidaj u'uhig (House Finch) D

ba'i chukul (Horned Lark) D

ba'ivchul (Belted Kingfisher) N

bichpoḍ (Abert's Towhee; Canyon Towhee) D

biit keishnam (Bendire's Thrasher) D

chechlogam (mythical?) N?

cheedagĭ u'uhig, tavir (parrot, Amazona?) (also *s-ñeokdam u'uhig*, syn) N

cheedagĭ vishag (Peregrine) Falcon (F Sp) D

chehegam (Ladder-backed Woodpecker) N

chehopĭ (Common Ground-Dove) (also *vuhigam*, syn) O

chev chiñkam (Crissal Thrasher) (also *keli huch*, syn) D

chevel u'uhig (Black-necked Stilt) D

chivi-chúuch, sivi-chúuch (Killdeer) N

chuággiakam vakoañ (Brown Pelican) (F Sp) D

chukuḍ (Great Horned Owl) N

chukuḍ ka'amaḍ (Long-eared Owl) (F Sp) D

chukukmal (Western [+ Cassin's?] Kingbird) N

duuk vaidam (Pima Bajo); *kurta*? (Mt. Pima) (Coppery-tailed [Elegant] Trogon) D?

ee'eḍ vahuḍam (Barn Owl) D

gihodag (Band-tailed Pigeon] N (or O?)

giidval, pl. *gigidval* (swallows, generic) N

giisobĭ (Verdin) N

go'ogov (Logger-head Shrike) N

gugú (Inca Dove) O

haiñ uushab, haiñ julshab (*unid. waterbird [Wood Stork or avocet?]) D

haupal (Red-tailed Hawk) N or ±

havañ (raven), *kokoñ* (myth) N

heet kawuḍk (Summer Tanager) D

hevachuḍ, var. *hevaichuḍ* (Blue Grosbeak) N

hevel moos (Say's Phoebe; Ash-throated Flycatcher) D

hialak (Snow Goose, all geese?) O

hikvig (Gila Woodpecker) N

hokkaḍ, var. *hogkaḍ* (Cactus Wren) N

hoohi (Mourning Dove) O

jujupul [*unid. waterbird] N

kaachkbak u'uhig (gull) (also *kaachk baasho u'uhig*, syn) D

kaachkĭ baasho u'uhig (gull) (also *kaachkbak u'uhig*, syn) D *kadgam* (Yellow-billed Cuckoo) N

kakaichu, kakachu (Gambel's Quail) O

keli huch (Crissal Thrasher) (also *chev chiñkam*, syn) D

kokoho (Burrowing Owl) O

kokon, koko'on (raven, Pima Bajo) (except Akimel and Tohono O'odham) N

kookoḍ (Sandhill Crane) N

koologam, koolo'ogam (Common [Nuttall's] Poorwill) N (?)

koomagĭ u'uhig (Le Conte's Thrasher) D

koomagĭ vakoañ (Great Blue Heron) (F Sp) D

kuḍat (Northern Flicker) N

kuḍvik, kuḍvich (Curve-billed Thrasher) N

kuigam (Phainopepla) D

kulisgam, kukulisgam [*unid. waterbird, Avocet?] N

kusijim (Crested Caracara) N

kuukvul (Western Screech-Owl) O (or ±)

ñepuḍ, ñepoḍ (Lesser Nighthawk) D

ñui (Turkey Vulture) N

ñuipa (Black Vulture) (F Sp) ±?

oám u'uhig (*unid. "yellow bird"; Yellow-breasted Chat?] D

okokoi (White-winged Dove) O

olas kakaichu (Montezuma, [Mearns'] Quail) syn. D+O

o'oḍopiwua (Lark Bunting) (also *s-baabanmakam*, syn) D

o"ok kakaichu (Montezuma [Mearns'] Quail) syn. D+O

s-chuk mo'okam giisobĭ (Black-tailed Gnatcatcher) (F Sp) D

s-hadam kakaichu (Scaled Quail) D

shalav, shadav (Black-throated Magpie-Jay [Pima Bajo]) N?

shashañ (Red-winged Blackbird; all blackbirds?) N

shuudagĭ maḍ (sandpiper, Least and Western mostly) D

shuug (Northern Mockingbird) N

sipuk, sipok (Northern Cardinal) N

sisikĭ (American Kestrel) N

taḍai (Greater Roadrunner) ±

tamtol (White-crowned Sparrow) N

tash cheenam vakaoñ (Green Heron) (F Sp) D

toha vakoañ (Snowy and Common Egret) (F Sp) D

tobavĭ (Harris' Hawk) N (?)

tosiv (Western Meadowlark) N

tova (Common Turkey) N

unmu'uhig, u'unam u'uhig ("enclosure bird" [jay or dream bird?]) D

vachpik (American Coot) D

vajukukĭ, vajakuk (Bullock's Oriole) N

vakav (Northern Harrier, Marsh Hawk) (also *s-heepĭ wuadam*) N

vakoañ (heron/egret) ±

vakoañ ba'ag (Osprey) (F Sp) D

vapkag, vapka'ig (ducks [and geese?]) ±

vavas (Rock [and Canyon?] Wren) ±

vipismal (hummingbird, mostly Costa's and Black-chinned) ±

vishag (Prairie Falcon) N

vuhigam, wuhigam (Common Ground-Dove) (also *chehopĭ*, syn) O

N = nonanalyzable lexeme
± = partially analyzable lexeme
O = onomatopoeic name
D = descriptive name
# = folk specific
syn = folk synonym or alternate name
*unid. = unidentified (see appendix F)

Appendix B

Scientific Names of Plants Mentioned in Text

Alkali Heliotrope, *Heliotropium curassavicum*

Amapa Colorada, *Tabebuia palmeri*

Amaranth, *Amaranthus palmeri*

Annual Sowthistle, *Sonchus oleraceus*

Arizona Cypress, *Cupressus arizonica*

Arizona Sycamore, *Plantanus wrightii*

Arizona Madrone, *Arbutus arizonica*

Arizona Rosewood, *Vauquelinia californica*

Arizona Walnut, *Juglans major*

Arrow-weed, *Pluchia sericea*

ash, *Fraximus* spp.

Athel Tamarisk, *Tamarix aphylla*

Banana Yucca, *Yucca baccata*

Batamote, *Baccharis salicifolia*

Bear-grass, *Nolina microcarpa*

Blue-green Paloverde, *Parkinsonia (Cercidium) florida*

Bonpland's Willow, *Salix bonplandiana*

Brazil, *Haematoxylon brasiletto*

Bristlegrass, *Setaria* spp.

bursages, *Ambrosia deltoidea, A. dumosa*

Bush Penstemon, *Keckiella antirrhinoides microphylla*

California Fan Palm, *Washingtonia filifera*

Candilillo, *Euphorbia collectioides*

Canyon Bursage, *Ambrosia ambrosiodes*

Cardón, *Pachycereus pringlei*

Carelessweed, *Amaranthus palmeri*

Carrizo, Common Reed, *Phragmites communis*

Catclaw, *Acacia greggii*

Chain-fruit Cholla, *Opuntia fulgida*

Chuparosa (Beloperone), *Justicia californica*

Coral-bean, *Erythrina flabelliformis*

Creosote Bush, *Larrea divaricata*

Crucifixion-thorn, *Castela emoryi*

day-flower, Yerba-del-pollo, *Commelina erecta*

Desert Agave, *Agave deserti*

Desert-broom, *Baccharis sarothroides*

Desert Elderberry (Mexican Elder), *Sambucus mexicana*

Desert Hackberry, *Celtis pallida*

Desert Ironwood, *Olneya tesota*

Desert-lavender, *Hyptis emoryi*

Desert Mistletoe, *Phoradendron californicum*

Desert Mulberry, *Morus microphylla*

Desert Saltbush, *Atriplex polycarpa*

Dove-weed, *Croton texensis*

Elephant Tree, *Bursera microphylla*

Four-winged saltbush, *Atriplex canescens*

Frémont's Cottonwood, *Populus fremontii*

Giant Reed, *Arundo donax*

Golden-flowered Agave, *Agave chrysantha*

Goodding's Willow, *Salix gooddingii*

Graythorn, *Ziziphus obtusifolia*

Greasewood, *Sarcobatus vermiculatus*

ground-cherry, *Physalis* cf. *maxima*, *P. acutifolia (wrightii)*

Guayacán, *Guaiacum coulteri*

Hetcho, *Pachycereus pecten-aboriginum*

Horned Pondweed, *Zannichellia palustris*

Horse-purslane, *Trianthema portulacastrum*

Jimsonweed, *Datura wrightii*

Jumete, *Euphorbia collectioides*

Little-leafed Paloverde, *Parkinsonia (Cercidium) parvifolia*

manzanita, *Arctostaphylos* spp.

Maoto Negro, *Acacia coulteri*

Mexican Paloverde, *Parkinsonia aculeata*

Mexican (or California) Poppy, *Eschscholzia mexicana*

Ocotillo, *Fouquieria splendens*

Organ-pipe Cactus, *Stenocereus thurberi*

Ojo de Chanate, *Rhynchosia precatoria*

Osage-orange, *Maclura pomifera*

Otate, *Arundinaria longifolia* [fide Pennington]

Otate, *Otatea acuminata* (probably bamboo)

Palmer's Agave, *Agave palmeri*

Palmer's Carelessweed, Palmer's Amaranth, *Amaranthus palmeri*

Palmilla, *Yucca elata*

Palo Chino, *Pithecellobium mexicanum*

Palo Colorado, *Caesalpinia platyloba*

paloverde, *Parkinsonia (Cercidium)* spp.

Pickleweed, *Allenrolfea occidentalis*

Pima pumpkin, squash, *Cucurbita argyrosperma*

Ponderosa Pine, *Pinus ponderosa*

Prickly Poppy, *Argemone* sp.

Purple Nut-sedge, *Cyperus rotundus*

Quail Brush, *Atriplex lentiformis*

Quail-plant, *Heliotropium curassavicum*

Sacaton Grass, *Sporobolus airoides*, *S. wrightii*

Sacred Datura, *Datura* spp.

Saguaro, *Carnegiea (Cereus) gigantea*

Salt-cedar, *Tamarisk ramosissima*

Screwbean (Mesquite), *Prosopis pubescens*

Seepweed, *Suaeda moquinii*

Seep-willow, *Baccharis salicifolia*

smartweed, *Polygonum* spp.

Soaptree Yucca, *Yucca elata*

Sotol, *Dasylirion wheeleri*

Sweet Bush, *Bebbia juncea*

Tansy-mustard, *Descurainia pinnata*

Tescalama, *Ficus petiolaris*

Tree Morning-glory, *Ipomoea arborescens*

Tree Ocotillo, *Fouquieria macdougalii*

Velvet Mesquite, *Prosopis velutina*

Verdolaga, *Portulaca oleracea*, *P. retusa*

Western Soapberry, *Sapindus saponaria*

White Ratany, *Krameria grayi*

Wild Sunflower, *Helianthus annuus*

wolfberry, *Lycium* spp.

Yellow Bird-of-paradise, *Caesalpinia gilliesii*

Yellow Mistletoe, *Phoradendron tomentosum*

Yerba-de-la-flecha, *Sapium biloculare*

Yerba Mansa, *Anemopsis californicus*

Appendix C

Scientific Names of Reptiles Mentioned in Text

Desert Iguana, *Dipsosaurus dorsalis*

Desert Spiny Lizard, *Sceloporus magister*

Desert Tortoise, *Gopherus agassizi*

horned lizard, *Phrynosoma* spp.

rattlesnakes, *Crotolus* spp.

Red Racer (Coachwhip, Whipsnake), *Masticophis flagellum*

toads, *Bufo* spp., *Scaphiopus couchi*

Western Whiptail Lizard, *Cnemidophorus tigris*

Appendix D

Scientific Names of Mammals Mentioned in Text

Antelope (Ground) Squirrel, *Ammospermophilus harrisii*

Antelope Jackrabbit, *Lepus alleni*

Badger, North American, *Taxidea taxus*

bats (all spp.), Chiroptera

Beaver, North American, *Castor canadensis*

Bobcat, Wildcat, *Felis (Lynx) rufus*

Black Bear, *Ursus americanus*

Black-tailed Jackrabbit, *Lepus californicus*

Coatimundi, *Nasua nasua*

cotton rat, *Sigmodon* spp.

Coyote, *Canis latrans*

Desert Bighorn (Sheep), *Ovis canadensis*

Desert Cottontail, *Sylvilagus auduboni*

Desert Shrew, *Notiosorex crawfordi*

dog, *Canis familiaris*

grasshopper mouse, *Onychomys* spp.

Gray Fox, *Urocyon cinereoargenteus*

Jaguar, *Felis onca*

Javelina (Collared Peccary), *Tayassu (Dicotyles) tajacu*

kangaroo rat, *Dipodomys* spp.

Mountain-lion, Cougar, Puma, *Felis concolor*

Mule Deer, *Odocoileus hemionus*

Muskrat, *Ondatura zibethicus*

packrat, *Neotoma* spp.

Porcupine, North American, *Erithizon dorsatum*

Pronghorn, *Antilocapra americana*

Raccoon, *Procyon lotor*

Ringtail, *Bassariscus astutus*

Round-tailed Ground Squirrel, *Spermophilus tereticaudus*

skunk spp., *Mephitis* spp., *Spilogale gracilis*

Valley Pocket Gopher, *Thomomys bottae*

white-footed (deer)mouse, *Peromyscus* spp.

White-tailed Deer, *Odocoileus virginianus*

Wolf, Gray, *Canis lupus*

Appendix E

Upper Piman Personal (Men's) Names Relating to Birds

Al Hohhoi, Little Mourning Dove

Ba'ag Mo'o, Eagle Head (Pima leader, mid-1800s)

Ba'ivchul, Belted Kingfisher

Chevel U'uhig, Black-necked Stilt (literally "Tall Bird")

Chivi-chuch, Killdeer

Chuk Hawañĭ, Black Raven

Chukuḍ Mo'o, Great Horned Owl's Head

Chukuḍ Naank, Great Horned Owl's Ears

Chukukmal, Western Kingbird

Havañĭ Kosh, Raven Nest (old storyteller fide Shaw 1968:28)

Havanĭ Mo'o, Raven Head (Javanimó, Gabanimo, historic Piman rebel leader)

Haupal Tonomkam, Thirsty Red-tailed Hawk (Pima name of Francisco Lucas, Maricopa)

Hikvig, Gila Woodpecker

Hoohi, Mourning Dove

Koomagĭ Shashañ, Gray Blackbird (Female Red-wing?)

Muhaḍagĭ Ba'ag, Greasy[-colored] Eagle

Ñui, Turkey Vulture

Oam Ba'ag, Brown Eagle

Okokoi, White-winged Dove

Paado, (Domestic?) Duck

Shashañ, Blackbird

Shashañ Jusip, Blackbird Joseph

Shuug, Mockingbird

Taḍaikaj Shalivim, "Roadrunner Pants"

Tosiw, Meadowlark

Towa Kusvokam, Turkey Neck–has

U'uhig Atkam (various ones including Joe Giff), Bird Buttocks–has

U'uhig O'odham (the author), Bird Man

Vavas, Rock Wren

Vishag Kavaḍkam, Falcon Shield–has

Vishag Voi'i, Prairie Falcon Flying

Appendix F

Unidentified Northern Piman Birds

Various Piman folk taxa remain unidentified. The information recalled and usually the sources are given here, sometimes with a best guess as to what the species might be. Some birds are listed here because I was never able to get a confirmation of the name from additional speakers.

akimel taak u'uhig

"Riverside bird" fide Sylvester Matthias (January 1986). "I don't remember what color. Maybe I did see one, but I didn't know the birds at that time."

chechlogam, chechevk, chechevk u'uhig

George Kyyitan (April 1988) said, "I don't know. It's some kind of bird. Emmett [White]'s grandfather is named like that."

I asked him, *"Chechlogam?"*

"Yeah."

"It's a bird?"

"It's a bird, I think. I think he sings that *chechlogam* [song]. They're like *chechevk, chechevk u'uhig*. It's a bird, but it's *way* up there, and it stays *way* up there in the air, all the time."

Leonard Pancott (October 1999) said, "*Chechlogam*, I've never seen it. The old roundup boss [used to] sing the *chechlogam* songs, so they call him that—Tom and Emmett's grandfather. They stay way up high. Lot of the birds never show up anymore. *Chechevk*, I've never seen. There's an old man who sings that *chechevk* songs. I've never known anybody that seen it. They might have seen it in their dreams."

The name is not from *chev*, tall; more likely the name is *chevagig, chechevagig*, cloudiness. George Kyyitan did not think that *chechlogam* was a variant of *koologam*, the Common Poorwill.

Father Antonine (1935:9) listed *chuechlogam* (no plural) as a gray bird, larger than a sparrow, placing it between *tamtol* (White-crowned Sparrow) and *ñepuḍ* (Lesser Nighthawk). This reference would indicate it is a real bird, not just mythological. Joe Giff said it was "kind of a fat, pretty bird." The name was not known to Vincent Josemaria or other Tohono O'odham I asked.

ge'eged kakachu

Delores Lewis of Ge O'idag Tohono O'odham village used this name for the Masked Bobwhite, identifying it in my elicitation notebook from a good painting of the male. He said they were on both sides of the International Border ("You can't stop them") and just on the other side of the Baboquivaris. He said he was unfamiliar with the Scaled Quail (a pale species from the area he described). If indeed a folk taxon, it would bring the total of folk species in the genus *kakachu* to four, matching the four biological species of the area.

haiñ uushab (haiñ julshab)

The first word of the name, *haiñ*, means "broken." Culver Cassa says *uushab* comes from *esh*, mouth, chin. When spoken, the second word becomes *jushab* in this environment.

Sylvester Matthias (June 1968) told me that these birds stand very tall. They were regularly seen every summer when the boys were out of school. There would be half a dozen at a time and would come into the field when a man was irrigating. Sylvester did not know if they were after the fish brought in with the irrigation water or the gophers or other rodents that were flooded out. The boys would try to get close to these interesting birds, but somehow, not noticeably, the birds would always keep at a considerable distance in the field. At other times, *haiñ uulshab* were seen along sandbars in the river or along the banks of sand at the edges.

At one time, he said they were the size of a Great Blue Heron. This detail suggests either Sandhill Crane ("broken" referring to the naked red crown patch) or (as Allan R. Phillips suggested to me) Wood Stork ("broken" alluding to the heavy, prominently decurved bill). The crane can be ruled out because of the season (the birds are not in southern Arizona during the summer) and because cranes already have a legitimate pan-Piman name, *kookoḍ*, grulla, based on the colonial Tepehuán and Névome vocabularies and other evidence (see Sandhill Crane species account). There are several references to a white neck. Avocets might be suggested by the black-and-white pattern and up-curved bill that appears "broken," but the stated size is off.

In a group discussion (January 1973), Sylvester said these birds have white necks. Others knew the name, if not the bird. However, David Brown recalled, "Most have white necks."

In a telephone conversation (21 June 1982), Sylvester said, "Stands on one leg. 'Stork'; dark gray, little smaller than a Great Blue Heron."

Sylvester (September 1990?) also said, "Something in the line of a heron. I never saw it [*sic*]. When St. John's [Indian School] school girls started wearing uniforms—black or navy blue—with white square at the back—World War I, when all schools turned into military schools and they start discipline and drilling—[they called the girls by this bird's name]."

Some of the information seems to be coming from faulty memory. The Wood Stork, *Mycteria americana*, has conspicuous black-and-white wings, unlike the White-faced Ibis *(Plegadis chihi)*, Sandhill Crane, or Great Blue Heron.

I suspect the bird is an American Avocet, *Recurvirostra americana*, a species found at various times throughout the year on the Gila River, most abundantly during its protracted migrations. Its neck is very pale gray during most of the time that it is seen in southern Arizona. It may even occasionally breed here (Monson and Phillips 1981; Witzeman, Demaree, and Radke 1997), at which time it has a pinkish buffy neck, hence David's remark. Most important, the very thin upcurved bill does indeed seem to be broken or at least unnaturally bent, and the color pattern of the back suggests a black jumper over a white blouse, hence the comparison with the girls' school uniform. Avocets often associate with Black-necked Stilts, a species that is identified in the River Pima lexicon.

hewel ch-eḍ u'uhig

Said by Saxton, Saxton, and Enos (1983:21) to be the Brown-headed Cowbird. I was unable to find anyone who recognized this name.

hohogĭ u'uhig

Identified by Mathiot (n.d.:502) as "Yaqui Quail." Possibly Masked Bobwhite? *Hohogi* is the name most Mountain Pima call their local Montezuma (Mearns') Quail. It is tantalizing that Mathiot turned up this name far to the north in the Santa Rosa area (Totoguañ dialect of Tohono O'odham).

hokimal u'uhig

This name, meaning "butterfly bird," was given in the Hohokimal Ñe'i (Butterfly Song) in Russell (1908:295). No one I talked to recognized it as a species. (See Western Kingbird account.)

jujupul (or chuchupul?)

Sylvester Matthias (December 1980) gave this information: "*S-oam* [yellow, orange, or brown], in owl family. Desert bird. Comes to water, like a swampy area round Maricopa Wells or Pima Butte somewhere. Either yellow or brown."

If he was correct about it being an owl, the Short-eared Owl, *Asio flammeus*, is a possibility, given the area and habitat (once grasslands and marshes). If he was incorrect, the Long-billed Curlew might be a candidate, given the color brown. These birds are rather common transients during their protracted migrations through the desert, and a few may overwinter.

kahagam

George Kyyitan said, "It's not the cuckoo, *kadgam*. *Kahagam* is a bluebird of some sort. I don't know what it looks like. I think it's just the same as those other bluebirds [Blue Grosbeaks], not big; just a small bird." When I asked him in what season it appeared, he replied, "I don't know just when it is. I think summer. I don't know for sure." When I asked about its color pattern, he compared it to a *ba'ivchul*, kingfisher. "That's what we had on our sticks, when we have those sticks and those feathers. It's got a white [below], what we got on our dancers. *Kadgam* [Yellow-billed Cuckoo], I think it's kind of a big bird. Arnold Juan's grandfather used to sing the *kahagam* song. A blue bird." Again George affirmed it is not the *kadgam*. "I used to know lots of them [*kahagam* songs], but now I know only three of them, *three* of them! The going in, coming out, and one in the middle."

I asked Sylvester Matthias if he had heard of the *kahagam* (7 October 1993). He said, "There *is* a *kahagam*. I don't know where it can be found and how does it look. There's a *kahagam* that my grandmother talked about."

Kadgam (Cadigum) is given in the Hayden-Smith version of the Creation Story (Hayden 1935:18) as a character's personal name. This name refers to the Yellow-billed Cuckoo (see species account). In working up this story, Bahr (in Bahr and associates 1994:114), confused the name with *kuhigam*, concluding that it might be the name of a leafy plant even though Hayden's text says clearly, "The man's name was Cadigum, a bird's name." For *kuhigam*, see the Phainopepla account.

koomagĭ shashañ

Mentioned by several Gila Pimas (Sylvester Matthias, Ruth Giff, and others). It is unclear if it is a species name or just a description of the female Red-winged Blackbird.

kulisgam, kukulisgam

George Kyyitan (21 April 1996) thought this bird might be an avocet or stilt. He mentioned this at the Sacaton sewage ponds where we were watching both species. George noted a place called "Kukulisgam ha Vo'oga—hard [bottom], where water stays all the time and these birds drink there; just about half an acre or an acre; too hard for water plants"—that he described as being sort of north of Bapchule and Sacaton, east or southeast of Chandler, by the reservation boundary line on Price Road. George explained the *vo'o* part of the place-name: "The water comes there when it rains; it ponded itself, natural pond. That's what is meant by saying *vo'o*, when the water comes in, just staying there, like that." The possessive marker *-gam* indicates that this pond is owned by these birds.

Sylvester Matthias (October 1993) said he had heard of this name from his grandmother and knows it is in the legend, but does not know the bird itself or where it nests or where it may be found, just the name.

Another possibility is the White-faced Ibis, a distinctive species that visits River Pima country but, as far as is known now, lacks a Piman name. The stilt and probably the avocet (see *haiñ uushab* given earlier in this appendix) are already named in the River Pima lexicon.

oám u'uhig, pl. o'am uu'uhig

The words mean "yellow bird." Although the name seems nonspecific, we see other situations where a color combined with the life-form word *u'uhig* specify a folk genus (*toha u'uhig*, egret, and *koomagĭ u'uhig*, Le Conte's Thrasher), just as in English (blackbird, bluebird). Shaw (1968:32–35) gave a story about Yellow Bird, but it is used as the proper name of a young Pima girl, not a bird.

Many people I asked, at least among River Pima, had heard the name, but did not know what bird it referred to. Sylvester Matthias (telephone conversation, 21 June 1982) was the only one to supply some particulars: "Yellow breast; kind of brownish head on down. When flies, opens tail, has yellow. Found wherever there is water and lots of trees. In trees. Yellow bill. Eats insects. Size of *shashañ* [Red-winged Blackbird] or larger."

Some of this description fits Yellow-breasted Chat, *Icteria virens*, curiously absent from the Pima lexicon. These vocal and conspicuous birds were once common in the desert riparian community and should have attracted Pima attention with their loud and varied repertoire. But the bill is black, not yellow, and there is no yellow in the tail. The Yellow Warbler, *Dendroica petechia*, another bird once breeding commonly in the cottonwood-willow canopy, does show extensive yellow in the spread tail. But it is less than half the size of a Red-wing and is clearly yellow above in the local breeding race.

Another possibility is a goldfinch. Both Grossmann's 1877 and Cook's late 1800 manuscripts list *su-ám u-hick* as "canary bird," a name sometimes used for goldfinches. The Lesser Goldfinch, *Carduelis psaltria*, is known to breed in Pima country where there is well-developed gallery forest (Rea 1983:251). But, like the warbler, goldfinches are very small birds.

pivikum (pivikam?)

Said to be a hawk (Hayden 1935:61, given as *pi vi cum*). The young boy of the story in the Hayden-Smith manuscript turned himself into this bird. He is associated with an old man called Tova Keli, Old Man Turkey, a "Hohokam" whom the boy kills. Bahr (in Bahr and associates 1994:269 n. *j*) says "Unknown, not a known hawk name." In a parallel story (Russell 1908:244–245), the boy turns first into a "little dove," then into a flock of hawks (the Pima not given for either). Word analysis seems unproductive (*vi'ikam*, orphan; *pi-*, negation). Mason (1917:329) listed *pichaag* (pl. *pipchag*), aguililla ("little eagle"), for the Tepecano. The biological referent is not known.

shashañĭ

Saxton, Saxton, and Enos (1983:52) offer the following breakdown:

shashañĭ—a blackbird
s-chuk shashañ—any black bird [*sic*]
s-oam shashañ—an oriole [!] [not a Yellow-headed Blackbird?]
s-wegĭ shashañ–Red-winged Blackbird [!]

This breakdown is not a River Pima dichotomy, nor did any Papago I consulted recognize these folk species.

shuudagĭ u'uhig

Given by Saxton, Saxton, and Enos (1983:54) as "a crane, heron, or other water bird." The validity of this name as a folk taxon in Papago dialects needs to be determined. Is it merely used when someone has forgotten the other proper folk generics? Or is it used as an umbrella term? What then would it include? In contrast to Akimel O'odham, Tohono O'odham taxonomy of waterbirds is greatly attenuated, even in English.

sudguskoko

Pennington (1980:192) gives this Pima Bajo name both for quelele (which should refer to the caracara but is used here for the White Ibis) and for pato negro and pato chocolate (coot? duck?). One of the four ducks in the Névome vocabulario, patos peruleros, is given as *sudaqui coconi*, a better transcription (Pennington 1979:90). The words *suudagi kokoni* literally mean "water raven," which I suspect is a cormorant, a bird seen regularly along the Río Yaqui in the vicinity of Ónavas. Here Pedro Estrella called the bird *s-tuk vakaik* or *vaka'ig*, or Corvejón in Spanish.

taatai a'anam

"Wings backwards," a mythical bird, fide Sylvester Matthias.

unmu'uhik

Sylvester Matthias knew the word but not the bird, thinking that it might be some "blue jay." "Gila River consultants relate that the Pima singers from the Salt River Reservation have a song about a bird that is blue, comes from the cold country, and 'lives in the spruce, but nobody knows what it is'" (Rea 1983:197).

Bahr and associates (1997:23) give *uun mu'uhig*. Bahr writes "I had thought this might be *Uam U'uhig*, 'Yellow Bird,' the goldfinch, according to Saxton and Saxton,

1969:167," but his singer, Lloyd Paul, says that the words are the name of a bird and nothing else. Paul apparently didn't know the bird either. "The string isn't analyzable into meaningful words."

Culver Cassa commented, "Nobody knows what it is, apparently, except the person who dreamed it. It may be only a dream bird rather than a biological species." He renders the name in full as *u'unam u'uhig*, "enclosure bird." The *unama* was a circular enclosure made of Arrowweeds thrust into the ground, forming a windbreak where women cooked (see Russell 1908:plate XXXVI, plate VIb). This enclosure is also knows as the *uksha*.

vachimukdam

Luciano Noriega spoke of a bird smaller than a duck (or coot) that dives under the water (a grebe?). *Vachmukdam, vachumdam, vachimukdam*, and perhaps *vachpickdam* (and even *vachpik*) are probably all names for grebes or coots, birds that dive.

vakakam or va'ukakam u'uhig

Sylvester Matthias (21 June 1982) said of this bird, "Sings in the morning where there are cliffs. (Where there are swallows.)" He gave no more description. Is this name associated with *vavas*, the Rock or Canyon Wren?

waakuk (vaakuk)

Given as Tropical Kingbird by Luciano Noriega at Quitovac, Sonora. He also called the Verdin *wakok*.

wachumdam

In their Papago dictionary, Saxton, Saxton, and Enos (1983:61) give this name for coot (American Coot), Mudhen, instead of the River Pima *vachpik*. Perhaps it is a legitimate variant. They also (page 62) list *wakaig* for the same referent. At Quitovac oasis, where coots are much in evidence, Hiá ch-eḍ O'odham Luciano Noriego told us they were called *vapkik*. Theodore Peters of Santa Cruz village on the Gila, originally settled by Sand Papago, called the coot *vachpikdam*, as did Leonard Pancott at Haashañ Keek (Sacaton Flats). (See discussion of *vachimukdam* in this appendix.)

wahum, chum wahum

These two names were given by Luciano Noriega (August 1981), a Hiá ch-eḍ O'odham of Quitovac. He identified pictures of the Short-eared Owl in my elicitation book as *wahum* and of the El Owl as *chum wahum*, a name that means "little *wahum*." But we did not have a complete contrast set because we were lacking the ubiquitous screech-owl. Neither Luciano nor Delores Lewis recognized the Ferruginous (Pygmy) Owl. Although virtually everyone was clear about the identifications of Barn Owl, Great Horned Owl, Burrowing Owl, and Western Screech-Owl, the two smallest owls eluded identification. The Elf Owl, a noisy and often common species in the breeding season, should have been well known to these Sonoran Desert peoples. *Chum wahum* is the closest I came to a name for it, but I was never able to verify this identification with another person or with actual encounters of the birds in the field.

Appendix G

Orthography: The Sounds of Akimel and Tohono O'odham

Symbol	Approximate Equivalent
a	f*a*ther
e [ɨ = barred i]	[not in English, similar to *e* in op*e*n with lips spread]
i	mach*i*ne (but usually = *i* as in w*i*n when syllable final)
o [= ô = aw]	*a*ll, *au*tumn, *aw*l, *aw*ful, s*o*ft, d*o*g, l*aw*, (sometimes) b*oa*t, n*o*te
u	br*u*te, f*oo*d, r*u*de
˘	nonvoiced or whispered vowel
repeated vowel without glottal [']	long vowel (duration)
w	*w*in (glide; semivowel)
v	*v*ineyard (voiced bilabial friative, similar to Spanish *v*aca)*
g	*g*o, e*gg* (voiced dorso-velar stop)
k	*k*ite, oa*k* (nonvoiced dorso-velar stop)
ŋ = manual = n + g	so*ng*, fi*ng*er (voiced nasal counterpart of *g*)
h	*h*ot
j	*j*ot
sh [= ṣ = š = x]	like *sh*ip but tongue curled (apicoalveolar fricative)
ch [ç = č]	*ch*ur*ch*
l [= ř]	closest to single flap *r* in Spanish, pe*r*o, pa*r*a, but never as English *l*ean or *r*ich (almost *d*)
ḍ	retroflex apico-alveolar *d*, like the *t* with a glottal stop in bu*t*
d	soft apicodental stop, similar to the *th* in *th*is
t	strong apicodental stop, like the *t* in *t*op

* Lacking in most Tohono O'odham dialects.

Notes

Chapter 4

1. Atran (1990:124–127) prefers the term *generic specific* instead of *folk genus* because this category corresponds more accurately with the correlation between folk taxa and scientific or Western taxonomy. The more widely used term, *folk genus*, is used in this volume.

2. A term preceded by an asterisk (for example, **va'konoi*, heron) is reconstructed from a word, in this case Proto-Pimic, based on modern Tepiman languages. Words that have cognates in several Piman languages I have termed *pan-Piman* (or, if attested in Tepehuán as well, *pan-Tepiman*). I have made no attempt at reconstructions, those few given being from Fowler 1983.

3. The word *u'uhig* seems to be the base variant, or at least that is what is heard among River Pima. Sylvester Matthias observed, "Another thing, in Papago there's no plural for *u'uhig*, but we have a plural, *uu'uhig*." For the Santa Rosa (Totoguañ) dialect of Papago, Mathiot (n.d.:502) recorded *u'uvhig*—that is, *u'uwhig*. Saxton, Saxton, and Enos (1983:72) list the variant *u'whig* in addition to *u'uhig*, also Tohono O'odham. All are Upper Piman forms. Modern Pima Bajo of the tierra caliente say *u'uvig* (Hale n.d.a; Rea notes). In colonial times, *aves* in común was given as *ubiga*—that is, *uviga*. A footnote in the Névome vocabulario (Pennington 1979:13), added in a different handwriting, gave *uugi*—that is *u'ugi*. Based on internal evidence, most of these latter notations were appended by someone working among what we would now call Mountain Pima. Today these pine-oak country Pimans still say *u'ugi* (pl. *u'hugi, uhugi*) (Shaul 1994:340). Pennington (1969:116) recorded *úrugi* as "bird" with modern Northern Tepehuán. Rinaldini's 1743 Southern Tepehuan vocabulario lists *urugui, uuggui*.

4. A number of birds considered here as resident (that is, present during both the breeding and wintering seasons) are technically not permanent residents in that they have a postbreeding exodus from the low hot country in summer (Costa's Hummingbird, White-throated Swift, Say's Phoebe, Phainopepla).

Some birds considered winter residents are actually much more dynamic, with different individuals (and often even different subspecific populations) moving through Piman country throughout the winter months, earlier ones moving south and later ones replacing them. Conspicuous examples are Dark-eyed Juncos, White-crowned Sparrows, and Lark Buntings.

Some birds are included in the breeding list because they formerly nested along the middle Gila: Green Heron, Great Blue Heron, ?Snowy Egret, Ferruginous Pygmy-Owl, Belted Kingfisher, ?Cliff Swallow, Black Phoebe, Blue Grosbeak,

Summer Tanager, Yellow Warbler, Yellow-breasted Chat. Excluded from the list are those breeding species that have only recently arrived: White-tailed Kite (*Elanus leucurus*), Mississippi Kite (*Ictinia mississippiensis*), Black-bellied Whistling-Duck (*Dendrocygna autumnalis*), Great-tailed Grackle, Red-eyed (Bronzed) Cowbird, and Pyrrhuloxia (though this latter species has probably a long association in Papago country). In most of these cases, the details of the particular species' breeding status can be found in Rea 1983.

Chapter 5

1. "*[N]amk, nanmek, nanammek*, Vt [to] meet; encounter and presumably gain spiritual power from (usually an animal)" (Saxton, Saxton, and Enos 1983:44; see also Mathiot n.d.:100). Mathiot lists *hevel namkam*, wind meeter, and *vuihom(i) namkam*, thunder meeter, as examples of specialists having encounters with inanimate (for Westerners) phenomena.

2. Bahr (in Bahr and associates 1994:321) frets over the fact that the animals were supposed to have already been made by this point in the Origin Story. Indeed, at first the Hayden-Smith narrative states that the poorwill, Verdin, and so forth were "made" by a *maakai*, shaman, for the specific purpose they were to serve in overcoming the original inhabitants of the land. But as the story unfolds, it becomes clearer that this is not creation for the occasion, but rather the *mamakai* have these specific animals as their guardian spirits, their helpers. In fact, at one point the narrative states, "And this medicine man had the power of a bluebird" (Hayden 1935:52; see also Bahr and associates 1994:243). In another case, the *maakai* whose help is sought is a *juḍum namkam*, "A man who has the power of a bear." This phenomenon is even explained by the Pima narrators:

> So they made camp there and they asked another medicine man to work for them who had the power of a bluebird, "hü wut jut nam kum" [*hevachuḍ namkam*, Blue Grosbeak meeter]. (Namkum means that a person sleeps at night and if this man is very interested in something such as birds or animals [*sic*], why, if that animal know[s] that he is interested in him the animal or bird comes to him in his dream and shows him what to do and how to act like this animal or bird. That's how the medicine people get their power, by learning from different things that they're interested in their dream; that's what namkum means. (Hayden 1935:55–56; see also Bahr and associates 1994:251)

3. The southern origin hypothesis has another interesting twist. As we saw earlier, historically, the youth seeking dreams and songs from some animal helper might engage in one of three power activities that would expose him to dreaming: a salt expedition, an eagle killing, or an enemy killing. The ocean (Gulf of California), the Golden Eagle, and the Apache (synonymous with *enemy*) would have been encountered as Pimans expanded northwestward from a southern ancestral home. In the south, the coastal plain was occupied by other tribes.

Chapter 6

1. The *theme* of this folk tale is generally agreed by mythologists to be post-European contact (a Cindarella story). However, the *setting* is thoroughly Puebloan, incorporating indigenous, not European, practices.

Species Accounts

1. Joseph Giff, a Piman scholar, is undoubtedly correct in calling attention to the distinction between the mythic characters Ñui and Oam Ñui. Ñui is one of the four primordial beings responsible for creation of the earth. Oam Ñui, Joe points out, is a character in Maricopa-derived, that is Yuman, stories. As Culver Cassa notes, Oam Ñui is associated with the shadow side of human nature: negativity, unhappiness, depression, violence, and murder. Some story tellers, even Piman ones, have conflated these two very different personages. This is particularly evident in the Creation Stories with strong Maricopa story lines (for example Densmore 1929, Herzog 1929).

2. Pima is more precise than English in calling this character Ba'ivchul Uvĭ. The term *uvĭ* indicates a marriageable or child-bearing age girl or woman between puberty and menopause. A pre-pubescent girl is *chehia* and a post-menopausal woman is *oks*. These distinctions have important social and ceremonial ramifications.

Bibliography

Alcock, John. 1990. Sonoran Desert Summer. University of Arizona Press, Tucson.

American Ornithologists' Union (AOU). 1957. Check-list of North American Birds. 5th ed. AOU, Baltimore.

American Ornithologists' Union (AOU). 1998. Check-list of North American Birds. 7th ed. AOU, Washington, D.C.

American Ornithologists' Union (AOU) Check-list Committee. 1973. Thirty-second Supplement to the American Ornithologists' Union Check-list of North American Birds. Auk 90:411–419.

American Ornithologists' Union (AOU) Check-list Committee. 1995. Fortieth Supplement to the American Ornithologists' Union Check-list of North American Birds. Auk 112:819–830.

Anderson, Walter L. 1978. Vocalizations of Scaled Quail. Condor 80:49–63.

Andrews, John P., and Todd W. Bostwick. 2000. Desert Farmers at the River's Edge: The Hohokam and Pueblo Grande. Pueblo Grande Museum Publications, Phoenix.

Antonine, Fr. See Willenbrink, Antonine.

Attí vocabulary. [1774]. Unpublished partial ms. M-M475, Bancroft Library, University of California, Berkeley.

Atran, Scott. 1990. Maya Folkbiological Taxonomy: Cognitive Universals and Cultural Particulars. Pp. 119–203 *in* Folkbiology, ed. Douglas L. Medin and Scott Atran. MIT Press, Cambridge, Mass.

Austin, George T., and W. Glen Bradley. 1969. Additional Responses of the Poor-will to Low Temperatures. Auk 86:717–725.

Bahr, Donald M. 1975. Pima and Papago Ritual Oratory: A Study of Three Texts. Indian Historian Press, San Francisco.

Bahr, Donald, Joseph Giff, and Manuel Havier. 1979. Piman Songs on Hunting. Journal of the Society for Ethnomusicology 23:245–296.

Bahr, Donald M., Juan Gregorio, David Lopez, and Albert Alvarez. 1974. Piman Shamanism and Staying Sickness *(Ká:cim Múmkidag)*. University of Arizona Press, Tucson.

Bahr, Donald M., and J. Richard Haefer. 1978. Song in Piman Curing. Ethnomusicology 22:89–122.

Bahr, Donald, Lloyd Paul, and Vincent Joseph. 1997. Ants and Orioles: Showing the Art of Pima Poetry. University of Utah Press, Salt Lake City.

Bahr, Donald, Juan Smith, William Smith Allison, and Julian Hayden. 1994. The Short, Swift Time of Gods on Earth: The Hohokam Chronicles. University of California Press, Berkeley.

Bailey, Florence Merriam. 1928. Birds of New Mexico. New Mexico Department of Game and Fish, State Game Protective Association, and Bureau of Biological Survey. Judd & Detweiler, Inc., Press, Washington, D.C.

Balfour, Bruce J. 1984. When a Plume Meant Aplomb. Westways 81(10):46–48.

Bancroft, Hubert H. 1889. The History of Arizona and New Mexico, 1530–1888. The History Company, San Francisco. Reprinted 1962, Horn and Wallace, Albuquerque.

Bandelier, A. F. 1892. Final Report of Investigations among the Indians of the Southwestern United States, Carried on Mainly in the Years from 1880–1885. Part II. Papers of the Archaeological Institute of America, American Series IV. Cambridge University Press, Cambridge, Mass.

Barney, James M. 1955. Ostrich Farming: Forgotten Industry of Salt River Valley. Sheriff Magazine (April–May):35–36.

Bartholomew, George A., Thomas R. Howell, and Thomas J. Cade. 1957. Torpidity in the White-throated Swift, Anna Hummingbird, and Poor-will. Condor 59:145–155.

Bartlett, Des, and Jen Bartlett. 1973. Beyond the North Wind with the Snow Goose. National Geographic 144:822–847.

Bartlett, John Russell. 1854. Personal Narrative of Explorations and Incidents in Texas, New Mexico, California, Sonora, and Chihuahua. 2 vols. D. Appleton Co., New York.

Bascom, B. W., Jr. 1965. Proto-Tepiman. Ph.D. diss., University of Washington.

Beaglehole, Ernest. 1936. Hopi Hunting and Hunting Ritual. Yale University Publications in Anthropology no. 4, Yale University Press, New Haven, Conn.

Beals, Ralph L. 1943. The Aboriginal Culture of the Cáhita Indians. Ibero-Americana no. 19, University of California Press, Berkeley.

Beers, Henry Putney. 1979. Spanish and Mexican Records of the American Southwest: A Bibliographical Guide to Archive and Manuscript Sources. University of Arizona Press, Tucson.

Bendire, Charles E. 1895. Life Histories of North American Birds. U.S. National Museum Special Bulletin no. 3, Washington, D.C.

Benedict, Ruth. 2001. O'odham Creation and Related Events: As Told to Ruth Benedict in 1927 in Prose, Oratory, and Song by the Pimas William Blackwater, Thomas Vanyiko, Clara Ahiel, William Stevens, Oliver Wellington, and Kisto. Ed. Donald Bahr. University of Arizona Press, Tucson.

Bent, Arthur Cleveland. 1938. Life Histories of North American Birds of Prey, Part Two. U.S. National Museum Bulletin no. 170, Washington, D.C.

Bent, Arthur Cleveland. 1940. Life Histories of North American Cuckoos, Goatsuckers, Hummingbirds, and Their Allies. U.S. National Museum Bulletin no. 176, Washington, D.C.

Berlin, Brent. 1972. Speculations on the Growth of Ethnobotanical Nomenclature. Language and Society 1:51–86.

Berlin, Brent. 1976. The Concept of Rank in Ethnobiological Classification: Some Evidence of Aguaruna Folk Botany. American Ethnologist 3:381–399.

Berlin, Brent. 1992. Ethnobiological Classification: Principles of Categorization of Plants and Animals in Traditional Societies. Princeton University Press, Princeton, N.J.

Berlin, Brent, Dennis Breedlove, and Peter H. Raven. 1973. General Principles of Classification and Nomenclature in Folk Biology. American Anthropologist 75:214–242.

Berlin, Brent, Dennis Breedlove, and Peter H. Raven. 1974. Principles of Tzeltal Plant Classification: An Introduction to the Botanical Ethnography of a Mayan-Speaking People of Highland Chiapas. Academic Press, New York.

Bierhorst, John. 1979. A Cry from the Earth: Music of the North American Indians. Four Winds Press, New York. Reissued 1992, Ancient City Press, Santa Fe.

Bierhorst, John. 1985. The Mythology of North America. Quill, William Morrow, New York.

Birkenstein, Lillian R., and Roy E. Tomlinson. 1981. Native Names of Mexican Birds. U.S. Department of Interior, Fish and Wildlife Service, Resource Publication no. 139, Washington, D.C.

Bourke, John G. 1889. Notes on the Cosmogony and Theogony of the Mojave Indians of the Rio Colorado, Arizona. Journal of American Folk-Lore 2:169–189.

Brambila, David. 1983. Diccionario Castellano/Raramuri (Tarahumar). Obra Nacional de la Buena Prensa, A.C., Mexico City.

Brand, Janette C., B. Janelle Snow, Gary P. Nabhan, and A. Stewart Truswell. 1990. Plasma Glucose and Insulin Response to Traditional Pima Indian Meals. American Journal of Clinical Nutrition 1:416–420.

Brandt, Herbert. 1951. Arizona and Its Bird Life: A Naturalist's Adventures with the Nesting Birds of the Deserts, Grasslands, Foothills, and Mountains of Southeastern Arizona. Privately published, Bird Research Foundation, Cleveland, Ohio.

Brauner, J. 1952. Reactions of Poor-wills to Light and Temperature. Condor 54:152–159.

Brewster, William. 1925. The Birds of the Lake Umbagog Region of Maine. Bulletin of the Museum of Comparative Zoology 66 (part 2):211–402.

Bright, William, comp. 1968. A Luiseño Dictionary. University of California Publications in Linguistics no. 51, University of California Press, Berkeley.

Brown, Cecil H. 1979. Folk Zoological Life-Forms: Their Universality and Growth. American Anthropologist 79:325–342.

Brown, Cecil H. 1981. More on Folk Zoological Life-Forms. American Anthropologist 83:398–401.

Brown, Cecil H. 1982. Folk Zoological Life-Forms and Linguistic Marking. Journal of Ethnobiology 2:95–112.

Brown, Cecil H. 1984. Language and Living Things: Uniformities in Folk Classification and Naming. Rutgers University Press, New Brunswick, N.J.

Brown, David E. 1973. Western Range Extensions of Scaled Quail, Montezuma Quail, and Coppery-tailed Trogon in Arizona. Western Birds 4:59–60.

Brown, David E. 1978. The Vegetation and Occurrence of Chaparral and Woodland Flora on Isolated Mountains within the Sonoran and Mohave Deserts in Arizona. Journal of the Arizona-Nevada Academy of Science 13:7–12.

Buelna, Eustaquio, ed. 1890. Arte de la lengua Cahita por un padre de la Compañía de Jesús. Reissued 1989, Siglo Veintiuno Editores, Mexico City.

Cade, Tom J. 1982. The Falcons of the World. Comstock, Cornell University Press, Ithaca, N.Y.

Campbell, Kenneth E., Jr., and E. P. Tonni. 1983. Size and Locomotion in Teratorns (Aves: Teratornithidae). Auk 100:390–403.

Campbell, Paul D. 1999. Survival Skills of Native California. Gibbs Smith, Layton, Utah.

Carter, George F. 1971. Pre-Columbian Chickens in America. Pp. 178–218 *in* Man across the Sea: Problems of Pre-Columbian Contacts, ed. Carroll L. Riley, J. Charles Kelley, Campbell W. Pennington, and Robert L. Rand. University of Texas Press, Austin.

Castetter, Edward F., and Willis H. Bell. 1942. Pima and Papago Indian Agriculture. Inter-Americana Studies I. University of New Mexico Press, Albuquerque. Reprinted 1980, AMS Press, New York.

Castetter, Edward F., and Ruth M. Underhill. 1935. The Ethnobiology of the Papago Indians. Ethnobiological Studies in the American Southwest no. 2. University of New Mexico Bulletin 275:3–84. Reissued 1978, AMS Press, New York.

Childs, Thomas. 1954. Sketch of the "Sand Indians." Ed. Henry F. Dobyns. Kiva 19:27–39.

Clark, William S. 1983. The Field Identification of North American Eagles. American Birds 37:822–826.

Colley, Charles C. 1979. Ostrich Farming: Arizona's Strangest Venture. Arizona Highways 55(5):40–45.

Cook, Charles H. [1943]. A List of Pima Words Compiled [Between 1870 and 1900] by C. H. Cook, Transcribed from the Original by Mrs. John W. Murphy, 1943. Manuscript, Arizona Historical Society Library, Tucson (copy, Gila River Indian Community Cultural Resources Library, Sacaton, Ariz.).

Cowen, Ron. 1990. Seeds of Protection: Ancestral Menus May Hold a Message for Diabetes-Prone Descendants. Science News 137:350–351.

Crosswhite, Frank S. 1980. The Annual Saguaro Harvest and Crop Cycle of the Papago, with Reference to Ecology and Symbolism. Desert Plants 2:3–61.

Curtin, Leonora S. M. 1949. By the Prophet of the Earth: Ethnobotany of the Pima. San Vincente Foundation, Santa Fe, New Mexico. Reissued 1984, with a new foreword by Gary Paul Nabhan, University of Arizona Press, Tucson.

Curtis, Edward S. 1908. The North American Indian. Vol. 2. Harvard University Press, Cambridge, Mass.

Cushing, Frank Hamilton. 1901. Zuñi Folk Tales. G. P. Putnam's Sons, New York. Reissued 1931, Alfred A. Knopf, New York; 1986, University of Arizona Press, Tucson.

Dawson, James W. 1998a. Great Horned Owl. Pp. 152–155 *in* The Raptors of Arizona, ed. Richard L. Glinski. University of Arizona Press, Tucson.

Dawson, James W. 1998b. Harris' Hawk. Pp. 77–81 *in* The Raptors of Arizona, ed. Richard L. Glinski. University of Arizona Press, Tucson.

Dawson, J. W., and R. W. Mannan. 1991a. Dominance Hierarchies and Helper Contributions in Harris' Hawks. Auk 108:649–660.

Dawson, J. W., and R. W. Mannan. 1991b. The Role of Territoriality in the Social Organization of Harris' Hawks. Auk 108:661–672.

Dawson, William Leon. 1923. The Birds of California. Students ed., vol. 1. South Moulton Company, San Diego.

Densmore, Frances. 1929. Papago Music. Bulletin of the Bureau of American Ethnography no. 90. Smithsonian Institution, Washington, D.C.

de Schauensee, Rodolphe Meyer, and William H. Phelps Jr. 1978. A Guide to the Birds of Venezuela. Princeton University Press, Princeton, N.J.

Di Peso, Charles C. 1974a. Bird Remains. Pp. 267–273 *in* Casas Grandes: A Fallen Trading Center of the Gran Chichimeca, vol. 8, ed. Charles C. Di Peso, John B. Rinaldo, and Gloria Fenner. Northland Press, Flagstaff, Ariz.

Di Peso, Charles C. 1974b. Commerce. Pp. 123–232 *in* Casas Grandes: A Fallen Trading Center of the Gran Chichimeca, vol. 8, ed. Charles C. Di Peso, John B. Rinaldo, and Gloria Fenner. Northland Press, Flagstaff, Ariz.

Dobyns, Henry F. 1978. Who Killed the Gila? Journal of Arizona History 19:17–30.

Dobyns, Henry F. 1981. From Fire to Flood: Historic Human Destruction of Sonoran Desert Riverine Oases. Anthropological Paper no. 20, Ballena Press, Soccoro, N.Mex.

Dodd, Norris L., and J. Robert Vahle. 1998. Osprey. Pp. 37–41 *in* The Raptors of Arizona, ed. Richard L. Glinski. University of Arizona Press, Tucson.

Dolores, Juan. 1923. Papago Nominal Stems. University of California Publications in American Archaeology and Ethnology 20:19–31.

Dolores, Juan. 1936. Papago Nicknames. Pp. 45–47 *in* Essays in Anthropology in Honor of Alfred Louis Kroeber, ed. J. Alden Mason. University of California Press, Berkeley.

Drucker, Philip. 1941. Culture Element Distributions: XVII, Yuman-Piman. University of California Anthropological Records 6(3): 91–203.

Dundes, Alan, ed. 1988. The Flood Myth. University of California Press, Berkeley.

Dunning, John B., Jr. 1993. CRC Handbook of Avian Body Masses. CRC Press, Boca Raton, Fla.

Edmonds, Margot, and Ella E. Clark. 1989. Voices of the Winds: Native American Legends. Facts on File, New York.

Edwards, Ernest P. 1972. A Field Guide to the Birds of Mexico. Ernest P. Edwards, Sweet Briar, Va.

Ellis, David H., and Gale Monson. 1989. A Historical Review of Peregrine Falcon Breeding and Summer Records for Arizona and the Navajo Indian Reservation. Southwestern Naturalist 34:565–569.

Emory, William H. 1848. Notes of a Military Reconnaissance, from Fort Leavenworth, in Missouri, to San Diego, in California, Including Parts of Arkansas, del Norte, and Gila Rivers. Made in 1846–1847. U.S. Congress, Senate, 30th Congr., 1st sess., Senate Executive Doc. no. 7 (serial no. 505), Washington, D.C.

Emslie, Stephen D. 1988a. An Early Condor-Like Vulture from North America. Auk 105:529–535.

Emslie, Stephen D. 1988b. The Fossil History and Phylogenetic Relationships of Condors (Ciconiiformes: Vulturidae) in the New World. Journal of Vertebrate Paleontology 8:212–228.

Evers, Larry, ed. 1980. The South Corner of Time: Hopi, Navajo, Papago, Yaqui Tribal Literature. Suntracks series, University of Arizona Press, Tucson.

Ezell, Paul H. 1961. The Hispanic Acculturation of the Gila River Pima. Memoir of the American Anthropological Association no. 90, Arlington, Va.

Ezell, Paul H. 1983. History of the Pima Indians. Pp. 149–160 *in* Handbook of North American Indians, vol. 10, Southwest, ed. Alfonso Ortiz. Smithsonian Institution, Washington, D.C.

Felger, Richard Stephen, and Mary Beck Moser. 1985. People of the Desert and Sea: Ethnobotany of the Seri Indians. University of Arizona Press, Tucson.

Ferg, Alan. 1985. Avifauna of the University Indian Ruin. Kiva 50:111–128.

Ferg, Alan, and Amadeo M. Rea. 1983. Prehistoric Bird Bone from the Big Ditch Site, Arizona. Journal of Ethnobiology 3:99–108.

Fewkes, J. Walter. 1900. Property-right in Eagles among the Hopi. American Anthropologist (n.s.) 2:690–707.

Fewkes, J. Walter. 1912. Casa Grande, Arizona. Pp. 25–179 *in* Annual Report of the Bureau of American Ethnology for the Years 1906–1907. Smithsonian Institution, Washington, D.C.

Font, Pedro. 1931. Anza's California Expeditions, vol. 4, Font's Complete Diary: A Chronicle of the Founding of San Francisco; Diary Kept by the Father Preacher Fray Pedro Font . . . during the Journey Which He Made to Monterey. Ed. and trans. Herbert E. Bolton. University of California Press, Berkeley.

Fontana, Bernard L. 1975. Introduction to the Re-edition. Pp. ix–xv *in* The Pima Indians, by Frank Russell. University of Arizona, Tucson (orig. pub. 1908).

Fontana, Bernard L. 1981. Of Earth and Little Rain: The Papago Indians. Northland Press, Flagstaff, Ariz.

Fontana, Bernard L. 1983. Pima and Papago: Introduction. Pp. 125–136 *in* Handbook of North American Indians, vol. 10, Southwest, ed. Alfonso Ortiz. Smithsonian Institution, Washington, D.C.

Forshaw, Joseph M. 1973. Parrots of the World. Lansdowne Press, Melbourne. Reissued 1977, T.F.H. Publications, Inc., Neptune, N.J.

Fowler, Catherine S. 1972. Some Ecological Clues to Proto-Numic Homelands. Pp. 105–121 *in* Great Basin Cultural Ecology: A Symposium, ed. Donald D. Fowler. Desert Research Institute Publications in the Social Sciences no. 8, Reno, Nev.

Fowler, Catherine S. 1983. Some Lexical Clues to Uto-Aztecan Prehistory. International Journal of American Linguistics 49:224–257.

Friedrich, P. 1970. Proto-Indo-European Trees: The Arboreal System of a Prehistoric People. University of Chicago Press, Chicago.

Garcés, Francisco. 1900. On the Trail of a Spanish Pioneer: The Diary and Itinerary of Francisco Garcés in His Travels Through Sonora, Arizona, and California, 1775–1776. 2 vols. Trans. and ed. Elliott Coues. Harper, N.Y.

Garcés, Francisco. 1965. A Record of Travel in Arizona and California, 1775–1776: A New Translation. Trans. and ed. John Galvin. John Howell, San Francisco.

Giff, Joseph. 1980. Pima Blue Swallow Songs of Gratitude. Pp. 127–139 *in* Speaking, Singing, and Teaching: A Multidisciplinary Approach to Language Variation, ed. Florence Barkin and Elizabeth Brandt. Proceedings of the 8th Annual Southwestern Areal Language and Linguistics Workshop. Swallow 8. Arizona State University Anthropological Research Papers no. 29, Tempe.

Gifford, E. W. 1933. Northeastern and Western Yavapai Myths. Journal of American Folk-Lore 46:347–415.

Gilman, M. French. 1909. Some Owls along the Gila River in Arizona. Condor 5:145–150.

Gilman, M. French. 1911. Doves on the Pima Reservation. Condor 13:51–56.

Gilman, M. French. 1914. Notes from Sacaton, Arizona. Condor 16:260–261.

Gilman, M. French. 1915a. A Forty Acre Bird Census at Sacaton, Arizona. Condor 17:86–90.

Gilman, M. French. 1915b. Woodpeckers of the Arizona Lowlands. Condor 17:151–163.

Gilman, M. French. 1930. Cacti as Nesting Sites.

Journal of the Cactus and Succulent Society of America 1:153–155.
Gilman, M. French. n.d. Birds of the Pima Indian Reservation in Arizona. Manuscript on typed index cards. A. R. Phillips archive, Museo de las Aves, Saltillo, Coahuila.
Ginzburg, Ralph. 1997. Wall Street Falcon. Natural History 106(9):52–53.
Glinski, Richard L. 1998a. Golden Eagle. Pp. 112–114 *in* The Raptors of Arizona, ed. Richard L. Glinski. University of Arizona Press, Tucson.
Glinski, Richard L. 1998b. Peregrine Falcon. Pp. 128–132 *in* The Raptors of Arizona, ed. Richard L. Glinski. University of Arizona Press, Tucson.
González, Luis. 1977. Etnología y misión en la Pimería Alta 1715–1740. Instituto Investigaciones Históricas. Universidad Nacional Autónoma de México, Mexico City.
Goulding, William R. n.d. Diary, 1849. Unpublished manuscript. Yale University Library, New Haven, Conn.
Griffin, John S. 1943. A Doctor Comes to California. California Historical Society, San Francisco.
Grossmann, F. E. 1873. The Pima Indians of Arizona. Pp. 407–419 *in* Annual Report of the Board of Regents Smithsonian Institution . . . for . . . 1871. Government Printing Office, Washington, D.C.
Grossmann [Grossman], Frederick E., comp. [1877]. English & Pima Dictionary. Typescript. Bureau of Ethnography, National Archives, Washington, D.C.
Gursky, Karl-Heinz. 1967. A Widespread Word for "Owl." International Journal of American Linguistics 33:328–329.
Hackenberg, Robert A. 1983. Pima and Papago Ecological Adaptations. Pp. 161–177 *in* Handbook of North American Indians, vol. 10, Southwest, ed. Alfonso Ortiz. Smithsonian Institution, Washington, D.C.
Hale, Kenneth L. 1964. Uncataloged Mountain Pima word list compiled from Andrés Duarte, from audio tapes. Arizona State Museum Archives, University of Arizona, Tucson.
Hale, Kenneth L. n.d.a. Breve vocabulario del idioma Pima de Ónavas. Unpublished manuscript compiled after summer of 1971. Arizona State Museum, University of Arizona, Tucson.
Hale, Kenneth L. n.d.b. Papago word list in shoe box, uncataloged. Arizona State Museum Archives, University of Arizona, Tucson.
Hancock, James, and Hugh Elliott. 1978. The Herons of the World. Harper and Row, New York.
Hargrave, Lyndon L. 1968. The Proper Way to Report Birds in Print. American Antiquity 33:384–385.
Hargrave, Lyndon L. 1970. Mexican Macaws: Comparative Osteology and Survey of Remains from the Southwest. University of Arizona Anthropological Papers no. 20, Tucson.
Harrington, John Peabody. 1908. A Yuma Account of Origins. Journal of American Folk-Lore 21:324–348.
Harris, Benjamin Butler. 1960. The Gila Trail: The Texas Argonauats and the California Gold Rush. Ed. Richard H. Dillon. University of Oklahoma Press, Norman.
Hatch, Daniel E. 1970. Energy Conserving and Heat Dissipating Mechanisms of the Turkey Vulture. Auk 87:111–124.
Hayden, Julian D. 1935. Pima Creation Myth, as Told by Juan Smith, Snaketown, Arizona. Unpublished manuscript. University Arizona Library, Tucson.
Hayden, Julian D. 1987. The Vikita Ceremony of the Papago. Journal of the Southwest 29:273–324.
Hayes, William D. 1957. Indian Tales of the Desert People. David McKay Company, New York.
Heermann, A. L. 1859. Report upon the Birds Collected on the Survey. Zoological Report Explorations and Surveys for a Railroad Route from the Mississippi River to the Pacific Ocean 10 (part 4), no. 1:9–20, no. 2:29–80. U.S. War Department, Washington, D.C.
Herzog, George. [1929]. Pima Origin Story. Unpublished manuscript, Pima and English translation, 1929 to early 1930s. Copy, Gila River Indian Community Cultural Resources Department Library, Sacaton, Ariz.
Herzog, George. 1936. A Comparison of Pueblo and Pima Musical Styles. Journal of American Folk-lore 49:283–417.
Hill, Gertrude. 1940a. Papago Legends from Santa Rosa, Arizona–I. The Man Transformed into an Eagle. Southwestern Lore 6:18–20.
Hill, Gertrude. 1940b. Papago Legends from Santa Rosa, Arizona–II. Hawk-Woman and the First Tobacco. Southwestern Lore 6:34–35.
Hill, W. W. 1936. Notes on Pima Land Law and Tenure. American Anthropologist (n.s.) 38:586–589.
Hinton, Thomas B. 1983. Southern Periphery: West. Pp. 315–328 *in* Handbook of North American Indians, vol. 10, Southwest, ed. Alfonso Ortiz. Smithsonian Institution, Washington, D.C.
Hoffmann, Ralph. 1927. Birds of the Pacific States. Houghton Mifflin, Cambridge, Mass.
Houston, David C. 1994. To the Vultures Belong the Spoils. Natural History 103:35–41.
Hrdlička, Aleš. 1906. Notes on the Pima of Arizona. American Anthropologist (n.s.) 8:39–46.
Hrdlička, Aleš. 1908. Physiological and Medical Observations among the Indians of Southwestern United States and Northern Mexico. Bureau of American Ethnology Bulletin 34:1–266.
Hubbard, John P. 1978. Revised Check-list of the Birds of New Mexico. New Mexico Ornithological Society Publication no. 6, Albuquerque.
Hunn, Eugene. 1975a. A Measure of the Degree of Correspondence of Folk to Scientific Biological Classification. American Ethnologist 2:309–327.
Hunn, Eugene. 1975b. Words for Owls in North American Indian Languages. International Journal of American Linguistics 41:237–239.
Hunn, Eugene. 1977. Tzeltal Folk Zoology: The Classification of Discontinuities in Nature. Academic Press, New York.
Hunn, Eugene. 1999. Size as Limiting the Recognition of Biodiversity in Folkbiological Classifications: One of Four Factors Governing the Cultural Recognition of Biological Taxa. Pp. 47–69 *in* Folkbiology, ed. Douglas L. Medin and Scott Atran. MIT Press, Cambridge, Mass.
Hunt, W. Grainger. 1998. Bald Eagle. Pp. 50–54 *in* The Raptors of Arizona, ed. Richard L. Glinski. University of Arizona Press, Tucson.
Jaeger, Edmund C. 1948. Does the Poor-will "Hibernate"? Condor 50: 45–46.
Jaeger, Edmund C. 1949. Further Observations on the Hibernation of the Poor-will. Condor 51:105–109.
Jaeger, Edmund C. 1953. Poorwill Sleeps Away the Winter. National Geographic 103:273–280.
Jensen, Allen Arthur. 1988. Sistemas indígenas de classificação de aves: Aspectos comparativos, ecológicos e evolutivos. Museu Paraense Emílio Goeldi, Belém, Brasil.
Johnsgard, Paul A. 1973. Grouse and Quails of North America. University of Nebraska, Lincoln.
Johnson, Jean B. n.d. Lower Pima Word Lists, uncataloged. Manuscript. Arizona State Museum Archives, University of Arizona, Tucson.
Jones, Richard D. 1971. The Wi'igita of Achi and Quitobac. Kiva 36:1–29.
Joseph, Alice, Rosamond B. Spicer, and Jane Chesky. 1949. The Desert People: A Study of the Papago Indians. University of Chicago Press, Chicago.
Kahl, Philip M., Jr. 1963. Thermoregulation in the Wood Stork, with Special Reference to the Role of the Legs. Physiological Zoology 36:141–151.
Karttunen, Frances. 1983. An Analytical Dictionary of Nahuatl. University of Texas Press, Austin.
Kelly, William H. 1977. Cocopa Ethnography. Anthropological Papers of the University of Arizona no. 29, Tucson.
Key, Harold. 1954. Vocabularies for Languages of the Uto-Aztecan Family. Microfilm collection, Manuscripts on Middle American Cultural Anthropology no. 38, University of Chicago Library, Chicago.
Kino, Eusebio Francisco. 1919. Kino's Historical Memoir of Pimería Alta: A Contemporary

Account of the Beginnings of California, Sonora, and Arizona by Father Eusebio Francisco Kino, S.J., Pioneer Missionary Explorer, Cartographer, and Ranchman, 1683–1711. Vol. 1. Ed. and trans. H. E. Bolton. A. H. Clark, Cleveland. Reissued 1948, University of California, Berkeley.

Kissell, Mary Lois. 1916. Basketry of the Papago and Pima. Anthropological Papers of the American Museum of Natural History 17:115–264. Reissued 1972, Rio Grande Press, Glorieta, N.Mex.

Knowler, William C., Peter H. Bennett, Richard F. Hamman, and Max Miller. 1978. Diabetes Incidence and Prevalence in Pima Indians: A 19-fold Greater Incidence Than in Rochester, Minnesota. American Journal of Epidemiology 108:497–505.

Knowler, William C., David J. Pettit, Peter H. Bennett, and Robert C. Williams. 1983. Diabetes Mellitus in the Pima Indians: Genetic and Evolutionary Considerations. American Journal of Physical Anthropology 62:107–114.

Knowler, W. C., D. J. Pettit, S. Lillioja, and R. G. Nelson. 1988. Genetic and Environmental Factors in the Development of Diabetes Mellitus in Pima Indians. Pp. 67–74 *in* Genetic Susceptibility to Environmental Factors: A Challenge for Public Intervention, ed. Ulf Smith, Sten Eriksson, and Folke Lindgärde. Almqvist and Wiksell International, Stockholm.

Knowler, William C., David J. Pettit, Mohammed F. Saad, and Peter H. Bennett. 1990. Diabetes Mellitus in the Pima Indians: Incidence, Risk Factors, and Pathogenesis. Diabetes/Metabolism Reviews 6:1–27.

Knowler, William C., David J. Pettit, Mohammed F. Saad, Marie A. Charles, Robert G. Nelson, Barbara V. Howard, Clifton Bogardus, and Peter H. Bennett. 1991. Obesity in the Pima Indians: Its Magnitude and Relationship with Diabetes. American Journal of Clinical Nutrition 53:1543S–1551S.

Knowler, William C., David J. Pettit, Peter J. Savage, and Peter H. Bennett. 1981. Diabetes Incidence in Pima Indians: Contributions of Obesity and Parental Diabetes. American Journal of Epidemiology 113:144–156.

König, V. C. 1982. Zur systematischen Stellung der Neuweltgeier (Cathartidae). Journal für Ornithologie 123:259–267.

Kroeber, Clifton B., and Bernard L. Fontana. 1986. Massacre on the Gila: An Account of the Last Major Battle Between American Indians, with Reflections on the Origin of War. University of Arizona Press, Tucson.

Kurath, William. n.d. Uncataloged cigar box of Papago words compiled before his death in March 1948. Arizona State Museum Archives, University of Arizona, Tucson.

Ladd, Edmund James. 1998. Ethno-ornithology of the Zuni. Pp. 119–137 *in* Stars above, Earth below: American Indians and Nature, ed. Marsha C. Bol. Roberts Rinehard, Niwot, Colo.

Lang, Richard E., and Arthur Harris. 1984. The Faunal Remains for Arroyo Hondo Pueblo, New Mexico. Archaeological Series, vol. 5. School of American Research Press, Santa Fe.

Leeming, David Adams. 1990. The World of Myth: An Anthology. Oxford University Press, New York.

Levy, Seymour H. 1961. The Caracara Nesting in Arizona. Auk 78:99.

Levy, Seymour H. 1998. Crested Caracara. Pp. 115–117 *in* The Raptors of Arizona, ed. Richard L. Glinski. University of Arizona Press, Tucson.

Ligon, J. David. 1967. Relationships of the Cathartid Vultures. Occasional Papers of the Museum of Zoology no. 651, University of Michigan, Ann Arbor.

Linares, Olga F. 1976. "Garden Hunting" in the American Tropics. Human Ecology 4:331–349.

Lionnet, Andrés. 1986. Un idioma extinto de Sonora: El Eudeve. Instituto de Investigaciones Anthropológicas. Serie Anthropología 60. Universidad Nacional Autónoma de México, Mexico City.

Lloyd, J. W. 1911. Aw-aw-tam Indian Nights. Lloyd Group, Westfield, N.J.

Lopez, Danny. 1980. O'odham Ha-ñeñei: Songs of the Papago People. Pp. 123–126 *in* The South Corner of Time: Hopi, Navajo, Papago, Yaqui Tribal Literature, ed. Larry Evers. Suntracks series. University of Arizona Press, Tucson.

Lumholtz, Carl S. 1912. New Trails in Mexico. An Account of One Year's Exploration in North-Western Sonora, Mexico, and South-Western Arizona, 1919–1910. Scribner's Sons, New York. Reissued 1971, Rio Grande Press, Glorieta, N.Mex.; 1990, University of Arizona Press, Tucson.

Macazaga Ordoño, César. 1985. Diccionario de zoologia nahuatl. Editorial Innovación, S.A., Mexico City.

Mader, William J. 1975. Biology of the Harris' Hawk in Southern Arizona. Living Bird 14:59–85.

Mader, William J. 1978. A Comparative Nesting Study of Red-tailed Hawks and Harris' Hawks in Southern Arizona. Auk 95:327–337.

Manje, Juan Mateo. 1954. Unknown Arizona and Sonora, 1693–1721, from the Francisco Fernández del Castillo [1926] Version of Luz de Tierra Incógnita. Trans. Harry J. Karns and associates. Arizona Silhouettes, Tucson.

Marshall, Joe T., Jr. 1955. Hibernation in Captive Goatsuckers. Condor 57:129–134.

Marshall, Joe T., Jr. 1967. Parallel Variation in North and Middle American Screech-Owls. Western Foundation of Vertebrate Zoology Monograph no. 1, Los Angeles.

Marshall, Joe. 1981. Family Strigidae: Typical Owls. Pp. 69–74 *in* Annotated Checklist of the Birds of Arizona, 2d ed., rev. and exp., ed. Gale Monson and Allan R. Phillips. University of Arizona Press, Tucson.

Martin, Alexander C., Herbert S. Zim, and Arnold L. Nelson. 1951. American Wildlife and Plants: A Guide to Wildlife Food Habits: The Use of Trees, Shrubs, Weeds, and Herbs by Birds and Mammals of the United States. McGraw-Hill, Columbus, Ohio. Reissued 1961, Dover, New York.

Mason, J. Alden. 1917. Tepecano, a Piman Language of Western Mexico. Annals of the New York Academy of Science 25:309–416.

Mason, J. Alden. 1920. The Papago Harvest Festival. American Anthropologist 22:13–25.

Mason, J. Alden. 1921. The Papago Migration Legend. Journal of American Folk-Lore 34:254–268.

Mason, J. Alden. 1950. The Language of the Papago of Arizona. University of Pennsylvania Museum Monograph no. 3, Philadelphia.

Mathiot, Madeleine. 1962. Noun Classes and Folk Taxonomy in Papago. American Anthropologist 64:340–350. Reprinted 1964, pp. 154–163 *in* Language in Culture and Society, ed. Dell Hymes. Harper and Row, New York.

Mathiot, Madeleine. 1973. A Dictionary of Papago Usage, vol. 1, B–K. Language Science Monographs no. 8(1). Indiana University Publications, Bloomington.

Mathiot, Madeleine. n.d. A Dictionary of Papago Usage, vol. 2, Ku-'u. Language Science Monographs no. 8(2). Indiana University Publications, Bloomington.

McAuliffe, Joseph R., and Paul Hendricks. 1988. Determinants of the Vertical Distribution of Woodpecker Nest Cavities in the Sahuaro Cactus. Condor 90:791–801.

McKusick, Charmion R. 1974. The Casas Grandes Avian Report. Pp. 273–284 *in* Casas Grandes: A Fallen Trading Center of the Gran Chichimeca, vol. 8, ed. Charles C. Di Peso, John B. Rinaldo, and Gloria Fenner. Northland Press, Flagstaff, Ariz.

McKusick, Charmion R. 1976. Appendix 6: Avifauna. Pp. 374–377 *in* The Hohokam: Desert Farmers and Craftsmen: Excavations at Snaketown, 1964–1965, by Emil W. Haury. University of Arizona Press, Tucson.

McKusick, Charmion R. 1980. Three Groups of Turkeys from Southwestern Archaeological Sites. Pp. 225–235 *in* Papers in Avian Paleontology Honoring Hildegard Howard, ed. Kenneth E. Campbell Jr. Natural History Museum of Los Angeles County Contributions in Science no. 330, Los Angeles.

McKusick, Charmion R. 1986. Southwest Indian Turkeys: Prehistory and Comparative Osteology. Privately published, Southwest Bird Laboratory, Globe, Ariz.

McKusick, Charmion R. 2001. Southwest Birds of Sacrifice. Arizona Archeologist no. 31, Arizona Archaeological Society, Tucson.

Mills, G. Scott. 1998. American Kestrel. Pp. 118–

120 *in* The Raptors of Arizona, ed. Richard L. Glinski. University of Arizona Press, Tucson.

Minnis, Paul E. 1985. Social Adaptation to Food Stress: A Prehistoric Southwestern Example. University of Chicago Press, Chicago.

Molina, Felipe S., Herminia Valenzuela, and David Leedom Shaul. 1999. Hippocrene Standard Dictionary Yoeme-English, English-Yoeme: With a Comprehensive Grammar of Yoeme Language. Hippocrene Books, New York.

Molina Molina, Flavio. 1972. Nombres indígenas de Sonora y su traducción al español. Impresora y Editorial S.A. de C.V., Hermosillo, Sonora.

Monson, Gale, and Allan R. Phillips. 1981. Annotated Checklist of the Birds of Arizona. 2d ed., rev. and exp. University of Arizona Press, Tucson.

Munn, Charles A. 1994. Macaws: Winged Rainbows. National Geographic 185:118–140.

Murray, Blanche K. 1939. Ostriches in Arizona. Arizona Highways 15(11):22–23, 27–28.

Nabhan, Gary Paul. 1982. The Desert Smells Like Rain: A Naturalist in Papago Indian Country. North Point Press, San Francisco. Reissued 1987, University of Arizona, Tucson.

Nabhan, Gary Paul. 1983. Papago Fields: Arid Lands Ethnobotany and Agricultural Ecology. Ph.D. diss., University of Arizona, Tucson.

Nabhan, Gary Paul. 1984. Soil Fertility Renewal and Water Harvesting in Sonoran Desert Agriculture: The Papago Example. Arid Lands Newsletter 20:20–28.

Nabhan, Gary Paul. 1997. Diabetes, Diet, and Native American Foraging Traditions. Pp. 197–206 *in* Cultures of Habitat: On Nature, Culture, and Story. Counterpoint, Washington, D.C.

Nabhan, Gary Paul. 2004. Cross-pollinations: The Marriage of Science and Poetry. Credo series. Milkweed Editions, Minneapolis.

Nabhan, Gary Paul, Wendy Hodgson, and Frances Fellows. 1989. A Meager Living on Lava and Sand? Hia Cheḍ O'odham Food Resources and Habitat Diversity in Oral and Documentary Histories. Journal of the Southwest 31:508–533.

Nabhan, Gary P., Amadeo M. Rea, Karen L. Reichhardt, Eric Mellink, and Charles F. Hutchinson. 1982. Papago (O'odham) Influences on Habitat and Biotic Diversity: Quitovac Oasis Ethnoecology. Journal of Ethnobiology 2:124–143. Reissued 2000, pp. 41–62 *in* Ethnobotany: A Reader, ed. Paul E. Minnis. University of Oklahoma Press, Norman.

National Geographic Society (U.S.). 1983. A Field Guide to the Birds of North America. Ed. Shirley L. Scott. National Geographic Book Service, Washington, D.C.

Nazarea, Virginia D. 1998. Cultural Memory and Biodiversity. University of Arizona Press, Tucson.

Neff, Mary L. 1912. Pima and Papago Legends. Journal of American Folk-lore 25:51–65.

Nelson, R. G., J. M. Newman, W. C. Knowler, M. L. Sievers, C. L. Kunzelman, D. J. Pettitt, C. D. Moffett, S. M. Teutsch, and P. H. Bennett. 1988. Incidence of End-stage Renal Disease in Type 2 (Non-insulin-dependent) Diabetes Mellitus in Pima Indians. Diabetologia 31:730–736.

Nelson, Richard K. 1983. Make Prayers to the Raven: A Koyukon View of the Northern Forest. University of Chicago Press, Chicago.

Nentvig, Juan. 1971. Descripción geográfica . . . de Sonora. Ed. Germán Viveros. Publicaciones del Archivo General de la Nación, 2d ser., no. 1, Mexico City.

Nentvig, Juan. 1980. Rudo Ensayo: A Description of Sonora and Arizona in 1764. Trans. and annotated by Alberto Francisco Pradeau and Robert R. Rasmussen. University of Arizona Press, Tucson.

Obregón, Baltasar de. 1928. Obregón's History of 16th Century Explorations in Western America. Trans. and ed. George P. Hammond and Agapito Rey. Wetzel, Los Angeles.

Och, Joseph. 1965. Missionary in Sonora: The Travel Reports of Joseph Och, S.J., 1755–1767. Trans. and ed. Theodore E. Treutlein. California Historical Society, San Francisco.

Ohmart, Robert D., and Robert C. Lasiewski. 1971. Roadrunners: Energy Conservation by Hypothermia and Absorption of Sunlight. Science 172:67–69.

Olson, Storrs L. 1979. Multiple Origins of the Ciconiiformes. Pp. 165–170 *in* Proceedings of the 1978 Conference of the Colonial Waterbird Group, ed. W. E. Southern. American Museum of Natural History, New York.

Olson, Storrs L. 1985. The Fossil Record of Birds. Pp. 79–238 *in* Avian Biology, vol. 8, ed. D. Farner, J. King, and K. Parkes. Academic Press, New York.

Opler, Morris E. 1936. A Summary of Jicarilla Apache Culture. American Anthropologist 38:202–223.

Parsons, Elsie Clews. 1928. Notes on the Pima, 1926. American Anthropologist (n.s.) 30:445–464.

Parsons, Elsie Clews. 1939. Pueblo Indian Religion. University of Chicago Press, Chicago. Reissued 1996, Bison Books, University of Nebraska Press, Lincoln.

Pennington, Campbell W. 1969. The Tepehuán of Chihuahua: Their Material Culture. University of Utah Press, Salt Lake City.

Pennington, Campbell W. 1979. Vocabulario en la lengua Névome: The Pima Bajo of Central Sonora, Mexico. Vol. 2. University of Utah Press, Salt Lake City.

Pennington, Campbell W. 1980. The Pima Bajo of Central Sonora, Mexico, vol. 1, The Material Culture. University of Utah Press, Salt Lake City.

Pennington, Campbell W., ed. 1981. Arte y vocabulario de la lengua Dohema, Heve o Eudeve: Manuscript of an Anonymous Eighteenth Century Author. Universidad Autónoma de México, Instituto de Investigaciones Filológicol, Mexico City.

Pennington, Campbell W. n.d. A Vocabulary Made at Yepáchic, Chihuahua, among the Pima Bajo (1968–1970), [and] A Vocabulary Made at Maicoba, Sonora, among the Pima Bajo (1968, 1970). Unpublished manuscripts in Pennington's and Rea's possession.

Pérez de Ribas, Andrés. 1944. Páginas para la historia de Sinaloa y Sonora [historia de los] triuunfos de nuestra Santa Fe entre gentes las más bárbaras y fieras del Nuevo Orbe. Vol 2. Editorial Layac, Mexico City.

Pérez de Ribas, Andrés. 1999. Historia de los triunfos de nuestra Santa Fe entre gentes las mas barbaras y fieras del nueve orbe. 1645. History of the Triumphs of Our Holy Faith Amongst the Most Barbarous and Fierce Peoples of the New World. Trans. Daniel T. Reff, Maureen Ahern, and Richard K. Danford. Annotated by Daniel T. Reff. University of Arizona Press, Tucson.

Peterson, Roger T. 1961. A Field Guide to Western Birds. Rev. ed. Houghton Mifflin, Boston.

Pfefferkorn, Ignaz. 1949. Sonora: A Description of the Province, vol. 12, Coronado Cuarto Centennial Publications, 1540–1940. Trans. and annotated by Theodore E. Treutlein. University of New Mexico Press, Albuquerque.

Phillips, Allan R. 1946. The Birds of Arizona. Ph.D. diss., Cornell University, Ithaca, N.Y. (Copy in author's possession.)

Phillips, Allan R. 1968. The Instability of the Distribution of Land Birds in the Southwest. Pp. 129–162 *in* Collected Papers in Honor of Lyndon Lane Hargrave, ed. Albert H. Schroeder. Papers of the Archaeological Society of New Mexico no. 1, Museum of New Mexico Press, Santa Fe.

Phillips, Allan R. 1986. The Known Birds of North and Middle America. Part 1. Hirundinidae to Mimidae; Certhiidae. Privately published, Denver, Colo.

Phillips, Allan, Joe Marshall, and Gale Monson. 1964. The Birds of Arizona. University of Arizona Press, Tucson.

Pierce, Fred J. 1921. The Meadowlark as a Conversationalist. Wilson Bulletin 33:154–155.

Pilcher, William W. 1967. Some Comments on the Folk Taxonomy of the Papago. American Anthropologist 69:204–208.

Randall, Robert. 1982. Peru's Pilgrimage to the Sky. National Geographic 162:60–69.

Ravussin, Eric, Mauro E. Valencia, Julian Esparza, Peter H. Bennett, and Leslie O. Schulz. 1994. Effects of a Traditional Lifestyle on Obesity in Pima Indians. Diabetes Care 17:1067–1073.

Rea, Amadeo M. 1973. The Scaled Quail *(Callipepla squamata)* of the Southwest: System-

atic and Historical Considerations. Condor 75:322–329.
Rea, Amadeo M. 1978. The Ecology of Pima Fields. Environment Southwest 484:1–6.
Rea, Amadeo M. 1980. Late Pleistocene and Holocene Turkeys in the Southwest. Pp. 209–224 *in* Papers in Avian Paleontology Honoring Hildegard Howard, ed. Kenneth E. Campbell Jr. Natural History Museum of Los Angeles County Contributions in Science no. 330, Los Angeles.
Rea, Amadeo M. 1981a. Avian Remains from Las Colinas. Pp. 297–302 *in* The 1968 Excavations at Mound 8 Las Colinas Ruins Group Phoenix, Ariz., ed. Laurens C. Hammack and Alan P. Sullivan. Arizona State Museum Archaeological Series no. 154, University of Arizona, Tucson.
Rea, Amadeo M. 1981b. Resource Utilization and Food Taboos of Sonoran Desert Peoples. Journal of Ethnobiology 1:69–83.
Rea, Amadeo M. 1983. Once a River: Bird Life and Habitat Changes on the Middle Gila. University of Arizona Press, Tucson.
Rea, Amadeo M. 1988. Habitat Restoration and Avian Recolonization from Wastewater on the Middle Gila River, Arizona. Pp. 1395–1405 *in* Arid Lands: Today and Tomorrow, ed. E. E. Whitehead, C. F. Hutchinson, B. N. Timmermann, and R. G. Varady. Westbutte Press, Boulder, Colo.
Rea, Amadeo M. 1997. At the Desert's Green Edge: An Ethnobotany of the Gila River Pima. University of Arizona Press, Tucson.
Rea, Amadeo M. 1998a. Corn Man and Tobacco Woman in Pima Cosmology. Pp. 211–227 *in* Stars above, Earth below: American Indians and Nature, ed. Marsha C. Bol. Roberts Rinehard, Niwot, Colo.
Rea, Amadeo M. 1998b. Folk Mammalogy of the Northern Pimans. University of Arizona Press, Tucson.
Rea, Amadeo M., Gary P. Nabhan, and Karen L. Reichhardt. 1983. Sonoran Desert Oases: Plants, Birds, and Native People. Environment Southwest 503:5–9.
Reichhardt, Karen L., Eric Mellink, Gary P. Nabhan, and Amadeo Rea. 1994. Habitat Heterogeneity and Biodiversity Associated with Indigenous Agriculture in the Sonoran Desert. Etnoecológica 2:21–34.
Rinaldini, Benito. 1743. Arte de la lengua tepeguana con vocabulario, confessionario y catechismo. Mexico. Reissued 1994, with a foreword by Javier Guerrero Romero, Consejo Nacional para la Cultura y las Artes, Gobierno del Estado de Durango, Mexico.
Rosenberg, Kenneth V., Robert D. Ohmart, William C. Hunter, and Bertin W. Anderson. 1991. Birds of the Lower Colorado River Valley. University of Arizona Press, Tucson.
Russell, Frank. 1908. The Pima Indians. Annual Report of the Bureau of American Ethnology 26:3–389. Reissued 1975, with introduction, citation sources, and bibliography by Bernard L. Fontana. University of Arizona Press, Tucson.
Russell, Priscilla N., and George C. West. 2003. Bird Traditions of the Lime Village Area Dena'ina: Upper Stony River Ethno-Ornithology. Alaska Native Knowledge Network, Center for Cross-Cultural Studies, University of Alaska, Fairbanks.
Russell, Stephen M., and Gale Monson. 1998. The Birds of Sonora. University of Arizona Press, Tucson.
Sada, Andrés M., Allan R. Phillips, and Mario A. Ramos. 1987. Nombres en Castellano para las aves mexicanas. Secunda impressión corregida. Instituto Nacional de Investigaciones sobre Recursos Bióticos, no. 17, Xalapa, Veracruz.
Savage, Candace. 1995. Bird Brains. Sierra Club Books, San Francisco.
Saxton, Dean F., and Lucille Saxton. 1969. Dictionary: Papago and Pima to English, English to Papago and Pima. O'odham–Mil-gahn, Mil-gahn–O'odham. University of Arizona Press, Tucson.
Saxton, Dean, and Lucille Saxton. 1973. O'othham Hoho'ok A'agitha: Legends and Lore of the Papago and Pima Indians. University of Arizona Press, Tucson.
Saxton, Dean, Lucille Saxton, and Susie Enos. 1983. Dictionary, Papago/Pima–English, O'othham–Mil-gahn; English–Papago/Pima, Mil-gahn–O'othham. 2d ed., rev. and exp. Ed. R. L. Cherry. University of Arizona Press, Tucson.
Schroeder, Albert H. 1968. Birds and Feathers in Documents Relating to Indians of the Southwest. Pp. 95–114 *in* Collected Papers in Honor of Lyndon Lane Hargrave, ed. Albert H. Schroeder. Papers of the Archaeological Society of New Mexico no. 1, Museum of New Mexico Press, Santa Fe.
Scott, W. E. D. 1886. On the Avi-fauna of Pinal County, with Remarks on Some Birds of Pima and Gila Counties, Arizona. With annotations by J. A. Allen. Auk 3:249–258, 383–389, 421–432.
Sedelmayr, Jacobo. 1939. Sedelmayr's Relación of 1746. Trans. Ronald L. Ives. Bulletin of the Bureau American Ethnology, no. 123; Anthropological Papers, no. 9:101–117.
Sedelmayr, Jacobo. 1955. Jacobo Sedelmayr, 1744–1751. Missionary, Frontiersman, Explorer in Arizona and Sonora. Four Original Manuscript Narratives [including 1744 draft of the relación]. Trans. and ed. P. M. Donne. Arizona Pioneers' Historical Society, Tucson.
Segesser, Philipp. 1945. Document: The Relation of Philipp Segesser. Introduced and trans. by P. M. Donne. Arizona Pioneers' Historical Society, Tucson.
Shaul, David Leedom. 1994. A Sketch of the Structure of Oob No'ok (Mountain Pima). Anthropological Linguistics 37:277–365.
Shaw, Anna Moore. 1968. Pima Indian Legends. University of Arizona Press, Tucson.
Shaw, Anna More. 1974. A Pima Past. University of Arizona Press, Tucson.
Sibley, Charles G., and Jon E. Ahlquist. 1986. Reconstructing Bird Phylogeny by Comparing DNAs. Scientific American 254:82–92.
Sibley, Charles G., and Jon E. Ahlquist. 1990. Phylogeny and Classification of Birds: A Study in Molecular Evolution. Yale University Press, New Haven, Conn.
Skinner, Charles. 1896. Myths and Legends of Our Own Land. 3 vols. J. P. Lippincott, Philadelphia. Reissued 1969, Singing Tree Press, Auburn, Calif.
Snyder, Helen A. 1998. Northern Harrier. Pp. 55–57 *in* The Raptors of Arizona, ed. Richard L. Glinski. University of Arizona Press, Tucson.
Snyder, Noel F. R., and Amadeo M. Rea. 1998. California Condor. Pp. 32–36 *in* The Raptors of Arizona, ed. Richard L. Glinski. University of Arizona Press, Tucson.
Sobarzo, Horacio. 1991. Vocabulario Sonorense. 3d ed. Instituto Sonorense de Cultura, Hermosillo, Sonora (orig. pub. 1966).
Spier, Leslie. 1928. Havasupai Ethnography. Anthropological Papers of the American Museum of Natural History 29(3):81–392.
Spier, Leslie. 1933. Yuman Tribes of the Gila River. University of Chicago Press, Chicago. Reissued 1978, Dover Press, New York.
Stager, Kenneth E. 1964. The Role of Olfaction in Food Location by the Turkey Vulture *(Cathartes aura)*. Los Angeles County Natural History Museum Contributions in Science no. 81, Los Angeles.
Steadman, David W. 1980. A Review of the Osteology and Paleontology of Turkeys (Aves: Meleagridinae). Pp. 131–207 *in* Papers in Avian Paleontology Honoring Hildegard Howard, ed. Kenneth E. Campbell Jr. Los Angeles County Natural History Museum Contributions in Science no. 330, Los Angeles.
Stewart, Kenneth M. 1947. Mohave Hunting. The Masterkey 21:80–84.
Swarth, Harry S. 1914. A Distributional List of the Birds of Arizona. Cooper Ornithological Club. Pacific Coast Avifauna no. 10, Albuquerque.
Swinburn, B. A., V. L. Boyce, R. N. Bergman, B. V. Howard, and C. Bogardus. 1991. Deterioration in Carbohydrate Metabolism and Lipoprotein Changes Induced by Modern, High Fat Diet in Pima Indians and Caucasians. Journal of Clinical Endocrinology and Metabolism 73:156–165.
Thorburg, Florence. 1953. Another Hibernating Poor-will. Condor 55:274.
Turner, Nancy J. 1988. "The Importance of a Rose": Evaluating the Cultural Significance of Plants in Thompson and Lillooet Interior Salish. American Anthropologist 90:272–290.
Turner, Raymond M., Janice E. Bowers, and

Tony L. Burgess. 1995. Sonoran Desert Plants: An Ecological Atlas. University of Arizona Press, Tucson.

Tyler, Hamilton A. 1979. Pueblo Birds and Myths. University of Oklahoma Press, Norman. Reissued 1991, Northland Press, Flagstaff, Ariz.

Underhill, Ruth M. 1936. The Autobiography of a Papago Woman. Memoirs of the American Anthropological Association no. 46, Arlington, Va.

Underhill, Ruth Murray. 1938. Singing for Power: The Song Magic of the Papago Indians of Southern Arizona. University of California Press, Berkeley. Reissued 1993, with a new foreword by Ofelia Zepeda, University of Arizona Press, Tucson.

Underhill, Ruth M. 1939. Social Organization of the Papago Indians. Columbia University Contributions in Anthropology no. 30, New York.

Underhill, Ruth M. 1940a. Hawk over Whirlpools. J. J. Augustin, New York.

Underhill, Ruth. 1940b. The Papago Indians of Arizona and Their Relatives the Pima. Bureau of Indian Affairs, Branch of Education. Sherman Pamphlets no. 3, Washington, D.C.

Underhill, Ruth M. 1946. Papago Indian Religion. Columbia University Press, New York.

Underhill, Ruth. 1951. People of the Crimson Evening. U.S. Bureau of Indian Affairs, Branch of Education. Haskell Press, Lawrence, Kans.

Underhill, Ruth M., Donald M. Bahr, Baptisto Lopez, Jose Pancho, and David Lopez. 1979. Rainhouse and Ocean: Speeches for the Papago Year. American Tribal Religions, vol. 4. Museum of Northern Arizona Press, Flagstaff.

van Rossem, Adrian J. 1945. A Distributional Survey of the Birds of Sonora, México. Occasional Papers of the Museum of Zoology, Louisiana State University, no. 21, Baton Rouge.

Velarde, Luis. 1931. Padre Luis Velarde's *Relación* of Pimería Alta. Ed. R. K. Wyllys. New Mexico Historical Review 6:111–157.

Velarde, Pablita. 1960. Old Father the Story Teller. Dale Stuart King, Globe, Ariz.

Voth, H. R. 1912. II. Notes on the Eagle Cult of the Hopi. Brief Miscellaneous Papers. Field Museum of Natural History, Publication no. 157, Anthropological Series, 11:102–109.

Walkinshaw, Lawrence [H.]. 1973. Cranes of the World. Winchester Press, New York.

Wallace, Michael P., and Stanley A. Temple. 1987. Competitive Interactions within and Between Species in a Guild of Avian Scavengers. Auk 104:290–295.

Webb, George. 1959. A Pima Remembers. University of Arizona Press, Tucson. Reissued 1994.

Willenbrink, Antonine. 1935. Notes on the Pima Indian Language. Franciscan Fathers of California, Santa Barbara.

Wilson, John P. n.d. Peoples of the Middle Gila: A Documentary History of the Pimas and Maricopas, 1500's–1945. Unpublished manuscript. Gila River Indian Community Cultural Resources Department Library, Sacaton, Ariz.

Winter, Joseph C. 1973. Cultural Modifications of the Gila Pima: A.D. 1697–A.D. 1846. Ethnohistory 20:67–77.

Witkowski, Stanley R., and Cecil H. Brown. 1983. Marking Reversals and Cultural Importance. Language 59:569–581.

Witzeman, Janet L., Salome R. Demaree, and Eleanor L. Radke. 1997. Birds of Phoenix and Maricopa County, Arizona. Maricopa Audubon Society, Phoenix.

Woodward, Arthur. 1933. A Man's Way. The Masterkey 7:165–167.

Wright, Harold Bell. 1929. Long Ago Told (Huh-Kew Ah-Kah): Legends of the Papago Indians. D. Appleton, New York.

Index

Numbers in **bold** indicate main entries for those terms